Computational Physics
Second Edition

This Second Edition has been fully updated. The wide range of topics covered in the First Edition has been extended with new chapters on finite element methods and lattice Boltzmann simulation. New sections have been added to the chapters on density functional theory, quantum molecular dynamics, Monte Carlo simulation and diagonalisation of one-dimensional quantum systems.

The book covers many different areas of physics research and different computational methodologies, with an emphasis on condensed matter physics and physical chemistry. It includes computational methods such as Monte Carlo and molecular dynamics, various electronic structure methodologies, methods for solving partial differential equations, and lattice gauge theory. Throughout the book, the relations between the methods used in different fields of physics are emphasised. Several new programs are described and these can be downloaded from www.cambridge.org/9781107677135

The book requires a background in elementary programming, numerical analysis and field theory, as well as undergraduate knowledge of condensed matter theory and statistical physics. It will be of interest to graduate students and researchers in theoretical, computational and experimental physics.

Jos Thijssen is a lecturer at the Kavli Institute of Nanoscience at Delft University of Technology.

COMPUTATIONAL PHYSICS

Second Edition

JOS THIJSSEN
Kavli Institute of Nanoscience, Delft University of Technology

CAMBRIDGE
UNIVERSITY PRESS

University Printing House, Cambridge CB2 8BS, United Kingdom

One Liberty Plaza, 20th Floor, New York, NY 10006, USA

477 Williamstown Road, Port Melbourne, VIC 3207, Australia

4843/24, 2nd Floor, Ansari Road, Daryaganj, Delhi - 110002, India

79 Anson Road, #06-04/06, Singapore 079906

Cambridge University Press is part of the University of Cambridge.

It furthers the University's mission by disseminating knowledge in the pursuit of education, learning and research at the highest international levels of excellence.

www.cambridge.org
Information on this title: www.cambridge.org/9781107677135

First published 1999
Reprinted 2003
Second edition 2007
3rd printing 2012
First paperback edition 2013

A catalogue record for this publication is available from the British Library

ISBN 978-0-521-83346-2 Hardback
ISBN 978-1-107-67713-5 Paperback

Contents

Preface to the first edition

This is a book on computational methods used in theoretical physics research, with an emphasis on condensed matter applications.

Computational physics is concerned with performing computer calculations and simulations for solving physical problems. Although computer memory and processor performance have increased dramatically over the last two decades, most physical problems are too complicated to be solved without approximations to the physics, quite apart from the approximations inherent in any numerical method. Therefore, most calculations done in computational physics involve some degree of approximation. In this book, emphasis is on the derivation of algorithms and the implementation of these: it is a book which tells you how methods work, why they work, and what the approximations are. It does not contain extensive discussions on results obtained for large classes of different physical systems.

This book is not elementary: the reader should have a background in basic undergraduate physics and in computing. Some background in numerical analysis is also helpful. On the other hand, the topics discussed are not treated in a comprehensive way; rather, this book hopefully bridges the gap between more elementary texts by Koonin, Gould and Giordano, and specialised monographs and review papers on the applications described. The fact that a wide range of topics is included has the advantage that the many similarities in the methods used in seemingly very different fields could be highlighted. Many important topics and applications are however not considered in this book – the material presented obviously reflects my own expertise and interest.

I hope that this book will be useful as a source for intermediate and advanced courses on the subject. I furthermore hope that it will be helpful for graduates and researchers who want to increase their knowledge of the field.

Some variation in the degree of difficulty is inherent to the topics addressed in this book. For example, in molecular dynamics, the equations of motion of a collection of particles are solved numerically, and as such it is a rather elementary subject. However, a careful analysis of the integration algorithms used, the problem of performing these simulations in different statistical ensembles, and the problem of treating long range forces with periodic boundary conditions, are much more difficult. Therefore, sections addressing advanced material are marked with an asterisk (*) – they can be skipped at first reading. Also, extensive theoretical derivations are sometimes moved to sections with asterisks, so that the reader who wants to write programs rather than go into the theory may use the results, taking the derivations for granted.

Aside from theoretical sections, implementations of algorithms are discussed, often in a step-by-step fashion, so that the reader can program the algorithms him- or herself. Suggestions for checking the program are included. In the exercises after each chapter, additional suggestions for programs are given, but there are also exercises in which the computer is not used. The computer exercises are marked by the symbol [**C**]; if the exercise is divided up into parts, this sign occurs before the parts in which a computer program is to be written (a problem marked with [**C**] may contain major parts which are to be done analytically). The programs are not easy to write – most of them took me a long time to complete! Some data-files and numerical routines can be found on www.cambridge.org/9780521833469.

The first person who suggested that I should write this book was Aloysio Janner. Thanks to the support and enthusiasm of my colleague and friend John Inglesfield in Nijmegen, I then started writing a proposal containing a draft of the first hundred pages. After we both moved to the University of Cardiff (UK), he also checked many chapters with painstaking precision, correcting the numerous errors, both in the physics and in the English; without his support, this book would probably never have been completed.

Bill Smith, from Daresbury Laboratories (UK), has checked the chapters on classical many-particle systems and Professor Konrad Singer those on quantum simulation methods. Simon Hands from the University of Swansea (UK) has read the chapter on lattice field theories, and Hubert Knops (University of Nijmegen, The Netherlands) those on statistical mechanics and transfer matrix calculations. Maziar Nekovee (Imperial College, London, UK) commented on the chapter on quantum Monte Carlo methods. I am very grateful for the numerous suggestions and corrections from them all. I am also indebted to Paul Hayman for helping me correcting the final version of the manuscript. Many errors in the book have been pointed out to me by colleagues and students. I thank Professor Ron Cohen in particular for spotting many mistakes and discussing several issues via email.

In writing this book, I have discovered that the acknowledgements to the author's family, often expressed in an apologetic tone as a result of the disruption caused by the writing process to family life, are too real to be disqualified as a cliché. My sons Maurice, Boudewijn and Arthur have in turn disrupted the process of writing in the most pleasant way possible, regularly asking me to show growing trees or fireworks on the screen of my PC, instead of the dull black-on-white text windows. Boudewijn and Maurice's professional imitation of their dad, tapping on the keyboard, and sideways reading formulae, is promising for the future.

It is to my wife Ellen that I dedicate this book, with gratitude for her patience, strength and everlasting support during the long, and sometimes difficult time in which the book came into being.

Preface to the second edition

Six years have passed since the first edition of this book appeared. In these years I have learned a lot more about computational physics – a process which will hopefully never stop. I learned from books and papers, but also from the excellent colleagues with whom I worked on teaching and research during this period. Some of this knowledge has found its place in this edition, which is a substantial extension of the first.

New topics include finite elements, lattice Boltzmann simulation and density matrix renormalisation group, and there are quite a few sections here and there in the book which either give a more in-depth treatment of the material than can be found in the first edition, or extensions to widen the view on the subject matter. Moreover I have tried to eliminate as many errors as possible, but I am afraid that it is difficult for me to beat the entropy of possible things which can go wrong in writing a book of over 650 pages.

In Delft, where I have now a position involving a substantial amount of teaching, I worked for several years in the computational physics group of Simon the Leeuw. I participated in an exciting and enjoyable effort: teaching in an international context. Together with Rajiv Kalia, from Louisana State, we let students from Delft collaborate with Louisiana students, having them do projects in the field of computational physics. Both Simon and Rajiv are experts in the field of molecular dynamics, and I learned a lot from them. Moreover, dealing with students and their questions has often forced me to deepen my knowledge in this field. Similar courses with Hiroshi Iyetomi from Niigata University in Japan, and now with Phil Duxbury at Michigan State have followed, and form my most enjoyable teaching experience. Much of the knowledge picked up in these courses has gone into the new material in this edition.

For one of the new parts of the book, the self-consistent pseudopotential and the Car–Parrinello program, I worked closely together with Erwin de Wolff for a few months. I am grateful for his support in this, and not least for his structured, neat way of tackling the problem.

Many students, university lecturers and researchers have shared their corrections on the text with me. I want to thank Ronald Cohen, Dominic Holland, Ari Harju,

John Mauro, Joachim Stolze and all the others whose names may have disappeared from my hard disks when moving to a new machine.

Preparing this edition in addition to the regular duties of a university position has turned out to be a demanding job, which has prevented me now and then from being a good husband and father. I thank Ellen and my sons Maurice, Boudewijn and Arthur for their patience and support, and express the hope that I will have more time for them in the future.

1

Introduction

1.1 Physics and computational physics

Solving a physical problem often amounts to solving an ordinary or partial differential equation. This is the case in classical mechanics, electrodynamics, quantum mechanics, fluid dynamics and so on. In statistical physics we must calculate sums or integrals over large numbers of degrees of freedom. Whatever type of problem we attack, it is very rare that analytical solutions are possible. In most cases we therefore resort to numerical calculations to obtain useful results. Computer performance has increased dramatically over the last few decades (see also Chapter 16) and we can solve complicated equations and evaluate large integrals in a reasonable amount of time.

Often we can apply numerical routines (found in software libraries for example) directly to the physical equations and obtain a solution. We shall see, however, that although computers have become very powerful, they are still unable to provide a solution to most problems without approximations to the physical equations. In this book, we shall focus on these approximations: that is, we shall concentrate on the development of computational methods (and also on their implementation into computer programs). In this introductory chapter we give a bird's-eye perspective of different fields of physics and the computational methods used to solve problems in these areas. We give examples of direct application of numerical methods but we also give brief and heuristic descriptions of the additional theoretical analysis and approximations necessary to obtain workable methods for more complicated problems which are described in more detail in the remainder of this book. The order adopted in the following sections differs somewhat from the order in which the material is treated in this book.

1.2 Classical mechanics and statistical mechanics

The motion of a point particle in one dimension subject to a force F depending on the particle's position x, and perhaps on the velocity $\dot{x}$ and on time t, is determined by Newton's equation of motion:

$$m\ddot{x}(t) = F[x(t), \dot{x}(t), t]. \tag{1.1}$$

The (double) dot denotes a (double) derivative with respect to time. A solution can be found for each set of initial conditions $x(t_0)$ and $\dot{x}(t_0)$ given at some time t_0. Analytical solutions exist for constant force, for the harmonic oscillator ($F = \kappa x^2/2$), and for a number of other cases. In Appendix A7.1 a simple numerical method for solving this equation is described and this can be applied straightforwardly to arbitrary forces and initial conditions.

Interesting and sometimes surprising physical phenomena can now be studied. As an example, consider the Duffing oscillator [1], with a force given by

$$F[x, \dot{x}, t] = -\gamma\dot{x} + 2ax - 4bx^3 + F_0 \cos(\omega t). \tag{1.2}$$

The first term on the right hand side represents a velocity-dependent friction; the second and third terms are the force a particle feels when it moves in a double potential well $bx^4 - ax^2$, and the last term is an external periodic force. An experimental realisation is a pendulum consisting of an iron ball suspended by a thin string, with two magnets below it. The pendulum and the magnets are placed on a table which is moved back and forth with frequency ω. The string and the air provide the frictional force, the two magnets together with gravity form some kind of double potential well, and, in the reference frame in which the pendulum is at rest, the periodic motion of the table is felt as a periodic force. It turns out that the Duffing oscillator exhibits *chaotic behaviour* for particular values of the parameters γ, a, b, F_0 and ω. This means that the motion itself looks irregular and that a very small change in the initial conditions will grow and result in a completely different motion. Figure 1.1 shows the behaviour of the Duffing oscillator for two nearly equal initial conditions, showing the sensitivity to these conditions. Over the past few decades, chaotic systems have been studied extensively. A system that often behaves chaotically is the weather: the difficulty in predicting the evolution of chaotic systems causes weather forecasts to be increasingly unreliable as they look further into the future, and occasionally to be dramatically wrong.

Another interesting problem is that of several particles, moving in three dimensions and subject to each other's gravitational interaction. Our Solar System is an example. For the simplest nontrivial case of three particles (for two particles, Newton has given the analytical solution), analytical solutions exist for particular configurations, but the general problem can only be solved numerically. This problem is called the *three-body problem* (N-body problem in general). The motion of satellites orbiting in space is calculated numerically using programs for the N-body problem, and the evolution of galaxies is calculated with similar programs using a large number of test particles (representing the stars). Millions of particles can

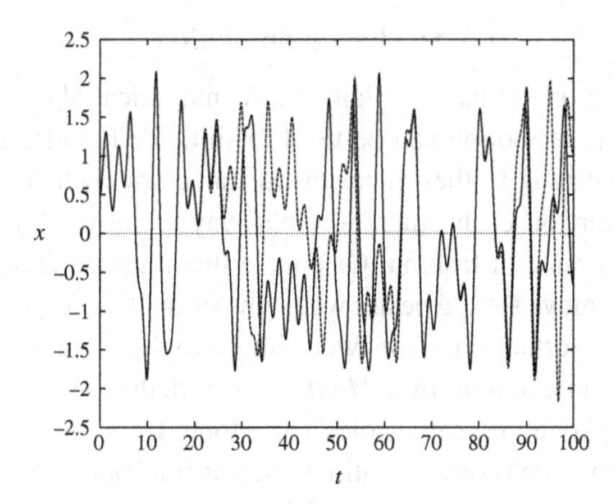

Figure 1.1. Solution of the Duffing oscillator. Parameters are $m = 1$, $a = 1/4$, $b = 1/2$, $F_0 = 2.0$, $\omega = 2.4$, $\gamma = 0.1$. Two solutions are shown: one with initial position $x_0 = 0.5$, the other with $x_0 = 0.5001$ ($\dot{x}_0 = 0$ in both cases). For these nearly equal initial conditions, the solutions soon become uncorrelated, showing the difficulty in predicting the time evolution of a chaotic system.

be treated using a combination of high-end computers and clever computational methods which will be considered in Chapter 8. Electrostatic forces are related to gravitational forces, as both the gravitational and the electrostatic (Coulomb) potential have a $1/r$ form. The difference between the two is that electrostatic forces can be repulsive or attractive, whereas gravitational forces are always attractive.

Neutral atoms interact via a different potential: they attract each other weakly through induced polarisation, unless they come too close – then the Pauli principle causes the electron clouds to repel each other. The problem of many interacting atoms and molecules is a very important subfield of computational physics: it is called *molecular dynamics*. In molecular dynamics, the equations of motion for the particles are solved straightforwardly using numerical algorithms similar to those with which a Duffing oscillator is analysed, the main difference being the larger number of degrees of freedom in molecular dynamics. The aim of molecular dynamics simulations is to predict the behaviour of gases, liquids and solids (and systems in other phases, like liquid crystals). An important result is the equation of state: this is the relation between temperature, number of particles, pressure and volume. Also, the microscopic structure as exhibited by the pair correlation function, which is experimentally accessible via neutron scattering, is an interesting property which can be determined in simulations. There are, however, many problems and pitfalls associated with computer simulations: the systems that can be simulated are always much smaller than realistic systems, and simulating a system at a predefined temperature or chemical potential is nontrivial. All these aspects will be considered in Chapter 8.

1.3 Stochastic simulations

In the previous section we have explained how numerical algorithms for solving Newton's equations of motion can be used to simulate liquids. The particles are moved around according to their mechanical trajectories which are governed by the forces they exert on each other. Another way of moving them around is to displace them in a random fashion. Of course this must be done in a controlled way, and not every move should be allowed, but we shall see in Chapter 10 that it is possible to obtain information in this way similar to that obtained from molecular dynamics. This is an example of a *Monte Carlo* method – procedures in which random numbers play an essential role. The Monte Carlo method is not suitable for studying dynamical physical quantities such as transport coefficients, as it uses artificial dynamics to simulate many-particle systems.

Random number generators can also be used in *direct simulations*: some process of which we do not know the details is replaced by a random generator. If you simulate a card game, for example, the cards are distributed among the players by using random numbers. An example of a direct simulation in physics is diffusion limited aggregation (DLA), which describes the growth of dendritic clusters (see Figure 1.2). Consider a square lattice in two dimensions. The sites of the lattice are either occupied or unoccupied. Initially, only one site in the centre is occupied. We release a random walker from the boundary of the lattice. The walker moves over the lattice in a stepwise fashion. At each step, the walker moves from a site to

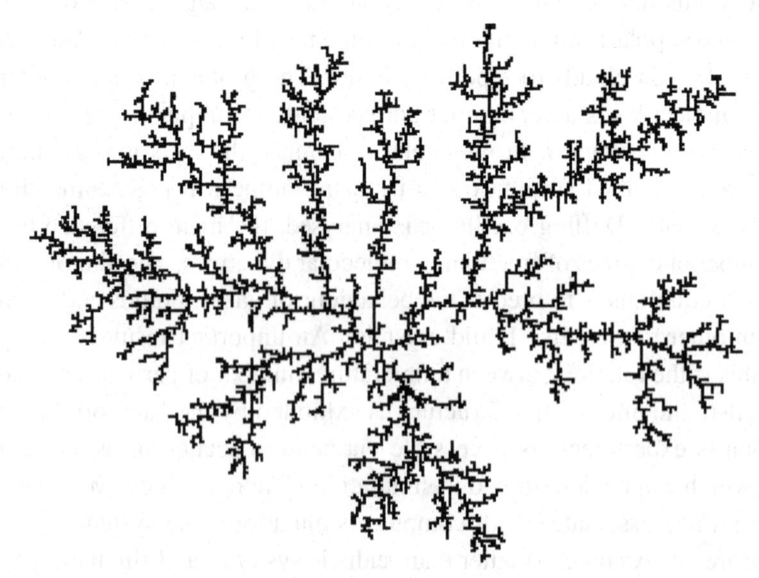

Figure 1.2. Dendritic cluster grown in a DLA simulation. The cluster consists of 9400 sites and it was grown on a 175×175 lattice.

one of its neighbour sites, which is chosen at random (there are four neighbours for each site in the interior of the lattice; the boundary sites have three neighbours, or two if they lie on a corner). If the walker arrives at a site neighbouring the occupied central site, it sticks there, so that a two-site cluster is formed. Then a new walker is released from the boundary. This walker also performs a random walk on the lattice until it arrives at a site neighbouring the cluster of two occupied sites, to form a three-site cluster, and so on. After a long time, a dendritic cluster is formed (see Figure 1.2), which shows a strong resemblance to actual dendrites formed in crystal growth, or by growing bacterial colonies [2], frost patterns on the window and numerous other physical phenomena.

This shows again that interesting physics can be studied by straightforward application of simple algorithms. In Chapter 10 we shall concentrate on the Monte Carlo method for studying many-particle systems at a predefined temperature, volume and particle number. This technique is less direct than DLA, and, just as in molecular dynamics, studying the system for different predefined parameters, such as chemical potential, and evaluating free energies are nontrivial aspects which need further theoretical consideration. The Monte Carlo method also enables us to analyse lattice spin models, which are important for studying magnetism and field theory (see below). These models cannot always be analysed using molecular dynamics methods, and Monte Carlo is often the only tool we have at our disposal in that case. There also exist alternative, more powerful techniques for simulating dendrite formation, but these are not treated in this book.

1.4 Electrodynamics and hydrodynamics

The equations of electrodynamics and hydrodynamics are partial differential equations. There exist numerical methods for solving these equations, but the problem is intrinsically demanding because the fields are continuous and an infinite number of variables is involved. The standard approach is to apply some sort of *discretisation* and consider the solution for the electric potential or for the flow field only on the points of the discrete grid, thus reducing the infinite number of variables to a finite number. Another method of solution consists of writing the field as a linear combination of smooth functions, such as plane waves, and solving for the best values of the expansion coefficients.

There exist several methods for solving partial differential equations: finite difference methods (FDM), finite element methods (FEM), Fourier transform methods and multigrid methods. These methods are also very often used in engineering problems, and are essentially the domain of numerical analysis. The finite element method is very versatile and therefore receives our particular attention in Chapter 13. The other methods can be found in Appendix A7.2.

1.5 Quantum mechanics

In quantum mechanics we regularly need to solve the Schrödinger equation for one or more particles. There is usually an external potential felt by the particles, and in addition there might be interactions between the particles. For a single particle moving in one dimension, the stationary form of the Schrödinger equation reduces to an ordinary differential equation, and techniques similar to those used in solving Newton's equations can be used. The main difference is that the stationary Schrödinger equation is an eigenvalue equation, and in the case of a discrete spectrum, the energy eigenvalue must be varied until the wave function is physically acceptable, which means that it matches some boundary conditions and is normalisable. Examples of this direct approach are discussed in Appendix A, in particular Problem A4.

In two and more dimensions, or if we have more than one particle, or if we want to solve the time-dependent Schrödinger equation, we must solve a partial differential equation. Sometimes, the particular geometry of the problem and the boundary conditions allow us to reduce the complexity of the problem and transform it into ordinary differential equations. This will be done in Chapter 2, where we shall study particles scattering off a spherically symmetric potential.

Among the most important quantum problems in physics is the behaviour of electrons moving in the field generated by nuclei, which occurs in atoms, molecules and solids. This problem is treated quite extensively in this book, but the methods we develop for it are also applied in nuclear physics. Solving the Schrödinger equation for one electron moving in the potential generated by the atomic static nuclei is already a difficult problem, as it involves solving a partial differential equation. Moreover, the potential is strong close to the nuclei and weak elsewhere, so the typical length scale of the wave function varies strongly through space. Therefore, discretisation methods must use grids which are finer close to the nuclei, rendering such methods difficult. The method of choice is, in fact, to expand the wave function as a linear combination of fixed basis functions that vary strongly close to the nuclei and are smooth elsewhere, and find the optimal values for the expansion coefficients. This is an example of the *variational method*, which will be discussed in Chapter 3. This application of the variational method leads to a matrix eigenvalue problem which can be solved very efficiently on a computer.

An extra complication arises when there are many (say N) electrons, interacting via the Coulomb potential, so that we must solve a partial differential equation in $3N$ dimensions. In addition to this we must realise that electrons are fermions and the many-electron wave function must therefore be antisymmetric with respect to exchange of any pair of electrons. Because of the large number of dimensions, solving the Schrödinger equation is not feasible using any of the standard numerical methods for solving partial differential equations, so we must make approximations. One approach is the Hartree–Fock (HF) method,developed in the early days of quantum mechanics, which takes into account the antisymmetry of the many-electron wave function. This leads to an independent particle picture, in which each electron moves in the potential generated by the nuclei plus an average potential generated by the other electrons. The latter depends on the electronic wave functions, and hence the problem must be solved *self-consistently* – in Chapter 4 we shall see how this is done. The HF method leads to wave functions that are fully antisymmetric, but contributions arising from the Coulomb interaction between the particles are taken into account in an approximate way, analogous to the way correlations are treated in the mean field approach in statistical mechanics.

Another approach to the quantum many-electron problem is given by *density functional theory* (DFT), which will be discussed in Chapter 5. This theory, which is in principle exact, can in practice only be used in conjunction with approximate schemes to be discussed in Chapter 5, the most important of which is the *local density approximation* (LDA). This also leads to an independent-particle Schrödinger equation, but in this case, the correlation effects resulting from the antisymmetry of the wave function are not incorporated exactly, leading to a small, unphysical interaction of an electron with itself (*self-interaction*). However, in contrast to Hartree–Fock,the approach does account (in an approximate way) for the dynamic correlation effects due to the electrons moving out of each other's way as a result of the Coulomb repulsion between them.

All these approaches lead in the end to a matrix eigenvalue problem, whose size depends on the number of electrons present in the system. The resulting solutions enable us to calculate total energies and excitation spectra which can be compared with experimental results.

1.6 Relations between quantum mechanics and classical statistical physics

In the previous two sections we have seen that problems in classical statistical mechanics can be studied with Monte Carlo techniques, using random numbers, and that the solution of quantum mechanical problems reduces to solving matrix eigenvalue problems. It turns out that quantum mechanics and classical statistical mechanics are related in their mathematical structure. Consider for example the partition function for a classical mechanics system at temperature T, with degrees of freedom denoted by the variable X and described by an energy function (that is, a classical Hamiltonian) $\mathcal{H}$:

$$Z_{\mathrm{Cl}} = \sum_X \mathrm{e}^{-\mathcal{H}(X)/(k_\mathrm{B}T)}, \tag{1.3}$$

and that of a quantum system with quantum Hamiltonian H:

$$Z_{QM} = \mathrm{Tr}(\mathrm{e}^{-H/(k_B T)}); \tag{1.4}$$

'Tr' denotes the trace of the operator following it. We will show in Chapter 12 that in the path-integral formalism, the second expression can be transformed into the same form as the first one. Also, there is a strong similarity between the exponent occurring in the quantum partition function and the quantum time-evolution operator $U(t) = \exp(-\mathrm{i}tH/\hbar)$, so solving the time evolution of a quantum system is equivalent to evaluating a classical or quantum partition function, the difference being an imaginary factor $\mathrm{i}t/\hbar$ replacing the real factor $1/(k_B T)$, and taking the trace in the case of the quantum partition function rather than a sum over states in the classical analogue.

These mathematical analogies suggest that numerical methods for either classical statistical mechanics or quantum mechanics are applicable in both fields. Indeed, in Chapter 11, we shall see that it is possible to analyse classical statistical spin problems on lattices by diagonalising large matrices. In Chapter 12, on the other hand, we shall use Monte Carlo methods for solving quantum problems. These methods enable us to treat the quantum many-particle problem without systematic approximations, because, as will be shown in Chapter 12, Monte Carlo techniques are very efficient for calculating integrals in many dimensions. This, as we have seen above, was precisely the problem arising in the solution of interacting many-particle systems.

1.7 Quantum molecular dynamics

Systems of many interacting atoms or molecules can be studied classically by solving Newton's equations of motion, as is done in molecular dynamics. Pair potentials are often used to describe the atomic interactions, and these can be found from quantum mechanical calculations, using Hartree–Fock,density functional theory or quantum Monte Carlo methods. In a dense system, the pair potential is inadequate as the interactions between two particles in the system are influenced by other particles. In order to incorporate these effects in a simulation, it would be necessary to calculate the forces from full electronic structure calculations for all configurations occurring in the simulation. Car and Parrinello have devised a clever way to calculate these forces as the calculation proceeds, by combining density functional theory with molecular dynamics methods.

In the Car–Parrinello approach, electron correlations are not treated exactly because of the reliance on LDA (see Section 1.5), but it will be clear that it is an important improvement on fully classical simulations where the interatomic interactions are described by a simple form, such as pair potentials. It is possible to include some damping mechanism in the equations of motion and then let the nuclei relax to their ground state positions, so that equilibrium configurations of molecules and solids can be determined (neglecting quantum fluctuations).

1.8 Quantum field theory

Quantum field theory provides a quantum description for fields: strings in one dimension, sheets in two dimensions, etc. Quantum field theory is also believed to describe elementary particles and their interactions. The best known example is quantum electrodynamics (QED) which gives a very accurate description of the interaction between charged spin-$1/2$ fermions (electrons) and electromagnetic fields. The results of QED are obtained using perturbation theory which works very well for this case, because the perturbative parameter remains small for all but the smallest length scales (at large length scales this is the fine structure constant).

In quantum chromodynamics (QCD), the theory which supposedly describes quarks bound together in a proton or neutron, the coupling constant grows large for large scales, and perturbation theory breaks down. One way to obtain useful results for this theory is to discretise space-time, and simulate the theory on this space-time lattice on a computer. This can be done using Monte Carlo or molecular dynamics techniques. The application of these techniques is far from easy as the QCD field theory is intrinsically complicated. A problem which needs to be addressed is efficiency, notably overcoming *critical slowing down*, which decreases the efficiency of simple Monte Carlo and molecular dynamics techniques for the cases which are of physical interest. The fact that quarks are fermions leads to additional complications.

QCD simulations relate quark masses to masses and interaction constants of hadrons (mesons, protons, neutrons).

1.9 About this book

In this book, the emphasis is on methods which do not merely involve straightforward application of numerical methods, and which are specific to problems studied in physics. In most cases, the theory is treated in some detail in order to exhibit clearly what the approximations are and why the methods work. However, some of this theoretical material can be skipped at first reading (this is the material in the sections marked with an asterisk *). Details on implementation are given for most of the methods described.

We start off with a chapter on quantum mechanical scattering theory. This is a rather straightforward application of numerical techniques, and is used as an illustration of solving a rather simple (not completely trivial) physical problem on a computer. The results of a sample program are compared with experiment. In Chapters 3 to 5 we discuss computational methods for the electronic structure: variational calculus, Hartree–Fock and density functional theory. We apply these methods to some simple systems: the hydrogen and the helium atoms, and the hydrogen molecule. We calculate the energies of these systems. Chapter 6 deals with solving the independent-particle Schrödinger equation in solids.

In Chapters 7 to 12 we describe molecular dynamics and Monte Carlo techniques for classical and quantum many-particle systems. Chapter 7 contains an overview of classical statistical mechanics, with emphasis on ensembles and on critical phenomena, which are also important for field theory, as discussed in Chapter 15. The molecular dynamics and Monte Carlo techniques are treated in Chapters 8 and 10. The standard example of a molecular liquid, argon, is analysed, but simulations for liquid nitrogen and for lattice spin systems (Ising model) are also discussed. Chapter 9 deals with the quantum molecular dynamics technique.

The relations between classical and statistical mechanics are exploited in Chapter 11 where the transfer matrix method for lattice spin systems is described. The next chapter deals with the application of Monte Carlo methods to quantum mechanics, and we revisit the helium atom which is now treated without Hartree–Fock or DFT approximations.

In Chapter 15 we consider numerical methods for field theory. Techniques for analysing the simplest interesting field theory, the scalar ϕ^4 theory, are studied, and methods for studying more complicated field theories (QED and QCD) are discussed. Because of the relation between statistical and quantum mechanics, some of the techniques discussed in this chapter are also relevant for classical statistical mechanics.

Finally, in Chapter 16 modern computer architectures are briefly considered and an example of a parallel algorithm for molecular dynamics is given.

The algorithms presented, and the programs to be written in the exercises, can be coded in different languages: C, C++, Java, Fortran 77, Fortran 90 etc. Also, an integrated scientific computer environment such as MatLab may be used. They all have their pluses and minuses: Fortran 77 allows for dirty programming, but is quite efficient, and the same holds for C; Fortran 90 is efficient and neat. MatLab is easy to use, but not as efficient as using a high-level programming language. Perhaps the most structured way of programming is by using the objected-oriented programming paradigm, as implemented in the langauges C++ and Java. For large and complex projects, these languages are unbeatable. However, for smaller jobs MatLab or Fortran 90 is usually sufficient. It is my experience that students relatively new to programming get their programs to run correctly most quickly when using Fortran 90 or MatLab.

Perhaps the most decisive criterion for choosing a particular language is whether you have experience with it. There is no emphasis on any of these languages in this book: they are all suitable for writing the numerical types of programs considered here. If asked for a recommendation, I would not hesitate to advocate Fortran 90 as the most suitable language for inexperienced programmers. Students with substantial programming skills are probably better off using C++ or Java.

It is hoped that in the future it might be easier to move around pieces of software and embed them in new programs, using graphical user interfaces (GUIs). Object-oriented programming (OOP) techniques will play a major role in this development. Ideally, the programs in this book could be built using a set of building blocks which are graphically connected by the programmer to form programs which include the numerical work along with a user-friendly input/output environment.

Exercises

1.1 [C] In this problem it is assumed that you have at your disposal a routine for solving an ordinary second order differential equation with given initial conditions (see also Appendix A7.1).

(a) Write a program for the Duffing oscillator [see Eqs. (1.1) and (1.2)] and study the motion for different sets of values of the parameters F_0, ω, γ (use $m = 1$, $a = 1/4$, $b = 1/2$) and initial conditions x_0 and $\dot{x}_0$.

(b) For the same values of the parameters, print out the values of x and $p = \dot{x}$ each time a period $T = 2\pi/\omega$ has elapsed. Plotting these points should yield a structure like Figure 1.3. The resulting curve is called the *strange attractor*.

It is possible to assign a dimension to such a structure. This is done by covering it with grids of different sizes, and counting the number of squares $N(b)$ needed to cover the structure with squares of size $b \times b$. This number then scales with b as

$$N(b) \propto b^{-D_\mathrm{f}}, \text{ small } b,$$

and the exponent D_f is then the dimension. This is not always an integer number, and if this is indeed not the case, we call the dimension, and the corresponding structure, *fractal*.

(c) Argue that the dimension of a line determined in this way is equal to one, and that that of a surface area is equal to two.

(d) Write a program to determine the fractal dimension of the strange attractor constructed above. This proceeds as follows. First, read all the points of the attractor into an array. The attractor lies in the square of side 6 centred at the origin. Divide this square up into $l \times l$ cells, where l is first taken to be 2, then 4, and so on up to $l = 128$. The side of the cells is then $b = 6/l$. A Boolean array of size at least 128 by 128 assumes the value `TRUE` for cells which contain a point of the attractor and `FALSE` if that is not the case. Fill this array by performing a loop over the points on the attractor, and then count the number of filled cells $N(b)$.

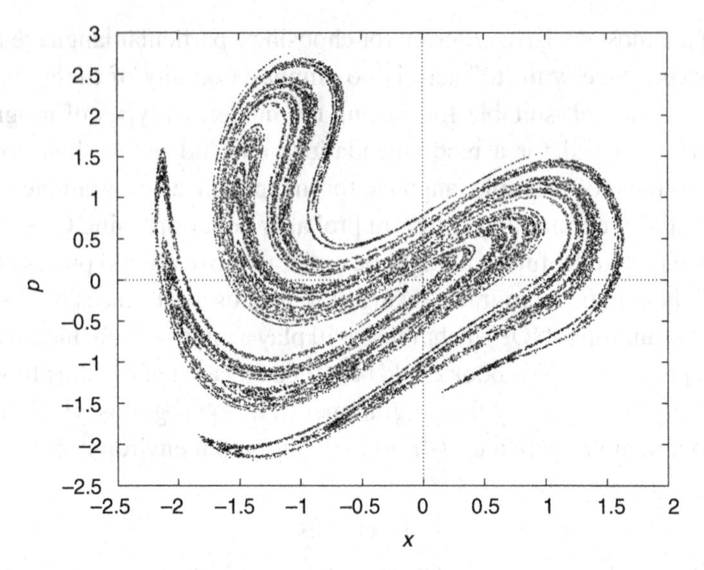

Figure 1.3. Strange attractor for the Duffing oscillator. Values of the parameters are $F_0 = 2.0$, $\omega = 2.4$, $\gamma = 0.1$. The initial conditions are $x_0 = 0.5$, $\dot{x}_0 = 0$.

The results $\log[N(b)]$ and $\log(b)$ should be written to a file. For an attractor of 25 000 points, the resulting points lie more or less on a straight line with slope $-D_f \approx -1.68$, for $2 \leq l \leq 7$.

1.2 [C] In this problem, we consider diffusion limited aggregation.

(a) Write a program for generating DLA clusters on a square lattice of size 150×150 (see Section 1.3). Generate a cluster of about 9000 sites, and write the sites occupied by this cluster to a file for viewing using a graphics program.

(b) Another definition of the fractal dimension (see Problem 1.1) is obtained by relating the number of sites N of the cluster to its *radius of gyration*, defined by

$$R_g = \frac{1}{N}\sum_{i=1}^{N}(\mathbf{r}_i - \mathbf{r}_0)^2,$$

where

$$\mathbf{r}_0 = \frac{1}{N}\sum_{i=1}^{N}\mathbf{r}_i$$

is the 'centre of mass' of the cluster. Show that the radius of gyration can be rewritten as

$$R_g = \frac{1}{N}\left(\sum_{i=1}^{N}\mathbf{r}_i^2\right) - \mathbf{r}_0^2.$$

Use this formula to calculate the radius of gyration after every 200 newly added sites, and write the values $\log(R_g)$, $\log(N)$ to a file. Plot this file and fit the results to a straight line. The slope of this line is then the fractal dimension which must be about 1.7.

References

[1] J. Awrejcewicz, 'On the occurrence of chaos in Duffing's oscillator,' *J. Sound Vibr.*, **108** (1986), 176–8.

[2] E. Ben-Jacob, I. Cohen, O. Sochet, *et al.* 'Complex bacterial patterns,' *Nature*, **373** (1986), 566–7.

2

Quantum scattering with a spherically symmetric potential

2.1 Introduction

In this chapter, we shall discuss quantum scattering with a spherically symmetric potential as a typical example of the problems studied in computational physics [1, 2]. Scattering experiments are perhaps the most important tool for obtaining detailed information on the structure of matter, in particular the interaction between particles. Examples of scattering techniques include neutron and X-ray scattering for liquids, atoms scattering from crystal surfaces and elementary particle collisions in accelerators. In most of these scattering experiments, a beam of incident particles hits a target which also consists of many particles. The distribution of scattered particles over the different directions is then measured, for different energies of the incident particles. This distribution is the result of many individual scattering events. Quantum mechanics enables us, in principle, to evaluate for an individual event the probabilities for the incident particles to be scattered off in different directions; and this probability is identified with the measured distribution.

Suppose we have an idea of what the potential between the particles involved in the scattering process might look like, for example from quantum mechanical energy calculations (programs for this purpose will be discussed in the next few chapters). We can then *parametrise* the interaction potential, i.e. we write it as an analytic expression involving a set of constants: the parameters. If we evaluate the scattering probability as a function of the scattering angle for different values of these parameters, and compare the results with experimental scattering data, we can find those parameter values for which the agreement between theory and experiment is optimal. Of course, it would be nice if we could evaluate the scattering potential directly from the scattering data (this is called the *inverse problem*), but this is unfortunately very difficult (if not impossible): many different interaction potentials can have similar scattering properties, as we shall see below.

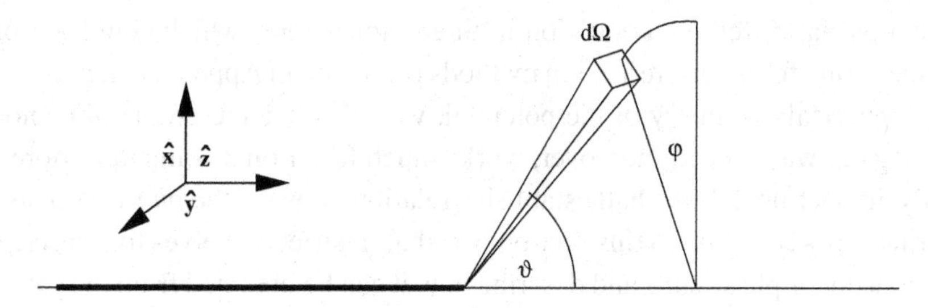

Figure 2.1. Geometry of a scattering process.

There might be many different motives for obtaining accurate interaction potentials. One is that we might use the interaction potential to make predictions about the behaviour of a system consisting of many interacting particles, such as a dense gas or a liquid. Methods for doing this will be discussed in Chapters 8 and 10.

Scattering may be *elastic* or *inelastic*. In the former case the energy is conserved, in the latter it disappears. This means that energy transfer takes place from the scattered particles to degrees of freedom which are not included explicitly in the system (inclusion of these degrees of freedom would cause the energy to be conserved). In this chapter we shall consider elastic scattering. We restrict ourselves furthermore to spherically symmetric interaction potentials. In Chapter 15 we shall briefly discuss scattering in the context of quantum field theory for elementary particles.

We analyse the scattering process of a particle incident on a scattering centre which is usually another particle.[1] We assume that we know the scattering potential, which is spherically symmetric so that it depends on the distance between the particle and the scattering centre only.

In an experiment, one typically measures the scattered flux, that is, the intensity of the outgoing beam for various directions which are denoted by the spatial angle $\Omega = (\theta, \varphi)$ as in Figure 2.1. The *differential cross section*, $\mathrm{d}\sigma(\Omega)/\mathrm{d}\Omega$, describes how these intensities are distributed over the various spatial angles Ω, and the integrated flux of the scattered particles is the *total cross section*, σ_{tot}. These experimental quantities are what we want to calculate.

The scattering process is described by the solutions of the single-particle Schrödinger equation involving the (reduced) mass m, the relative coordinate $\mathbf{r}$ and the interaction potential V between the particle and the interaction centre:

$$\left[-\frac{\hbar^2}{2m}\nabla^2 + V(r)\right]\psi(\mathbf{r}) = E\psi(\mathbf{r}). \tag{2.1}$$

[1] Every two-particle collision can be transformed into a single scattering problem involving the relative position; in the transformed problem the incoming particle has the reduced mass $m = m_1 m_2/(m_1 + m_2)$.

This is a partial differential equation in three dimensions, which could be solved using the 'brute force' discretisation methods presented in Appendix A, but exploiting the spherical symmetry of the potential, we can solve the problem in another, more elegant, way which, moreover, works much faster on a computer. More specifically, in Section 2.3 we shall establish a relation between the *phase shift* and the scattering cross sections. In this section, we shall restrict ourselves to a description of the concept of phase shift and describe how it can be obtained from the solutions of the radial Schrödinger equation. The expressions for the scattering cross sections will then be used to build the computer program which is described in Section 2.2.

For the potential $V(r)$ we make the assumption that it vanishes for r larger than a certain value $r_{\max}$. If we are dealing with an asymptotically decaying potential, we neglect contributions from the potential beyond the range $r_{\max}$, which must be chosen suitably, or treat the tail in a perturbative manner as described in Problem 2.2.

For a spherically symmetric potential, the solution of the Schrödinger equation can always be written as

$$\psi(\mathbf{r}) = \sum_{l=0}^{\infty} \sum_{m=-l}^{l} A_{lm} \frac{u_l(r)}{r} Y_l^m(\theta, \varphi) \tag{2.2}$$

where u_l satisfies the radial Schrödinger equation:

$$\left\{ \frac{\hbar^2}{2m} \frac{\mathrm{d}^2}{\mathrm{d}r^2} + \left[E - V(r) - \frac{\hbar^2 l(l+1)}{2mr^2} \right] \right\} u_l(r) = 0. \tag{2.3}$$

Figure 2.2 shows the solution of the radial Schrödinger equation with $l = 0$ for a square well potential for various well depths – our discussion applies also to nonzero values of l. Outside the well, the solution u_l can be written as a linear combination of the two independent solutions j_l and n_l, the regular and irregular spherical Bessel functions. We write this linear combination in the particular form

$$u_l(r > r_{\max}) \propto kr[\cos \delta_l j_l(kr) - \sin \delta_l n_l(kr)]; \tag{2.4}$$

$$k = \sqrt{2mE}/\hbar.$$

Here $r_{\max}$ is the radius of the well, and δ_l is determined via a matching procedure at the well boundary. The motivation for writing u_l in this form follows from the asymptotic expansion for the spherical Bessel functions:

$$krj_l(kr) \approx \sin(kr - l\pi/2) \tag{2.5a}$$

$$krn_l(kr) \approx -\cos(kr - l\pi/2) \tag{2.5b}$$

which can be used to rewrite (2.4) as

$$u_l(r) \propto \sin(kr - l\pi/2 + \delta_l), \quad \text{large } r. \tag{2.6}$$

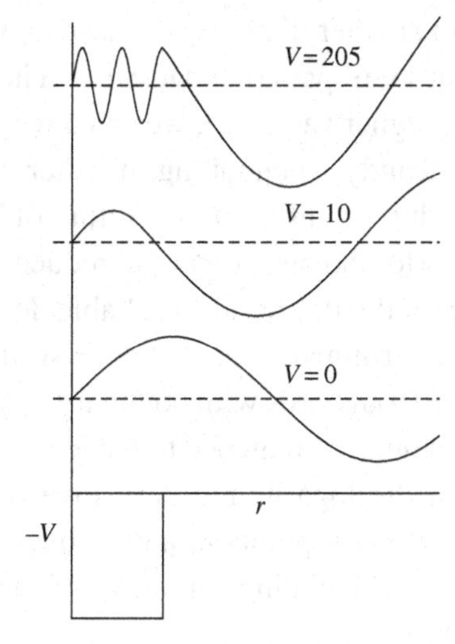

Figure 2.2. The radial wave functions for $l = 0$ for various square well potential depths.

We see that u_l approaches a sine-wave form for large r and the phase of this wave is determined by δ_l, hence the name 'phase shift' for δ_l (for $l = 0$, u_l is a sine wave for all $r > r_{\text{max}}$).

The phase shift as a function of energy and l contains all the information about the scattering properties of the potential. In particular, the phase shift enables us to calculate the scattering cross sections and this will be done in Section 2.3; here we simply quote the results. The differential cross section is given in terms of the phase shift by

$$\frac{\mathrm{d}\sigma}{\mathrm{d}\Omega} = \frac{1}{k^2}\left|\sum_{l=0}^{\infty}(2l+1)\mathrm{e}^{\mathrm{i}\delta_l}\sin(\delta_l)P_l(\cos\theta)\right|^2 \tag{2.7}$$

and for the total cross section we find

$$\sigma_{\text{tot}} = 2\pi\int \mathrm{d}\theta \sin\theta \frac{\mathrm{d}\sigma}{\mathrm{d}\Omega}(\theta) = \frac{4\pi}{k^2}\sum_{l=0}^{\infty}(2l+1)\sin^2\delta_l. \tag{2.8}$$

Summarising the analysis up to this point, we see that the potential determines the phase shift through the solution of the Schrödinger equation for $r < r_{\text{max}}$. The phase shift acts as an intermediate object between the interaction potential and the experimental scattering cross sections, as the latter can be determined from it.

Unfortunately, the expressions (2.7) and (2.8) contain sums over an infinite number of terms – hence they cannot be evaluated on the computer exactly. However, there is a physical argument for cutting off these sums. Classically, only those waves with an angular momentum smaller than $\hbar l_{\max} = \hbar k r_{\max}$ will 'feel' the potential – particles with higher l-values will pass by unaffected. Therefore we can safely cut off the sums at a somewhat higher value of l; we can always check whether the results obtained change significantly when taking more terms into account. We shall frequently encounter procedures similar to the cutting off described here. It is the art of computational physics to find clever ways to reduce infinite problems to ones which fit into the computer and still provide a reliable description.

How is the phase shift determined in practice? First, the Schrödinger equation must be integrated from $r = 0$ outwards with boundary condition $u_l(r = 0) = 0$. At $r_{\max}$, the numerical solution must be matched to the form (2.4) to fix δ_l. The matching can be done either via the logarithmic derivative or using the value of the numerical solution at two different points r_1 and r_2 beyond $r_{\max}$. We will use the latter method in order to avoid calculating derivatives. From (2.4) it follows directly that the phase shift is given by

$$\tan \delta_l = \frac{K j_l^{(1)} - j_l^{(2)}}{K n_l^{(1)} - n_l^{(2)}} \quad \text{with} \tag{2.9a}$$

$$K = \frac{r_1 u_l^{(2)}}{r_2 u_l^{(1)}}. \tag{2.9b}$$

In this equation, $j_l^{(1)}$ stands for $j_l(kr_1)$ etc.

2.2 A program for calculating cross sections

In this section we describe the construction of a program for calculating cross sections for a particular scattering problem: hydrogen atoms scattered off (much heavier) krypton atoms. Both atoms are considered as single particles and their structure (nucleus and electrons) is not explicitly taken into account. After completion, we are able to compare the results with experimental data. The program described here closely follows the work of Toennies *et al.* who carried out various atomic collisions experimentally and modelled the results using a similar computer program [3].

The program is built up in several steps.

- First, the integration method for solving the radial Schrödinger equation is programmed. Various numerical methods can be used; we consider in particular Numerov's method (see Appendix A7.1).
- Second, we need routines yielding spherical Bessel functions in order to determine the phase shift via the matching procedure Eq. (2.9a). If we want to

calculate differential cross sections, we need Legendre polynomials too. In Appendix A2, iterative methods for evaluating special functions are discussed.

- Finally, we complete the program with a routine for calculating the cross sections from the phase shifts.

2.2.1 Numerov's algorithm for the radial Schrödinger equation

The radial Schrödinger equation is given in Eq. (2.3). We define

$$F(l, r, E) = V(r) + \frac{\hbar^2 l(l+1)}{2mr^2} - E \tag{2.10}$$

so that the radial Schrödinger equation now reads:

$$\frac{\hbar^2}{2m}\frac{\mathrm{d}^2}{\mathrm{d}r^2}u(r) = F(l, r, E)u(r). \tag{2.11}$$

Units are chosen in which $\hbar^2/(2m)$ assumes a reasonable value, that is, not extremely large and not extremely small (see below). You can choose a library routine for integrating this equation but if you prefer to write one yourself, Numerov's method is a good choice because it combines the simplicity of a regular mesh with good efficiency. The Runge–Kutta method can be used if you want to have the freedom of varying the integration step when the potential changes rapidly (see Problem 2.1).

Numerov's algorithm is described in Appendix A7.1. It makes use of the special structure of this equation to solve it with an error of order h^6 (h is the discretisation interval) using only a three-point method. For $\hbar^2/2m \equiv 1$ it reads:

$$w(r+h) = 2w(r) - w(r-h) + h^2 F(l, r, E)u(r) \tag{2.12}$$

and

$$u(r) = \left[1 - \frac{h^2}{12}F(l, r, E)\right]^{-1} w(r). \tag{2.13}$$

It is useful to keep several things in mind when coding this algorithm.

- The function $F(l, r, E)$, consisting of the energy, potential and centrifugal barrier, given in Eq. (2.10), is coded into a function `F(L,R,E)`, with `L` an integer and `R` and `E` being real variables.
- As you can see from Eq. (2.9a), the value of the wave function is needed for two values of the radial coordinate r, both beyond r_{max}. We can take r_1 equal to the first integration point beyond r_{max} (if the grid constant h for the integration fits an integer number of times into r_{max}, it is natural to take $r_1 = r_{max}$). The value of r_2 is larger than r_1 and it is advisable to take it roughly half a wavelength beyond the latter. The wavelength is given by $\lambda = 2\pi/k = 2\pi\hbar/\sqrt{2mE}$. As

both r_1 and r_2 are equal to an integer times the integration step h (they will in general not differ by exactly half a wavelength) the precise values of r_1 and r_2 are determined in the routine and output to the appropriate routine parameters.

- The starting value at $r = 0$ is given by $u(r = 0) = 0$. We do not know the value of the derivative, which determines the normalisation of the resulting function – this normalisation can be determined afterward. We take $u_l(0) = 0$ and $u_l(h) = h^{l+1}$ (h is the integration step), which is the asymptotic approximation for u_l near the origin for a *regular* potential (for the H–Kr interaction potential which diverges strongly near the origin, we must use a different boundary condition as we shall see below).

PROGRAMMING EXERCISE

Write a code for the Numerov algorithm. The input parameters to the routine must include the integration step h, the radial quantum number l, the energy E and the radial coordinate r_{max}; on output it yields the coordinates r_1 and r_2 and the values of the wave function $u_l(r_1)$ and $u_l(r_2)$.

When building a program of some complexity, it is very important to build it up step by step and to check every routine extensively. Comparison with analytical solutions is then of prime importance. We now describe several checks that should be performed after completion of the Numerov routine (it is also sensible to test a library routine).

Check 1 The numerical solutions can be compared with analytical solutions for the case of the three-dimensional harmonic oscillator. Bound states occur for energies $E = \hbar\omega(n + 3/2), n = 0, 1, 2, \ldots$ It is convenient in this case to choose units such that $\hbar^2/2m = 1$. Taking $V(r) = r^2$, we have $\hbar\omega = 2$ and the lowest state occurs for $l = 0$ with energy $E = 3.0$, with eigenfunction $Ar\exp(-r^2/2)$, A being some constant. Using $E = 3.0$ in our numerical integration routine should give us this solution with $A = \exp(h^2/2)$ for the starting conditions described above. Check this for r-values up to r_2.

Check 2 The integration method has an error of $\mathcal{O}(h^6)$ (where $\mathcal{O}$ indicates 'order'). The error found at the end of a finite interval then turns out to be less than $\mathcal{O}(h^4)$ (see Problem A3). This can be checked by comparing the numerical solution for the harmonic oscillator with the exact one. Carry out this comparison for several values of N, for example $N = 4, 8, 16, \ldots$ For N large enough, the difference between the exact and the numerical solution should decrease for each new value of N by a factor of at least 16. If your program does not yield this behaviour, there must be an error in the code!

We shall now turn to the H–Kr interaction. The two-atom interaction potential for atoms is often modelled by the so-called Lennard–Jones (LJ) potential, which has the following form:

$$V_{\mathrm{LJ}}(r) = \varepsilon \left[\left(\frac{\rho}{r} \right)^{12} - 2 \left(\frac{\rho}{r} \right)^{6} \right]. \tag{2.14}$$

This form of potential contains two parameters, ε and ρ, and for H–Kr the best values for these are

$$\varepsilon = 5.9\,\mathrm{meV} \quad \text{and} \quad \rho = 3.57\,\text{Å}. \tag{2.15}$$

Note that the energies are given in *milli*-electronvolts! In units of meV and ρ for energy and distance respectively, the factor $2m/\hbar^2$ is equal to about $6.12\,\mathrm{meV}^{-1}\rho^{-2}$. The potential used by Toennies *et al.* [3] included small corrections to the Lennard–Jones shape.

For the Lennard–Jones potential the integration of the radial Schrödinger equation gives problems for small r because of the $1/r^{12}$ divergence at the origin. We avoid integrating in this region and start at a nonzero radius r_{min} where we use the analytic approximation of the solution for small r to find the starting values of the numerical solution. For $r < r_{\mathrm{min}}$, the term $1/r^{12}$ dominates the other terms in the potential and the energy, so that the Schrödinger equation reduces to

$$\frac{\mathrm{d}^2 u}{\mathrm{d}r^2} = \varepsilon\alpha \frac{1}{r^{12}} u(r) \tag{2.16}$$

with $\alpha = 6.12$. The solution of this equation is given by

$$u(r) = \exp(-Cr^{-5}) \tag{2.17}$$

with $C = \sqrt{\varepsilon\alpha/25}$. This fixes the starting values of the numerical solution at r_{min} which should be chosen such that it can safely be assumed that the $1/r^{12}$ dominates the remaining terms in the potential; typical values for the starting value of r lie between 0.5ρ and 0.8ρ (the minimum of the Lennard–Jones potential is found at $r = 2$). Note that Eq. (2.17) provides the starting value and derivative of the wavefunction u at the starting point. In Appendix A7.1 a procedure is described by which two consecutive values can then be found which, when used as the starting values of the Numerov method, provide a solution with the proper accuracy. This will not be the case when two consecutive points are simply set to the solution Eq. (2.17), as this is not an exact solution to either the continuum differential equation or to its discrete (Numerov) form.

You can adapt your program to the problem at hand by simply changing the function $F(l, r, E)$ to contain the Lennard–Jones potential and by implementing the boundary conditions as described. As a check, you can verify that the solution does not become enormously large or remain very small.

2.2.2 The spherical Bessel functions

For the present problem, you need only the first six spherical Bessel functions j_l and n_l, and you can type in the explicit expressions directly. If you want a general routine for the spherical Bessel functions, however, you can use the recursive procedures described in Appendix A (see also Problem A1). Although upward recursion can be unstable for j_l (see Appendix A), this is not noticeable for the small l values (up to $l = 6$) that we need and you can safely use the simple upward recursion for *both* n_l and j_l (or use a library routine).

PROGRAMMING EXERCISE

Write routines for generating the values of the spherical Bessel functions j_l and n_l. On input, the values of l and the argument x are specified and on output the value of the appropriate Bessel function is obtained.

Check 3 If your program is correct, it should yield the values for j_5 and n_5 given in Problem A1.

2.2.3 Putting the pieces together: results

To obtain the scattering cross sections, some extra routines must be added to the program. First of all, the phase shift must be extracted from the values r_1, $u(r_1)$ and r_2, $u(r_2)$. This is straightforward using Eq. (2.9a). The total cross section can then readily be calculated using Eq. (2.8). The choice of r_{max} must be made carefully, preferably keeping the error of the same order as the $\mathcal{O}(h^6)$ error of the Numerov routine (or the error of your library routine). In Problem 2.2 it is shown that the deviation in the phase shift caused by cutting off the potential at r_{max} is given by

$$\Delta\delta_l = -\frac{2m}{\hbar^2}k\int_{r_{\text{max}}}^{\infty} j_l^2(kr)V_{\text{LJ}}(r)r^2\text{d}r \tag{2.18}$$

and this formula can be used to estimate the resulting error in the phase shift or to improve the value found for it with a potential cut-off beyond r_{max}. A good value is $r_{\text{max}} \approx 5\rho$.

For the determination of the differential cross section you will need additional routines for the Legendre polynomials.[2] In the following we shall only describe results for the total cross section.

PROGRAMMING EXERCISE

Add the necessary routines to the ones you have written so far and combine them into a program for calculating the total cross section.

[2] These can be generated using the recursion relation $(l+1)P_{l+1}(x) = (2l+1)xP_l(x) - lP_{l-1}(x)$.

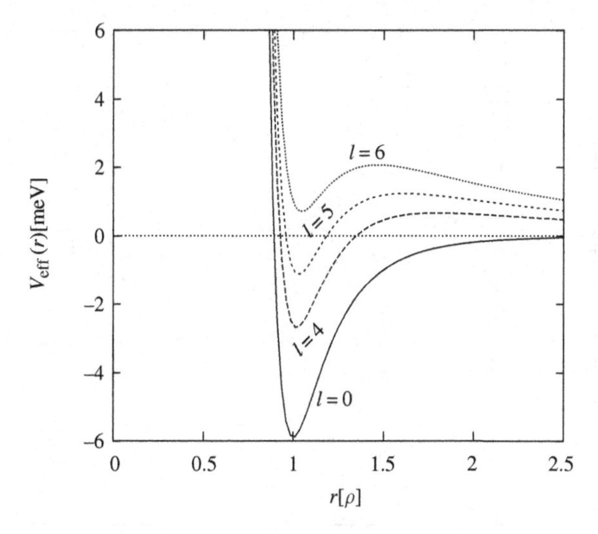

Figure 2.3. The effective potential for the Lennard–Jones interaction for various l-values.

A computer program similar to the one described here was used by Toennies *et al.* [3] to compare the results of scattering experiments with theory. The experiment consisted of the bombardment of krypton atoms with hydrogen atoms. Figure 2.3 shows the Lennard–Jones interaction potential plus the centrifugal barrier $l(l+1)/r^2$ of the radial Schrödinger equation. For higher l-values, the potential consists essentially of a hard core, a well and a barrier which is caused by the $1/r^2$ centrifugal term in the Schrödinger equation. In such a potential, quasi-bound states are possible. These are states which would be genuine bound states for a potential for which the barrier does not drop to zero for larger values of r, but remains at its maximum height. You can imagine the following to happen when a particle is injected into the potential at precisely this energy: it tunnels through the barrier, remains in the well for a relatively long time, and then tunnels outward through the barrier in an arbitrary direction because it has 'forgotten' its original direction. In wave-like terms, the particle resonates in the well, and this state decays after a relatively long time. This phenomenon is called 'scattering resonance'. This means that particles injected at this energy are strongly scattered and this shows up as a peak in the total cross section.

Such peaks can be seen in Figure 2.4, which shows the total cross section as a function of the energy calculated with a program as described above. The peaks are due to $l = 4$, $l = 5$ and $l = 6$ scattering, with energies increasing with l. Figure 2.5 finally shows the experimental results for the total cross section for H–Kr. We see that the agreement is excellent.

You should be able now to reproduce the data of Figure 2.4 with your program.

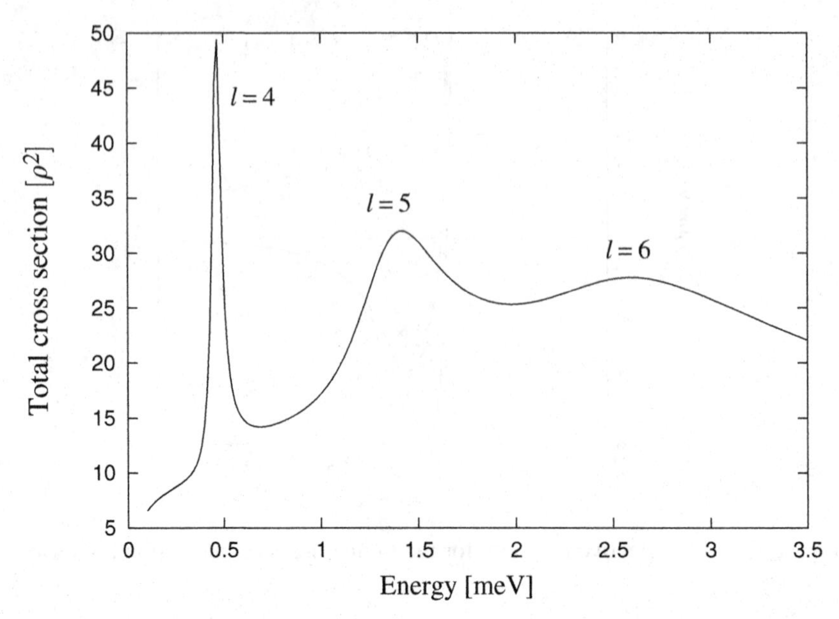

Figure 2.4. The total cross section shown as function of the energy for a Lennard–Jones potential modelling the H–Kr system. Peaks correspond to the resonant scattering states. The total cross section is expressed in terms of the range ρ of the Lennard–Jones potential.

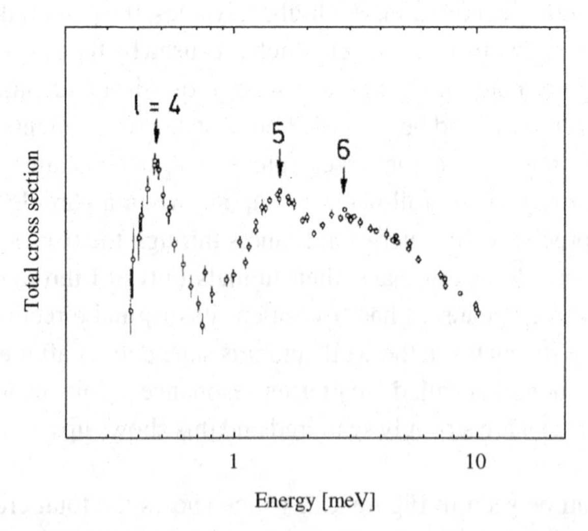

Figure 2.5. Experimental results as obtained by Toennies *et al.* [3] for the total cross section (arbitrary units) of the scattering of hydrogen atoms by krypton atoms as function of centre of mass energy.

*2.3 Calculation of scattering cross sections

In this section we derive Eqs. (2.7) and (2.8). At a large distance from the scattering centre we can make an *Ansatz* for the wave function. This consists of the incoming beam and a scattered wave:

$$\psi(\mathbf{r}) \propto \mathrm{e}^{\mathrm{i}\mathbf{k}\cdot\mathbf{r}} + f(\theta)\frac{\mathrm{e}^{ikr}}{r}. \tag{2.19}$$

Here, θ is the angle between the incoming beam and the line passing through $\mathbf{r}$ and the scattering centre. The function f does not depend on the azimuthal angle φ because the incoming wave has azimuthal symmetry, and the spherically symmetric potential will not generate $m \neq 0$ contributions to the scattered wave. $f(\theta)$ is called the scattering amplitude. From the *Ansatz* it follows that the differential cross section is given directly by the square of this amplitude:

$$\frac{\mathrm{d}\sigma}{\mathrm{d}\Omega} = |f(\theta)|^2 \tag{2.20}$$

with the appropriate normalisation (see for example Ref. [1]).

Beyond r_{max}, the solution can also be written in the form (2.2) leaving out all $m \neq 0$ contributions because of the azimuthal symmetry:

$$\psi(\mathbf{r}) = \sum_{l=0}^{\infty} A_l \frac{u_l(r)}{r} P_l(\cos\theta) \tag{2.21}$$

where we have used the fact that $Y_0^l(\theta, \phi)$ is proportional to $P_l(\cos\theta)$. Because the potential vanishes in the region $r > r_{\mathrm{max}}$, the solution $u_l(r)/r$ is given by the linear combination of the regular and irregular spherical Bessel functions, and as we have seen this reduces for large r to

$$u_l(r) \approx \sin\left(kr - \frac{l\pi}{2} + \delta_l\right). \tag{2.22}$$

We want to derive the scattering amplitude $f(\theta)$ by equating the expressions (2.19) and (2.21) for the wave function. For large r we obtain, using (2.22):

$$\sum_{l=0}^{\infty} A_l \left[\frac{\sin(kr - l\pi/2 + \delta_l)}{kr}\right] P_l(\cos\theta) = \mathrm{e}^{\mathrm{i}\mathbf{k}\cdot\mathbf{r}} + f(\theta)\frac{\mathrm{e}^{ikr}}{r}. \tag{2.23}$$

We write the right hand side of this equation as an expansion similar to that in the left hand side, using the following expression for a plane wave [4]

$$\mathrm{e}^{\mathrm{i}\mathbf{k}\cdot\mathbf{r}} = \sum_{l=0}^{\infty} (2l+1) i^l j_l(kr) P_l(\cos\theta). \tag{2.24}$$

$f(\theta)$ can also be written as an expansion in Legendre polynomials:

$$f(\theta) = \sum_{l=0}^{\infty} f_l P_l(\cos\theta), \tag{2.25}$$

so that we obtain:

$$\sum_{l=0}^{\infty} A_l \left[\frac{\sin(kr - l\pi/2 + \delta_l)}{kr}\right] P_l(\cos\theta)$$
$$= \sum_{l=0}^{\infty} \left[(2l+1)i^l j_l(kr) + f_l \frac{e^{ikr}}{r}\right] P_l(\cos\theta). \tag{2.26}$$

If we substitute the asymptotic form (2.5a) of j_l in the right hand side, we find:

$$\sum_{l=0}^{\infty} A_l \left[\frac{\sin(kr - l\pi/2 + \delta_l)}{kr}\right] P_l(\cos\theta)$$
$$= \frac{1}{r}\sum_{l=0}^{\infty} \left[\frac{2l+1}{2ik}(-)^{l+1} e^{-ikr} + \left(f_l + \frac{2l+1}{2ik}\right) e^{ikr}\right] P_l(\cos\theta). \tag{2.27}$$

Both the left and the right hand sides of (2.27) contain incoming and outgoing spherical waves (the occurrence of incoming spherical waves does not violate causality: they arise from the incoming plane wave). For each l, the prefactors of the incoming and outgoing waves should be equal on both sides in (2.27). This condition leads to

$$A_l = (2l+1)e^{i\delta_l} i^l \tag{2.28}$$

and

$$f_l = \frac{2l+1}{k} e^{i\delta_l} \sin(\delta_l). \tag{2.29}$$

Using (2.20), (2.25), and (2.29), we can write down an expression for the differential cross section in terms of the phase shifts δ_l:

$$\frac{d\sigma}{d\Omega} = \frac{1}{k^2}\left|\sum_{l=0}^{\infty}(2l+1)e^{i\delta_l}\sin(\delta_l)P_l(\cos\theta)\right|^2. \tag{2.30}$$

For the total cross section we find, using the orthonormality relations of the Legendre polynomials:

$$\sigma_{\text{tot}} = 2\pi \int d\theta \sin\theta \frac{d\sigma}{d\Omega}(\theta) = \frac{4\pi}{k^2}\sum_{l=0}^{\infty}(2l+1)\sin^2\delta_l. \tag{2.31}$$

Exercises

2.1 [C] Try using the Runge–Kutta method with an adaptive time step to integrate the radial Schrödinger equation in the program of Section 2.2, keeping the estimated error fixed as described in Appendix A7.1. What is the advantage of this method over Numerov's method for this particular case?

2.2 [C] Consider two radial potentials V_1 and V_2 and the solutions $u_l^{(1)}$ and $u_l^{(2)}$ to the radial Schrödinger equation for these two potentials (at the same energy):

$$\left[\frac{\hbar^2}{2m}\frac{\mathrm{d}^2}{\mathrm{d}r^2} + \left(E - V_1(r) - \frac{\hbar^2 l(l+1)}{2mr^2}\right)\right] u_l^{(1)}(r) = 0$$

$$\left[\frac{\hbar^2}{2m}\frac{\mathrm{d}^2}{\mathrm{d}r^2} + \left(E - V_2(r) - \frac{\hbar^2 l(l+1)}{2mr^2}\right)\right] u_l^{(2)}(r) = 0.$$

(a) Show that by multiplying the first equation from the left by $u_l^{(2)}(r)$ and the second one from the left by $u_l^{(1)}(r)$ and then subtracting, it follows that:

$$\int_0^L \mathrm{d}r u_l^{(2)}(r)[V_1(r) - V_2(r)]u_l^{(1)}(r) = \frac{\hbar^2}{2m}\left[u_l^{(2)}(L)\frac{\partial u_l^{(1)}(L)}{\partial r} - u_l^{(1)}(L)\frac{\partial u_l^{(2)}(L)}{\partial r}\right].$$

(b) If $V_i \to 0$ for large r, then both solutions are given for large r by $\sin[kr - (l\pi/2) + \delta_l^{(i)}]/k$. Show that from this it follows that:

$$\int_0^\infty \mathrm{d}r u_l^{(2)}(r)[V_1(r) - V_2(r)]u_l^{(1)}(r) = \frac{\hbar^2}{2mk}\sin(\delta_l^{(2)} - \delta_l^{(1)}).$$

Now take $V_1 \equiv 0$ and $V_2 \equiv V$ small everywhere. In that case, $u_l^{(1)}$ and $u_l^{(2)}$ on the left hand side can both be approximated by $rj_l(kr)$, so that we obtain:

$$\delta_l \approx -\frac{2mk}{\hbar^2}\int_0^\infty \mathrm{d}r\, r^2 j_l^2(kr)V(r).$$

This is the *Born approximation* for the phase shift. This approximation works well for potentials that are small with respect to the energy.

(c) [C] Write a (very simple) routine for calculating this integral (or use a library routine). Of course, it is sufficient to carry out the integration up to r_{max} since beyond that range $V \equiv 0$. Compare the Born approximation with the solution of the program developed in the previous problem. For the potential, take a weak Gaussian well:

$$V(r) = -A\exp[-(r-1)^2], \quad x < r_{\mathrm{max}}$$

and

$$V(r) = 0, \quad x \geq r_{\mathrm{max}}.$$

with $A = 0.01$ and r_{max} chosen suitably. Result?

(d) Now consider the analysis of items (a) and (b) where V_1 is the Lennard–Jones potential without cut-off and V_2 with cut-off. Show that the phase shift for the

Lennard–Jones potential without cut-off is given by the phase shift for the potential with cut-off plus a correction given by:

$$\Delta\delta_l = \frac{2m}{\hbar^2} k \int_{r_{\max}}^{\infty} j_l^2(kr) V_{\mathrm{LJ}}(r) r^2 \mathrm{d}r.$$

References

[1] A. Messiah, *Quantum Mechanics*, vols. 1 and 2. Amsterdam, North-Holland, 1961.

[2] S. E. Koonin, *Computational Physics*. Reading, Benjamin/Cummings, 1986.

[3] J. P. Toennies, W. Welz, and G. Wolf, 'Molecular beam scattering studies of orbiting resonances and the determination of Van der Waals potentials for H–He, Ar, Kr, and Xe and for H_2–Ar, Kr and Xe,' *J. Chem. Phys.*, **71** (1979), 614–42.

[4] M. Abramowitz and I. A. Stegun, *Handbook of Mathematical Functions*. Washington DC, National Bureau of Standards, 1964.

3

The variational method for the Schrödinger equation

3.1 Variational calculus

Quantum systems are governed by the Schrödinger equation. In particular, the solutions to the stationary form of this equation determine many physical properties of the system at hand. The stationary Schrödinger equation can be solved analytically in a very restricted number of cases – examples include the free particle, the harmonic oscillator and the hydrogen atom. In most cases we must resort to computers to determine the solutions. It is of course possible to integrate the Schrödinger equation using discretisation methods – see the different methods in Appendix A7.2 – but in most realistic electronic structure calculations we would need huge numbers of grid points, leading to high computer time and memory requirements. The variational method on the other hand enables us to solve the Schrödinger equation much more efficiently in many cases. In the next few chapters, which deal with electronic structure calculations, we shall make frequent use of the variational method described in this chapter.

In the variational method, the possible solutions are restricted to a subspace of the Hilbert space, and in this subspace we seek the best possible solution (below we shall define what is to be understood by the 'best' solution). To see how this works, we first show that the stationary Schrödinger equation can be derived by a stationarity condition of the functional:

$$E[\psi] = \frac{\int \mathrm{d}X \psi^*(X) H \psi(X)}{\int \mathrm{d}X \psi^*(X) \psi(X)} = \frac{\langle \psi | H | \psi \rangle}{\langle \psi | \psi \rangle} \tag{3.1}$$

which is recognised as the expectation value of the energy for a stationary state ψ (to keep the analysis general, we are not specific about the form of the generalised coordinate X – it may include the space and spin coordinates of a collection of particles). The stationary states of this energy-functional are defined by postulating that if such a state is changed by an arbitrary but small amount $\delta\psi$, the corresponding change in E vanishes to first order:

$$\delta E \equiv 0. \tag{3.2}$$

Defining

$$\begin{aligned} P &= \langle\psi|H|\psi\rangle \quad \text{and} \\ Q &= \langle\psi|\psi\rangle, \end{aligned} \tag{3.3}$$

we can write the change δE in the energy to first order in $\delta\psi$ as

$$\begin{aligned} \delta E &= \frac{\langle\psi+\delta\psi|H|\psi+\delta\psi\rangle}{\langle\psi+\delta\psi|\psi+\delta\psi\rangle} - \frac{\langle\psi|H|\psi\rangle}{\langle\psi|\psi\rangle} \\ &\approx \frac{\langle\delta\psi|H|\psi\rangle - (P/Q)\langle\delta\psi|\psi\rangle}{Q} + \frac{\langle\psi|H|\delta\psi\rangle - (P/Q)\langle\psi|\delta\psi\rangle}{Q}. \end{aligned} \tag{3.4}$$

As this should vanish for an *arbitrary* but small change in ψ, we find, using $E = P/Q$:

$$H\psi = E\psi, \tag{3.5}$$

together with the Hermitian conjugate of this equation, which is equivalent.

In variational calculus, stationary states of the energy-functional are found *within* a subspace of the Hilbert space. An important example is linear variational calculus, in which the subspace is spanned by a set of basis vectors $|\chi_p\rangle$, $p = 1, \ldots, N$. We take these to be orthonormal at first, that is,

$$\langle\chi_p|\chi_q\rangle = \delta_{pq}, \tag{3.6}$$

where δ_{pq} is the Kronecker delta-function which is 0 unless $p = q$, and in that case, it is 1.

For a state

$$|\psi\rangle = \sum_p C_p|\chi_p\rangle, \tag{3.7}$$

the energy-functional is given by

$$E = \frac{\sum_{p,q=1}^{N} C_p^* C_q H_{pq}}{\sum_{p,q=1}^{N} C_p^* C_q \delta_{pq}} \tag{3.8}$$

with

$$H_{pq} = \langle\chi_p|H|\chi_q\rangle. \tag{3.9}$$

The stationary states follow from the condition that the derivative of this functional with respect to the C_p vanishes, which leads to

$$\sum_{q=1}^{N} (H_{pq} - E\delta_{pq})C_q = 0 \quad \text{for } p = 1, \ldots, N. \tag{3.10}$$

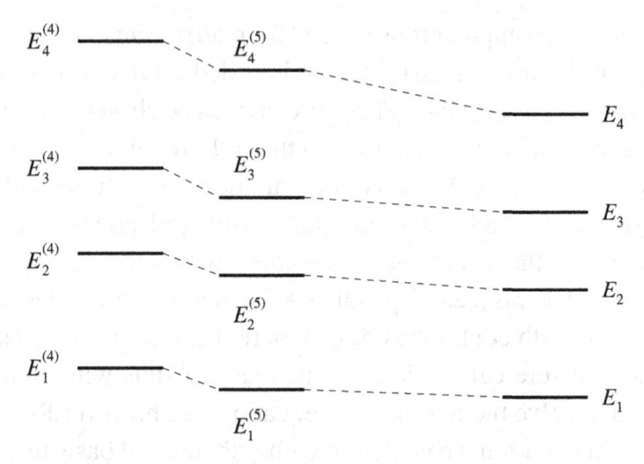

Figure 3.1. The behaviour of the spectrum of Eq. (3.11) with increasing basis set size in linear variational calculus. The upper index is the number of states in the basis set, and the lower index labels the spectral levels.

Equation (3.10) is an eigenvalue problem which can be written in matrix notation:

$$\mathbf{HC} = E\mathbf{C}. \tag{3.11}$$

This is the Schrödinger equation, formulated for a finite, orthonormal basis.

Although in principle it is possible to use nonlinear parametrisations of the wave function, linear parametrisations are used in the large majority of cases because of the simplicity of the resulting method, allowing for numerical matrix diagonalisation techniques, discussed in Appendix A7.2, to be used. The lowest eigenvalue of (3.11) is always higher than or equal to the ground state energy of Eq. (3.5), as the ground state is the minimal value assumed by the energy-functional in the full Hilbert space. If we restrict ourselves to a part of this space, then the minimum value of the energy-functional must always be higher than or equal to the ground state of the full Hilbert space. Including more basis functions into our set, the subspace becomes larger, and consequently the minimum of the energy-functional will decrease (or stay the same). For the specific case of linear variational calculus, this result can be generalised to higher stationary states: they are always higher than the equivalent solution to the full problem, but approximate the latter better with increasing basis set size (see Problem 3.1). The behaviour of the spectrum found by solving (3.11) with increasing basis size is depicted in Figure 3.1.

We note here that it is possible to formulate the standard discretisation methods such as the finite difference method of Appendix A7.2 as linear variational methods with an additional nonvariational approximation caused by the discretised representation of the kinetic energy operator. These methods are usually considered as separate: the term variational calculus implies continuous (and often analytic) basis functions. Because the computer time needed for matrix diagonalisation scales with the third power of the linear matrix size (it is called a $\mathcal{O}(N^3)$ process), the basis should be kept as small as possible. Therefore, it must be chosen carefully: it should be possible to approximate the solutions to the full problem with a small number of basis states. The fact that the basis in (continuous) variational calculus can be chosen to be so much smaller than the number of grid points in a finite difference approach implies that even though the latter can be solved using special $\mathcal{O}(N)$ methods for sparse systems (see Appendix A8.2), they are still far less efficient than variational methods with continuous basis functions in most cases. This is why, in most electronic structure calculations, variational calculus with continuous basis functions is used to solve the Schrödinger equation; see however Refs. [1] and [2].

An example of a variational calculation with orthonormal basis functions will be considered in Problem 3.4. We now describe how to proceed when the basis consists of nonorthonormal basis functions, as is often the case in practical calculations. In that case, we must reformulate (3.11), taking care of the fact that the *overlap matrix* **S**, whose elements S_{pq} are given by

$$S_{pq} = \langle \chi_p | \chi_q \rangle \tag{3.12}$$

is not the unit matrix. This means that in Eq. (3.8) the matrix elements δ_{pq} of the unit matrix, occurring in the denominator, have to be replaced by S_{pq}, and we obtain

$$\mathbf{HC} = E\mathbf{SC}. \tag{3.13}$$

This looks like an ordinary eigenvalue equation, the only difference being the matrix **S** in the right hand side. It is called a *generalised eigenvalue equation* and there exist computer programs for solving such a problem. The numerical method used in such programs is described in Section 3.3.

3.2 Examples of variational calculations

In this section, we describe two quantum mechanical problems and the computer programs that can solve these problems numerically by a variational calculation. In both cases, we must solve a generalised matrix eigenvalue problem (3.13).

You can find a description of the method for diagonalising a symmetric matrix in Appendix A8.2, and the method for solving the generalised eigenvalue problem is considered in Section 3.3; see also problem 3.3. It is not advisable to program the matrix diagonalisation routine yourself; numerous routines can be found on the internet. Solving the generalised eigenvalue problem is not so difficult if you have a matrix diagonalisation routine at your disposal. It is easy to find such a routine on the network (it is part of the LAPACK library, which is part of the ATLAS numerical library; these can be found in the NETLIB repository). In the following we shall assume that we have such programs available.

3.2.1 The infinitely deep potential well

The potential well with infinite barriers is given by:

$$V(x) = \begin{cases} \infty & \text{for } |x| > |a| \\ 0 & \text{for } |x| \leq |a| \end{cases} \tag{3.14}$$

and it forces the wave function to vanish at the boundaries of the well ($x = \pm a$). The exact solution for this problem is known and treated in every textbook on quantum mechanics [3, 4]. Here we discuss a linear variational approach to be compared with the exact solution. We take $a = 1$ and use natural units such that $\hbar^2/2m = 1$.

As basis functions we take simple polynomials that vanish on the boundaries of the well:

$$\psi_n(x) = x^n(x-1)(x+1), \ n = 0, 1, 2, \ldots \tag{3.15}$$

The reason for choosing this particular form of basis functions is that the relevant matrix elements can easily be calculated analytically. We start with the matrix elements of the overlap matrix, defined by

$$S_{mn} = \langle \psi_n | \psi_m \rangle = \int_{-1}^{1} \psi_n(x) \psi_m(x) \mathrm{d}x. \tag{3.16}$$

Working out the integral gives

$$S_{mn} = \frac{2}{n+m+5} - \frac{4}{n+m+3} + \frac{2}{n+m+1} \tag{3.17}$$

for $n + m$ even; otherwise $S_{mn} = 0$.

We can also calculate the Hamilton matrix elements, and you can check that they are given by:

$$\begin{aligned} H_{mn} = \langle \psi_n | p^2 | \psi_m \rangle &= \int_{-1}^{1} \psi_n(x) \left(-\frac{\mathrm{d}^2}{\mathrm{d}X^2} \right) \psi_m(x) \mathrm{d}x \\ &= -8 \left[\frac{1 - m - n - 2mn}{(m+n+3)(m+n+1)(m+n-1)} \right] \end{aligned} \tag{3.18}$$

for $m + n$ even, otherwise $H_{mn} = 0$.

PROGRAMMING EXERCISE

Write a computer program in which you fill the overlap and Hamilton matrix for this problem. Use standard software to solve the generalised eigenvalue problem.

Table 3.1. *Energy levels of the infinitely deep potential well.*

$N = 5$	$N = 8$	$N = 12$	$N = 16$	Exact
2.4674	2.4674	2.4674	2.4674	2.4674
9.8754	9.8696	9.8696	9.8696	9.8696
22.2934	22.2074	22.2066	22.2066	22.2066
50.1246	39.4892	39.4784	39.4784	39.4784
87.7392	63.6045	61.6862	61.6850	61.6850

The first four columns show the variational energy levels for various numbers of basis states N. The last column shows the exact values. The exact levels are approached from above as in Figure 3.1.

Check Compare the results with the analytic solutions. These are given by

$$\psi_n(x) = \begin{cases} \cos(k_n x) & n \text{ odd} \\ \sin(k_n x) & n \text{ even and positive} \end{cases} \tag{3.19}$$

with $k_n = n\pi/2$, $n = 1, 2, \ldots$, and the corresponding energies are given by

$$E_n = k_n^2 = \frac{n^2\pi^2}{4}. \tag{3.20}$$

For each eigenvector $\mathbf{C}$, the function $\sum_{p=1}^{N} C_p \chi_p(x)$ should approximate an eigenfunction (3.19). They can be compared by displaying both graphically. Carry out the comparison for various numbers of basis states. The variational levels are shown in Table 3.1, together with the analytical results.

3.2.2 Variational calculation for the hydrogen atom

As we shall see in the next two chapters, one of the main problems of electronic structure calculations is the treatment of the electron–electron interactions. Here we develop a program for solving the Schrödinger equation for an electron in a hydrogen atom for which the many-electron problem does not arise, so that a direct variational treatment of the problem is possible which can be compared with the analytical solution [3, 4].

The program described here is the first in a series leading to a program for calculating the electronic structure of the hydrogen molecule. The extension to the H_2^+ ion can be found in the next chapter in Problem 4.8 and a program for the hydrogen molecule is considered in Problem 4.12.

The electronic Schrödinger equation for the hydrogen atom reads:

$$\left[-\frac{\hbar^2}{2m}\nabla^2 - \frac{1}{4\pi\epsilon_0}\frac{1}{r}\right]\psi(\mathbf{r}) = E\psi(\mathbf{r}) \tag{3.21}$$

where the second term in the square brackets is the Coulomb attraction potential of the nucleus. The mass m is the reduced mass of the proton–electron system which is approximately equal to the electron mass. The ground state is found at energy

$$E = -\frac{m}{\hbar^2}\left(\frac{e^2}{4\pi\epsilon_0}\right)^2 \approx -13.6058\ \text{eV} \tag{3.22}$$

and the wave function is given by

$$\psi(\mathbf{r}) = \frac{2}{a_0^{3/2}} e^{-r/a_0} \tag{3.23}$$

in which a_0 is the Bohr radius,

$$a_0 = \frac{4\pi\epsilon_0\hbar^2}{me^2} \approx 0.529\,18\ \text{Å}. \tag{3.24}$$

In computer programming, it is convenient to use units such that equations take on a simple form, involving only coefficients of order 1. Standard units in electronic structure physics are so-called *atomic units*:the unit of distance is the Bohr radius a_0, masses are expressed in the electron mass m_e and the charge is measured in unit charges (e). The energy is finally given in 'hartrees' (E_H), given by $m_e c^2 \alpha^2$ (α is the fine-structure constant and m_e is the electron mass) which is roughly equal to 27.212 eV. In these units, the Schrödinger equation for the hydrogen atom assumes the following simple form:

$$\left[-\frac{1}{2}\nabla^2 - \frac{1}{r}\right]\psi(\mathbf{r}) = E\psi(\mathbf{r}). \tag{3.25}$$

We try to approximate the ground state energy and wave function of the hydrogen atom in a linear variational procedure. We use Gaussian basis functions which will be discussed extensively in the next chapter (Section 4.6.2). For the ground state, we only need angular momentum $l = 0$ functions (s-functions), which have the form:

$$\chi_p(r) = e^{-\alpha_p r^2} \tag{3.26}$$

centred on the nucleus (which is thus placed at the origin). We have to specify the values of the exponents α; these are kept fixed in our program. Optimal values for these exponents have previously been found by solving the *nonlinear* variational problem including the linear coefficients C_p *and* the exponents α [5]. We shall use these values of the exponents in the program:

$$\begin{aligned}\alpha_1 &= 13.007\,73\\ \alpha_2 &= 1.962\,079\\ \alpha_3 &= 0.444\,529\\ \alpha_4 &= 0.121\,949\,2.\end{aligned} \tag{3.27}$$

If the program works correctly, it should yield a value close to the exact ground state energy $-1/2\, E_{\rm H}$ (which is equal to -13.6058 eV).

It remains to determine the linear coefficients C_p in a computer program which solves the generalised eigenvalue problem, just as in Section 3.2.1:

$$\mathbf{HC} = E\mathbf{SC}. \tag{3.28}$$

It is not so difficult to show that the elements of the overlap matrix **S**, the kinetic energy matrix **T** and the Coulomb matrix **A** are given by:

$$\begin{aligned}
S_{pq} &= \int d^3r\, e^{-\alpha_p r^2} e^{-\alpha_q r^2} = \left(\frac{\pi}{\alpha_p + \alpha_q}\right)^{3/2};\\
T_{pq} &= -\frac{1}{2}\int d^3r\, e^{-\alpha_p r^2} \nabla^2 e^{-\alpha_q r^2} = 3\frac{\alpha_p \alpha_q \pi^{3/2}}{(\alpha_p + \alpha_q)^{5/2}};\\
A_{pq} &= -\int d^3r\, e^{-\alpha_p r^2} \frac{1}{r} e^{-\alpha_q r^2} = -\frac{2\pi}{\alpha_p + \alpha_q}.
\end{aligned} \tag{3.29}$$

See also Section 4.8. Using these expressions, you can fill the overlap and the Hamilton matrix. Since both matrices are symmetric, it is clear that only the upper (or the lower) triangular part (including the diagonal) has to be calculated; the other elements follow from the symmetry.

PROGRAMMING EXERCISE

Write a program in which the relevant matrices are filled and which solves the generalised eigenvalue problem for the variational calculation.

Check 1 Fortunately, we again have an exact answer for the ground state energy: this should be equal to -0.5 hartree $= 13.6058$ eV, and, if your program contains no errors, you should find $-0.499\,278$ hartree, which is amazingly good if you realise that only four functions have been taken into account.

Check 2 The solution of the eigenvalue problem not only yields the eigenvalues (energies) but also the eigenvectors. Use these to draw the variational ground state wave function and compare with the exact form (3.23). (See also Figure 4.3.)

*3.3 Solution of the generalised eigenvalue problem

It is possible to transform (3.13) into an ordinary eigenvalue equation by performing a basis transformation which brings **S** to unit form. Suppose we have found a matrix **V** which transforms **S** to the unit matrix:

$$\mathbf{V}^\dagger \mathbf{S} \mathbf{V} = \mathbf{I}. \tag{3.30}$$

Then we can rewrite (3.13) as

$$\mathbf{V}^\dagger \mathbf{H} \mathbf{V} \mathbf{V}^{-1} \mathbf{C} = E \mathbf{V}^\dagger \mathbf{S} \mathbf{V} \mathbf{V}^{-1} \mathbf{C} \tag{3.31}$$

and, defining

$$\mathbf{C}' = \mathbf{V}^{-1} \mathbf{C} \tag{3.32}$$

and

$$\mathbf{H}' = \mathbf{V}^\dagger \mathbf{H} \mathbf{V}, \tag{3.33}$$

we obtain

$$\mathbf{H}' \mathbf{C}' = E \mathbf{C}'. \tag{3.34}$$

This is an ordinary eigenvalue problem which we can solve for $\mathbf{C}'$ and E, and then we can find the eigenvector $\mathbf{C}$ of the original problem as $\mathbf{V}\mathbf{C}'$.

The problem remains of finding a matrix $\mathbf{V}$ which brings $\mathbf{S}$ to unit form according to (3.30). This matrix can be found if we have a unitary matrix $\mathbf{U}$ which diagonalises $\mathbf{S}$:

$$\mathbf{U}^\dagger \mathbf{S} \mathbf{U} = \mathbf{s} \tag{3.35}$$

with $\mathbf{s}$ the diagonalised form of $\mathbf{S}$. In fact, the matrix $\mathbf{U}$ is automatically generated when diagonalising $\mathbf{S}$ by a Givens–Householder *QR* procedure (see Appendix A8.2). From the fact that $\mathbf{S}$ is an overlap matrix, defined by (3.12), it follows directly that the eigenvalues of $\mathbf{S}$ are positive (see Problem 3.2). Therefore, it is possible to define the inverse square root of $\mathbf{s}$: it is the matrix containing the inverse of the square root of the eigenvalues of $\mathbf{S}$ on the diagonal. Choosing the matrix $\mathbf{V}$ as $\mathbf{U}\mathbf{s}^{-1/2}$, we obtain

$$\mathbf{V}^\dagger \mathbf{S} \mathbf{V} = \mathbf{s}^{-1/2} \mathbf{U}^\dagger \mathbf{S} \mathbf{U} \mathbf{s}^{-1/2} = \mathbf{I} \tag{3.36}$$

so the matrix $\mathbf{V}$ indeed has the desired property.

*3.4 Perturbation theory and variational calculus

In 1951, Löwdin [6] devised a method in which, in addition to a standard basis set A, a number of extra basis states (B) is taken into account in a perturbative manner, thus allowing for huge basis sets to be used without excessive demands on computer time and memory. The size of the matrix to be diagonalised in this method is equal to the number of basis states in the restricted set A; the remaining states are taken into account in constructing this matrix. A disadvantage is that the latter depends on the energy (which is obviously not known at the beginning), but, as we shall see, this does not prevent the method from being useful in many cases.

We start with an orthonormal basis, which could be a set of plane waves. The basis is partitioned into the two sets A and B, and for the plane wave example, A will contain the slowly varying waves and B those with shorter wavelength. We shall use the following notation: n and m label the states in A, α and β label the states in B, and p and q label the states in both sets. Furthermore we define

$$H'_{pq} = H_{pq}(1 - \delta_{pq}), \tag{3.37}$$

that is, H' is H with the diagonal elements set to 0. Now we can write Eq. (3.11) as

$$(E - H_{pp})C_p = \sum_{n \epsilon A} H'_{pn} C_n + \sum_{\alpha \epsilon B} H'_{p\alpha} C_\alpha. \tag{3.38}$$

If we define

$$h'_{pn} = H'_{pn}/(E - H_{pp}), \tag{3.39}$$

and similarly for $h'_{p\alpha}$, then we can write Eq. (3.38) as

$$C_p = \sum_{n \epsilon A} h'_{pn} C_n + \sum_{\alpha \epsilon B} h'_{p\alpha} C_\alpha. \tag{3.40}$$

Using this expression to rewrite C_α in the second term of the right hand side, we obtain

$$\begin{aligned} C_p &= \sum_{n \epsilon A} h'_{pn} C_n + \sum_{\alpha \epsilon B} h'_{p\alpha} \left(\sum_{n \epsilon A} h'_{\alpha n} C_n + \sum_{\beta \epsilon B} h'_{\alpha\beta} C_\beta \right) \\ &= \sum_{n \epsilon A} \left(h'_{pn} + \sum_{\alpha \epsilon B} h'_{p\alpha} h'_{\alpha n} \right) C_n + \sum_{\alpha \epsilon B} \sum_{\beta \epsilon B} h'_{p\alpha} h'_{\alpha\beta} C_\beta. \end{aligned} \tag{3.41}$$

After using (3.40) again to re-express C_β and repeating this procedure over and over, we arrive at

$$C_p = \sum_{n \epsilon A} \left(h'_{pn} + \sum_{\alpha \epsilon B} h'_{p\alpha} h'_{\alpha n} + \sum_{\alpha, \beta \epsilon B} h'_{p\alpha} h'_{\alpha\beta} h'_{\beta n} + \cdots \right) C_n. \tag{3.42}$$

We now introduce the following notation:

$$U^A_{pn} = H_{pn} + \sum_{\alpha \epsilon B} \frac{H'_{p\alpha} H'_{\alpha n}}{E - H_{\alpha\alpha}} + \sum_{\alpha\beta \epsilon B} \frac{H'_{p\alpha} H'_{\alpha\beta} H'_{\beta n}}{(E - H_{\alpha\alpha})(E - H_{\beta\beta})} + \cdots \tag{3.43}$$

Then (3.42) transforms into

$$C_p = \sum_{n \epsilon A} \frac{U^A_{pn} - H_{pn}\delta_{pn}}{E - H_{pp}} C_n. \tag{3.44}$$

Choosing p in A (and calling it m), (3.44) becomes

$$(E - H_{mm})C_m = \sum_{n \epsilon A} U^A_{mn} C_n - H_{mm} C_m, \tag{3.45}$$

so

$$\mathbf{UC} = E\mathbf{C}. \tag{3.46}$$

This equation is similar to (3.11), except that **H** is replaced by **U**. Notice that **U** depends on the energy which remains to be calculated, which makes Eq. (3.46) rather difficult to solve. In practice, a fixed value for E is chosen somewhere in the region for which we want accurate results. For electrons in a solid, this might be the region around the Fermi energy, since the states with these energies determine many physical properties.

The convergence of the expansion for **U**, Eq. (3.44), depends on the matrix elements $h'_{p\alpha}$ and $h'_{\alpha\beta}$, which should be small. Cutting off after the first term yields

$$U^A_{mn} = H_{mn} + \sum_{\alpha \epsilon B} \frac{H'_{m\alpha} H'_{\alpha n}}{E - H_{\alpha\alpha}}. \tag{3.47}$$

Löwdin perturbation theory is used mostly in this form.

It is not a priori clear that the elements $h'_{p\alpha}$ and $h'_{\alpha\beta}$ are small. However, keeping in mind a plane wave basis set, if we have a potential that varies substantially slower than the states in set B, these numbers will indeed be small as the H'_{pn} are small, so in that case the method will improve the efficiency of the diagonalisation process. The Löwdin method is frequently used in pseudopotential methods for electrons in solids which will be discussed in Chapter 6.

Exercises

3.1 MacDonald's theorem states that, in linear variational calculus, not only the variational ground state but also the higher variational eigenvectors have eigenvalues that are higher than the corresponding eigenvalues of the full problem.

Consider an Hermitian operator $\mathcal{H}$ and its variational matrix representation **H** defined by

$$H_{pq} = \langle \chi_p | \mathcal{H} | \chi_q \rangle.$$

χ_p are the basis vectors of the linear variational calculus. They form a finite set.

We shall denote the eigenvectors of $\mathcal{H}$ by ϕ_k and the corresponding eigenvalues by λ_k; Φ_k are the eigenvectors of **H** with eigenvalues Λ_k. They are all ordered, i.e. ϕ_0 corresponds to the lowest eigenvalue and so on, and similarly for the Φ_k.

(a) Write Φ_0 as an expansion in the complete set ϕ_k in order to show that

$$\frac{\langle \Phi_0 | \mathcal{H} | \Phi_0 \rangle}{\langle \Phi_0 | \Phi_0 \rangle} = \Lambda_0 \geq \lambda_0.$$

(b) Suppose Φ'_1 is a vector perpendicular to ϕ_0. Show that

$$\frac{\langle \Phi'_1 | \mathcal{H} | \Phi'_1 \rangle}{\langle \Phi'_1 | \Phi'_1 \rangle} \geq \lambda_1.$$

(Note that, in general, the lowest-but-one variational eigenstate Φ_1 is not perpendicular to ϕ_0 so this result does not guarantee $\Lambda_1 \geq \lambda_1$.)

(c) Consider a vector $\Phi_1' = \alpha\Phi_0 + \beta\Phi_1$ which is perpendicular to ϕ_0. From (b) it is clear that $\langle\Phi_1'|H|\Phi_1'\rangle/\langle\Phi_1'|\Phi_1'\rangle \geq \lambda_1$. Show that

$$\frac{\langle\Phi_1'|\mathcal{H}|\Phi_1'\rangle}{\langle\Phi_1'|\Phi_1'\rangle} = \frac{|\alpha|^2\Lambda_0 + |\beta|^2\Lambda_1}{|\alpha|^2 + |\beta|^2}$$

and that from this it follows that $\Lambda_1 \geq \lambda_1$. This result can be generalised for higher states.

3.2 The overlap matrix **S** is defined as

$$S_{pq} = \langle\chi_p|\chi_q\rangle.$$

Consider a vector ψ that can be expanded in the basis χ_p as:

$$\psi = \sum_p C_p\chi_p.$$

(a) Suppose ψ is normalised. Show that **C** then satisfies:

$$\sum_{pq} C_p^* S_{pq} C_q = 1.$$

(b) Show that the eigenvalues of **S** are positive.

3.3 [**C**] In this problem, it is assumed that a routine for diagonalising a real, symmetric matrix is available.

(a) [**C**] Using a library routine for diagonalising a real, symmetric matrix, write a routine which, given the overlap matrix **S**, generates a matrix **V** which brings **S** to unit form:

$$\mathbf{V}^\dagger\mathbf{S}\mathbf{V} = \mathbf{I}.$$

(b) [**C**] Write a routine which uses the matrix **V** to produce the solutions (eigenvectors and eigenvalues) to the generalised eigenvalue problem:

$$\mathbf{HC} = E\mathbf{SC}.$$

The resulting routines can be used in the programs of Sections 3.2.1 and 3.2.2.

3.4 [**C**] The potential for a finite well is given by

$$V(x) = \begin{cases} 0 & \text{for } |x| > |a| \\ -V_0 & \text{for } |x| \leq |a| \end{cases}$$

In this problem, we determine the bound solutions to the Schrödinger equation using plane waves on the interval $(-L, +L)$ as basis functions:

$$\psi_n(x) = 1/\sqrt{2L}\, e^{ik_n x}$$

with

$$k_n = \pm\frac{n\pi}{L}, \quad n = 0, 1, \ldots$$

It is important to note that, apart from the approximation involved in having a finite basis set, there is another one connected with the periodicity imposed by the specific form of the basis functions on the finite interval $(-L, L)$. In this problem, we use units such that the factor $\hbar^2/2m$ assumes the value 1.

(a) Show that the relevant matrix elements are given by

$$S_{mn} = \delta_{mn}$$
$$\langle \psi_n | p^2 | \psi_m \rangle = -k_n^2 \delta_{nm} \quad \text{and}$$
$$\langle \psi_n | V | \psi_m \rangle = -\frac{V_0}{L} \frac{\sin(k_m - k_n)a}{k_m - k_n} \quad \text{for } n \neq m$$
$$\langle \psi_n | V | \psi_n \rangle = -\frac{V_0}{L} a$$

The stationary states in an even potential (i.e. $V(x) = V(-x)$) have either positive or negative parity [3]. From this it follows that if we use a basis $1/\sqrt{L} \cos k_n x$ (and $1/\sqrt{2L}$ for $n = 0$), we shall find the even stationary states, and if we take the basis functions $1/\sqrt{L} \sin k_n x$, only the odd states. It is of course less time-consuming to diagonalise two $N \times N$ matrices than a single $2N \times 2N$, knowing that matrix diagonalisation scales with N^3.

(b) Show that the matrix elements with the cosine basis read

$$S_{mn} = \delta_{mn}$$
$$\langle \psi_n | p^2 | \psi_m \rangle = -k_n^2 \delta_{nm} \quad \text{and}$$
$$\langle \psi_n | V | \psi_m \rangle = -\frac{V_0}{L} \left[\frac{\sin(k_m - k_n)a}{k_m - k_n} + \frac{\sin(k_m + k_n)a}{k_m + k_n} \right]$$
$$\text{for } n \neq m$$
$$\langle \psi_n | V | \psi_n \rangle = -\frac{V_0}{L} \left[a + \frac{\sin(2k_n a)}{2k_n} \right] \quad \text{for } n \neq 0$$
$$\langle \psi_0 | V | \psi_0 \rangle = -\frac{V_0}{L} a \quad \text{for } n = 0$$

In the sine-basis, the last terms in the third and fourth expressions occur with a minus sign.

(c) **[C]** Write a computer program for determining the spectrum. Compare the results with those of the direct calculation (which, for $V_0 = 1$ and $a = 1$, yields a ground state energy $E \approx -0.4538$).

As you will note, for many values of A, V_0, L and N, the variational ground state energy lies below the exact ground state energy number. Explain why this happens.

References

[1] L.-W. Wang and A. Zunger, 'Electronic-structure pseudopotential calculations of large (approximate-to-1000 atoms) Si quantum dots,' *J. Phys. Chem.*, **98** (1994), 2158–65.

[2] J. M. Thijssen and J. E. Inglesfield, 'Embedding muffin tins into a finite difference grid,' *Europhys. Lett.*, **27** (1994), 65–70.
[3] A. Messiah, *Quantum Mechanics*, vols. 1 and 2. Amsterdam, North-Holland, 1961.
[4] S. Gasiorowicz, *Quantum Physics*. New York, John Wiley, 1974.
[5] R. Ditchfield, W. J. Hehre, and J. A. Pople, 'Self-consistent molecular orbital methods. VI. Energy optimised Gaussian atomic orbitals,' *J. Chem. Phys.*, **52** (1970), 5001–7.
[6] P.-O. Löwdin, 'A note on the quantum mechanical perturbation theory,' *J. Chem. Phys.*, **19** (1951), 1396–401.

4

The Hartree–Fock method

4.1 Introduction

Here and in the following chapter we treat two different approaches to the many-electron problem: the Hartree–Fock theory and the density functional theory. Both theories are simplifications of the full problem of many electrons moving in a potential field. In fact, the physical systems we want to study, such as atoms, molecules and solids, consist not only of electrons but also of nuclei, and each of these particles moves in the field generated by the others. A first approximation is to consider the nuclei as being fixed, and to solve the Schrödinger equation for the electronic system in the field of the static nuclei. This approach, called the Born–Oppenheimer approximation, is justified by the nuclei being much heavier than the electrons so that they move at much slower speeds. It remains then to solve for the electronic structure.

The Hartree–Fock methodcan be viewed as a variational method in which the wave functions of the many-electron system have the form of an antisymmetrised product of one-electron wave functions (the antisymmetrisation is necessary because of the fermion character of the electrons). This restriction leads to an effective Schrödinger equation for the individual one-electron wave functions (called *orbitals*) with a potential determined by the orbitals occupied by the other electrons. This coupling between the orbitals via the potentials causes the resulting equations to become nonlinear in the orbitals, and the solution must be found iteratively in a self-consistency procedure. The Hartree–Fock (HF)procedure is close in spirit to the mean-field aproach in statistical mechanics.

We shall see that in this variational approach, correlations between the electrons are neglected to some extent. In particular, the Coulomb repulsion between the electrons is represented in an averaged way. However, the effective interaction caused by the fact that the electrons are fermions, obeying Pauli's principle, and hence avoid each other if they have the same spin, is accurately included in the HF approach. There exist several methods that improve on the approximations made in the HF method.

The Hartree–Fock approach is very popular among chemists, and it has also been applied to solids. In this chapter, we give an introduction to the Hartree–Fock method and apply it to simple two-electron systems: the helium atom and the hydrogen molecule. We describe the Born–Oppenheimer approach and independent particle approaches (of which HF is an example) in a bit more detail in the next section. In Section 4.3 we then derive the Hartree method for a two-electron system (the helium atom). In Section 4.3.2, a program for calculating the ground state of the helium atom is described.

In Sections 4.4 and 4.5 the HF method for systems containing more than two electrons is described in detail, and in Section 4.6 the basis functions used for molecular systems are described. In sections 4.7 and 4.8 some details concerning the implementation of the HF method are considered. In Section 4.9, results of the HF method are presented, and in Section 4.10 the configuration interaction (CI) method, which improves on the HF method is described.

4.2 The Born–Oppenheimer approximation and the independent-particle method

The Hamiltonian of a system consisting of N electrons and K nuclei with charges Z_n reads

$$H = \sum_{i=1}^{N} \frac{p_i^2}{2m} + \sum_{n=1}^{K} \frac{P_n^2}{2M_n} + \frac{1}{4\pi\epsilon_0}\frac{1}{2} \sum_{i,j=1;i\neq j}^{N} \frac{e^2}{|\mathbf{r}_i - \mathbf{r}_j|}$$
$$- \frac{1}{4\pi\epsilon_0} \sum_{n=1}^{K}\sum_{i=1}^{N} \frac{Z_n e^2}{|\mathbf{r}_i - \mathbf{R}_n|} + \frac{1}{4\pi\epsilon_0}\frac{1}{2} \sum_{n,n'=1;n\neq n'}^{K} \frac{Z_n Z_{n'} e^2}{|\mathbf{R}_n - \mathbf{R}_{n'}|}. \tag{4.1}$$

The index i refers to the electrons and n to the nuclei, m is the electron mass, and M_n are the masses of the different nuclei. The first two terms represent the kinetic energies of the electrons and nuclei respectively; the third term represents the Coulomb repulsion between the electrons and the fourth term the Coulomb attraction between electrons and nuclei. Finally, the last term contains the Coulomb repulsion between the nuclei. The wave function of this system depends on the positions $\mathbf{r}_i$ and $\mathbf{R}_n$ of the electrons and nuclei respectively. This Hamiltonian looks quite complicated, and in fact it turns out that if the number of electrons and nuclei is not extremely small (typically smaller than four), it is impossible to solve the stationary Schrödinger equation for this Hamiltonian directly on even the largest and fastest computer available.

Therefore, important approximations must be made, and a first step consists of separating the degrees of freedom connected with the motion of the nuclei from those of the electrons. This procedure is known as the Born–Oppenheimer approximation [1] and its justification resides in the fact that the nuclei are much heavier than the electrons (the mass of a proton or neutron is about 1835 times as large as the electron mass) so it is intuitively clear that the nuclei move much more slowly than the electrons. The latter will then be able to adapt themselves to the current configuration of nuclei. This approach results also from formal calculations (see Problem 4.9), and leads to a Hamiltonian for the electrons in the field generated by a static configuration of nuclei, and a separate Schrödinger equation for the nuclei in which the electronic energy enters as a potential. The Born–Oppenheimer Hamiltonian for the electrons reads

$$H_{\rm BO} = \sum_{i=1}^{N} \frac{p_i^2}{2m} + \frac{1}{2}\frac{1}{4\pi\epsilon_0}\sum_{i,j=1;i\neq j}^{N} \frac{e^2}{|\mathbf{r}_i - \mathbf{r}_j|} - \frac{1}{4\pi\epsilon_0}\sum_{n=1}^{K}\sum_{i=1}^{N} \frac{Z_n e^2}{|\mathbf{r}_i - \mathbf{R}_n|}. \quad (4.2)$$

The total energy is the sum of the energy of the electrons and the energy resulting from the Schrödinger equation satisfied by the nuclei. In a further approximation, the motion of the nuclei is neglected and only the electrostatic energy of the nuclei should be added to the energy of the electrons to arrive at the total energy. The positions of the nuclei can be varied in order to find the minimum of this energy, that is, the ground state of the whole system (within the Born–Oppenheimer approximation with static nuclei). In this procedure, the nuclei are treated on a classical footing since their ground state is determined as the minimum of their potential energy, neglecting quantum fluctuations.[1]

Even with the positions of the nuclei kept fixed, the problem of solving for the electronic wave functions using the Hamiltonian (4.2) remains intractable, even on a computer, since too many degrees of freedom are involved. It is the second term containing the interactions between the electrons that makes the problem so difficult. If this term were not present, we would be dealing with a sum of one-electron Hamiltonians which can be solved relatively easily. There exist several ways of approximating the eigenfunctions of the Hamiltonian (4.2). In these approaches, the many-electron problem is reduced to an uncoupled problem in which the interaction of one electron with the remaining ones is incorporated in an averaged way into a potential felt by the electron.

[1] Vibrational modes of the nuclei can, however, be treated after expanding the total energy in deviations of the nuclear degrees of freedom from the ground state configuration. A transformation to normal modes then gives us a system consisting of independent harmonic oscillators.

The resulting uncoupled or *independent-particle* (IP) Hamiltonian has the form

$$H_{\mathrm{IP}} = \sum_{i=1}^{N} \left[\frac{p_i^2}{2m} + V(\mathbf{r}_i) \right]. \tag{4.3}$$

$V(\mathbf{r})$ is a potential depending on the positions $\mathbf{R}_i$ of the nuclei. As we shall see, its form can be quite complicated; in particular, V depends on the wave function ψ on which the IP Hamiltonian is acting. Moreover, V is often a nonlocal operator which means that the value of $V\psi$, evaluated at position $\mathbf{r}$, is determined by the values of ψ at other positions $\mathbf{r}' \neq \mathbf{r}$, and V depends on the energy in some approaches. These complications are the price we have to pay for an independent electron picture.

In the remaining sections of this chapter we shall study the Hartree–Fock approximation and in the next chapter we shall discuss the density functional theory. We start by considering the the helium atom to illustrate the general techniques which will be developed in later sections.

4.3 The helium atom

4.3.1 Self-consistency

In this section, we find an approximate independent-particle Hamiltonian (4.3) for the helium atom within the Born–Oppenheimer approximation by restricting the electronic wave function to a simple form. The coordinates of the wave function are $\mathbf{x}_1$ and $\mathbf{x}_2$, which are combined position and spin coordinates: $\mathbf{x}_i = (\mathbf{r}_i, s_i)$. As electrons are fermions, the wave function must be antisymmetric in the two coordinates $\mathbf{x}_1$ and $\mathbf{x}_2$ (more details concerning antisymmetry and fermions will be given in Section 4.4). We use the following antisymmetric trial wave function for the ground state:

$$\Psi(\mathbf{r}_1, s_1; \mathbf{r}_2, s_2) = \phi(\mathbf{r}_1)\phi(\mathbf{r}_2)\frac{1}{\sqrt{2}}[\alpha(s_1)\beta(s_2) - \alpha(s_2)\beta(s_1)], \tag{4.4}$$

where $\alpha(s)$ denotes the spin-up and $\beta(s)$ the spin-down wave function and ϕ is an orbital – a function depending on a single spatial coordinate – which is shared by the two electrons.

The Born–Oppenheimer Hamiltonian (4.2) for the helium atom reads

$$H_{\mathrm{BO}} = -\frac{1}{2}\nabla_1^2 - \frac{1}{2}\nabla_2^2 + \frac{1}{|\mathbf{r}_1 - \mathbf{r}_2|} - \frac{2}{r_1} - \frac{2}{r_2}, \tag{4.5}$$

where we have used atomic units introduced in Section 3.2.2. We now let this Hamiltonian act on the wave function (4.4). Since the Hamiltonian does not act on the spin, the spin-dependent part drops out on the left and right hand side of the

Schrödinger equation and we are left with:[2]

$$\left[-\frac{1}{2}\nabla_1{}^2 - \frac{1}{2}\nabla_2{}^2 - \frac{2}{r_1} - \frac{2}{r_2} + \frac{1}{|\mathbf{r}_1 - \mathbf{r}_2|}\right]\phi(\mathbf{r}_1)\phi(\mathbf{r}_2) = E\phi(\mathbf{r}_1)\phi(\mathbf{r}_2). \quad (4.6)$$

In order to arrive at a simpler equation we remove the $\mathbf{r}_2$-dependence by multiplying both sides from the left by $\phi^*(\mathbf{r}_2)$ and by integrating over $\mathbf{r}_2$. We then arrive at

$$\left[-\frac{1}{2}\nabla_1{}^2 - \frac{2}{r_1} + \int \mathrm{d}^3r_2\, |\phi(\mathbf{r}_2)|^2 \frac{1}{|\mathbf{r}_1 - \mathbf{r}_2|}\right]\phi(\mathbf{r}_1) = E'\phi(\mathbf{r}_1), \quad (4.7)$$

where several integrals yielding a constant (i.e. not dependent on $\mathbf{r}_1$) are absorbed in E'. The third term on the left hand side is recognised as the Coulomb energy of particle 1 in the electric field generated by the charge density of particle 2. To obtain this equation we have used the fact that ϕ is normalised to unity and this normalisation is from now on implicitly assumed for ϕ as occurring in the integral on the left hand side of (4.7). The effective Hamiltonian acting on the orbital of particle 1 has the independent particle form of Eq. (4.3). A remarkable feature is the dependence of the potential on the wave function we are searching for.

Equation (4.7) has the form of a *self-consistency* problem: ϕ is the solution to the Schrödinger equation but the latter is determined by ϕ itself. To solve an equation of this type, one starts with some trial ground state solution $\phi^{(0)}$ which is used in constructing the potential. Solving the Schrödinger equation with this potential, we obtain a new ground state $\phi^{(1)}$ which is used in turn to build a new potential. This procedure is repeated until the ground state $\phi^{(i)}$ and the corresponding energy $E^{(i)}$ of the Schrödinger equation at step i do not deviate appreciably from those in the previous step (if convergence does not occur, we must use some tricks to be discussed in Section 4.7).

The wave function we have used is called *uncorrelated* because of the fact that the probability $P(\mathbf{r}_1, \mathbf{r}_2)$ for finding an electron at $\mathbf{r}_1$ and another one at $\mathbf{r}_2$ is uncorrelated, i.e. it can be written as a product of two one-electron probabilities:

$$P(\mathbf{r}_1, \mathbf{r}_2) = p(\mathbf{r}_1)p(\mathbf{r}_2). \quad (4.8)$$

This does not mean that the electrons do not feel each other: in the determination of the spatial function ϕ, the interaction term $1/|\mathbf{r}_1 - \mathbf{r}_2|$ has been taken into account. But this interaction has been taken into account in an averaged way: it is not the *actual* position of $\mathbf{r}_2$ that determines the wave function for electron 1, but the *average* charge distribution of electron 2. This approach bears much relation to the mean field theory approach in statistical mechanics.

[2] This equation cannot be satisfied exactly with the form of trial function chosen, as the left hand side depends on $\mathbf{r}_1 - \mathbf{r}_2$ whereas the right hand side does not. We are, however, after the optimal wave function within the set of functions of the form (4.4) in a variational sense, along the lines of the previous chapter, but we want to avoid the complications of carrying out the variational procedure formally. This will be done in Section 4.5.2 for arbitrary numbers of electrons.

The neglect of correlations sometimes leads to unphysical results. An example is found in the dissociation of the hydrogen molecule. Suppose the nuclei are placed at positions $\mathbf{R}_A$ and $\mathbf{R}_B$ and we approximate the one-electron orbitals by spherically symmetric (1s) basis orbitals centred on the two nuclei: $u(\mathbf{r} - \mathbf{R}_A)$ and $u(\mathbf{r} - \mathbf{R}_B)$. Because of the symmetry of the hydrogen molecule, the ground state orbital solution of the independent particle Hamiltonian is given by the symmetric combination of these two basis orbitals:

$$\phi(\mathbf{r}) = u(\mathbf{r} - \mathbf{R}_A) + u(\mathbf{r} - \mathbf{R}_B). \tag{4.9}$$

The total wave function, which contains the product $\phi(\mathbf{r}_1)\phi(\mathbf{r}_2)$ therefore contains ionic terms in which both electrons sit on the same nucleus. This is not so disastrous if the two nuclei are close, but if we separate them, these terms should not be present: they contain the wrong physics and they result in a serious over-estimation of the energy. Physically, this is caused by the fact that electron 1 in our present approximation feels the potential resulting from the *average* charge distribution of electron 2, which is symmetrically distributed over the two nuclei, and thus it ends up on A or B with equal probability. If electron 1 was to feel the *actual* potential caused by electron 2, it would end up on a different nucleus from electron 2. A better description of the state would therefore be

$$\psi(\mathbf{r}_1; \mathbf{r}_2) = \tfrac{1}{2}[u(\mathbf{r}_1 - \mathbf{R}_A)u(\mathbf{r}_2 - \mathbf{R}_B) + u(\mathbf{r}_2 - \mathbf{R}_A)u(\mathbf{r}_1 - \mathbf{R}_B)], \tag{4.10}$$

which must then be multiplied by an antisymmetric spin wave function. This wave function is however not of the form (4.4).

The fact that the spatial part of the wave function is equal for the two electrons is specific for the case of two electrons: the antisymmetry is taken care of by the spin part of the wave function. If there are more than two electrons, the situation becomes more complicated and requires much more bookkeeping; this case will be treated in Section 4.5. Neglecting the antisymmetry requirement, one can, however, generalise the results obtained for two electrons to systems with more electrons. Writing the wave function as a product of spin-orbitals $\psi_k(\mathbf{x})$ (spin-orbitals are functions depending on the spatial and spin coordinates of one electron), the following equation for these spin-orbitals is obtained:

$$\left[-\frac{1}{2}\nabla^2 - \sum_n \frac{Z_n}{|\mathbf{r} - \mathbf{R}_n|} + \sum_{l=1}^{N} \int \mathrm{d}x' |\psi_l(\mathbf{x}')|^2 \frac{1}{|\mathbf{r} - \mathbf{r}'|}\right] \psi_k(\mathbf{x}) = E'\psi_k(\mathbf{x}). \tag{4.11}$$

Here k and l label the spin-orbitals; $\int \mathrm{d}x'$ denotes a sum over the spin s' and an integral over the spatial coordinate $\mathbf{r}'$: $\int \mathrm{d}x' = \sum_{s'} \int \mathrm{d}^3 r'$. As the Hamiltonian does not act on the spin-dependent part of the spin-orbitals, ψ_k can be written as a product of a spatial orbital with a one-electron spin wave function. In the last term on the left hand side we recognise the potential resulting from a charge distribution caused by all the electrons; it is called the *Hartree potential.* There is something unphysical about this term: it contains a coupling between orbital k and itself, since this orbital is included in the electron density, even though an electron clearly does not interact with itself. This can be remedied by excluding k from the sum over l in the Hartree term, but then every orbital feels a different potential. In the next subsection, we shall see that this problem is automatically solved in the Hartree–Fock theory which takes the antisymmetry of the many-electron wave function fully into account. Note that in our discussion of the helium case, we have already taken the self-interaction into account because the electron–electron interaction is half the size of that in (4.11) (after summation over the spin in this equation). Equation (4.11) was derived in 1927 by Hartree [2]; it neglects exchange as well as other correlations.

Before studying the problem of more electrons with an antisymmetric wave function, we shall now describe the construction of a program for actually calculating the solution of Eq. (4.7).

4.3.2 A program for calculating the helium ground state

In this section we construct a program for calculating the ground state energy and wave function for the helium atom. In the previous section we have restricted the form of the wave function to be uncorrelated; here we restrict it even further by writing it as a linear combination of four fixed, real basis functions in the same way as in Section 3.2.2. Let us first consider the form assumed by the Schrödinger equation for the independent particle formulation, Eq. (4.7). The parametrisation

$$\phi(\mathbf{r}) = \sum_{p=1}^{4} C_p \chi_p(\mathbf{r}) \tag{4.12}$$

leads directly to

$$\left[-\frac{1}{2}\nabla_1^2 - \frac{2}{r_1} + \sum_{r,s=1}^{4} C_r C_s \int d^3r_2\, \chi_r(\mathbf{r}_2)\chi_s(\mathbf{r}_2)\frac{1}{|\mathbf{r}_1 - \mathbf{r}_2|}\right] \sum_{q=1}^{4} C_q \chi_q(\mathbf{r}_1)$$
$$= E' \sum_{q=1}^{4} C_q \chi_q(\mathbf{r}_1). \tag{4.13}$$

Note that the C_p are real as the functions $\chi_p(\mathbf{r})$ are real. From now on we implicitly assume sums over indices $p, q, \ldots$ to run from 1 to the number of basis functions N, which is 4 in our case. Multiplying Eq. (4.13) from the left by $\chi_p(\mathbf{r}_1)$ and integrating over $\mathbf{r}_1$ leads to

$$\sum_{pq}\Big(h_{pq}+\sum_{rs}C_rC_sQ_{prqs}\Big)C_q=E'\sum_{pq}S_{pq}C_q \tag{4.14}$$

with

$$h_{pq}=\left\langle\chi_p\left|-\frac{1}{2}\nabla^2-\frac{2}{r}\right|\chi_q\right\rangle; \tag{4.15a}$$

$$Q_{prqs}=\int \mathrm{d}^3r_1\mathrm{d}^3r_2\,\chi_p(\mathbf{r}_1)\chi_r(\mathbf{r}_2)\frac{1}{|\mathbf{r}_1-\mathbf{r}_2|}\chi_q(\mathbf{r}_1)\chi_s(\mathbf{r}_2)\quad\text{and} \tag{4.15b}$$

$$S_{pq}=\langle\chi_p|\chi_q\rangle. \tag{4.15c}$$

Unfortunately, (4.14) is not a generalised eigenvalue equation because of the presence of the variables C_r and C_s between the brackets on the left hand side. However, if we carry out the self-consistency iteration process as indicated in the previous section, the C_r and C_s are kept *fixed*, and finding the C_q in (4.14) reduces to solving a generalised eigenvalue equation. We then replace C_r, C_s by the solution found and start the same procedure again.

The matrix elements from (4.15) remain to be found. We shall use Gaussian $l=0$ basis functions (s-functions), just as in the case of the hydrogen atom (see Section 3.2.2). Of course, the optimal exponents α_p occurring in the Gaussian s-basis functions χ_p,

$$\chi_p(\mathbf{r})=\mathrm{e}^{-\alpha_p r^2}, \tag{4.16}$$

are different from those of the hydrogen atom. Again, rather than solve the non-linear variational problem, which involves not only the prefactors C_p but also the exponents α_p as parameters of the wave function, we shall take the optimal values calculated from a different program which we do not go into here. They are

$$\alpha_1=0.298\,073$$
$$\alpha_2=1.242\,567$$
$$\alpha_2=5.782\,948,$$
$$\alpha_4=38.474\,970.$$

The matrix elements of the kinetic and the Coulomb energy are similar to those calculated for the hydrogen atom (see Eq. (3.29)), except for an extra factor of 2 in the nuclear attraction (due to the nuclear charge). In Section 4.8, the matrix element Q_{prqs} will be calculated; the result is given by

$$Q_{prqs}=\frac{2\pi^{5/2}}{(\alpha_p+\alpha_q)(\alpha_r+\alpha_s)\sqrt{\alpha_p+\alpha_q+\alpha_r+\alpha_s}}. \tag{4.17}$$

The program is constructed as follows.

- First, the 4×4 matrices h_{pq}, S_{pq} and the $4 \times 4 \times 4 \times 4$ array Q_{prqs} are calculated.
- Then initial values for C_p are chosen; they can, for example, all be taken to be equal (of course, you are free to choose other initial values – for this simple system most initial values will converge to the correct answer).
- These C-values are used for constructing the matrix F_{pq} given by

$$F_{pq} = h_{pq} + \sum_{rs} Q_{prqs} C_r C_s. \tag{4.18}$$

It should be kept in mind that the vector $\mathbf{C}$ should always be normalised to unity via the overlap matrix *before* inserting it into Eq. (4.18):

$$\sum_{p,q=1}^{4} C_p S_{pq} C_q = 1 \tag{4.19}$$

(see Problem 3.2).

- Now the generalised eigenvalue problem

$$\mathbf{FC} = E'\mathbf{SC} \tag{4.20}$$

is solved. For the ground state, the vector $\mathbf{C}$ is the one corresponding to the *lowest* eigenvalue.
- The energy for the state found is not simply given by E' as follows from the derivation of the self-consistent Schrödinger equation, Eq. (4.7). The ground state energy can be found by evaluating the expectation value of the Hamiltonian for the ground state just obtained:

$$E_G = 2\sum_{pq} C_p C_q h_{pq} + \sum_{pqrs} Q_{prqs} C_p C_q C_r C_s, \tag{4.21}$$

where the (normalised) eigenvector $\mathbf{C}$ results from the last diagonalisation of $\mathbf{F}$.
- The solution $\mathbf{C}$ of the generalised eigenvalue problem (4.20) is then used to build the matrix $\mathbf{F}$ again and so on.

PROGRAMMING EXERCISE

Write a program for calculating the ground state wave function of the helium atom.

Check 1 If your program is correct, the resulting ground state energy should be equal to $-2.855\,160\,38$ a.u. (remember that the atomic energy unit is the Hartree, see Section 3.2.2). The effect of using a small basis set can be judged by comparing with the value -2.8616 a.u. resulting from calculations using continuum integration techniques within the framework of the present calculation as described in Chapter 5. The effect of neglecting correlations in our approach results in the deviation from the exact value -2.903 a.u. (very accurate calculations can be performed for systems containing small numbers of electrons [3]).

4.4 Many-electron systems and the Slater determinant

In the helium problem, we could make use of the fact that in the ground state the required antisymmetry is taken care of by the spin part of the wave function, which drops out of the Schrödinger equation. If it is not the ground state we are after, or if more than two electrons are involved, antisymmetry requirements affect the orbital part of the wave function, and in the next two sections we shall consider a more general approach to an independent electron Hamiltonian, taking this antisymmetry into account. In the present section we consider a particular class of antisymmetric many-electron wave functions and in the next section we shall derive the equations obeyed by them.

When considering a many-electron problem, it must be remembered that electrons are identical particles. This is reflected in the form of the Hamiltonian: for example in (4.2), interchanging electrons i and j does not change the Hamiltonian and the same holds for the independent particle Hamiltonian (4.3). We say that the Hamiltonian commutes with the particle-exchange operator, P_{ij}. This operator acts on a many-electron state and it has the effect of interchanging the coordinates of particles i and j. For an N-particle state[3] Ψ

$$\mathbf{P}_{ij}\Psi(\mathbf{x}_1, \mathbf{x}_2, \ldots, \mathbf{x}_i, \ldots, \mathbf{x}_j, \ldots, \mathbf{x}_N) = \Psi(\mathbf{x}_1, \mathbf{x}_2, \ldots, \mathbf{x}_j, \ldots, \mathbf{x}_i, \ldots, \mathbf{x}_N). \quad (4.22)$$

In this equation, $\mathbf{x}_i$ is again the combined spin and orbital coordinate:

$$\mathbf{x}_i = (\mathbf{r}_i, s_i). \quad (4.23)$$

As P_{ij} is an Hermitian operator which commutes with the Hamiltonian, the eigenstates of the Hamiltonian are simultaneous eigenstates of P_{ij} with real eigenvalues. Furthermore, as $P_{ij}^2 = 1$ (interchanging a pair twice in a state brings the state back to its original form), its eigenvalue is either $+1$ or -1. It is an experimental fact that for particles with half-integer spin (fermions) the eigenvalue of the permutation operator is always -1, and for particles with integer spin (bosons) it is always $+1$. In the first case, the wave function is antisymmetric with respect to particle exchange and in the second case it is symmetric with respect to this operation. As electrons have spin-1/2, the wave function of a many-electron system is antisymmetric with respect to particle exchange.

Let us forget about antisymmetry for a moment. For the case of an independent-particle Hamiltonian, which is a sum of one-electron Hamiltonians as in (4.3), we can write the solution of the Schrödinger equation as a product of one-electron

[3] In order to clarify the role of the coordinate s_i, we note that for a single electron the wave function can be written as a *two-spinor*, that is, a two-dimensional vector, and s, which is a two-valued coordinate, selects one component of this spinor. When dealing with more particles (N), the two-spinors combine into a 2^N-dimensional one and the combined coordinates $s_1, \ldots, s_N$ select a component of this large spinor (which depends on the positions $\mathbf{r}_i$).

states:

$$\Psi(\mathbf{x}_1, \ldots, \mathbf{x}_N) = \psi_1(\mathbf{x}_1) \cdots \psi_N(\mathbf{x}_N). \tag{4.24}$$

The one-electron states ψ_k are eigenstates of the one-particle Hamiltonian, so they are orthogonal. The probability density for finding the particles with specific values $\mathbf{x}_1, \ldots, \mathbf{x}_N$ of their coordinates is given by

$$\rho(\mathbf{x}_1, \mathbf{x}_2, \ldots, \mathbf{x}_N) = |\psi_1(\mathbf{x}_1)|^2 |\psi_2(\mathbf{x}_2)|^2 \cdots |\psi_N(\mathbf{x}_N)|^2, \tag{4.25}$$

which is just the product of the one-electron probability densities. Such a probability distribution is called *uncorrelated*, and therefore we will use the term 'uncorrelated' for the wave function in (4.24) too.

Of course, the same state as (4.24) but with the spin-orbitals permuted, is a solution too, as are linear combinations of several such states. But we require antisymmetric states, and an antisymmetric linear combination of a minimal number of terms of the form (4.24) is given by

$$\Psi_{\text{AS}}(\mathbf{x}_1, \ldots, \mathbf{x}_N) = \frac{1}{\sqrt{N!}} \sum_{\mathbf{P}} \epsilon_P \mathbf{P} \psi_1(\mathbf{x}_1) \cdots \psi_N(\mathbf{x}_N). \tag{4.26}$$

P is a permutation operator which permutes the *coordinates* of the spin-orbitals only, and not their labels (if **P** acted on the latter too, it would have no effect at all!); alternatively, one could have **P** acting on the labels only, the choice is merely a matter of convention. The above-mentioned exchange operator is an example of this type of operator. In (4.26), all permutations are summed over and the states are multiplied by the sign ϵ_P of the permutation (the sign is $+1$ or -1 according to whether the permutation can be written as product of an even or an odd number of pair interchanges respectively).

We can write (4.26) in the form of a *Slater determinant*:

$$\Psi_{\text{AS}}(\mathbf{x}_1, \ldots, \mathbf{x}_N) = \frac{1}{\sqrt{N!}} \begin{vmatrix} \psi_1(\mathbf{x}_1) & \psi_2(\mathbf{x}_1) & \cdots & \psi_N(\mathbf{x}_1) \\ \psi_1(\mathbf{x}_2) & \psi_2(\mathbf{x}_2) & \cdots & \psi_N(\mathbf{x}_2) \\ \vdots & \vdots & & \vdots \\ \psi_1(\mathbf{x}_N) & \psi_2(\mathbf{x}_N) & \cdots & \psi_N(\mathbf{x}_N) \end{vmatrix}. \tag{4.27}$$

It is important to note that after this antisymmetrisation procedure the electrons are correlated. To see this, consider the probability density of finding one electron with coordinates $\mathbf{x}_1$ and another with $\mathbf{x}_2$:

$$\begin{aligned} \rho(\mathbf{x}_1, \mathbf{x}_2) &= \int \mathrm{d}x_3 \cdots \mathrm{d}x_N |\Psi_{\text{AS}}(\mathbf{x}_1, \ldots, \mathbf{x}_N)|^2 \\ &= \frac{1}{N(N-1)} \sum_{k,l} [|\psi_k(\mathbf{x}_1)|^2 |\psi_l(\mathbf{x}_2)|^2 - \psi_k^*(\mathbf{x}_1)\psi_k(\mathbf{x}_2)\psi_l^*(\mathbf{x}_2)\psi_l(\mathbf{x}_1)]. \end{aligned} \tag{4.28}$$

To find the probability of finding two electrons at positions $\mathbf{r}_1$ and $\mathbf{r}_2$,we must sum over the spin variables:

$$\rho(\mathbf{r}_1, \mathbf{r}_2) = \sum_{s_1, s_2} \rho(\mathbf{x}_1, \mathbf{x}_2). \tag{4.29}$$

For spin-orbitals that can be written as a product of a spatial orbital and a one-particle spin wave function, it is seen that for ψ_k and ψ_l having opposite spin, the second term vanishes and therefore opposite spin-orbitals are still uncorrelated (the first term of (4.28) obviously describes uncorrelated probabilities) but for equal spins, the two terms cancel when $\mathbf{r}_1 = \mathbf{r}_2$, so we see that electron pairs with parallel spin are kept apart. Every electron is surrounded by an 'exchange hole' [4] in which other electrons having the same spin are hardly found. Comparing (4.28) with the uncorrelated form (4.25), we see that exchange introduces correlation effects. However, the term 'correlation effects' is usually reserved for all correlations *apart from* exchange.

It is possible to construct Slater determinants from general spin-orbitals, i.e. not necessarily eigenstates of the one-electron Hamiltonian. It is even possible to take these spin-orbitals to be nonorthogonal. However, if there is overlap between two such spin-orbitals, this drops out in constructing the Slater determinant. Therefore we shall take the spin-orbitals from which the Slater determinant is constructed to be orthonormal.

Single Slater determinants form a basis in the space of all antisymmetric wave functions. In Section 4.10, we shall describe a method in which this fact is used to take correlations into account.

4.5 Self-consistency and exchange: Hartree–Fock theory

4.5.1 The Hartree–Fock equations – physical picture

Fock extended the Hartree equation (4.11) by taking antisymmetry into account. We first give the result which is known as the Hartree–Fock equation; the full derivation is given in Section 4.5.2 [5, 6]:

$$\mathcal{F}\psi_k = \epsilon_k \psi_k \text{ with} \tag{4.30}$$

$$\mathcal{F}\psi_k = \left[-\frac{1}{2}\nabla^2 - \sum_n \frac{Z_n}{|\mathbf{r} - \mathbf{R}_n|} \right] \psi_k(\mathbf{x}) + \sum_{l=1}^{N} \int \mathrm{d}x' |\psi_l(\mathbf{x}')|^2 \frac{1}{|\mathbf{r} - \mathbf{r}'|} \psi_k(\mathbf{x})$$
$$- \sum_{l=1}^{N} \int \mathrm{d}x' \psi_l^*(\mathbf{x}') \frac{1}{|\mathbf{r} - \mathbf{r}'|} \psi_k(\mathbf{x}') \psi_l(\mathbf{x}). \tag{4.31}$$

The operator $\mathcal{F}$ is called the *Fock operator*. The first three terms on the right hand side are the same as in those appearing in the Hartree equation. The fourth term is

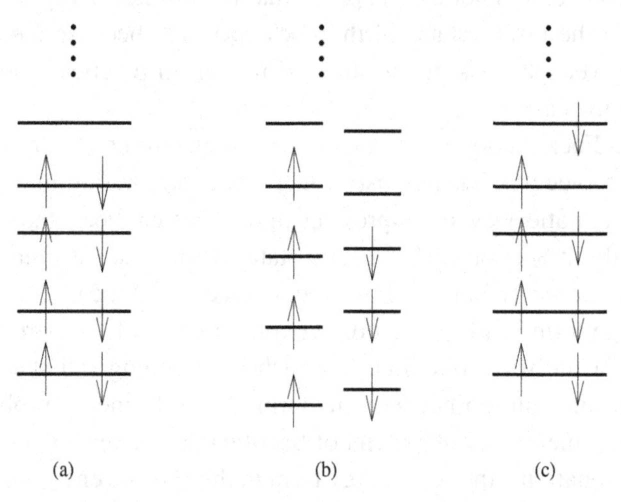

Figure 4.1. The Hartree–Fock spectrum. The figure shows how the levels are filled for (a) the ground state of an even number of electrons, (b) the ground state of an odd number of electrons and (c) an excited state in the spectrum of (a). Note that the spectrum in (c) does not correspond to the ground state; see Section 4.5.3. Instead it corresponds to the restricted approximation, in which the same set of energy levels is available for electrons with both spins.

the same as the third, with two spin-orbital labels k and l interchanged and a minus sign in front resulting from the antisymmetry of the wave function – it is called the exchange term. Note that this term is nonlocal: it is an operator acting on ψ_k, but its value at $\mathbf{r}$ is determined by the value assumed by ψ_k at all possible positions $\mathbf{r}'$.

A subtlety is that the eigenvalues ϵ_k of the Fock operator are not the energies of single electron orbitals, although they are related to the total energy by

$$E = \frac{1}{2} \sum_k [\epsilon_k + \langle \psi_k | h | \psi_k \rangle]. \tag{4.32}$$

In Section 4.5.3 we shall see that the individual levels ϵ_k can be related to excitation energies within some approximation.

It is clear that (4.31) is a nonlinear equation, which must be solved by a self-consistency iterative procedure analogously to the previous section. Sometimes the name 'self-consistent field theory' (SCF) is used for this type of approach. The self-consistency procedure is carried out as follows. Solving (4.31) yields an infinite spectrum. To find the ground state, we must take the lowest N eigenstates of this spectrum as the spin-orbitals of the electrons. These are the ψ_l which are then used to build the new Fock operator which is diagonalised again and the procedure is repeated over and over until convergence is achieved. Figure 4.1(a) and (b) gives a schematic representation of the Hartree–Fock spectrum and shows how the levels are filled. Of course, it is not clear a priori that the lowest energy of the system is found by filling the lowest states of the Fock spectrum because the energy is not simply a sum over the Fock eigenvalues. However, in practical applications this turns out to be the case.

The Hartree–Fock theory is the cornerstone of electronic structure calculations for atoms and molecules. There exists a method, configuration interaction, which provides a systematic way of improving upon Hartree–Fock theory; it will be described briefly in Section 4.10. In solid state physics, density functional theory is used mostly instead of Hartree–Fock theory (see Chapter 5).

The exchange term in (4.31) is a direct consequence of the particle exchange-antisymmetry of the wave function. It vanishes for orthogonal states k and l, so pairs with opposite spin do not feel this term. The self-energy problem with the Hartree potential mentioned at the end of Section 4.3.1 appears to be solved in the Hartree–Fock equations: the self-energy term in the Hartree energy is cancelled by the exchange contribution as a result of the antisymmetry.

The exchange contribution *lowers* the Coulomb interaction between the electrons, which can be viewed as a consequence of the fact that exchange keeps electrons with the same spin apart; see the discussion below Eq. (4.28). The dependence of this change in Coulomb energy on the electron density can be estimated using a simple classical argument. Suppose that in an electron gas with average density n, each electron occupies a volume which is not accessible to other electrons with like spin. This volume can be approximated by a sphere with radius $r_{\mathrm{c}} \propto n^{-1/3}$. Comparing the Coulomb interaction per volume for such a system with one in which the electrons are distributed homogeneously throughout space, we obtain

$$\Delta E_{\mathrm{C}} \approx n^2 \left[\int_{r_{\mathrm{c}}}^{\infty} r^2 \mathrm{d}r \frac{1}{r} - \int_0^{\infty} r^2 \mathrm{d}r \frac{1}{r} \right] = -n^2 \int_0^{r_{\mathrm{c}}} r^2 \mathrm{d}r \frac{1}{r} \propto -n^2 r_c^2 \propto n^{4/3}. \tag{4.33}$$

One of the two factors n in front of the integral comes from the average density seen by one electron, and the second factor counts how many electrons per volume experience this change in electrostatic energy. The $n^{4/3}$ dependence of the exchange contribution is also found in more sophisticated derivations [7] and we shall meet it again when discussing the local density approximation in the next chapter.

*4.5.2 Derivation of the Hartree–Fock equations

The derivation of the Fock equation consists of performing a variational calculation for the Schrödinger equation, where the subspace to which we shall confine ourselves is the space of all single Slater determinants like Eq. (4.27). We must therefore calculate the expectation value of the energy for an arbitrary Slater determinant using the Born–Oppenheimer Hamiltonian and then minimise the result with respect to the spin-orbitals in the determinant.

We write the Hamiltonian as follows:

$$H = \sum_i h(i) + \frac{1}{2} \sum_{i,j;\, i \neq j} g(i,j) \quad \text{with}$$

$$g(i,j) = \frac{1}{|\mathbf{r}_i - \mathbf{r}_j|} \quad \text{and} \tag{4.34}$$

$$h(i) = -\frac{1}{2}\nabla_i^2 - \sum_n \frac{Z_n}{|\mathbf{r}_i - \mathbf{R}_n|}.$$

$h(i)$ depends on $\mathbf{r}_i$ only and $g(i,j)$ on $\mathbf{r}_i$ and $\mathbf{r}_j$. Writing the Slater determinant ψ as a sum of products of spin-orbitals and using the orthonormality of the latter, it can easily be verified that this determinant is normalised, and for the matrix element of the one-electron part of the Hamiltonian, we find (see Problem 4.3)

$$\left\langle \Psi_{\text{AS}} \left| \sum_i h(i) \right| \Psi_{\text{AS}} \right\rangle = N \frac{(N-1)!}{N!} \sum_k \langle \psi_k | h | \psi_k \rangle$$

$$= \sum_k \langle \psi_k | h | \psi_k \rangle = \sum_k \int \mathrm{d}x\, \psi_k^*(\mathbf{x}) h(\mathbf{r}) \psi_k(\mathbf{x}). \tag{4.35}$$

By $\int \mathrm{d}x$ we denote an integral over the spatial coordinates and a sum over the spin-degrees of freedom as usual.

The matrix element of the two-electron term $g(i,j)$ for a Slater determinant not only gives a nonzero contribution when the spin-orbitals in the left and right hand sides of the inner product occur in the same order, but also for k and l interchanged on one side (the derivation is treated in Problem 4.3):

$$\left\langle \Psi_{\text{AS}} \left| \sum_{i,j} g(i,j) \right| \Psi_{\text{AS}} \right\rangle = \sum_{k,l} \langle \psi_k \psi_l | g | \psi_k \psi_l \rangle - \sum_{k,l} \langle \psi_k \psi_l | g | \psi_l \psi_k \rangle. \tag{4.36}$$

In this equation, the following notation is used:

$$\langle \psi_k \psi_l | g | \psi_m \psi_n \rangle = \int \mathrm{d}x_1 \mathrm{d}x_2 \psi_k^*(\mathbf{x}_1) \psi_l^*(\mathbf{x}_2) \frac{1}{|\mathbf{r}_1 - \mathbf{r}_2|} \psi_m(\mathbf{x}_1) \psi_n(\mathbf{x}_2). \tag{4.37}$$

In summary, we obtain for the expectation value of the energy:

$$E = \sum_k \langle \psi_k | h | \psi_k \rangle + \frac{1}{2} \sum_{kl} [\langle \psi_k \psi_l | g | \psi_k \psi_l \rangle - \langle \psi_k \psi_l | g | \psi_l \psi_k \rangle]. \tag{4.38}$$

We now define the operators

$$J_k(\mathbf{x})\psi(\mathbf{x}) = \int \psi_k^*(\mathbf{x}')\frac{1}{r_{12}}\psi_k(\mathbf{x}')\psi(\mathbf{x})\,\mathrm{d}x' \quad \text{and} \tag{4.39a}$$

$$K_k(\mathbf{x})\psi(\mathbf{x}) = \int \psi_k^*(\mathbf{x}')\frac{1}{r_{12}}\psi(\mathbf{x}')\psi_k(\mathbf{x})\,\mathrm{d}x' \tag{4.39b}$$

and furthermore

$$J = \sum_k J_k; \quad K = \sum_k K_k. \tag{4.40}$$

J is called the *Coulomb* operator and K the *exchange* operator as it can be obtained from the Coulomb operator by interchanging the two rightmost spin-orbitals. In terms of these operators, we can write the energy as

$$E = \sum_k \left\langle \psi_k \left| h + \frac{1}{2}(J - K) \right| \psi_k \right\rangle. \tag{4.41}$$

This is the energy-functional for a Slater determinant. We determine the minimum of this functional as a function of the spin-orbitals ψ_k, and the spin-orbitals for which this minimum is assumed give us the many-electron ground state. Notice however that the variation in the spin-orbitals ψ_k is not completely arbitrary, but should respect the orthonormality relation:

$$\langle \psi_k | \psi_l \rangle = \delta_{kl}. \tag{4.42}$$

This implies that we have a minimisation problem with constraints, which can be solved using the Lagrange multiplier theorem. Note that there are only $N(N+1)/2$ independent constraints as $\langle \psi_k | \psi_l \rangle = \langle \psi_l | \psi_k \rangle^*$. Using the Lagrange multipliers Λ_{kl} for the constraints (4.42), we have

$$\delta E - \sum_{kl} \Lambda_{kl}[\langle \delta\psi_k | \psi_l \rangle + \langle \psi_k | \delta\psi_l \rangle] = 0 \tag{4.43}$$

with

$$\begin{aligned}
\delta E = & \sum_k \langle \delta\psi_k | h | \psi_k \rangle + \text{complex conj.} \\
& + \frac{1}{2}\sum_{kl} (\langle \delta\psi_k \psi_l | g | \psi_k \psi_l \rangle + \langle \psi_l \delta\psi_k | g | \psi_l \psi_k \rangle \\
& - \langle \delta\psi_k \psi_l | g | \psi_l \psi_k \rangle - \langle \psi_l \delta\psi_k | g | \psi_k \psi_l \rangle) + \text{complex conj.} \\
= & \sum_k \langle \delta\psi_k | h | \psi_k \rangle + \text{complex conj.} \\
& + \sum_{kl} (\langle \delta\psi_k \psi_l | g | \psi_k \psi_l \rangle - \langle \delta\psi_k \psi_l | g | \psi_l \psi_k \rangle) + \text{complex conj.}
\end{aligned} \tag{4.44}$$

where in the second step the following symmetry property of the two-electron matrix elements is used:

$$\langle\psi_k\psi_l|g|\psi_m\psi_n\rangle = \langle\psi_l\psi_k|g|\psi_n\psi_m\rangle. \tag{4.45}$$

Note furthermore that because of the symmetry of the constraint equations, we must have $\Lambda_{kl} = \Lambda^*_{lk}$. Eq. (4.44) can be rewritten as

$$\delta E = \sum_k \langle\delta\psi_k|\mathcal{F}|\psi_k\rangle + \langle\psi_k|\mathcal{F}|\delta\psi_k\rangle \tag{4.46}$$

with

$$\mathcal{F} = h + J - K. \tag{4.47}$$

The Hermitian operator $\mathcal{F}$ is the Fock operator, now formulated in terms of the operators J and K. It is important to note that in this equation, J and K occur with the same prefactor as h, in contrast to Eq. (4.41) in which both J and K have a factor $1/2$ compared with h. This extra factor is caused by the presence of two spin-orbitals in the expressions for J and K which yield extra terms in the derivative of the energy. The matrix elements of the Fock operator with respect to the spin-orbitals ψ_k are

$$\langle\psi_k|\mathcal{F}|\psi_l\rangle = h_{kl} + \sum_{k'}[\langle\psi_k\psi_{k'}|g|\psi_l\psi_{k'}\rangle - \langle\psi_k\psi_{k'}|g|\psi_{k'}\psi_l\rangle]. \tag{4.48}$$

We finally arrive at the equation

$$\langle\delta\psi_k|\mathcal{F}|\psi_k\rangle + \langle\psi_k|\mathcal{F}|\delta\psi_k\rangle - \sum_l \Lambda_{kl}(\langle\delta\psi_k|\psi_l\rangle + \langle\psi_l|\delta\psi_k\rangle) = 0 \tag{4.49}$$

and since $\delta\psi$ is small but arbitrary, this, with $\Lambda_{kl} = \Lambda^*_{lk}$, leads to

$$\mathcal{F}\psi_k = \sum_l \Lambda_{kl}\psi_l. \tag{4.50}$$

The Lagrange parameters Λ_{kl} in this equation cannot be chosen freely: they must be such that the solutions ψ_k form an orthonormal set. An obvious solution of the above equation is found by taking the ψ_k as the eigenvectors of the Fock operators with eigenvalues ϵ_k, and $\Lambda_{kl} = \epsilon_k\delta_{kl}$:

$$\mathcal{F}\psi_k = \epsilon_k\psi_k. \tag{4.51}$$

This equation is the same as (4.31), presented at the beginning of the previous subsection. We can find other solutions to the general Fock equation (4.50) by transforming the set of eigenstates $\{\psi_k\}$ according to a unitary transformation, defined by a (unitary) matrix U:

$$\psi'_k = \sum_l U_{kl}\psi_l. \tag{4.52}$$

The resulting states ψ'_k then form an orthonormal set, satisfying (4.50) with

$$\Lambda_{kl} = \sum_{lm} U_{km}\epsilon_m U^\dagger_{ml}. \tag{4.53}$$

In fact, a unitary transformation of the set $\{\psi_k\}$ leaves the Slater determinant unchanged (see Problem 4.7).

Equation (4.51) has the form of an ordinary Schrödinger equation although the eigenvalues ϵ_k are identified as Lagrange multipliers rather than as energies; nevertheless they are often called 'orbital energies'. From (4.51) and (4.38) it can be seen that the energy is related to the parameters ϵ_k by

$$E = \frac{1}{2}\sum_k [\epsilon_k + \langle\psi_k|h|\psi_k\rangle] = \sum_k \left[\epsilon_k - \frac{1}{2}\langle k|J - K|k\rangle\right]. \tag{4.54}$$

The second form shows how the Coulomb and exchange contribution must be subtracted from the sum of the Fock levels to avoid counting the two-electron integrals twice.

In the previous section we have already seen how the self-consistency procedure for solving the resulting equations is carried out.

4.5.3 *Koopman's theorem*

If we were to calculate an excited state, we would have to take the lowest $N - 1$ spin-orbitals from the Fock spectrum and one excited spin-orbital for example (see Fig. 4.1c), and carry out the self-consistency procedure for this configuration. The resulting eigenstates will differ from the corresponding eigenstates in the ground state. If we assume that the states do not vary appreciably when constructing the Slater determinant from excited spin-orbitals instead of the ground state ones, we can predict excitation energies from a ground state calculation. It turns out that – within the approximation that the spin-orbitals are those of the ground state – the difference between the sums of the eigenvalues, ϵ_k, of the ground state and excited state configuration, is equal to the real energy difference; see Problem 4.6. This is known as *Koopman's theorem*. This is not really a theorem but a way of approximating excitation energies which turns out to work well for many systems. For further reading, see Refs. [5, 6, 8].

4.6 Basis functions

In the derivation leading to (4.51) (or (4.31)), the possible solutions of the Schrödinger equation were restricted to the space of single Slater determinants. To solve the resulting eigenvalue equation, another variational principle in the same spirit as in the previous chapter and in Section 4.3.2 can be used, that is, expanding the spin-orbitals ψ_k as linear combinations of a finite number of basis states χ_p:

$$\psi_k(\mathbf{x}) = \sum_{p=1}^{M} C_{pk}\chi_p(\mathbf{x}). \tag{4.55}$$

Then (4.51) assumes a matrix form

$$\mathbf{FC}_k = \epsilon_k \mathbf{SC}_k \tag{4.56}$$

where $\mathbf{S}$ is the overlap matrix for the basis used.

In the next subsection we shall consider how spin and orbital parts are combined in the basis sets and in Section 4.6.2 we shall discuss the form of the orbital basis functions.

4.6.1 Closed- and open-shell systems

In a *closed-shell* system, the levels are occupied by two electrons with opposite spin whereas in an *open-shell* system there are partially filled levels containing only one electron. If the number of electrons is even, the system does not necessarily have to be closed-shell since there may be degenerate levels (apart from spin-degeneracy) each containing one electron – or we might be considering an excited state in which an electron is pushed up to a higher level. If the number of electrons is odd, the system will always be open-shell.

Consider the addition of an electron to a closed-shell system. The new electron will interact differently with the spin-up and -down electrons present in the system, as exchange is felt by parallel spin pairs only. Therefore, if the levels of the system without the extra electron are spin-up and -down degenerate, they will now split into two sublevels with different orbital dependence, the lower sublevel having the same spin as the new electron.[4] We see that the spin-up and -down degeneracy of the levels of a closed-shell system is lifted in the open-shell case.

It is important to note that even when the number of electrons is even, the unrestricted solution may be different from the restricted one. To see this, consider again the discussion of the hydrogen molecule in Section 4.3.1. A possible description of the state within the unrestricted scheme is

$$\psi(\mathbf{x}_1, \mathbf{x}_2) = \frac{1}{\sqrt{2}}[u(\mathbf{r}_1 - \mathbf{R}_A)\alpha(s_1)u(\mathbf{r}_2 - \mathbf{R}_B)\beta(s_2) \\ - u(\mathbf{r}_2 - \mathbf{R}_A)\alpha(s_2)u(\mathbf{r}_1 - \mathbf{R}_B)\beta(s_1)]. \tag{4.57}$$

[4] An exception to this rule occurs when the Coulomb interaction between the degenerate levels and the new electron vanishes as a result of symmetry.

This describes two electrons located at different nuclei, which is correct for large nuclear separation, but this is not an eigenstate of the total spin operator. When the nuclei are separated, the state crosses over from a restricted to an unrestricted one. The distance at which this happens is the Coulson–Fisher point [9].

In a closed-shell system the $2N$ orbitals can be grouped in pairs with the same orbital dependence but with opposite spin, thus reflecting the spin-degeneracy:

$$\{\psi_{2k-1}(\mathbf{x}), \psi_{2k}(\mathbf{x})\} = \{\phi_k(\mathbf{r})\alpha(s), \phi_k(\mathbf{r})\beta(s)\}, \quad k = 1, \ldots, N. \tag{4.58}$$

The $\phi_k(\mathbf{r})$ are the spatial orbitals and $\alpha(s)$, $\beta(s)$ are the up and down spin-states respectively. For an open-shell system, such pairing does not occur for all levels, and to obtain accurate results, it is necessary to allow for a different orbital dependence for each spin-orbital in most cases. Even for an open-shell system it is possible to impose the restriction (4.58) on the spin-orbitals, neglecting the splitting of the latter, but the results will be less accurate in that case. Calculations with the spin-orbitals paired as in (4.58) are called restricted Hartree–Fock (RHF) and those in which all spin-orbitals are allowed to have a different spatial dependence are called unrestricted Hartree–Fock (UHF). UHF eigenstates are usually inconvenient because they are not eigenstates of the total spin-operator, as can easily be verified by combining two different orbitals with a spin-up and -down function respectively. On the other hand, the energy is more accurate.

We shall now rewrite the Hartree–Fock equations for RHF using the special structure of the set of spin-orbitals given in (4.58). As we have seen in the previous section, the general form of the Fock operator is

$$\mathcal{F} = h + J - K \tag{4.59}$$

with

$$\begin{aligned} J(\mathbf{x})\psi(\mathbf{x}) &= \sum_l \int \mathrm{d}x'\ \psi_l^*(\mathbf{x}')\psi_l(\mathbf{x}')\frac{1}{r_{12}}\psi(\mathbf{x}); \\ K(\mathbf{x})\psi(\mathbf{x}) &= \sum_l \int \mathrm{d}x'\ \psi_l^*(\mathbf{x}')\psi(\mathbf{x}')\frac{1}{r_{12}}\psi_l(\mathbf{x}). \end{aligned} \tag{4.60}$$

The sum over l is over all occupied Fock levels. As the Fock operator depends explicitly on the spatial coordinate only (there is an implicit spin-dependence via the spin-orbitals occurring in the Coulomb and exchange operators), it is possible to eliminate the spin degrees of freedom by summing over them and find an operator acting only on the spatial orbitals $\phi(\mathbf{r})$. The uncoupled single-particle Hamiltonian h remains the same since it contains neither explicit nor implicit spin-dependence, and from (4.60) it is seen that the Coulomb and exchange operators, written in terms of the orbital parts only, read

$$\tilde{J}(\mathbf{r})\phi(\mathbf{r}) = 2\sum_l \int d^3r' \phi_l^*(\mathbf{r}')\phi_l(\mathbf{r}')\frac{1}{|\mathbf{r}'-\mathbf{r}|}\phi(\mathbf{r}) \quad \text{and}$$
$$\tilde{K}(\mathbf{r})\phi(\mathbf{r}) = \sum_l \int d^3r' \phi_l^*(\mathbf{r}')\phi(\mathbf{r}')\frac{1}{|\mathbf{r}'-\mathbf{r}|}\phi_l(\mathbf{r}). \tag{4.61}$$

In contrast with Eq. (4.60), the sums over l run over half the number of electrons because the spin degrees of freedom have been summed over. The Fock operator now becomes

$$\tilde{\mathcal{F}}(\mathbf{r}) = h(\mathbf{r}) + 2\tilde{J}(\mathbf{r}) - \tilde{K}(\mathbf{r}). \tag{4.62}$$

From now on, we shall only use this spatial form of the Fock operator and drop the tilde from the operators in (4.61) and (4.62). The corresponding expression for the energy is found analogously and is given by

$$E_g = 2\sum_k \langle\phi_k|h|\phi_k\rangle + \sum_k (2\langle\phi_k|J|\phi_k\rangle - \langle\phi_k|K|\phi_k\rangle). \tag{4.63}$$

It is possible to solve the Fock equation using a finite basis set, in the same spirit as the helium calculation of Section 4.3.2. The spin part of the basis functions is simply $\alpha(s)$ or $\beta(s)$ (spin-up and -down respectively) and the orbital part $\chi_p(\mathbf{r})$ needs to be specified – this will be done in the next section. For a given basis $\chi_p(\mathbf{r})$, we obtain the following matrix equation, which is known as the *Roothaan equation*:

$$\mathbf{FC}_k = \epsilon\mathbf{SC}_k, \tag{4.64}$$

similar to (4.51), but now $\mathbf{S}$ is the overlap matrix for the *orbital* basis $\chi_p(\mathbf{r})$ and the matrix $\mathbf{F}$ is given by

$$F_{pq} = h_{pq} + \sum_k\sum_{rs} C_{rk}^* C_{sk}(2\langle pr|g|qs\rangle - \langle pr|g|sq\rangle) \tag{4.65}$$

where

$$h_{pq} = \langle p|h|q\rangle = \int d^3r\, \chi_p^*(\mathbf{r})\left[-\frac{1}{2}\nabla^2 - \sum_n \frac{Z_n}{|\mathbf{R}_n - \mathbf{r}|}\right]\chi_q(\mathbf{r}), \tag{4.66}$$

and

$$\langle pr|g|qs\rangle = \int d^3r_1 d^3r_2 \chi_p^*(\mathbf{r}_1)\chi_r^*(\mathbf{r}_2)\frac{1}{|\mathbf{r}_1 - \mathbf{r}_2|}\chi_q(\mathbf{r}_1)\chi_s(\mathbf{r}_2). \tag{4.67}$$

k labels the orbitals ϕ_k and p, q, r and s label the basis functions. Generally, sums over labels k and l run over the occupied orbitals, and sums over p, q, r, s run over the functions in the basis set.

It is convenient to introduce the density matrix which (for RHF) is defined as

$$P_{pq} = 2 \sum_k C_{pk} C^*_{qk}. \tag{4.68}$$

This is the matrix representation of the operator

$$\rho = 2 \sum_k |\phi_k\rangle \langle \phi_k| \tag{4.69}$$

which is recognised as the usual definition of the density matrix in quantum theory (the factor 2 is due to the spin). Using (4.68), the Fock matrix can be written as

$$F_{pq} = h_{pq} + \frac{1}{2} \sum_{rs} P_{sr} [2\langle pr|g|qs\rangle - \langle pr|g|sq\rangle], \tag{4.70}$$

and the energy is given by

$$E = \sum_{pq} P_{pq} h_{pq} + \frac{1}{2} \sum_{pqrs} P_{pq} P_{sr} \left[\langle pr|g|qs\rangle - \frac{1}{2} \langle pr|g|sq\rangle \right]. \tag{4.71}$$

For the UHF case, it is convenient to define an orbital basis $\chi_p(\mathbf{r})$ and the spin-up orbitals are now represented by the vector $\mathbf{C}^+$ and the spin-down ones by $\mathbf{C}^-$. Using these vectors to reformulate the Hartree–Fock equations, the so-called Pople–Nesbet equations are obtained:

$$\begin{aligned} \mathbf{F}^+ \mathbf{C}^+ &= \epsilon^+ \mathbf{S} \mathbf{C}^+ \\ \mathbf{F}^- \mathbf{C}^- &= \epsilon^- \mathbf{S} \mathbf{C}^- \end{aligned} \tag{4.72}$$

with

$$\begin{aligned} F^{\pm}_{pq} = h_{pq} &+ \sum_{k_\pm} \sum_{rs} C^{\pm *}_{rk_\pm} C^{\pm}_{sk_\pm} [\langle pr|g|qs\rangle - \langle pr|g|sq\rangle] \\ &+ \sum_{k_\mp} \sum_{rs} C^{\mp *}_{rk_\mp} C^{\mp}_{sk_\mp} \langle pr|g|qs\rangle. \end{aligned} \tag{4.73}$$

In practice, real orbital basis functions are used, so that complex conjugates can be removed from the C_{pk} in Eqs. (4.65), (4.68) and (4.73). In the following, we shall restrict ourselves to RHF.

4.6.2 Basis functions: STO and GTO

In this subsection, we discuss the basis functions used in the atomic and molecular Hartree–Fock programs. As already noted in Chapter 3, the basis must be chosen carefully: the matrix diagonalisation we must perform scales with the third power of the number of basis functions, so a small basis set is desirable which is able to model the exact solutions to the Fock equations accurately. A molecule consists of atoms which, in isolation, have a number of atomic orbitals occupied by the electrons. If we put the atoms together in a molecule, the orbitals with low energies will be slightly perturbed by the new environment and the valence electrons will now orbit around more than one nucleus, thus binding the molecule together. In the molecule, the electrons now occupy *molecular orbitals* (MO). In constructing a basis, it turns out to be efficient to start from the atomic orbitals. The one-electron wave functions that can be constructed from these orbitals are linear combinations of atomic orbitals (LCAO). The solutions to the HF equations, which have the form

$$\phi_k(\mathbf{r}) = \sum_p C_{pk}\chi_p(\mathbf{r}), \tag{4.74}$$

are the molecular orbitals, written in LCAO form.

Analytic forms of the atomic orbitals are only known for the hydrogen atom but they can be used for more general systems. The orbitals of the hydrogen atom have the following form:

$$\chi(\mathbf{r}) = f_{n-1}(r) r^{-l} P_l(x, y, z) \mathrm{e}^{-r/n}, \tag{4.75}$$

where l is the angular momentum quantum number, P_l is a polynomial in x, y and z of degree l containing the angular dependence; $f_{n-1}(r)$ is a polynomial in r of degree $n-1$; n is an integer (r is expressed in atomic units a_0). This leads to the following general form of atomic orbital basis functions:

$$\chi_\zeta(\mathbf{r}) = r^m P_l(x, y, z) \mathrm{e}^{-\zeta|\mathbf{r}-\mathbf{R}_A|} \tag{4.76}$$

which is centred around a nucleus located at $\mathbf{R}_A$. Functions of this form are called Slater type orbitals (STO). The parameter ζ, defining the range of the orbital, remains to be determined; P_l is taken the same as for the hydrogen atom. For *atomic* Hartree–Fock calculations, this basis yields accurate results with a restricted basis set size. However, in molecular calculations, integrals involving products of two and four basis functions centred at different nuclei are needed, and these are hard to calculate since the product of the two exponentials,

$$\mathrm{e}^{-\zeta_1|\mathbf{r}-\mathbf{R}_A|}\mathrm{e}^{-\zeta_2|\mathbf{r}-\mathbf{R}_B|}, \tag{4.77}$$

is a complicated expression in $\mathbf{r}$. A solution might be to evaluate these integrals numerically, but for a large basis set this is impractical.

Another basis set which avoids this problem, was proposed in 1950 by Boys [10], who replaced the simple exponential in (4.76) by a Gaussian function:

$$\chi_\alpha(\mathbf{r}) = P_M(x, y, z)\mathrm{e}^{-\alpha(\mathbf{r}-\mathbf{R}_A)^2}. \tag{4.78}$$

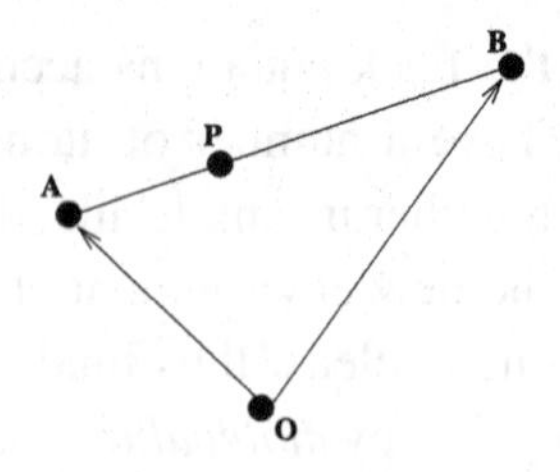

Figure 4.2. Positions in the Gaussian product theorem (4.79).

These functions are called *primitive basis functions* for reasons which will be explained below. These Gaussian type orbitals (GTO) have the nice property that the product of two such functions centred on different nuclei again has a Gaussian form as in (4.78):

$$P_M(x,y,z)\mathrm{e}^{-\alpha(\mathbf{r}-\mathbf{R}_A)^2} Q_N(x,y,z)\mathrm{e}^{-\beta(\mathbf{r}-\mathbf{R}_B)^2} = R_{N+M}(x,y,z)\mathrm{e}^{-(\alpha+\beta)(\mathbf{r}-\mathbf{R}_P)^2}. \tag{4.79}$$

Here, $\mathbf{R}_P$ is the 'centre of mass' of the two points $\mathbf{R}_A$ and $\mathbf{R}_B$ with masses α and β:

$$\mathbf{R}_P = \frac{\alpha\mathbf{R}_A + \beta\mathbf{R}_B}{\alpha+\beta} \tag{4.80}$$

(see also Figure 4.2), and R_{M+N} is a polynomial of degree $M+N$. Equation (4.79) is easy to prove; it is known as the 'Gaussian product theorem'. This theorem makes it possible for the integrals involved in the Hartree–Fock equations to be either calculated analytically or reduced to an expression suitable for fast evaluation on a computer. In Section 4.8 we shall derive some of these integrals.

The polynomials P_M in Eq. (4.78) contain the angular-dependent part of the orbitals, which is given by the spherical harmonics $Y^l_m(\theta,\phi)$. For $l = 0$, these functions are spherically symmetric (no angular dependence) – hence a 1s-orbital (having no nodes) is given as

$$\chi^{(\mathrm{s})}_\alpha(\mathbf{r}) = \mathrm{e}^{-\alpha(\mathbf{r}-\mathbf{R}_A)^2}. \tag{4.81}$$

Note that we need not normalise our basis functions: the overlap matrix will ensure proper normalisation of the final molecular orbitals. For $l = 1$, the L_z-quantum number m can take on the three different values 1, 0 and -1, so there are three p-orbitals, and an explicit GTO representation is

$$\chi^{(\mathrm{p}_x)}_\alpha(\mathbf{r}) = x\mathrm{e}^{-\alpha(\mathbf{r}-\mathbf{R}_A)^2}, \tag{4.82}$$

and similarly for p_y and p_z. Proceeding in the same fashion for $l = 2$, we find six quadratic factors x^2, y^2, z^2, xy, yz and xz before the Gaussian exponential, but there are only five d-states! This paradox is solved by noticing that the linear combination

$$(x^2 + y^2 + z^2)\mathrm{e}^{-\alpha(\mathbf{r}-\mathbf{R}_A)^2} \tag{4.83}$$

has the symmetry of an s-orbital, and therefore, instead of x^2, y^2 and z^2, only the orbitals $x^2 - y^2$ and $3z^2 - r^2$ are used (any independent combination is allowed), thus arriving at five d-states.

GTOs are widely used for molecular calculations, and from now on we shall restrict ourselves to basis functions of this form. The simplest basis consists of one GTO per atomic orbital; it is called a *minimal basis*. The parameter α in the exponent must somehow be chosen such that the GTO fits the atomic orbital in an optimal way. However, since there is only one parameter to be fitted in the minimal basis, this will give poor results, and in general more GTOs per atomic orbital are used. This means that a 1s basis orbital is now given as a linear combination of Gaussian functions:

$$\sum_p D_p \mathrm{e}^{-\alpha_p r^2}. \tag{4.84}$$

As the parameters α_p occur in the exponent, determination of the best combination (D_p, α_p) is a nonlinear optimisation problem. We shall not go into details of solving such a problem (see Ref. [11]), but discuss the criterion according to which the best values for (D_p, α_p) are selected. A first approach is to take Hartree–Fock orbitals resulting from an *atomic* calculation, perhaps determined using Slater type orbitals, and to fit the form (4.84) to these orbitals. A second way is to perform the *atomic* Hartree–Fock calculation using Gaussian primitive basis functions and determine the optimal set as a solution to the nonlinear variational problem in the space given by (4.84).

Suppose we have determined the optimal set (D_p, α_p), then there are in principle two options for constructing the basis set. The first option is to incorporate for each exponential parameter α_p, the Gaussian function

$$\mathrm{e}^{-\alpha_p(\mathbf{r}-\mathbf{R}_A)^2} \tag{4.85}$$

into the basis, that is, the values of the D_p-parameters are relaxed since the prefactors of the Gaussian primitive functions can vary at will with this basis. A second approach is to take the linear combination (4.84) with the optimal set of (D_p, α_p) as a *single* basis function and add it to the basis set, i.e. keeping the D_p fixed as well as the α_p. If we have optimised the solution (4.84) using four primitive functions, the second option yields a basis four times smaller than the first but, because of its lack of flexibility (remember the D_p are kept fixed), it will result in lower accuracy. The procedure of taking fixed linear combinations of Gaussian primitive functions as a single basis function is called *contraction*; the basis set is called a *contracted set*. The difference between the GTOs from which the basis functions are constructed

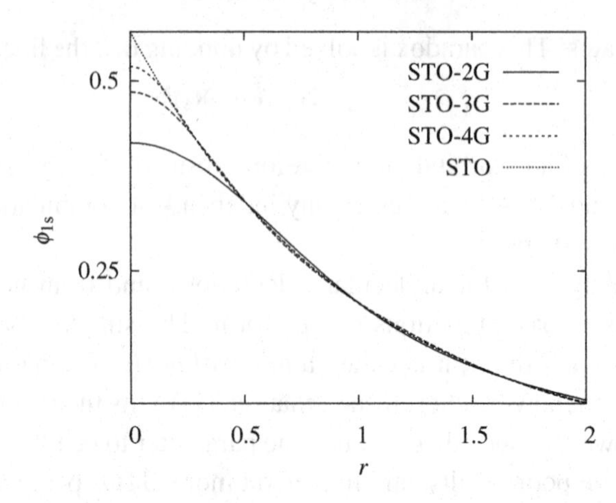

Figure 4.3. Approximation of a 1s Slater orbital (STO) by STO-2G up to STO-4G basis functions.

and the basis functions themselves is the reason why the GTOs were called *primitive* basis functions above: they are used to build contracted basis functions. If all N primitive basis functions are contracted into one normal basis function, the basis set is called a STO-NG basis, denoting that N GTOs have been used to fit a Slater type orbital. Figure 4.3 shows how a Slater orbital is approximated by STO-NG basis functions for $N = 2, 3, 4$. Note that deviations from the exact Slater orbital are only noticeable for small r, and these are usually suppressed, as in volume integrals the integrands are weighted with a factor r^2.

In most basis sets, an intermediate approach is taken, in which the set of primitive Gaussians is split into two (or more) parts. The primitive Gaussian functions in each such part are contracted into one basis function. A common example is the STO-31 basis set, in which the three basis functions with shortest range (i.e. highest α_p) are contracted (using the corresponding D_ps) into one basis function and the remaining one (with the longest range) is taken as a second basis function. The reason for splitting the basis in this way is that the perturbation of the orbitals by the environment will affect primarily the long-range tails, and leave the short-range part essentially unchanged.

There exist many other basis sets, like split-basis sets, in which from every primitive Gaussian two new ones are constructed, one with a range slightly shorter than in the original function (α slightly larger) and another with a slightly longer range (α slightly smaller), so that it is possible for an orbital in a molecule to assume a slightly contracted or expanded form with respect to the atomic orbital. For more details concerning the various basis sets, see Ref. [12].

An important consideration is the symmetry of the orbitals which should be taken into account. As an example, consider a HF calculation for H_2. The starting point for this calculation is the atomic orbitals of the ground state of the isolated H-atom. In the latter, only the (spherically symmetric) 1s orbitals are filled, and in the H_2 molecule, these orbitals will merge into a single molecular orbital which is given by the sum of the two atomic orbitals plus a correction containing a substantial contribution from the atomic p_z-orbitals (the axis connecting the two nuclei is taken to be the z-axis). This shows that it is sensible to include these p_z-orbitals in the basis, even though they are not occupied in the ground state of the isolated atom. Such basis states, which are included in the basis set to make it possible for the basis to represent the polarisation of the atom by its anisotropic environment, are called *polarisation orbitals*. When calculating (dipole, quadrupole, ...) moments by switching on an electric or magnetic field and studying the response of the orbitals to this field, it is essential to include such states into the basis.

4.7 The structure of a Hartree–Fock computer program

In this section we describe the structure of a typical computer program for solving the Roothaan equations. The program for the helium atom as described in Section 4.3.2 contained most of the features present in Hartree–Fock programs already – the treatment given here is a generalisation for arbitrary molecules.

As the most time-consuming steps in this program involve the two-electron integrals, we consider these in some detail in the next subsection. The general scheme of the HF program is then given in Section 4.7.2.

4.7.1 The two-electron integrals

The two-electron integrals are the quantities

$$\langle pr|g|qs\rangle = \int d^3r d^3r' \chi_p(\mathbf{r})\chi_r(\mathbf{r}') \frac{1}{|\mathbf{r}-\mathbf{r}'|} \chi_q(\mathbf{r})\chi_s(\mathbf{r}'). \tag{4.86}$$

(Do not confuse the *label* r with the orbital coordinate $\mathbf{r}$!) The two-electron matrix elements obey the following symmetry relations:

$$p \leftrightarrow q \quad r \leftrightarrow s \quad p, q \leftrightarrow r, s. \tag{4.87}$$

This implies that, starting with K basis functions, there are roughly $K^4/8$ different two-electron matrix elements, since each of the symmetries in (4.87) allows the reduction of the number of different matrix elements to be stored by a factor of about 2.

The subset of two-electron matrix elements to be calculated can be selected in the following way. Because p and q can be interchanged, we can take $p \geq q$. As the pair p, q can be interchanged with the pair r, s, we may also take $p \geq r$. The range of s depends on whether $p = r$ or not. If $p \neq r$, we have $s \leq r$, but for $p = r$, the fact that p, q can be interchanged with r, s means that the range of s can be restricted to $s \leq q$. All in all, a loop over the different two-electron matrix elements, using K basis vectors and taking into account all symmetries, reads:

```
FOR p = 1 TO K DO
   FOR q = 1 TO p DO
      FOR r = 1 TO p − 1 DO
         FOR s = 1 TO r DO
            ......
            ......
         END FOR
      END FOR
      r = p
      FOR s = 1 TO q DO
         ......
         ......
      END FOR
   END FOR
END FOR
```

Suppose we are dealing with a molecule consisting of 6 atoms, each having 12 basis functions, then $K^4/8$ is about 3.4 million. It is not always possible to keep all these numbers in core-memory, in which case they must be stored on disk. In earlier days (before about 1975) this was done by storing them in a prescribed order in a one-dimensional array whose index could be converted into the numbers p, q, r and s. This was inefficient, however, since because of symmetry, many matrix elements may vanish or be equal, or matrix elements may vanish because they involve basis orbitals lying far apart, i.e. having negligible overlap, so nowadays only the non-negligible matrix elements are stored together with the values of the corresponding indices p, q, r and s.

4.7.2 General scheme of the HF program

First, we give an outline and then fill in the details [6, 13].

- Input data;
- Determine matrices;
- Bring overlap matrix to unit form;
- Make a first guess for the density matrix;
- REPEAT
  - Calculate Coulomb and exchange contributions to the Fock matrix;
  - Add these terms to the uncoupled one-electron Hamiltonian h_{pq};

- Diagonalise the Fock matrix;
- Construct a new density matrix **P** from the eigenvectors found;

• UNTIL converged;
• Output.

Input data: The user provides the coordinates of the positions of the nuclei $\mathbf{R}_n$, the atomic numbers Z_n, the total number of electrons N and a basis set to be used (most programs have several basis sets built into them).

Determine matrices: All matrices that do not depend on the eigenvectors $\mathbf{C}_k$ can be determined here: the overlap matrix S_{pq}, the uncoupled one-electron Hamiltonian h_{pq}, and the two-electron integrals $\langle pr|g|qs\rangle$.

Bring overlap matrix to unit form: This is the procedure described in Section 3.3. It results in a matrix **V** defining the basis transformation which brings **S** to unit form.

Make a first guess for the density matrix: It is possible to take $P_{pq} = 0$ as a first guess: this implies that the electrons feel the nuclei but not each other. It is then expected that the HF self-consistency procedure will converge with this choice. More elaborate guesses for P can be used in order to increase the probability of convergence, but we shall not go into this here.

Calculate Coulomb and exchange contributions to the Fock matrix: *This is the most time-consuming step in the program!*
The exchange and Coulomb contributions are stored in a matrix **G** according to the following formula (see also (4.70)):

$$G_{pq} = \sum_{rs} P_{rs} \left[\langle pr|g|qs\rangle - \frac{1}{2}\langle pr|g|sq\rangle \right]. \tag{4.88}$$

This is done in a loop of the type displayed in the previous subsection. For each combination of the indices p, q, r and s occuring in this loop it must be checked to which elements of **G** the corresponding two-electron matrix element contributes. This, however, depends on which of the indices coincide. When all four indices are different for example, one obtains the following contributions to the matrix **G**:

$$\begin{aligned}
G_{pq} &= G_{pq} + 2\langle pr|g|qs\rangle P_{rs} \\
G_{rs} &= G_{rs} + 2\langle pr|g|qs\rangle P_{pq} \\
G_{ps} &= G_{ps} - 1/2\langle pr|g|qs\rangle P_{qr} \\
G_{qr} &= G_{qr} - 1/2\langle pr|g|qs\rangle P_{ps} \\
G_{pr} &= G_{pr} - 1/2\langle pr|g|qs\rangle P_{qs} \\
G_{qs} &= G_{qs} - 1/2\langle pr|g|qs\rangle P_{pr}.
\end{aligned} \tag{4.89}$$

Counting all different configurations of the indices, like $p = r \neq q, s$ and $q \neq s$, and so on, 14 different cases are found, and for each of these, the analogue of (4.89) has to be worked out. We shall not do this here but leave it as an exercise to the reader.

Add G and h: The Fock matrix is simply the sum of the matrix **G** which was discussed above, and the uncoupled one-electron matrix **h**:

$$F_{pq} = h_{pq} + G_{pq}. \tag{4.90}$$

Diagonalise the Fock matrix: To diagonalise **F**, we transform the Roothaan equation (4.64) with the help of the matrix **V** to an ordinary eigenvalue as in Section 3.3:

$$\mathbf{F}' = \mathbf{V}^\dagger \mathbf{F} \mathbf{V}, \tag{4.91}$$

and then we must solve

$$\mathbf{F}'\mathbf{C}' = \epsilon \mathbf{C}' \tag{4.92}$$

and transform the eigenvectors $\mathbf{C}'$ back to the original ones:

$$\mathbf{C} = \mathbf{V}\mathbf{C}'. \tag{4.93}$$

For diagonalising the Fock matrix, we may use the Givens–Householder *QR* method discussed in Section 7.2.1 of Appendix A, which can be found in the Lapack/Atlas library.

Construct a new density matrix P: From the eigenvectors **C** found in diagonalising the Fock matrix, a new density matrix can be constructed which is the ingredient of the Fock matrix in the next iteration. It is important to realise that the vectors $\mathbf{C}_k$ used in constructing the density matrix should be normalised according to

$$1 = \langle \psi_k | \psi_k \rangle = \sum_{pq} C_{pk} \langle p | q \rangle C_{qk} = \mathbf{C}_k \mathbf{S} \mathbf{C}_k. \tag{4.94}$$

Converged: The criterion for convergence of the iterative process is the amount by which the Fock levels and/or the basis functions change from one iteration to the next. A typical criterion is that the Fock levels have converged to within a small margin, such as 10^{-6} a.u. Sometimes, however, the process does not converge but ends up oscillating between two or more configurations, or it diverges. This is often because that the initial guess for P was too far off the converged value. It might then help to use *mixing*: if the changes in the density matrix from one iteration to another are too large, one takes for the density matrix in the next iteration a weighted average of the last and the previous density matrix:

$$P_{pq}^{\text{new}} = \alpha P_{pq}^{\text{last}} + (1 - \alpha) P_{pq}^{\text{previous}}, \quad 0 < \alpha < 1. \tag{4.95}$$

The convergence can finally be enhanced by extrapolating the values of the density matrix. Various extrapolation schemes are used, one of the most popular being Pulay's DIIS scheme (see Section 9.4) [14].

Output: Output of the program are the Fock levels and corresponding eigenstates. From these data, the total energy can be determined:

$$E = \frac{1}{2}\left[\sum_{rs} h_{rs}P_{rs} + \sum_k \epsilon_k\right] + E_{\text{nucl}}. \tag{4.96}$$

ϵ_k are the Fock levels and E_{nucl} represents the electrostatic nuclear repulsion energy which is determined by the nuclear charges Z_n and positions R_n. Invoking Koopman's theorem (see Section 4.5.3), the Fock levels may be interpreted as electron removal or addition energies.

In Problem 4.12, the hydrogen molecule is treated again in Hartree–Fock rather than in Hartree theory, as is done in Problem 4.9.

*4.8 Integrals involving Gaussian functions

In this section, we describe some simple calculations of integrals involving two or four GTOs. We restrict ourselves to 1s-functions; for matrix elements involving higher l-values, see Refs. [15, 16]. As noticed already in Section 4.6.2, the central result which is used in these calculations is the Gaussian product theorem: denoting the Gaussian function $\exp(-\alpha|\mathbf{r} - \mathbf{R}_A|^2)$ by $g_{1s,\alpha}(\mathbf{r} - \mathbf{R}_A)$, we have:

$$g_{1s,\alpha}(\mathbf{r} - \mathbf{R}_A)g_{1s,\beta}(\mathbf{r} - \mathbf{R}_B) = Kg_{1s,\gamma}(\mathbf{r} - \mathbf{R}_P), \tag{4.97}$$

with

$$\begin{aligned} K &= \exp\left[-\frac{\alpha\beta}{\alpha+\beta}|\mathbf{R}_A - \mathbf{R}_B|^2\right] \\ \gamma &= \alpha + \beta \\ \mathbf{R}_P &= \frac{\alpha\mathbf{R}_A + \beta\mathbf{R}_B}{\alpha+\beta}. \end{aligned} \tag{4.98}$$

From now on, we shall use the Dirac notation:

$$g_{1s,\alpha}(\mathbf{r} - \mathbf{R}_A) = |1s, \alpha, A\rangle. \tag{4.99}$$

The overlap integral: The overlap matrix for two 1s-functions can be calculated directly using (4.97):

$$\begin{aligned} \langle 1s, \alpha, A|1s, \beta, B\rangle &= 4\pi \int \mathrm{d}r\, r^2 K \mathrm{e}^{-\gamma r^2} \\ &= \left(\frac{\pi}{\alpha+\beta}\right)^{3/2} \exp\left[-\frac{\alpha\beta}{\alpha+\beta}|\mathbf{R}_A - \mathbf{R}_B|^2\right]. \end{aligned} \tag{4.100}$$

The kinetic integral: This is given by

$$\langle 1s, \alpha, A| - \nabla^2|1s, \beta, B\rangle = \int d^3r\, \nabla e^{-\alpha(\mathbf{r}-\mathbf{R}_A)^2} \nabla e^{-\beta(\mathbf{r}-\mathbf{R}_B)^2} \tag{4.101}$$

where we have used Green's theorem. Working out the gradients and using the Gaussian product theorem, we arrive at

$$\langle 1s, \alpha, A| - \nabla^2|1s, \beta, B\rangle = 4\alpha\beta \int d^3r(\mathbf{r} - \mathbf{R}_A)(\mathbf{r} - \mathbf{R}_B)K e^{-\gamma(\mathbf{r}-\mathbf{R}_P)^2} \tag{4.102}$$

with γ, K and $\mathbf{R}_P$ given above. Substituting $\mathbf{u} = \mathbf{r} - \mathbf{R}_P$ and using the fact that integrals antisymmetric in $\mathbf{u}$ vanish, we arrive at:

$$\begin{aligned} &\langle 1s, \alpha, A| - \nabla^2|1s, \beta, B\rangle \\ &\quad = 16\alpha\beta K\pi \left[\int_0^\infty du\, u^4 e^{-\gamma u} + (\mathbf{R}_P - \mathbf{R}_A) \cdot (\mathbf{R}_P - \mathbf{R}_B) \int_0^\infty du\, u^2 e^{-\gamma u} \right] \\ &\quad = \frac{\alpha\beta}{\alpha+\beta} \left[6 - 4\frac{\alpha\beta}{\alpha+\beta} |\mathbf{R}_A - \mathbf{R}_B|^2 \right] \\ &\quad \times \left(\frac{\pi}{\alpha+\beta} \right)^{3/2} \exp[-\alpha\beta/(\alpha+\beta)|\mathbf{R}_A - \mathbf{R}_B|^2]. \end{aligned} \tag{4.103}$$

The nuclear attraction integral: This integral is more difficult than the previous ones, because after applying the Gaussian product theorem we are still left with the $1/r$ Coulomb term of the nucleus (whose position does in general not coincide with the centre of the orbital). To reduce the integral to a simpler form, we use Fourier transforms:

$$\hat{f}(\mathbf{k}) = \int d^3r\, f(\mathbf{r}) e^{-i\mathbf{k}\cdot\mathbf{r}}. \tag{4.104}$$

The inverse transformation is given by

$$f(\mathbf{r}) = (2\pi)^{-3} \int d^3k\, \hat{f}(\mathbf{k}) e^{i\mathbf{k}\cdot\mathbf{r}}. \tag{4.105}$$

The Dirac delta-function can be written as

$$\delta(\mathbf{r}) = (2\pi)^{-3} \int d^3k\; e^{i\mathbf{k}\cdot\mathbf{r}}. \tag{4.106}$$

The Coulomb integral is given by

$$\langle 1s, \alpha, A| - Z/r_C|1s, \beta, B\rangle = -Z \int d^3r\, K e^{-\gamma|\mathbf{r}-\mathbf{R}_P|^2} |\mathbf{r} - \mathbf{R}_C|^{-1}. \tag{4.107}$$

The Fourier transform of $1/r$ is $4\pi/k^2$, as can be seen for example by Fourier transforming the Poisson equation

$$-\nabla^2 \frac{1}{r} = 4\pi\delta(r). \tag{4.108}$$

Furthermore, the Fourier transform of $\exp(-\gamma r^2)$ is $(\pi/\gamma)^{3/2}\exp(-k^2/4\gamma)$, so substituting these transforms into (4.107), we obtain

$$\langle 1\mathrm{s},\alpha,A|-Z/r_C|1\mathrm{s},\beta,B\rangle$$
$$= -Z(2\pi)^{-6}\left(\frac{\pi}{\gamma}\right)^{3/2}\int \mathrm{d}^3r\,\mathrm{d}^3k_1\,\mathrm{d}^3k_2\;K\mathrm{e}^{-k_1^2/(4\gamma)}\mathrm{e}^{\mathrm{i}\mathbf{k}_1\cdot(\mathbf{r}-\mathbf{R}_P)}$$
$$\times\, 4\pi k_2^{-2}\mathrm{e}^{\mathrm{i}\mathbf{k}_2\cdot(\mathbf{r}-\mathbf{R}_C)}. \tag{4.109}$$

In this equation, the expression (4.106) for the delta-function for $\mathbf{k}_1+\mathbf{k}_2$ is recognised, and this transforms the integral into

$$\langle 1\mathrm{s},\alpha,A|-Z/r_C|1\mathrm{s},\beta,B\rangle$$
$$= -ZK(2\pi^2)^{-1}\left(\frac{\pi}{\gamma}\right)^{3/2}\int \mathrm{d}^3k\;\mathrm{e}^{-k^2/(4\gamma)}k^{-2}\mathrm{e}^{-\mathrm{i}\mathbf{k}\cdot(\mathbf{R}_P-\mathbf{R}_C)}. \tag{4.110}$$

Integrating over the angular variables leads to

$$\langle 1\mathrm{s},\alpha,A|-Z/r_C|1\mathrm{s},\beta,B\rangle = \mathcal{N}\int_0^\infty \mathrm{d}k\;\mathrm{e}^{-k^2/(4\gamma)}1/k\,\sin(k|\mathbf{R}_P-\mathbf{R}_C|);$$
$$\mathcal{N} = -2ZK(\pi|\mathbf{R}_P-\mathbf{R}_C|)^{-1}(\pi/\gamma)^{3/2}. \tag{4.111}$$

The integral (without $\mathcal{N}$) can be rewritten as

$$I(x) \equiv \frac{1}{2}\int_0^x \mathrm{d}y\int_{-\infty}^{\infty}\mathrm{d}k\;\mathrm{e}^{-k^2/(4\gamma)}\cos(ky) \tag{4.112}$$

with $x=|\mathbf{R}_P-\mathbf{R}_C|$. The integral over k is easy, and the result is

$$I(x) = \frac{1}{2}\sqrt{\pi/\gamma}\int_0^x \mathrm{d}y\;\mathrm{e}^{-\gamma y^2}. \tag{4.113}$$

So, finally, we have

$$\langle 1\mathrm{s},\alpha,A|-Z/r_C|1\mathrm{s},\beta,B\rangle$$
$$= -2\pi KZ\gamma^{-1}(\gamma^{1/2}|\mathbf{R}_P-\mathbf{R}_C|)^{-1}\int_0^{\gamma^{1/2}|\mathbf{R}_P-\mathbf{R}_C|}\mathrm{d}y\;\mathrm{e}^{-y^2},$$

and, using the definition

$$F_0(t) = t^{-1/2}\int_0^{t^{1/2}}\mathrm{d}y\;\mathrm{e}^{-y^2}, \tag{4.114}$$

the final result can be rewritten as

$$\langle 1\mathrm{s},\alpha,A|-Z/r_C|1\mathrm{s},\beta,B\rangle$$
$$= -2\pi Z(\alpha+\beta)^{-1}\exp[-\alpha\beta/(\alpha+\beta)|\mathbf{R}_A-\mathbf{R}_B|^2]$$
$$\times\, F_0[(\alpha+\beta)|\mathbf{R}_P-\mathbf{R}_C|^2]. \tag{4.115}$$

The function $F_0(t)$ can be evaluated using the error function erf, which is available in most high-level programming languages as an intrinsic funtion. The error function is defined by

$$\mathrm{erf}(x) = \frac{2}{\sqrt{\pi}} \int_0^x \mathrm{d}x' \mathrm{e}^{-x'^2}. \tag{4.116}$$

If your compiler does not have the error function as an intrinsic, you can calculate it using a recursive procedure. The function $F_0(t)$ is then considered as one in a series of functions defined as

$$F_m(t) = \int_0^1 \exp(-ts^2)s^{2m}\mathrm{d}s. \tag{4.117}$$

The following recursion relation for $F_m(t)$ can easily be derived via partial integration:

$$F_m(x) = \frac{\mathrm{e}^{-x} + 2xF_{m+1}(x)}{2m+1}. \tag{4.118}$$

This recursion is stable only if performed downward (see Appendix A2).
The two-electron integral: This has the form:

$$\begin{aligned} &\langle 1\mathrm{s}, \alpha, A; 1\mathrm{s}, \beta, B|g|1\mathrm{s}, \gamma, C; 1\mathrm{s}, \delta, D\rangle \\ &= \int \mathrm{d}^3r_1 \mathrm{d}^3r_2\, \mathrm{e}^{-\alpha|\mathbf{r}_1-\mathbf{R}_A|^2} \mathrm{e}^{-\beta|\mathbf{r}_2-\mathbf{R}_B|^2} \frac{1}{r_{12}} \mathrm{e}^{-\gamma|\mathbf{r}_1-\mathbf{R}_C|^2} \mathrm{e}^{-\delta|\mathbf{r}_2-\mathbf{R}_D|^2}. \end{aligned} \tag{4.119}$$

Using the Gaussian product theorem, we can write the Gaussian functions depending on $\mathbf{r}_1$ and $\mathbf{r}_2$ as new 1s-functions with exponential parameters ρ and σ and centres $\mathbf{R}_P$ (of $\mathbf{R}_A$ and $\mathbf{R}_C$) and $\mathbf{R}_Q$ (of $\mathbf{R}_B$ and $\mathbf{R}_D$):

$$\begin{aligned} &\langle 1\mathrm{s}, \alpha, A; 1\mathrm{s}, \beta, B|g|1\mathrm{s}, \gamma, C; 1\mathrm{s}, \delta, D\rangle \\ &= \exp[-\alpha\gamma/(\alpha+\gamma)|\mathbf{R}_A - \mathbf{R}_C|^2 - \beta\delta/(\beta+\delta)|\mathbf{R}_B - \mathbf{R}_D|^2] \\ &\times \int \mathrm{d}^3r_1 \mathrm{d}^3r_2 \mathrm{e}^{-\rho|\mathbf{r}_1-\mathbf{R}_P|^2} \frac{1}{r_{12}} \mathrm{e}^{-\sigma|\mathbf{r}_2-\mathbf{R}_Q|^2}. \end{aligned} \tag{4.120}$$

Calling $\mathcal{M}$ the factor in front of the integral, and replacing the Gaussian exponentials in the integrand by their Fourier transforms, and similarly for the $1/r_{12}$ term, we obtain

$$\begin{aligned} &\langle 1\mathrm{s}, \alpha, A; 1\mathrm{s}, \beta, B|g|1\mathrm{s}, \gamma, C; 1\mathrm{s}, \delta, D\rangle \\ &= \mathcal{M}(2\pi)^{-9} \int \mathrm{d}^3r_1\, \mathrm{d}^3r_2\, \mathrm{d}^3k_1\, \mathrm{d}^3k_2\, \mathrm{d}^3k_3\, (\pi/\rho)^{3/2} \mathrm{e}^{-k_1^2/4\rho} \mathrm{e}^{\mathrm{i}\mathbf{k}_1\cdot(\mathbf{r}_1-\mathbf{R}_P)} \\ &\times 4\pi k_2^{-2} \mathrm{e}^{\mathrm{i}\mathbf{k}_2\cdot(\mathbf{r}_1-\mathbf{r}_2)} (\pi/\sigma)^{3/2} \mathrm{e}^{-k_3^2/4\sigma} \mathrm{e}^{\mathrm{i}\mathbf{k}_3\cdot(\mathbf{r}_2-\mathbf{R}_Q)}. \end{aligned} \tag{4.121}$$

Table 4.1. *Bond lengths in atomic units for three different molecules. Hartree–Fock (HF) and experimental results are shown.*

Molecule	HF	Expt.
H_2	1.385	1.401
N_2	2.013	2.074
CO	1.914	1.943

Data taken from Ref. [6].

The integrals over $\mathbf{r}_1$ and $\mathbf{r}_2$ yield two delta-functions in $\mathbf{k}_1$ and $\mathbf{k}_2$, and Eq. (4.121) transforms into

$$\langle 1s, \alpha, A; 1s, \beta, B|g|1s, \gamma, C; 1s, \delta, D\rangle$$
$$= 4\pi \mathcal{M}(2\pi)^{-3}(\pi^2/\rho\sigma)^{3/2} \int d^3k\, k^{-2} e^{-k^2(\rho+\sigma)/(4\rho\sigma)} e^{i\mathbf{k}(\mathbf{R}_P-\mathbf{R}_Q)}. \quad (4.122)$$

We have already encountered this integral in the previous subsection. The final result is now

$$\langle 1s, \alpha, A; 1s, \beta, B|g|1s, \gamma, C; 1s, \delta, D\rangle$$
$$= \frac{2\pi^{(5/2)}}{(\alpha+\gamma)(\beta+\delta)(\alpha+\beta+\gamma+\delta)^{1/2}}$$
$$\times \exp[-\alpha\gamma/(\alpha+\gamma)|\mathbf{R}_A - \mathbf{R}_C|^2 - \beta\delta/(\beta+\delta)|\mathbf{R}_B - \mathbf{R}_D|^2]$$
$$\times F_0\left[\frac{(\alpha+\gamma)(\beta+\delta)}{(\alpha+\beta+\gamma+\delta)}|\mathbf{R}_P - \mathbf{R}_Q|^2\right]. \quad (4.123)$$

4.9 Applications and results

After having considered the implementation of the Hartree–Fock in a computer program, we now present some results of HF calculations for simple molecules [6, 17]. As the HF calculations yield an energy for a static configuration of nuclei, it is possible to find the stable configuration by varying the positions of the nuclei and calculating the corresponding energies – the lowest energy corresponds to the stable configuration. In this way, the equilibrium bond lengths of diatomic molecules can be determined. In Table 4.1, HF results are shown for these bond lengths, together with experimental results. The table shows good agreement between the two. The same holds for bond angles, given in Table 4.2. It is also possible to calculate the

Table 4.2. *Bond angles in degrees for* H_2O *and* NH_3. *The angles are those of the* H–O–H *and* H–N–H *chains respectively. Hartree–Fock (HF) and experimental results are shown.*

Molecule	HF	Expt.
H_2O	107.1	104.5
NH_3	108.9	106.7

Data taken from Ref. [17].

Table 4.3. *Dissociation energies in atomic units for* LiF *and* NaBr. *Hartree–Fock (HF) and experimental results are shown.*

Molecule	HF	Expt
LiF	0.2938	0.2934
NaBr	0.1978	0.2069

Data taken from Ref. [17].

energies needed to dissociate diatomic molecules (see Table 4.3) and again good agreement is found with experiment.

Koopman's theorem can be used to calculate ionisation potentials, that is, the minimum energy needed to remove an electron from the molecule. Comparing the results in Table 4.4 for the ionisation potentials calculated via Koopman's theorem with those of the previous tables, it is seen that the approximations involved in this 'theorem' are more severe than those of the Hartree–Fock theory, although agreement with experiment is still reasonable.

*4.10 Improving upon the Hartree–Fock approximation

The Hartree–Fock approximation sometimes yields unsatisfactory results. This is of course due to Coulomb correlations not taken into account in the Hartree–Fock formalism. There exists a systematic way to improve on Hartree–Fock by constructing a many-electron state as a linear combination of Slater determinants (remember the Slater determinants span the N-electron Hilbert space of many-electron wave

Table 4.4. *Ionisation potentials in atomic units for various molecules. Results obtained via Koopman's theorem and experimental results are shown.*

Molecule	Koopman	Expt
H_2	0.595	0.584
CO	0.550	0.510
H_2O	0.507	0.463

Data taken from Ref. [6].

functions as mentioned at the end of Section 4.4). These determinants are constructed from the ground state by excitation: the first determinant is the Hartree–Fock ground state and the second one is the first excited state (within the spectrum determined self-consistently for the ground state) and so on. The matrix elements of the Hamiltonian between these Slater determinants are calculated and the resulting Hamilton matrix (which has a dimension equal to the number of Slater determinants taken into account) is diagonalised. The resulting state is then a linear combination of Slater determinants

$$\Psi(\mathbf{x}_1, \ldots, \mathbf{x}_N) = \sum_n \alpha_n \Psi_{\mathrm{AS}}^{(n)}(\mathbf{x}_1, \ldots, \mathbf{x}_N) \tag{4.124}$$

and its energy will be lower than the Hartree–Fock ground state energy. This is a time-consuming procedure so that for systems containing many electrons, only a limited number of determinants can be taken into account. This is the configuration interaction (CI) method. In simple systems, for which this method allows very high accuracy to be achieved, excellent agreement with experimental results can be obtained. The CI method is in principle exact (within the Born–Oppenheimer approximation), but since for a finite basis set the Fock spectrum is finite, only a finite (though large) number of Slater determinants is possible within that basis set. A CI procedure in which all possible Slater determinants possible within a chosen basis set are taken into account is called 'full CI'. For most systems, full CI is impossible because of the large number of Slater determinants needed, but it is sometimes possible to obtain an estimate for the full CI result by extrapolating results for larger and larger numbers of Slater determinants.

As an illustration, we show bond lengths and correlation energies for H_2 and H_2O in Tables 4.5 and 4.6. The correlation energy is defined as the difference between the Hartree–Fock and the exact energy. For small systems such as H_2, the electronic structure can be calculated taking the electron correlation fully into account (but

Table 4.5. *Correlation energies in atomic units for* H_2 *and* H_2O.

Molecule	CI	Exact
H_2	−0.039 69	−0.0409
H_2O	−0.298	−0.37

Data taken from Ref. [6].

Table 4.6. *Bond lengths in atomic units for* H_2 *and* H_2O.

Molecule	HF	CI	Exact	Experiment
H_2	1.385	1.396	1.401	
H_2O	1.776	1.800		1.809

Data taken from Ref. [6]. Exact results were obtained by variational calculus [18]. Experimental results are from Ref. [19].

within the Born–Oppenheimer approximation) by a variational method using basis functions depending on the positions of both electrons [18]. The CI results are excellent for both cases.

In CI, the spin-orbitals from which the Slater determinants are built are the eigenstates of the Fock operator as determined self-consistently for the ground state. As only a restricted number of determinants can be taken into account, the dimension of the subspace spanned by the Slater determinants is quite limited. If, within these Slater determinants, the orbitals are allowed to vary by relaxing the ground state coefficients of the basis functions, this subspace can be increased considerably. In the so-called multi-configuration self-consistent field theory (MCSCF), this process is carried out, but because of the large amount of variation possible this leads to a huge numerical problem. Finally, perturbative analysis allows correlation effects to be taken into account [5, 6].

Exercises

4.1 In this problem we show that the large masses of the nuclei compared with those of the electrons lead to the Born–Oppenheimer approximation.

The wave function Ψ of a collection of electrons and nuclei depends on the positions $\mathbf{R}_n$ of the nuclei and $\mathbf{r}_i$ of the electrons (we neglect the spin-degrees of freedom). For this function we make the following *Ansatz*:

$$\Psi(\mathbf{R}_n, \mathbf{r}_i) = \chi(\mathbf{R}_n)\Phi(\mathbf{r}_i)$$

with $\Phi(\mathbf{r}_i)$ an eigenstate with eigenvalue $E_{\rm el}$ of the N-electron 'Born–Oppenheimer Hamiltonian' Eq. (4.2), which in atomic units reads:

$$H_{\rm BO} = \sum_{i=1}^{N} -\frac{1}{2}\nabla_i^2 + \frac{1}{2}\sum_{i,j=1;i\neq j}^{N} \frac{1}{|\mathbf{r}_i - \mathbf{r}_j|} - \sum_{n=1}^{K}\sum_{i=1}^{N} \frac{Z_n}{|\mathbf{r}_i - \mathbf{R}_n|}.$$

It is clear that Φ and $E_{\rm el}$ depend on the nuclear positions $\mathbf{R}_n$, since the Born–Oppenheimer Hamiltonian does.

Show that substitution of this *Ansatz* into the full Hamiltonian, Eq. (4.1), leads to:

$$\Phi(\mathbf{r})\left[\sum_{n=1}^{K} -\frac{1}{2M_n}\nabla_n^2 + E_{\rm el} + \sum_{n,n'=1;n\neq n'}^{K} \frac{Z_n Z_{n'}}{|\mathbf{R}_n - \mathbf{R}_{n'}|}\right]\chi(\mathbf{R}_n)$$

$$-\chi(\mathbf{R}_n)\sum_{n=1}^{K}\frac{1}{2M_n}\nabla_n^2\Phi(\mathbf{r}_i) - \sum_{n=1}^{K}\frac{1}{M_n}\nabla_n\chi(\mathbf{R}_n)\cdot\nabla_n\Phi(\mathbf{r}_i) = E\chi(\mathbf{R}_n)\Phi(\mathbf{r}_i),$$

so that neglecting the last two terms on the left hand side of this equation, we arrive at a Schrödinger equation for the nuclei which contains the electronic degrees of freedom via the electronic energy $E_{\rm el}$ only:

$$\left[\sum_{n=1}^{K} -\frac{1}{2M_n}\nabla_n^2 + E_{\rm el} + \sum_{n,n'=1;n\neq n'}^{K} \frac{Z_n Z_{n'}}{|\mathbf{R}_n - \mathbf{R}_{n'}|}\right]\chi(\mathbf{R}_n) = E_{\rm nuc}\chi(\mathbf{R}_n).$$

The fact that the term $(1/2M_n)\nabla_n^2\Phi(\mathbf{r}_i)$ can be neglected can be understood by realising that it is $1/M_n$ times the variation of the kinetic energy of the electrons with the positions of the nuclei. Of course, the core electrons have large kinetic energy, but they feel almost exclusively their own nucleus, hence their kinetic energy is insensitive to variations in the relative nuclear positions. The valence electrons have smaller kinetic energies, so the variation of this energy with nuclear positions will be small too. In a solid, deleting the term $(1/M_n)\nabla_n\chi(\mathbf{R}_n)\cdot\nabla_n\Phi(\mathbf{r}_i)$ means that electron–phonon couplings are neglected, so that some physical phenomena cannot be treated in calculations involving the Born–Oppenheimer approach, although these effects can often be studied perturbatively.

4.2 For a two-electron system, the wave function can be written as

$$\Psi(\mathbf{x}_1, \mathbf{x}_2) = \Phi(\mathbf{r}_1, \mathbf{r}_2)\cdot\chi(s_1, s_2).$$

Because the wave function Ψ is antisymmetric under particle exchange, we may take Φ symmetric in 1 and 2 and χ antisymmetric, or vice versa.

We construct the functions Φ and χ from the orthonormal spatial orbitals $\phi_1(\mathbf{r})$, $\phi_2(\mathbf{r})$ and the spin-up and -down functions $\alpha(s)$ and $\beta(s)$ respectively.

(a) Write down the antisymmetric wave functions that can be constructed in this way (there are six of them).
(b) Write down all possible Slater determinants that can be built from the one-electron spin-orbitals consisting of a product of one of the orbitals ϕ_1 and ϕ_2 and a spin-up or -down spinor (you will find six of these determinants too).
(c) Express the wave functions of part (a) of this problem in those of (b).

4.3 Consider a Slater determinant

$$\Psi_{\mathrm{AS}}(\mathbf{x}_1, \ldots, \mathbf{x}_N) = \frac{1}{\sqrt{N!}} \begin{vmatrix} \psi_1(\mathbf{x}_1) & \psi_2(\mathbf{x}_1) & \cdots & \psi_N(\mathbf{x}_1) \\ \psi_1(\mathbf{x}_2) & \psi_2(\mathbf{x}_2) & \cdots & \psi_N(\mathbf{x}_2) \\ \vdots & \vdots & & \vdots \\ \psi_1(\mathbf{x}_N) & \psi_2(\mathbf{x}_N) & \cdots & \psi_N(\mathbf{x}_N) \end{vmatrix}$$
$$= \frac{1}{\sqrt{N!}} \sum_{\mathbf{P}} \epsilon_P \mathbf{P} \psi_1(\mathbf{x}_1) \ldots \psi_N(\mathbf{x}_N).$$

The spin-orbitals $\psi_k(\mathbf{x})$ are orthonormal.

(a) Show that the Slater determinant is normalised, by considering the inner product of two arbitrary terms occurring in the sum of the Slater determinant and then summing over all possible pairs of such terms.
(b) Show in the same way that the density of electrons with coordinates $\mathbf{x}$, given by:

$$n(\mathbf{x}) = N \int \mathrm{d}x_2 \ldots \mathrm{d}x_N |\Psi_{\mathrm{AS}}(\mathbf{x}, \mathbf{x}_2, \ldots, \mathbf{x}_N)|^2,$$

can be written in terms of the ψ_k as:

$$n(\mathbf{x}) = \sum_k |\psi_k(\mathbf{x})|^2.$$

Suppose all spin-orbitals can be written as the product of a normalised orbital and a normalised one-particle spinor, what is then the spatial charge density of the electrons (i.e. regardless of the spin)?
(c) Derive Eqs. (4.35) and (4.36) using the methods employed in (a) and (b).

4.4 Consider the helium atom with two electrons having the *same* spin, represented by the spinor $\alpha(s)$.

(a) Give the form of the two-electron wave function, expressed in orthonormal spatial orbitals ϕ_1 and ϕ_2.
(b) Write down the Schrödinger equation for this system.
(c) Give an expression for the expectation value of the energy in the orbitals ϕ_i.

4.5 The Hartree–Fock analysis can be performed not only for fermions, but also for bosons. Consider a system consisting of N spin-0 particles in one dimension having spin-orbital coordinates x_i (for spin-0 particles, only the orbital coordinate matters).

The bosons interact via a δ-function potential:

$$H = -\sum_{i=1}^{N} \frac{\partial^2}{\partial x_i^2} + g \sum_{i>j}^{N} \delta(x_i - x_j).$$

This means that the particles feel each other only if they are at the same position: they experience an infinitely large attraction with an infinitely short range. Although this problem can be solved exactly, we consider the Hartree–Fock approximation here.

A boson wave function Ψ is symmetric under particle exchange. This means that *all* the particles are in the same orbital ϕ.

(a) Show that the kinetic term of the Hamiltonian has the form:

$$\left\langle \Psi \left| -\sum_{i=1}^{N} \frac{\partial^2}{\partial x_i^2} \right| \Psi \right\rangle = -N \left\langle \phi \left| \frac{\mathrm{d}^2}{\mathrm{d}x^2} \right| \phi \right\rangle.$$

(b) Show that the expectation value of the interaction potential is given by

$$\left\langle \Psi \left| g \sum_{i<j} \delta(x_i - x_j) \right| \Psi \right\rangle = \frac{1}{2} N(N-1) \int \mathrm{d}x |\phi(x)|^4.$$

(c) Show, by minimisation of the energy-functional

$$\langle E \rangle = \frac{\langle \Psi | H | \Psi \rangle}{\langle \Psi | \Psi \rangle}$$

with respect to ϕ, that the Hartree–Fock equation reads

$$\left[-\frac{\partial^2}{\partial x^2} - g(N-1)|\phi(x)|^2 \right] \phi(x) = \epsilon \phi(x).$$

(d) The solution to this last equation is found as

$$\phi(x) = \frac{\left[(1/8)g(N-1)\right]^{1/2}}{\cosh\left[(1/4)g(N-1)x\right]};$$

$$\epsilon = \frac{1}{16} g^2 (N-1)^2.$$

Show that this function indeed satisfies the Hartree–Fock equation.

4.6 Consider the Fock spectrum of a many-electron system. In the ground state, the N electrons fill the lowest N levels of this spectrum. Consider the same spectrum, but now with one electron removed from it. This means that the system has been ionised. Show from the expressions for the Fock operator and the energy in terms of the spin-orbitals, Eqs. (4.41) and (4.48), that the ionisation energy is equal to the difference between the sum over the occupied Fock levels ϵ_k in the ground state and the same sum for the ionised state.

It is then clear that the same holds for adding an electron to the ground state, and therefore for moving an electron from level a to level b (by first removing the electron from level a and then adding one in level b).

4.7 Consider a Slater determinant constructed from spin-orbitals ψ_k, $k = 1, \ldots, N$. A unitary transformation transforms the spin-orbitals ψ_k into new ones, which we denote by ψ'_k:

$$\psi'_k = \sum_{l=1}^{N} U_{kl}\psi_l.$$

U_{kl} are the elements of the unitary matrix $\mathbf{U}$.

(a) Show that the new basis is orthonormal.

(b) Show that the Slater determinant constructed from the new spin-orbitals can be written as the determinant of the product of the matrix

$$\frac{1}{\sqrt{N!}}\begin{pmatrix} \psi_1(\mathbf{x}_1) & \psi_2(\mathbf{x}_1) & \cdots & \psi_N(\mathbf{x}_1) \\ \psi_1(\mathbf{x}_2) & \psi_2(\mathbf{x}_2) & \cdots & \psi_N(\mathbf{x}_2) \\ \vdots & \vdots & & \vdots \\ \psi_1(\mathbf{x}_N) & \psi_2(\mathbf{x}_N) & \cdots & \psi_N(\mathbf{x}_N) \end{pmatrix}$$

and the matrix $\mathbf{U}$.

(c) Show that the Slater determinant built from ψ_k and that built from ψ'_k are equal up to a complex constant of absolute value unity.

4.8 [C] In this problem, the program for calculating the electronic structure of the hydrogen atom (see Section 3.2.2), is extended to the H_2^+ ion. The H_2^+ ion contains only one electron and the problem is therefore essentially the same as that of Section 3.2.2, the difference being a second nucleus at a distance R from the first one. The global structure of the program is therefore the same, the main difference being that the basis now consists of eight functions: four functions centred around each nucleus. Therefore, all matrices now have dimension 8.

It is important to note that for basis functions centred around one of the two nuclei, the Coulomb attraction of the other nucleus is still important, as is immediately clear from the expression of the Coulomb matrix $\mathbf{A}$:

$$A_{pq} = \int \mathrm{d}^3r\, \chi_p(\mathbf{r}) \left(\frac{1}{|\mathbf{r} - \mathbf{R}_A|} + \frac{1}{|\mathbf{r} - \mathbf{R}_B|} \right) \chi_q(\mathbf{r})$$

where $\mathbf{R}_A$ and $\mathbf{R}_B$ denote the positions of the two nuclei. The integrals for calculating the matrix elements of the various operators can be found in Section 4.8. Write a program to determine the ground state of the H_2^+ ion.

For a distance $1a_0$ between the nuclei, the program should yield an energy (without the Coulomb repulsion between the nuclei) equal to $-1.442\,455$ a.u.

4.9 [C] In this problem, the program developed in the previous problem is extended along the lines of the helium ground state calculation of Section 4.3.2 in order to calculate the electronic structure of the hydrogen molecule. This means that a second electron is added to the ionic hydrogen system and we must solve the Hartree equation for a finite basis, Eq. (4.14), self-consistently analogous to the helium calculation.

The matrix to be diagonalised is given by

$$F_{pq} = h_{pq} + \sum_{rs} C_r C_s Q_{prqs}$$

with

$$Q_{prqs} = \langle pr|g|qs\rangle = \int \mathrm{d}^3r_1\, \mathrm{d}^3r_2\, \phi_p(\mathbf{r}_1)\phi_r(\mathbf{r}_2)\frac{1}{r_{12}}\phi_q(\mathbf{r}_1)\phi_s(\mathbf{r}_2).$$

As we are dealing with two electrons only, we do not have to sum over the different orbitals k.

In Problem 4.8 you have already programmed the expressions for the matrix elements h_{pq}, so these pose no problems. The two-electron matrix elements Q_{prqs} are rather complicated; they are given in Section 4.8. The resulting expressions can be written in the form:

$$\langle \chi_p\chi_r|g|\chi_q\chi_s\rangle = 2\sqrt{\frac{AB}{\pi(A+B)}}S_{pq}S_{sr}F_0(t).$$

Here, t is defined as

$$t = \frac{(\alpha_p+\alpha_q)(\alpha_r+\alpha_s)}{(\alpha_p+\alpha_q+\alpha_r+\alpha_s)}|\mathbf{R}_A - \mathbf{R}_B|^2;$$

$$\mathbf{R}_A = \frac{\alpha_p\mathbf{R}_p + \alpha_q\mathbf{R}_q}{\alpha_p+\alpha_q}$$

$$\mathbf{R}_B = \frac{\alpha_r\mathbf{R}_r + \alpha_s\mathbf{R}_s}{\alpha_r+\alpha_s},$$

and A and B as

$$A = \alpha_p + \alpha_q$$

$$B = \alpha_r + \alpha_s.$$

S_{pq} is the overlap matrix.

You can now use these matrix elements in a program which has the same structure as that of the helium atom.

Check: for a distance 1 a.u. between the atoms, one finds for the ground state vector $\mathbf{C}$:

(0.092 561 548 6, 0.165 180 118, 0.120 122 665, 0.021 154 565 7,
0.092 561 548 6, 0.165 180 118, 0.120 122 665, 0.021 154 565 7)
and an energy $-1.078\,547\,61$ (nuclear repulsion $+1$ included!).

4.10 In a restricted Hartree–Fock (RHF) calculation using a finite basis $\chi_p(\mathbf{r})$, the kinetic and Coulomb integrals must be calculated. We can gain speed by using the symmetry of these matrices, for example

$$\langle\chi_p|\nabla^2|\chi_q\rangle = \langle\chi_q|\nabla^2|\chi_p\rangle.$$

We do not assume other symmetries to be present in the system.

(a) Suppose the basis contains M basis functions, at least how many of these matrix elements must be calculated?

(b) How many two-electron matrix elements

$$\langle pr|g|qs\rangle$$

must be calculated?

Now suppose that the molecule for which we are performing the calculation consists of three identical atoms, located on an equilateral triangle:

On every atom, we have M basis functions which we denote by $\chi_p^A(\mathbf{r})$ etc., $p = 1, \ldots, M$. The basis functions on the different atoms have the same form:

$$\chi_p^A(\mathbf{r}) = \chi_p(\mathbf{r} - \mathbf{R}_A),$$

$$\chi_p^B(\mathbf{r}) = \chi_p(\mathbf{r} - \mathbf{R}_B)$$

and similarly for C.

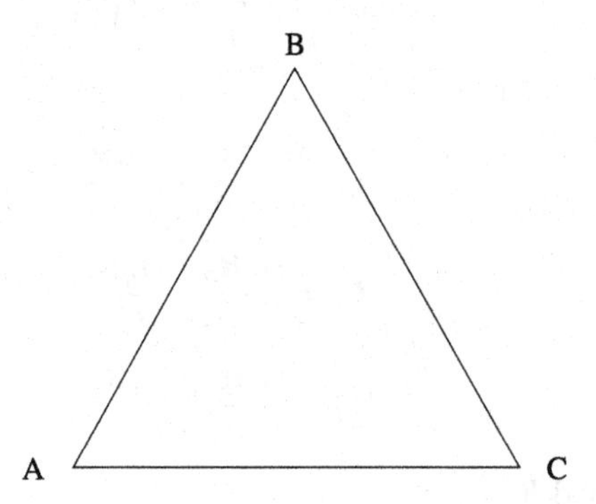

(c) How many different kinetic and Coulomb matrix elements must be calculated in this case?

(d) How many different two-electron matrix elements must be calculated?

(e) Suppose now that M is very large. What is then the gain in speed when using all symmetries of the matrix elements instead of using no symmetry at all?

4.11 Suppose a Hartree–Fock calculation is carried out for a linear chain of K identical atoms and N electrons, where K is a large number. The distance between two successive atoms is a. For each atom, the same set of M basis functions is used. We assume that the overlap between two wave functions centred around two atoms at a distance larger than pa (p is some integer) vanishes.

(a) How many elements of the Hamilton matrix are nonzero (for large K)?

(b) How many nonzero two-electron matrix elements $\langle pr|g|qs\rangle$ do we have (for large K)?

4.12 [C] In this problem, we modify the hydrogen calculation as carried out in Problem 4.9 to a Hartree–Fock calculation – remember that in the previous version we solved the Hartree equation and not the Hartree–Fock equation. Furthermore, we consider exploiting the symmetry in order to speed up the calculation. As we have mentioned in Section 4.3.1, the Hartree approach is good enough for the ground state of a two-electron system because the two electrons are described by the *same* orbital: antisymmetry is taken into account via the spins which are opposite. The present program should therefore yield the same result as the previous one but have the structure of the programs dealing with more electrons or excited states.

The main difference from the previous version is that the array Q is replaced by $\tilde{Q}$ in this program, defined in terms of the Q_{prqs} as follows:

$$\tilde{Q}_{pqrs} = 2Q_{prqs} - Q_{prsq}.$$

(a) [C] First of all, you can exploit the symmetry in the calculation of S_{pq} and h_{pq}: both matrices are symmetric, so you only need to calculate the upper or lower triangular elements. Implement this in your program.

(b) [C] In order to calculate the matrix Q_{prqs} fast, you can restrict the indices to $p \geq q$, $q \geq r$ and if $p = r$, $q \geq s$, otherwise $r \geq s$, see Section 4.7. For each set of p, q, r, s in these ranges, seven other Q-matrix elements having the same values can then be found because of symmetry. These are: Q_{qrps}, Q_{psqr}, Q_{qspr}, Q_{rpsq}, Q_{sprq}, Q_{rqsp} and Q_{sqrp}.

(c) [C] Now use the matrix elements $\tilde{Q}_{pqrs}$ instead of Q_{prqs} and check that you obtain the same results as in the previous program.

(d) The fact that using $\tilde{Q}$ instead of Q leads to the same results is quite surprising, since it is a different Fock matrix we are diagonalising (you can check this by printing out the Fock matrices of the old and the present program). Show that if the vector **C** has converged, the results are equivalent (hint: by inspection of the energy, rather than the Fock matrix).

(e) Exploit the full symmetry of the two-electron matrix elements by distinguishing all possible cases for p, q, r, s being different or equal. Formulate the equations analogous to Eq. (4.89) for all these cases (there are 14 of them).

(f) [C] Implement these equations into your program.

References

[1] B. T. Sutcliffe, 'Fundamentals of computational quantum chemistry,' in *Computational Techniques in Quantum Chemistry and Molecular Physics* (ed. G. H. F. Diercksen, B. T. Sutcliffe, and A. Veillard), Proceedings of the NATO ASI held at Ramsau, Germany, Boston, Reidel, 1975, p. 1.

[2] D. R. Hartree, *Proc. Camb. Phil. Soc.*, **24** (1928), 89.

[3] C.-O. Almbladh and A. C. Pedroza, 'Density-functional exchange-correlation potentials and orbital eigenvalues for light atoms,' *Phys. Rev. A*, **29** (1984), 2322–30.

[4] J. C. Slater, *Quantum Theory of Molecules and Solids*, vol. IV. New York, McGraw-Hill, 1982.

[5] R. McWeeny, *Methods of Molecular Quantum Mechanics*, 2nd edn. New York, Academic Press, 1989.

[6] A. Szabo and N. S. Ostlund, *Modern Quantum Chemistry*. London, Macmillan, 1982.

[7] N. W. Ashcroft and N. D. Mermin, *Solid State Physics*. New York, Holt, Reinhart and Winston, 1976.

[8] W. J. Hehre, L. Radom, P. R. Schleyer, and J. A. Pople, *Ab Initio Molecular Orbital Theory*. New York, John Wiley, 1986.
[9] C. A. Coulson and I. Fisher, 'Notes on the molecular orbital treatment of the hydrogen molecule,' *Philos. Mag.*, **40** (1949), 386–93.
[10] S. F. Boys, 'A general method of calculation for the stationary states of any molecular system,' *Proc. Roy. Soc. (London)*, **200** (1950), 542–54.
[11] W. H. Press, S. A. Teukolsky, W. T. Vetterling, and B. P. Flannery, *Numerical Recipes*, 2nd edn. Cambridge, Cambridge University Press, 1992.
[12] R. Poirer, R. Kari, and I. G. Czismadia, *Handbook of Gaussian Basis Sets*. Amsterdam, Elsevier, 1985.
[13] A. Veillard, 'The logic of SCF procedures,' in *Computational Techniques in Quantum Chemistry and Molecular Physics* (ed. G. H. F. Diercksen, B. T. Sutcliffe, and A. Veillard), Proceedings of the NATO ASI held at Ramsau, Germany, Boston, Reidel, 1975, p. 201.
[14] P. Pulay, 'Improved SCF convergence acceleration,' *J. Comp. Chem.*, **3** (1982), 556–60.
[15] E. Clementi and D. R. Davis, 'Electronic structure of large molecular systems,' *J. Comp. Phys.*, **2** (1967), 223–44.
[16] V. R. Saunders, 'An Introduction to Molecular Integral Evaluation,' in *Computational Techniques in Quantum Chemistry and Molecular Physics* (ed. G. H. F. Diercksen, B. T. Sutcliffe, and A. Veillard), Proceedings of the NATO ASI held at Ramsau, Germany, Boston, Reidel, 1975, pp. 347–92.
[17] A. Hinchliffe, *Computational Quantum Chemistry*. Chichester, John Wiley & Sons, 1988.
[18] W. Kolos and L. Wolniewicz, 'Improved theoretical ground-state energy of the hydrogen molecule,' *J. Chem. Phys.*, **49** (1968), 404–10.
[19] B. J. Rosenberg and I. Shavitt, '*Ab initio* SCF and CI studies on the ground state of the water molecule. I. Comparison of CGTO and STO basis sets near the Hartree–Fock limit,' *J. Chem. Phys.*, **63** (1975), 2162–74.

5

Density functional theory

5.1 Introduction

In the previous chapter we saw how the many-electron problem can be treated in the Hartree–Fock formalism in which the solution of the many-body Schrödinger equation is written in the form of a Slater determinant. The resulting HF equations depend on the occupied electron orbitals, which enter these equations in a nonlocal way. The nonlocal potential of Hartree–Fock is difficult to apply in extended systems, and for this reason there have been relatively few applications to solids; see however Ref. [1].

Most electronic structure calculations for solids are based on density functional theory (DFT), which results from the work of Hohenberg, Kohn and Sham [2, 3]. This approach has also become popular for atoms and molecules. In the density functional theory, the electronic orbitals are solutions to a Schrödinger equation which depends on the electron density rather than on the individual electron orbitals. However, the dependence of the one-particle Hamiltonian on this density is in principle nonlocal. Often, this Hamiltonian is taken to depend on the *local* value of the density only – this is the *local density approximation* (LDA). In the vast majority of DFT electronic structure calculations for solids, this approximation is adopted. It is, however, also applied to atomic and molecular systems [4].

In this chapter we describe the density functional method for electronic structure calculations. In the present section, the physical interpretation of the density functional equations is first described and the formal derivations are given. In the next section the local density approximation is considered. An application to the ground state of the helium atom will be described in some detail in Section 5.5. Finally, some results obtained using density functional theory will be discussed in Section 5.5.3. For further reading, there are many reviews and books available; see for example Refs. [4–7].

5.1.1 Density functional theory: physical picture

In density functional theory, an effective independent-particle Hamiltonian is arrived at, leading to the following Schrödinger equation for one-electron spin-orbitals:

$$\left[-\frac{1}{2}\nabla^2 - \sum_n \frac{Z_n}{|\mathbf{r}-\mathbf{R}_n|} + \int d^3r'\, n(\mathbf{r}')\frac{1}{|\mathbf{r}-\mathbf{r}'|} + V_{\rm xc}[n](\mathbf{r})\right]\psi_k(\mathbf{r}) = \varepsilon_k\psi_k(\mathbf{r}). \tag{5.1}$$

The first three terms in the left hand side of this equation are exactly the same as those of Hartree–Fock, Eq. (4.31), namely the kinetic energy, the electrostatic interaction between the electrons and the nuclei, and the electrostatic energy of the electron in the field generated by the total electron density $n(\mathbf{r})$. The fourth term contains the many-body effects, lumped together in an exchange-correlation potential. The main result of density functional theory is that there exists a form of this potential, depending only on the electron density $n(\mathbf{r})$, that yields the *exact* ground state energy and density. Unfortunately, this exact form is not known, but there exist several approximations to it, as we shall see in Sections 5.2 and 5.3. The dependence of the independent-particle Hamiltonian on the density only is in striking contrast with Hartree–Fock theory, where the Hamiltonian depends on the individual orbitals. The solutions of Eq. (5.1) must be self-consistent in the density, which is given by

$$n(\mathbf{r}) = \sum_{k=1}^{N} |\psi_k(\mathbf{r})|^2, \tag{5.2}$$

where the sum is over the N spin-orbitals ψ_k having the lowest eigenvalues ε_k in (5.1), and N is the number of electrons in the system.

The total energy of the many-electron system is given by

$$E = \sum_{k=1}^{N} \varepsilon_k - \frac{1}{2}\int d^3r\, d^3r'\, n(\mathbf{r})\frac{1}{|\mathbf{r}-\mathbf{r}'|}n(\mathbf{r}') + E_{\rm xc}[n] - \int d^3r\, V_{\rm xc}[n](\mathbf{r})n(\mathbf{r}) \tag{5.3}$$

where the parameters ε_k are the eigenvalues occurring in Eq. (5.1) and $E_{\rm xc}$ is the exchange correlation energy. The exchange correlation potential $V_{\rm xc}[n]$ which occurs in (5.1) is the functional derivative of this energy with respect to the density:

$$V_{\rm xc}[n](\mathbf{r}) = \frac{\delta}{\delta n(\mathbf{r})}E_{\rm xc}[n]. \tag{5.4}$$

Although the energy parameters ε_k are not, strictly speaking, one-electron energies they are often treated as such for comparison with spectroscopy experiments according to an extended version of Koopman's theorem (see Problem 5.4). The wave functions ψ_k also have no individual meaning but are used to construct the total charge density. This again contrasts with Hartree–Fock where the one-electron spin-orbitals have a definite interpretation: they are the constituents of the many-electron wave function.

Equations (5.1) and (5.2) are solved in an iterative self-consistency loop, which is started by choosing an initial density $n(\mathbf{r})$, constructing the Schrödinger equation (5.1) from it, solving the latter and calculating the resulting density from (5.2). Then a new Schrödinger equation is constructed and so on, until the density does not change appreciably any more.

In both DFT and Hartree–Fock theory, the electrons move in a background composed of the Hartree and external potentials. In addition to this, the exchange term in Hartree–Fock accounts for the fact that electrons with parallel spin avoid each other as a result of the exclusion principle (exchange hole). Opposite spin pairs do not feel this interaction. In DFT, the exchange-correlation potential includes not only exchange effects but also correlation effects due to the Coulomb repulsion between the electrons (*dynamic correlation* effects). In HF, the exchange interaction is treated exactly, but dynamic correlations are neglected. DFT is in principle exact, but we do not know the exact form of the exchange correlation potential – both exchange and dynamic correlation effects are in practice treated approximately.

It is essential that the exchange correlation energy is given in terms of the electron density only and contains no explicit dependence on the external potential (in our case the potential due to the atomic nuclei). As we shall see in Section 5.2, a local approximation for the exchange correlation energy occuring in the DFT equation (5.1) is usually made, thereby simplifying the implementation significantly with respect to Hartree–Fock with its complicated nonlocal exchange term.

*5.1.2 Density functional formalism and derivation of the Kohn–Sham equations

For a many-electron system, the Hamiltonian is given by

$$H = \sum_i \left[-\frac{1}{2}\nabla_i^2 + V_{\text{ext}}(\mathbf{r}_i) \right] + \frac{1}{2} \sum_{\substack{i,j;\\ i \neq j}} \frac{1}{|\mathbf{r}_i - \mathbf{r}_j|}. \tag{5.5}$$

V_{ext} is an external potential which, in the systems of interest to us, is the Coulomb attraction by the static nuclei.

In Chapter 3 we have seen how the ground state can be found by varying the energy-functional with respect to the wave function. Now consider carrying out this variational procedure in two stages: first – for a given electron density – minimise with respect to the wave functions consistent with this density, and then minimise with respect to the density. Denoting by $\min_{\Psi|n}$ a minimisation with respect to the wave functions Ψ which are consistent with the density $n(\mathbf{r})$, we can write

$$E[n] = \min_{\Psi|n} \langle \Psi | H | \Psi \rangle \tag{5.6}$$

and it will be clear that the ground state of the many-electron Hamiltonian can be found by minimising the functional $E[n]$ with respect to the density, subject to the constraint

$$\int \mathrm{d}^3 r \; n(\mathbf{r}) = N \tag{5.7}$$

where N is the total number of electrons.

Now consider a separation of the Hamiltonian into the Hamiltonian H_0 of the homogeneous electron gas (with external potential $V_{\text{ext}} \equiv 0$), and the external potential:

$$H = H_0 + V_{\text{ext}}(\mathbf{r}). \tag{5.8}$$

In this case we can write $E[n]$ as

$$E[n] = \min_{\Psi|n} \left[\langle \Psi | H_0 | \Psi \rangle + \int \mathrm{d}^3 r \; V_{\text{ext}}(\mathbf{r}) n(\mathbf{r}) \right]. \tag{5.9}$$

If we minimise the term in square brackets for a given density $n(\mathbf{r})$, the second term is a constant so that we do not have to include it in the minimisation:

$$E[n] = \min_{\Psi|n} \left[\langle \Psi | H_0 | \Psi \rangle \right] + \int \mathrm{d}^3 r \; V_{\text{ext}}(\mathbf{r}) n(\mathbf{r}). \tag{5.10}$$

Writing

$$F[n] = \min_{\Psi|n} \left[\langle \Psi | H_0 | \Psi \rangle \right] \tag{5.11}$$

we see that $E[n]$ can be written as

$$E[n] = F[n] + \int \mathrm{d}^3 r \; V_{\text{ext}}(\mathbf{r}) n(\mathbf{r}) \tag{5.12}$$

and $F[n]$ obviously does not depend on the external potential. We shall now use these general statements to treat our problem of interacting electrons in an external potential. Summarising the results obtained so far, we see that:

- The ground state density can be obtained by minimising the energy-functional (5.6).
- If we split the Hamiltonian H into a homogeneous one, H_0, and the external potential, the energy-functional can be split into a part $F[n]$, which is defined in (5.11) and which is independent of the external potential, and the functional $\int \mathrm{d}^3 r \; V_{\text{ext}}(\mathbf{r}) n(\mathbf{r})$.

The problem with treating the many-electron system lies in the electron–electron interaction. In fact, for both interacting and noninteracting electron systems the form of the functional $E[n]$ is unknown, but the ground state energy for noninteracting electrons can be solved for trivially, and we can use this to tackle the problem of interacting electrons. In the noninteracting case, $E[n]$ has a kinetic contribution and a contribution from the external potential V_{ext}:

$$E[n] = T[n] + \int \mathrm{d}^3 r \, n(\mathbf{r}) V_{\text{ext}}(\mathbf{r}). \tag{5.13}$$

Variation of E with respect to the density leads to the following equation:

$$\frac{\delta T[n]}{\delta n(\mathbf{r})} + V_{\text{ext}}(\mathbf{r}) = \lambda n(\mathbf{r}), \tag{5.14}$$

where λ is the Lagrange parameter associated with the restriction of the density to yield the correct total number of electrons, N. The form of $T[n]$ is unknown, but we know that the ground state of the system can be written as a Slater determinant with spin-orbitals satisfying the single-particle Schrödinger equation:

$$\left[-\tfrac{1}{2}\nabla^2 + V_{\text{ext}}(\mathbf{r})\right] \psi_k(\mathbf{r}) = \varepsilon_k \psi_k(\mathbf{r}). \tag{5.15}$$

The ground state density is then given by

$$n(\mathbf{r}) = \sum_{k=1}^{N} |\psi_k(\mathbf{r})|^2 \tag{5.16}$$

where the spin-orbitals ψ_k are supposed to be normalised so that the density satisfies the correct normalisation to the number of particles N. Using the above analysis, and taking $T[n]$ for the functional $F[n]$, we immediately see that the kinetic energy-functional T is independent of the potential V_{ext}. Summarising, we have:

- The energy-functional of a noninteracting electron gas can be split into a kinetic functional $T[n]$, and a functional representing the interaction with the external potential, $\int \mathrm{d}^3 r \, V_{\text{ext}}(\mathbf{r}) n(\mathbf{r})$. The kinetic functional does not depend on the external potential.
- The exact solution of the noninteracting electron gas is given in terms of the eigenfunction solutions of the independent-particle Hamiltonian; see Eq. (5.15).

The energy-functional for a many-electron system with electronic interactions included can be written in the form

$$\begin{aligned} E[n] = T[n] &+ \int \mathrm{d}^3 r \, n(\mathbf{r}) V_{\text{ext}}(\mathbf{r}) \\ &+ \frac{1}{2} \int \mathrm{d}^3 r \int \mathrm{d}^3 r' \, n(\mathbf{r}') \frac{1}{|\mathbf{r} - \mathbf{r}'|} n(\mathbf{r}) + E_{\text{xc}}[n], \end{aligned} \tag{5.17}$$

where the last term, the exchange correlation energy, contains, by definition, all the contributions not taken into account by the first three terms which represent the kinetic energy-functional of the *noninteracting* electron gas, the external and the

Hartree energy respectively. It is important to note that we have made no approximations so far but moved all the unknown correlations into $E_{\rm xc}$, which depends on the density n rather than on the explicit form of the wave function because all the other terms in (5.17) depend on the density. For the interacting electron gas it is not clear that the kinetic energy and the electron–electron interaction can be written as a sum of two terms depending on the density only; therefore the kinetic functional for *noninteracting* electrons, which depends only on the density, has been split off and the remaining part of the kinetic energy has been moved into $E_{\rm xc}$. Varying this equation with respect to the density, we obtain

$$\frac{\delta T[n]}{\delta n(\mathbf{r})} + \frac{\delta E_{\rm xc}[n]}{\delta n(\mathbf{r})} + \int d^3r'\, n(\mathbf{r}') \frac{1}{|\mathbf{r}-\mathbf{r}'|} + V_{\rm ext}(\mathbf{r}) = \lambda n(\mathbf{r}). \tag{5.18}$$

This equation has the same form as (5.14), the only difference being the potential replaced by a more complicated one, the 'effective potential':

$$V_{\rm eff}(\mathbf{r}) = V_{\rm ext}(\mathbf{r}) + \frac{\delta E_{\rm xc}[n]}{\delta n(\mathbf{r})} + \int d^3r'\, n(\mathbf{r}') \frac{1}{|\mathbf{r}-\mathbf{r}'|}. \tag{5.19}$$

The analogue of Eq. (5.15) now becomes

$$\left[-\frac{1}{2}\nabla^2 + V_{\rm eff}(\mathbf{r})\right] \psi_k(\mathbf{r}) = \varepsilon_k \psi_k(\mathbf{r}). \tag{5.20}$$

Comparing Eqs. (5.20) and (5.17), we see that adding the eigenvalues ε_k of the occupied states does not lead to the total energy as the Hartree energy is overestimated by a factor of 2, and there is a further difference in the exchange correlation term, so that we have altogether:

$$E = \sum_{k=1}^{N} \varepsilon_k - \frac{1}{2}\int d^3r\, d^3r'\, n(\mathbf{r}) \frac{1}{|\mathbf{r}-\mathbf{r}'|} n(\mathbf{r}') + E_{\rm xc}[n] - \int d^3r\, V_{\rm xc}[n(\mathbf{r})] n(\mathbf{r}). \tag{5.21}$$

where $V_{\rm xc}$ is defined in (5.4). The density functional procedure is now given by Eqs. (5.16), (5.19), (5.20) and (5.21). These equations were first derived by Kohn and Sham [3].

We have already mentioned that the exact form of the exchange correlation potential is not known. This energy is a functional of the density, but there may be an additional explicit dependence on the external potential. Such a dependence would imply that each physical system has its own particular exchange correlation energy-functional. That the exchange correlation potential does not have such a dependence follows immediately from the argument given at the beginning of this section (Eqs. (5.8–5.12)) by separating the external potential off the Hamiltonian and taking the remaining contributions to the energy-functional for $F[n]$. This shows that the exact exchange correlation potential, which should work for *all* materials, is simply a functional of the density. In practice we have to use approximations for E_{xc}, as the exact form of this functional is unknown, and our approximation might be better for some materials than for others. The final conclusion can then be formulated as follows:

- If we split the energy-functional according to (5.17), the term $E_{\text{xc}}[n]$ into which we have moved all the terms we do not have under control, is independent of the external potential.
- The minimisation problem of the energy-functional can be carried out using the Kohn-Sham equations (5.20) together with the constraint (5.16).

5.2 The local density approximation

The difference between the Hartree–Fock and density functional approximation is the replacement of the HF exchange term by the exchange correlation energy E_{xc} which is a functional of the density. The exchange correlation potential is a functional derivative of the exchange correlation energy with respect to the local density and for a homogeneous electron gas this will depend on the value of the electron density. For a nonhomogeneous system, the value of the exchange correlation potential at the point $\mathbf{r}$ depends not only on the value of the density at $\mathbf{r}$ but also on its variation close to $\mathbf{r}$, and it can therefore be written as an expansion in the gradients to arbitrary order of the density:

$$V_{\text{xc}}[n](\mathbf{r}) = V_{\text{xc}}[n(\mathbf{r}), \nabla n(\mathbf{r}), \nabla(\nabla n(\mathbf{r})), \ldots]. \tag{5.22}$$

Apart from the fact that the exact form of the energy-functional is unknown, inclusion of density gradients makes the solution of the DFT equations rather difficult, and usually the *Ansatz* is made that the exchange correlation energy leads to an exchange correlation potential depending on the value of the density in $\mathbf{r}$ only and not on its gradients – this is the *local density approximation* (LDA):

$$E_{\text{xc}} = \int \mathrm{d}^3 r \; \varepsilon_{\text{xc}}[n(\mathbf{r})] n(\mathbf{r}) \tag{5.23}$$

where $\varepsilon_{\text{xc}}[n]$ is the exchange correlation energy per particle of an homogeneous electron gas at density n. The local density approximation is exact for an homogeneous electron gas, so it works well for systems in which the electron density does not vary too rapidly. We shall briefly discuss the various forms used for the exchange correlation energy density in the local density approximation, $\varepsilon_{\text{xc}}[n(\mathbf{r})]$, and refer to the literature for more details [4, 8, 9].

The exchange effects (denoted by the subscript 'x') are usually included in a term based on calculations for the homogeneous electron gas [10] giving the following form for the exchange energy in density functional theory:

$$\varepsilon_{\text{x}}[n(\mathbf{r})] = \text{Const.} \times n^{1/3}(\mathbf{r}) \tag{5.24}$$

which we have already encountered at the end of Section 4.5.1. The value for the constant is found as $-(3/4)(3/\pi)^{1/3}$.

For open-shell systems the spin-up and -down densities n_+ and n_- are usually taken into account as two independent densities in the exchange correlation energy according to a natural extension of the DFT formalism [4]. In local density approximation (now called local spin density approximation), the exchange is given as

$$E_{\mathrm{x}}[n_+, n_-] = -\text{Const.} \int \mathrm{d}^3r \, [n_+^{4/3}(\mathbf{r}) + n_-^{4/3}(\mathbf{r})], \tag{5.25}$$

with Const. $= (3/2)(3/4\pi)^{1/3}$ in accordance with the closed-shell prefactor in (5.24), as can be checked by putting $n_+ = n_- = n/2$. As is to be expected for an exchange coupling, this expression contains interactions between parallel spin pairs only.

In addition to exchange, there is a contribution from the dynamic correlation effects (due to the Coulomb interaction between the electrons) present in the exchange correlation potential, and several local density parametrisations of this interaction have been proposed. A successful one is a parametrised version of the correlation energy obtained in quantum Monte Carlo simulations of the homogeneous electron gas at different densities [11, 9]. Other parametrisations have been presented by von Barth and Hedin [12], and Gunnarson and Lundqvist [13]. These dynamic correlations represent couplings between both parallel and opposite spin pairs.

In calculations of the electronic structure, the DFT–LDA approach has turned out very successfully. In some systems, however, it leads to noticeable deviations or even failures – for examples some stable negative ions such as H^-, O^- and F^- are predicted to be unstable. Many improvements on LDA have therefore been proposed. One of these consists of including gradients of the density in the exchange correlation functional (we will come back to this in the second part of the next section), whose form is motivated by calculations taking many-electron effects into account [8].

Another approach focuses on the self-interaction present in the Hartree energy which contains Coulomb couplings between an electron and its own charge distribution. This overestimation of the electron–electron interaction should be cancelled by the exchange correlation term, which – in LDA – succeeds only partially (although in the hydrogen atom for example, 95% of the self-interaction is cancelled by the exchange correlation). It is possible to add these corrections afterward to the exchange correlation potential [9], but this introduces a dependence of the exchange correlation on the individual orbitals, ψ_k, instead of a dependence on the density only. Both the gradient-correction and self-interaction methods lead to important improvements in calculations of physical properties [4].

5.3 Exchange and correlation: a closer look

5.3.1 The adiabatic theorem and the normalisation conditions

In this section we consider exchange and correlation in more detail. We shall take into account the spin as well as the spatial coordinates. All spin-space coordinates $(\mathbf{r}_1, s_1; \ldots \mathbf{r}_N, s_N)$ are denoted by X. Let us first consider the exact energy-functional (of the spin-orbitals):

$$E_{\text{exact}} = \int \Psi_{\text{AS}}^*(X)\left(-\frac{1}{2}\sum_i \nabla_i^2 + V_{\text{ext}} + V_{ee}\right)\Psi_{\text{AS}}(X)\text{d}X. \tag{5.26}$$

Here, $\Psi_{\text{AS}}(X)$ is a wave function which is antisymmetric in the $x_i = (\mathbf{r}_i, s_i)$, but not necessarily a Slater determinant. We compare the exact energy with the Kohn–Sham functional (which should also be exact for the correct exchange-correlation functional):

$$E_{\text{KS}} = -\sum_k \int \psi_k^*(x)\frac{1}{2}\nabla_k^2\psi_k(x)\text{d}x + \sum \int n(\mathbf{r})V_{\text{ext}}(\mathbf{r})\text{d}x$$
$$+\frac{1}{2}\int n(\mathbf{r})\frac{1}{|\mathbf{r}-\mathbf{r}'|}n(\mathbf{r}')\,\text{d}^3r\,\text{d}^3r' + E_{\text{xc}}[n]. \tag{5.27}$$

The terms related to $V_{\text{ext}}(\mathbf{r})$ are the same in both cases: the exchange and correlation term E_{xc} makes up for the difference in the kinetic energies and the difference between the exact Coulomb interaction and the Hartree approximation in the Kohn–Sham scheme.

We now try to connect the exact form to the Kohn–Sham picture in order to pinpoint this difference better. This is done in the *adiabatic connection* procedure, which works as follows. We first introduce a tunable electron–electron interaction

$$V_{\text{c},\lambda} = \sum_{ij}\frac{\lambda}{|\mathbf{r}_i - \mathbf{r}_j'|} = \lambda V_{\text{c}}, \tag{5.28}$$

where the subscript C stands for Coulomb and where V_{c} is identified with $V_{\text{c},\lambda=1}$.

Just as we did in Section 5.1.2, we split the many-body Hamiltonian into that of an homogeneous electron gas with interaction V_λ and the external potential:

$$H_\lambda = H_{0,\lambda} + \sum_i V_{\text{ext}}(\mathbf{r}_i) = (T + V_{\text{c},\lambda}) + \sum_i V_{\text{ext}}(\mathbf{r}_i), \tag{5.29}$$

and note that for *fixed* densities $n(\mathbf{r})$, the last term will always give the same contribution to the energy. Indeed, we minimise this Hamiltonian for such a fixed density:

$$E_\lambda[n] = \min_{\psi|n}\langle\Psi|H_{0,\lambda}|\Psi\rangle + \int V_{\text{ext}}(\mathbf{r})n(\mathbf{r})\text{d}^3r = F_\lambda[n] + \int V_{\text{ext}}(\mathbf{r})n(\mathbf{r})\text{d}^3r, \tag{5.30}$$

where we have used the definition

$$F_\lambda[n] = \min_{\psi|n} \langle \Psi | H_{0,\lambda} | \Psi \rangle \tag{5.31}$$

The minimisation is carried out on the set of wave functions compatible with the given densities $n(\mathbf{r})$.

We now need a theorem that plays an important role in the quantum molecular dynamics method (see Chapter 9): the *Hellmann–Feynman theorem*. Here we shall prove this theorem for the simple case in which we have a Hamiltonian depending on a single parameter λ. The theorem tells us how the energy eigenvalues of a Hamiltonian H_λ, depending on a parameter λ, vary with λ. Differentiating the Schrödinger equation

$$H_\lambda |\psi_\lambda\rangle = E_\lambda |\psi_\lambda\rangle \tag{5.32}$$

with respect to λ we obtain (the prime indicates derivative with respect to λ):

$$H'_\lambda |\psi_\lambda\rangle + H_\lambda |\psi'_\lambda\rangle = E'_\lambda |\psi_\lambda\rangle + E_\lambda |\psi'_\lambda\rangle. \tag{5.33}$$

Taking the inner product from the left with $\langle \psi_\lambda |$ and using the Hermitian conjugate of (5.32), we see that

$$\frac{\mathrm{d}E_\lambda}{\mathrm{d}\lambda} = \frac{\langle \psi_\lambda | \mathrm{d}H_\lambda / \mathrm{d}\lambda | \psi_\lambda \rangle}{\langle \psi_\lambda | \psi_\lambda \rangle}. \tag{5.34}$$

We can write the derivative of F_λ from the Hellmann–Feynman theorem, by realising that, since $|\psi_\lambda\rangle$ is the variational ground state of F_λ, it must be the lowest eigenstate of $H_{0,\lambda}$. We then obtain

$$\frac{\mathrm{d}F_\lambda}{\mathrm{d}\lambda} = \langle \Psi_\lambda | V_\mathrm{c} | \Psi_\lambda \rangle. \tag{5.35}$$

From this, and from the fact that for $\lambda = 0$ we have a trivial, noninteracting electron gas, we have

$$F_{\lambda=1}[n] = T_\mathrm{KS}[n] + \int_0^1 \langle \psi_\lambda | V_\mathrm{c} | \Psi_\lambda \rangle \mathrm{d}\lambda. \tag{5.36}$$

We now find the exchange correlation potential as the difference between the interacting and noninteracting electron gas including the Hartree energy E_H:

$$\begin{aligned} E_\mathrm{xc} &= F_{\lambda=1}[n] - T_{KS}[n] - \sum \frac{1}{2} \int n(\mathbf{r}) \frac{1}{|\mathbf{r} - \mathbf{r}'|} n(\mathbf{r}')\, \mathrm{d}^3 r\, \mathrm{d}^3 r' \\ &= \int_0^1 \langle \psi_\lambda | V_\mathrm{c} | \Psi_\lambda \rangle \mathrm{d}\lambda - E_\mathrm{H}. \end{aligned} \tag{5.37}$$

The main point of the derivation is that in (5.36), which holds for the interacting gas, the kinetic energy is that of the noninteracting gas; therefore, we find the exchange correlation correction only in terms of the Coulomb interaction. For $\lambda = 0$, the XC correction term is nonzero as the Hartree energy does not take the antisymmetry of the full many-body wave function into account: it is the exchange-only part of the correction.

There is another fruitful way of looking at the exchange correlation term, which is related to the discussion in Section 5.1.2. There we considered the probability density for finding the particles 1 and 2 with coordinates x and x' respectively:

$$P(x, x') = \int |\Psi(x, x', x_3, \ldots, x_N)|^2 \mathrm{d}x_3 \ldots \mathrm{d}x_N. \tag{5.38}$$

We now use this definition for a general wave function (not necessarily a Slater determinant).

The single-particle density is given as

$$n(x) = N \int |\Psi(x, x_2, \ldots, x_N)|^2 \mathrm{d}x_2 \ldots \mathrm{d}x_N. \tag{5.39}$$

Integrating the single-particle density gives the number of particles:

$$\int n(x)\mathrm{d}x = N. \tag{5.40}$$

From the definition of $n(x)$ we immediately see that

$$N \int P(x, x')\mathrm{d}x' = n(x). \tag{5.41}$$

The reason for introducing these quantities is that they give insight in the exchange and correlation energies. To see this, consider the Coulomb energy:

$$E_{\mathrm{c}} = \frac{N(N-1)}{2} \int P(x, x') \frac{1}{|\mathbf{r} - \mathbf{r}'|} \,\mathrm{d}x\,\mathrm{d}x' \tag{5.42}$$

(the prefactor counts the number of particle pairs). We now define the *exchange correlation hole*, $n_{\mathrm{xc}}(x, x')$, through

$$N(N-1)P(x, x') = n(x)n(x') + n(x)n_{xc}(x, x'). \tag{5.43}$$

The exchange correlation hole indicates how the actual distribution of a second electron, given a first electron at x, deviates from the average density. Then we can write

$$E_{\mathrm{c}} = E_{\mathrm{H}} + \frac{1}{2} \int n(x)n_{\mathrm{xc}}(x, x') \frac{1}{|\mathbf{r} - \mathbf{r}'|} \,\mathrm{d}x\,\mathrm{d}x'. \tag{5.44}$$

Note that the second term can be identified with the exchange correlation energy.

The most important consequence of this is that we can derive some properties of the exchange correlation hole, which any exchange correlation energy should satisfy. The first of these properties follows from the normalisation of P:

$$\int P(x, x')\,\mathrm{d}x\,\mathrm{d}x' = 1 \tag{5.45}$$

which follows directly from the normalisation of the wave function. Furthermore,

$$\int n(x)\mathrm{d}x = N \tag{5.46}$$

for the same reason. Integrating Eq. (5.43) now over $\mathrm{d}x'$ (this actually denotes an integration over $\mathbf{r}'$ and a sum over s'), we obtain the result

$$(N-1)n(x) = Nn(x) + n(x)\int n_{\mathrm{xc}}(x,x')\,\mathrm{d}x'; \tag{5.47}$$

in other words,

$$\int n_{\mathrm{xc}}(x,x')\mathrm{d}x' = -1. \tag{5.48}$$

Realising that the second term in Eq. (5.44) is the exchange correlation correction to the Coulomb energy, we see that this correction can be described in terms of a charge distribution which carries a positive unit charge: this is the exchange correlation hole n_{xc}.

Now let us return to the Hartree–Fock approximation. There we considered a Slater determinant containing all exchange effects. If we apply the above analysis to a Slater determinant we obtain an exchange hole (Coulomb correlations are absent in this case) which integrates up to a charge -1 (that is, a positive hole), *irrespective of the strength* λ *of the Coulomb interaction*. Therefore we conclude that the exchange hole adds up to -1 and, supposing that the exchange correlation hole is the sum of an exchange and a correlation contribution, the correlation hole must add up to 0.

Let us summarise the results obtained so far. The first is that we can remove the exchange and correlation contribution to the kinetic energy from the description by applying the adiabatic connection formula. The price we have to pay is that we have to integrate the Coulombic term due to exchange and correlation over the interaction strength λ. The second result is that this contribution can be described in terms of an exchange and a correlation hole, the first of which integrates up to -1 and the second integrates to 0.

5.3.2 The generalised gradient approximation

We can now understand the success of LDA: the exchange and correlation holes are taken from very accurate quantum Monte Carlo results for the homogeneous electron gas and therefore they satisfy the two normalisation conditions for exchange and correlation just described.

We can now also describe how a gradient expansion can be constructed: we must take into account isotropy conditions and then make sure that the exchange and the correlation hole satisfy their respective normalisation conditions. This scheme has been carried out by several groups, and some well known functionals are those of Perdew and Wang of 1986 [14, 15] and 1991 [16] (respectively PW86 and PW91), and of Becke [17], Lee, Yang and Parr [18] (LYP) and Perdew, Burke and Enzerhof [19, 20]. These exchange correlation functionals go by the name of generalised gradient approximations (GGAs).

In general, GGA improves on LDA for the quantities which are already successfully treated in LDA: total energies and hence binding energies, bond lengths and angles. Ionisation energies based on Kohn–Sham energy eigenvalues are approximately the same as for LDA. In general, LDA tends to over-estimate the correlation energy and underestimates the exchange energy; these are remedied to some extent in GGA, but as the two corrections are opposite, the net effect is not too spectacular. That does not mean that the improvement is not important: the GGA gives a more accurate description of the many-body electron system than LDA.

One major deficiency which is shared by GGA and LDA is the fact that the exchange correlation correction does not cancel the self-interaction present in the Hartree energy. This in particular affects the interpretation of the highest Kohn Sham energy as the ionisation energy of the system (see also Section 5.4.1).

5.3.3 Exact exchange

The problem with the known exchange functionals which are given as explicit functions of the density is the presence of self-interaction terms, a feature that was absent in the Hartree–Fock theory. It is possible to include the HF exchange term in the exchange correlation functional. This is justified in the so-called *optimized potential method* [21, 22] which leads to a Kohn–Sham picture where the exchange correlation functional is allowed to depend explicitly on the orbitals rather than on the density. The advantage of having no self-interaction left is counteracted by a less favourable scaling behaviour: just as in the HF theory, we must calculate and sum over two-electron integrals which makes this method rather time-consuming.

Finally, hybrid functionals combine exact exchange with traditional functionals.

5.4 Beyond DFT: one- and two-particle excitations

5.4.1 One-particle theories: ionisation and electron addition energies

The DFT is designed to yield correct ground state energies for a many-body system. It is, however, not justifiable to interpret Kohn–Sham energies as energy levels which can be detected in a spectroscopy experiment. An exception must be made for the highest occupied level, which gives the correct ionisation potential in exact DFT. To see that this is indeed the case, note that if one of the electrons (we take this to be electron number N) of a neutral system is moved very far away from all the nuclei in the system, the *exact* ground state wave function for the N electrons can be written as

$$\psi_N(\mathbf{r}_1, \ldots, \mathbf{r}_N) = \psi_{N-1}(\mathbf{r}_1, \ldots, \mathbf{r}_{N-1})\varphi(\mathbf{r}_N), \tag{5.49}$$

where this form is justified by the notion that at the large distance between particle N and its partners, any correlation between them has disappeared. Note that $\psi_{N-1}(\mathbf{r}_1, \ldots, \mathbf{r}_{N-1})$ is the normalised ground state wave function of the $N - 1$ particles close to the nuclei, as the perturbation due to particle N can be neglected.

The Hamiltonian for the N-particle system can be written as [23]

$$H(N) = H(N-1) + \frac{p_N^2}{2m} + V_{\text{ext}}(\mathbf{r}_N) + \sum_{j=1}^{N-1} \frac{1}{|\mathbf{r}_j - \mathbf{r}_N|}, \tag{5.50}$$

where $H(N-1)$ is the $(N-1)$-particle Hamiltonian. Writing up the Schrödinger equation for the N electrons, using the wave function (5.49) and using the fact that the first term on the right hand side of that equation represents the $(N-1)$-particle ground state, we obtain an equation for φ:

$$\left[E_{N-1}^{\text{GS}} + \frac{p_N^2}{2m} + V_{\text{ext}}(\mathbf{r}_N) + \left\langle \psi_{N-1} \left| \sum_{j=1}^{N-1} \frac{1}{|\mathbf{r}_j - \mathbf{r}_N|} \right| \psi_{N-1} \right\rangle \right] \varphi(\mathbf{r}_N) = E_N^{\text{GS}} \varphi(\mathbf{r}_N). \tag{5.51}$$

The asymptotic (large r) behaviour of this equation is exactly the same as that of the Kohn–Sham equation (which also describes an electron far away from a localised charge distribution with net charge $+1$), and this can only be the case when the 'energy' eigenvalue of the Kohn–Sham equation is the same as $E_N^{\text{GS}} - E_{N-1}^{\text{GS}}$ [24].

This is a very interesting result when it is combined with Janak's theorem [25] which says that the highest occupied orbital energy gives the chemical potential (see Problem 5.4). In DFT, we can fill the orbitals partially by calculating the density with a fractional filling factor f_j:

$$n(\mathbf{r}) = \sum_j f_j |\psi(\mathbf{r})|^2. \tag{5.52}$$

This shows that we can really take an infinitesimal differential of the total energy (by varying f_N) with respect to the charge in the highest occupied orbital, which is the proper definition of the chemical potential. From the fact that the chemical potential and the ionisation energy are both given as the highest occupied Kohn–Sham eigenvalue, we see that the discrete derivative of the total energy with respect to the charge in the highest occupied level must be equal to the continuous derivative.

Perdew *et al.* [26] have argued that the derivative is constant for any fractional occupation of the highest occupied level, but their reasoning can be criticised because they impose this property in their form of the energy-functional (based on a density operator form), which need not describe the pure-state functional of DFT [27].

The property we have just derived – the Kohn–Sham energy of the highest occupied level gives us the ionisation energy – is satisfied very well for extended systems, but poorly for molecules, where the highest occupied Kohn–Sham energy (called the highest occupied molecular orbital, or HOMO) is generally found a few eV *above* the experimental value. Hartree–Fock usually gives a much better value. The reason why DFT fails so badly in practice lies in the poor asymptotic behaviour of the available exchange-correlation potentials, which, among other imperfections, do not cancel the self-interaction and hence give an incorrect asymptotic form of the Kohn–Sham potential. In our derivation, this asymptotic behaviour played a crucial role. The fact that we do not have the exact exchange correlation functional at our disposal therefore is a serious handicap in describing the spectra of atoms and molecules.

There is a way around this: given the fact that DFT is very good at calculating ground state energies, we can perform two calculations: one for N electrons, and one for $N - 1$ (for the electron addition energies, the second calculation would be performed for $N + 1$ electrons). The difference in the total energies then gives the ionisation (or electron addition) energy. Instead of these two energies, it is also possible to do one calculation at half filling of the highest occupied (or lowest unoccupied) level. The Kohn–Sham energy of that level is the derivative of the total energy with respect to the charge, so that we can predict the ionisation energy from

$$E^{\mathrm{GS}}(N) - E^{\mathrm{GS}}(N-1) = \left[\frac{\partial E_{\mathrm{tot}}}{\partial N}\right]_{N+1/2} = \varepsilon_N^{\mathrm{KS}}(N+1/2). \tag{5.53}$$

A similar procedure gives the electron addition energy. This method is known as *delta-SCF*.

5.4.2 General theories for excitation energies

Looking at what causes a system which is in the ground state to go to an excited state, we conclude that there should always be some time-dependent perturbation to the Hamiltonian which is responsible for this. Therefore, we should consider the response of the system to an external, time-dependent perturbation. The standard approach is to consider the response to a monochromatic perturbation with frequence ω. However, if the response of the system to a perturbation is faster than the typical period of the perturbation, we may consider a time-independent approach.

An electron which has been excited to a higher energy level will return to its ground state orbital after some time. This finite lifetime gives rise to a finite width of the energy spectrum, according to the time–energy uncertainty relation. Therefore, we can no longer speak of a discrete energy level, but we can still find a fingerprint of the spectrum in quantities such as the macroscopic dielectric function, which is the long-wavelength limit of the microscopic function $\varepsilon(\mathbf{r}, \mathbf{r}, \omega)$. This will exhibit peaks as a function of ω whose centres can be viewed as energy levels, and the widths as lifetimes.

Experimentally, excited states are studied by using spectroscopy techniques. In *direct* photo-emission, an incident photon excites an electron to sufficiently high energy that it can leave the system (ionisation). In *inverse* photo-emission we send an electron into the material to occupy an unoccupied state, causing emission of a photon whose energy is detected (electron addition). In absorption spectroscopy, the electron or photon that is sent into the system is also detected when it leaves the system. It may meanwhile have changed its energy by interaction with another electron in the material which is excited to a higher state.

The first two processes, ionisation and electron addition, are called *one-particle* processes; the third is a *two-particle* process. The two-particle character arises because, when an electron is excited in a system, it leaves a (positive) hole behind. If the electron remains in the system, it interacts with the hole, and in particular it may form an *exciton*: a bound state of the particle–hole system. We shall briefly describe the analysis for one-particle processes, and then review two-particle methods.

In the previous subsection, the problem of finding the ionisation energy was addressed. In general, when performing spectroscopy experiments, levels other than the highest one may be excited. Of course, one could try to use a generalised delta-SCF procedure for these, but this is difficult because for a band structure calculation, we would need many calculations as each $\mathbf{k}$-vector in the Brillouin zone has its own particular excitation. Another problem is that for a band state, DFT differs essentially from Hartree–Fock, which allows for calculating excited states: the HF orbital energies can be interpreted as excitation energies according to Koopman's theorem (which only holds for the highest band in DFT, see above). This theorem is based on the assumption that the orbitals do not relax when the configuration changes by emptying full, or filling empty levels. This approximation fails miserably in solids, for example in diamond, where the band gap is in HF predicted to be 15 eV, more than twice the experimental gap of about 7 eV.

What is missing from the description of a ground state system, is the fact that an electron added to the system does not feel the pure Coulomb interaction from the ground state charge distribution: the resident electrons will re-order in the presence of the visiting electron, and tend to *screen* the effect of the Coulomb interaction. A many-body theory for band structure takes these effects into account; HF and DFT do not.

Such a many-body theory was formulated by Hedin in 1965 [28], for reviews see Refs. [29, 30]. We shall not go into the details of the many-body theory behind this approach, but consider a particular, relatively simple form, the COHSEX approximation in some detail (we shall explain the name COHSEX below). This approximation can be understood quite well without going into the formal theory.

Suppose we put an extra unit charge into the system. This charge will occupy some state with orbital wave function $\psi(\mathbf{r})$. We could describe the interaction between this electron and the resident electrons by Hartree–Fock terms, i.e. a Coulomb interaction and an exchange interaction. However, although exchange is treated correctly (apart from neglecting screening effects, see below), the Coulomb interaction will push the resident electrons away from the visitor and thereby lower the interaction between visitor and residents.

Let us first neglect screening. The electrostatic energy is then given by

$$E_{\mathrm{ES}}(\mathbf{r}) = \int n(\mathbf{r}')v(\mathbf{r},\mathbf{r}')|\psi(\mathbf{r})|^2 \mathrm{d}^3r\,\mathrm{d}^3r', \tag{5.54}$$

where the interaction $v(\mathbf{r},\mathbf{r}')$ is the 'bare' Coulomb interaction potential $v(\mathbf{r},\mathbf{r}') = 1/|\mathbf{r}-\mathbf{r}'|$. Screening can be viewed from two different standpoints. The first is to consider the change Δn in charge distribution due to the presence of the new electron. The second view is to take for the potential felt by this electron a screened potential $w(r)$ which falls off more rapidly than the bare potential. Obviously, the two are connected.

For the correction to the energy due to the change $\Delta n(\mathbf{r})$ in the charge distribution we write:

$$\Delta E = \int \Delta n(\mathbf{r}')v(\mathbf{r},\mathbf{r}')|\psi(\mathbf{r})|^2 \mathrm{d}^3r\,\mathrm{d}^3r'. \tag{5.55}$$

However, this result is wrong, because the response Δn to the test charge is proportional to that charge! Therefore, if we integrate the energy up over the extra charge put into the system, we get a prefactor of $1/2$:

$$\Delta E = \frac{1}{2}\int \Delta n(\mathbf{r}')v(\mathbf{r},\mathbf{r}')|\psi(\mathbf{r})|^2\, \mathrm{d}^3r\,\mathrm{d}^3r'. \tag{5.56}$$

In order to get a handle on the screened potential w, we note that it is defined as the potential measured at $\mathbf{r}'$ given the fact that there is a test point charge at $\mathbf{r}$. We therefore may write:

$$\begin{aligned} w(\mathbf{r}',\mathbf{r}) &= \int \delta(\mathbf{r}-\mathbf{r}'')v(\mathbf{r}'',\mathbf{r}')\mathrm{d}^3r'' + \int \Delta n(\mathbf{r}''|\mathbf{r})v(\mathbf{r}'',\mathbf{r}')\mathrm{d}^3r'' \\ &= v(\mathbf{r},\mathbf{r}') + \int \Delta n(\mathbf{r}''|\mathbf{r})v(\mathbf{r}'',\mathbf{r}')\,\mathrm{d}^3r''. \end{aligned} \tag{5.57}$$

Here, $\Delta n(\mathbf{r}'|\mathbf{r})$ is the change in the charge density at $\mathbf{r}'$ due to a unit test charge placed at $\mathbf{r}$. The induced charge density $\Delta n(\mathbf{r})$ due to a charge distribution $|\psi(\mathbf{r})|^2$

is given as the integral of the induced charge $\Delta n(\mathbf{r}''|\mathbf{r})$ over $\mathbf{r}$, weighted by $|\psi(\mathbf{r})|^2$. Putting these results back into Eq. (5.56) leads to the formal expression

$$\Delta E = \frac{1}{2}\int \delta(\mathbf{r}-\mathbf{r}')[w(\mathbf{r},\mathbf{r}') - v(\mathbf{r},\mathbf{r}')]|\psi(\mathbf{r})|^2\,\mathrm{d}^3r\,\mathrm{d}^3r'. \tag{5.58}$$

If we take the functional derivative of this expression with respect to $\psi(\mathbf{r})$, we obtain a term

$$V\psi(\mathbf{r}) = \frac{1}{2}\int \delta(\mathbf{r}-\mathbf{r}')[w(\mathbf{r},\mathbf{r}') - v(\mathbf{r},\mathbf{r}')]\,\mathrm{d}^3r'\,\psi(\mathbf{r}) \tag{5.59}$$

in the one-particle Schrödinger equation. Exchange is already treated correctly, so in the exchange term, we can simply replace the bare Coulomb interaction by the screened interaction. We see that the correction boils down to taking into account the COulomb Hole and Screened EXchange – hence the name COHSEX.

Now there is still something missing: we do not know the screened interaction potential $w(\mathbf{r},\mathbf{r}')$. This can however be found in a so-called random phase approximation (RPA) scheme, which is based on perturbation theory. It works as follows. We place a test charge at position $\mathbf{r}$. As we have seen above, this test charge generates a change Δn of the resident charge distribution, and the bare potential $v(\mathbf{r},\mathbf{r}')$ is replaced by the screened potential $w(\mathbf{r},\mathbf{r}')$. The relation between the two is usually formulated in terms of the dielectric constant. This is defined as:

$$v(\mathbf{r},\mathbf{r}') = \int \varepsilon(\mathbf{r}',\mathbf{r}'')w(\mathbf{r},\mathbf{r}'')\,\mathrm{d}^3r''. \tag{5.60}$$

We can therefore write for the screening correction

$$\begin{aligned} w(\mathbf{r},\mathbf{r}') - v(\mathbf{r},\mathbf{r}') &= \int [\delta(\mathbf{r}',\mathbf{r}'') - \varepsilon(\mathbf{r}',\mathbf{r}'')]w(\mathbf{r},\mathbf{r}'')\,\mathrm{d}^3r'' \\ &= \int \Delta n(\mathbf{r}''|\mathbf{r})v(\mathbf{r}',\mathbf{r}'')\mathrm{d}^3r'', \end{aligned} \tag{5.61}$$

where $\Delta n(\mathbf{r}''|\mathbf{r})$ denotes a change of the density at $\mathbf{r}''$ due to a unit point charge being placed at $\mathbf{r}$.

Now we view the effect of this point charge at $\mathbf{r}$ as a perturbing potential w. The lowest order correction to the occupied orbital j in stationary perturbation theory is given by [31]

$$\Delta\psi_j(\mathbf{r}') = \sum_{k\ \text{unocc.}} \frac{\langle\psi_k|w|\psi_j\rangle}{E_j - E_k}\psi_k(\mathbf{r}'). \tag{5.62}$$

The total change in the density is therefore given by

$$\Delta n(\mathbf{r}') = 2 \sum_{j\text{ occ.}} \psi_j^*(\mathbf{r}')\Delta\psi_j(\mathbf{r}')$$
$$= 2{\sum_{j,k}}' \int \frac{\psi_j^*(\mathbf{r}')\psi_k(\mathbf{r}')\psi_k^*(\mathbf{r}'')\psi_j(\mathbf{r}'')}{E_j - E_k} w(\mathbf{r}', \mathbf{r})\, \mathrm{d}^3 r'', \tag{5.63}$$

where the prime with the sum indicates that the index j runs over occupied, and k over unoccupied levels. Putting this back into the rightmost term of Eq. (5.61) and using the equality between the second and third expression in this equation yields

$$\varepsilon(\mathbf{r}', \mathbf{r}) - \delta(\mathbf{r}', \mathbf{r}) = 2{\sum_{j,k}}' \int \frac{\psi_j(\mathbf{r})\psi_k^*(\mathbf{r})\psi_k^*(\mathbf{r}'')\psi_j(\mathbf{r}'')}{(E_j - E_k)|\mathbf{r} - \mathbf{r}'|} \mathrm{d}^3 r''. \tag{5.64}$$

In this derivation we have assumed that the effects of the new electron on the resident one can be completely described in terms of the Hartree term in the Hamiltonian. This is known as the *random phase approximation* [32].

Hybertsen and Louie have implemented the full GW approximation into an LDA framework [33, 34], and obtained energy spectra with excellent agreement with experiment. The static COHSEX approximation is only a first step in this procedure. It is possible to replace the relation (5.60) by a local one:

$$v(\mathbf{r}, \mathbf{r}') = w(\mathbf{r}, \mathbf{r}')\varepsilon(\mathbf{r}, \mathbf{r}'), \tag{5.65}$$

which is sometimes done for convenience. The detailed structure overlooked in this approximation is denoted as *local field effects*. From the work of Hybertsen and Louie it is clear that local field effects have a major impact on the energy spectra.

We see that a particle which is added to the system will influence the behaviour of the other particles. If we could switch off the interaction between the particles, the newly added particles would occupy sharp energy levels, and the new particle on its own would completely determine the new level. Landau [35] analysed the many-body behaviour of liquid helium-3 and argued that if we had a knob with which we could tune the interactions, the spectrum would change in a *continuous* way. That is, for no interaction, the spectrum consists of a series of delta-functions, which start to broaden and shift when the interactions are switched on. The corresponding excitations involve, as we have seen, the presence of a new particle (or, in the case of two-particle problems, a particle occupying a new state), accompanied by a slight change of the orbitals of the other particles. This excitation is called a *quasi-particle*. Quasi-particle excitations can be analysed in terms of many-body Green's functions [36].

5.4.3 Two-particle effects

A two-particle description within the many-body theory of Green's functions can be formulated: it is known as the *Bethe–Salpeter* theory. Implementation of this is possible but generally demanding – for a review see Ref. [37]. Another approach which potentially describes any type of excitation of a many-body system in the presence of a time-dependent field is *time-dependent density functional theory* (TDDFT) [38–40]. The formalism of this theory is analogous to that of plain DFT, and the analogue of the DFT Hohenberg–Kohn theorem in TDDFT is the *Runge–Gross* theorem. This reads:

> Two densities $\rho(\mathbf{r}, t)$ and $\rho'(\mathbf{r}, t)$ evolving from a common initial state $\psi(\mathbf{R}, t = 0)$ $[\mathbf{R} = (\mathbf{r}_1, \mathbf{r}_2, \ldots, \mathbf{r}_N)]$ under the influence of two external potentials $v(\mathbf{r}, t)$ and $v'(\mathbf{r}, t)$ are always different provided these potentials differ by more than a purely time-dependent function
> $$v'(\mathbf{r}, t) \neq v(\mathbf{r}, t) + c(t). \tag{5.66}$$

The presence of the uniform function $c(t)$ in this last condition is related to the 'gauge invariance': multiplying $\psi(\mathbf{r}, t)$ by a factor $\exp[-\mathrm{i}C(t)/\hbar]$ solves the time-dependent Schrödinger equation with a potential shifted by $c(t) = \dot{C}(t)$. This is easily verified.

A time-dependent Kohn–Sham formulation can be derived from this theorem. This formulation gives the time-evolution of single-particle orbitals which generate the same density as the full many-body problem. These orbitals evolve according to a time-dependent Schrödinger equation:

$$\mathrm{i}\frac{\partial \psi_k(\mathbf{r}, t)}{\partial t} = \left[-\frac{1}{2}\nabla^2 + V_{\text{ext}}(\mathbf{r}, t) + V_{\text{H}}(\mathbf{r}, t) + V_{\text{xc}}(\mathbf{r}, t)\right]\psi_k(\mathbf{r}, t) \text{ for } k = 1, \ldots, N. \tag{5.67}$$

The density is now time-dependent: it is as usual given by

$$n(\mathbf{r}, t) = \sum_{k=1}^{N} |\psi_k(\mathbf{r}, t)|^2. \tag{5.68}$$

The Hartree and exchange-correlation potentials V_{H} and V_{xc} are defined in terms of the time-dependent density using the same expressions as in static DFT. Note, however, that an exchange-correlation potential that works in static DFT is not guaranteed to work in TDDFT. In fact this is the greatest weakness of TDDFT at this moment: it is as yet unclear which are the reliable approximations to this potential.

Technically, the solution of the time-dependent Kohn–Sham equations can be carried out in a Crank–Nicholson or in a split-operator scheme (see Appendix 7.2). The application of these schemes is however slightly tricky [41]. The reason is that in a proper Crank–Nicholson scheme, we use the Hamiltonian operator evaluated at $t + h/2$, where h is the time step in going from t to $t + h$. As the Hamiltonian depends on the solutions $\psi_k(t + h/2)$ which are not yet known, we must first estimate $\psi_k(t + h)$ using H evaluated at t ($\hbar = 1$; do not confuse with the time step h):

$$\tilde{\psi}_k(t + h) = \frac{1 - \mathrm{i}hH(t)/2}{1 + \mathrm{i}hH(t)/2}\psi_k(t). \tag{5.69}$$

Using the $\tilde{\psi}_k(t + h)$, we evaluate $\tilde{H}(t + h)$. Then we again perform a Crank–Nicholson step where we use the mean of $H(t)$ and $\tilde{H}(t + h)$.

In the split-operator scheme, the solution to the fact that the orbitals are unknown at $t + h/2$ can be solved in an elegant way [41]. The scheme brings us from t to $t + h$ by applying the following operation:

$$\psi_k(t + h) = \exp(-\mathrm{i}T/2)\exp[-\mathrm{i}V(t + h/2)]\exp(-\mathrm{i}T/2)\psi_k(t), \tag{5.70}$$

where T is the kinetic, and V the potential energy. In order to perform this step, we need to Fourier-transform back and forth between the momentum and direct-space representations where the operators occurring in the exponentials are diagonal:

$$\begin{aligned} &\psi_k(\mathbf{r}, t) \xrightarrow{\text{FFT}} \psi_k(\mathbf{p}, t) \xrightarrow{\times \exp[-\mathrm{i}T/2)} \psi'_k(\mathbf{p}, t) \xrightarrow{\text{FFT}} \psi'_k(\mathbf{r}, t) \xrightarrow{\times \exp[-\mathrm{i}V(t+h/2)]} \\ &\psi''_k(\mathbf{r}, t) \xrightarrow{\text{FFT}} \psi''(\mathbf{p}, t) \xrightarrow{\times \exp(-\mathrm{i}T/2)} \psi'''_k(\mathbf{p}, t) \xrightarrow{\text{FFT}} \psi'''_k(\mathbf{r}, t) = \psi_k(t + h). \end{aligned} \tag{5.71}$$

The nice property of applying the second operator ($\times \exp[-\mathrm{i}V(t + h/2)]$) is that it does *not* change the density, as it represents just a phase factor in real space. Therefore we can just take the orbitals ψ'_k to evaluate the potential in this procedure. This implies that we already have ψ_k at the half-integer steps at our disposal. Furthermore, we can glue the last stage of this procedure onto the first stage of the next step, at the expense of not having the ψ_k at our disposal at the integer time steps.

A particularly nice sample application of TDDFT is the description of *higher harmonic generation* in helium [42], which describes the generation of higher harmonics in the response to monochromatic light of high intensity [43]. Generally, TDDFT is a very useful tool for calculating dynamic response functions (frequency-dependent polarisabilities) [44].

5.5 A density functional program for the helium atom

In this section we describe the construction of a program for the calculation of the ground state of the helium atom within the local density approximation. As the two electrons occupy the 1s-orbital, the density and hence the Hartree potential are radially symmetric and we exploit this symmetry in spatial integrations. Instead of using basis functions, we solve the radial Schrödinger equation directly, just as we have done in the first chapter for calculating scattering cross sections. The program is set up in three steps. First, we use a simple integration algorithm and combine this with an interpolation routine in order to find the stationary states of hydrogen-like atoms. Second, a routine for obtaining the Hartree potential from the (radial) electronic density is added. At this point the results should compare with those obtained in the previous chapter using Gaussian basis functions. Finally, the exchange correlation potential is added and we have a fully self-consistent local density program.

5.5.1 Solving the radial equation

To solve the radial Schrödinger equation you can use the Numerov algorithm which is discussed in Appendix 7.1 and which has been used for the scattering program in Chapter 2. However, we also have to solve other differential equations and integrals, and in order to avoid complications we shall not require the $\mathcal{O}(h^6)$ accuracy of Numerov's algorithm – hence you can also use the simpler Verlet–Stoermer algorithm of Appendix A7.1. It is of course possible and recommended to use library routines throughout the program. For integrating the radial Schrödinger equation, a nonuniform grid is often used which is dense near the nucleus where the Coulomb potential diverges (see Problem 5.1). For the hydrogen atom, the radial equation for $l = 0$ reads (in atomic units)

$$\left[-\frac{1}{2}\nabla^2 - \frac{1}{r}\right]u(r) = Eu(r) \tag{5.72}$$

where $u(r)$ is given as $rR(r)$, $R(r)$ being the radial wave function. For the hydrogen atom we know that the solution for the ground state is given by $E = -0.5$ a.u. and $u(r) \propto re^{-r}$, and this enables us to test our programs. The energy eigenvalues can be found by integrating the radial Schrödinger equation from some large radius r_{max} inward to $r = 0$ and checking whether the solution vanishes there. The procedure is analogous to that described in Problem A4. You should first check whether for $E = -0.5$ a.u. the radial solution does indeed vanish at $r = 0$. Note that for the regular solutions [$u(0) = 0$] we are looking for, the divergence of the potential near the origin causes no problems in the integration routine as long as it is not evaluated at $r = 0$.

For the starting values at r_{max} you can substitute the exact values $u(r_{\text{max}}) = r_{\text{max}} \exp(-r_{\text{max}})$ and similarly for $u(r_{\text{max}} - h)$, but it is also possible to take $u(r_{\text{max}})$ equal to 0 and $u(r_{\text{max}} - h)$ equal to a very small value. It is interesting to play around varying the starting conditions and the value of r_{max} in order to get a feeling for how the accuracy is affected by these.

To arrive at a program which determines the spectrum for you, you must couple the integration routine to a root-finding scheme and apply it to the value of u at the origin. Although it is in principle possible to solve for the energy derivative of u alongside the determination of u itself, we assume here that the integration routine does not provide energy derivatives. Therefore a library root-finding routine must not use the derivative and the same holds for one you write yourself. In the latter case, the secant method is appropriate; see Appendix A3. You will have to supply the boundaries between which the root must lie when using the program.

PROGRAMMING EXERCISE

Combine the integration routine and the root-finding routine into a method for finding the $l = 0$ states of a radial potential.

Check Test your program for the hydrogen atom.

5.5.2 Including the Hartree potential

We now describe an extension of the hydrogen program to the helium case, which implies having a nuclear potential $-2/r$ in the Hamiltonian and requires some treatment of the electron–electron interaction. In this section we take the latter into account in the same way as in Section 4.3.2, that is by a so-called Hartree potential which is the electrostatic potential generated by the charge distribution following from the wave function. This potential is given by

$$V_{\rm H}(\mathbf{r}) = \int {\rm d}^3 r' \; n_{\rm s}(\mathbf{r}') \frac{1}{|\mathbf{r} - \mathbf{r}'|}. \tag{5.73}$$

Here, $n_{\rm s}$ stands for the density of a *single* orbital – the total charge density is twice as large as a result of summation over the spin. The proper Hartree potential is therefore twice as large, but half of it consists of the self-interaction which we have subtracted off because this can easily be done for the helium case (see also the end of Section 4.3.1). Rather than solving for this potential by calculating the integral (5.73) directly, we shall find it by solving Poisson's equation:

$$\nabla^2 V_{\rm H}(\mathbf{r}) = -4\pi n_{\rm s}(\mathbf{r}). \tag{5.74}$$

Using the radial symmetry of the density and defining $U(r) = rV_{\rm H}(r)$, this equation reduces to the form

$$\frac{{\rm d}^2}{{\rm d}r^2} U(r) = -4\pi r n_{\rm s}(r). \tag{5.75}$$

This is an ordinary second order differential equation which can be solved again using Verlet's algorithm (or a library routine). Note that it is necessary to normalise the radial wave function before integrating Poisson's equation! If you take for the normalisation

$$\int \mathrm{d}r\, u^2(r) = \int \mathrm{d}r\, r^2 R^2(r) = 1, \tag{5.76}$$

you have already included a factor 4π into the density (arising from the angular integrations) and the factor 4π in Poisson's equation drops out:

$$U''(r) = -\frac{u^2(r)}{r}. \tag{5.77}$$

We shall use the normalisation (5.76) throughout this section.

The solution of Eq. (5.77) contains two integration constants which have to be fixed by the boundary conditions. We take $U(0) = 0$ as the first boundary condition. Elementary electrostatics then leads to the second condition

$$V'_H(r_{\max}) = \frac{q_{\max}}{r^2_{\max}}, \tag{5.78}$$

where $q_{\max}$ is the electron charge contained in a sphere of radius $r_{\max}$:

$$q_{\max} = \int_0^{r_{\max}} \mathrm{d}r\, u^2(r). \tag{5.79}$$

For large $r_{\max}$, $q_{\max}$ is the total electron charge. In that case we see from the asymptotic form of (5.78), using the fact that $U(r_{\max})$ is now constant as a function of $r_{\max}$), that $U(r_{\max}) = q_{\max}$. When carrying out the numerical integration, we take for the first starting condition $U(0) = 0$. The second starting condition, for $U(h)$, is not known at the beginning – we take $U(h) = h$. As the solution $U(r) = \alpha r$ solves the homogeneous differential equation, $U''(r) = 0$, we can add this solution to the numerical solution found, with α taken such as to satisfy the end condition $U(r_{\max}) = q_{\max}$, without violating the starting condition $U(0) = 0$.

PROGRAMMING EXERCISE

Add an extra integration to your program which solves Eq. (5.77).

It is useful to check for correctness by using the hydrogen atom as an example. The normalised ground state density (in the sense of (5.76)), found at $E = -0.5$ a.u., is $4\mathrm{e}^{-2r}$ and we must solve

$$U''(r) = -4r\mathrm{e}^{-2r}, \tag{5.80}$$

with the boundary conditions $U(0) = 0$, $U(\infty) = 1$, so

$$U(r) = -(r + 1)\mathrm{e}^{-2r} + 1. \tag{5.81}$$

Check Check whether your program produces these results

The next step is to make the program self-consistent. This is done by adding the Hartree potential to the nuclear potential and solving for the eigenstate again. You repeat this process until the energy does not change appreciably between subsequent steps. The total energy is given by

$$E = 2\varepsilon - \int \mathrm{d}r\, V_{\mathrm{H}}(r)u^2(r). \tag{5.82}$$

The Hartree correction arises because the Hartree energy is quadratic in the density.

Check Try to reproduce the results for the helium Hartree–Fock calculation in Section 4.3.2. In fact, the present method is more accurate as the wave functions are not restricted to linear combinations of four Gaussians. For an integration step $h = 0.01$ (in the Verlet algorithm) you will find for the eigenvalue of the radial Schrödinger equation the value −0.923 a.u. and for the Hartree correction 1.0155 a.u., so that the total energy amounts to $E = -2.861$ a.u., in good agreement with the result obtained in the previous chapter. The experimental value is −2.903 a.u.

5.5.3 The local density exchange potential

The aim of the exercise has not yet been achieved: we must calculate the energy and eigenvalues in the density functional formalism within the local density approximation. Remember that in density functional theory, the density that gives rise to the Hartree potential is the *full* density $n(\mathbf{r})$, i.e. the density of the two electrons, and in the previous section we have subtracted off the self-interaction contribution, leading to a reduction by a factor of 2 of the Hartree potential. Multiplying the Hartree potential by a factor of 2 in the previous program yields very poor results and therefore we hope that the exchange potential will correct for the self-interaction. As we have noted above, a popular form of the local density exchange potential is the one based on a treatment of the exchange hole in a homogeneous electron gas and is given by

$$V_{\mathrm{x}}(\mathbf{r}) = \text{Const.} \times n^{1/3}(\mathbf{r}) \tag{5.83}$$

where the constant is given as

$$\text{Const.} = -\left(\frac{3}{\pi}\right)^{1/3}. \tag{5.84}$$

Here, again, the *full* density is to be taken in the right hand side of (5.83) and this is twice the single electron density arising from the radial Schrödinger equation, since we have two electrons. Therefore, in terms of the radial eigenfunctions u

normalised as in (5.76), our exchange potential reads

$$V_X(\mathbf{r}) = -\left[\frac{3u^2(r)}{2\pi^2 r^2}\right]^{1/3} \tag{5.85}$$

which, for the s-states under consideration, depends only on the radial coordinate r. The total energy is given by

$$E = 2\varepsilon - \int \mathrm{d}r\, V_\mathrm{H}(r)u^2(r) + \frac{1}{2}\int \mathrm{d}r\, u^2(r)V_X(r). \tag{5.86}$$

The extension of your program to a local density version is now straightforward: instead of adding only the Hartree potential to the nuclear attraction, you take twice this potential and add the exchange potential to it. The self-consistency loop remains unaltered.

PROGRAMMING EXERCISE

Extend your Hartree–Fock program to include the exchange potential.

Check If your program is correct, it should give the following values for the energies: $\varepsilon = -0.52$ and $E = -2.72$ a.u.

Obviously the result is inferior to Hartree–Fock as the exchange potential is included only in an approximate way. Improvement is possible by considering an exchange correlation potential based on an interpolation of quantum Monte Carlo results by Ceperley and Alder [45], and it yields a ground state energy of $E = -2.83$ a.u. [9] which is an important improvement with respect to -2.72, although it is still worse than the HF result of -2.86 a.u. Implementation of this is straightforward and will be done in Problem 5.6.

5.6 Applications and results

In numerous calculations for atoms, molecules and solids the DFT–LDA approach has been very successful. In this section we quote some results which have been taken from the review by Jones and Gunnarson [4].

The original applications were to the ground state properties of solids, and some typical results are shown in Table 5.1. Binding energies for atoms and molecules are often better than HF (Table 5.3); total energies are close to but a bit worse than HF (Table 5.2). Interpretation of the Kohn–Sham eigenvalues as excitation energies works surprisingly well in many solids, where the energy bands frequently agree with those measured in photo-emission for example (see Problem 5.4 and

Table 5.1. *Lattice constants and cohesive energies for diamond,* Si *and* Ge. *Atomic units are used.*

	Lattice constant		Cohesion energy	
	DFT	Expt	DFT	Expt
Diamond	6.807	6.740	7.58	7.37
Si	10.30	10.26	4.84	4.64
Ge	10.69	10.68	4.02	3.85

Data taken from Ref. [4].

Table 5.2. *Energies in a.u. for various atoms.*

Atom	HF	DFT	Expt.
Li	−7.433	−7.353	−7.479
C	−37.702	−37.479	−37.858
O	−74.858	−74.532	−75.113

Data taken from Ref. [4]

Table 5.3. *Binding energies in a.u. for diatomic molecules.*

Atom	HF	DFT	Expt.
H_2	3.64	4.91	4.75
C_2	0.79	7.19	6.32
O_2	1.28	7.54	5.22

Data taken from Ref. [4].

Section 5.3). But we should be cautious about interpreting ψ_k and ε_k as anything other than auxiliary quantities for constructing the ground state energy and density as explained extensively in that section. There are several examples where interpretation of ε_k as excitation energies goes drastically wrong: band gaps in semiconductors and insulators are almost invariably too small, and ionisation energies for atoms and molecules are usually way too small. The inclusion of self-interaction corrections, mentioned in the previous subsection, gives better results for these gaps, but remember that these corrections introduce dependence of the Hamiltonian on individual orbitals instead of the density only and are therefore incompatible with DFT. The best approach is to use many-body theories for calculating actual excitation energies.

Exercises

5.1 [C] Instead of the regular grid which was used in the helium program of Section 5.5, it is better to use a grid with a step size which grows from a very small value near the nucleus to larger values in the valence region, because the wave function will oscillate more rapidly near the nucleus as a result of the deep Coulomb potential. Consider a grid with grid points given by the following formula:

$$r_j = r_\text{p}[\exp(j\delta) - 1], \quad j = 0, 1, \ldots, j_\text{max}.$$

The grid point with $j = 0$ coincides with the nucleus and the grid runs up to a radius r_max which fixes the value of the prefactor r_p to

$$r_\text{p} = r_\text{max}/[\exp(j_\text{max}\delta) - 1].$$

The grid is defined by the number of grid points j_max, by the outermost point r_max and by the parameter δ which determines how much the grid constant near the nucleus differs from that near r_max. All these three values must be specified and then the prefactor r_p can be determined.

(a) Show that, in terms of j, the radial Schrödinger equation

$$\frac{\mathrm{d}^2}{\mathrm{d}r^2}u(r) = [V(r) - E]u(r)$$

transforms into

$$\frac{\mathrm{d}^2}{\mathrm{d}j^2}u(j) - \delta\frac{\mathrm{d}}{\mathrm{d}j}u(j) = r_\text{p}^2\delta^2\mathrm{e}^{2j\delta}[V(j) - E]u(j),$$

where $u(j) = u(r_j)$.

(b) Write a general integral $\int_0^{r\text{max}} f(r)\mathrm{d}r$ as an integral over j.

(c) [C] Transform all integrals and differential equation methods in the density functional program to the nonhomogeneous grid defined above. Compare the accuracies of the two versions.

(d) Show that the first derivative occurring in the radial Schrödinger equation in terms of j above can be transformed away by writing $u(j) = v(j)\exp(j\delta/2)$. Show that the resulting equation for v reads

$$\frac{\mathrm{d}^2}{\mathrm{d}j^2}v(j) - \frac{\delta^2}{4}v(j) = r_\text{p}^2\delta^2\mathrm{e}^{2j\delta}[V(j) - E]v(j).$$

(e) [C] Numerov's algorithm (see Appendix A7.1) can be used for solving this differential equation. Try this out for the ground state of the hydrogen atom and show that the numerical error scales as $1/N^4$ as is expected (see Problem A3). Note that when the number of points is doubled, δ should be decreased by a factor of 2.

5.2 [C] The Hartree energy

$$E_{\mathrm{H}} = \frac{1}{2}\int \mathrm{d}^3r\, \mathrm{d}^3r' \,\frac{n(\mathbf{r}')n(\mathbf{r})}{|\mathbf{r}-\mathbf{r}'|}$$

overestimates the classical electrostatic energy of the electrons because it includes interactions of the electrons with themselves – these are the so-called self-interactions. In Hartree–Fock theory, this spurious effect is cancelled by the exchange energy. In density functional theory, the exchange correlation energy does not ensure this cancellation a priori and we can only hope that it cancels the self-interaction as much as possible. To see to what extent the exchange correlation potential succeeds in doing so, we consider the hydrogen atom in DFT (of course, DFT was designed for many-electron systems, but its use is not a priori restricted to systems containing more than one electron). In the hydrogen atom, we find a nonvanishing Hartree and exchange correlation energy, which can easily be evaluated with our DFT program for helium.

Change the nuclear charge back to $Z = 1$ and make sure that the density used in the Hartree and exchange correlation energies is evaluated for the single electron (i.e. not multiplied by 2 as in the helium case). Evaluate both energies for the exact solution of the hydrogen atom.

You should find that the exchange correlation energy compensates about 80% of the self-interaction. For better exchange correlation energies, a value of 96% can be found – see the following problem and Ref. [9]. See also Ref. [46] for more examples.

5.3 [C] The Slater exchange potential

$$V_{\mathrm{X}}(\mathbf{r}) = -\left(\frac{3}{\pi}\right)^{1/3} n^{1/3}(\mathbf{r})$$

is based on the exchange energy in a homogeneous electron gas [47]. It is quite a crude approximation, and a refinement can be made using quantum Monte Carlo results obtained by Ceperley and Alder [9, 11, 45]. This leads to a parametrised correlation energy which should be *added to* the Slater term given above. The parametrisation is given in terms of the parameter r_s which is related to the density n according to

$$n = \frac{3}{4\pi r_s^3}.$$

The parametrisation is split into two parts: $r_s \geq 1$ and $r_s < 1$. We need an expression for the correlation energy parameter ε_{c} defined by

$$E_{\mathrm{c}} = \int \mathrm{d}^3r\; n(\mathbf{r})\varepsilon_{\mathrm{c}}(n)n(\mathbf{r}).$$

(a) Show that from this an expression for the correlation potential V_{c} can be derived according to

$$V_{\mathrm{c}}(r_s) = \left(1 - \frac{r_s}{3}\frac{\mathrm{d}}{\mathrm{d}r_s}\right)\varepsilon_{\mathrm{c}}(r_s).$$

Table 5.4. *Parameters for correlation energy*

	Unpolarised	Polarised
A	0.0311	0.01555
B	−0.048	−0.0269
C	0.0020	0.0014
D	−0.0116	−0.0108
γ	−0.1423	−0.0843
β_1	1.0529	1.3981
β_2	0.3334	0.2611

(b) [C] A parametrised form of ε_c is given by the following expressions. For $r_s \geq 1$ we have

$$\varepsilon_c = \gamma/(1 + \beta_1\sqrt{r_s} + \beta_2 r_s)$$

and for $r_s > 1$

$$\varepsilon_c = A \ln r_s + B + C r_s \ln r_s + D r_s.$$

From this, we obtain the following expressions for the correlation potential:

$$V_c(r_s) = \varepsilon_c \frac{1 + 7/6\beta_1\sqrt{r_s} + \beta_2 r_s}{1 + \beta_1\sqrt{r_s} + \beta_2 r_s}$$

for $r_s \geq 1$ and

$$V_c(r_s) = A \ln r_s + B - A/3 + \frac{2}{3} C r_s \ln r_s + (2D - C) r_s/3.$$

The values of the parameters A, B etc. depend on whether we are dealing with the polarised (all spins same z component) or unpolarised case. For both cases, the values are given in Table 5.4.

Use this parametrisation in your helium density functional theory program (unpolarised). You should find an energy $E = -2.83$ atomic units,to be compared with -2.72 without this correction.

(c) [C] Use the polarised parametrisation for the hydrogen program of the previous problem. You should find an energy $E = -0.478$ a.u.

(d) [C] It is also possible to combine the self-energy correction with the correlation energy. You should consult the paper by Perdew and Zunger [9], if you intend to do this. This results in an energy $E = -2.918$ a.u., which is only 0.015 a.u. off the experimental value.

5.4 In this problem, we consider a generalisation of Koopman's theorem (see Section 4.5.3) to the density functional formalism. To this end, we consider the spectrum $\{\varepsilon_i\}$ and the corresponding eigenstates of the Kohn–Sham Hamiltonian. We consider the chemical potential, which is found by removing a small amount of charge from the system. In practice this means that the highest level (which is level N) is not fully occupied. We usually calculate the density according to

$$n(\mathbf{r}) = \sum_{i=1}^{N} f_i |\psi_N(\mathbf{r})|^2.$$

The change in the density is realised by reducing the value f_N slightly:

$$f_N \to f_N - \delta f_N.$$

This induces a change in the density

$$\delta n(\mathbf{r}) = \delta f_N |\psi_N(\mathbf{r})|^2.$$

The total energy is calculated according to:

$$E(N) = \sum_{i=1}^{N} f_i \varepsilon_i - \int \frac{1}{2} \mathrm{d}^3 r \, \mathrm{d}^3 r' \, \frac{n(\mathbf{r}) n(\mathbf{r}')}{|\mathbf{r} - \mathbf{r}'|} - \int \mathrm{d}^3 r \, n(\mathbf{r}) V_{\mathrm{xc}}(\mathbf{r}) + E_{\mathrm{xc}}[n].$$

The levels ε_i arise from taking the matrix elements $\langle \psi_i | H(N) | \psi_i \rangle$. As a result of the change in density, both the Hamiltonian occurring in these matrix elements and the remaining terms in the energy expression change.

We therefore have three contributions to the change in the total energy. First, the factor f_N in the sum over the energy levels changes; second, the potential for which the levels are calculated changes slightly; and third, the correction terms in the expression for the energy change.

Show that, to linear order in $\delta n(\mathbf{r})$, the combined effect of the change in the Hamiltonian matrix elements is precisely compensated by the change in the remaining terms in the energy expression so that we obtain

$$E(N) - E(N - \delta f_N) = \varepsilon_N \delta f_N$$

Hint: the change in the exchange correlation energy $E_{\mathrm{xc}}[n]$ is given by the expression

$$\delta E_{\mathrm{xc}}[n] = \int \mathrm{d}^3 r \, \delta n(\mathbf{r}) \frac{\delta E_{\mathrm{xc}}[n]}{\delta n(\mathbf{r})} = \int \mathrm{d}^3 r \, V_{\mathrm{xc}}[n](\mathbf{r}) \delta n(\mathbf{r}).$$

This proves Janak's theorem [25].

References

[1] C. Pisany, R. Dovea, and C. Roetti, *Hartree–Fock Ab-initio Treatment of Crystalline Systems*. Berlin, Springer, 1988.

[2] P. Hohenberg and W. Kohn, 'Inhomogeneous electron gas,' *Phys. Rev.*, **136** (1964), B864–71.

[3] W. Kohn and L. J. Sham, 'Self-consistent equations including exchange and correlation effects,' *Phys. Rev.*, **140** (1965), A1133.

[4] R. O. Jones and O. Gunnarsson, 'The density functional formalism, its applications and prospects,' *Rev. Mod. Phys.*, **61** (1989), 689–746.

[5] S. Lundqvist and N. March, *Theory of the Inhomogeneous Electron Gas*. New York, Plenum, 1983.

[6] P. Phariseau and W. M. Temmerman, *The Electronic Structure of Complex Systems*, NATO ASI series B. New York, Plenum, 1984.
[7] R. M. Martin, *Electronic Structure*. Cambridge, Cambridge University Press, 2004.
[8] D. C. Langreth and M. J. Mehl, 'Easily implementable nonlocal exchange-correlation energy-functional,' *Phys. Rev. Lett.* **47** (1981), 446–50.
[9] J. P. Perdew and A. Zunger, 'Self-interaction correction to density-functional approximations for many-electron systems,' *Phys. Rev. B*, **23** (1981), 5048–79.
[10] J. C. Slater, *Quantum Theory of Molecules and Solids*, vol. IV. New York, McGraw-Hill, 1982.
[11] D. M. Ceperley, 'Ground state of the fermion one-component plasma – a Monte Carlo study in two and three dimensions,' *Phys. Rev. B*, **18** (1978), 3126–38.
[12] U. von Barth and L. Hedin, 'A local exchange-correlation potential for the spin-polarized case: I,' *J. Phys. C*, **5** (1972), 1629–42.
[13] O. Gunnarson and B. I. Lundqvist, 'Exchange and correlation in atoms, molecules and solids by the spin-density-functional formalism,' *Phys. Rev. B*, **13** (1976), 4274–98.
[14] J. P. Perdew and Y. Wang, 'Accurate and simple density functional for the electronic exchange energy: Generalized gradient approximation,' *Phys. Rev. B*, **33** (1986), 8800–2.
[15] J. P. Perdew, 'Density-functional approximation for the correlation energy of the inhomogeneous electron gas,' *Phys. Rev. B*, **33** (1986), 8822–4.
[16] Y. Wang and J. P. Perdew, 'Correlation hole of the spin-polarized electron gas, with exact small-wave-vector and high-density scaling,' *Phys. Rev. B*, **44** (1991), 13298–307.
[17] A. D. Becke, 'Density functional exchange energy approximation with correct asymptotic behaviour,' *Phys. Rev. A*, **38** (1988), 3098–100.
[18] C. Lee, W. Yang, and R. G. Parr, 'Development of the Colle–Salvetti correlation-energy formula into a functional of the electron density,' *Phys. Rev. B*, **37** (1988), 785–9.
[19] J. P. Perdew, K. Burke, and M. Enzerhof, 'Generalized gradient approximation made simple,' *Phys. Rev. Lett.*, **77** (1996), 3865–86.
[20] J. P. Perdew, K. Burke, and M. Enzerhof, 'Generalized gradient approximation made simple (Erratum),' *Phys. Rev. Lett.*, **78** (1997), 1396.
[21] R. T. Sharp and G. K. Horton, 'A variational approach to the unipotential many-electron problem,' *Phys. Rev.*, **90** (1953), 317.
[22] J. D. Talman and W. F. Shadwick, 'Optimized effective atomic central potential,' *Phys. Rev. A*, **14** (1976), 36–40.
[23] J. D. Talman and W. F. Shadwick, 'Asymptotic behavior of atomic and molecular wave functions,' *Proc. Natl. Acad. Sci*, **77** (1980), 4403–6.
[24] M. Levy, J. P. Perdew, and V. Shani, 'Exact differential equation for the density and ionization energy of a many-particle system,' *Phys. Rev. A*, **30** (1984), 2745–8.
[25] J. F. Janak, 'Proof that $\partial E/\partial n_i = \varepsilon$ in density-functional theory,' *Phys. Rev. B*, **18** (1978), 7165–8.
[26] J. P. Perdew, R. G. Par, M. Levy, and J. L. Balduz, 'Density-functional theory for fractional particle number: derivative discontinuities of the energy,' *Phys. Rev. Lett.*, **49** (1982), 1691–4.
[27] J. F. Janak, 'Significance of the highest occupied Kohn–Sham eigenvalue,' *Phys. Rev. B*, **56** (1997), 12042–5.
[28] L. Hedin, 'New method for calculating the one-particle Green's function with application to the electron-gas problem,' *Phys. Rev.*, **139** (1965), A796–A823.
[29] F. Aryasetiawan and O. Gunnarsson, 'The GW method,' *Rep. Prog. Phys.*, **61** (1998), 237–312.
[30] W. G. Aulbur, L. Jönsson, and J. W. Wilkins, 'Quasiparticle calculations in solids,' in *Solid State Physics*, vol. 54 (H. Ehrenreich and F. Spaepen, eds.). San Diego, Academic Press, 2000, pp. 1–218.

[31] C. Cohen-Tannoudji, B. Diu, and F. Laloë, *Quantum Mechanics*, vols. 1 and 2. New York/Paris, John Wiley/Hermann, 1977.

[32] D. Pines, *Elementary Excitations in Solids*. New York, Wiley, 1964.

[33] M. S. Hybertsen and S. G. Louie, 'First-principles theory of quasiparticles: calculation of band gaps in semiconductors and insulators,' *Phys. Rev. Lett.*, **55** (1985), 1418–21.

[34] M. S. Hybertsen and S. G. Louie, 'Electron correlation in semiconductors and insulators: Band gaps and quasiparticle energies,' *Phys. Rev. B*, **34** (1986), 5390.

[35] L. D. Landau, 'Theory of the Fermi liquid,' *Soviet Physics JETP*, **3** (1957), 920–5.

[36] G. D. Mahan, *Many-Particle Physics*, 3rd edn. New York, Kluwer Academic/Plenum Press, 2000.

[37] G. Onida, L. Reining, and A. Rubio, 'Electronic excitations: density functional versus many-body Green's function approaches,' *Rev. Mod. Phys.*, **74** (2002), 601–59.

[38] E. Runge and E. K. U. Gross, 'Density-functional theory for time-dependent systems,' *Phys. Rev. Lett.*, **52** (1984), 997–1000.

[39] E. K. U. Gross, J. F. Dobson, and M. Petersilka, 'Density functional theory of time-independent phenomena,' in *Topics in Current Chemistry: Density Functional Theory* (R. F. Nalewajski, ed.), Heidelberg, Springer, 1996, pp. 81–172.

[40] R. van Leeuwen, 'Key concepts in time-dependent density-functional theory,' *Int. J. Mod. Phys. B*, **15** (2001), 1969–2023.

[41] H. Appel and E. K. U. Gross, 'Static and time-dependent many-body effects via density-functional theory,' in *Quantum Simulations of Complex Many-Body Systems: From Theory to Algorithms; Kerkrade, The Netherlands* (J. Grotendorst, D. Marx, and A. Muramatsu, eds.), Jülich, John von Neumann Institute for Computing, 2002, pp. 255–68.

[42] J. F. Ward and G. H. C. New, 'Optical third-harmonic generation in gases by a focused laser beam,' *Phys. Rev.*, **185** (1969), 57–72.

[43] S. Erhard and E. K. U. Gross, 'High harmonic generation in hydrogen and helium atoms subject to one- and twocolor laser pulses,' in *Multiphoton Processes 1996* (P. Lambropoulus and H. Walther, eds.). Bristol, Institute of Physics, 1997, pp. 37–46.

[44] S. J. A. van Gisbergen, J. M. Pacheco, and E. J. Baerends, 'Influence of the exchange-correlation potential in density-functional calculations on polarizabilities and absorption spectra of alkali-metal clusters,' *Phys. Rev. A*, **63** (2001), 063201.

[45] D. M. Ceperley and B. J. Alder, 'Ground state of the electron gas by a stochastic method,' *Phys. Rev. Lett.*, **45** (1980), 566–9.

[46] C.-O. Almbladh and A. C. Pedroza, 'Density-functional exchange-correlation potentials and orbital eigenvalues for light atoms,' *Phys. Rev. A*, **29** (1984), 2322–30.

[47] N. W. Ashcroft and N. D. Mermin, *Solid State Physics*. New York, Holt, Reinhart and Winston, 1976.

6

Solving the Schrödinger equation in periodic solids

In the previous chapter we encountered density functional theory (DFT) which is extensively used for calculating the electronic structure of periodic solids. Aside from DFT, carefully designed potentials often allow accurate electronic structures to be obtained by simply solving the Schrödinger equation without going through the self-consistency machinery of DFT. In both approaches it is necessary to solve the Schrödinger equation and the present chapter focuses on this problem, although some comments on implementing a DFT self-consistency loop will be made.

The large number of electrons contained in a macroscopic crystal prohibits a direct solution of the Schrödinger equation for such a system. Fortunately, the solid has periodic symmetry in the bulk, and this can be exploited to reduce the size of the problem significantly, using *Bloch's theorem*, which enables us to replace the problem of solving the Schrödinger equation for an infinite periodic solid by that of solving the Schrödinger equation in a unit cell with a series of different boundary conditions – the so-called *Bloch boundary conditions*. Having done this, there remains the problem that close to the nuclei the potential diverges, whereas it is weak when we are not too close to any of the nuclei (interstitial region). We can take advantage of the fact that the potential is approximately spherically symmetric close to the nuclei, but further away the periodicity of the crystal becomes noticeable. These two different symmetries render the solution of the Schrödinger equation in periodic solids difficult. In this chapter we consider an example of an electronic structure method, the augmented plane wave (APW) method, which uses a spatial decomposition of the wave functions: close to the nuclei they are solutions to a spherical potential, and in the interstitial region they are plane waves satisfying the appropriate Bloch boundary conditions.

It is possible to avoid the problem of the deep potential altogether by replacing it by a weaker one, which leaves the interesting physical properties unchanged. This is done in the pseudopotential method which we shall also discuss in this chapter.

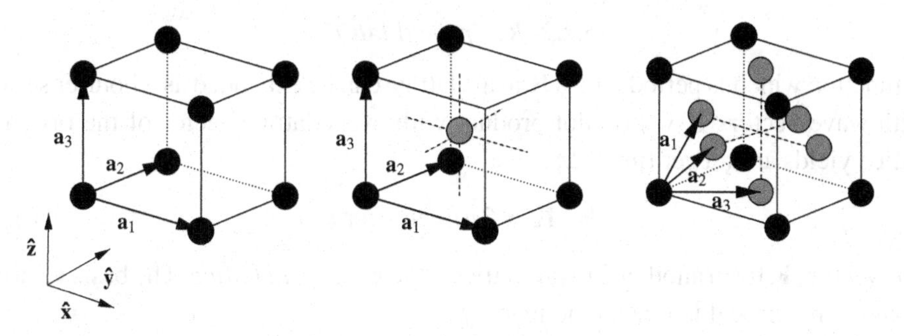

Figure 6.1. Lattice structure of the simple cubic (left), body-centred cubic (middle) and face-centred cubic (left) lattices with basis vectors.

Before going into these methods, we start with a brief review of the theory of electronic structure of solids. For further reading concerning the material in the first three sections of this chapter, we refer to general books on solid state physics [1, 2].

An excellent reference for computational band structures is the book by R. M. Martin [3].

6.1 Introduction: definitions

6.1.1 Crystal lattices

We consider crystals in which the atomic nuclei are perfectly ordered in a periodic lattice. Such a lattice, a so-called *Bravais* lattice, is defined by three basis vectors. The lattice sites $\mathbf{R}$ are given by the integer linear combinations of the basis vectors:

$$\mathbf{R} = \sum_{i=1}^{3} n_i \mathbf{a}_i, \quad n_i \text{ integer}. \tag{6.1}$$

Each cell may contain one or more nuclei – in the latter case we speak of a lattice with a basis. The periodicity implies that the arrangement of these nuclei must be the same within each cell of the lattice. Of course, in reality solids will only be approximately periodic: thermal vibrations and imperfections will destroy perfect periodicity and moreover, periodicity is destroyed at the crystal surface. Nevertheless, the infinite, perfectly periodic lattice is usually used for calculating electronic structure because periodicity facilitates calculations and a crystal usually contains large regions in which the structure is periodic to an excellent approximation.

Three common crystal structures, the simple cubic (sc), body-centred cubic (bcc) and face-centred cubic (fcc) structures, are shown in Figure 6.1.

6.1.2 Reciprocal lattice

A function which is periodic on a Bravais lattice can be expanded as a Fourier series with wave vectors $\mathbf{K}$ whose dot product with every lattice vector of the original lattice yields an integer times 2π:

$$\mathbf{R} \cdot \mathbf{K} = 2\pi n, \text{ integer } n. \tag{6.2}$$

The vectors $\mathbf{K}$ form another Bravais lattice – the *reciprocal lattice*. The basis vectors $\mathbf{b}_j$ of the reciprocal lattice are defined by

$$\mathbf{a}_i \cdot \mathbf{b}_j = 2\pi \delta_{ij} \tag{6.3}$$

and an explicit expression for the $\mathbf{b}_j$ is

$$\mathbf{b}_j = 2\pi \, \varepsilon_{jkl} \frac{\mathbf{a}_k \times \mathbf{a}_l}{\mathbf{a}_1 \cdot (\mathbf{a}_2 \times \mathbf{a}_3)}. \tag{6.4}$$

ε_{jkl} is the Lévi–Civita tensor, which is $+1$ for jkl an even permutation of $(1, 2, 3)$, and -1 for odd permutations.

In the reciprocal lattice, the first Brillouin zone is defined as the volume in reciprocal space consisting of the points that are closer to the origin than to any other reciprocal lattice point. A general wave vector $\mathbf{q}$ is usually decomposed into a vector $\mathbf{k}$ of the first Brillouin zone and a vector $\mathbf{K}$ of the reciprocal lattice:

$$\mathbf{q} = \mathbf{k} + \mathbf{K}. \tag{6.5}$$

For a finite *rectangular* lattice of size $L_x \times L_y \times L_z$, the allowed wave vectors $\mathbf{q}$ to be used for expanding functions defined on the lattice are restricted by the boundary conditions. A convenient choice is periodic boundary conditions in which functions are taken periodic within the volume of $L_x \times L_y \times L_z$. In that case, vectors in reciprocal space run over the following values:

$$\mathbf{q} = 2\pi \left(\frac{n_x}{L_x}, \frac{n_y}{L_y}, \frac{n_z}{L_z} \right) \tag{6.6}$$

with integer n_x, n_y and n_z.

6.2 Band structures and Bloch's theorem

We know that the energy spectra of electrons in atoms are discrete. If we place two identical atoms at a very large distance from each other, their atomic energy levels will remain unchanged. Electrons can occupy the atomic levels on either of both atoms and this results in a double degeneracy. On moving the atoms closer together, this degeneracy will be lifted and each level splits into two; the closer we move the atoms together, the stronger this splitting. Suppose we play the same game with three instead of two atoms: then the atomic levels split into three different ones, and so on. A solid consists of an infinite number of atoms moved close together and therefore each atomic level splits into an infinite number, forming a *band*. It is our aim to calculate these bands.

We shall now prove the famous Bloch theorem which says that the eigenstates of the Hamiltonian with a periodic potential are the same in the lattice cells located at $\mathbf{R}_i$ and $\mathbf{R}_j$ up to a phase factor $\exp[i\mathbf{q}\cdot(\mathbf{R}_i - \mathbf{R}_j)]$ for some reciprocal vector $\mathbf{q}$. We shall see that a consequence of this theorem is that the energy spectra are indeed composed of bands.

We write the Schrödinger equation in reciprocal space. The potential V is periodic and it can therefore be expanded as a Fourier sum over reciprocal lattice vectors $\mathbf{K}$:

$$V(\mathbf{r}) = \sum_{\mathbf{K}} e^{i\mathbf{K}\cdot\mathbf{r}} V_{\mathbf{K}}. \tag{6.7}$$

An arbitrary wave function ψ can expanded as a Fourier series with wave vectors $\mathbf{q}$ allowed by the periodic boundary conditions (6.6):[1]

$$\psi(\mathbf{r}) = \sum_{\mathbf{q}} e^{i\mathbf{q}\cdot\mathbf{r}} C_{\mathbf{q}}. \tag{6.8}$$

Writing $\mathbf{q} = \mathbf{k} + \mathbf{K}$, the Schrödinger equation reads (in atomic units)

$$\left[\frac{1}{2}(\mathbf{k}+\mathbf{K})^2 - \varepsilon\right] C_{\mathbf{k}+\mathbf{K}} + \sum_{\mathbf{K}'} V_{\mathbf{K}-\mathbf{K}'} C_{\mathbf{k}+\mathbf{K}'} = 0. \tag{6.9}$$

This equation holds for each vector $\mathbf{k}$ in the first Brillouin zone: in the equation, wave vectors $\mathbf{k}+\mathbf{K}$ and $\mathbf{k}+\mathbf{K}'$ are coupled by the term with the sum over $\mathbf{K}'$, but no coupling occurs between $\mathbf{k}+\mathbf{K}$ and $\mathbf{k}'+\mathbf{K}'$ for different $\mathbf{k}$ and $\mathbf{k}'$. Therefore, for each $\mathbf{k}$ we can, in principle, solve the eigenvalue equation (6.9) and obtain the energy eigenvalues ε and eigenvectors $\mathbf{C}_{\mathbf{k}}$ with components $C_{\mathbf{k}+\mathbf{K}}$, leading to wave functions of the form (see (6.8))

$$\psi_{\mathbf{k}}(\mathbf{r}) = e^{i\mathbf{k}\cdot\mathbf{r}} \left(\sum_{K} C_{\mathbf{k}+\mathbf{K}} e^{i\mathbf{K}\cdot\mathbf{r}} \right). \tag{6.10}$$

The eigenvalues form a discrete spectrum for each $\mathbf{k}$. The levels vary with $\mathbf{k}$ and therefore give rise to energy bands. Equation (6.9) yields an infinite spectrum for each $\mathbf{k}$. We might attach a label n, running over the spectral levels, alongside the label $\mathbf{k}$, to the energy level ε: $\varepsilon = \varepsilon_{n\mathbf{k}}$.

We can rewrite (6.10) in a more transparent form. To this end we note that the expression in brackets in this equation is a periodic function in $\mathbf{r}$. Denoting this periodic function by $u_{\mathbf{k}}(\mathbf{r})$, we obtain

$$\psi_{\mathbf{k}}(\mathbf{r}) = e^{i\mathbf{k}\cdot\mathbf{r}} u_{\mathbf{k}}(\mathbf{r}). \tag{6.11}$$

[1] For an infinite solid, the sum over $\mathbf{q}$ becomes an integral.

The eigenstates of the Hamiltonian can thus be written in the form of a plane wave times a periodic function. Equivalently, evaluating such a wave function at two positions, separated by a lattice vector $\mathbf{R}$, yields a difference of a phase factor $e^{i\mathbf{k}\cdot\mathbf{R}}$, according to our previous formulation of Bloch's theorem.

Electronic structure methods for periodic crystals are usually formulated in reciprocal space, by solving an equation like (6.9) in which the basis functions are plane waves labelled by reciprocal space vectors. In fact, Bloch's theorem allows us to solve for the full electronic structure in real space by considering only one cell of the lattice for each $\mathbf{k}$ and applying boundary conditions to the cell as dictated by the Bloch condition (6.11). In particular, each facet of the unit cell boundary has a 'partner' facet which is found by translating the facet over a lattice vector $\mathbf{R}$. The solutions to the Schrödinger equation should on both facets be equal up to factor $\exp(i\mathbf{k}\cdot\mathbf{R})$. These boundary conditions determine the solutions inside the cell completely. We see that we can try to solve the Schrödinger equation either in reciprocal space or in real space. For nonperiodic systems, real-space methods enjoy an increasing popularity [4–7].

6.3 Approximations

The Schrödinger equation for an electron in a crystal can be solved in two limiting cases: the nearly free electron approximation, in which the potential is considered to be weak everywhere, and the tight-binding approximation in which it is assumed that the states are tightly bound to the nuclei. Both methods aim to reduce the difficulty of the band structure problem and to increase the understanding of band structures by relating them to those of two different systems which we can easily describe and understand: free electrons and electrons in single-atom orbitals. The tight-binding method has led to many computational applications. We shall apply it to graphene and carbon nanotubes.

6.3.1 The nearly free electron approximation

It is possible to solve Eq. (6.9) if the potential is small, by using perturbative methods. This is called the nearly free electron (NFE) approximation. You might consider this to be inappropriate as the Coulomb potential is certainly not small near the nuclei. Surprisingly, the NFE bands closely resemble those of aluminium, for example! We shall see later on that the pseudopotential formalism provides an explanation for this.

The main results of the NFE are that the bands are perturbed by an amount which is quadratic in the size of the weak potential V except close to *Bragg planes*

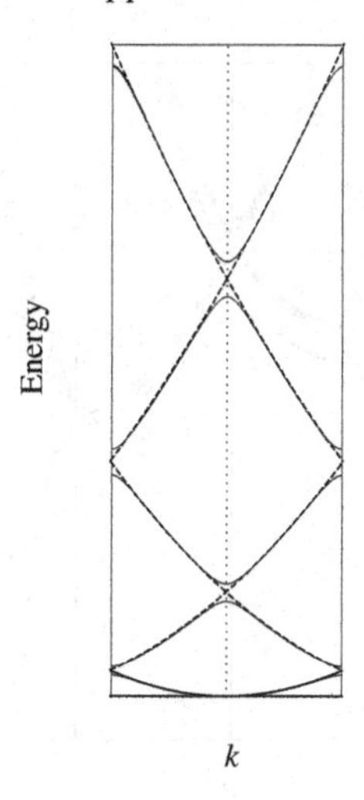

Figure 6.2. Nearly free electron spectrum for a periodic potential in one dimension.

consisting of reciprocal points $\mathbf{q}$ which satisfy

$$|\mathbf{q}| = |\mathbf{K} - \mathbf{q}|, \tag{6.12}$$

where $\mathbf{K}$ is a reciprocal lattice vector. At a Bragg plane, a band gap of size $2|V_{\mathbf{q}}|$ opens up. Figure (6.2) gives the resulting bands for the one-dimensional case. Figure (6.3) shows how well the bands in aluminium resemble the free electron bands.

6.3.2 The tight-binding approximation

The tight-binding (TB) approximation will be discussed in more detail as it is an important way for performing electronic structure calculations with many atoms in the unit cell. It naturally comes about when considering states which are tightly bound to the nuclei. The method is essentially a linear combination of atomic orbitals (LCAO) type of approach, in which the atomic states are used as basis orbitals. Let us denote these states, which are assumed to be available from some atomic electronic structure calculation, by $u_p(\mathbf{r}-\mathbf{R})$, in which the index p labels the levels of an atom located at $\mathbf{R}$. From these states, we build Bloch basis functions as

$$\phi_{p,\mathbf{k}}(\mathbf{r}) = \frac{1}{\sqrt{N}} \sum_{\mathbf{R}} \mathrm{e}^{\mathrm{i}\mathbf{k}\cdot\mathbf{R}} u_p(\mathbf{r} - \mathbf{R}), \tag{6.13}$$

and a general Bloch state is a linear combination of these:

$$\phi_{\mathbf{k}}(\mathbf{r}) = \sum_p C_p(\mathbf{k}) \phi_{p,\mathbf{k}}(\mathbf{r}). \tag{6.14}$$

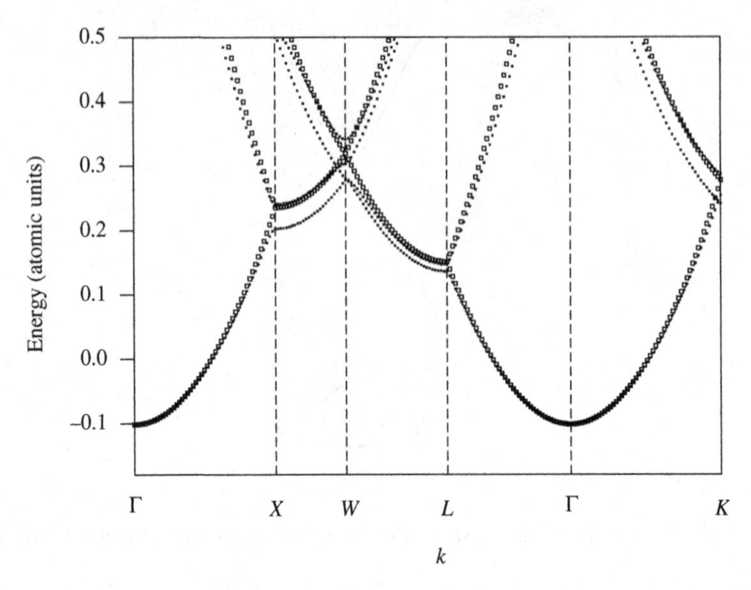

Figure 6.3. Band structure of aluminium. Also shown (open squares) is the free electron result. X, W etc. are special points in the Brillouin zone (see Section 6.5).

The coefficients $C_p(\mathbf{k})$ can be found from the variational principle. Applying the techniques of Chapter 3, we see that we must solve the generalised eigenvalue problem

$$\mathbf{HC}(\mathbf{k}) = E\mathbf{SC}(\mathbf{k}) \tag{6.15}$$

for the vector $\mathbf{C}(\mathbf{k})$ with components $C_p(\mathbf{k})$. The matrix elements of the Hamiltonian $\mathbf{H}$ and of the overlap matrix $\mathbf{S}$ (which depend on $\mathbf{k}$) are given by

$$H_{pq} = \langle \phi_{p,\mathbf{k}} | H | \phi_{q,\mathbf{k}} \rangle \text{ and} \tag{6.16a}$$

$$S_{pq} = \langle \phi_{p,\mathbf{k}} | \phi_{q,\mathbf{k}} \rangle. \tag{6.16b}$$

Writing out the matrix elements using (6.13) and the lattice periodicity, we obtain the following expressions:

$$H_{pq} = \sum_{\mathbf{R}} \mathrm{e}^{\mathrm{i}\mathbf{k}\cdot\mathbf{R}} \int \mathrm{d}^3 r \, u_p(\mathbf{r} - \mathbf{R}) H u_q(\mathbf{r}) \tag{6.17a}$$

$$S_{pq} = \sum_{\mathbf{R}} \mathrm{e}^{\mathrm{i}\mathbf{k}\cdot\mathbf{R}} \int \mathrm{d}^3 r \, u_p(\mathbf{r} - \mathbf{R}) S u_q(\mathbf{r}). \tag{6.17b}$$

As the states $u_p(\mathbf{r})$ are rather strongly localised near the nuclei, they will have virtually no overlap when centred on atoms lying far apart. This restricts the sums in (6.17) to only the first few shells of neighbouring atoms, sometimes only nearest neighbours. The numerical solution of the generalised eigenvalue problem $\mathbf{HC} = E\mathbf{SC}$ is treated in Chapter 3.

Of course, we may relax the condition that we consider atomic orbitals as basis functions, and extend the method to allow for arbitrary, but still localised, basis functions. This even works for the valence orbitals in metals, although in that case relatively many neighbours have to be coupled, so that the approach pays off only for large unit cells. We may use the tight-binding approach for fixed interatomic distances – for a tight-binding method in which the atoms are allowed to move, thereby requiring varying distances, see Chapter 9.

The tight-binding method comes in two flavours. The first is the *semi-empirical* TB method, in which only a few valence orbital basis functions are used. Their couplings are restricted to nearest neighbour atoms and the value of the couplings are fitted to either experimental data such as band gaps and band widths, or to similar data obtained using more sophisticated band structure calculations. Once satisfactory values have been obtained for the TB couplings, more complicated structures may be considered which are beyond reach of self-consistent DFT calculations.

We may also be more ambitious and use more TB parameters which are fitted to DFT Hamiltonians. This is particulary useful when the TB Hamiltonians were obtained using localised basis functions, such as Gaussian or Slater orbitals (see Chapter 4 for a dicussion of these basis sets). For DFT calculations, Slater type orbitals are becoming increasingly popular, as the reason for choosing Gaussians in Hartree–Fock calculations, i.e. the fact that integrals can be evaluated analytically, ceases to be relevant in DFT with its highly nonlinear exchange correlation potential. Therefore, the Hamiltonian naturally has a tight-binding form, and this means that it is *sparse*, that is, a small minority of the elements of the Hamiltonian are nonzero. Such a Hamiltonian allows for iterative methods to be used.

In the next subsection we consider an appealing application of the tight-binding method: graphene and carbon nanotubes.

6.3.3 Tight-binding calculation for graphene and carbon nanotubes

In this subsection, we calculate the band structure of a carbon nanotube within the tight-binding approximation. It is a very instructive exercise which is strongly recommended as an introduction to band structure calculations. Here we follow the discussion in Refs. [3, 8].

We assume that only elements of the Hamiltonian and overlap matrix coupling two atomic orbitals are relevant, and that three-point terms do not occur. This is the so-called *Slater–Koster* approximation [9]. We shall first apply this to graphene, which is a sheet consisting of carbon atoms ordered within a hexagonal lattice. This is not a Bravais lattice, but it can be described as a triangular Bravais lattice with a

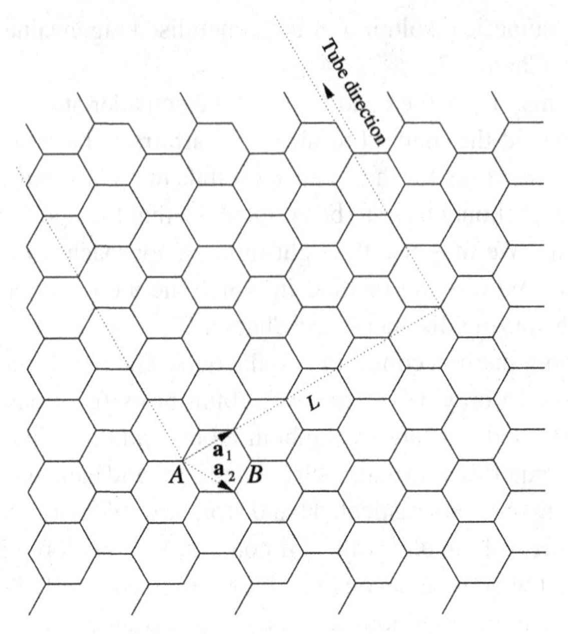

Figure 6.4. Hexagonal lattice of the graphene sheet with basis vectors $\mathbf{a}_1$ and $\mathbf{a}_2$ indicated. The zigzag nanotube is also indicated.

basis consisting of two atoms, A and B (see Figure 6.4). The two basis vectors are

$$\mathbf{a}_1 = a\left(\frac{1}{2}\sqrt{3}, \frac{1}{2}\right); \quad \mathbf{a}_2 = a\left(\frac{1}{2}\sqrt{3}, -\frac{1}{2}\right). \tag{6.18}$$

where the lattice constant $a = 2.461$ Å. We only include nearest neighbour interactions.

The most relevant part of the band structure is the valence band – it turns out that this is formed by the π-orbitals which are built from the p_z-atomic orbitals. This leads us to taking only a single orbital per atom into account. Furthermore we keep only nearest neighbour matrix elements in the tight-binding matrices. We see from Figure 6.4 that an A-atom has nearest neighbours of the type B only.

The essential idea of using Bloch's theorem in calculating the band structure is to reduce the entire problem to that of the unit cell, which contains only two orbitals: the p_z orbitals of A and B. We must therefore calculate $H_{AA}(\mathbf{k}) = H_{BB}(\mathbf{k})$, $S_{AA}(\mathbf{k}) = S_{BB}(\mathbf{k})$, $H_{AB}(\mathbf{k}) = H^*_{BA}(\mathbf{k})$ and $S_{AB}(\mathbf{k}) = S^*_{BA}(\mathbf{k})$ *for each Bloch vector* $\mathbf{k}$. The spectrum is then given by the equation:

$$\begin{pmatrix} H_{AA}(\mathbf{k}) & H_{AB}(\mathbf{k}) \\ H_{BA}(\mathbf{k}) & H_{BB}(\mathbf{k}) \end{pmatrix} \begin{pmatrix} \psi_A(\mathbf{k}) \\ \psi_B(\mathbf{k}) \end{pmatrix} = E(\mathbf{k}) \begin{pmatrix} S_{AA}(\mathbf{k}) & S_{AB}(\mathbf{k}) \\ S_{BA}(\mathbf{k}) & S_{BB}(\mathbf{k}) \end{pmatrix} \begin{pmatrix} \psi_A(\mathbf{k}) \\ \psi_B(\mathbf{k}) \end{pmatrix}. \tag{6.19}$$

It follows that the energies $E(\mathbf{k})$ are given as

$$E_{\pm}(\mathbf{k}) = \frac{-(-2E_0 + E_1) \pm \sqrt{(-2E_0 + E_1)^2 - 4E_2E_3}}{2E_3}, \tag{6.20}$$

where

$$E_0 = H_{AA}S_{AA}; \qquad E_1 = S_{AB}H^*_{AB} + H_{AB}S^*_{AB};$$
$$E_2 = H^2_{AA} - H_{AB}H^*_{AB}; \qquad E_3 = S^2_{AA} - S_{AB}S^*_{AB}.$$

The matrix element $H_{AA}(\mathbf{k})$ must be a real constant which does not depend on $\mathbf{k}$ and the same holds for $S_{AA}(\mathbf{k})$. We take the first to be 0 (with a suitable shift of the energy scale) and the second is 1 because of the normalisation of the orbitals.

You may verify that the off-diagonal elements have the form

$$H_{AB} = \gamma_0[\exp(\mathrm{i}\mathbf{k} \cdot \mathbf{R}_1) + \exp(\mathrm{i}\mathbf{k} \cdot \mathbf{R}_2) + \exp(\mathrm{i}\mathbf{k} \cdot \mathbf{R}_3)] \tag{6.21}$$

with γ_0 a real constant independent of $\mathbf{k}$. The vectors $\mathbf{R}_\mathrm{i}$ connect A to its three neighbours. For S_{AB} we find the same form but we call the constant s_0.

After some calculation, the energies are found as

$$E_{\pm}(\mathbf{k}) = \frac{\varepsilon_{2p} \mp \gamma_0\sqrt{f(\mathbf{k})}}{1 \mp s_0\sqrt{f(\mathbf{k})}}, \tag{6.22}$$

where

$$f(\mathbf{k}) = 3 + 2\cos(k_1) + 2\cos(k_2) + 2\cos(k_1 - k_2); \tag{6.23}$$

$$k_1 = \mathbf{k} \cdot \mathbf{a}_1, \quad k_2 = \mathbf{k} \cdot \mathbf{a}_2 \text{ etc.} \tag{6.24}$$

In the Brillouin zone, the point Γ is identified with $\mathbf{k} = (0, 0)$, and the point K is a vector of length $4\pi/3a$ in the direction of $\mathbf{a}_2$; M is the point $2\pi/\sqrt{3}(1, 0)$. We can now plot the bands (i.e. the values $E_{\pm}(\mathbf{k})$ between M, Γ and K. The result is shown in Figure 6.5. Note that we find two energy values per atom. As we know that there should only be a single electron per p_z orbital, the Fermi energy must be the highest negative energy. We see that graphene is a metal: the bands touch each other precisely at the Fermi energy, so infinitesimal excitations are possible which yield nonzero momentum.

Now we turn to carbon nanotubes. These are graphene sheets rolled into a cylindrical form. There are many ways in which the two long ends can be glued together. These ways correspond to different strips which can be drawn on this sheet such that the two sides of the strip can be connected together smoothly. This is possible when the vector running perpendicularly across the sheet is an integer linear combination of the basis vectors $\mathbf{a}_1$ and $\mathbf{a}_2$. This is also indicated in Figure 6.4.

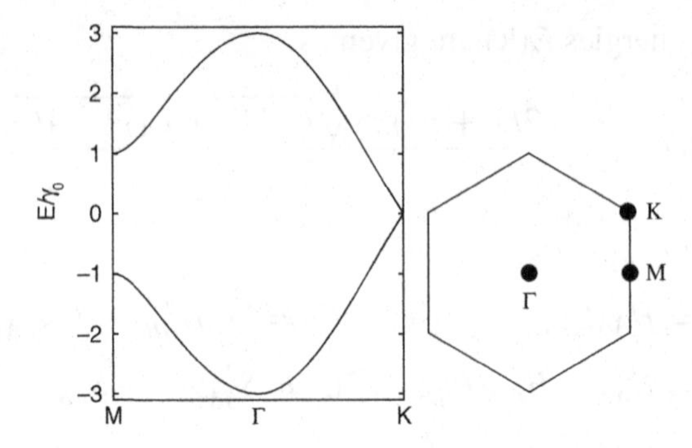

Figure 6.5. Tight-binding band structure of graphene. The points of the Brillouin zone are shown on the right hand side.

We speak now of an (n, m) nanotube where n and m are the integer coefficients. We restrict ourselves here to the case where $m = 0$. Such tubes are called 'zigzag' tubes because a circle running around the tube consists of a zigzag structure of nearest neighbour bonds.

A carbon nanotube is a one-dimensional object – therefore the states are labelled by a Bloch vector *along* the tube. We neglect effects due to the curvature of the sheet, which alter the interactions. For large tubes (i.e. tubes with a large diameter) this is a good approximation. Across the tube, the wave function *must be periodic*. The difference from a periodic cell in a periodic crystal must be emphasised here. In a crystal, the potential and the density are periodic, but the wave function (in general) is not. In the present case the wave function must match onto itself across the tube – hence it is really periodic. This implies that the transverse component of the wave vector must be $2\pi j/L$ for a tube circumference L and integer j. For each n we find an energy value, that is, for each fixed longitudinal k-vector we find a discrete energy spectrum.

For this case, the period along the tube is $a\sqrt{3}$. This means that the longitudinal Brillouin zone runs up to $k = \pi/(a\sqrt{3})$. This point is denoted as X. The transverse period is given as $L = na$. In order to calculate the band structure, we perform a loop over the longitudinal k-vector. For each such vector we run over the possible transverse k-vectors (values $2\pi j/L$ where j lies between 0 and n). We calculate the two energies in (6.20), and plot these as a function of the longitudinal k. The result is shown in Figure 6.6 for a tube with an odd and an even number of orbitals. Note the difference between the two: one is a metal, the other an insulator. In reality, the even tube has a small gap due to the curvature with respect to the graphene case. Tubes of another type, the so-called arm-chair tubes, characterised by $m = n$, are always metallic. The reader is invited to investigate that case.

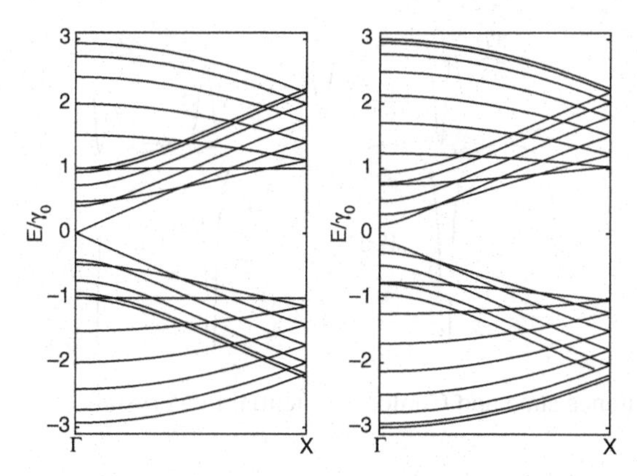

Figure 6.6. Energy bands for a (12, 0) and a (13, 0) tube. In reality, both have an energy gap. That of the even tube is very small. It is absent in the present analysis owing to the neglect of tube curvature.

6.4 Band structure methods and basis functions

Many band structure methods exist and they all have their own particular features. A distinction can be made between *ab initio* methods, which use no experimental input, and *semi-empirical* methods, which do. The latter should not be considered as mere fitting procedures: by fitting a few numbers to a few experimental data, many new results may be predicted, such as full band structures. Moreover, the power of *ab initio* methods should not be exaggerated: there are always approximations inherent to them, in particular the reliance on the Born–Oppenheimer approximation separating the electronic and nuclear motion, and often the local density approximation for exchange and correlation.

In *ab initio* methods, the potential is usually determined self-consistently with the electron density according to the DFT scheme. Some methods, however, solve the Schrödinger equation for a *given*, cleverly determined potential designed to give reliable results. The latter approach is particularly useful for nonperiodic systems where many atoms must be treated in the calculation.

In a general electronic structure calculation scheme we must give the basis functions a good deal of attention since we know that by cleverly choosing the basis states we can reduce their number, which has a huge impact on the computer time needed as the latter is dominated by the $\mathcal{O}(N^3)$ matrix diagonalisation.

Two remarks concerning the potential in a periodic solid are important in this respect. First, the potential grows very large near the nuclei whereas it is (relatively) small in the interstitial region. There is no sharp boundary between the two regions, but it is related to the distance from the nucleus where the atomic wave function

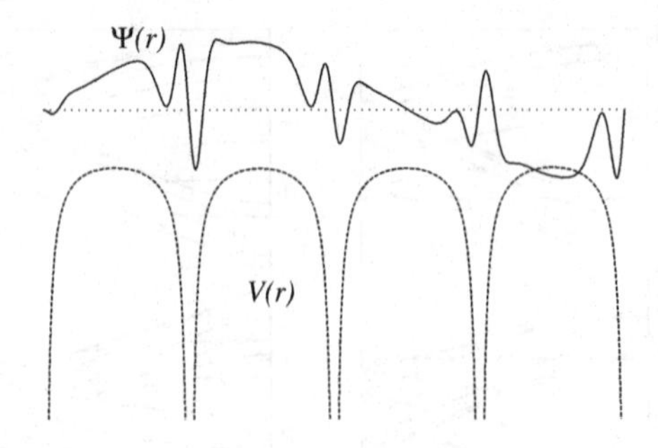

Figure 6.7. Valence state and Coulomb potential in a crystal.

becomes small. Second, the potential is approximately spherically symmetric near the nuclei, whereas at larger distances the crystal symmetry dominates.

Consider the valence state shown in Figure 6.7, together with the potential. Close to the nucleus, the valence state feels the strong spherically symmetric Coulomb potential and it will oscillate rapidly in this region. In between two nuclei, the potential is relatively weak and the orbital will oscillate slowly. The shape of this valence function can also be explained in a different way. Suppose that close to the nucleus, where the potential is essentially spherically symmetric, the valence wave function has s-symmetry (angular momentum quantum number $l = 0$), i.e. no angular dependence. There might also be core states with the same symmetry. As the states must be mutually orthogonal, the valence states must oscillate rapidly in order to be orthogonal to the lower states. A good basis set must be able to approximate such a shape using a limited number of basis functions.

For each Bloch wave vector, we need a basis set satisfying the appropriate Bloch condition. The most convenient Bloch basis set consists of plane waves:

$$\psi^{\text{PW}}_{\mathbf{k}+\mathbf{K}}(\mathbf{r}) = \exp[\mathrm{i}(\mathbf{k}+\mathbf{K})\cdot\mathbf{r}]. \tag{6.25}$$

For a fixed Bloch vector $\mathbf{k}$ in the first Brillouin zone, each reciprocal lattice vector $\mathbf{K}$ defines a Bloch basis function for $\mathbf{k}$. If we take a sufficient number of such basis functions into account, we can match any Bloch function for a Bloch vector $\mathbf{k}$. However, Figure 6.7 suggests that we would need a huge number of plane waves to match our Bloch states because of the rapid oscillations near the nuclei. This is indeed the case: the classic example is aluminium for which it is estimated that 10^6 plane waves are necessary to describe the valence states properly [10]. Although plane waves allow efficient numerical techniques to be used, this number is very high compared with other basis sets, which yield satisfactory results with of the order of only 100 functions. Plane waves can only be used after cleverly transforming away the rapid oscillations near the nuclei, as in pseudopotential methods (Section 6.7).

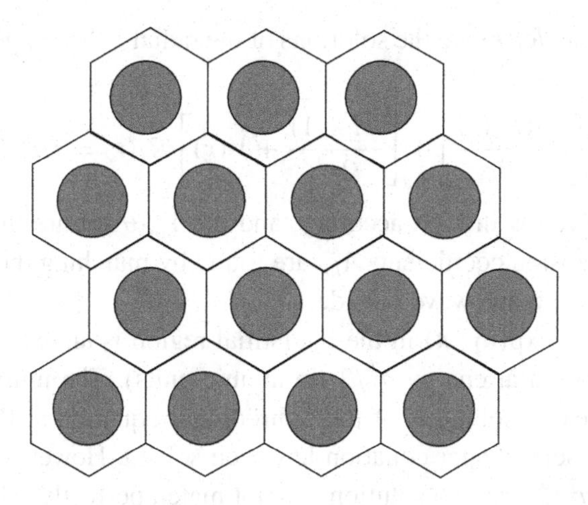

Figure 6.8. The muffin tin approximation.

In the next sections, we consider the augmented plane wave (APW) method and its linearised version, and construct a program for calculating the band structure of copper. In Section 6.7, the pseudopotential method will be considered with a few applications.

6.5 Augmented plane wave methods

6.5.1 Plane waves and augmentation

As we have just indicated, the problem in constructing basis sets starting from plane waves lies in the core region. Close to the nucleus, the potential is approximately spherically symmetric and this symmetry should be exploited in constructing the basis functions. In the augmented plane wave (APW) [11, 12] method, the nuclei are surrounded by spheres in which the potential is considered to be spherically symmetric and outside which the potential is constant, and usually taken to be zero. From a two-dimensional picture (Figure 6.8) it is clear why this approximation is known as the 'muffin tin' approximation.

Outside the muffin tin spheres, the basis functions are simple plane waves $e^{i\mathbf{q}\cdot\mathbf{r}}$. Inside the spheres, they are linear combinations of the solutions to the Schrödinger equation, evaluated at a predefined energy E. These linear combinations can be written as

$$\sum_{l=0}^{\infty}\sum_{m=-l}^{l} A_{lm}\mathcal{R}_l(r)Y_m^l(\theta,\varphi) \tag{6.26}$$

where the functions $\mathcal{R}_l(r)$ are the solutions of the radial Schrödinger equation with energy E:

$$-\frac{1}{2r^2}\frac{\mathrm{d}}{\mathrm{d}r}\left[r^2\frac{\mathrm{d}\mathcal{R}_l(r)}{\mathrm{d}r}\right]+\left[\frac{l(l+1)}{2r^2}+V(r)\right]\mathcal{R}_l(r)=E\mathcal{R}_l(r) \tag{6.27}$$

which can be solved with high accuracy, and the $Y_m^l(\theta,\varphi)$ are the spherical harmonics. The expansion coefficients A_{lm} are found by matching the solution inside the muffin tin to the plane wave outside.

The Bloch wave $\exp(\mathrm{i}\mathbf{q}\cdot\mathbf{x})$ in the interstitial region is an exact solution to the Schrödinger equation at energy $q^2/2$ (in atomic units). The muffin tin solutions are numerically exact solutions of the Schrödinger equation at the energy E for which the radial Schrödinger equation has been solved. However, if we take this energy equal to $q^2/2$, the two solutions do not match perfectly. The reason is that the *general* solution in the interstitial region for some energy $E = q^2/2$ includes *all* wave vectors $\mathbf{q}$ with the same length. In the Bloch solution we take only one of these. If we want to solve the Schrödinger equation inside the muffin tins with the boundary condition imposed by this single plane wave solution then we would have to include solutions that diverge at the nucleus, which is physically not allowed.

An APW basis function contains a muffin tin solution with a definite energy E, and a Bloch wave $\exp(\mathrm{i}\mathbf{q}\cdot\mathbf{x})$ in the interstitial region. It turns out to be possible to match the amplitude of the wave function across the muffin tin sphere boundary. In order to carry out the matching procedure at the muffin tin boundary we expand the plane wave in spherical harmonics [13]:

$$\exp(\mathrm{i}\mathbf{q}\cdot\mathbf{r})=4\pi\sum_{l=0}^{\infty}\sum_{m=-l}^{l}\mathrm{i}^l j_l(qr)Y_m^{l\,*}(\theta_{\mathbf{q}},\varphi_{\mathbf{q}})Y_m^l(\theta,\varphi) \tag{6.28}$$

where r, θ and φ are the polar coordinates of $\mathbf{r}$ and q, $\theta_{\mathbf{q}}$ and $\varphi_{\mathbf{q}}$ those of $\mathbf{q}$. To keep the problem tractable, we cut all expansions in lm off at a finite value for l:

$$\sum_{l=0}^{\infty}\sum_{m=-l}^{l}\rightarrow\sum_{l=0}^{l_{\max}}\sum_{m=-l}^{l} \tag{6.29}$$

From now on, we shall denote these sums by $\sum_{lm}$.

The matching condition implies that the coefficients of the Y_m^l must be equal for both parts of the basis function, (6.26) and (6.28), as the Y_m^l form an orthogonal set over the spherical coordinates. This condition fixes the coefficients A_{lm}, and we arrive at

$$\psi_{\mathbf{q}}^{\mathrm{APW}}(\mathbf{r})=4\pi\sum_{lm}\mathrm{i}^l\left[\frac{j_l(qR)}{\mathcal{R}_l(R)}\right]\mathcal{R}_l(r)Y_m^{l\,*}(\theta_{\mathbf{q}},\varphi_{\mathbf{q}})Y_m^l(\theta,\varphi) \tag{6.30}$$

for the APW basis function inside the sphere.

Summarising the results so far, we can say that in the APW method the wave function is approximated in the interstitial region by plane waves, whereas in the core region the rapid oscillations are automatically incorporated via direct integration of the Schrödinger equation. The basis functions are continuous at the sphere boundaries, but their derivative is not. The APW functions are not exact solutions to the Schrödinger equation, but they are appropriate basis functions for expanding the actual wave function:

$$\psi_{\mathbf{k}}(\mathbf{r}) = \sum_{\mathbf{K}} C_{\mathbf{K}} \psi^{\text{APW}}_{\mathbf{k}+\mathbf{K}}(\mathbf{r}). \tag{6.31}$$

The muffin tin parts of the ψ_{APW} in this expansion are all evaluated at the same energy E. The coefficients $C_{\mathbf{K}}$ are given by the lowest energy solution of the generalised eigenvalue equation:

$$\mathbf{HC} = E\mathbf{SC} \tag{6.32}$$

where the matrix elements of $\mathbf{H}$ and $\mathbf{S}$ are given by quite complicated expressions. In the resulting solution, the mismatch in the derivative across the sphere boundary is minimised.

Before giving the matrix elements of the Hamiltonian and the overlap matrix, we must point out that (6.32) differs from usual generalised eigenvalue equations in that the matrix elements of the Hamiltonian *depend on energy*. This dependence is caused by the fact that they are calculated as matrix elements of energy-dependent wave functions (remember the radial wave functions depend on the energy). Straightforward application of the matrix methods for generalised eigenvalue problems is therefore impossible.

In order to obtain the spectrum, we rewrite Eq. (6.32) in the following form:

$$(\mathcal{H} - E)\mathbf{C} = 0 \tag{6.33}$$

where $\mathcal{H} = \mathbf{H} - E\mathbf{S} + E\mathbf{I}$ ($\mathbf{I}$ is the unit matrix). Although the form (6.33) suggests that we are dealing with an ordinary eigenvalue problem, this is not the case: the overlap matrix has been moved into $\mathcal{H}$. To find the eigenvalues we calculate the determinant $|\mathcal{H} - E\mathbf{I}|$ on a fine energy mesh and see where the zeroes are. It is sometimes possible to use root-finding algorithms (see Appendix A3) to locate the zeroes of the determinant, but these often fail because the energy levels may be calculated along symmetry lines in the Brillouin zone and in that case the determinant may vary quadratically about the zero, so that it becomes impossible to locate this point by detecting a change of sign (this problem does not arise in Green's function approaches, where one evaluates the Green's function at a definite energy).

For crystals with one atom per unit cell, the matrix elements of the Hamiltonian for APW basis functions with wave vectors $\mathbf{q}_i = \mathbf{k} + \mathbf{K}_i$ and $\mathbf{q}_j = \mathbf{k} + \mathbf{K}_j$ are given by

$$\mathcal{H}_{ij} = \langle \mathbf{k} + \mathbf{K}_i | \mathcal{H} | \mathbf{k} + \mathbf{K}_j \rangle = -EA_{ij} + B_{ij} + \sum_{l=0}^{l_{\max}} C_{ijl} \frac{\mathcal{R}'_l(R)}{\mathcal{R}_l(R)}. \tag{6.34}$$

In this expression, $\mathcal{R}'_l(R)$ is $d\mathcal{R}_l(r)/dr$, evaluated at the muffin tin boundary $r = R$. The coefficients A_{ij}, B_{ij} and C_{ijl} are given by

$$A_{ij} = \frac{-4\pi R^2}{\Omega} \frac{j_1(|\mathbf{K}_i - \mathbf{K}_j|R)}{|\mathbf{K}_i - \mathbf{K}_j|} + \delta_{ij},$$

$$B_{ij} = \frac{1}{2} A_{ij} (\mathbf{q}_i \cdot \mathbf{q}_j) \text{ and}$$

$$C_{ijl} = (2l+1) \frac{2\pi R^2}{\Omega} P_l \left(\frac{\mathbf{q}_i \cdot \mathbf{q}_j}{q_i q_j} \right) j_l(q_i R) j_l(q_j R). \tag{6.35}$$

Here, Ω is the volume of the unit cell. Note that there is no divergence in the expression for the matrix elements A_{ij} if $|\mathbf{K}_i - \mathbf{K}_j|$ vanishes, as $j_1(x) \to x/3$ for small x.

In addition to the inconvenient energy dependence of the Hamiltonian, another problem arises from the occurrence of $\mathcal{R}_l(R)$ in the denominator in (6.34). For energies for which the radial solution $\mathcal{R}_l$ happens to be zero or nearly zero on the border of the muffin tin spheres, the matrix elements become very large or even diverge, which may cause numerical instabilities. In the linearised APW (LAPW) method, an energy-independent Hamiltonian is used in which the radial solution does not occur in a denominator, and therefore both problems of the APW method are avoided. The APW method is hardly used now because of the energy-dependence problem. The reasons we have treated it here are that it is conceptually simple and that the principle of this method lies at the basis of many other methods. Further details on the APW method can be found in Refs. [14–16].

6.5.2 An APW program for the band structure of copper

Copper has atomic number $Z = 29$. We consider only the valence states. The core states are two s- and two p-states, and theeleven valence electrons occupy the third s-level and the first d-levels. Its crystal structure is fcc (Figure 6.1) with lattice constant $a = 6.822$ a.u. The unit cell volume Ω is equal to $3a^3/4$. The reciprocal lattice is a bcc lattice with basis vectors

$$\mathbf{b}_1 = \frac{2\pi}{a} \begin{pmatrix} -1 \\ 1 \\ 1 \end{pmatrix}, \quad \mathbf{b}_2 = \frac{2\pi}{a} \begin{pmatrix} 1 \\ -1 \\ 1 \end{pmatrix}, \quad \mathbf{b}_3 = \frac{2\pi}{a} \begin{pmatrix} 1 \\ 1 \\ -1 \end{pmatrix}. \tag{6.36}$$

The Brillouin zone of this lattice is shown in Figure 6.9.

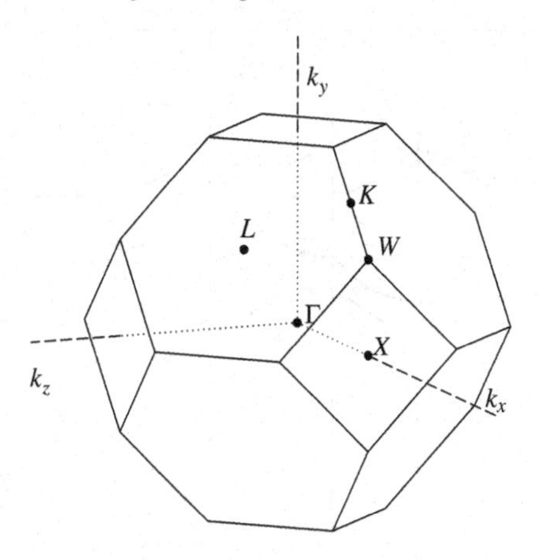

Figure 6.9. Brillouin zone of the fcc lattice.

For a given vector $\mathbf{k}$ in the Brillouin zone, we construct the APW basis vectors as $\mathbf{q} = \mathbf{k} + \mathbf{K}$. The norm of a reciprocal lattice vector $\mathbf{K} = l\mathbf{b}_1 + m\mathbf{b}_2 + n\mathbf{b}_3$ is given by

$$|\mathbf{K}| = \frac{2\pi}{a}\sqrt{3l^2 + 3m^2 + 3n^2 - 2lm - 2nl - 2nm}. \tag{6.37}$$

We take a set of reciprocal lattice vectors with norm smaller than some cut-off and it turns out that the sizes of these sets are 1, 9, 15, 27 etc. A good basis set size to start with is 27, but you might eventually do calculations using 113 basis vectors, for example. The set of such reciprocal lattice vectors is easy to generate by considering all vectors with l, m and n between say -6 and 6 and neglecting all those with norm beyond some cut-off.

The program must contain loops over sets of k-points in the Brillouin zone between for example Γ and X in Figure 6.9. The locations of the various points indicated in Figure 6.9, expressed in cartesian coordinates, are

$$\begin{aligned} \Gamma &= \frac{2\pi}{a}\begin{pmatrix}0\\0\\0\end{pmatrix}, \quad X = \frac{2\pi}{a}\begin{pmatrix}1\\0\\0\end{pmatrix}, \quad K = \frac{2\pi}{a}\begin{pmatrix}3/4\\3/4\\0\end{pmatrix}, \\ W &= \frac{2\pi}{a}\begin{pmatrix}1\\1/2\\0\end{pmatrix}, \quad L = \frac{2\pi}{a}\begin{pmatrix}1/2\\1/2\\1/2\end{pmatrix}. \end{aligned} \tag{6.38}$$

For each $\mathbf{k}$, the matrix elements A_{ij}, B_{ij} and C_{ijl} in (6.35) are to be determined. Good values for the cut-off angular momentum are $l_{\max} = 3$ or 4. Then, for any energy E, the matrix elements of $\mathcal{H}$ according to (6.34) can be found by first solving the radial Schrödinger equation numerically from $r = 0$ to $r = R$ and then using the

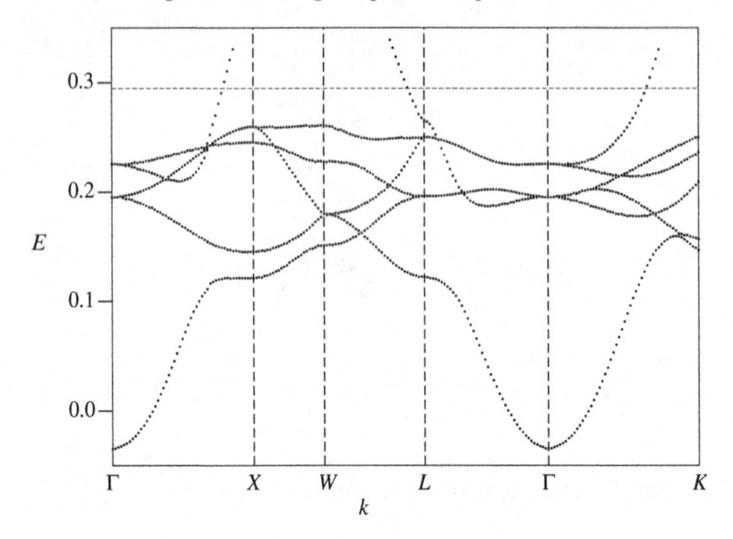

Figure 6.10. Band structure of fcc copper. The Fermi energy is shown as a horizontal dashed line.

quotient $\mathcal{R}_l'(R)/\mathcal{R}_l(R)$ as obtained from this solution in (6.34). Our program will not be self-consistent as we shall use a reasonable one-electron potential.[2]

It is best to use some numerical routine for calculating the determinant. If such a routine is not available, you can bring your matrix to an upper-triangular form as described in Appendix A8 and multiply the diagonal elements of the resulting upper triangular matrix to obtain the determinant.

If you have a routine at your disposal which can calculate the determinant for each k-vector and for any energy, the last step is to calculate the eigenvalues (energies) at some **k**-point. This is a very difficult step and you are advised not to put too much effort into finding an optimal solution for this. The problem is that often the determinant may not change sign at a doubly degenerate level, and energy levels may be extremely close. Finally, changes of sign may occur across a singularity. A highly inefficient but fool-proof method is to calculate the determinant for a large amount of closely spaced energies containing the relevant part of the spectrum (to be read off from Figure 6.10) and then scan the results for changes of sign or near-zeroes. It is certainly not advisable to try writing a routine which finds all the energy eigenvalues automatically using clever root-finding algorithms.

[2] The potential can be found on www.cambridge.org/9780521833469. There are in fact two files in which the potential is given on different grids: the first one is a uniform grid and the second an exponential grid considered in Problem 5.1. Details concerning the integration of the Schrödinger equation on the latter are to be found in this problem.

PROGRAMMING EXERCISE

Write a program for calculating the determinant $|\mathcal{H} - E|$.

Check Check that the determinant vanishes near the values which you can read off from Figure 6.10 for a few points in the Brillouin zone.

The Fermi level for the potential supplied lies approximately at 0.29 a.u., so one conclusion you can draw from the resulting band structure is that copper is a conductor as the Fermi energy does not lie in the energy gap.

You will by now have appreciated why people have tried to avoid energy-dependent Hamiltonians. In the next section we shall describe the linearised APW (LAPW) method which is based on the APW method, but avoids the problems associated with the latter.

6.6 The linearised APW (LAPW) method

A naive way of avoiding the energy-dependence problem in APW calculations would be to use a fixed 'pivot' energy for which the basis functions are calculated and to use these for a range of energies around the pivot energy. If the form of the basis functions inside the muffin tin varies rapidly with energy (and this turns out to be often the case) this will lead to unsatisfactory results.

The idea of the LAPW method [17, 18] is to use a set of pivot energies for which not only the solution to the radial Schrödinger equation is taken into account in constructing the basis set, but also its energy derivative. This means that the new basis set should be adequate for a range of energies around the pivot energy in which the radial basis functions can be reasonably approximated by an energy linearisation:

$$\mathcal{R}(r, E) = \mathcal{R}(r, E_p) + (E - E_p)\dot{\mathcal{R}}(r, E_p). \tag{6.39}$$

Here, and in the remainder of this section, the dot stands for the energy derivative, as opposed to the prime, which is used for the radial derivative – for any differentiable function $f(r, E)$:

$$\dot{f}(r, E) = \frac{\partial}{\partial E} f(r, E) \quad \text{and} \tag{6.40a}$$

$$f'(r, E) = \frac{\partial}{\partial r} f(r, E). \tag{6.40b}$$

The energy derivatives of the radial solution within the muffin tins are used alongside the radial solutions themselves to match onto the plane wave outside the spheres. Note that the APW Hamiltonian depends on energy only via the radial solutions $\mathcal{R}_l$, so if we take these solutions and their energy derivatives $\dot{\mathcal{R}}_l$ at a *fixed* energy into account, we have eliminated all energy dependence from the Hamiltonian.

In comparison with the APW method, we have twice as many radial functions inside the muffin tin sphere, $\mathcal{R}_l$ and $\dot{\mathcal{R}}_l$, and we can match not only the value but also the derivative of the plane wave $\exp(i\mathbf{q}\cdot\mathbf{r})$ across the sphere boundary. We write the wave function inside as the expansion

$$\Phi_{\mathbf{k}+\mathbf{K}}(\mathbf{r}) = \sum_{lm} [A_{lm}\mathcal{R}_l(r; E_p) + B_{lm}\dot{\mathcal{R}}_l(r; E_p)] Y^l_m(\theta, \phi) \tag{6.41}$$

and the numbers A_{lm} and B_{lm} are fixed by the matching condition. There is no energy dependence of the wave functions, and they are smooth across the sphere boundary, but the price which is paid for this is giving up the exactness of the solution inside the sphere for the range of energies we consider.

We end up with a generalised eigenvalue problem with energy-independent overlap and Hamiltonian matrices. These matrices are reliable for energies in some range around the pivot energy. It turns out that the resulting wave functions have an inaccuracy of $(E - E_p)^2$ as a result of the linearisation and that the energy eigenvalues deviate as $(E - E_p)^4$ from those evaluated at the correct energy – see Ref. [18].

The expressions for the matrix elements are again quite complicated. They depend on the normalisations for $\mathcal{R}_l$ and $\dot{\mathcal{R}}_l$ which will be specified below. For the coefficients A_{lm} and B_{lm}, the matching conditions lead (with $\mathbf{q} = \mathbf{k} + \mathbf{K}$, $\mathbf{q}' = \mathbf{k} + \mathbf{K}'$) to:

$$A_{lm}(\mathbf{q}) = 4\pi R^2 \mathrm{i}^l \Omega^{-1/2} Y^{l\,*}_m(\theta_q, \phi_q) a_l; \tag{6.42a}$$

$$a_l = j'_l(qR)\dot{\mathcal{R}}_l(R) - j_l(qR)\dot{\mathcal{R}}'_l(R); \tag{6.42b}$$

$$B_{lm}(\mathbf{q}) = 4\pi R^2 \mathrm{i}^l \Omega^{-1/2} Y^{l\,*}_m(\theta_q, \phi_q) b_l; \tag{6.42c}$$

$$b_l = j_l(qR)\mathcal{R}'_l(R) - j'_l(qR)\mathcal{R}_l(R). \tag{6.42d}$$

The matrix elements of the overlap matrix and the Hamiltonian can now be calculated straightforwardly – the result for the overlap matrix is [18]

$$S_{\mathbf{K},\mathbf{K}'} = U(\mathbf{K} - \mathbf{K}') + \frac{4\pi R^4}{\Omega} \sum_l (2l+1) P_l(\hat{\mathbf{q}} \cdot \hat{\mathbf{q}}') s^l_{\mathbf{K},\mathbf{K}'} \quad \text{with} \tag{6.43a}$$

$$s^l_{\mathbf{K},\mathbf{K}'} = a_l(q)a_l(q') + b_l(q)b_l(q')N_l \quad \text{and} \tag{6.43b}$$

$$U(\mathbf{K}) = \delta_{\mathbf{K},\mathbf{0}} - \frac{4\pi R^2}{\Omega} \frac{j_l(KR)}{K}. \tag{6.43c}$$

Here, N_l is the norm of the energy derivative inside the muffin tin (see below). The Hamiltonian is given by

$$H_{\mathbf{K},\mathbf{K}'} = (\mathbf{q} \cdot \mathbf{q}') U(\mathbf{K} - \mathbf{K}') + \frac{4\pi R^2}{\Omega} \sum_l (2l+1) P_l (E_l s^l_{\mathbf{K},\mathbf{K}'} + \gamma_l) \tag{6.44a}$$

with

$$\gamma_l = \mathcal{R}'_l(R)\dot{\mathcal{R}}_l(R)[j'_l(qR)j_l(q'R) + j_l(qR)j'_l(q'R)]$$
$$- [\mathcal{R}'_l(R)\dot{\mathcal{R}}'_l(R)j_l(qR)j_l(q'R) + \mathcal{R}_l(R)\dot{\mathcal{R}}_l(R)j'_l(qR)j'_l(q'R)]. \tag{6.44b}$$

We see that a pleasing feature of these expressions is that we do not get APW-type numerical inaccuracies due to radial solutions vanishing at the muffin tin radius and occurring in the denominator of the expressions for the matrix elements.

Finally, we must find out how the energy derivative of the solution of the radial Schrödinger equation, $\dot{\mathcal{R}}_l$, can be calculated. By differentiating the radial Schrödinger equation

$$(H - E)\mathcal{R}_l(r; E) = 0 \tag{6.45}$$

with respect to E, we find that $\dot{\mathcal{R}}_l$ satisfies the following differential equation:

$$(H - E)\dot{\mathcal{R}}_l(r; E) = \mathcal{R}_l(r; E). \tag{6.46}$$

This second order inhomogeneous differential equation needs two conditions to fix the solution. The first condition is that $\dot{\mathcal{R}}_l$ (like $\mathcal{R}_l$) is regular at the origin which leaves the freedom of adding $\alpha\mathcal{R}_l(r)$ to it, for arbitrary α ($\mathcal{R}_l$ is the solution of the homogeneous equation). The number α is fixed by the requirement that $\mathcal{R}_l$ is normalised:

$$\int_0^R r^2\mathcal{R}_l^2(r; E)\mathrm{d}r = 1 \tag{6.47}$$

which, after differentiation with respect to E, leads to

$$\int_0^R r^2\mathcal{R}_l(r)\dot{\mathcal{R}}_l(r)\mathrm{d}r = 0 \tag{6.48}$$

i.e. $\mathcal{R}_l$ and $\dot{\mathcal{R}}_l$ are orthogonal. The norm of $\dot{\mathcal{R}}_l$,

$$N_l = \int_0^R r^2|\dot{\mathcal{R}}_l(r)|^2\mathrm{d}r, \tag{6.49}$$

which occurs in the definition of the overlap matrix, is therefore in general not equal to one. It can be shown that the normalisation condition (6.47) leads to the following boundary condition at the muffin tin bounday ($r = R$):

$$R^2[\mathcal{R}'_l(R)\dot{\mathcal{R}}_l(R) - \mathcal{R}(R)\dot{\mathcal{R}}'_l(R)] = 1; \tag{6.50}$$

see Problem 6.5.

The interested reader might try to write a program for calculating the band structure of copper using this linearisation technique. The determination of the eigenvalues will be found much more easily than in the case of the APW calculation, as the Hamiltonian is energy-independent.

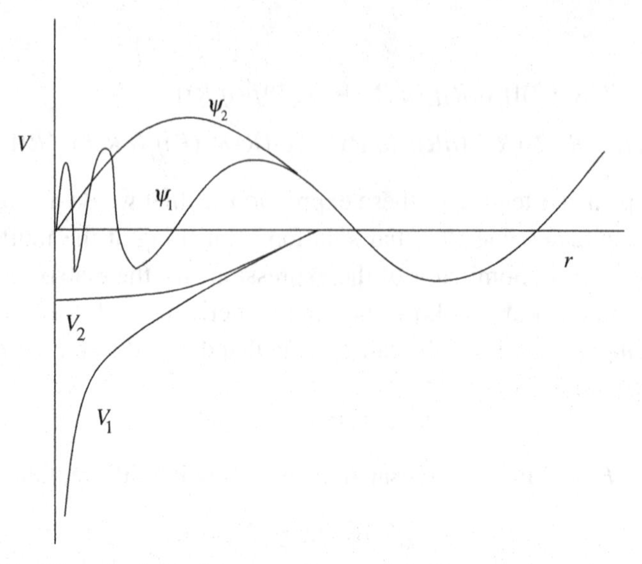

Figure 6.11. The principle of the pseudopotential. The wave functions of the full potential (Ψ_1) and of the pseudopotential (Ψ_2) are equal beyond some radius.

6.7 The pseudopotential method

We have already seen that the main problem in calculating band structures is the deep Coulomb potential giving rise to rapid oscillations close to the nuclei. In the pseudopotential method this problem is cleverly transformed away by choosing a potential which is weak. It is not immediately obvious that this is possible: after all, the solutions of the problem with a weak and with a deep potential can hardly describe the same system! The point is that the pseudopotential does not aim at describing accurately what happens in the core region, but it focuses on the valence region. A weak potential might give results that outside the core region are the same as those of the full potential.

In order to obtain a better understanding of this, we must return to Chapter 2, where the concept of phase shift was discussed. The phase shift uniquely determines the scattering properties of a potential – indeed, we seek a weak pseudopotential that scatters the valence electrons in the same way as the full potential, so that the solution beyond the core region is the same for both potentials. An important point is that we can add an integer times π to the phase shift without changing the solution outside the core region, and there exist therefore many different potentials yielding the same valence wave function. To put it another way: if we make the potential within the core region deeper and deeper, the phase shift will increase steadily, but an increase by π does not affect the solution outside. The pseudopotential is a weak potential which gives the same phase shift (modulo π) as the full potential and hence the same solution outside the core region. The principle is shown in

Figure 6.11 which shows two different potentials and their solutions (for the same energy). These solutions differ strongly within the core region but they coincide in the valence region.

What the pseudopotential does is to remove nodes from the core region of the valence wave function while leaving it unchanged in the valence region. The nodes in the core region are necessary in order to make the valence wave functions orthogonal to the core states. If there are no core states for a given l, the valence wave function is nodeless and the pseudopotential method is less effective. Such is the case in 3d transition metals, such as copper.

The phase shift depends on the angular momentum l and on the energy. A pseudopotential that gives the correct phase shift will therefore also depend on these quantities. The energy dependence is particularly inconvenient, as we have seen in the discussion of the APW method. In Section 6.7.2 we shall see that this dependence disappears automatically when solving another problem associated with the pseudopotential: that of the distribution of the charge inside and outside the core region. More details will be given in that section, and we restrict ourselves here to energy-independent pseudopotentials.

There is a considerable freedom in choosing the pseudopotential as it only has to yield the correct phase shift outside the core region, and several simple parametrised forms of pseudopotentials have been proposed. These are fitted either to experimental data for the material in question (the *semi-empirical* approach), or to data obtained using *ab initio* methods for ions and atoms of the same material, obtained using full-potential calculations.

We give two examples of pseudopotentials.

- The Ashcroft empty-core pseudopotential [19]:

$$V(r) = \begin{cases} -\dfrac{Ze}{r}, & r > r_c \\ 0 & r < r_c \end{cases}. \tag{6.51}$$

  Z is the valence of the ion and there is only one parameter to be adjusted: the cut-off length r_c. Although its simplictity is very attractive, this potential does not perform very well for wide energy ranges, although it reproduces some material properties reasonably well.
- The Fourier-component parametrisation

$$V(r) = \sum\nolimits'_{\mathbf{K}} V_{\mathbf{K}} e^{i\mathbf{K}\cdot\mathbf{R}}. \tag{6.52}$$

  where the sum $\sum'$ is over a limited set of $\mathbf{K}$-vectors. This parametrisation is convenient for the plane wave basis set which is (nearly) always used in pseudopotential calculations. In the next subsection we shall use this form of the pseudopotential in a band structure program for silicon.

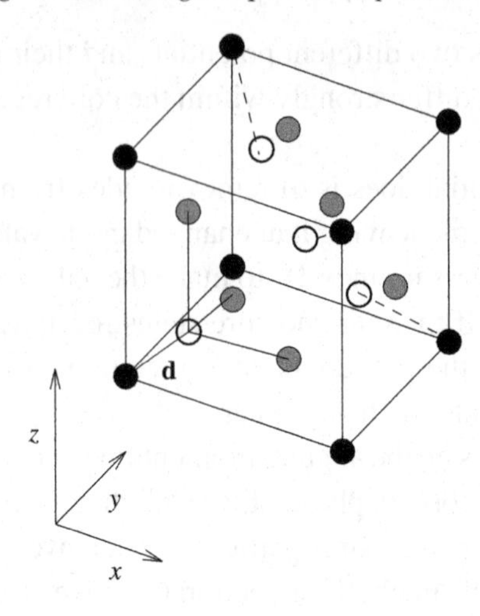

Figure 6.12. The diamond structure.

There are numerous review articles on the pseudopotenial method and readers who are interested in the subject are referred to those by Heine [10], Brust [20] and Pickett [21].

6.7.1 A pseudopotential band structure program for silicon

In this section, the construction of a pseudopotential program for silicon is described. For details, the review by Brust [20] and the paper by Chelikowsky and Cohen [22] may be consulted.

Silicon is considered here in the diamond structure which is a fcc lattice with, at each lattice point, two atoms at relative positions $\pm 1/8(\mathbf{a}_1 + \mathbf{a}_2 + \mathbf{a}_3)$ (see Figure 6.12). We have already described the fcc crystal structure and the special points in the first Brillouin zone in Section 6.5. The lattice constant is $5.43a_0$. The pseudopotential is given in the convenient form of a few Fourier components (see above). We restrict the number of coefficients by assuming the pseudopotential to be a repetition of spherically symmetric potentials in cells surrounding the atoms, which leads to the following form of the Fourier components of the pseudopotential V_{ps} arising from a single atom per cell:

$$V_{\text{ps}}^{(\text{at})}(K) = \frac{1}{V_{\text{cell}}} \int_{\text{cell}} \mathrm{d}^3 r \, V_{\text{ps}}^{(\text{at})}(r) \mathrm{e}^{-\mathrm{i}\mathbf{K}\cdot\mathbf{r}}. \tag{6.53}$$

The Fourier components depend only on the length of the wave vector $\mathbf{K}$, and this property reduces the number of independent Fourier coefficients.

Another reduction comes about when calculating the Fourier component of the sum of the potentials arising from the two atoms at positions

$\mathbf{d}_{1,2} = \pm 1/8(\mathbf{a}_1 + \mathbf{a}_2 + \mathbf{a}_3)$ relative to the lattice points:

$$V_{\rm ps}^{\rm (tot)}(\mathbf{K}) = V_{\rm ps}^{\rm (at)}(K)(e^{i\mathbf{K}\cdot\mathbf{d}_1} + e^{i\mathbf{K}\cdot\mathbf{d}_2}). \tag{6.54}$$

The sum of the exponentials on the right hand side is known as the *structure factor*. Therefore, we find for a vector $\mathbf{K} = \sum_{i=1}^{3} n_i \mathbf{b}_i$ that

$$V_{\rm ps}^{\rm (tot)}(\mathbf{K}) = \cos[\pi(n_1 + n_2 + n_3)/4] V_{\rm ps}^{\rm (at)}(K). \tag{6.55}$$

It follows immediately that the pseudopotential components vanish if the sum of the n_i is an odd multiple of 2: the structure factor causes extinction of certain wave vectors. Furthermore, we can choose the component $V_{\rm ps}^{\rm (tot)}(\mathbf{K} = \mathbf{0})$ to be equal to zero, as this induces a mere shift in the energy offset. Collecting all bits and pieces, we are left with the following values of $|\mathbf{K}| = K$ for which the pseudopotential does not vanish (apart from a factor $4\pi^2/a^2$):

$$K^2 = 3, 8, 11, \ldots \tag{6.56}$$

and only these first three components are taken into account in the pseudopotential. This means that only three numbers are to be fitted and the whole band structure follows from them.

We shall not carry out the fitting procedure but quote the resulting values for the potential from literature, resulting from a fit to optical transition energies [22] – they read (in atomic units):

$$V_{\rm ps}^{\rm (tot)}(\sqrt{3}) = -0.1121;\ V_{\rm ps}^{\rm (tot)}(\sqrt{8}) = 0.0276;\ V_{\rm ps}^{\rm (tot)}(\sqrt{11}) = 0.0362. \tag{6.57}$$

The matrix element of the pseudopotential Hamiltonian for plane waves $\mathbf{k} + \mathbf{K}$ and $\mathbf{k} + \mathbf{K}'$ is given by ($\Delta\mathbf{K} = \mathbf{K} - \mathbf{K}'$)

$$H_{\mathbf{K},\mathbf{K}'} = \frac{1}{2}|\mathbf{k} + \mathbf{K}|^2 \delta_{\mathbf{K},\mathbf{K}'} + V(|\Delta\mathbf{K}|) \cos\left[(\Delta K_1 + \Delta K_2 + \Delta K_3)\frac{\pi}{4}\right]. \tag{6.58}$$

The diagonalisation of the resulting matrix is straightforward: the plane waves are orthogonal and hence no overlap matrix has to be taken into account. You can use basis sets of size 9, 15, 27 etc., just as in the APW case. In Figure 6.13 the band structure is represented for a basis with 113 states.

The band structure in Figure 6.13 matches the results of calculations using more sophisticated methods very well, which is remarkable if you note that only three free parameters enter into the potential. The fact that the band gap comes out well is not surprising since it has been used in the fitting procedure. It turns out to be 1.17 eV, and you might compare this with $k_{\rm B}T$ at room temperature in order to estimate the fraction of electrons excited into the conduction band using the Fermi–Dirac distribution.

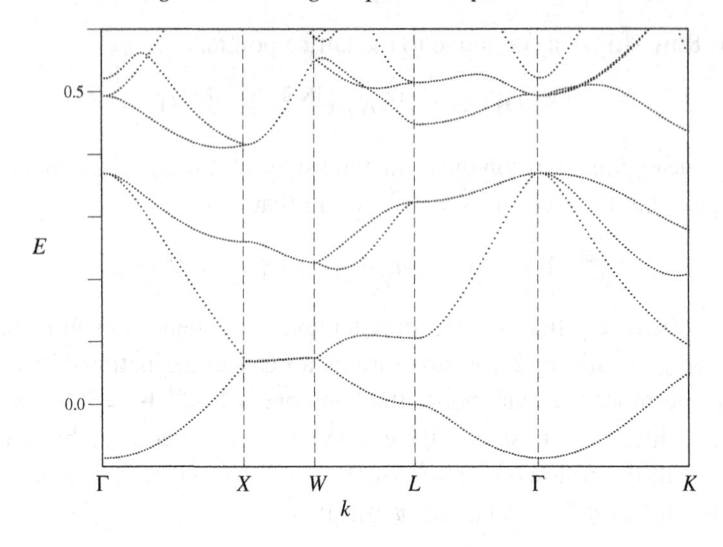

Figure 6.13. Band structure of silicon.

6.7.2 Accurate energy-independent pseudopotentials

Suppose we have a pseudopotential that gives exactly the same phase shift as the full potential. In the valence region, the wave functions have the same shape as for the full potential, but their normalisation may differ: the wave functions in the valence region for the two potentials are the same only up to a scaling factor. The point is that if two normalised wave functions differ within the core region while being similar (up to a multiplication constant) in the valence region, their respective charges will be distributed differently among core and valence regions. The resulting charge difference is called *orthogonality hole* and one should correct for it, for example by rescaling the full pseudo-wave function.

It turns out that the normalisation of the states is related to energy dependence of the pseudopotential. It can be shown (see Problem 6.1) that for the *full* potential, the charge inside a sphere around the nucleus with radius R_c carried by the solution ψ of the Schrödinger equation evaluated at energy E is related to the energy derivative of the wave function at R_c:

$$\int_{\text{core}} \mathrm{d}^3 r |\psi(\mathbf{r})|^2 = -\frac{1}{2} \int \mathrm{d}\Omega\, R_c^2\, \psi(\mathbf{r}) \left.\frac{\partial^2 \psi(\mathbf{r})}{\partial r \partial E}\right|_{r=R_c} \tag{6.59}$$

where the integration is carried out over the spherical angles, $\mathrm{d}\Omega = \mathrm{d}\cos\vartheta\, \mathrm{d}\varphi$. This is another instance of the relation between norm and energy derivative which was previously encountered in connection with the energy derivatives of the solution of the radial equation in the LAPW wave functions in Section 6.6.

If we now consider an energy-dependent pseudopotential V_{ps} with eigenstate ϕ, we obtain, aside from the surface integral on the right hand side, an integral over the energy derivative of V_{ps}:

$$\int_{\text{core}} \mathrm{d}^3 r |\phi(\mathbf{r})|^2 = -\frac{1}{2} \int \mathrm{d}\Omega R_c^2 \, \phi(\mathbf{r}) \left. \frac{\partial^2 \phi(\mathbf{r})}{\partial r \partial E} \right|_{r=R_c} + \int_{\text{core}} \mathrm{d}^3 r \frac{\partial V(\mathbf{r}, E)}{\partial E} |\phi(\mathbf{r})|^2. \tag{6.60}$$

The first term on the right hand side is equal to the right hand side in (6.59) if we fix the amplitude of ϕ to be equal to that of ψ at R_c. Therefore, if both solutions have the same amount of charge inside the core region, the second term on the right hand side must vanish, which implies that the pseudopotential is independent of energy. This means that if we have solved the orthogonality hole problem, we have obtained an energy-independent pseudopotential, so that we have solved two problems at once.

Bachelet, Hamann and Schlüter [23] have constructed accurate norm-conserving pseudopotentials and we refer to their paper and to the review article by Pickett [21] for further details. Goedecker, Teter and Hutter [24] have developed a particularly convenient type of pseudopotential which, being based on Gaussian functions, allows for analytic Fourier transforms. We shall use this pseudopotential in the following section in the construction of a fully self-consistent pseudopotential program.

6.7.3 Building a self-consistent pseudopotential program

The construction of a fully self-consistent pseudopotential program is quite elaborate – we shall therefore restrict ourselves to the case of a cubic unit cell and instead of summing over all points in the Brillouin zone, we shall only consider the Γ-point (i.e. reciprocal vector $\mathbf{K} = \mathbf{0}$). This restriction is often applied when dealing with molecules: the cell is taken big enough to ensure that the electron density vanishes near the cell boundary. This choice therefore renders our program more suitable for molecular systems or clusters than for periodic solids. However, the method for periodic systems uses very similar techniques, and the interested reader is invited to extend his or her program to that case.

It is important to build up the program in a step by step fashion and check each step very carefully. The steps are described below. We closely follow the setup of the CPMD program, described in the review paper by Marx and Hutter [25]. For more details concerning the program and further background, that paper should be consulted.

We start with some remarks concerning definitions and conventions relating to Fourier transforms and choice of basis functions. For simplicity, we restrict ourselves to cubic unit cells. Although all quantities are expanded in a basis of plane waves with wave vectors on a grid in reciprocal space, we cannot always use periodicity on this grid. Consider for example the potential of a nucleus or ion core, located at position $\mathbf{r}_0$ in the unit cell. We expand this potential using a grid of wave vectors

$$\mathbf{K} = \frac{2\pi}{L}(n_x, n_y, n_z) \tag{6.61}$$

with n_x running from 1 to N, L/N being the resolution in real space. If we represent the potential in *real* space on the real-space grid, its Fourier transform is periodic in K-space, with a period $2\pi N/L$ in each Cartesian direction. Usually we are given the (Fourier transform of the) pseudopotential for an ion core located at the origin. If the atom is actually located at some place $\mathbf{r}_0$, the Fourier transfrom acquires an extra structure factor $\exp(-\mathrm{i}\mathbf{K}\cdot\mathbf{r}_0)$. If $\mathbf{r}_0$ does not lie on the real-space grid, this structure factor is nonperiodic in reciprocal space!

Another example of a nonperiodic operator in reciprocal space is the kinetic energy, for which we know the Fourier transform of the operator in continuum space:

$$T = \frac{\hbar^2 K^2}{2m} \tag{6.62}$$

(in atomic units, this reduces to $K^2/2$). This expression is cut off at some maximum wave vector beyond which the components of the orbitals are supposed to be very small. Note that we do not use the periodic discrete form of the kinetic energy (with Fourier transform $3 - \cos(K_x a) - \cos(K_y a) - \cos(K_z a)$; $a = L/N$); the form (6.61) is a more accurate representation.

As a basis, we use Fourier waves $\mathrm{e}^{\mathrm{i}\mathbf{K}\cdot\mathbf{r}}/\sqrt{\Omega}$ (remember that Ω is volume of the unit cell). An orbital $\phi^{(j)}$ is then expanded in these basis vectors as

$$\phi^{(j)}(\mathbf{r}) = \frac{1}{\sqrt{\Omega}} \sum_{\mathbf{K}} c_{\mathbf{K}}^{(j)} \mathrm{e}^{\mathrm{i}\mathbf{K}\cdot\mathbf{r}}. \tag{6.63}$$

The coefficients $c_{\mathbf{K}}^{(j)}$ come out of a diagonalisation routine and are usually normalised according to

$$\sum_{\mathbf{K}} |c_{\mathbf{K}}^{(j)}|^2 = 1. \tag{6.64}$$

Therefore we have:

$$\sum_{\mathbf{r}} |\phi^{(j)}(\mathbf{r})|^2 = \frac{1}{\Omega} \sum_{\mathbf{r}} \sum_{\mathbf{K},\mathbf{K}'} c_{\mathbf{K}}^{(j)} c_{\mathbf{K}'}^{(j)*} \mathrm{e}^{\mathrm{i}(\mathbf{K}-\mathbf{K}')\cdot\mathbf{r}} = \frac{N^3}{\Omega}. \tag{6.65}$$

The density due to all orbitals is therefore given by

$$n(\mathbf{r}) = \sum_j f_j |\phi^{(j)}(\mathbf{r})|^2, \tag{6.66}$$

where f_j is the *occupancy*, which is usually the Fermi–Dirac function with an additional factor of 2 in the closed-shell case. To check that the normalisation is indeed correct, we calculate the total charge for the closed-shell system at $T = 0$:

$$\int n(\mathbf{r})\, \mathrm{d}^3r = 2\sum_{j_{\rm occ}} |\phi^{(j)}(\mathbf{r})|^2\, \mathrm{d}^3r = 2\frac{\Omega}{N^3}\sum_{j_{\rm occ}}\sum_{\mathbf{r}} |\phi^{(j)}(\mathbf{r})|^2 = N_{\rm el} \tag{6.67}$$

where $N_{\rm el}$ is the number of electrons (the factor 2 in front of the second and third expressions is due to the spin degeneracy). Note that the prefactor Ω/N^3 results from the transition of the integral to a sum.

We can formulate Fourier transforms *in continuous, real space* by writing operators and vectors with respect to the basis functions $\exp(\mathrm{i}\mathbf{K}\cdot\mathbf{r})/\sqrt{\Omega}$. In that case, the discrete representation converges to the continuum one for fine grids, and there is no ambiguity concerning the prefactors in the Fourier transforms (powers of Ω). The only exception is the density, which is not a vector or operator in Hilbert space, and we have therefore some freedom in the definition of its Fourier transform. We adopt the convention usually taken in this field, writing

$$n(\mathbf{r}) = \sum_{\mathbf{K}} n(\mathbf{K})\mathrm{e}^{\mathrm{i}\mathbf{K}\cdot\mathbf{r}}. \tag{6.68}$$

It then follows from Eqs. (6.63) and (6.66) that

$$n(\mathbf{K}) = \frac{1}{\Omega}\int n(\mathbf{r})\mathrm{e}^{-\mathrm{i}\mathbf{K}\cdot\mathbf{r}}\, \mathrm{d}^3r = \frac{1}{\Omega}\sum_j f_j \sum_{\mathbf{K}'} c^{(j)}_{\mathbf{K}+\mathbf{K}'}c^{*(j)}_{\mathbf{K}'}. \tag{6.69}$$

An important issue concerns the truncation of the sums over $\mathbf{K}$. The point is that the potential is expressed in terms of the *density*, and not of the wave function. Now suppose that we can safely assume that the wave function vanishes for K-vectors beyond some maximum value $K_{\rm max}$. In that case, from working out the density in real space,

$$\begin{aligned} n(\mathbf{r}) &= \sum_j f_j|\phi_j(\mathbf{r})|^2 = \sum_j f_j \sum_{\mathbf{K},\mathbf{K}'} c^{(j)}_{\mathbf{K}}\mathrm{e}^{\mathrm{i}\mathbf{K}\cdot\mathbf{r}}c^{*(j)}_{\mathbf{K}'}\mathrm{e}^{-\mathrm{i}\mathbf{K}'\cdot\mathbf{r}} \\ &= \frac{1}{\Omega}\sum_j f_j \sum_{\mathbf{K},\mathbf{K}'} c^{(j)}_{\mathbf{K}+\mathbf{K}'}c^{*(j)}_{\mathbf{K}'}\mathrm{e}^{\mathrm{i}\mathbf{K}\cdot\mathbf{r}} = \sum_{\mathbf{K}} n(\mathbf{K})\mathrm{e}^{\mathrm{i}\mathbf{K}\cdot\mathbf{r}}, \end{aligned} \tag{6.70}$$

we see that $n(\mathbf{K})$ contains contributions $\mathbf{K} - \mathbf{K}'$ running up to $2K_{\rm max}$! Therefore, the potential also contains nonzero components for K up to $2K_{\rm max}$. To see that these terms occur in the Hamiltonian matrix, we consider now a *local* potential: this is a potential which depends only on $\mathbf{r}$. Fourier transforming leads to a potential $V_{\mathbf{K},\mathbf{K}'}$ in reciprocal space which is translationally invariant:

$$V_{\mathbf{K},\mathbf{K}'} = V(\mathbf{K} - \mathbf{K}'). \tag{6.71}$$

Even if both $|\mathbf{K}|$ and $|\mathbf{K}'|$ are smaller than K_{max}, their difference can attain lengths up to $2K_{\text{max}}$!

A two-dimensional representation of the situation is depicted in Figure 6.14, where the sphere of radius K_{max} is indicated as a dashed circle, called $C1$. The kinetic energy is evaluated only for values of $\mathbf{K}$ within this circle. When evaluating the kinetic energy, only the points lying inside $C1$ must be taken into account. However, the Fourier transorm of the density, which satisfies periodicity in K-space, has nonzero components for points inside the bigger circle $C2$.

Now let us again consider the contribution to the Hamiltonian of a *local potential*. The matrix elements are given by

$$V_{\mathbf{K},\mathbf{K}'} = \frac{1}{\Omega}\int \mathrm{e}^{-\mathrm{i}(\mathbf{K}-\mathbf{K}')\cdot\mathbf{r}} V(\mathbf{r})\,\mathrm{d}^3 r \approx \frac{1}{N^3}\sum_{\mathbf{r}} \mathrm{e}^{-\mathrm{i}(\mathbf{K}-\mathbf{K}')\cdot\mathbf{r}} V(\mathbf{r}) = V(\mathbf{K}-\mathbf{K}'). \tag{6.72}$$

The last expression on the right hand side is the discrete Fourier transform of the potential in real space. We regularly must perform Fourier transforms, for which we use the FFT algorithm (see Appendix A9). There exist packages containing FFT routines, and we mention here the FFTW package (`http://www.fftw.org`). Often, these packages have their own type definitions for real and complex numbers, which should then be used throughout your program.

The simplest possible nontrivial case is the one with seven wave vectors in $C1$: one in the origin and two along each of the three Cartesian axes. These are used for the Hamiltonian and the wave functions. The wave vectors for which the density is evaluated run over a grid with a linear size at least four times as large as the cut-off wave vector K_{max} (see Figure 6.14), which would suggest a unit cell with a side of four grid points. We take the grid size one larger (i.e. a $5 \times 5 \times 5$ grid) in order to avoid the coincidence of point pairs like $(2, 0, 0)$ and $(-2, 0, 0)$ for functions or operators which are nonperiodic in reciprocal space (such as the pseudopotential of an ion which is not located at a real-space grid point, see above).

6.7.4 Free particle in a box

We start by considering a free particle in the box. The Hamiltonian only contains the kinetic term:

$$H_{\mathbf{K},\mathbf{K}'} = \frac{K^2}{2}\delta(\mathbf{K}-\mathbf{K}'). \tag{6.73}$$

For a box of size $L \times L \times L$, the seven vectors $\mathbf{K}$ we take into account are the null vector and the vectors with size $2\pi/L$ along the positive and negative Cartesian axes. The eigenvalues are therefore equal to 0 (with multiplicity 1) and $2\pi^2/L^2$ (with multiplicity 6). If we put four electrons in the box, the total energy is given by

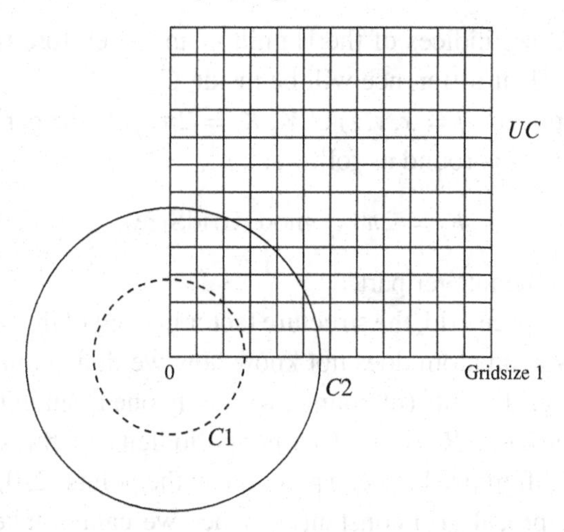

Figure 6.14. Two-dimensional representation of the reciprocal grid. The dashed sphere $C1$ contains the wave vectors of the wavefunctions – its periodic images are also shown as solid circles. The bigger circle $C2$ contains the reciprocal wave vectors for representing the electron density, and we need all those points for accurately constructing the Hamiltonian. The cell UC is a unit cell of the reciprocal lattice.

$4\pi^2/L^2$, as two of these have energy 0 and the other two are divided over the second level. In fact, in each of the six degenerate levels, we should put 1/3 electron because of symmetry. This is an example of fractional filling resulting from degeneracy.

PROGRAMMING EXERCISE

Write a program which diagonalises the Hamiltonian for a particle confined to a box.

Check For the density in a box of size 5 a.u. we find in this case that the density is homogeneous and equals $0.032 = 4/125$. This is not surprising as we put four electrons in a box of volume $5^3 = 125$.

6.7.5 Adding a pseudopotential

The pseudopotential is part of the total potential felt by the electrons. The pseudopotential consists of a local and a nonlocal part. A *local* potential can be evaluated as in Eq. (6.71). The local pseudopotential potential depends only on $\mathbf{r} - \mathbf{r}_0$, where $\mathbf{r}_0$ is the centre of an atom. We have:

$$V_{\mathbf{K},\mathbf{K}'} = V(\mathbf{K} - \mathbf{K}') = \mathrm{e}^{-\mathrm{i}(\mathbf{K}-\mathbf{K}')\cdot\mathbf{r}_0} \int \mathrm{e}^{-\mathrm{i}(\mathbf{K}-\mathbf{K}')\cdot\mathbf{r}} V(\mathbf{r})\, \mathrm{d}^3 r. \tag{6.74}$$

Note that **K** and **K**′ are indices of the Hamiltonian – therefore they lie inside the $C1$ in Figure 6.14. Their difference will be inside $C2$.

For the j-th component ($j = x, y, z$) of **K**, $K_j = 2\pi n_j/L$, the periodic image lying on the grid in Fig. 6.14 is found as follows:

$$K_j = 2\pi(n_j \mod \text{GridSize}), \tag{6.75}$$

where [...] denotes the integer part.

For $\mathbf{r}_0$ on the real-space grid, the structure factor in front of the integral is periodic in K-space. However, an atom does not know how we define our grid and may be located in between grid points (of course, with only one atom in the cell, we could always move the atom to $\mathbf{R} = 0$). This nonperiodicity of the structure factor is responsible for the difference between potential at the points $(2, 0, 0)$ and $(-2, 0, 0)$ (in units of the reciprocal grid constant) – hence we cannot take a grid size of 4 units, but instead need a $5 \times 5 \times 5$ grid, as mentioned above.

We now give a specific form of the pseudopotential. We shall use the Goedecker–Teter–Hutter (GTH) potential described in refs. [24] and [26]. This potential for a core, located at the origin, with s and p electrons has the form:[3]

$$V(\mathbf{r}, \mathbf{r}') = V_{\text{core}}(\mathbf{r}) + V_{\text{loc}}(\mathbf{r})\delta(\mathbf{r} - \mathbf{r}') + V_{\text{nonloc}}(\mathbf{r}, \mathbf{r}') \tag{6.76}$$

with

$$V_{\text{core}} = -\frac{Z_{\text{eff}}}{r}\text{erf}\left(\frac{r}{\sqrt{2}\xi}\right); \tag{6.77a}$$

$$V_{\text{loc}}(\mathbf{r}) = \exp[-(r/\xi)^2/2] \times [C_1 + C_2(r/\xi)^2], \tag{6.77b}$$

and

$$V_{\text{nonloc}}(\mathbf{r}, \mathbf{r}') = \sum_{i=1}^{2} Y_0^0(\hat{\mathbf{r}}) p_i^0(r) h_i^0 p_i^0(r') Y_0^{0*}(\hat{\mathbf{r}}') + \sum_{m=1,0,-1} Y_m^1(\hat{\mathbf{r}}) p_1^1(r) h_1^1 p_1^1(r') Y_m^{1*}(\hat{\mathbf{r}}') \tag{6.78}$$

In these expressions, ξ, C_i, h_i^l are parameters, and the p_i^l are the functions

$$p_1^l(r) = \sqrt{2}\frac{r^l e^{-(1/2)(r/r_l)^2}}{r^{l+3/2}\sqrt{\Gamma(l+3/2)}} \text{and} \tag{6.79}$$

$$p_2^l(r) = \sqrt{2}\frac{r^{l+2} e^{-(1/2)(r/r_l)^2}}{r^{l+7/2}\sqrt{\Gamma(l+7/2)}} \tag{6.80}$$

[3] The form given here is somewhat simpler than the full GTH potential. For the atoms considered here, however, the present form is sufficient.

Table 6.1. *Parameters for the GTH pseudopotential*

	Hydrogen	Silicon
ξ	0.2	0.44
C_1	−4.0663326	−6.9136286
C_2	0.6778322D0	0.0
$r_{l=0}$	–	0.4243338
$h_1^{l=0}$	–	3.2081318
$h_2^{l=0}$	–	2.5888808
$r_{l=1}$	–	0.4853587
$h_2^{l=0}$	–	2.6562230

Source: [24, 26]
Only values for hydrogen and silicon are listed.

The Gamma-function in the denominators ensures proper normalisation:

$$\int p_i^l(r) p_i^l(r) r^2 \, \mathrm{d}r = 1. \tag{6.81}$$

Let us spend a few moments studying this potential. The very first term, $-Z_{\mathrm{eff}}/r \operatorname{erf}(r/2\xi)$, is the Coulomb potential of a Gaussian charge distribution with total charge Z_{eff} and width ξ: for large arguments, that is, far from the ion core, the error function erf tends to 1. The remaining terms are short-ranged and allow therefore for refinement of the shape of the radial charge distribution. The nonlocal term is, as usual, a projection onto the different l subspaces. For a complete list of pseudopotential parameters, we refer to Refs. [24] and [26]; here we give those for hydrogen and silicon – see Table 6.1.

The Fourier transform of the GTH potential can be calculated analytically, yielding the following closed forms.

$$V_{\mathrm{core}}(\mathbf{K}) = -4\pi \frac{Z_{\mathrm{eff}}}{\Omega} \frac{\mathrm{e}^{-(K\xi)^2/2}}{K^2}; \tag{6.82}$$

$$V_{\mathrm{loc}}(\mathbf{K}) = \sqrt{(2\pi)^3} \frac{\xi^3}{\Omega} \mathrm{e}^{-(K\xi)^2/2} \{C_1 + C_2[3 - (K\xi)^2]\} \tag{6.83}$$

and

$$V_{\mathrm{nonloc}}(\mathbf{K}, \mathbf{K}') = \sum_{i=1}^{2} Y_0^0(\hat{\mathbf{K}}) p_i^0(K) h_i^0 p_i^0(K') Y_0^{0*}(\hat{\mathbf{K}}') - \sum_{m=1,0,-1} Y_m^1(\hat{\mathbf{K}}) p_1^1(K) h_1^1 p_1^1(K') Y_m^{1*}(\hat{\mathbf{K}}'). \tag{6.84}$$

The projector functions have the form p_i^l:

$$p_1^0 = \frac{1}{\sqrt{\Omega}} 4r_s\sqrt{2r_s}\pi^{5/4}\mathrm{e}^{-(Kr_s)^2/2}, \tag{6.85a}$$

$$p_2^0 = \frac{1}{\sqrt{\Omega}} 8r_s\sqrt{\frac{2r_s}{15}}\pi^{5/4}\mathrm{e}^{-(Kr_s)^2/2}[3-(Kr_s)^2], \text{ and} \tag{6.85b}$$

$$p_1^1(K) = \frac{1}{\Omega} 8r_1^2\sqrt{\frac{r_1}{3}}\pi^{5/4}\mathrm{e}^{-(Kr_s)^2/2}K. \tag{6.85c}$$

This pseudopotential can be directly incorporated into the Kohn–Sham Hamiltonian. It does not depend on the density, so if we simply want to calulate the energies and eigenfunctions of a particle moving in the pseudopotential, we just have to diagonalise the Hamiltonian which consists of the kinetic energy plus pseudopotential. A self-consistency cycle is not necessary.

The Fourier transform of the local part of the pseudopotential for a core located at $\mathbf{R}_n$ must be multiplied by $\exp(-\mathrm{i}\mathbf{K}\cdot\mathbf{R}_n)$. The nonlocal part must be multiplied by $\exp[\mathrm{i}(\mathbf{K}-\mathbf{K}')\cdot\mathbf{R}_n]$.

Check Doing this for a cubic cell with an edge length of 5 a.u. containing one hydrogen-core and an energy cut-off of $(1/2)K_{\text{max}}^2 = 1.3$, we obtain the eigenvalues −0.03572203 (once), 0.68175686 (once), 0.80555307 (three times) and 0.83735807 (twice). If we fill all seven levels, the density should be 0.05600 on any real-space grid point. Try this first for a hydrogen at the origin, and then some arbitrary position within the cell.

6.7.6 Exchange-correlation and Hartree potentials

The exchange-correlation and Hartree potentials are density-dependent; therefore, including them makes a self-consistency cycle necessary. We shall first consider the general problem of including density-dependent potentials into the problem. First, we must have the density at our disposal. After diagonalising the Hamiltonian, we calculate the Fourier transforms $\phi^{(j)}(\mathbf{r})$ of the eigenfunctions $c_{\mathbf{K}}^{(j)}$ as in (6.63). Then we calculate the density on all real-space grid points inside the cell according to (6.66). Finally, we calculate the density in reciprocal space according to

$$n(\mathbf{K}) = \frac{1}{N^3}\sum_{\mathbf{r}} n(\mathbf{r})\mathrm{e}^{-\mathrm{i}\mathbf{K}\cdot\mathbf{r}}. \tag{6.86}$$

For the exchange-correlation potential, we use the GTH parametrisation of the pseudopotential of Perdew and Wang [27]. This is a form of Padé approximant:

$$\varepsilon_{\text{xc}} = -\frac{\sum_{i=1}^4 a_i r_s^{i-1}}{\sum_{i=1}^4 b_i r_s^i}. \tag{6.87}$$

Table 6.2. *Parameters for the GTH parametrisation of the exchange-correlation energy.*

a_1	0.4581652932831429	b_1	1.0
a_2	2.217058676663745	b_2	4.504130959426697
a_3	0.7405551735357053	b_3	1.110667363742916
a_4	0.01968227878617998	b_4	0.02359291751427506

Here, $r_s = [3/(4\pi n)]^{1/3}$ is the radius of the spherical volume per atom; the numbers a_i and b_i are given in Table 6.2. The exchange-correlation potential is given as the derivative of the energy with respect to n. It must be calculated in real space, where it is periodic (as it depends on the density, which is periodic), and then Fourier-transformed so that it can be added to the (K-space) Hamiltonian. The procedure is therefore to first fill a grid with the values of $V_{\text{xc}}(\mathbf{r})$. This is then Fourier-transformed to $V_{\text{xc}}(\mathbf{K})$. Then, the contribution $V_{\text{xc}}(\mathbf{K}, \mathbf{K}')$, where $\mathbf{K}$ and $\mathbf{K}'$ lie inside the circle $C1$ of Figure 6.14, is found by first translating $\mathbf{K} - \mathbf{K}'$ to a point inside the unit cell UC of the reciprocal grid, and then taking for $V_{\text{xc}}(\mathbf{K}, \mathbf{K}')$ the Fourier-transformed exchange-correlation potential at that reciprocal grid point.

The Hartree potential

$$V_{\text{H}}(\mathbf{r}) = \int \frac{n(\mathbf{r}')}{|\mathbf{r} - \mathbf{r}'|} \mathrm{d}^3 r' \tag{6.88}$$

can be Fourier-transformed to give

$$V_{\text{H}}(\mathbf{K}, \mathbf{K}') = V_{\text{H}}(\mathbf{K} - \mathbf{K}') = \frac{4\pi}{|\mathbf{K} - \mathbf{K}'|^2} n(\mathbf{K} - \mathbf{K}'). \tag{6.89}$$

For the density, the difference $\mathbf{K} - \mathbf{K}'$ has to be translated to lie inside the unit cell UC (see Figure 6.13), just as in the case of the exchange correlation potential. For the denominator, we simply take the norm of the smallest periodic image of the difference $\mathbf{K} - \mathbf{K}'$.

Check If we incorporate both the exchange-correlation and the Hartree potentials, we have a complete self-consistent pseudopotential Kohn–Sham program. For the calculation with one hydrogen atom in a cubic cell of size 5 and a cut-off of 1.3 atomic units, we obtain the energy spectrum:

−0.468131; 0.249348; 0.373144; 0.373144;

0.373144; 0.404949; 0.404949.

6.7.7 Evaluating the energy

If you have obtained the correct spectrum, the density should necessarily be correct too. One major task remains, however: evaluating the total energy. The energy can be evaluated either by adding all the Kohn–Sham eigenvalues and subtracting the appropriate corrections as in Eq. (5.3), or by using the Kohn–Sham eigenfunctions to evaluate all the contributions to the energy as in (5.17) one by one. We take the second approach. First of all, the *kinetic energy* is given by

$$E_{\text{kin}} = \sum_j f(E_j) \sum_{\mathbf{K}} |c^{(j)}|^2 K^2/2. \tag{6.90}$$

The exchange-correlation energy is evaluated in real space:

$$E_{\text{xc}} = \sum_{\mathbf{r}} \varepsilon_{\text{xc}}(\mathbf{r}) n(\mathbf{r}), \tag{6.91}$$

where the sum is over the real-space lattice points, or in reciprocal space, where it reads:

$$E_{\text{xc}} = \Omega \sum_{\mathbf{K}} \varepsilon(\mathbf{K}) n^*(\mathbf{K}). \tag{6.92}$$

Following Marx and Hutter [25], we combine the electrostatic contributions from the electrons and the ion cores. Remember that the core part of the pseudopotential,

$$V_{\text{core}} = -\frac{Z_n}{|\mathbf{r}-\mathbf{R}_n|} \text{erf}\left(\frac{|\mathbf{r}-\mathbf{R}_n|}{\sqrt{2}\xi}\right); \tag{6.93}$$

derives from a Gaussian charge distribution:

$$n_{\text{core}}(\mathbf{r}) = -\frac{Z_n}{(\sqrt{2}\xi_n)^3} \pi^{-3/2} \exp\left[-\frac{1}{2}\left(\frac{\mathbf{r}-\mathbf{R}_n}{\xi}\right)^2\right]. \tag{6.94}$$

The Fourier transform of the core density is

$$n_{\text{core}}(\mathbf{K}) = -\frac{Z_n}{\Omega} \exp\left[-\frac{1}{2}(\xi K)^2\right] e^{-i\mathbf{K}\cdot\mathbf{R}_n}. \tag{6.95}$$

For the total charge density we have

$$n_{\text{tot}}(\mathbf{K}) = n_{\text{el}}(\mathbf{K}) + n_{\text{core}}(\mathbf{K}). \tag{6.96}$$

The electrostatic energy resulting from the total charge density is

$$E_{\text{ES}} = \frac{1}{2} \int \frac{n_{\text{el}}(\mathbf{r}) n_{\text{el}}(\mathbf{r}')}{|\mathbf{r}-\mathbf{r}'|} \, \mathrm{d}^3r \, \mathrm{d}^3r' + \int \frac{n_{\text{core}}(\mathbf{r}) n_{\text{el}}(\mathbf{r}')}{|\mathbf{r}-\mathbf{r}'|} \, \mathrm{d}^3r \, \mathrm{d}^3r' + \frac{1}{2} \sum_{n \neq n'} \frac{Z_n Z'_n}{|\mathbf{R}_n - \mathbf{R}'_n|}, \tag{6.97}$$

where the first term is the Hartree electrostatic energy due to the electrons, the second term is the interaction between the electrons and the core, and the last term is the core–core interaction.

This last term causes problems as we must sum it over all periodic images of the unit cell for which we are performing the calculations (for a discussion concerning convergence of this type of expression, see Section 8.7.1). These problems can be avoided by replacing the last term by the electrostatic interaction of the core charges. This is done by adding and subtracting a term

$$E_{cc} = \frac{1}{2}\int \frac{n_{core}(\mathbf{r})n_{core}(\mathbf{r}')}{|\mathbf{r}-\mathbf{r}'|}\,d^3r\,d^3r' \tag{6.98}$$

to the expression for the total energy, Eq. (6.97). The added term, together with the first two terms in that equation, yields a contribution

$$\frac{1}{2}\int \frac{n_{tot}(\mathbf{r})n_{tot}(\mathbf{r}')}{|\mathbf{r}-\mathbf{r}'|}\,d^3r\,d^3r' \tag{6.99}$$

The remaining terms can be written as a convergent sum [25], leading to

$$E_{ES} = \frac{1}{2}\int \frac{n_{tot}(\mathbf{r})n_{tot}(\mathbf{r}')}{|\mathbf{r}-\mathbf{r}'|}\,d^3r d^3r' + \frac{1}{2}\sum_{n,n'} \frac{Z_n Z_{n'}}{|\mathbf{R}_n - \mathbf{R}_{n'}|}\operatorname{erfc}\left[\frac{|\mathbf{R}_n - \mathbf{R}_{n'}|}{\sqrt{2(\xi_n^2 + \xi_{n'}^2)}}\right] - \sum_n \frac{Z_n^2}{2\sqrt{\pi}\xi_n}. \tag{6.100}$$

The second term on the right hand side is due to the overlap of the core distributions, and the third term corrects for the self-energy (that is, the energy of a core with itself). Both of these are contained in the first term.

For periodic boundaries, we can reformulate this expression in Fourier space, where it reads:

$$E_{ES} = 2\pi\Omega \sum_{\mathbf{K}\neq\mathbf{0}} \frac{|n_{tot}(\mathbf{K})|^2}{K^2} + E_{ovrl} - E_{self}, \tag{6.101}$$

where $n_{tot}(\mathbf{K})$ is given above (Eq. (6.95)) and where

$$E_{ovrl} = \sum_{\mathbf{L}}\sum_{n,n'}{}' \frac{Z_n Z_{n'}}{|\mathbf{R}_n - \mathbf{R}_{n'} - \mathbf{L}|}\operatorname{erfc}\left[\frac{|\mathbf{R}_n - \mathbf{R}_{n'} - \mathbf{L}|}{\sqrt{2(\xi_n^2 + \xi_{n'}^2)}}\right], \tag{6.102}$$

where $\mathbf{L}$ is an integer linear combination of the sides of the unit cell, the second sum is restricted to $n < n'$ for $\mathbf{L} = \mathbf{0}$, and

$$E_{self} = \sum_n \frac{Z_n^2}{2\sqrt{\pi}\xi_n}. \tag{6.103}$$

Table 6.3. *Contributions to electronic energy of hydrogen atom.*

Contribution	*Value*
Kinetic	0.159 230 87
Short range part of pseudopotential	−0.021 083 18
Local pseudopotential	−0.244 116 10
Exchange correlation	−0.210 599 25
Hartree energy	0.024 351 53
Nonlocal pseudopotential	0.000 000 00
Local core energy	1.12582002
Self-energy	0.92508958
Electrostatic overlap	0.000 000 00
Total energy	−0.557 835 94

Finally, we must include the energy contributions due to the pseudopotential. These do not depend on the charge distribution and they contain the local and the nonlocal terms. The local contribution is easily evaluated using

$$E_{\text{local}} = \int \sum_n V_{\text{local},n}(\mathbf{r} - \mathbf{R}_n) n(\mathbf{r}) \, \mathrm{d}^3 r$$
$$= \Omega \sum_n \sum_{\mathbf{K}} V_{\text{local},n}(\mathbf{K}) \mathrm{e}^{-\mathrm{i}\mathbf{K}\cdot\mathbf{R}_n} n^*(\mathbf{K}). \tag{6.104}$$

where n runs over the atoms in the cell. The nonlocal energy reads:

$$E_{\text{nonlocal}} = \sum_j f_j \sum_n \sum_{l,m\varepsilon n} (F^n_{jlm})^* h^n_{lm} F^n_{jlm} \tag{6.105}$$

where $l, m\varepsilon n$ denotes the orbital with quantum numbers l, m belonging to atom number n, and

$$F^n_{jlm} = \sum_{\mathbf{K}} \mathrm{e}^{-\mathrm{i}\mathbf{K}\cdot\mathbf{R}_n} c_j^*(\mathbf{K}) Y_{lm}(\hat{\mathbf{K}}) p^l_m(K). \tag{6.106}$$

Now that you have everything in place, you can calculate the electronic energy of the hydrogen atom. It is built up from the contributions shown in Table 6.3.

6.8 Extracting information from band structures

Apart from ground state energies, from which cohesion energies and lattice spacings can be determined, and those energy levels that can be measured directly using spectroscopy experiments, it is useful to determine the density of states, $n(E)$, which can also be determined experimentally. This is defined as the number of levels between E and $E + \mathrm{d}E$, divided by $\mathrm{d}E$. Another quantity of interest is the charge density, which is needed for calculating the Hartree and exchange and correlation potentials in the DFT self-consistency loop. The charge density is given by

$$n(\mathbf{r}) = \sum_{\mathbf{k},n}{}' |\psi_{\mathbf{k},n}(\mathbf{r})|^2 = \int_{-\infty}^{E_{\mathrm{F}}} \mathrm{d}E \; n(\mathbf{r}, E), \tag{6.107}$$

where the sum in the second expression is over the occupied levels, i.e. those with energy below the Fermi energy E_{F}.

The charge density can also be found from an integration over the energy of the *local density of states*, which is defined as the charge density resulting exclusively from states at energy E. An elegant way of finding this quantity using Green's functions is described in Problem 6.3. Such an approach is necessary when the total charge of the system is not known, as is the case for a small system coupled to a large reservoir which determines the chemical potential: for a metal, this is (to very good approximation) the Fermi energy of the large system. This Fermi energy is defined with respect to the vacuum energy as the *work function*: the energy needed to remove an electron from the large system. There exist Green's function methods in which the small system (e.g. an atom or a molecule) is coupled to the surface Green's function of a metal. The electronic structure can then still be determined in a self-consistency loop, in which the charge density is determined from the Green's function of the combined system plus reservoir rather than from the eigenstates of the Hamiltonian [28].

To find physical properties or quantities, we often must perform an integration over the Brillouin zone, as the vectors (together with the band labels) in this zone are quantum numbers of the stationary states. Taking the crystal symmetry into account, these integrations only need to be carried out in the 'irreducible wedge' of the Brillouin zone: this can be used to fill the whole Brillouin zone by crystal symmetry transformations. For example, in a two-dimensional lattice having the symmetry of the square, the Brillouin zone is also a square, but to integrate quantity over the Brillouin zone, an integration over a wedge of area 1/8 of the whole square needs to be carried out. For the Brillouin zone of the fcc lattice in Figure 6.9, this irreducible wedge is the volume bounded by the labelled points.

There exist many different methods for performing Brillouin zone integration [21]. The most popular methods are those using *special points* [29, 30] and tetrahedron methods. In the latter, (part of) the Brillouin zone is divided up into tetrahedra, in each of which either a linear or a quadratic approximation of the function to be integrated is made. For calculating the density of states, the quadratic works very well since it is capable of reproducing all known Van Hove singularities [31–34].

6.9 Some additional remarks

In this chapter we have described how the nonrelativistic Schrödinger equation can be solved efficiently in a solid. The core electrons near heavy nuclei move at speeds where relativistic effects become significant, although they are still small, so that relativistic corrections must be included. This can be done in a perturbative way, and the resulting equation for the radial part $\mathcal{R}_{nl}$ of the wave function reads:

$$\left[-\frac{1}{2M}\frac{1}{r^2}\frac{\mathrm{d}}{\mathrm{d}r}\left(r^2\frac{\mathrm{d}^2}{\mathrm{d}r^2}\right)+\frac{l(l+1)}{2Mr^2}+V(r)-\frac{V'(r)}{4M^2c^2}\frac{\mathrm{d}}{\mathrm{d}r}\right]\mathcal{R}_{nl}(r)=E\mathcal{R}_{nl}(r). \tag{6.108}$$

$V'(r)$ is the derivative of the potential V, and M is given in terms of the electron rest mass m, the energy E and the potential as:

$$M(r)=m+\frac{1}{2c^2}[E-V(r)]. \tag{6.109}$$

This equation is derived from the Dirac equation; see for example Ref. [13].

Solving the Schrödinger equation is only one step in a DFT self-consistency equation. Having found the density as described in the previous section, we must calculate the Hartree potential by solving Poisson's equation:

$$\nabla^2 V_{\mathrm{H}}(r)=-4\pi n(\mathbf{r}). \tag{6.110}$$

Solving this equation in a pseudopotential method with a plane wave basis is not so difficult, as the Laplace operator ∇^2 has the diagonal form $k_r^2\delta_{rs}$ in reciprocal space. For muffin tins, most of the codes use a method developed by Weinert [35]. In this method we obtain an expansion of the potential in spherical harmonics. Note that in the APW method considered above, we use the spherical average of the full potential. We shall only briefly discuss the two main ideas upon which Weinert's method is based.

First of all, inside the muffin tins, the charge density and potential are expanded in spherical harmonics. The radial part of the Hartree potential can then be found by integration of a radial differential equation, as was done for the $l = 0$ case in the local density program for helium – see Section 5.5. The problem then remains of finding the solution outside the muffin tins, which is determined by the charge density in and outside the muffin tin. It seems a good idea to solve this problem in reciprocal space because of the Laplace operator being diagonal there. However, a huge number of plane waves would be necessary for obtaining an accurate solution, as the charge inside the muffin tin contains rapid oscillations (it is constructed from the wave functions which, as we have seen, vary rapidly close to the nucleus). The second ingredient of Weinert's method is the replacement of the charge density inside the spheres by a weaker one, just as in the replacement of the full potential by a pseudopotential. The new, weak charge density is called *pseudo-charge* density. That this replacement is possible can be seen by realising that the effect of a muffin tin charge distribution can be formulated in terms of the multipole moments of the charge density, and many different charge densities give the same multipole moments. Note that this justification is also analogous to the pseudopotential method, where the fact that many different potentials yield the same phase shifts justifies the replacement of the full potential by a pseudopotential. For details we refer to Weinert's paper.

6.10 Other band methods

There are numerous band structure methods [1, 10, 36]), and we have considered only two illustrative examples in this chapter. Another important approach is the Korringa–Kohn–Rostocker (KKR) method, [37–39] based on a scattering approach with a muffin tin form of potential. It leads to a matrix whose size is equal to the number of different states used in the muffin tins.

Linearising the KKR method, one obtains the linear muffin tin orbital (LMTO) method with localised, energy-independent wave functions which are centred at each atom – see Refs. [40] and [41].

Exercises

6.1 In this exercise we want to establish the relation between the energy derivative and the charge of a core wave function, Eq. (6.59). Our derivation will not rely on a spherical shape for the core region. We use the normalisation convention that the value of the wave function at the boundary of the core region is equal to some fixed number, so that we have

$$\frac{\partial \psi(\mathbf{r}_s)}{\partial E} = \dot{\psi}(\mathbf{r}_s) = 0$$

where $\mathbf{r}_s$ lies at the core boundary.

(a) Starting from the Schrödinger equation, derive an equation satisfied by $\dot{\psi}$. Note that we use the full potential, which does not depend on energy.

(b) Green's theorem applied to the core region for two arbitrary functions, ψ_1 and ψ_2, reads

$$\int_{\text{core}} d^3r[\psi_1(\mathbf{r})\nabla^2\psi_2(\mathbf{r}) - \psi_2(\mathbf{r})\nabla^2\psi_1(\mathbf{r})]$$
$$= \int_{\text{shell}} d^2a[\psi_1(\mathbf{a})\hat{\mathbf{n}} \cdot \nabla\psi_2(\mathbf{a}) - \psi_2(\mathbf{a})\hat{\mathbf{n}} \cdot \nabla\psi_1(\mathbf{a})]$$

where the integral on the right hand side is a surface integral over the boundary of the core region and $\hat{\mathbf{n}}$ is a normal vector pointing out of the core boundary. Apply this theorem to ψ and its energy derivative and use the normalisation convention to show that

$$\int_{\text{core}} \mathrm{d}^3 r \psi^2(\mathbf{r}) = -\frac{1}{2}\int_{\text{shell}} \mathrm{d}^2 a \psi(\mathbf{a})\hat{\mathbf{n}} \cdot \nabla\dot{\psi}(\mathbf{a}).$$

6.2 [C] Consider the following periodic potential in one dimension.

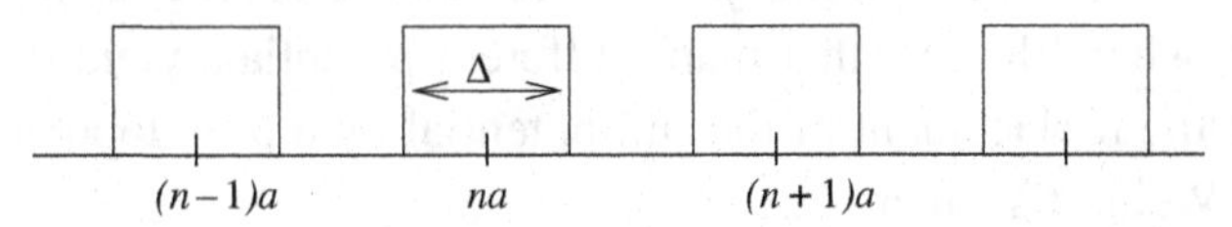

The height of the barriers is V_0. The solution of the Schrödinger equation in between two barriers at $(n-1)a$ and na can be written as

$$\psi(x) = A_n \mathrm{e}^{\mathrm{i}q(x-na)} + B_n \mathrm{e}^{-\mathrm{i}q(x-na)}$$

with $q = \sqrt{2E}$. Assume that the energy we are interested in is higher than the barrier height V_0. On the nth barrier, the solution is written as

$$\psi(x) = C_n \mathrm{e}^{\mathrm{i}\kappa(x-na)} + D_n \mathrm{e}^{-\mathrm{i}\kappa(x-na)}$$

with $\kappa = \sqrt{2(E - V_0)}$.

The values of A_n and B_n in neighbouring interstitial regions are connected through the so-called 'transfer matrix':

$$\begin{pmatrix} A_{n+1} \\ B_{n+1} \end{pmatrix} = T(E) \begin{pmatrix} A_n \\ B_n \end{pmatrix}.$$

T is a 2×2 matrix which depends on energy.

(a) Show that the transfer matrix is given by

$$T = \frac{q}{4\kappa} \begin{pmatrix} T_{11} & T_{12} \\ T_{21} & T_{22} \end{pmatrix},$$

with

$$T_{11} = \mathrm{e}^{\mathrm{i}q(a-\Delta)} \left[\mathrm{e}^{\mathrm{i}\kappa\Delta} \left(1 + \frac{\kappa}{q}\right)^2 - \mathrm{e}^{-\mathrm{i}\kappa\Delta} \left(1 - \frac{\kappa}{q}\right)^2 \right],$$

$$T_{12} = -2\mathrm{i}\mathrm{e}^{\mathrm{i}qa} \left(1 - \frac{\kappa^2}{q^2}\right) \sin(\kappa\Delta);$$

and

$$T_{22} = T_{11}^*,$$

$$T_{21} = T_{12}^*.$$

Show that the product of the two eigenvalues of this matrix is equal to 1. Hence these eigenvalues can either be written as $\mathrm{e}^{\pm ik}$ (or as $\mathrm{e}^{\pm\alpha}$, real α). From Bloch's theorem we know that the solutions can be labelled by a wave vector $\tilde{q}$ which is not necessarily equal to q, and that these solutions can be written as a periodic function times $e^{i\tilde{q}x}$. In our case this implies that

$$\begin{pmatrix} A_{n+1} \\ B_{n+1} \end{pmatrix} = e^{i\tilde{q}a} \begin{pmatrix} A_n \\ B_n \end{pmatrix}$$

and therefore

$$T \begin{pmatrix} A_n \\ B_n \end{pmatrix} = e^{i\tilde{q}a} \begin{pmatrix} A_n \\ B_n \end{pmatrix}.$$

This equation defines the band spectrum of the system. It is now easier to find the vector $\tilde{q}$ as a function of energy than vice versa: the above-mentioned eigenvalues (which depend on energy via q and κ) must be equal to $e^{i\tilde{q}a}$.

(b) [C] Write a simple computer program to determine the spectrum. In an APW approach, the wave function outside the barriers is written as $e^{iq_m x}$, where $q_m = k + 2\pi m/a$ and $-\pi/a < k < \pi/a$ (m is integer). It is now convenient to confine ourselves to the unit cell $[-a/2, a/2]$ and to use Bloch boundary conditions on that cell. For a Bloch state χ_m:

$$\chi_m(-a/2) = \chi(a/2)e^{-ika}.$$

For each q_r, the value of the wave function outside and inside the barrier can be matched at the boundaries of the barrier. Show that C_m and D_m are given by

$$C_m = \frac{\sin[(\kappa + q_m)\Delta/2]}{\sin(\kappa\Delta)},$$

$$D_m = \frac{\sin[(\kappa - q_m)\Delta/2]}{\sin(\kappa\Delta)}.$$

In the APW method, the coefficients b_m of the expansion

$$\psi(x) = \sum_m b_m \chi_m(x)$$

are found by solving the generalised eigenvalue problem

$$\mathbf{Hb} = E\mathbf{Sb}$$

in which the matrix S is given by

$$\begin{aligned} S_{ml} &= \int_{-a/2}^{-\Delta/2} e^{-iq_m x} e^{iq_l x} dx + \int_{\Delta/2}^{a/2} e^{-iq_m x} e^{iq_l x} dx \\ &+ \int_{-\Delta/2}^{\Delta/2} [C_m^* e^{-i\kappa_m x} + D_m^* e^{i\kappa_m x}][C_l e^{i\kappa_n x} + D_l e^{-i\kappa_l x}] dx \\ &= S_{ml}^{\text{int}} + S_{ml}^{\text{ext}} \end{aligned}$$

where we have split the expression for S into an integration over the interior of the barriers and the part outside.

(c) Show that

$$S^{\text{ext}}_{ml} = \begin{cases} a - \Delta & \text{if } m = l \\ \dfrac{-2}{q_m - q_l} \sin \dfrac{(q_m - q_l)\Delta}{2} & \text{otherwise} \end{cases};$$

and

$$S^{\text{int}}_{ml} = C_m C_n + D_m D_n + (C_m D_n + C_n D_m)\frac{\kappa\Delta}{\kappa}.$$

(d) Show that the Hamiltonian matrix is given as

$$H_{ml} = H^{\text{int}}_{mn} + \frac{1}{2} q_m q_n S^{\text{ext}}_{mn} + \partial H_{mn}$$

where

$$H^{\text{int}}_{ml} = -\kappa \sin(\kappa\Delta)(C_m D_n + C_n D_m) + \kappa^2 \Delta (C_m C_n + D_m D_n),$$

and where $\partial \mathbf{H}$ is the matrix due to the jumps in derivatives across the barrier boundaries. Show that $\partial \mathbf{H}$ is given by

$$\begin{aligned} \partial H_{mn} &= -q_n \sin[(q_n - q_m)\Delta/2] \\ &\quad - C_m \kappa \sin[(\kappa - q_n)\Delta/2] - D_m \kappa \sin[(\kappa + q_n)\Delta/2]. \end{aligned}$$

(e) Write a program in which the matrices H_{ml} and S_{ml} are filled and find the zeroes of the determinant $|\mathbf{H} - E\mathbf{S}|$ for various k. Compare the results with the numerically exact ones, resulting from the previous program.

6.3 In this problem we consider the determination of the local charge density using the Green's function. The Green's function for a Hamiltonian's H is defined as

$$(H - E)G(\mathbf{r}, \mathbf{r}'; E) = \delta(\mathbf{r}, \mathbf{r}').$$

(a) Show that G can be written as

$$G(\mathbf{r}, \mathbf{r}'; z) = \sum_{n=1}^{\infty} \psi_n(\mathbf{r}) \frac{1}{z - E_n} \psi_n(\mathbf{r}').$$

(b) Show that the electron density (charge density) can be found as

$$n(r) = \frac{1}{2\pi i} \int_\Gamma G(\mathbf{r}, \mathbf{r}; z) \mathrm{d}z,$$

where Γ is a closed contour in the complex plane which contains all the *occupied* energy levels (these of course all lie on the real axis).

6.4 [C] As plane waves form an orthogonal basis, it is possible to use the Lowdin perturbation method discussed in Section 3.4. Write an extension to your pseudopotential program to incorporate large lattice vectors into the Hamiltonian in a perturbative manner. Compare the results with those of the direct diagonalisation.

6.5 In this problem we derive the normalisation condition (6.50) from the normalisation (6.47) of the radial solution inside the muffin tin.

(a) Show that the normalisation condition (6.47) can be rewritten as

$$\langle \mathcal{R}_l | H - E | \dot{\mathcal{R}}_l \rangle = 1.$$

(b) Use this result, together with the fact that $\mathcal{R}_l$ is an eigenfunction of H inside the muffin tin with eigenvalue E, and partial differentiation, to derive Eq. (6.50).

References

[1] N. W. Ashcroft and N. D. Mermin, *Solid State Physics*. New York, Holt, Reinhart and Winston, 1976.

[2] C. Kittel, *Introduction to Solid State Physics*, 6th edn. New York, John Wiley, 1973.

[3] R. M. Martin, *Electronic Structure*. Cambridge, Cambridge University Press, 2004.

[4] J. M. Thijssen and J. E. Inglesfield, 'Embedding muffin tins into a finite difference grid,' *Europhys. Lett.*, **27** (1994), 65–70.

[5] S. Baroni and P. Giannozzi, 'Towards very large-scale electronic structure calculations,' *Europhys. Lett.*, **17** (1992), 547–52.

[6] L.-W. Wang and A. Zunger, 'Electronic-structure pseudopotential calculations of large (approximate-to-1000 atoms) Si quantum dots,' *J. Phys. Chem.*, **98** (1994), 2158–65.

[7] T. L. Beck, 'Real-space mesh techniques in density-functional theory,' *Rev. Mod. Phys.*, **72** (2000), 1041–80.

[8] S. Reich, J. Maultzsch, C. Thomsen, and P. Ordéjon, 'Tight-binding description of graphene,' *Phys. Rev. B*, **66** (2002), 035412.

[9] J. C. Slater and G. F. Koster, 'Simplified LCAO method for the periodic potential problem,' *Phys. Rev.*, **94** (1954), 1498–524.

[10] V. Heine, 'The pseudopotential concept,' in *Solid State Physics* (F. Seitz and D. Turnbull, eds.), vol. 24. New York, Academic Press, 1970, p. 1.

[11] J. C. Slater, 'Wave functions in a periodic potential,' *Phys. Rev.*, **51** (1937), 846–51.

[12] J. C. Slater, 'An augmented plane wave method for the periodic potential problem,' *Phys. Rev.*, **92** (1953), 603–8.

[13] A. Messiah, *Quantum Mechanics*, vols. 1 and 2. Amsterdam, North-Holland, 1961.

[14] L. F. Mattheis, J. H. Wood, and A. C. Switendick, 'A procedure for calculating electronic energy bands using symmetrised augmented planes,' in *Methods in Computational Physics*, vol. 8. New York, Academic Press, 1968, pp. 63–147.

[15] T. Loucks, *Augmented Plane Wave Method*. New York, Benjamin, 1967.

[16] J. Callaway, *Quantum Theory of the Solid State*. 2nd edn. San Diego, Academic Press, 1991.

[17] O. K. Andersen, 'Linear methods in band theory,' *Phys. Rev. B*, **12** (1975), 3060–83.

[18] D. D. Koelling and G. O. Arbman, 'Use of the energy derivative of the radial solution in an augmented plane wave method: application to copper,' *J. Phys. F*, **5** (1975), 2041–54.

[19] N. W. Ashcroft and D. C. Lang, 'Compressibility and binding energy of the simple metals,' *Phys. Rev.*, **155** (1967), 682–4.

[20] D. Brust, 'The pseudopotential method and the single-particle electronic excitation spectra of crystals,' *Methods in Computational Physics*, vol. 8. pp. 33–61, New York, Academic Press, 1968, pp. 33–61.

[21] W. E. Pickett, 'Pseudopotential methods in condensed matter applications,' *Comp. Phys. Rep.*, **9** (1989), 115–97.

[22] J. R. Chelikowsky and M. L. Cohen, 'Electronic structure of silicon,' *Phys. Rev. B*, **10** (1974), 5095–107.

[23] G. B. Bachelet, D. R. Hamann, and M. Schlüter, 'Pseudopotentials that work,' *Phys. Rev. B*, **26** (1982), 4199–288.

[24] S. Goedecker, M. Teter, and J. Hutter, 'Separable dual space Gaussian pseudopotentials,' *Phys. Rev. B*, **54** (1996), 1703–10.

[25] D. Marx and J. Hutter, 'Ab initio molecular dynamics: theory and implementations,' in *Modern Methods and Algorithms of Quantum Chemistry, NIC Series*, vol. 1. Jülich, John von Neumann Institute for Computing, 2000, pp. 301–449.

[26] C. Hartwigsen, S. Goedecker, and J. Hutter, 'Relativistic separable dual-space Gaussian pseudopotentials from H to Rn,' *Phys. Rev. B*, **58** (1998), 3641–62.

[27] J. P. Perdew and Y. Wang, 'Accurate and simple analytic representation of the electron-gas correlation energy,' *Phys. Rev. B*, **45** (1992), 13244–9.

[28] N. D. Lang and A. R. Williams, 'Theory of atomic chemisorption on simple metals,' *Phys. Rev. B*, **18** (1978), 616–36.

[29] D. J. Chadi and M. L. Cohen, 'Special points in the Brillouin zone,' *Phys. Rev. B*, **8** (1973), 5747–5753.

[30] H. J. Monkhorst and J. D. Pack, 'Special points for Brillouin-zone integrations,' *Phys. Rev. B*, **13** (1976), 5188–192.

[31] M. S. Methfessel, M. H. Boon, and F. M. Müller, 'Analytic-quadratic method of calculating the density of states,' *J. Phys. C*, **16** (1983), L949–54.

[32] M. S. Methfessel, M. H. Boon, and F. M. Müller, 'Singular integrals over the Brillouin zone: the analytic-quadratic method for the density of states,' *J. Phys. C*, **19** (1986), 5337–64.

[33] M. S. Methfessel, M. H. Boon, and F. M. Müller, 'Singular integrals over the Brillouin zone: inclusion of k-dependent matrix elements,' *J. Phys. C*, **20** (1987), 1069–77.

[34] G. Wiesenecker and E. J. Baerends, 'Analytic quadratic integration over the two-dimensional Brillouin zone,' *J. Phys. C*, **21** (1988), 4263–83.

[35] M. Weinert, 'Solution of Poisson equation – beyond Ewald-type methods,' *J. Math. Phys.*, **22** (1981), 2433–9.

[36] R. Zeller, 'Band structure methods,' in *Unoccupied Electron States* (J. E. Inglesfield and J. Fuggle, eds.), Heidelberg, Springer, 1991, pp. 25–49.

[37] J. Korringa, 'On the calculation of the energy of a Bloch wave in a metal,' *Physica*, **13** (1947), 392–400.

[38] W. Kohn and N. Rostocker, 'Solution of the Schrödinger equation in periodic lattices with an application to metallic lithium,' *Phys. Rev.*, **94** (1954), 1111–20.

[39] A. R. Williams, S. M. Hu, and D. W. Jepsen, 'Recent developments in KKR theory,' in *Computational Methods in Band Theory* (P. M. Marcus, J. F. Janak, and A. R. Williams, eds.), New York, Plenum, 1971, p. 157.

[40] O. K. Andersen, O. Jepsen, and M. Sob, 'Linearised band structure methods,' in *Electronic Band Structure and its Applications* (M. Yussouff, ed.), *Lecture Notes in Physics*, vol. 283, Heidelberg-Berlin, Springer, 1987, ch. 1, pp. 1–57.

[41] H. L. Skriver, *The LMTO Method*. New York, Springer, 1984.

7

Classical equilibrium statistical mechanics

7.1 Basic theory

In this chapter we briefly review the theory of classical statistical mechanics with emphasis on those issues which are relevant to computer simulations. We shall assume that the reader has some background in thermodynamics and statistical mechanics; for further reading, numerous textbooks are available [1–8].

Statistical mechanics concerns the study of systems with many (in principle infinitely many) degrees of freedom. The degrees of freedom are usually the positions and momenta of particles, or magnetic moments ('spins'). We restrict ourselves to classical systems for which all degrees of freedom commute. The space spanned by the degrees of freedom is called *phase space* – every point in phase space represents a particular configuration of the system. In the course of time, the system follows a path in phase space, determined by the equations of motion. We are obviously not interested in the values of all these degrees of freedom as a function of time: only the time averages of physical quantities such as pressure are measurable. This is because our measuring devices (thermometers, barometers) respond relatively slowly; hence they give a time average of the physical quantity of interest. However, even if we could perform an instantaneous measurement of some quantity we would find a result very close to the time average of that quantity as a result of the law of large numbers, which teaches us that if a quantity is composed of N uncorrelated contributions, fluctuations in that quantity are of order $1/\sqrt{N}$. This implies that for typical macroscopic physical quantities (such as the temperature of your cup of tea) for which N is of the order of 10^{24}, the fluctuations are as small as $\sim 10^{-12}$ if we neglect correlations. If correlations extend over ~ 100 particles, the number of uncorrelated contributions is $\sim 10^{24}/100 = 10^{22}$, so the fluctuations remain extremely small.

Computer simulations always sample relatively few degrees of freedom, since only a restricted amount of data can be stored in memory: system sizes in simulations are always much smaller than those of experimental systems.[1] Furthermore, a time average of a physical quantity A is given by

$$\bar{A} = \lim_{T\to\infty} \frac{1}{T} \int_0^T A(t)\mathrm{d}t, \tag{7.1}$$

and we want to obtain results in a finite amount of time! In a molecular dynamics simulation (see Chapter 8), the typical simulation time is of the order of 10^{-9}–10^{-6} seconds, far below the time in which most measuring devices sample physical quantities. The results of such simulations can only be representative if the spatial correlations extend over ranges smaller than the system size and if the correlation time of the system is smaller than the simulation time. Sometimes it is possible to extract useful information from simulations of systems with a size much smaller than the correlation length by extrapolation – this is done in the finite-size scaling method which will be discussed in Section 7.3.2. In this chapter, we shall almost exclusively be concerned with systems in equilibrium.

7.1.1 Ensembles

If a system is thermally and mechanically insulated, the internal energy will remain unchanged in the course of time. If the system is not insulated, it will eventually take on the temperature of its surroundings (we assume that the surroundings have a constant temperature). Such physical quantities, which are either kept fixed or whose average value is controlled externally, are called *system parameters*. Different experimental circumstances correspond to different parameters being kept fixed. In the theory of statistical physics, these cases correspond to different *ensembles*. We shall see that adapting the simulation techniques for classical many-particle systems (Monte Carlo and molecular dynamics) to these experimental situations is a nontrivial problem – that is why we consider the ensemble theory in some detail in this section.

The fundamental postulate, or assumption, of statistical mechanics pertains to systems with fixed energy E, volume V and particle number N (in magnetic systems, instead of the volume V, the external magnetic field H is kept constant). The fundamental postulate says that all states accessible to the system and having a prescribed energy, volume and number of particles are equally likely to be visited in the course of time (the ergodic hypothesis). This leads to an identification of the time average $\bar{A}$ (7.1) of the physical quantity A with a uniform average over all accessible states – the latter is denoted as $\langle A \rangle$. Denoting the states by X,

[1] A notable exception is formed by the so-called mesoscopic systems which contain typically 10^2 to 10^5 particles.

we have

$$\langle A \rangle = \frac{\sum_{\{X|E\}} A(X)}{\sum_{\{X|E\}}} = \frac{\sum_X A(X)\delta[\mathcal{H}(X) - E]}{\sum_X \delta[\mathcal{H}(X) - E]} = \bar{A}. \tag{7.2}$$

$\mathcal{H}(X)$ is the Hamiltonian which gives the energy for a point X in phase space. The denominator ensures proper normalisation. The sum $\sum_{\{X|E\}}$ denotes a sum over all states X with a fixed energy E; in the unrestricted sums the delta-function takes care of the restriction to the states with energy E (the restriction to a specific volume and particle number is tacitly assumed). In the case of continuous degrees of freedom, the sums will generally be replaced by integrals. In the case of a monatomic liquid consisting of N moving particles with spherically symmetric interactions, for example, the sum is replaced by the following integral over the positions $\mathbf{r}_i$ and momenta $\mathbf{p}_i$ of the particles:

$$\sum_X \to \left(\frac{1}{h}\right)^{3N} \int_V d^3r_1 d^3r_2 \ldots d^3r_N \int d^3p_1 d^3p_2 \ldots d^3p_N \tag{7.3}$$

where h is Planck's constant. The average (7.2) is called the *ensemble average* and the set of states under consideration (fixed N, V and E) is called the *microcanonical ensemble* or (NVE) ensemble (the (NHE) ensemble in the magnetic case). From now on, the volume V of a system of moving particles can be replaced by the external magnetic field H for magnetic systems unless stated otherwise.

The denominator in (7.2) counts the number of states with the prescribed energy. In fact, quantum mechanics imposes a way of counting which for the case of identical particles is quite different from the classical procedure: as the particles are indistinguishable, configurations that can be obtained from each other by permuting the particles should be counted only once. This implies that the sum in the denominator of (7.2) should be divided by $N!$.[2] The number of states with energy E is then given by

$$\Omega(N, V, E) = \frac{1}{N!} \sum_X \delta[\mathcal{H}(X) - E] \tag{7.4}$$

(for mixtures, the factor $N!$ is replaced by the product $N_1!N_2!\ldots$, where the subscripts label the different species). The *entropy* is defined in terms of $\Omega(N, V, E)$ as

$$S(N, V, E) = k_B \ln \Omega(N, V, E) \tag{7.5}$$

where k_B is Boltzmann's constant. The quantum counting factor $N!$ is necessary in order to make the entropy thus defined an extensive variable, i.e. a variable that scales linearly with system size. The thermodynamic quantities temperature T,

[2] This only holds for systems in which there is at most one particle per quantum state. Properly taking into account more particles per state leads to quantum statistical distributions.

chemical potential μ and pressure P are given as derivatives of the entropy with respect to the system parameters:

$$T = \left(\frac{\partial S}{\partial E}\right)^{-1}_{N,V} \qquad \mu = -T\left(\frac{\partial S}{\partial N}\right)_{E,V} \qquad P = T\left(\frac{\partial S}{\partial V}\right)_{E,N} \tag{7.6}$$

as can be readily seen from the first law of thermodynamics:[3]

$$\mathrm{d}E = T\mathrm{d}S - P\mathrm{d}V + \mu\mathrm{d}N. \tag{7.7}$$

In experimental situations, it is often the temperature that is kept constant and not the energy (for the latter to be constant, the system must be insulated thermally and mechanically). In order to achieve constant temperature, the system under consideration is coupled to a heat bath, a much larger system with which it can exchange heat. It turns out that a time average for the system under consideration is equal to a weighted average over states with fixed volume and particle number (the energy is no longer restricted); the weighting factor is the so-called *Boltzmann factor* $\exp[-\mathcal{H}(X)/(k_B T)]$. Writing $\beta = 1/(k_B T)$, we have

$$\langle A\rangle_{NVT} = \frac{1}{N!Z}\sum_X A(X)\mathrm{e}^{-\beta\mathcal{H}(X)}; \tag{7.8a}$$

$$Z(N, V, T) = \frac{1}{N!}\sum_X \mathrm{e}^{-\beta\mathcal{H}(X)}. \tag{7.8b}$$

The factor Z ensures proper normalisation. It is called the *partition function* and it is related to the free energy F:

$$F = -k_B T \ln Z(N, V, T) \tag{7.9}$$

which, in terms of thermodynamic quantities, is given by

$$F = E - TS. \tag{7.10}$$

In equilibrium, the free energy assumes its minimum under the constraint of fixed volume and particle number. The average in (7.8) is called the *canonical ensemble average* or (NVT) ensemble average. Note that the partition function can be written as a sum over sets of states with fixed energy:

$$Z(N, V, T) = \sum_E \mathrm{e}^{-\beta E}\Omega(N, V, E), \tag{7.11}$$

where $\Omega(N, V, E)$ is the number of states with energy E as defined already in the microcanonical ensemble. The number of states $\Omega(N, V, E)$ is a rapidly increasing function of E and the Boltzmann distribution is a rapidly decreasing function of E.

[3] Often, the first law is stated without including changes in particle number $\mathrm{d}N$.

The product of the two functions peaks sharply at some value $\bar{E}$ and the system will be found to have an energy very close to this value most of the time. This suggests that there is in practice not much difference between the canonical and the microcanonical system in which the energy is kept rigorously fixed at $\bar{E}$. This is a manifestation of the so-called ensemble equivalence: because of the law of large numbers, measurable physical quantities exhibit very small fluctuations – hence fixing them to their average value leaves the system essentially unchanged. For finite systems, the differences between the ensembles increase with decreasing system size.

Using the definition of the entropy (7.5), we may write (7.11) as

$$Z(N,V,T) = \sum_E e^{-\beta(E-TS)} = \sum_E e^{-\beta F_E}, \tag{7.12}$$

where F_E is the free energy $E - TS$ with S evaluated in the microcanonical ensemble with energy E, and we see that the sum is indeed dominated by the states for which the free energy is minimal.

Again using the first law of thermodynamics, (7.7), we can derive the following thermodynamic quantities from the free energy:

$$\mu = \left(\frac{\partial F}{\partial N}\right)_{V,T} \quad P = -\left(\frac{\partial F}{\partial V}\right)_{N,T} \quad S = -\left(\frac{\partial F}{\partial T}\right)_{V,N}. \tag{7.13}$$

If the pressure P is kept constant and not the volume, as in a cylinder closed by a movable piston, we obtain an average over the isothermal-isobaric or (NPT) ensemble:

$$\langle A \rangle_{NPT} = \frac{1}{N!Q} \int dV\, e^{-\beta PV} \sum_X e^{-\beta \mathcal{H}(X)} A(X); \tag{7.14a}$$

$$Q(N,P,T) = \int dV\, e^{-\beta PV} \frac{1}{N!} \sum_X e^{-\beta \mathcal{H}(X)} = \int dV\, e^{-\beta PV} Z(N,V,T), \tag{7.14b}$$

where $Q(N,P,T)$ is again called the partition function. We see that Q is related to the canonical partition function Z in much the same way as Z was related to the function Ω in the microcanonical ensemble – see Eq. (7.11). Q is related to the Gibbs free energy or Gibbs potential G:

$$G = -k_B T \ln Q(N,P,T). \tag{7.15}$$

G can be expressed in terms of thermodynamic quantities as

$$G = E - TS + PV, \tag{7.16}$$

and it assumes its mimimum value when the system has reached equilibrium under the condition of fixed temperature and pressure. For magnetic systems, the role of the pressure P is taken over by the total magnetic moment M. The other relevant thermodynamic quantities follow from the definition of $G(N, P, T)$:

$$\mu = \left(\frac{\partial G}{\partial N}\right)_{P,T} \quad V = \left(\frac{\partial G}{\partial P}\right)_{N,T} \quad S = -\left(\frac{\partial G}{\partial T}\right)_{P,N}. \tag{7.17}$$

If the volume is again fixed, but the number of particles is allowed to vary, we obtain the *grand canonical ensemble* average:

$$\langle A \rangle = \frac{1}{Z_G} \sum_N e^{\beta \mu N} \frac{1}{N!} \sum_X e^{-\beta \mathcal{H}(X)} A(X) \tag{7.18a}$$

$$Z_G(\mu, V, T) = \sum_N e^{\beta \mu N} \frac{1}{N!} \sum_X e^{-\beta \mathcal{H}(X)}. \tag{7.18b}$$

Here, μ is the chemical potential for the addition or removal of a particle. $Z_G(\mu, V, T)$ should not be confused with the canonical partition function $Z(N, V, T)$; it can be expressed in terms of the latter as

$$Z_G(\mu, V, T) = \sum_N e^{\beta \mu N} Z(N, V, T). \tag{7.19}$$

Z_G defines the *grand canonical potential* Ω_G, analogous to similar definitions for the other ensembles:

$$\Omega_G(\mu, V, T) = -k_B T \ln Z_G(\mu, V, T). \tag{7.20}$$

In equilibrium, this potential assumes its minimum value for fixed μ, T and V. From the definition of Z_G and from the expression for the average values in the grand canonical ensemble, it follows that

$$\Omega_G(\mu, V, T) = F - \mu N. \tag{7.21}$$

The internal energy can be written in terms of the variables S, V and N and it satisfies the Gibbs–Duhem equation [4]

$$E(S, V, N) = TS - PV + \mu N \tag{7.22}$$

so that we have

$$\Omega_G(\mu, V, T) = -PV. \tag{7.23}$$

From the grand canonical potential we can derive thermodynamic quantities:

$$N = -\left(\frac{\partial \Omega_G}{\partial \mu}\right)_{V,T} \quad P = -\left(\frac{\partial \Omega_G}{\partial V}\right)_{\mu,T} \quad S = -\left(\frac{\partial \Omega_G}{\partial T}\right)_{V,\mu}. \tag{7.24}$$

Expectation values of thermodynamic quantities are calculated either as ensemble averages or as integrals over phase space. As an example of an ensemble average, consider the internal energy. The expectation value of this quantity in the canonical ensemble is given by

$$\langle E\rangle_{NVT} = \frac{\sum_X \mathrm{e}^{-\beta\mathcal{H}(X)}\mathcal{H}(X)}{\sum_X \mathrm{e}^{-\beta\mathcal{H}(X)}} \tag{7.25}$$

and from this it is readily seen that

$$\langle E\rangle_{NVT} = -\frac{\partial \ln Z}{\partial\beta}. \tag{7.26}$$

The specifc heat at constant volume C_V is defined as

$$C_V = \left(\frac{\partial E}{\partial T}\right)_{N,V} \tag{7.27}$$

and it can therefore be related to the root mean square (rms) fluctuation of the energy:

$$\begin{aligned} C_V &= \frac{1}{k_B T^2}\frac{\partial^2 \ln Z}{\partial\beta^2} \\ &= \frac{1}{k_B T^2}\left[\frac{\sum_X \mathrm{e}^{-\beta\mathcal{H}(X)}\mathcal{H}^2(X)}{\sum_X \mathrm{e}^{-\beta\mathcal{H}(X)}} - \left(\frac{\sum_X \mathrm{e}^{-\beta\mathcal{H}(X)}\mathcal{H}(X)}{\sum_X \mathrm{e}^{-\beta\mathcal{H}(X)}}\right)^2\right] \\ &= \frac{1}{k_B T^2}(\langle E^2\rangle_{NVT} - \langle E\rangle_{NVT}^2). \end{aligned} \tag{7.28}$$

Information about the microscopic properties of the system is given by correlation functions, which can sometimes be measured experimentally, for example through neutron scattering experiments [9]. In the next section we shall encounter several examples of correlation functions.

In later chapters, we shall describe the molecular dynamics and Monte Carlo simulation methods, which enable us to evaluate ensemble averages of different physical quantities expressed in terms of the system coordinates. Such ensemble averages are called *mechanical averages*. Free energies and chemical potentials are not directly given as mechanical averages but as phase space integrals. Integrals over phase space cannot be estimated directly in simulations, but fortunately differences between free energies at two different temperatures can be formulated as ensemble averages. Suppose, for example, that we know the free energy of system at a temperature T, and we would like to know it at a different temperature T'. The difference $\beta F(T) - \beta' F(T')$ is then found as

$$\begin{aligned} \exp[\beta F(\beta) - \beta' F(\beta')] &= \frac{Z(\beta')}{Z(\beta)} \\ &= \frac{\sum_X \exp[-\beta'\mathcal{H}(X)]}{\sum_X \exp[-\beta\mathcal{H}(X)]} = \langle \exp[(-\beta' + \beta)\mathcal{H}]\rangle_\beta \end{aligned} \tag{7.29}$$

where $\langle \cdots \rangle_\beta$ denotes a canonical ensemble average evaluated at inverse temperature β. Determination of this expectation value in a simulation suffers from bad statistics.

The reason is that in these simulations the system is pushed into a narrow region around a hypersurface in phase space where the configurational energy is equal to its average value, say $\bar{E}$, at temperature β. In Eq. (7.29), we want to probe the region where the configurational energy is equal to its average $\bar{E}'$ at temperature β' – hence this region will only be probed correctly if β and β' are fairly close, so that the hypersurface with configurational energy $\bar{E}'$ lies within the narrow region around the $\bar{E}$-hypersurface probed by the phase space integral. If this is not the case, simulations can be performed for a number of temperatures between T and T'; the resulting free energy differences are then added to find the desired free energy difference. Such is frequently done, although a slightly more subtle approach is used in practice [10].

Another approach is to integrate the free energy numerically from one value of the volume or temperature to another (*thermodynamic integration*). According to Eqs. (7.13) and (7.26), we have [10]

$$F(T, V_1) = F(T, V_0) - \int_{V_0}^{V_1} P(T, V) \, \mathrm{d}V \tag{7.30a}$$

$$\frac{F(T_1, V)}{T_1} = \frac{F(T_0, V)}{T_0} + \int_{T_0}^{T_1} \frac{E(T, V)}{T^2} \, \mathrm{d}T. \tag{7.30b}$$

This method can be used to calculate energy differences between systems at different temperatures or with different volumes. Integration over a particular path in phase space can be performed by carrying out simulations for a number of points on that path in order to determine $\langle P \rangle$ or $\langle E \rangle$ and then performing a numerical integration of (7.30). It is advisable to choose these points in accordance with the Gauss–Legendre integration scheme – see Appendix A6. At a phase transition (see Section 7.3), the free energy does not behave smoothly as a function of the system parameters. Either the path must circumvent the transition line, or two integrations must be performed, one for each phase, with starting points corresponding to appropriate reference systems for which the free energy is known, for example at zero or infinite temperature.

In Chapter 10 we shall consider additional methods for calculating free energies and chemical potentials. For a review of free energy calculation methods see Ref. [10].

7.2 Examples of statistical models; phase transitions

7.2.1 Molecular systems

A *model* is defined by its degrees of freedom and by the Hamiltonian which assigns an energy to every possible state of the system – that is, a specific set of values of the degrees of freedom. If we consider, for example, a system consisting of N identical point particles, the degrees of freedom are given by all positions $\mathbf{r}_i$ and all momenta $\mathbf{p}_i$, $i = 1, \dots, N$ of the particles. We shall denote the full sets of positions and momenta by R and P, respectively. The Hamiltonian $\mathcal{H}$ is given as

$$\mathcal{H}(R,P) = \sum_{i=1}^{N} \frac{p_i^2}{2m} + V_N(R). \tag{7.31}$$

$V_N(R)$ denotes the total potential energy of all the particles with positions given by the $3N$-coordinate R. In simulations one often uses an approximation in which $V_N(R)$ is written as a sum over pair potentials:

$$V_N(R) = \frac{1}{2} \sum_{\substack{i,j \\ i \neq j}}^{N} V_2(|\mathbf{r}_i - \mathbf{r}_j|), \tag{7.32}$$

where the sum is over all pairs i,j, except those with $i = j$. The factor $1/2$ compensates the double counting of pairs in the sum. Pair potentials are so popular because usually the evaluation of all forces or all potentials is the most time-consuming part of the program, and the time needed for this calculation increases rapidly with the number of particles involved in the interaction. For pair potentials, for example, there are $N(N-1)/2$ interactions, for three-particle interactions we would have $\mathcal{O}(N^3)$ contributions etc.

A Lennard–Jones parametrisation for the pair potential is often adopted:

$$V_{\mathrm{LJ}}(r) = 4\varepsilon \left[\left(\frac{\sigma}{r}\right)^{12} - \left(\frac{\sigma}{r}\right)^{6} \right]. \tag{7.33}$$

Such a potential has already been used in Chapter 2 for describing the interaction between a hydrogen and a krypton atom.[4] The $1/r^6$ tail is based on polarisation effects of the interacting atoms and the $1/r^{12}$ repulsive is chosen for numerical convenience. For argon, the Lennard–Jones description has been quite successful [11]; it has been applied to the solid, liquid and gas phases.

The canonical partition function Z is given as

$$Z(N,V,T) = \frac{1}{h^{3N} N!} \int_V \mathrm{d}^{3N}R \; \mathrm{d}^{3N}P \; \exp\left[-\beta \left(\sum_{i=1}^{N} \frac{p_i^2}{2m} + V_N(R) \right) \right]. \tag{7.34}$$

Irrespective of the form of V_N, we can perform the (Gaussian) integration over the momenta since they do not couple with the spatial coordinates, and we find

$$Z(N,V,T) = \frac{1}{N!} \left(\frac{2m\pi}{\beta h^2} \right)^{3N/2} \int_V \mathrm{d}^{3N}R \; \exp[-\beta V_N(R)]. \tag{7.35}$$

[4] Note that this form deviates from that given in Chapter 2. The present form is common in molecular dynamics.

For systems consisting of rigid polyatomic molecules, the interaction potential is usually taken to be the sum of atomic pair potentials, aside from rigidity constraints. A tantalising problem is the satisfactory description of water in simulations using *ab initio* interaction potentials [12].

Macroscopic quantities such as pressure, specific heat, etc, can be determined relatively easily from simulations and can be compared with experimental results. They give global information concerning the state of the system. The pressure can be found in a simulation using the virial theorem [13]:

$$\frac{\beta P}{n} = 1 - \frac{\beta}{3N}\left\langle \sum_{i=1}^{N} \mathbf{r}_i \nabla_i V_N(R) \right\rangle \tag{7.36}$$

where $\langle\cdots\rangle$ denotes the usual ensemble average, but in a dynamic system the time average can be used instead.

The specific heat at constant volume can easily be calculated in the canonical ensemble using Eq. (7.28), which relates this quantity to the fluctuation of the total energy. However, in the microcanonical ensemble, the total energy is fixed, so its fluctuation vanishes at all times. Fortunately, it can be calculated from the fluctuation of the kinetic energy from a formula derived by Lebowitz [14]:

$$\frac{\langle \delta K^2 \rangle}{\langle K \rangle^2} = \frac{2}{3N}\left(1 - \frac{3N}{2C_V}\right). \tag{7.37}$$

More detailed information can experimentally be obtained via X-ray and neutron scattering experiments. In particular, several correlation functions can be measured experimentally and they can also be determined in simulations. The static pair correlation function $g(\mathbf{r}, \mathbf{r}')$ is proportional to the probability of finding a particle at $\mathbf{r}$ and simultaneously one at $\mathbf{r}'$. In the canonical ensemble, it is given by the following expression:

$$g(\mathbf{r}, \mathbf{r}') = V^2 \frac{1}{N! h^{3N} Z} \int_V \mathrm{d}^3 r_3 \cdots \mathrm{d}^3 r_N \exp[-\beta V_N(\mathbf{r}, \mathbf{r}', \mathbf{r}_3, \ldots, \mathbf{r}_N]. \tag{7.38}$$

For a homogeneous system, this function depends on $\Delta\mathbf{r} = \mathbf{r} - \mathbf{r}'$ only and therefore for large N it can be written as

$$g(\Delta\mathbf{r}) = \frac{V}{N(N-1)}\left\langle \int \mathrm{d}^3 r' \sum_{\substack{i,j \\ i \neq j}}^{N} \delta(\mathbf{r}' - \mathbf{r}_i)\delta(\mathbf{r}' + \Delta\mathbf{r} - \mathbf{r}_j) \right\rangle. \tag{7.39}$$

For large $\Delta\mathbf{r}$, the correlation function tends to 1, and often the 'bare' correlation function $h(\Delta\mathbf{r})$, which is defined as $h(\Delta\mathbf{r}) = \mathrm{g}(\Delta\mathbf{r}) - 1$, is used instead.

The pair correlation function contains information concerning the local structure of the fluid. For an isotropic, homogeneous system, the pair correlation function depends only on the distance $\Delta r = |\mathbf{r} - \mathbf{r}'|$. Suppose we were to sit somewhere

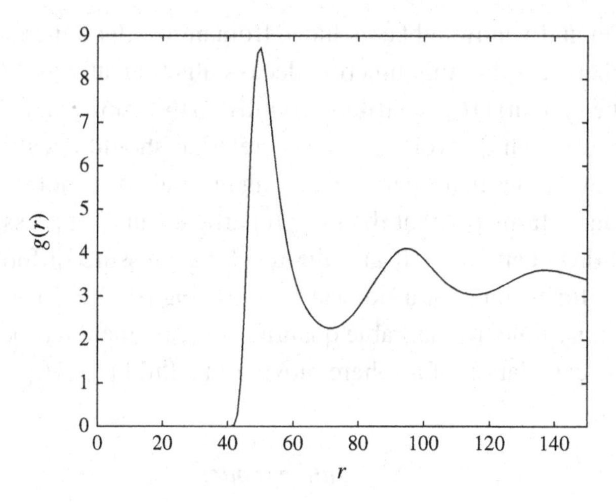

Figure 7.1. The pair correlation function of argon at its triple point.

in the fluid and watch the surroundings for some time, then, on average, we would see a homogeneous structure. If we were to move along with a particular particle, however, and watch the scenery from this particle, we would find no particles close to us because of the strong short-range repulsion. Then we have an increase in density due to a layer of particles surrounding our particle, followed by a drop in density marking the boundary between this layer and a second layer, and so on. Because of the fluctuations, the layer structure becomes more and more diffuse for increasing distances and the correlation function will approach a constant value at large distances. A typical example of a pair distribution function in a fluid is shown in Figure 7.1. For a discussion on the experimental determination of static and dynamic correlation functions, see Ref. [13].

Another important correlation function is the velocity autocorrelation function, which is a function of time. It is the expectation value of the dot product of the velocity of a particular particle ('tagged particle') at time 0 with the velocity of the same particle at time t:

$$c_{v_i}(t) = \langle \mathbf{v}_i(0) \cdot \mathbf{v}_i(t) \rangle \tag{7.40}$$

for an arbitrary particle i. For a homogeneous system this is independent of i. Since this correlation function is a dynamic quantity, it cannot be found as an ensemble average, as the latter is suitable for evaluation of averages of static quantities only. For identical particles, the velocity autocorrelation function is usually evaluated as a combined time average and an average over the N particles in equilibrium:

$$c_v(t) = \frac{1}{N} \lim_{T\to\infty} \sum_{i=1}^{N} \frac{1}{T} \int_0^T \mathrm{d}t' \mathbf{v}_i(t') \cdot \mathbf{v}_i(t'+t). \tag{7.41}$$

In 1970, Alder and Wainwright concluded from molecular dynamics simulations for the hard sphere gas that this function decays algebraically as $1/t^{D/2}$ (D is the dimension of the system) [16], in striking contrast to the 'molecular chaos' assumption according to which the velocity autocorrelation should decay exponentially. The long time tail implies that a particle moving in a fluid does not so easily 'forget' its initial motion. It turns out that the tagged particle causes a pressure rise ahead and a pressure drop behind itself and the resulting pressure difference produces vortices (in two dimensions) or a sideways vortex ring (if $D = 3$) and these persist for a relatively long time. Remarkable quantitative agreement has been found with a hydrodynamic calculation of a sphere moving in a fluid [15, 16].

7.2.2 Lattice models

Another model is a 'magnetic' one: the famous Ising model [17, 18]. The quotes are put around the qualification 'magnetic' to indicate that the model does not describe magnetic systems satisfactorily; it does however give a good description of atoms adsorbed on surfaces and of two-component alloys. Furthermore, the Ising model is an example of a lattice field theory (lattice field theories will be discussed in Chapter 15). Last but not least: the two-dimensional Ising model on a square lattice was the first model that was found to exhibit a genuine phase transition and was solved exactly [18, 19, 20].

The Ising model is defined on a lattice and we shall confine ourselves to the two-dimensional version on a square lattice of size $L \times L$ (in the thermodynamic limit L goes to infinity). The lattice sites are labelled by a single index i, and with $\langle i,j \rangle$ we denote a pair of neighbouring sites, where it is assumed that the spins on the top row of the lattice are connected to the corresponding ones on the bottom row and similarly for the left and right columns of sites (periodic boundary conditions; see Figure 7.2). On each site i, a 'spin' s_i is located. This can assume two different values, which we shall take to be $+1$ and -1. The spins are the degrees of freedom, and the Hamiltonian assigns an energy to each configuration $\{s_i\}$ of the spins according to

$$\mathcal{H}\{s_i\} = -J \sum_{\langle i,j \rangle} s_i s_j - H \sum_i s_i. \tag{7.42}$$

J is a coupling constant. It couples only nearest neighbour spins: the first sum is over nearest neighbour pairs on the lattice (taking periodic boundary conditions into account). For positive J, the coupling term favours like nearest neighbour pairs as this lowers the total energy: each spin wants to be surrounded by like spins on neighbouring sites – this case is called ferromagnetic. For negative J-values the model is called antiferromagnetic. The second term favours the spins to have a sign equal to that of the external magnetic field H. The partition function of the Ising

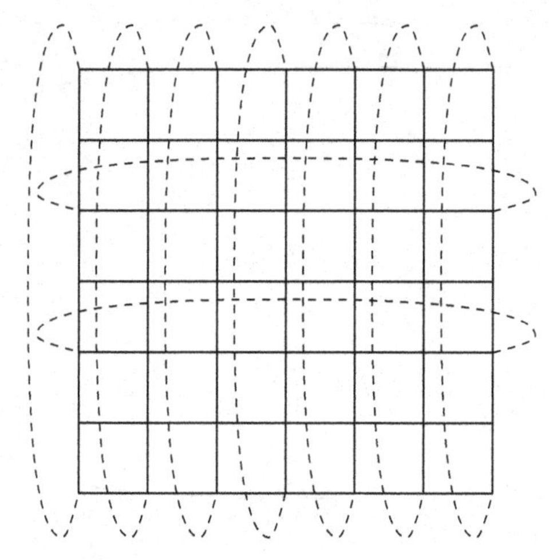

Figure 7.2. Periodic boundary conditions on the square lattice. All sites on the left column are coupled to their counterparts on the right column, but only two of these couplings are shown.

model is given by

$$Z = \sum_{\{s_i\}} \exp\left[\beta J \sum_{\langle i,j \rangle} s_i s_j + \beta H \sum_i s_i\right]. \tag{7.43}$$

Notice that the model is defined without any reference to dynamics. Dynamical Ising models have been formulated [21] and these reflect somehow the behaviour of real systems, but their form is not imposed by physical laws.

An interesting case is zero external magnetic field ($H = 0$), for which the model has been solved analytically. The Hamiltonian is then invariant with respect to global spin reversal. At absolute zero temperature, $\beta \to \infty$, either of two configurations, with all spins $+$ or all spins $-$, are allowed. Suppose we start off with all spins $+$. We are interested in the behaviour of the average value of the spins, which we shall call *magnetisation* and which is denoted m. Flipping a spin with four equal nearest neighbours induces a penalty via the Boltzmann factor being reduced by a factor $e^{-8\beta J}$ (remember the Boltzmann factor gives the weight, i.e. the probability of occurrence in a time sequence) and for low temperature, as β is still large, a particular spin turning over is therefore a very rare event. The relative occurrence of a configuration with an *arbitrary* single spin turned over with respect to one in which all spins are equal is given by $L^2 e^{-8\beta J}$. If we raise the temperature, the probability of having one or more spins turned over increases and therefore the magnetisation decreases (in absolute value). What will happen to the magnetisation when increasing the temperature further? Let us first consider $T \to \infty$, or $\beta = 0$. In that case all configurations have the same Boltzmann factor of 1 and the coupling

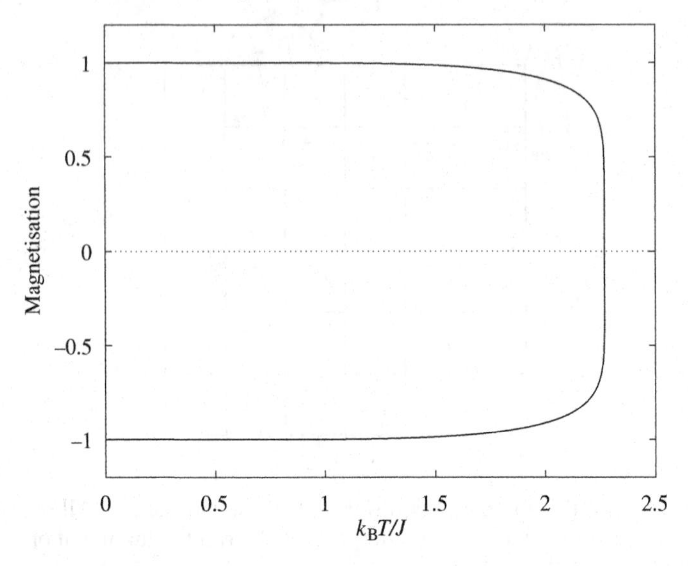

Figure 7.3. Phase diagram of the Ising model. There are two branches, one with negative and one with positive magnetisation, corresponding to the spin-reversal symmetry present in the model.

between the spins is no longer noticeable. Therefore, each spin will assume values $+1$ and -1 with equal probability and the average magnetisation will vanish. Two scenarios are possible for intermediate temperatures: either the magnetisation will decay asymptotically with increasing temperature, or it will vanish at some finite temperature. If the latter happens, we shall see a nonanalytic behaviour in the magnetisation curve, which seems highly improbable as the Hamiltonian depends analytically on all spins. Indeed for *finite* systems, all physical variables are analytic functions of the system parameters, but for $N \to \infty$, nonanalytic behaviour might show up. This is precisely what happens! The magnetisation for the infinite system vanishes at a finite temperature T_c given by $J/k_B T_c \approx 0.44$ and this phenomenon is called a *phase transition* [18, 19]. For reasons to be explained below, this phase transition is often called 'second order', 'critical' or 'continuous'. Figure 7.3 shows the (m, T) phase diagram for zero magnetic field. Two branches are shown, one for a system starting off with negative, and the other with positive magnetisation.

The behaviour of the Ising ferromagnet may be described in terms of the balance between entropy and energy. There is only one state with lowest energy (if we restrict ourselves to positive magnetisation at low temperatures, see below), L^2 states with one spin flipped, $L^2(L^2 - 1)/2$ states with two spins flipped and so on: the number of states increases rapidly with energy. It also increases rapidly with decreasing magnetisation for similar reasons. Therefore, there exist a huge number of disordered (zero magnetisation) states, having a relatively small Boltzmann factor, and a small number of ordered states, with a large Boltzmann factor. The Boltzmann effect is reduced by increasing the temperature. At the point where the numeric abundance (entropy effect) of the disordered states compensates for the Boltzmann effect, energy and entropy of the domain walls separating the spin-up from the spin-down phases are said to be in balance – this is the critical point, where the average magnetisation reaches zero.

This entropy–energy balance can be quantified using an argument given by Peierls [5]. A domain wall of length N, separating a $+$ from a $-$ region, represents an energy penalty of $2JN$, since each pair of opposite spins on both sides of the wall carries an energy J, as opposed to equal neighbouring spins representing an energy $-J$. We can estimate the number of possible domain wall configurations by realising that at each segment (a unit step of the interface) a domain wall has the option of turning left or right, or continuing straight on, leading to three possibilities. However, a domain wall cannot intersect itself, so at some segments only two of the three options are allowed. Therefore the number of domain wall configurations lies between 2^N and 3^N, and we have for the entropy S:

$$k_B T \ln 2^N < S < k_B T \ln 3^N. \tag{7.44}$$

The point where energy and entropy are in balance satisfies

$$k_B TN \ln 2 < 2NJ < k_B TN \ln 3, \tag{7.45}$$

which leads to $\ln 2 < 2J/(k_B T) < \ln 3$, or $0.3466 < J/(k_B T) < 0.549$, to be compared with the exact value $J/(k_B T) \approx 0.44$.

A remark is appropriate here. The picture sketched so far is a dynamic one: we start off with a particular state (all spins $+$) and consider what happens when the temperature is increased. According to the postulate of statistical mechanics, average values of physical quantities are given by ensemble averages, and we see immediately that the average magnetisation is always zero, as the Hamiltonian is symmetric with respect to flipping all spins. It is, however, believed that in any realistic system the spins turn over one after another, or perhaps in small groups at a time. Turning over the magnetisation requires a large number of spin flips and the occurrence of a domain wall between two regions of different spin with a length of the order of the linear system size. The probability of this happening is exceedingly small and the system will never enter the opposite magnetisation phase. This implies ergodicity violation since not all configurations are accessible to the system. A nice way to get round this violation is to switch on a small but positive magnetic field H which causes a difference between the energy of the positive and negative magnetisation phase by an amount $2HL^2$, and therefore the negative magnetisation phase no longer contributes to ensemble averages. After the calculation has been completed, the limit $H \to 0$ is taken. It is to be noted that for a finite external magnetic field the phase transition disappears.[5]

7.3 Phase transitions

7.3.1 First order and continuous phase transitions

As we have already seen in Section 7.2, phase transitions may occur in thermodynamic systems. These transitions can be of two different types, first order and second order. The latter are also called critical or continuous transitions. In this section we consider phase transitions in more detail, with emphasis on phenomena and techniques which are of interest in numerical simulations. In particular we discuss the finite-size scaling technique for studying second order transitions in simulations. The description here is short and simplified and for more detailed accounts the reader is referred to the books by Plischke and Bergersen [5], Reichl [3], Pathria [22], Le Bellac [8] and the various volumes in the Domb and Green/Lebowitz series [23].

The state of a system is usually characterised by a particular value of a physical quantity which is called the *order parameter*. This order parameter is used to distinguish between different phases. In the case of a gas–liquid transition at fixed pressure and temperature, it is the density which plays the role of the order parameter and the transition to the gas phase is indeed characterised by the density greatly decreasing. In magnetic systems, with the magnetic field and the temperature as system parameters, the order parameter is the magnetisation m which distinguishes the magnetic ($m \neq 0$) from the nonmagnetic ($m = 0$) phase and which, as we have seen above, is continuous at the zero-field Ising phase transition (the point where it vanishes) but has a discontinuous derivative. The order parameter is a derivative of the free energy (the density is expressed in terms of the volume, which is a derivative with respect to pressure, and magnetisation is a derivative with respect to magnetic field) and therefore a *jump* in the order parameter means a discontinuity in a first derivative of the free energy – hence the name 'first order' for this type of transition. If the order parameter is continuous at the phase transition, we speak of a continuous, critical or second order transition. In fact, the discontinuity shows up 'before the second derivative', as the free energy generally behaves as a broken power of one of the external parameters, $f \sim (K - K_c)^\alpha$, where K is the external parameter which assumes the value K_c at the critical point, and α lies between 1 and 2.

As we have seen in Section 7.1, any system in equilibrium is characterised by some free energy assuming its minimum for given values of the system parameters,

[5] Switching from a positive magnetic field to a negative one induces a change in sign of the magnetisation m if $T < T_C$. This is a first order phase transition, induced by the magnetic field instead of the temperature.

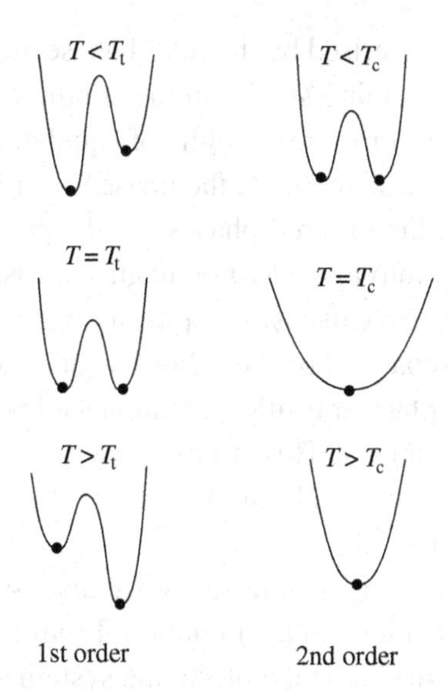

Figure 7.4. Typical behaviour of the free energy as a function of the order parameter and temperature. The left hand side corresponds to the first order case, with transitions temperature T_{f}, and the right hand side to the continuous case, with critical transition temperature T_{c}.

and for this minimum the order parameter assumes a particular value. It is possible to define a free energy for any fixed value of the order parameter by calculating the partition function for exclusively those configurations that have the prescribed value of the order parameter. As an example, we can define the free energy, $F(m)$, for the Ising model with fixed magnetic field in terms of a partition function, $Z(m)$, defined as

$$Z(m) = \sum_{\{s_i\}} \mathrm{e}^{-\beta\mathcal{H}} \delta\Big(\sum_i s_i - L^d m\Big) \tag{7.46a}$$

$$F(m) = -k_{\mathrm{B}} T \ln Z(m), \tag{7.46b}$$

where d is the dimension of the system. Note the delta-function in the definition of $Z(m)$ restricting the sum to configurations with a fixed magnetisation m. It is instructive to consider how this free energy as a function of the order parameter changes with an external parameter (the temperature for example) across the transition for the two different types of phase transitions. Typical examples are shown in Figure 7.4.

The equilibrium situation is characterised by the minimum of the free energy. If we imagine the leftmost minimum in the first order case to correspond to the liquid phase and the right hand one to the gas phase, we see that, away from the transition temperature, one of the two phases is stable and the other one metastable.

The phase transition is characterised by the liquid phase going from stable to metastable and the gas phase becoming stable. In the continuous case (right hand side of Figure 7.4), there are two (or more) minima of equal depth, corresponding to as many ordered phases, and these merge at the phase transition into one, disordered phase; in the Ising model, the ordered phases are the positive and negative magnetisation phases, merging into a single, nonmagnetic, disordered phase. Close to the phase transition the system can easily hop from one (weakly) ordered phase to another, as the phases are separated by weak barriers and therefore fluctuations will increase considerably: the phase transition is announced before it actually happens by an increase in the fluctuations. This is unlike the first order case, in which the order parameter jumps from one well into the other without this being announced by an increase in the fluctuations.

Before focusing on second order transitions, we discuss some problems related to detecting first order transitions in a simulation. From Figure 7.4 it is seen that, in order for the actual transition to take place, the system should overcome a free-energy barrier, and obviously the higher the barrier the longer the time needed for this to happen. In the short time over which a typical system can be simulated, it will not be able to overcome the barrier at or near the first order transition and we shall observe a strong hysteresis: if, in the case of a liquid–gas transition, the system is cooled down from the gas phase, it will remain in that phase well below the transition temperature before it will actually decide to condense into the liquid phase. On the other hand, if a fluid is heated, it will remain in the fluid state above the transition temperature for quite some time before it enters the gas phase. In order to determine the transition temperature it is necessary to obtain the free energy for both phases so that the transition can be determined as the point where they become equal. However, as mentioned already in Section 7.1, the free energy cannot be extracted straightforwardly from molecular dynamics or Monte Carlo simulations, and the special techniques mentioned there and those to be discussed in Chapter 10 must be applied. In transfer matrix calculations (see Chapter 11), the free energy is directly obtainable but this method is restricted to lattice spin models. Panagiotopoulos [24, 25] has developed a method in which two phases of a molecular system can coexist by adjusting their chemical potentials by the exchange of particles – see Section 10.4.3.

*7.3.2 Critical phase transitions and finite-size scaling

Critical phase transitions are characterised by the disappearance of order caused by different ordered phases merging into one disordered phase at the transition. In contrast to first order transitions, critical phase transitions are ‘announced’ by an important increase in the fluctuations. The Ising model on a square lattice described above is an ideal model for visualising what is going on close to a second order phase transition.

An interesting object in connection with phase transitions is the pair correlation function. As the Ising model in itself is not dynamic, only the static correlation function is relevant. It is given by

$$\tilde{g}(m,n) = \langle s_m s_n \rangle = \frac{1}{Z} \sum_{\{s_i\}} s_m s_n \exp\Big[\beta J \sum_{\langle ij \rangle} s_i s_j + \beta H \sum_i s_i\Big]. \tag{7.47}$$

Instead of the pair correlation function defined in (7.47), the 'bare' correlation function is usually considered:

$$g(i,j) = \tilde{g}(i,j) - \langle s_i \rangle^2 \tag{7.48}$$

which decays to zero if i and j are far apart. The physical meaning of the bare pair correlation function is similar to that defined above for molecular systems. Suppose we sit on a site i, then $g(i,j)$ gives us the probability of finding the same spin value on site j in excess of the average spin on the lattice. The correlation function defined here obviously depends on the relative orientation of i and j because the lattice is anisotropic. However, for large distances this dependence is weak and the pair correlation function will depend only on the distance r_{ij} between i and j. The decay of the bare correlation function below the transition temperature is given by

$$g(r) \propto \mathrm{e}^{-r/\xi}, \quad \text{large } r. \tag{7.49}$$

ξ is called the *correlation length*: it sets the scale over which each spin has a significant probability of finding like spins in excess of the average probability. One can alternatively interpret ξ as a measure of the average linear size of the domains containing minority spins. If we approach the transition temperature, more and more spins turn over. Below the transition temperature, the system consists of a connected domain (the 'sea') of majority spins containing 'islands' of minority spin. When approaching the transition temperature, the islands increase in size and at T_c they must grow into a connected land cluster which extends through the whole system in order to equal the surface of the sea, which also extends through the whole system. For higher temperature the system is like a patchwork of unconnected domains of finite size. The picture described here implies that at the transition the correlation length will become of the order of the system size. Indeed, it turns out that at the critical phase transition the correlation length diverges and the physical picture [26] is that of huge droplets of one spin containing smaller droplets of the other spin containing still smaller droplets of the first spin and so on. This suggests that the system is self-similar for a large range of different length scales: if we zoomed in on part of a large Ising lattice at the phase transition, we would notice that the resulting picture is essentially indistinguishable from the one presented by the lattice as a whole: the differences only show up at the smallest scales, i.e. comparable to the lattice constant. This scale invariance is exploited in renormalisation theory [27, 28] which has led to a qualitative and quantitative understanding of critical phase transitions.[6]

One of the consequences of the scale invariance at the critical phase transition is that the form of the correlation function should be scale invariant, that is, it should be essentially invariant under a scale transformation with scaling factor b, and it follows from renormalisation theory that at the transition, g transforms under a rescaling as

$$g(r) = b^{2(d-y)} g(rb) \tag{7.50}$$

(d is the system dimension). From this, the form of g is found as

$$g(r) = \frac{\text{Constant}}{r^{2(d-y)}}. \tag{7.51}$$

The exponent y is called the *critical exponent*. It turns out that this exponent is *universal*: if we change details in the Hamiltonian, for instance by adding next nearest neighbour interactions to it, the temperature at which the transition takes place will change, but the critical exponent y will remain exactly the same. Systems which are related through such 'irrelevant' changes in the Hamiltonian are said to belong to the same *universality class*. If the changes to the Hamiltonian are too drastic, however, like changing the number of possible states of a spin (for example 3 or 4 instead of 2 in the Ising model), or if we add strong next-nearest neighbour interactions with a sign opposite to the nearest neighbour ones, the critical behaviour will change: we cross over to a different universality class.

It should be noted that the spin pair-correlation function is not the only correlation function of interest. Other correlation functions can be defined, which we shall not go into, but it is important that these give rise to new exponents. Different correlation functions may have the same exponent, or their exponents may be linearly dependent. The set of independent exponents defines the universality class. In the case of the Ising model this set contains two exponents, the 'magnetic' one, y_H, which we have encountered above, and the 'thermal' exponent y_T (which is related to a different correlation function).

The critical exponents not only show up in correlation functions, they also describe the behaviour of thermodynamic quantities close to the transition. For example, in magnetic systems, the magnetic susceptibility χ_m, defined as

$$\chi_m = \left(\frac{\partial m}{\partial H}\right)_T, \tag{7.52}$$

[6] More recently, the more extended conformal symmetry has been exploited in a similar way to the scale invariance alone. Conformal field theory has turned out a very powerful tool to study phase transitions in two-dimensional systems [29–31].

exhibits a singularity near the phase transition:

$$\chi_m(T) \propto |T - T_c|^{-\gamma} \tag{7.53}$$

where γ is also called the 'critical exponent'; its value is related to the y-exponents by $\gamma = (-d + 2y_H)/y_T$. For the specific heat c_H, the correlation length ξ and the magnetisation m we have similar critical exponents:

$$\begin{aligned} c_H(T) &\propto |T - T_c|^{-\alpha} \\ \xi(T) &\propto |T - T_c|^{-\nu} \\ m(T) &\propto (-T + T_c)^{\beta}; \quad T < T_c \end{aligned} \tag{7.54a}$$

and, moreover, we have an exponent for the behaviour of the magnetisation with varying small magnetic field at the transition temperature:

$$m(H, T_c) = H^{1/\delta}. \tag{7.55}$$

For the case of the two-dimensional Ising model on a square lattice, we know the values of the exponents from the exact solution:

$$\begin{aligned} &\alpha = 0, \quad \beta = 1/8, \quad \gamma = 7/4, \\ &\delta = 15, \quad \nu = 1. \end{aligned} \tag{7.56}$$

The value 0 of the exponent α denotes a logarithmic divergence:

$$c_H \propto \ln |T - T_c|. \tag{7.57}$$

The fact that there are only two y-exponents and the fact that the five exponents expressing the divergence of the thermodynamic quantities are expressed in terms of these indicates that there must exist relations between the exponents α, β etc. These relations are called *scaling laws* – examples are:

$$\alpha + 2\beta + \gamma = 2 \quad \text{and} \tag{7.58a}$$

$$2 - \alpha = d\nu, \tag{7.58b}$$

with d the dimension of the system. The Ising exponents listed above do indeed satisfy these scaling laws.

In dynamical versions of the Ising model, the relaxation time also diverges with a critical exponent. The correlation time is the time scale over which a physical quantity A relaxes towards its equilibrium value $\overline{A}$ – it is defined by[7]

$$\tau = \frac{\int_0^\infty t[A(t) - \overline{A}]\mathrm{d}t}{\int_0^\infty [A(t) - \overline{A}]\mathrm{d}t}. \tag{7.59}$$

[7] In Section 7.4 we shall give another definition of the correlation time which describes the decay of the time correlation function rather than that of the quantity A itself.

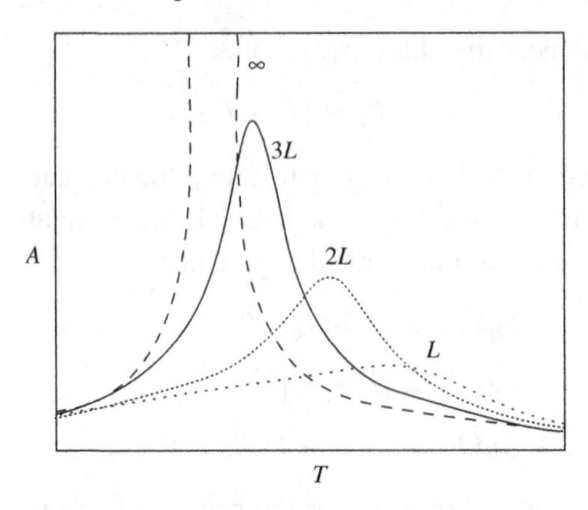

Figure 7.5. Typical behaviour of a physical quantity A vs temperature close to the critical point for various system sizes.

At the critical point the correlation time diverges according to

$$\tau = \xi^z. \tag{7.60}$$

This divergence implies that close to the critical point the simulation time needed to obtain reliable estimates for physical quantities increases dramatically. This phenomenon is called *critical slowing down.* For most models with a Hamiltonian containing only short-range couplings, the value of the exponent z is close to 2. For the Ising model in two dimensions, the dynamic critical exponent has been determined numerically – its value is $z \approx 2.125$ [32].

For systems far from the critical point, the correlation length is small, and it is easy to simulate systems that are considerably larger than the correlation length. The values of physical quantities measured will then converge rapidly to those of the infinite system. Close to the critical point, however, the correlation length of the infinite system might exceed the size of the simulated system; hence the system size will set the scale over which correlations can extend. This part of the phase diagram is called the *finite-size scaling region.* It turns out that it is possible to extract information concerning the critical exponents from the behaviour of physical quantities with varying system size close to the critical point. Of course, for a finite system, the partition function and hence the thermodynamic quantities are smooth functions of the system parameters, so the divergences of the critical point are absent. However, we can still see a signature of these divergences in the occurrence of peaks, which in the scaling region ($\xi \gg L$) become higher and narrower with increasing system size. Also, the location of the peak may be shifted with respect to the location of the critical point. The general behaviour is shown in Figure 7.5. These characteristics of the peak shape as a function of temperature are described in terms of additional exponents, the so-called *finite-size scaling exponents*:

- The shift in the position of the maximum with respect to the critical temperature is described by

$$T_c(L) - T_c(\infty) \propto L^{-\lambda}. \tag{7.61}$$

- The width of the peak scales as

$$\Delta T(L) \propto L^{-\Theta}. \tag{7.62}$$

- The peak height grows with the system size as

$$A_{\max}(L) \propto L^{\sigma_m}. \tag{7.63}$$

The behaviour of a system is determined by two length scales: L/a and ξ/a, with ξ the correlation length of the infinite system, which in the finite-size scaling region is larger than the linear system size L. As in the critical region, the fluctuations determining the behaviour of the system extend over large length scales; physical properties should be independent of a. This leaves L/ξ as the only possible parameter in the system and this leads to the so-called finite-size scaling *Ansatz*. Defining

$$\varepsilon \equiv \frac{T - T_c}{T_c}, \tag{7.64}$$

we can formulate the finite-size scaling *Ansatz* as follows:

$$\frac{A_L(\varepsilon)}{A_\infty(\varepsilon)} = f\left[\frac{L}{\xi_\infty(\varepsilon)}\right]. \tag{7.65}$$

Suppose the exponent of the critical divergence of the quantity A is σ:

$$A_\infty \propto \varepsilon^{-\sigma}. \tag{7.66}$$

Using, moreover, the scaling form of the correlation length $\xi \propto \varepsilon^{-\nu}$, we can write the scaling *Ansatz* as

$$A_L(\varepsilon) = \varepsilon^{-\sigma} f(L\varepsilon^\nu) \tag{7.67}$$

which can be reformulated as

$$A_L(\varepsilon) = L^{\sigma/\nu}\phi(L^{1/\nu}\varepsilon) \tag{7.68}$$

where we have replaced the scaling function, f, by another one, ϕ, by extracting a factor $(L\varepsilon^\nu)^{\sigma/\nu}$ from f and then writing the remaining function in terms of $(L\varepsilon^\nu)^{1/\nu}$ rather than $(L\varepsilon^\nu)$. Obviously, $\phi(x)$ will have a maximum $\phi_{\max}$ for some value $x = x_{\max}$ with a peak width Δx. From Eq. (7.68) we then see immediately that:

- The peak height scales as $L^{\sigma/\nu}$, hence $\sigma_m = \sigma/\nu$.
- The peak position scales as $L^{-1/\nu}$, hence $\lambda = 1/\nu$.
- The peak width also scales as $L^{-1/\nu}$, hence $\Theta = 1/\nu$.

These are the finite-size scaling laws for any thermodynamic quantity which diverges at the critical point as a power law. We see that if we monitor the peak height, position and width as a function of system size, we can extract the correlation length exponent ν and the exponent σ associated with A from the resulting data.

In reality this approach poses difficulties as the fluctuations increase near the critical point and hence the time needed to obtain reliable values for the physical quantities measured also increases. This increase is stronger when the system size increases – hence calculations for larger systems require more time, not only because more computational effort is used per time step for a larger system, but also because we need to generate more and more configurations in order to obtain reliable results. An extra complication is that the fluctuations are not only huge, but they correlate over increasing time scales, and the simulation time must be at least a few times the relaxation time in order to obtain reliable estimates for the physical quantities. In Chapter 15 we shall discuss various methods for reducing the dynamic exponent z in Monte Carlo type simulations.

We have presented only the most elementary results of the finite-size scaling analysis and the interested reader is invited to consult more specialised literature. There exists a nice collection of key papers on the field [33] and a recent volume on finite-size scaling [34].

7.4 Determination of averages in simulations

In Chapters 8 and 10 we shall encounter two simulation methods for classical many-particle systems: the molecular dynamics (MD) method and the Monte Carlo (MC) method. During a simulation of a many-particle system using either of these methods, we can monitor various physical quantities and determine their expectation values as averages over the configurations generated in the simulation. We denote such averages as 'time averages' although the word time does not necessarily denote physical time. For a physical quantity A, the time average is

$$\overline{A} = \frac{1}{M}\sum_{n=1}^{M} A_n. \tag{7.69}$$

If the system size and the simulation time are large enough, these averages will be very close to the averages in a macroscopic experimental system. Usually, the system sizes and simulation times that can be achieved are limited and it is important to find an estimate of the error bounds associated with the measured average. These are related to the standard deviation σ of the physical quantity A:

$$\sigma^2 = \langle A^2 \rangle - \langle A \rangle^2. \tag{7.70}$$

The ensemble average $\langle \cdots \rangle$ is an average over many independent simulations.

We can estimate the standard deviation as a time average:

$$\sigma^2 = \overline{A^2} - \overline{A}^2. \tag{7.71}$$

For a long enough simulation this reduces to the ensemble average, and the expectation value of this estimate becomes independent of the simulation time. Equation (7.71) estimates the standard deviation irrespective of time correlations between subsequent samples generated by the simulation. However, the *standard deviation of the mean value* of A calculated over M samples generated by the simulation, i.e. the statistical error, depends on the number of *independent* samples generated in the simulation, and this is the total number of samples divided by the correlation 'time' τ, measured in simulation steps.

In order to study the standard deviation of the mean (the statistical error), we first analyse the time correlations. These manifest themselves in the time correlation function:

$$c_{AA}(k) = \big\langle (A_n - \langle A_n \rangle)(A_{n+k} - \langle A_{n+k} \rangle) \big\rangle = \langle A_n A_{n+k} \rangle - \langle A_n \rangle^2. \tag{7.72}$$

Note that the right hand side of this expression does not depend on n because of time translation symmetry. For $k = 0$ this function is equal to σ^2, and time correlations manifest themselves in this function assuming nonzero values for $k \neq 0$. The time correlation function can be used to determine the *integrated correlation time* τ, defined as

$$\tau = \frac{1}{2} \sum_{n=-\infty}^{\infty} \frac{c_{AA}(n)}{c_{AA}(0)} \tag{7.73}$$

where the factor $1/2$ in front of the sum is chosen to guarantee that for a correlation function of the form $\exp(-|t|/\tau)$ with $\tau \gg 1$, the correlation time is equal to τ. Note that this definition of the time correlation is different from that given in Eq. (7.59). The current one is more useful as it can be determined throughout the simulation, and not only at the beginning when the quantity A decays to its equilibrium value. A third definition is the *exponential correlation time* τ_{exp}:

$$\tau_{\text{exp}} = -t/\ln\left|\frac{c_{AA}(t)}{c_{AA}(0)}\right|, \quad \text{large } t. \tag{7.74}$$

This quantity is the slowest decay time with which the system relaxes towards equilibrium (such as happens at the start of a simulation when the system is not yet in equilibrium), and it is in general not equal to the integrated correlation time.

Now let us return to the standard deviation of the mean value of A as determined in a simulation generating M configurations (with time correlations). It is easy to see that the standard deviation in the mean, ε, is given by

$$\varepsilon^2 = \left\langle \frac{1}{M^2} \sum_{n,m=1}^{M} A_n A_m \right\rangle - \left(\left\langle \frac{1}{M} \sum_{n=1}^{M} A_n \right\rangle \right)^2 = \frac{1}{M^2} \sum_{n,m=1}^{M} c_{AA}(n-m). \tag{7.75}$$

If we define $l = n - m$, then this can be rewritten as

$$\varepsilon^2 = \frac{1}{M^2} \sum_{n=1}^{M} \sum_{l=n-1}^{n-M} c_{AA}(l). \tag{7.76}$$

The lowest and highest values taken on by l are $-(M-1)$ and $M-1$ respectively, and some fixed value of l between these two boundaries occurs $M - |l|$ times. This leads to the expression

$$\varepsilon^2 = \frac{1}{M} \sum_{l=-(M-1)}^{M-1} \left(1 - \frac{|l|}{M}\right) c_{AA}(l) \xrightarrow{\text{large } M} 2\frac{\tau}{M} c_{AA}(0) = 2\frac{\tau}{M}\sigma^2. \tag{7.77}$$

We see that time correlations cause the error ε to be multiplied by a factor of $\sqrt{2\tau}$ with respect to the uncorrelated case. The obvious procedure for determining the statistical error is first to estimate the standard deviation and the correlation time, using (7.71) and (7.73) respectively, and then calculate the error using (7.77).

In practice, however, a simpler method is preferred. The values of the physical quantities are recorded in a file. Then the data sequence is chopped into a number of blocks of equal size which is larger than the correlation time. We calculate the averages of A within each block. For blocks of size m, the jth block average is then given as

$$\overline{A}_j = \frac{1}{m} \sum_{k=jm+1}^{m(j+1)} A_k. \tag{7.78}$$

The averages of the physical quantities in different blocks are uncorrelated and the error can be determined as the standard deviation of the uncorrelated block averages. This method should yield errors which are independent of the block size provided the latter is larger than the correlation time and sufficiently small to have enough blocks to calculate the standard deviation reliably. This method is called *data-blocking*.

Exercises

7.1 In this problem we analyse the relation between the differential scattering cross section for elastic X-ray scattering by a collection of particles and the structure factor in more detail. Consider an incoming X-ray with wave vector $\mathbf{k}_0$, which is scattered into $\mathbf{k}_1$ by particle number j at $\mathbf{r}_j$ at time t'. When the wave 'hits' particle j at time t', its phase factor is given by

$$e^{i\mathbf{k}_0\mathbf{r}_j - i\omega t'}.$$

(a) Give the phase of the scattered wave when it arrives at the detector located at $\mathbf{r}$ at time t.

(b) We assume that the incoming rays have intensity I_0. Show that the average total intensity of waves with wave vector $\mathbf{k}_1$ arriving at the detector is given by

$$I(\mathbf{k}_1, \mathbf{r}) = I_0 \left\langle \sum_{l,j=1}^{N} e^{i\Delta\mathbf{k}(\mathbf{r}_l - \mathbf{r}_j)} \right\rangle$$

with $\Delta\mathbf{k} = \mathbf{k}_1 - \mathbf{k}_0$.

(c) Show that this expression is equal to $I_0 N S(\Delta\mathbf{k})$, where S is the static structure factor, defined in terms of the correlation function g as

$$S(\mathbf{k}) = 1 + n \int d^3r\, g(\mathbf{r}) e^{i\mathbf{k}\mathbf{r}}.$$

(n is the particle density N/V.)

7.2 The magnetic susceptibility of the Ising model on an $L \times L$ square lattice is defined by $\chi = \partial m/\partial H$, where m is the magnetisation and h the magnetic field.

(a) Show that the magnetic susceptibility can be written as

$$\chi = \frac{1}{L^2 k_B T} \sum_{i,j} (\langle s_i s_j \rangle - \langle s_i \rangle^2).$$

(b) A scaling exponent η associated with the magnetic correlation function (see Eq. (7.48)) is defined by

$$g(r) \propto r^{2-d-\eta}.$$

Assuming that close to the critical point this form extends to a distance ξ, where ξ is the correlation length, find the following scaling relation between γ, η and ν:

$$\gamma = \nu(2 - \eta).$$

References

[1] T. L. Hill, *Statistical Mechanics*. New York, McGraw-Hill, 1956.

[2] K. Huang, *Statistical Mechanics*, 2nd edn. New York, John Wiley, 1987.

[3] L. E. Reichl, *Equilibrium Statistical Mechanics*. Englewood Cliffs, NJ, Prentice-Hall, 1989.

[4] F. Reiff, *Fundamentals of Statistical and Thermal Physics*. Kogakusha, McGraw-Hill, 1965.

[5] M. Plischke and H. Bergersen, *Equilibrium Statistical Physics*. Englewood Cliffs, NJ, Prentice-Hall, 1989.

[6] J. Yeomans, *Equilibrium Statistical Mechanics*. Oxford, Oxford University Press, 1989.

[7] D. Chandler, *Introduction to Modern Statistical Mechanics*. New York, Oxford University Press, 1987.

[8] M. Le Bellac, F. Mortessagne, and G. G. Batrouni, *Equilibrium and Non-equilibrium Statistical Thermodynamics*. Cambridge, Cambridge University Press, 2004.

[9] S. W. Lovesey, *Theory of Neutron Scattering from Condensed Matter*, vols 1 and 2. Oxford, Clarendon Press, 1984.

[10] D. Frenkel, 'Free energy computation and first-order phase transitions,' in *Molecular Dynamics Simulation of Statistical Mechanical Systems* (G. Ciccotti and W. G. Hoover, eds.), *Proceedings of the International School of Physics "Enrico Fermi", Varenna 1985*, vol. 97. Amsterdam, North-Holland, 1986, pp. 151–188.

[11] A. Rahman, 'Correlations in the motion of atoms in liquid argon,' *Phys. Rev.*, **136A** (1964) 405–11.

[12] A. Rahman and F. Stillinger, 'Molecular dynamics study of liquid water,' *J. Chem. Phys.*, **55** (1971) 3336–59.

[13] J. P. Hansen and I. R. McDonald, *Theory of Simple Liquids*, 2nd edn. New York, Academic Press, 1986.

[14] J. L. Lebowitz, J. K. Percus, and L. Verlet, 'Ensemble dependence of fluctuations with application to machine calculations,' *Phys. Rev.*, **253** (1967), 250–4.

[15] B. J. Alder and T. E. Wainwright, 'Enhancement of diffusion by vortex-like motion of classical hard particles,' *J. Phys. Soc. Japan Suppl.*, **26** (1969), 267–9.

[16] B. J. Alder and T. E. Wainwright, 'Decay of the velocity autocorrelation function,' *Phys. Rev. A*, **1** (1970), 18–21.

[17] M. E. Fisher, 'The theory of equilibrium critical phenomena,' *Rep. Prog. Phys.*, **30** (1967), 615–730.

[18] L. Onsager, 'Crystal statistics. I. A two-dimensional model with an order–disorder transition,' *Phys. Rev.*, **65** (1944), 117–49.

[19] T. D. Schultz, D. C. Mattis, and E. H. Lieb, 'Two-dimensional Ising model as a soluble problem of many fermions,' *Rev. Mod. Phys.*, **36** (1964), 856–71.

[20] R. J. Baxter, *Exactly Solved Models in Statistical Mechanics*. London, Academic Press, 1982.

[21] R. J. Glauber, 'Time-dependent statistics of the Ising model,' *J. Math. Phys*, **4** (1963), 294–307.

[22] R. Pathria, *Statistical Mechanics*, 2nd edn. Oxford, Butterworth–Heinemann, 1996.

[23] C. Domb and M. S. Green (vols. 1–7)/C. Domb and J. L. Lebowitz (vols. 7–19), *Phase Transitions and Critical Phenomena*. New York, Academic Press, 1972–2000.

[24] A. Z. Panagiotopoulos, 'Direct determination of phase coexistence properties of fluids by Monte Carlo simulation in a new ensemble,' *Mol. Phys.*, **61** (1987), 813–26.

[25] A. Z. Panagiotopoulos, N. Quirke, and D. J. Tildesley, 'Phase-equilibria by simulation in the Gibbs ensemble – alternative derivation, generalization and application to mixture and membrane equilibria,' *Mol. Phys.*, **63** (1988), 527–45.

[26] L. Kadanoff, 'Scaling laws for Ising models near T_c,' *Physics*, **2** (1966), 263–72.

[27] C. Domb and M. S. Green, eds., *Phase Transitions and Critical Phenomena*, vol. 6. New York, Academic Press, 1976.

[28] S.-K. Ma, *Modern Theory of Critical Phenomena*. New York, Benjamin, 1976.

[29] A. A. Belavin, A. M. Polyakov, and A. B. Zamolodchikov, 'Infinite conformal symmetry in two dimensional quantum field theory,' *Nucl. Phys. B*, **241** (1984), 333–80.

[30] A. A. Belavin, A. M. Polyakov, and A. B. Zamolodchikov, 'Infinite conformal symmetry of critical fluctuations in two dimensions,' *J. Stat. Phys.*, **34** (1984), 763–74.

[31] J. Cardy, 'Conformal Invariance,' *Phase Transitions and Critical Phenomena*, (C. Domb and J. L. Lebauitz, eds). London, Academic Press, 1987, ch. 2, vol. 11.

[32] S. Tang and D. P. Landau, 'Monte Carlo study of dynamic universality in two-dimensional Potts models,' *Phys. Rev. B*, **36** (1985), 567–73.

[33] J. Cardy, ed., *Finite Size Scaling*. Amsterdam, North-Holland, 1988.

[34] V. Privman, ed., *Finite Size Scaling and Numerical Simulation of Statistical Systems*. Singapore, World Scientific, 1988.

8

Molecular dynamics simulations

8.1 Introduction

In the previous chapter we saw that the experimental values of physical quantities of a many-particle system can be found as an ensemble average. Experimental systems are so large that it is impossible to determine this ensemble average by summing over all the accessible states in a computer. There exist essentially two methods for determining these physical quantities as statistical averages over a restricted set of states: the molecular dynamics and Monte Carlo methods. Imagine that we have a random sample of, say, 10^7 configurations of the system which are all compatible with the values of the system parameters. For such a large number we expect averages of physical quantities over the sample to be rather close to the ensemble average. It is unfortunately impossible to generate such a random sample; however, we can generate a sample consisting of a large number of configurations which are determined successively from each other and are hence correlated. This is done in the molecular dynamics and Monte Carlo methods. The latter will be described in Chapter 10.

Molecular dynamics is a widely used method for studying classical many-particle systems. It consists essentially of integrating the equations of motion of the system numerically. It can therefore be viewed as a simulation of the system as it develops over a period of time. The system moves in phase space along its physical trajectory as determined by the equations of motion, whereas in the Monte Carlo method it follows a (directed) random walk. The great advantage of the MD method is that it not only provides a way to evaluate expectation values of static physical quantities; dynamical phenomena, such as transport of heat or charge, or relaxation of systems far from equilibrium can also be studied.

In this section we discuss the general principles of the molecular dynamics method. In the following sections more details will be given and special techniques will be discussed. There exists a vast research literature on this subject and there are some review papers and books [1–5].

Consider a collection of N classical particles in a rectangular volume $L_1 \times L_2 \times L_3$. The particles interact with each other, and for simplicity we shall assume that the interaction force can be written as a sum over pair forces, $\mathbf{F}(r)$, whose magnitude depends only on the distance, r, between the particle pairs and which is directed between them (see also the previous chapter). In that case the internal force (i.e. the force due to interactions between the particles) acting on particle number i is given as

$$\mathbf{F}_i(R) = \sum_{\substack{j=1,N;\\ j\neq i}} F(|\mathbf{r}_i - \mathbf{r}_j|)\hat{\mathbf{r}}_{ij}. \tag{8.1}$$

R denotes the position coordinates $\mathbf{r}_i$ of all particles in the notation introduced in Section 7.2.1 (P denotes the momenta); $\hat{\mathbf{r}}_{ij}$ is a unit vector directed along $\mathbf{r}_j - \mathbf{r}_i$, pointing from particle i to particle j. In experimental situations there will be external forces in addition to the internal ones – examples are gravitational forces and forces due to the presence of boundaries. Neglecting these forces for the moment, we can use (8.1) in the equations of motion:

$$\frac{\mathrm{d}^2\mathbf{r}_i(t)}{\mathrm{d}t^2} = \frac{\mathbf{F}_i(R)}{m_i} \tag{8.2}$$

in which m_i is the mass of particle i. In this chapter we take the particles identical unless stated otherwise. Molecular dynamics is the simulation technique in which the equations (8.2) are solved numerically for a large collection of particles.

The solutions of the equations of motion describe the time evolution of a real system although obviously the molecular dynamics approach is approximate for the following reasons.

- First of all, instead of a quantum mechanical treatment we restrict ourselves to a classical description for the sake of simplicity. In Chapter 9, we shall describe a method in which ideas of the density functional description for quantum many-particle systems (Chapter 5) are combined with the classical molecular dynamics approach. The importance of the quantum effects depends strongly on the particular type of system considered and on the physical parameters (temperature, density ...).
- The forces between the particles are not known exactly: quantum mechanical calculations from which they can be determined are subject to systematic errors as a result of the neglect of correlation effects, as we have seen in previous chapters. Usually these forces are given in a parametrised form, and the parameters are determined either by *ab initio* calculations or by fitting the results of simulations to experimental data. There exist systems for which the forces are known to high precision, such as systems consisting of stars and galaxies at large mutual distances and at nonrelativistic velocities where the interaction is largely dominated by Newton's gravitational $1/r^2$ force.

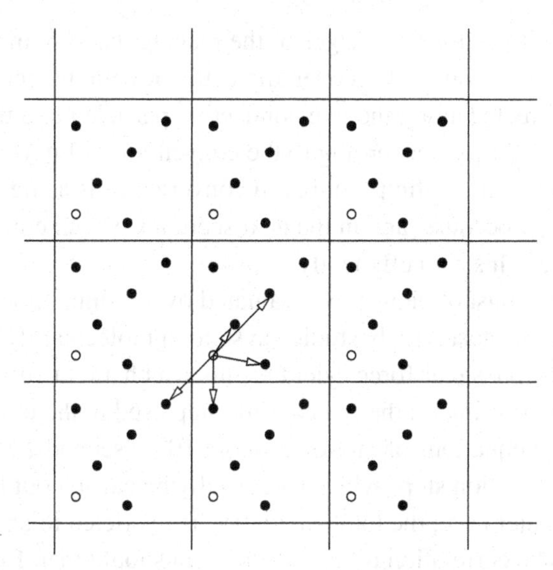

Figure 8.1. Periodic boundary conditions for molecular dynamics. Each particle interacts not only with every other particle in the system but also with all other particles in the copies of the system. The arrows from the white particle point to the nearest copies of the other particles in the system.

- Another approximation is inherent to most computer simulations aiming at a description of the real world: the system sizes in such simulations are much smaller than those of experimental systems. In the limit where the correlation length is much smaller than the system size this does not matter too much, and in the opposite regime, in which the correlation length exceeds the system size we can use the finite-size scaling methods discussed in Chapter 7 in order to extrapolate results for physical quantities in the finite system to those of the infinite system (although second order transitions are seldom studied in molecular dynamics because of the heavy demands on computing resources). The finiteness of the system size is felt through the presence of the boundary. The convention adopted in the vast majority of molecular simulations is to use periodic boundary conditions (PBC) as it is assumed that for these boundary conditions the behaviour of the system is most similar to that of a system of the same size embedded in an infinite system. In fact, with periodic boundary conditions the system of interest is surrounded by similar systems with exactly the same configuration of particles at any time (see Figure 8.1). The interaction between two particles i and j is then given by the following expression:

$$\mathbf{F}_{\text{PBC}}(\mathbf{r}_i - \mathbf{r}_j) = \sum_{\mathbf{n}} \mathbf{F}\left(\left|\mathbf{r}_i - \mathbf{r}_j + \sum_{\mu=1}^{3} \mathbf{L}_\mu n_\mu\right|\right) \tag{8.3}$$

where $\mathbf{L}_\mu$ are vectors along the edges of the rectangular system volume and the first sum on the right hand side is over all vectors $\mathbf{n}$ with integer coefficients n_μ. The force $\mathbf{F}$ is directed along the line connecting particle i and the image particle $\mathbf{r}_j - \sum_{\mu=1}^{3} \mathbf{L}_\mu n_\mu$ according to the convention of Eq. (8.1). Of course, calculating terms of this infinite sum until convergence is achieved is a time-consuming procedure, and in the next section we shall consider techniques for approximating this sum efficiently.

- The time average must obviously be evaluated over a finite time. For liquid argon, which is the most widely studied system in molecular dynamics because simple Lennard–Jones pair forces yield results which are in very good agreement with experiment, the typical time step used in the numerical integration of the equations of motion is about 10^{-14} seconds, which means that for the $\sim 10^5$ integration steps which can usually be carried out in a reasonable amount of computer time, the total simulation is restricted to about 10^{-9} seconds. The correlation time of the system should therefore be much smaller than this. There is also a limitation in time because of the finite size of the system. This might in principle become noticeable when the particles have travelled on average more than half the linear system size, but in practice such effects occur at much longer time scales, of the order of the *recurrence time*, the time after which the system returns to the initial configuration (in continuum mechanics, this is called the *Poincaré time*).
- The numerical integration algorithm is not infinitely accurate. This forces us to make some optimum choice between speed and accuracy: the larger the integration time step, the more inaccurate the results of the simulation. In fact, the system will follow a trajectory in phase space which deviates from the trajectory it would follow in reality. The effect on the physical quantities as measured in the simulation is of course related to this deviation in the course of time.

We may summarise by saying that MD is – in principle – a direct simulation of a many-particle system but we have seen that, just as with any computational technique in physics, MD simulations must be carried out with considerable care. It is furthermore advisable to carry out reference tests for systems for which exact results exist or for which there is an extensive literature for comparison.

8.2 Molecular dynamics at constant energy

In the previous section we sketched the molecular dynamics method briefly for the simplest case in which the equations of motion for a collection of particles are solved for forces depending on the relative positions of the particles only. In that case energy and momentum are conserved.[1] Trivially, the particle number and system volume are

[1] The angular momentum is not conserved because of the periodic boundary conditions breaking the spherical symmetry of the interactions.

conserved too, so the time averages of physical quantities obtained by this type of simulation are equivalent to averages in the microcanonical or (*NVE*) ensemble. In this section we describe the microcanonical MD method in more detail.

The algorithm of a standard MD simulation for studying systems in equilibrium is the following:

- Initialise;
- Start simulation and let the system reach equilibrium;
- Continue simulation and store results.

We will now describe these main steps in more detail.

The number of particles and the form of the interaction are specified. The temperature is usually of greater interest than the total energy of the system and is therefore usually specified as an input parameter. We shall see below how the system can be pushed toward the desired temperature.

Initialise: The particles are assigned positions and momenta. If a Lennard–Jones potential is used, the positions are usually chosen as the sites of a Bravais-fcc lattice, which is the ground state configuration of the noble gases like argon (although the Lennard–Jones system is hexagonal close-packed in the ground state [6]). The fcc lattice contains four particles per unit cell, and for a cubic volume the system contains therefore $4M^3$ particles, $M = 1, 2, \ldots$ This is the reason why MD simulations with Lennard–Jones interactions are often carried out with particle numbers 108, 256, 500, 864, ...

The velocities are drawn from a Maxwell distribution with the specified temperature. This is done by drawing the x, y and z velocity components for each particle from a Gaussian distribution; for the x-component of the velocity this distribution is $\exp[-mv_x^2/(2k_BT)]$. In Appendix B3 it is described how random numbers with a Gaussian distribution can be generated. After generating the momenta, the total momentum is made equal to zero by calculating the average momentum $\bar{\mathbf{p}}$ per particle, and then subtracting an amount $\bar{\mathbf{p}}$ from all the individual momenta $\mathbf{p}_i$.

Start simulation and let the system reach equilibrium: The particles being released from fcc lattice positions, the system is generally not in equilibrium and during the initial phase of the simulation it is given the opportunity to relax. We now describe how the integration of the equations of motion is carried out and how the forces are evaluated. Finally we shall explain how in this initial phase the desired temperature is arrived at.

Numerical algorithms for molecular dynamics will be considered in detail in Section 8.4. Suffice it here to mention briefly the most widely used algorithm which is simple and reliable at the same time – the Verlet algorithm (see also Appendix A7.1). The standard form of the Verlet algorithm for the integration of the equation of motion of a single particle subject to a force $\mathbf{F}$ depending only on the position of the particle reads

$$\mathbf{r}(t+h) = 2\mathbf{r}(t) - \mathbf{r}(t-h) + h^2\mathbf{F}[\mathbf{r}(t)]/m \tag{8.4}$$

where $\mathbf{r}(t)$ is the position of the particle at time $t = nh$ (h is the time step; n is an integer). From now on we choose units such that $m = 1$. The error per time step is of order h^4 and a worst case estimate for the error over a fixed time interval containing many time steps is of order h^2 (see Problem A3). To start up the algorithm we need the positions of the particles at two subsequent time steps. As we have only the initial ($t = 0$) positions and velocity at our disposal, the positions at $t = h$ are calculated as

$$\mathbf{r}(h) = \mathbf{r}(0) + h\mathbf{v}(0) + \frac{h^2}{2}\mathbf{F}[\mathbf{r}(t = 0)] \quad (m \equiv 1), \tag{8.5}$$

with an error of order h^3.

During the integration, the velocities can be calculated as

$$\mathbf{v}(t) = \frac{\mathbf{r}(t + h) - \mathbf{r}(t - h)}{2h} + \mathcal{O}(h^2). \tag{8.6}$$

When using periodic boundary conditions in the simulation, we must check for each particle whether it has left the simulation cell in the last integration step. If this is the case, the particle is translated back over a lattice vector $\mathbf{L}_\mu$ to keep it inside the cell (we shall see below that this procedure facilitates the common procedure for evaluating the forces with periodic boundary conditions). The velocity must obviously be determined before such a translation.

There exist two alternative formulations of the Verlet algorithm, which are exactly equivalent to it in exact arithmetic but which are less susceptible to errors resulting from finite numerical precision in the computer than the original version. The first of these, the *leap-frog* form, introduces the velocities at time steps precisely in between those at which the positions are evaluated:

$$\mathbf{v}(t + h/2) = \mathbf{v}(t - h/2) + h\mathbf{F}[\mathbf{r}(t)], \tag{8.7a}$$

$$\mathbf{r}(t + h) = \mathbf{r}(t) + h\mathbf{v}(t + h/2). \tag{8.7b}$$

These steps are then repeated over and over. Note that they must always be applied in the given order: the second step uses $\mathbf{v}(t + h/2)$ which is calculated in the first step.

Another form is the so-called velocity-Verlet algorithm [7] which is also more stable than the original Verlet form and which, via the definition

$$\mathbf{v}(t) = \frac{\mathbf{r}(t + h) - \mathbf{r}(t - h)}{2h} \tag{8.8}$$

evaluates velocities and positions at the same time instances:

$$\mathbf{r}(t + h) = \mathbf{r}(t) + h\mathbf{v}(t) + h^2\mathbf{F}(t)/2, \tag{8.9a}$$

$$\mathbf{v}(t + h) = \mathbf{v}(t) + h[\mathbf{F}(t + h) + \mathbf{F}(t)]/2. \tag{8.9b}$$

This form is most convenient because it is very stable with respect to errors due to finite precision arithmetic, and it does not require additional calculations in order to find the velocities. It should be noted that all formulations have essentially the same memory requirements. It may seem that, as this algorithm needs *two* forces in the second step, we need two arrays for these, one containing $\mathbf{F}(t)$ and the other $\mathbf{F}(t+h)$. However, the following form of the algorithm is exactly equivalent and avoids the need for two force arrays:

$$\tilde{\mathbf{v}}(t) = \mathbf{v}(t) + h\mathbf{F}(t)/2, \tag{8.10a}$$

$$\mathbf{r}(t+h) = \mathbf{r}(t) + h\tilde{\mathbf{v}}(t), \tag{8.10b}$$

$$\mathbf{v}(t+h) = \tilde{\mathbf{v}}(t) + h\mathbf{F}(t+h)/2. \tag{8.10c}$$

The new force $\mathbf{F}(t+h)$ is calculated between the second and third step.

The force acting on particle i results from the interaction forces between this particle and all the other particles in the system – usually pair-wise interactions are used. The calculation of the forces therefore takes a relatively long time as this requires $\mathcal{O}(N^2)$ steps. A problem in the evaluation of the force arises from the assumption of periodic boundary conditions. These imply that the system is surrounded by an infinite number of copies with exactly the same configuration as in Figure 8.1. A particle therefore interacts not only with each partner j in the system cell we are considering but also with the images of particle j in all the copies of the system. This means that in principle an infinite number of interactions has to be summed over. In many cases, the force decays rapidly with distance, and in that case remote particle copies will not contribute significantly to the force. If the force between the particles can safely be neglected beyond separations of half the linear system size, the force evaluation can be carried out efficiently by taking into account, for each particle in the system, only the interactions with the nearest copy of each of the remaining particles (see Figure 8.1): each infinite sum over all the copies is replaced by a single term! This is the *minimum image convention.* In formula, for a cubic system cell the minimum image convention reads

$$r_{ij}^{\min} = \min_{\mathbf{n}} |\mathbf{r}_i - \mathbf{r}_j + n_\mu \mathbf{L}_\mu| \tag{8.11}$$

with the same notation as in Eq. (8.3), but where the components of n_μ assume the values $0, \pm 1$, provided all the particles are kept within the system cell, by translating them back if they leave this cell. The potential is no longer analytic in this convention, but discontinuities will obviously be unimportant if the potential is small beyond half the linear system size.

Often it is possible to cut the interactions off at a distance $r_{\text{cut-off}}$ smaller than half the linear system size without introducing significant errors. In that case the forces do not have to be calculated for all pairs. However, all pairs must be considered to check whether their separation is larger than $r_{\text{cut-off}}$. In the same paper in which he introduced the midpoint integration algorithm into MD, Verlet [8] proposed keeping a list of particle pairs whose separation lies within some maximum distance r_{max} and updating this list at intervals of a fixed number of steps – this number lies typically between 10 and 20. The radius r_{max} is taken larger than $r_{\text{cut-off}}$ and must be chosen such that between two table updates it is unlikely for a pair not in the list to come closer than $r_{\text{cut-off}}$. If both distances are chosen carefully, the accuracy can remain very high and the increase in efficiency is of the order of a factor of 10 (the typical relative accuracy in macroscopic quantities in a MD simulation is of order 10^{-4}).

There exists another method for keeping track of which pairs are within a certain distance of each other: the *linked-cell method.* In this method, the system is divided up into (rectangular) cells. Each cell is characterized by its integer coordinates IX,IY,IZ in the grid of cells. The cell size is chosen larger than the interaction range which is about the size of $r_{\text{max}} > r_{\text{cut-off}}$ in the Verlet method. If we wanted a list of particles for each cell, we could simply restrict the interactions to particle pairs in the same, or in neighbouring cells. However, as particles will leave and enter the cells, the bookkeeping of these lists becomes a bit cumbersome. This bookkeeping can however be done very efficient by using a list of particle indices. The procedure is reminiscent of the use of pointers in a linked list. We need two ingredients: we must have a routine which generates a sort of table containing information about which particle is in what cell, and we need to organise the force calculation such that it uses this information.

To be specific, let us assume that there are $M \times M \times M$ cells. The particles are numbered 1 through N, so each particle has a definite index. We use an integer array called 'Header' which is of size $M \times M \times M$: Header(IX,IY,IZ) tells us the *highest* particle index to be found in cell IX,IY,IZ. We also introduce an integer array 'Link' which is of size N. The arrays Header and Link are filled in the following code: dimension header(M,M,M), link(N)

```
Set Header (IX,IY,IZ) to 0
Set Link(I) to 0
FOR I = 1,N DO
    IX = int(M*x(I)/L)+1
    IY = int(M*y(I)/L)+1
    IZ = int(M*z(I)/L)+1
    link(i) = header(IX,IY,IZ)
    header(IX,IY,IZ) = I
END FOR
```

Now, Header contains the highest index present in all cells. Furthermore, for particle I, Link(I) is another particle *in the same cell.* To find all particles in cell IX, IY, IZ, we look at Header(IX,IY,IZ) and then move down from particle I to the following by taking for the next particle the value Link(I). Using this in the force calculation leads to the pseudocode:

```
FOR all cells with indices (IX,IY,IZ) DO
{Fill the list xt, yt and zt with the particles of the central cell}
    icnt = 0;
    j = Header(IX,IY,IZ);
    WHILE (j>0) DO
        j = link(j);
        icnt = icnt + 1;
        xt(icnt) = x(j); yt(icnt) = y(j); zt(icnt) = z(j);
        LocNum = icnt;
    END WHILE
{Now, LocNum is the number of particles in the central cell}
    FOR half of the neighbouring cells DO
        Find particles in the same way as central cell
        and append them to the list xt, yt, zt;
    END FOR
    Calculate Lennard–Jones forces between all particles in the central cell;
    Calculate Lennard–Jones forces between particles in central and
      neighbouring cells;
END FOR
```

Note that we loop over only *half* the number of neighbouring cells in order to avoid double counting of particle pairs. The cell method is less efficient than the neighbour list method as the blocks containing possible interaction candidates for each particle substantially bigger than the spheres of the neighbour list. The advantage of the present method lies in its suitability for parallel computing (see Chapter 16).

Cutting off the force violates energy conservation although the effect is small if the cut-off radius is chosen suitably. To avoid this violation, the pair potential $U(r)$ can be shifted so that it becomes continuous at $r_{\text{cut-off}}$. The shifted potential can be written in terms of the original one as

$$U_{\text{shift}}(r) = U(r) - U(r_{\text{cut-off}}). \tag{8.12}$$

The force is not affected by this shift; it remains discontinuous at the cut-off and this gives rise to inaccuracies in the integration. Applying a shift in the force in addition to the shift in the potential yields [9, 10]

$$U_{\text{force shift}}(r) = U(r) - U(r_{\text{cut-off}}) - \frac{\mathrm{d}}{\mathrm{d}r}U(r_{\text{cut-off}})(r - r_{\text{cut-off}}) \tag{8.13}$$

and now the force and the potential are continuous. These adjustments to the potential can be compensated for by thermodynamic perturbation theory (see Ref. [11]).

Electric and gravitational forces decay as $1/r$ and cannot be truncated beyond a finite range without introducing important errors. These systems will be treated in Section 8.7.

The time needed to reach equilibrium depends on how far the initial configuration was from equilibrium, and on the relaxation time (see Section 7.4). To check whether equilibrium has been reached, it is best to monitor several physical quantities such as kinetic energy and pressure, and see whether they have levelled down. This can be judged after completing the simulation by plotting out the values of these physical quantities as a function of time. It is therefore convenient to save all these values on disk during the simulation and analyse the results afterwards. It is also possible to measure correlation times along the lines of Section 7.4, and let the system relax for a period of, for example, twice the longest correlation time measured.

A complication is that we want to study the system at a predefined temperature rather than at a predefined total energy because temperature is easily measurable and controllable in experimental situations. Unfortunately, we can hardly forecast the final temperature of the system from the initial configuration. To arrive at the desired value of the temperature, we rescale the velocities of the particles a number of times during the equilibration phase with a uniform scaling factor λ according to

$$\mathbf{v}_i(t) \rightarrow \lambda \mathbf{v}_i(t) \tag{8.14}$$

for all the particles $i = 1, \ldots, N$. The scaling factor λ is chosen such as to arrive at the desired temperature T_D after rescaling:

$$\lambda = \sqrt{\frac{(N-1)3k_\mathrm{B}T_\mathrm{D}}{\sum_{i=1}^{N} mv_i^2}}. \tag{8.15}$$

Note the factor $N - 1$ in the numerator of the square root: the kinetic energy is composed of the kinetic energies associated with the *independent* velocities, but as for interparticle interactions with PBC the total force vanishes, the total momentum is conserved and hence the number of independent velocity components is reduced by 3. This argument is rather heuristic and not entirely correct. We shall give a more rigorous treatment of the temperature calculation in Section 10.7.

After a rescaling the temperature of the system will drift away but this drift will become less and less important when the system approaches equilibrium. After a number of rescalings, the temperature then fluctuates around an equilibrium value. Now the 'production phase', during which data can be extracted from the simulation, begins.

Continue simulation and determine physical quantities: Integration of the equations of motion proceeds as described above. In this part of the simulation, the actual determination of the static and dynamic physical quantities takes place. We determine the expectation value of a static physical quantity as a time average according to

$$\bar{A} = \frac{1}{n - n_0} \sum_{\nu > n_0}^{n} A_\nu. \tag{8.16}$$

The indices ν label the n time steps of the numerical integration, and the first n_0 steps have been carried out during the equilibration. For determination of errors in the measured physical quantities, see the discussion in Section 7.4.

Difficulties in the determination of physical quantities may arise when the parameters are such that the system is close to a first or second order phase transition (see the previous chapter): in the first order case, the system might be 'trapped' in a metastable state and in the second order case, the correlation time might diverge for large system sizes.

In the previous chapter we have already considered some of the quantities of interest. In the case of a microcanonical simulation, we are usually interested in the temperature and pressure. Determination of these quantities enables us to determine the *equation of state*, a relation between pressure and temperature, and the system parameters – particle number, volume and energy (*NVE*). This relation is hard to establish analytically, although various approximate analytical techniques for this purpose exist: cluster expansions, Percus–Yevick approximation, etc. [11].

The pair correlation function is useful not only for studying the details of the system but also to obtain accurate values for the macrosopic quantities such as the potential energy and pressure, as we shall see below. The correlation function is determined by keeping a histogram which contains for every interval $[i\Delta r, (i+1)\Delta r]$ the number of pairs $n(r)$ with separation within that range. The list can be updated when the pair list for the force evaluation is updated. The correlation function is found in terms of $n(r)$ as

$$g(r) = \frac{2V}{N(N-1)} \left[\frac{\langle n(r) \rangle}{4\pi r^2 \Delta r} \right]. \tag{8.17}$$

Similar expressions can be found for time-dependent correlation functions – see Refs. [2] and [11].

If the force has been cut off during the simulation, the calculation of average values involving the potential U requires some care. Consider for example the potential energy itself. This is calculated at each step taking only the pairs with separation within the minimum cut-off distance into account; taking all pairs into account would imply losing the efficiency gained by cutting off the potential. The neglect of the tail of the potential can be corrected for by using the pair correlation function beyond $r_{\text{cut-off}}$:

$$\langle U\rangle = \langle U\rangle_{\text{cut-off}} + 2\pi\frac{N(N-1)}{V}\int_{r_{\text{cut-off}}}^{\infty} r^2\mathrm{d}r\; U(r)g(r) \tag{8.18}$$

where $\langle\cdots\rangle_{\text{cut-off}}$ is the average restricted to pairs with separation smaller than $r_{\text{cut-off}}$. Of course, we can determine the correlation function for r up to half the linear system size only because of periodic boundary conditions. Verlet [12] has used the Percus–Yevick approximation to extrapolate g beyond this range. Often g is simply approximated by its asymptotic value $g(r) \equiv 1$ for large r.

Similarly, the virial equation is corrected for the potential tail:

$$\frac{P}{nk_{\mathrm{B}}T} = 1 - \frac{1}{3Nk_{\mathrm{B}}T}\left\langle\sum_i\sum_{j>i} r_{ij}\frac{\partial U(R)}{\partial r_{ij}}\right\rangle_{\text{cut-off}} - \frac{2\pi N}{3k_{\mathrm{B}}TV}\int_{r_{\text{cut-off}}}^{\infty} r^3\frac{\partial U(r)}{\partial r}g(r)\mathrm{d}r, \tag{8.19}$$

where $g(r)$ can also be replaced by 1.

The specific heat can be calculated from Lebowitz's formula, see Eq. (7.37).

8.3 A molecular dynamics simulation program for argon

In the previous section we described the structure of a MD program and here we give some further details related to the actual implementation. The program simulates the behaviour of argon. In 1964, Rahman [13] published a paper on the properties of liquid argon – the first MD simulation involving particles with smoothly varying potentials. Previous work by Alder and Wainwright [14] was on hard sphere fluids. Rahman's work was later refined and extended by Verlet [8] who introduced several features that are still used, as we have seen in the previous section.

The Lennard–Jones pair potential turns out to give excellent results for argon:

$$U(r) = 4\varepsilon\left[\left(\frac{\sigma}{r}\right)^{12} - \left(\frac{\sigma}{r}\right)^{6}\right]. \tag{8.20}$$

The optimal values for the parameters ε and σ are $\varepsilon/k_{\mathrm{B}} = 119.8$ K and $\sigma = 3.405$ Å respectively.

In the initialisation routine, the positions of a face centred cubic lattice are generated. For an $L\times L\times L$ system containing $4M^3$ particles, the fcc lattice constant a is $a = L/M$. It may be safer to put the particles not exactly on the boundary facets of the system because as a result of rounding errors it might not always be clear whether they belong to the system under consideration or a neighbouring copy.

The procedure in Appendix B3 for generating random numbers with a Gaussian distribution should be used in order to generate momenta according to a Maxwell distribution. First generate all the momenta with some arbitrary distribution width. Then calculate the total momentum $\mathbf{p}_{\text{tot}}$ and subtract a momentum $\bar{\mathbf{p}} = \mathbf{p}_{\text{tot}}/N$ from each of the momenta in order to make the total momentum zero. Now the kinetic energy is calculated and then all momenta are rescaled to arrive at the desired kinetic energy.

When calculating the forces, the minimum image convention should be adopted. It is advisable to start without using a neighbour list. For the minimum image convention it should be checked for each pair (i, j) whether the difference of the x-components $x_i - x_j$ is larger or smaller than $L/2$ in absolute value. If it is larger, then an amount L should be added to or subtracted from this difference to transform it to a value which is smaller than $L/2$ (in absolute value). In many codes, this translation is implemented as follows:

$$x \to x - \text{nint}(x/L) * L, \tag{8.21}$$

where nint denotes the nearest integer of the argument. This procedure is then repeated for the y- and z-components. Potential and force may be adjusted according to Eqs (8.12) and (8.13).

The equations of motion are solved using the leap-frog or the velocity form of the Verlet algorithm. A good value for the time step is 10^{-14} s which in units of $(m\sigma^2/\varepsilon)^{1/2}$ is equal to about 0.004. Using the argon mass as the unit of mass, σ as the unit of distance and $\tau = (m\sigma^2/\varepsilon)^{1/2}$ as the unit of time, the x-component of the force acting on particle i resulting from the interaction with particle j is given by

$$F_x^{ij} = (x_i - x_j)(48r_{ij}^{-14} - 24r_{ij}^{-8}) \tag{8.22}$$

with similar expressions for the y- and z-components.

After each step in the Verlet/leap-frog algorithm, each particle should be checked to see whether it has left the volume. If this is the case, it should be translated over a distance $\pm L$ along one or more of the Cartesian axes in order to bring it back into the system in accordance with the periodic boundary conditions.

During equilibration, the velocities (momenta) should be rescaled at regular intervals. The user might specify the duration of this phase and the interval between momentum rescalings.

During the production phase, the following quantities should be stored in a file at each time step: the kinetic energy, potential energy, and the virial

$$\sum_{ij} r_{ij} F(r_{ij}). \tag{8.23}$$

Furthermore, the program should keep a histogram-array containing the numbers of pairs found with a separation between r and $r + \Delta$ for, say, $\Delta = L/200$ from which in the end the correlation function can be read off.

Table 8.1. *Molecular dynamics data for thermodynamic quantities of the Lennard–Jones liquid.*

$\rho(1/\sigma^3)$	$T_0(\varepsilon/k_B)$	T	$\beta P/\rho$	$U(\varepsilon)$
0.88	1.0	0.990 (2)	2.98 (2)	−5.704 (1)
0.80	1.0	1.010 (2)	1.31 (2)	−5.271 (1)
0.70	1.0	1.014 (2)	0.06 (4) (5)	−4.662 (1)

T_0 is the desired temperature; T is the temperature as determined from the simulation; ρ is the density: $\rho = N/V$. All values are in reduced units.

PROGRAMMING EXERCISE

Write a program that simulates the behaviour of a Lennard–Jones liquid with the proper argon parameters given above.

Check 1 To check the program, you can use small particle numbers, such as 32 or 108. Check whether the program is time-reversible by integrating for some time (without rescaling) and then reversing velocities. The system should then return to its initial configuration (graphical display of the system might be helpful).

Check 2 The definite check is to compare your results for argon with literature. A good value for the equilibration time is 10.0 τ and rescalings could take place after every 10 or 20 time steps. A sufficiently long simulation time to obtain accurate results is 20.0 τ (remember the time step is 0.004 τ). In Table 8.1 you can find a few values for the potential energy and pressure for different temperatures. Note that the average temperature in your simulation will not be precisely equal to the desired value. In Figure 7.1, the pair correlation function for $\rho = N/V = 1.06$ and $T = 0.827$ is shown.

It is interesting to study the specific heat (Eq. (7.37)) in the solid and in the gas phase. You may compare the behaviour with that of an ideal gas, $c_V = 3k_B/T$ per particle, and for a harmonic solid, $c_v = 3k_BT$ per particle (this is the Dulong–Petit law).

Note that phase transitions are difficult to locate, as there is a strong hysteresis in the physical quantities there. It is however interesting to obtain information about the different phases. For $T = 1$, $\rho = 0.8$ the argon Lennard–Jones system is found in the liquid phase, and for $\rho = 1.2$ and $T = 0.5$ in the solid phase. The gas phase is found for example with $\rho = 0.3$ and and $T = 3.0$. It is very instructive to plot the correlation function for the three phases and explain how they look. Another interesting exercise is to calculate the diffusion constant by plotting the displacement as a function of time averaged over all particles. For times smaller than the typical collision time (time of free flight), you should find

$$\langle x^2 \rangle \propto t^2, \tag{8.24}$$

and this crosses over to diffusive behaviour

$$\langle x^2 \rangle = Dt, \tag{8.25}$$

with D the diffusion constant. In the solid phase, the diffusion constant is 0. In the gas phase, the diffusive behaviour sets in at later times than in the fluid.

If the program works properly, keeping a Verlet neighbour list as discussed in the previous section can be implemented. Verlet [8] used $r_{\text{cut-off}} = 2.5\sigma$ and $r_{\max} = 3.3\sigma$. A more detailed analysis of the increase in efficiency for various values of $r_{\max}$ with $r_{\text{cut-off}} = 2.5\sigma$ shows that $r_{\max} = 3.0\sigma$ with the neighbour list updated once every 25 integration steps is indeed most efficient [2, 15].

PROGRAMMING EXERCISE

Implement the neighbourlist in your program and check whether the results remain essentially the same. Determine the increase in efficiency.

8.4 Integration methods: symplectic integrators

There exist many algorithms for integrating ordinary differential equations, and a few of these are described in Appendix A. In this section, we consider the particular case of numerically integrating the equations of motion for a dynamical system described by a time-independent Hamiltonian, of which the classical many-particle system at constant energy is an example. Throughout this section we consider the equation of motion for a single particle in one dimension. The discussion is easily generalised to more particles in more dimensions.

The Verlet algorithm is the most popular algorithm for molecular dynamics and we shall consider it in more detail in the next subsection. Before doing so, we describe a few criteria which were formulated by Berendsen and van Gunsteren [16] for integration methods for molecular dynamics. First of all, *accuracy* is an important criterion: it tells us to which power of the time step the numerical trajectory will deviate from the exact one after one integration step (see also Appendix A). Note that the prefactor of this may diverge if the algorithm is unstable (e.g. close to a singularity of the trajectory). The accuracy is the criterion that is usually considered in numerical analysis in connection with integration methods.

Two further criteria are related to the behaviour of the energy and other conserved quantities of a mechanical system which are related to symmetries of the interactions. Along the exact trajectory, energy is conserved as a result of the time-translation invariance of the Hamiltonian, but the energy of the numerical trajectory will deviate from the initial value and this deviation can be characterised by its *drift*, a steady increase or decrease, and the *noise*, fluctuations on top of the drift. Drift is obviously most undesirable. In microcanonical MD we want to sample the points in phase space with a given energy; these points form a hypersurface in phase space – the so-called *energy surface*. If the system drifts away steadily from this plane it is obviously not in equilibrium.

It is very important to distinguish in all these cases between two sources of error: those resulting from the numerical integration method as opposed to those resulting from finite precision arithmetic, inherent to computers. For example, we shall see below that the Verlet algorithm is not susceptible to energy drift in exact arithmetic. Drift will however occur in practice as a result of finite precision of computer arithmetic, and although different formulations of the Verlet algorithm have different susceptibility to this kind of drift, this depends also on the particular way in which numbers are rounded off in the computer.

Recently, there has been much interest in *symplectic integrators*. After considering the Verlet algorithm in some detail, we shall describe the concept of symplecticity[2] and its relevance to numerical integration methods.

8.4.1 The Verlet algorithm revisited

Properties of the Verlet algorithm

In this section we treat the Verlet algorithm

$$x(t+h) = 2x(t) - x(t-h) + h^2 F[x(t)] \tag{8.26}$$

in more detail with emphasis on issues which are relevant to MD. A derivation of this algorithm can be found in Appendix A7.1. The error per integration step is of the order h^4. Note that we take the mass of the particle(s) involved equal to 1. Unless stated otherwise, we analyse the one-dimensional single-particle version of the algorithm. The momenta are usually determined as

$$p(t) = [x(t+h) - x(t-h)]/(2h) + \mathcal{O}(h^2). \tag{8.27}$$

Note that there is no need for a more accurate formula, as the accumulated error in the positions after many steps is also of order h^2. We shall check this below, using also a more accurate expression for the momenta [16]:

$$p(t) = [x(t+h) - x(t-h)]/(2h) - \frac{h}{12}\{F[x(t+h)] - F[x(t-h)]\} + \mathcal{O}(h^3). \tag{8.28}$$

This form can be derived by subtracting the Taylor expansions for $x(t+h)$ and $x(t-h)$ about t, and approximating $\mathrm{d}F[x(t)]/\mathrm{d}t$ by $\{F[x(t+h)] - F[x(t-h)]\}/h$.

[2] Some authors use the term 'symplecticness' instead of 'symplecticity'.

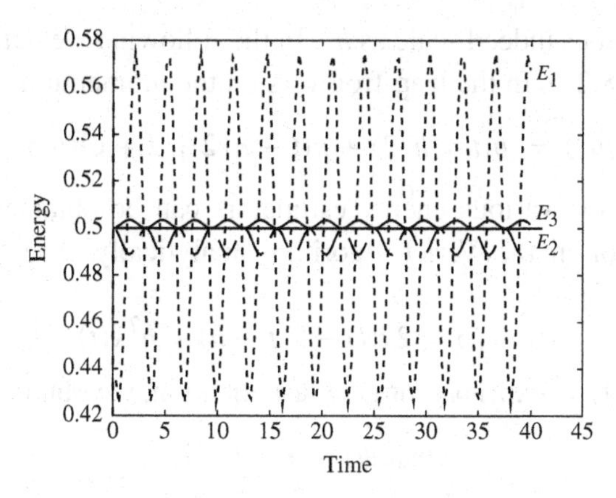

Figure 8.2. The energy of the harmonic oscillator determined using the various velocity estimators described in the text. E_1 is the energy using (8.29), E_2 uses (8.27) and E_3 was calculated using (8.28).

In the leap-frog version, we have the velocities at our disposal for times halfway between those at which the positions are given:

$$p(t + h/2) = [x(t + h) - x(t)]/h + \mathcal{O}(h^2). \tag{8.29}$$

Each of the expressions (8.27–8.29) for the momentum gives rise to a different expression for the energy.

We first analyse the different ways of calculating the total energy for the simple case of the one-dimensional harmonic oscillator

$$\mathcal{H} = (p^2 + x^2)/2 \tag{8.30}$$

and we can use any of the formulae (8.27–8.29) for the momentum. In Figure 8.2 the different energy estimators are shown as a function of time for the harmonic oscillator which is integrated using the Verlet algorithm with a time step $h = 0.3$. This is to be compared with the period $T = 2\pi$ of the motion $x(t) = \cos(t)$ (for appropriate initial conditions). It is seen that the leap-frog energy estimator is an order of magnitude worse than the other two. This is not surprising, since the fact that the velocity is not calculated at the same time instants as the position results in deviation of the energy from the continuum value of order h instead of h^2 when using (8.27). The energy estimator using third order momenta according to (8.28) is better than the second order form. Note that the error in the position accumulates in time to give $\mathcal{O}(h^2)$ (see Problem A3), so that there is no point in calculating the momenta with a higher order of accuracy, as this will not yield an order of magnitude improvement. The fact that the error for the third order estimator is about a factor of 3 better than that of the second order one for the harmonic oscillator does not therefore indicate a systematic trend. More importantly, the error in *both* estimators

(8.27) and (8.28) does indeed scale as h^2. In the following we determine momenta according to Eq. (8.27). In the leap-frog version the momentum estimator is

$$p(t) = [p(t + h/2) + p(t - h/2)]/2 + \mathcal{O}(h^2). \tag{8.31}$$

The results for the various energy estimators can be obtained by solving the harmonic oscillator in the Verlet algorithm analytically. The 'Verlet harmonic oscillator' reads

$$x(t + h) = 2x(t) - x(t - h) - h^2 x(t). \tag{8.32}$$

If we substitute $x(t) = \exp(\mathrm{i}\omega t)$ into the last equation, we obtain

$$\cos(\omega h) = 1 - h^2/2 \tag{8.33}$$

and this defines a frequency ω differing by an amount of order h^2 from the angular frequency $\omega = 1$ of the exact solution. The difference between the numerical and the exact solution will therefore show a slow beat.

A striking property of the energy determined from the Verlet/leap-frog solution is that it does not show any drift in the total energy (in exact arithmetic). This stability follows directly from the fact that the Verlet algorithm is time-reversible, which excludes steady increase or decrease of the energy for periodic motion. In a molecular dynamics simulation, however, the integration time, which is the duration of the simulation, is much smaller than the period of the system, which is the *Poincaré time*, that is the time after which the system returns to its starting configuration. The error in the energy might therefore grow steadily during the simulation. It turns out, however, that the deviation of the energy remains bounded in this case also, as the Verlet algorithm possesses an additional symmetry, called *symplecticity*. Symplecticity will be described in detail in Section 8.4.2. Here we briefly describe what the consequences of symplecticity are for an integration algorithm. Symplecticity gives rise to conserved quantities, and in particular, it can be shown that a discrete analogue of the total energy is rigorously conserved (in exact arithmetic) [17]. It turns out that this discrete energy deviates from the continuum energy at most an amount of order h^k, for some positive integer k. Therefore, the energy cannot drift away arbitrarily and it follows that the noise remains bounded.

To illustrate this point we return to the harmonic oscillator. In this particular case we can actually determine the conserved discrete energy. In the leap-frog formulation:

$$p(t + h/2) = p(t - h/2) - hx(t); \tag{8.34a}$$

$$x(t + h) = x(t) + hp(t + h/2), \tag{8.34b}$$

it is equal to [18]

$$\mathcal{H}_\mathrm{D} = \tfrac{1}{2}[p(t - h/2)^2 + x(t)^2 - hp(t - h/2)x(t)]. \tag{8.35}$$

The fact that this quantity is conserved can also be checked directly using (8.34b). This energy is equal to $1/2-h^2/8$ for the solution $\cos(\omega t)$ with ω given in Eq. (8.33). For general potentials, the discrete energy is not known.

As mentioned before, the absence of drift in the energy in the case of the harmonic oscillator can be explained by the time-reversibility of the Verlet algorithm, and comparisons with Runge–Kutta integrators for example, which are in general not time-reversible for potentials such as the harmonic oscillator, do not convincingly demonstrate the necessity for using a symplectic algorithm. Symplecticity does however impose a restriction on the noise, but time-reversibility does not.

Symplectic integrators are generally recommended for integrating dynamical systems because they generate solutions with the same geometric properties in phase space as the solutions of the continuum dynamical system. The fact that the deviation of the energy is always bounded is a pleasant property of symplectic integrators. Symplectic integrators are considered in more detail in Section 8.4.2.

Finite precision of computer arithmetic obviously does not respect the symplectic geometry in phase space. Hockney and Eastwood observed that when numbers are rounded off properly in the computer, the system tends to heat up because the rounding effects can be viewed as small random forces acting on the particles [19]. If real numbers are systematically truncated to finite precision numbers, the system cools down slowly. Both effects are clearly signs of nonsymplectic behaviour.

Several classes of symplectic integrators with explicit formulas for different orders of accuracy have been found. Runge–Kutta–Nystrom integrators (not to be confused with ordinary Runge–Kutta algorithms) have been studied by Okunbor and Skeel [20]. Yoshida [21] and Forest [22] have considered Lie-integrators. Their approach follows rather naturally from the structure of the symplectic group, as we shall see in Section 8.4.2.[3]

Let us make an inventory of relevant symmetry properties of integrators. First of all, time-reversibility is important. If it is present in the equations of motion, as is usually the case in MD, it is natural to require it in the integration method. Another symmetry is phase space conservation. This is a property of the trajectories of the continuum equations of motion – this property is given by Liouville's theorem – and it is useful to have our numerical trajectories obeying this condition too (note that time-reversibility by itself does not guarantee phase space conservation). The most detailed symmetry requirement is symplecticity, which will be considered in greater detail below (Section 8.4.2). This incorporates phase space conservation and conservation of a number of conserved quantities, the so-called *Poincaré invariants*. The symplectic symmetry properties can also be formulated in geometrical terms

[3] Gear algorithms [16, 23, 24] have been fashionable for MD simulations. These are predictor–corrector algorithms requiring only one force evaluation per time step. Gear algorithms are not symplectic and they are becoming less popular for that reason.

as we shall see below. Most important for molecular dynamics is the property that the total energy fluctuates within a narrow range around the exact one. Some comparison has been carried out between nonsymplectic phase space conserving and symplectic integrators [25], and this gave no indication of the superiority of symplectic integrators above merely phase-space conserving ones. As symplectic integrators are not more expensive to use than nonsymplectic time-reversible ones, their use is recommended as the safest option. Investigating the merits of the various classes of integration methods for microcanonical MD is a fruitful area for future research.

Frictional forces

Later we shall encounter extensions of the standard MD method where a frictional force is acting on the particles along the direction of the velocity. The Verlet algorithm can be generalised to include such frictional forces and we describe this extension for the one-dimensional case which can easily be generalised to more dimensions. The continuum equation of motion is

$$\ddot{x} = F(x) - \gamma \dot{x}, \tag{8.36}$$

and expanding $x(h)$ and $x(-h)$ around $t = 0$ in the usual way (see Appendix A7.1) gives

$$x(h) = x(0) + h\dot{x}(0) + h^2[-\gamma\dot{x}(0) + F(0)]/2 + h^3\dddot{x}(0)/6 + \mathcal{O}(h^4) \tag{8.37a}$$

$$x(-h) = x(0) - h\dot{x}(0) + h^2[-\gamma\dot{x}(0) + F(0)]/2 - h^3\dddot{x}(0)/6 + \mathcal{O}(h^4). \tag{8.37b}$$

Addition then leads to

$$x(h) = 2x(0) - x(-h) + h^2[-\gamma\dot{x}(0) + F(0)] + \mathcal{O}(h^4) \tag{8.38}$$

where $\dot{x}(0)$ remains to be evaluated. If we write

$$\dot{x}(0) = [x(h) - x(-h)]/(2h) + \mathcal{O}(h^2), \tag{8.39}$$

and substitute this into (8.38), we obtain

$$(1 + \gamma h/2)x(h) = 2x(0) - (1 - \gamma h/2)x(-h) + h^2 F(0) + \mathcal{O}(h^4). \tag{8.40}$$

A leap-frog version of the same algorithm is

$$x(h) = x(0) + hp(h/2); \tag{8.41a}$$

$$p(h/2) = \frac{(1 - \gamma h/2)p(-h/2) + hF(0)}{1 + \gamma h/2}. \tag{8.41b}$$

If the mass m is not equal to unity, the factors $1 \pm \gamma h/2$ are replaced by $1 \pm \gamma h/(2m)$.

It is often useful to simulate the system with a prescribed temperature rather than at constant energy. In Section 8.5 we shall discuss a constant-temperature MD method in which a time-dependent friction parameter occurs, obeying a first order differential equation:

$$\ddot{x}(t) = -\gamma(t)\dot{x}(t) + F[x(t)] \tag{8.42a}$$

$$\dot{\gamma}(t) = g[\dot{x}(t)]. \tag{8.42b}$$

The solution can conveniently be presented in the leap-frog formulation. As the momentum is given at half-integer time steps in this formulation, we can solve for γ in the following way:

$$\gamma(h) = \gamma(0) + hg[p(h/2)] + \mathcal{O}(h^2), \tag{8.43}$$

and this is to be combined with Eqs. (8.41). Velocity-Verlet formulations (Eqs. (8.9)) for equations of motions including friction terms can be found straightforwardly. This is left as an exercise to the reader – see also Ref. [26].

*8.4.2 Symplectic geometry; symplectic integrators

In recent years, major improvement has been achieved in understanding the merits of the various methods for integrating equations of motion which can be derived from a Hamiltonian. This development started in the early 1980s with the observations made independently by Ruth [27] and Feng [28] that methods for solving Hamiltonian equations of motion should preserve the geometrical structure of the continuum solution in phase space. This geometry is the so-called *symplectic geometry*. Below we shall explain what this geometry is about, and what the properties of symplectic integrators are. In Section 8.4.3 we shall see how symplectic integrators can be constructed. We restrict ourselves again to a two-dimensional phase space (one particle moving in one dimension) spanned by the coordinates p and x, but it should be realised that the analysis is trivially generalised to arbitrary numbers of particles in higher dimensional space with phase space points $(\mathbf{p}_1, \ldots, \mathbf{p}_m, \mathbf{r}_1, \ldots, \mathbf{r}_m)$.[4] The equations of motion for the particle are derived from a Hamiltonian which for a particle moving in a potential (in the absence of constraints) reads

$$\mathcal{H}(p, x) = \frac{p^2}{2} + V(x). \tag{8.44}$$

The Hamilton equations of motion are then given as

$$\dot{p} = -\frac{\partial \mathcal{H}(p, x)}{\partial x} \tag{8.45a}$$

$$\dot{x} = \frac{\partial \mathcal{H}(p, x)}{\partial p} \tag{8.45b}$$

[4] Although we use the notation $\mathbf{r}_i$ for the coordinates, they may be generalised coordinates.

It is convenient to introduce the combined momentum–position coordinate $z = (p, x)$, in terms of which the equations of motion read

$$\dot{z} = J\nabla\mathcal{H}(z) \tag{8.46}$$

where J is the matrix

$$J = \begin{pmatrix} 0 & -1 \\ 1 & 0 \end{pmatrix} \tag{8.47}$$

and $\nabla\mathcal{H}(z) = (\partial\mathcal{H}(z)/\partial p, \partial\mathcal{H}(z)/\partial x)$.[5]

Expanding the equation of motion (8.46) to first order, we obtain the time evolution of the point z to a new point in phase space:

$$z(t+h) = z(t) + hJ\nabla_z\mathcal{H}[z(t)]. \tag{8.49}$$

The exact solution of the equations of motion can formally be written as

$$z(t) = \exp(tJ\nabla_z\mathcal{H})[z(0)] \tag{8.50}$$

where the exponent is to be read as a series expansion of the operator $tJ\nabla_z\mathcal{H}$. This can be verified by substituting Eq. (8.50) into (8.46). This is a one-parameter family of mappings with the time t as the continuous parameter. The first order approximation to (8.50) coincides with (8.49).

Now consider a small region in phase space located at $z = (p, x)$ and spanned by the infinitesimal vectors δz^a and δz^b. The area δA of this region can be evaluated as the cross product of δz^a and δz^b which can be rewritten as[6]

$$\delta A = \delta z^a \times \delta z^b = \delta z^a \cdot (J\delta z^b). \tag{8.51}$$

It is now easy to see that the mapping (8.50) preserves the area δA. It is sufficient to show that its time derivative vanishes for $t = 0$, as for later times the analysis can be translated to this case. We have

$$\begin{aligned} \left.\frac{\mathrm{d}\delta A}{\mathrm{d}t}\right|_{t=0} &= \frac{\mathrm{d}}{\mathrm{d}t}\{[\mathrm{e}^{tJ\nabla_z\mathcal{H}}(\delta z^a)] \cdot [J\mathrm{e}^{tJ\nabla_z\mathcal{H}}(\delta z^b)]\}_{t=0} \\ &= [J\nabla_z\mathcal{H}(\delta z^a)] \cdot (J\delta z^b) + (\delta z^a) \cdot [JJ\nabla_z\mathcal{H}(\delta z^b)]. \end{aligned} \tag{8.52}$$

We can find $\mathcal{H}(\delta z^{a,b})$ using a first order Taylor expansion:

$$\mathcal{H}(\delta z^a) = \mathcal{H}(z + \delta z^a) - \mathcal{H}(z) = \delta z^a \cdot \nabla_z\mathcal{H}(z), \tag{8.53}$$

[5] In more than one dimension, the vector z is defined as $(p_1, \ldots, p_N, x_1, \ldots, x_N)$, and the matrix J reads in that case

$$J = \begin{pmatrix} 0 & -I \\ I & 0 \end{pmatrix} \tag{8.48}$$

where I is the $N \times N$ unit matrix.

[6] Note that the area can be negative: it is an *oriented* area. In the language of differential geometry this area is called a *two-form*.

and similar for $\mathcal{H}(\delta z^b)$. This leads to the form

$$\left.\frac{d\delta A}{dt}\right|_{t=0} = -(L^T\delta z^a)\cdot(J\delta z^b) - (\delta z^a)\cdot(JL^T\delta z^b) \tag{8.54}$$

where L is the Jacobian matrix of the operator $J\nabla_z\mathcal{H}$:

$$L_{ij} = \sum_k J_{ik}[\partial^2\mathcal{H}(z)/\partial z_k\partial z_j] = \begin{pmatrix} -\mathcal{H}_{px} & -\mathcal{H}_{xx} \\ \mathcal{H}_{pp} & \mathcal{H}_{px} \end{pmatrix}. \tag{8.55}$$

Here $\mathcal{H}_{xx}$ denotes the second partial derivative with respect to x etcetera. It is easy to see that the matrix L satisfies

$$L^{\mathrm{T}}J + JL = 0, \tag{8.56}$$

where L^{T} is the transpose of L, and hence from (8.54) the area δA is indeed conserved.

We can now define symplecticity in mathematical terms. The Jacobi matrix S of the mapping $\exp(tJ\nabla H)$ is given as $S = \exp(tL)$. This matrix satisfies the relation:

$$S^{\mathrm{T}}JS = J. \tag{8.57}$$

Matrices satisfying this requirement are called *symplectic*. They form a Lie group whose Lie algebra is formed by the matrices L satisfying (8.56). General nonlinear operators are symplectic if their Jacobi matrix is symplectic.

In more than two dimensions the above analysis can be generalised for *any* pair of canonical variables p_i, x_i – we say that phase space area is conserved for any pair of one-dimensional conjugate variables p_i, x_i. The conservation law can be formulated in an integral form [29]; this is depicted in Figure 8.3. In this picture the three axes correspond to p, x and t. If we consider the time evolution of the points lying on a closed loop in the p, x plane, we obtain a tube which represents the flow in phase space. The area conservation theorem says that *any* loop around the tube encloses the same area $\oint p dx$. In fact, there exists a similar conservation law for volumes enclosed by the areas of pairs of canonical variables: these volumes are called the *Poincaré invariants*. For the particular case of the volume enclosed by areas of *all* the pairs of canonical variables, we recover Liouville's theorem which says that the volume in phase space is conserved. Phase space volume conservation is equivalent to the Jacobi determinant of the time evolution operator in phase space being equal to 1 (or -1 if the orientation is not preserved). For two-dimensional matrices, the Jacobi determinant being equal to 1 is equivalent to symplecticity as can easily be checked from (8.57). This is also obvious from the geometric representation in Figure 8.3. For systems with a higher-dimensional phase space, however, the symplectic symmetry is a more detailed requirement than mere phase space conservation.

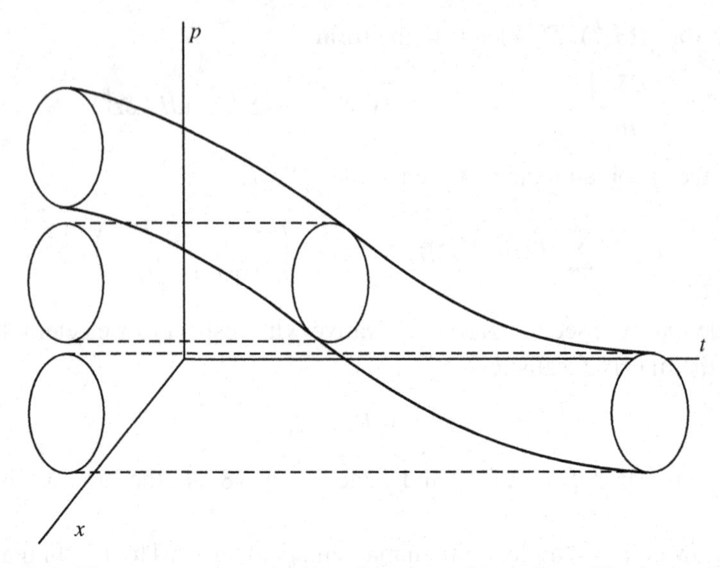

Figure 8.3. The area conservation law for a symplectic flow. The integral $\oint p\mathrm{d}x$ for any loop around the tube representing the flow of a closed loop in the p, x plane remains constant. This integral represents the area of the projection of the loop onto the xp plane. Note that the loops do not necessarily lie on a plane of constant time.

We have seen that symplecticity is a symmetry of Hamiltonian mechanics in continuum time; now we consider numerical integration methods for Hamiltonian systems (discrete time). As mentioned above, it is not clear whether full symplecticity is necessary for a reliable description of the dynamics of a system by a numerical integration. However, it will be clear that the preservation of the symmetries present in continuum time mechanics is the most reliable option. The fact mentioned above, that symplecticity implies conservation of the discrete version of the total energy, is an additional feature in favour of symplectic integrators for studying dynamical systems.

It should be noted that symplecticity does not guarantee time reversibility or vice versa. Time reversibility shows up as the Hamiltonian being invariant when replacing p by $-p$, and a Hamiltonian containing odd powers of p might still be symplectic.

*8.4.3 Derivation of symplectic integrators

The first symplectic integrators were found by requiring that an integrator of some particular form be symplectic. The complexity of the resulting algebraic equations for the parameters in the integration scheme was found to increase dramatically with increasing order of the integrator. Later Yoshida [21] and Forest [22] developed a different scheme for finding symplectic integrators, and in this section we follow their analysis.

Consider a Hamiltonian of the simple form:

$$\mathcal{H} = T(p) + U(x) \tag{8.58}$$

(we still restrict ourselves to a particle in one dimension – results are easily generalised). In terms of the variable $z = (p, x)$ the equations of motion read

$$\begin{aligned}\frac{\mathrm{d}z}{\mathrm{d}t} &= \left(-\frac{\partial \mathcal{H}}{\partial x}, \frac{\partial \mathcal{H}}{\partial p}\right) = \left(-\frac{\partial U(x)}{\partial x}, \frac{\partial T(p)}{\partial p}\right)\\ &= J\nabla\mathcal{H}(z) \equiv \tilde{T}(z) + \tilde{U}(z),\end{aligned} \tag{8.59}$$

where in the last expression the operator $J\nabla\mathcal{H}$, which acts on $z = (p, x)$, is split into the contributions from the kinetic and potential energy respectively:

$$\tilde{T}(z) = \left(0, \frac{\partial T(p)}{\partial p}\right) \tag{8.60a}$$

$$\tilde{U}(z) = \left(-\frac{\partial U(x)}{\partial x}, 0\right). \tag{8.60b}$$

$\tilde{T}$ and $\tilde{U}$ are therefore also operators which map a point $z = (p, x)$ in phase space onto another point in phase space.

As we have seen in the previous section, the exact solution of (8.59) is given as

$$z(t) = \exp(tJ\nabla H)[z(0)] = \exp[t(\tilde{T} + \tilde{U})][z(0)]. \tag{8.61}$$

The term $\exp(tJ\nabla H)$ is a time evolution operator. It is a symplectic operator, as are $\exp(t\hat{T})$ and $\exp(t\hat{U})$ since these can both be derived from a Hamiltonian (for a free particle and a particle with infinite mass respectively).

An nth *order integrator* for time step h is now defined by a set of numbers a_k, b_k, $k = 1, \ldots, m$, such that

$$\prod_{k=1}^{m} \exp(a_k h\tilde{T}) \exp(b_k h\tilde{U}) = \exp(hJ\nabla H) + \mathcal{O}(h^{n+1}). \tag{8.62}$$

Since the operators $\exp(a_k h\tilde{T})$ and $\exp(b_k h\tilde{U})$ are symplectic, the integrator (8.62) is symplectic too. The difference between the integrator and the exact evolution operator can be expressed in Campbell–Baker–Hausdorff (CBH) commutators: if $\mathrm{e}^C = \mathrm{e}^A\mathrm{e}^B$ then

$$C = A + B + [A, B]/2 + ([A, [A, B]] + [B, [B, A]])/12 + \cdots \tag{8.63}$$

where the dots represent higher order commutators. This formula can be derived by writing $\exp(tA)\exp(tB) = \exp[t(A + B) + \Delta]$, expanding the operator Δ in powers of t and equating equal powers of t on the left and right hand sides of the equality [30]. Applying this formula with $A = h\tilde{T}$ and $B = h\tilde{U}$ to increasing orders of commutators, we find

$$\exp(hJ\nabla H) = \exp(h\tilde{T})\exp(h\tilde{U}) + \mathcal{O}(h^2) \tag{8.64a}$$

$$\exp(hJ\nabla H) = \exp(h\tilde{T}/2)\exp(h\tilde{U})\exp(h\tilde{T}/2) + \mathcal{O}(h^3) \tag{8.64b}$$

etc.,

but the extra terms are often tedious to find. As $\tilde{T}$ and $\tilde{U}$ appear in the exponent, these expressions do not seem very useful. However, as it follows directly from Eq. (8.60) that applying $\tilde{T}$ and $\tilde{U}$ more than once gives zero, we have simply

$$\exp(ah\tilde{T}) = 1 + ah\tilde{T} \tag{8.65}$$

and similarly for $\exp(bh\tilde{U})$. Therefore, the first order integrator is

$$p(t+h) = p(t) - h\{\partial U[x(t)]/\partial x\}; \tag{8.66a}$$

$$x(t+h) = x(t) + h\{\partial T[p(t+h)]/\partial p\} \tag{8.66b}$$

which is recognised as the Verlet algorithm (although with a less accurate definition of the momentum).

The second order integrator is given by

$$p(t+h/2) = p(t) - h\{\partial\tilde{U}[x(t)]/\partial x\}/2; \tag{8.67a}$$

$$x(t+h) = x(t) + h\{\partial\tilde{T}[p(t+h/2)]/\partial p\}; \tag{8.67b}$$

$$p(t+h) = p(t+h/2) - h\{\partial\tilde{U}[x(t+h)]/\partial x\}/2. \tag{8.67c}$$

Applying this algorithm successively, the first and third step can be merged into one, and we obtain precisely the Verlet algorithm in leap-frog form with a third order error in the time step h. This error seems puzzling since we know that the Verlet algorithm gives positions with an error of order h^4 and momenta with an error of order h^2. The solution to this paradox lies in the interpretation of the variable p. If at time t, $p(t)$ is the continuous time derivative of the continuum solution $x(t)$, the above algorithm gives us $x(t+h)$ and $p(t+h)$ both with error h^3. If however $p(t)$ is defined as $[x(t+h) - x(t-h)]/(2h)$, the algorithm is equivalent to the velocity-Verlet algorithm and hence gives the positions $x(t+h)$ with an error of order h^4 and $p(t+h)$ is according to its definition given with a h^2 error. The way in which initial conditions are given defines which case we are in.

Finding higher order algorithms is nontrivial as we do not know the form of the higher order expansion terms of the operators $\exp(h\tilde{T})$ and $\exp(h\tilde{U})$. However, Yoshida [21] proposed writing the fourth order integrator in the following form:

$$S_2(\alpha h)S_2(\beta h)S_2(\alpha h) \tag{8.68}$$

where S_2 is the second order integrator, and he fixed α and β by the requirement that the resulting expression is equal to the continuum operator to fourth order. Higher order integrators were found similarly. The general result can be written as

$$\begin{aligned}
&\text{for } k = 1 \text{ to } n \text{ do}\\
&\qquad x^{(k)} = x^{(k-1)} - ha_k \partial T[p^{(k-1)}]/\partial p \\
&\qquad p^{(k)} = p^{(k-1)} - hb_k \partial U[x^{(k)}]/\partial x\\
&\text{end}
\end{aligned} \tag{8.69}$$

and the numbers a_k and b_k can be found in Yoshida's paper. For the fourth order case, they read

$$a_1 = a_4 = 1/[2(2 - 2^{1/3})]; \quad a_2 = a_3 = (1 - 2^{1/3})a_1 \tag{8.70a}$$

$$b_1 = b_3 = 2a_2; \quad b_2 = -2^{1/3}b_1; \quad b_4 = 0. \tag{8.70b}$$

From Yoshida's derivation it follows that there exists a conserved quantity which acts as the analog of the energy. The integrator is certainly not the same as the exact time evolution operator, but it deviates from the latter only by a small amount. Writing the integrator $S(h)$ as

$$S(h) = \exp(hA_\mathrm{D}) \tag{8.71}$$

we have a new operator A_D which deviates from the continuum operator A only by an amount of order h^{n+1}, as the difference can be written as a sum of higher order CBH commutators. It will be shown in Problem 8.9 that for an operator of the form $\exp(tA_\mathrm{D})$ which is symplectic for all t, there exists a Hamiltonian $\mathcal{H}_\mathrm{D}$ which is the analogue of the Hamiltonian in the continuum time evolution. This means that, if we know $\mathcal{H}_\mathrm{D}$ (which is usually impossible to find, except for the trivial case of the harmonic oscillator), we could either use the integrator (8.71) to give us the image at time h, or the continuum solution of the dynamical system with Hamiltonian $\mathcal{H}_\mathrm{D}$ for $t = h$: both mappings would give identical results. The Hamiltonian $\mathcal{H}_\mathrm{D}(z)$ is therefore a conserved quantity of the integrator, and it differs from the energy by an amount of order h^{n+1}. The existence of such a conserved quantity is also discussed in Refs. [17, 18, 31].

8.5 Molecular dynamics methods for different ensembles

8.5.1 Constant temperature

In experimental situations the total energy is often not a control variable as usually the temperature of the system is kept constant. We know that in the infinite system the temperature is proportional to the average kinetic energy per degree of freedom with proportionality constant $k_B/2$, and therefore this quantity is used in MD to calculate the temperature, even though the system is finite (see Section 10.7 for a discussion on temperature for a finite system). As the total energy remains constant in the straightforward implementation of the molecular dynamics paradigm as presented in the previous sections, the question arises how we can perform MD simulations at constant temperature or pressure. We start with a brief overview of the various techniques which have been developed for keeping the temperature constant. Then we shall discuss the most successful one, the Nosé–Hoover method, in greater detail.

Overview of constant temperature methods

Experience from real life is a useful guide to understanding procedures for keeping the temperature at a constant value. In real systems, the temperature is usually kept constant by letting the system under consideration exchange heat with a much larger system in equilibrium – the heat bath. The latter has a definite temperature (it is in equilibrium) and the smaller system that we consider will assume the same temperature, as it has a negligible influence on the heat bath. Microscopically the heat exchange takes place through collisions of the particles in the system with the particles of the wall that separates the system from the heat bath. If, for example, the temperature of the heat bath is much higher than that of the system under consideration, the system particles will on average increase their kinetic energy considerably in each such collision. Through collisions with their partners in the system, the extra kinetic energy spreads through the system, and this process continues until the system has attained the temperature of the heat bath.

In a simulation we must therefore allow for heat flow from and to the system in order to keep it at the desired temperature. Ideally, such a heat exchange leads to a distribution ρ of configurations according to the canonical ensemble, irrespective of the number of particles:

$$\rho(R, P) = e^{-\mathcal{H}(R,P)/(k_B T)}, \tag{8.72}$$

but some of the methods described below yield distributions differing from this by a correction of order $1/N^k$, $k > 0$. In comparison with the experimental situation, we are not confined to allowing heat exchange only with particles at the boundary: any particle in the system can be coupled to the heat bath.

Several canonical MD methods have been developed in the past. In 1980 Andersen [32] devised a method in which the temperature is kept constant by replacing every so often the velocity of a randomly chosen particle by a velocity drawn from a Maxwell distribution with the desired temperature. This method is closest to the experimental situation: the velocity alterations mimic particle collisions with the walls. The rate at which particles should undergo these changes in velocity influences the equilibration time and the kinetic energy fluctuations. If the rate is high, equilibration will proceed quickly, but as the velocity updates are uncorrelated, they will destroy the long time tail of the velocity autocorrelation function. Moreover, the system will then essentially perform a random walk through phase space, which means that it moves relatively slowly. If on the other hand the rate is low, the equilibration will be very slow. The rate $R_{\text{collisions}}$ for which wall collisions are best mimicked by Andersen's procedure is of the order of

$$R_{\text{collisions}} \sim \frac{\kappa}{k_B n^{1/3} N^{2/3}} \tag{8.73}$$

where κ is the thermal conductivity of the system, and n, N the particle density and number respectively [32] (see Problem 8.9). Andersen's method leads to a canonical distribution for all N. The proof of this statement needs some theory concerning Markov chains and is therefore postponed to Section 15.4.3, where we consider the application of this method to lattice field theories.

For evaluating equilibrium expectation values for time- and momentum-independent quantities, the full canonical distribution (8.72) is not required: a canonical distribution in the positional coordinates

$$\rho(R) = e^{-U(R)/(k_B T)} \tag{8.74}$$

is sufficient since the momentum part can be integrated out for momentum-independent expectation values. For a sufficiently large system the total kinetic energy of a canonical system will evolve towards its equilibrium value $3Nk_BT/2$ and fluctuations around this value are very small. We might therefore force the kinetic energy to have a value exactly equal to the one corresponding to the desired temperature. This means that we replace the narrow distribution of the kinetic energy by a delta-function

$$\rho(E_{\text{kin}}) \to \delta[E_{\text{kin}} - 3(N-1)k_BT/2]. \tag{8.75}$$

The simplest way of achieving this is by applying a simple velocity rescaling procedure as described in the previous section (Eqs. (8.14) and (8.15)) after *every* integration step rather than occasionally:

$$p_i \to p_i \sqrt{\frac{3/2(N-1)k_BT}{E_{\text{kin}}}}. \tag{8.76}$$

This method can also be derived by imposing a constant kinetic energy via a Lagrange multiplier term added to the Lagrangian of the isolated system [33]. It turns out [34] that this velocity rescaling procedure induces deviations from the canonical distribution of order $1/\sqrt{N}$, where N is the number of particles.

Apart from the rescaling method, which is rather *ad hoc*, there have been attempts to introduce the coupling via an extra force acting on the particles with the purpose of keeping the temperature constant. This force assumes the form of a friction proportional to the velocity of the particles, as this is the most direct way to affect velocities and hence the kinetic energy:

$$m\ddot{\mathbf{r}}_i = \mathbf{F}_i(R) - \zeta(R,\dot{R})\dot{\mathbf{r}}_i. \tag{8.77}$$

The parameter ζ acts as a friction parameter which is the same for all particles and which will be negative if heat is to be added and positive if heat must be drained from the system. Various forms for ζ have been used, and as a first example we consider [33, 35]

$$\zeta(R,\dot{R}) = \frac{\mathrm{d}V(R)/\mathrm{d}t}{\sum_i \dot{\mathbf{r}}_i^2}. \tag{8.78}$$

This force keeps the kinetic energy $K = m\sum_i v_i^2/2$ constant as can be seen using (8.77). From this equation, we obtain

$$\frac{\partial K}{\partial t} \propto \sum_i \mathbf{v}_i\dot{\mathbf{v}}_i = -\sum_i \mathbf{v}_i[\nabla_i V(R) - \zeta(R,\dot{R})\mathbf{v}_i] = \frac{\mathrm{d}V}{\mathrm{d}t} - \sum_i \dot{\mathbf{r}}_i^2\zeta(R,\dot{R}) = 0. \tag{8.79}$$

It can be shown [34] that for finite systems the resulting distribution is purely canonical (without $1/N^k$ corrections) in the restricted sense, i.e. in the coordinate part only.

Another form of the friction parameter ζ was proposed by Berendsen *et al.* [36] This now has the form $\zeta = \gamma(1 - T_\mathrm{D}/T)$ with constant γ, T is the actual temperature $T = \sum_i mv_i^2/(3k_\mathrm{B})$, and T_D is the desired temperature. It can be shown that the temperature decays to the desired temperature exponentially with time at rate given by the coefficient γ. However, this method is not time reversible; moreover, it can be shown that the Nosé method (see below) is the only method with a single friction parameter which gives a full canonical distribution [37], so Berendsen's method cannot have this property. Berendsen's method can be related to a Langevin description of thermal coupling, in the sense that the time evolution of the temperature for a Langevin system (see Section 8.8) can be shown to be equivalent to that of a system with a coupling via ζ as given by Berendsen.

Nosé's method in the formulation by Hoover [37] uses yet another friction parameter ζ which is now determined by a differential equation:

$$\frac{\mathrm{d}\zeta}{\mathrm{d}t} = \left(\sum_i v_i^2 - 3Nk_\mathrm{B}T_\mathrm{D}\right)/Q \tag{8.80}$$

where Q is a parameter which has to be chosen with some care (see below) [38]. This way of keeping the temperature constant yields the canonical distribution for positions and momenta, as will be shown in the next subsection.

The Nosé and the Andersen methods yield precise canonical distributions for position and momentum coordinates. They still have important disadvantages, however.

In the Andersen method, it is not always clear at which rate the velocities are to be altered and it has been found [39, 40] that the temperature sometimes levels down at the wrong value. The Nosé–Hoover thermostat suffers from similar problems. In this method, the coupling constant Q in Eq. (8.80) between the heat bath and the system must be chosen – this coupling constant is the analogue of the velocity alteration rate in the Andersen method. It turns out [38] that for a Lennard–Jones fluid at high temperatures, the canonical distribution comes out well, but if the temperature is lowered [26], the temperature starts oscillating with an amplitude much larger than the standard deviation expected in the canonical ensemble. It can also occur that such oscillations are much smaller than the expected standard deviation, but in this case the fluctuations on top of this oscillatory behaviour are much smaller than in the canonical ensemble. Martyna *et al.* [41] have devised a variant of the Nosé–Hoover thermostat which is believed to alleviate these problems to some extent. Although the difficulties with these constant temperature approaches are very serious, they have received rather little attention to date. It should be clear that it must always be checked explicitly whether the temperature shows unusual behaviour; in particular, it should not exhibit systematic oscillations, and the standard deviation for N particles in D dimensions should satisfy

$$\overline{\Delta T} = \sqrt{\frac{2}{ND}}\,\overline{T} \tag{8.81}$$

where $\overline{\Delta T}$ is the width of the temperature distribution and $\overline{T}$ is the mean value [26]. This equation follows directly from the Boltzmann distribution.

*Derivation of the Nosé–Hoover thermostat

In this section we shall discuss Nosé's approach [34, 42], in which the heat bath is explicitly introduced into the system in the form of a single degree of freedom s. The Hamiltonian of the total (extended) system is given as

$$\mathcal{H}(P, R, p_s, s) = \sum_i \frac{\mathbf{p}_i^2}{2ms^2} + \frac{1}{2}\sum_{ij,i\neq j} U(\mathbf{r}_i - \mathbf{r}_j) + \frac{p_s^2}{2Q} + gkT\ln(s). \tag{8.82}$$

g is the number of independent momentum-degrees of freedom of the system (see below), and R and P represent all the coordinates $\mathbf{r}_i$ and $\mathbf{p}_i$ as usual. The physical quantities R, P and t (time) are virtual variables – they are related to real variables R', P' and t' via $R' = R$, $P' = P/s$ and $t' = \int^t d\tau/s$. With these definitions we have for the real variables $P' = dQ'/dt'$.

First we derive the equations of motion in the usual way:

$$\frac{d\mathbf{r}_i}{dt} = \frac{\partial \mathcal{H}}{\partial \mathbf{p}_i} = \frac{\mathbf{p}_i}{ms^2} \tag{8.83a}$$

$$\frac{ds}{dt} = \frac{\partial \mathcal{H}}{\partial p_s} = \frac{p_s}{Q} \tag{8.83b}$$

$$\frac{d\mathbf{p}_i}{dt} = -\frac{\partial \mathcal{H}}{\partial \mathbf{r}_i} = -\nabla_i U(R) = -\sum_{i<j} \nabla_i U(\mathbf{r}_i - \mathbf{r}_j) \tag{8.83c}$$

$$\frac{dp_s}{dt} = -\frac{\partial \mathcal{H}}{\partial s} = \left(\frac{\sum_i p_i^2}{ms^2} - gk_B T \right) \Big/ s. \tag{8.83d}$$

We have used the notation $\partial\mathcal{H}/\partial\mathbf{p}_i = \nabla_{\mathbf{p}_i}\mathcal{H}$, etc. The partition function of the total system (i.e. including heat bath degree of freedom s) is given by the expression:

$$Z = \frac{1}{N!} \int dp_s \int ds \int dP \int dR \\ \times \delta\left(\sum_i \frac{p_i^2}{2ms^2} + \frac{1}{2}\sum_{ij,i\neq j} U(r_{ij}) + \frac{p_s^2}{2Q} + gk_B T \ln(s) - E \right). \tag{8.84}$$

Integrations $\int dR$ and $\int dP$ are over all position and momentum degrees of freedom. We now rescale the momenta $\mathbf{p}_i$:

$$\frac{\mathbf{p}_i}{s} = \mathbf{p}_i', \tag{8.85}$$

so that we can rewrite the partition function as

$$Z = \frac{1}{N!} \int dp_s \int ds \int dP' \int dR \\ \times s^{3N} \delta\left(\sum_i \frac{p_i'^2}{2m} + \frac{1}{2}\sum_{ij,i\neq j} U(r_{ij}) + \frac{p_s^2}{2Q} + gk_B T \ln(s) - E \right). \tag{8.86}$$

We define the Hamiltonian $\mathcal{H}_0$ in terms of R and P' as

$$\mathcal{H}_0 = \sum_i \frac{p_i'^2}{2m} + \frac{1}{2}\sum_{ij,i\neq j} U(r_{ij}). \tag{8.87}$$

Furthermore we use the relation $\delta[f(s)] = \delta(s - s_0)/f'(s)$ with $f(s_0) = 0$ and set $g = 3N + 1$, so that we can rewrite Eq. (8.86) as

$$\begin{aligned} Z &= \frac{1}{N!}\int \mathrm{d}p_s \int \mathrm{d}s \int \mathrm{d}P' \int \mathrm{d}R \frac{s^{3N+1}}{(3N+1)k_\mathrm{B}T} \\ &\quad \times \delta\left(s - \exp\left[-\frac{\mathcal{H}_0(P', R) + p_s^2/2Q - E}{(3N+1)k_\mathrm{B}T}\right]\right) \\ &= \frac{1}{(3N+1)k_\mathrm{B}T}\frac{1}{N!}\int \mathrm{d}p_s \int \mathrm{d}P' \int \mathrm{d}R \exp\left[-\frac{\mathcal{H}_0(P', R) + p_s^2/2Q - E}{k_\mathrm{B}T}\right]. \end{aligned} \tag{8.88}$$

The dependence on p_s is simply Gaussian and integrating over this coordinate we obtain

$$Z = \frac{1}{3N+1}\left(\frac{2\pi Q}{k_\mathrm{B}T}\right)^{1/2} \exp(E/k_\mathrm{B}T) Z_\mathrm{c} \tag{8.89}$$

where Z_c is the canonical partition function:

$$Z_\mathrm{c} = \frac{1}{N!}\int \mathrm{d}P' \int \mathrm{d}R \exp[-\mathcal{H}_0(P', R)/k_\mathrm{B}T]. \tag{8.90}$$

It follows that the expectation value of a quantity A which depends on R and P is given by

$$\langle A(P/s, R)\rangle = \langle A(P', R)\rangle_\mathrm{c} \tag{8.91}$$

where $\langle \cdots \rangle_\mathrm{c}$ denotes an average in the canonical ensemble. The ergodic hypothesis relates this ensemble average to a virtual-time average.

The Lagrangian equations of motion for the $\mathbf{r}_i$ can be obtained by eliminating the momenta from (8.83a):

$$\frac{\mathrm{d}^2\mathbf{r}_i}{\mathrm{d}t^2} = -\frac{1}{ms^2}\nabla_i V(R) - \frac{2}{s}\frac{\mathrm{d}\mathbf{r}_i}{\mathrm{d}t}\frac{\mathrm{d}s}{\mathrm{d}t}. \tag{8.92}$$

In this equation the ordinary force term is recognised with a factor $1/s^2$ in front and with an additional friction term describing the coupling to the heat bath. The factor $1/s^2$ is consistent with the relation between real and virtual-time $\mathrm{d}t' = \mathrm{d}t/s$ given above. Together with the definitions $P' = P/s$ and $p_s' = p_s/s$, this leads to the equations of motion in real variables:

$$\frac{\mathrm{d}\mathbf{r}_i'}{\mathrm{d}t'} = \frac{\mathbf{p}_i'}{m} \tag{8.93a}$$

$$\frac{\mathrm{d}\mathbf{p}_i'}{\mathrm{d}t'} = -\nabla_i V(R) - s p_s' \mathbf{p}_i'/Q \tag{8.93b}$$

$$\frac{\mathrm{d}s}{\mathrm{d}t'} = s'^2 p_s'/Q \tag{8.93c}$$

$$\frac{\mathrm{d}p_s'}{\mathrm{d}t'} = \left(\sum_i p_i'^2/m - gk_\mathrm{B}T\right)/s - s^2 p_s'^2/Q. \tag{8.93d}$$

Although these equations are equivalent to the equations for the virtual variables, there is a slight complication in the evaluation of averages. The point is that we have used the ergodic theorem for the canonical Hamiltonian expressed in virtual variables (P, R, t, s, p_s) in order to relate *virtual-time* averages to ensemble averages. The real time steps however are not equidistant and time averaging in real time is therefore not equivalent to averaging in virtual-time. Fortunately the two can be related. Expressing the real time t' as an integral over virtual-time τ according to $t' = \int_0^t \mathrm{d}\tau/s$ we obtain

$$\begin{aligned}
\lim_{t'\to\infty} \frac{1}{t'} \int_0^{t'} A(P/s, R)\mathrm{d}\tau' &= \lim_{t'\to\infty} \frac{t}{t'}\frac{1}{t} \int_0^t A(P/s, R)\mathrm{d}\tau/s \\
&= \left[\lim_{t'\to\infty} \frac{1}{t} \int_0^t A(P/s, R)\mathrm{d}\tau/s\right] \Big/ \left(\lim_{t'\to\infty} \frac{1}{t} \int_0^t \mathrm{d}\tau/s\right) \\
&= \langle A(P/s, R)/s\rangle/\langle 1/s\rangle.
\end{aligned} \tag{8.94}$$

It can be verified (see Problem 8.9) that the latter expression coincides with the canonical ensemble average if we put g equal to $3N$ instead of $3N + 1$. This means that if we carry out the simulation using Eqs. (8.93) with $g = 3N$, real-time averages are equivalent to canonical averages.

Hoover [37] showed that by defining $\zeta = sp_s'/Q$, Eqs. (8.93) can be reduced to the simpler form

$$\frac{\mathrm{d}\mathbf{r}_i'}{\mathrm{d}t'} = \frac{\mathbf{p}_i'}{m}; \quad \frac{\mathrm{d}\mathbf{p}_i'}{\mathrm{d}t'} = \mathbf{F}_i - \zeta \mathbf{p}_i'; \tag{8.95}$$

$$\frac{\mathrm{d}\zeta}{\mathrm{d}t'} = \left(\frac{\sum_i p_i'^2}{m} - gk_\mathrm{B}T\right)/Q, \tag{8.96}$$

and taking g equal to the number of degrees of freedom, i.e. $3N$, he was able to show that the distribution $f(P, R, \zeta)$ is phase space conserving, i.e. it satisfies Liouville's equation.

The disadvantage of the real-time equations is that they are not Hamiltonian, in the sense that they cannot be derived from a Hamiltonian. Although this might not seem to be a problem, we prefer Hamiltonian equations of motion as they allow for stable (symplectic) integration methods as discussed in Section 8.4. Winkler *et al.* [43] have formulated canonical equations of motion in real-time but these are subject to severe numerical problems when integrating the equations of motion for large systems.

8.5.2 Keeping the pressure constant

In experimental situations not only the temperature is kept under control but also the pressure. The partition function for the (NpT)-ensemble is given as

$$Q(N,p,T) = \int \mathrm{d}V\, \mathrm{e}^{-\beta pV} \frac{1}{N!} \int \mathrm{d}R\, \mathrm{d}P\, \mathrm{e}^{-\beta \mathcal{H}(R,P)} = \int \mathrm{d}V\, \mathrm{e}^{-\beta pV} Z_{\mathrm{c}}(N,V,T) \tag{8.97}$$

(see Chapter 7). We use a lower-case p for the pressure in order to avoid confusion with the total momentum coordinate P. We now describe the scheme which is commonly adopted for keeping the pressure constant but do not go into too much detail as the analysis follows the same lines as the Nosé–Hoover thermostat, and refer to the literature for details [32, 34, 37].

Andersen first presented this scheme. He proposed incorporating the volume into the equations of motion as a dynamical variable and scaled the spatial coordinates back to a unit volume:

$$\mathbf{r}_i' = \mathbf{r}_i V^{1/3}, \tag{8.98}$$

where again the prime denotes the real coordinate – unprimed coordinates are those of the virtual system. Moreover

$$\mathbf{p}_i' = \mathbf{p}_i/(sV^{1/3}). \tag{8.99}$$

The canonical Hamiltonian is extended by *two* variables, the volume V and the canonical momentum p_V which can be thought of as the momentum of a piston closing the system.[7] The Hamiltonian has an extra 'potential energy' term pV and a 'kinetic' term $p_V^2/2W$ (W is the 'mass' of the piston, and p_V its momentum):

$$\mathcal{H}(P,R,p_s,s,p_V,V) = \sum_i \frac{\mathbf{p}_i^2}{2mV^{2/3}s^2} + \frac{1}{2}\sum_{ij,i\neq j} U[V^{1/3}R] + \frac{p_s^2}{2Q} + gkT\ln(s) + pV + p_V^2/2W. \tag{8.100}$$

[7] Note that the system expands and contracts isotropically, so instead of a piston, the whole system boundary moves.

The equations of motion now read:

$$\frac{\mathrm{d}\mathbf{r}}{\mathrm{d}t} = \frac{\partial \mathcal{H}}{\partial \mathbf{p}_i} = \frac{\mathbf{p}_i}{mV^{2/3}s^2} \tag{8.101a}$$

$$\frac{\mathrm{d}s}{\mathrm{d}t} = \frac{\partial \mathcal{H}}{\partial p_s} = \frac{p_s}{Q} \tag{8.101b}$$

$$\frac{\mathrm{d}\mathbf{p}_i}{\mathrm{d}t} = -\frac{\partial \mathcal{H}}{\partial \mathbf{r}_i} = -\nabla_i U(V^{1/3}R) \tag{8.101c}$$

$$\frac{\mathrm{d}p_s}{\mathrm{d}t} = -\frac{\partial \mathcal{H}}{\partial s} = \left(\frac{\sum_i p_i^2}{mV^{2/3}s^2} - gk_\mathrm{B}T \right) / s \tag{8.101d}$$

$$\frac{\mathrm{d}V}{\mathrm{d}t} = \frac{\partial \mathcal{H}}{\partial p_V} = \frac{p_V}{W} \tag{8.101e}$$

$$\frac{\mathrm{d}p_V}{\mathrm{d}t} = -\frac{\partial \mathcal{H}}{\partial V} = \left(\frac{\sum_i p_i^2}{mV^{2/3}s^2} - \sum_i \nabla_i U(V^{1/3}R) \cdot \mathbf{r}_i \right) / (3V) - p. \tag{8.101f}$$

It can be shown in the same way as in the thermostat case that the distribution of configurations corresponds to that of the (NpT) ensemble:

$$\rho(P', R', V) = V^N \exp\{-[\mathcal{H}_0(P', R') + pV]/k_\mathrm{B}T\}. \tag{8.102}$$

Hoover [37] proposed similar equations of motion which conserve phase space, but they differ from this distribution by an extra factor V in front of the exponent [44].

The method described is restricted to isotropic volume changes and can therefore not be used for studying structural phase transitions in solids. A method which allows for anisotropic volume changes while keeping the pressure constant was developed by Parrinello and Rahman [45].

8.6 Molecular systems

8.6.1 Molecular degrees of freedom

Interactions in molecular systems can be divided into intramolecular and intermolecular ones. The latter are often taken to be atom-pair interactions similar to those considered in the previous sections. The intramolecular interactions (i.e. the interactions between the atoms of one molecule) are determined by chemical bonds, so not only are they strong compared with the intermolecular interactions (between atoms of different molecules) but they also include orientational dependencies. We now briefly describe the intramolecular degrees of freedom and interactions (see also Figure 8.4).

First of all, the chemical bonds can stretch. The interaction associated with this degree of freedom is usually described in the form of a harmonic potential for the

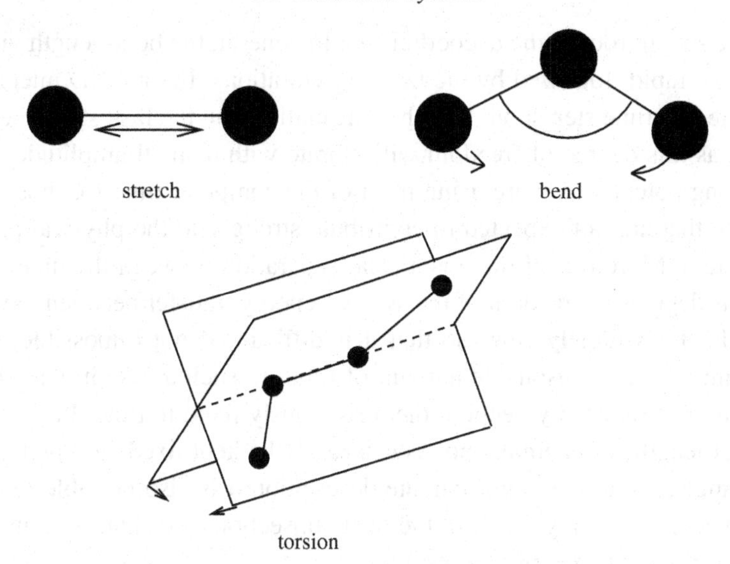

Figure 8.4. Different types of motion of atoms within a molecule.

bond length l:

$$V_{\text{stretch}}(l) = \tfrac{1}{2}\alpha_{\text{S}}(l - l_0)^2 \tag{8.103}$$

where l_0 is the equilibrium bond length.

Now consider three atoms bonded in a chain-like configuration A–B–C. This chain is characterised by a bending or *valence angle* φ which varies around an equilibrium value φ_0 and the potential is described in terms of a cosine:

$$V_{\text{valence}}(\varphi) = -\alpha_{\text{B}}[\cos(\varphi - \varphi_0) + \cos(\varphi + \varphi_0)] \tag{8.104}$$

where the equivalence of the angles φ_0 and $-\varphi_0$ is taken into account. Often, a harmonic approximation $\cos(\varphi) \approx 1 - \varphi^2/2$, valid for small angles, is used.

Finally there is an interaction associated with chain configurations of four atoms A–B–C–D. The plane through the first three atoms A, B, C does not coincide in general with that through B, C and D. The *torsion* interaction is similar to the bend interaction, but the angle (called *dihedral* angle), denoted by ϑ, is now that between these two planes:

$$V_{\text{torsion}}(\vartheta) = -\alpha_{\text{T}}[\cos(\vartheta - \vartheta_0) + \cos(\vartheta + \vartheta_0)]. \tag{8.105}$$

This interaction is also often replaced by its harmonic approximation. Other interactions and more complicated forms of these potentials can be used – we have only listed the simplest ones.

Characteristic vibrations associated with the different degrees of freedom distinguished here can be derived from the harmonic interactions – the frequencies vary as the square root of the α-coefficients. In general, the bond length vibrations are the most rapid, followed by the bending vibrations. For an MD integration to be accurate, the time step should be chosen smaller than the fastest degree of freedom. But as this degree of freedom will vibrate with a small amplitude, because of the strong potential, we are using most of the computer time for those parts of the motion that are not expected to contribute strongly to the physical properties of the system. Moreover, if there is a clear separation between the time scales of the various degrees of freedom of the system, energy transfer between the fast and slow modes is extremely slow, so that it is difficult, if not impossible, to reach equilibrium within a reasonable amount of time. In such a case it is advisable to 'freeze' the fast modes by keeping them rigorously fixed in time. In practice this means that lengths of chemical bonds can safely be kept fixed, and perhaps some bending angles. In a more approximate description it is also possible to consider entire molecules as being rigid. In the next subsections we shall describe how to deal with rigid and partly rigid molecules.

8.6.2 Rigid molecules

We consider molecules which can be treated as rigid bodies whose motion consists of translations of the centre of mass and rotations around this point. The forces acting between two rigid molecules are usually composed of atomic pair interactions between atoms belonging to the two different molecules.[8] The total force acting on a molecule determines the translational motion and the torque determines the rotational motion. In the next subsection, we shall describe a direct formulation of the equations of motion of a simple rigid molecule – the nitrogen molecule. In the following subsection we shall then describe a different approach in which rigidity is enforced through constraints added to the Lagrangian.

Direct approach for the rigid nitrogen molecule

As a simple example consider the nitrogen molecule, N_2. This consists of two nitrogen atoms, each of mass $m \approx 14$ atomic mass units (a.m.u.) and whose separation d is kept fixed in the rigid approximation. The coordinates of the molecule are the three coordinates of the centre of mass and the two coordinates defining its orientation. The latter can be polar angles but here we shall characterise the orientation of the molecule by a unit direction vector $\hat{\mathbf{n}}$, pointing from atom 1 to atom 2 (see Figure 8.5).

The motion of the centre of mass of the molecule is determined by the total force $\mathbf{F}_{\text{tot}}$ acting on a particular molecule. This force is the sum of all the forces between

[8] Sometimes, off-centre interactions (i.e. not centred on the atomic positions) are taken into account too but we shall not consider these.

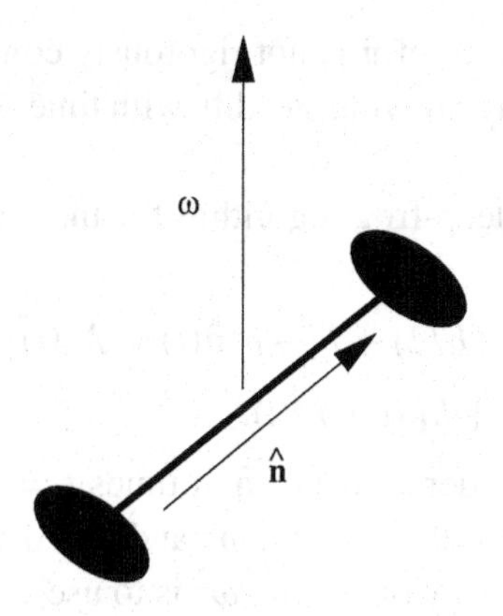

Figure 8.5. The nitrogen molecule. $\hat{\mathbf{n}}$ is a unit vector, $\boldsymbol{\omega}$ is the rotation vector.

each of the two atoms in the molecule and the atoms of the remaining molecules. The atomic forces can be modelled by a Lennard–Jones interaction with the appropriate atomic nitrogen parameters $\sigma = 3.31$ Å, $\varepsilon/k_B = 37.3$ K and $d = 0.3296\sigma$ [46]. The equation of motion for the centre of mass $\mathbf{R}_{CM}$ is then

$$\ddot{\mathbf{R}}_{CM} = \mathbf{F}_{tot}, \tag{8.106}$$

which can be solved in exactly the same way as in an ordinary MD simulation.

The motion of the orientation vector $\hat{\mathbf{n}}$ is determined by the torque $\mathcal{N}$ with respect to the centre of the molecule, which is given in terms of the forces $\mathbf{F}^{(1)}$ and $\mathbf{F}^{(2)}$ acting on atoms 1 and 2 respectively:

$$\mathcal{N} = (d/2)\hat{\mathbf{n}} \times (\mathbf{F}^{(1)} - \mathbf{F}^{(2)}). \tag{8.107}$$

The torque changes the angular momentum $\mathbf{L}$ of the molecule. This is equal to $I\boldsymbol{\omega}$, where I is the moment of inertia $md^2/2$ and $\boldsymbol{\omega}$ is the angular frequency vector whose norm is the angular frequency and whose direction is the axis around which the rotation takes place (see Figure 8.5). Note that $\mathcal{N}$ is not necessarily parallel to $\boldsymbol{\omega}$. The equation of motion for the angular momentum is $\dot{\mathbf{L}} = \mathcal{N}$ or

$$I\dot{\boldsymbol{\omega}} = \mathcal{N}. \tag{8.108}$$

The angular frequency $\boldsymbol{\omega}$ is in turn related to the time derivative of the direction vector $\hat{\mathbf{n}}$:

$$\dot{\hat{n}} = \boldsymbol{\omega} \times \hat{\mathbf{n}}. \tag{8.109}$$

Combining Eqs. (8.108) and (8.109) leads to

$$\ddot{\hat{n}} = \boldsymbol{\omega} \times (\boldsymbol{\omega} \times \hat{\mathbf{n}}) + \mathcal{N} \times \hat{\mathbf{n}}/I = -\omega^2\hat{\mathbf{n}} + \mathcal{N} \times \hat{\mathbf{n}}/I. \tag{8.110}$$

This equation of motion leaves the norm of the direction vector $\hat{\mathbf{n}}$ invariant, as it should – this follows directly from (8.109). In a numerical integration of the equations of motion the norm of $\hat{\mathbf{n}}$ is not rigorously conserved; it can suffer from numerical errors which may growing steadily with time. We shall now see how this can be avoided.

Let us write down the leap-frog algorithm for the equation of motion (8.110) for $\hat{\mathbf{n}}$:

$$\mathbf{p}(t + h/2) = \mathbf{p}(t - h/2) + h[-\omega^2\hat{\mathbf{n}}(t) + \mathcal{N}(t) \times \hat{\mathbf{n}}(t)/I] \tag{8.111a}$$

$$\hat{\mathbf{n}}(t + h) = \hat{\mathbf{n}}(t) + h\mathbf{p}(t + h/2). \tag{8.111b}$$

Here $\mathbf{p}$ represents the time-derivative of $\hat{\mathbf{n}}$ at times $t = (n + 1/2)h$. The problem with this algorithm is that it depends on ω^2 and this depends in turn on the time derivative $\dot{\hat{n}}$. A convenient way of finding ω^2 is to use

$$\mathbf{p}(t - h/2) = \mathbf{p}(t) - \frac{h}{2}(-\omega^2\hat{\mathbf{n}} + \mathcal{N} \times \hat{\mathbf{n}}/I) + \mathcal{O}(h^2), \tag{8.112}$$

so that, using $\hat{\mathbf{n}}(t) \cdot \mathbf{p}(t) = 0$, we obtain

$$2\mathbf{p}(t - h/2) \cdot \hat{\mathbf{n}}(t) = h\omega^2 + \mathcal{O}(h^2). \tag{8.113}$$

Calling the left hand side of this equation λ, we have [2, 47]

$$\lambda = 2\mathbf{p}(t - h/2) \cdot \hat{\mathbf{n}} \tag{8.114a}$$

$$\mathbf{p}(t + h/2) = \mathbf{p}(t - h/2) + h[\hat{\mathbf{n}}(t) \times \mathcal{N}(t)/I - \lambda\hat{\mathbf{n}}(t)] \tag{8.114b}$$

$$\hat{\mathbf{n}}(t + h) = \hat{\mathbf{n}}(t) + h\mathbf{p}(t + h/2). \tag{8.114c}$$

The continuum equations of motion guaranteed conservation of the norm of the unit vector $\hat{\mathbf{n}}$. The leap-frog algorithm will enforce this normalisation only up to an error of h^3. It is therefore sensible to normalise $\hat{\mathbf{n}}$ after every time step – the parameter λ can then be viewed as the Lagrange multiplier associated with the constraint $|\hat{\mathbf{n}}|^2 = 1$. In the next section we shall discuss a simpler method for simulating liquid nitrogen, using constraints in a different way.

For general molecules, we have an extra degree of freedom: the angle of rotation around a molecular axis. This is the third Euler angle, which is usually denoted as γ. The straightforward procedure for solving the equations of motion is to calculate the principal angular velocity $\boldsymbol{\omega}$ in terms of the time derivatives of the Euler angles. The Euler equation of motion gives the rate of change in $\boldsymbol{\omega}$ in terms of the torque. The time derivatives of the Euler angles can then be found again from $\boldsymbol{\omega}$, and these can be used to calculate the new atomic positions. There is however a problem when the Euler angle $\theta = 0$, as in that case the transformation from $\boldsymbol{\omega}$ to the time derivatives of the Euler angles becomes singular. Evans has discussed this problem and has presented methods to avoid the instability resulting from this singularity [48]. The most efficient one is to use the quaternion representation, in which the orientation of the molecule is defined in terms of a four-dimensional unit vector rather than three Euler angles. This method was implemented by Evans and Murad [49].

Enforcing rigidity via constraints

Another method for treating rigid molecules is by imposing holonomic constraints, i.e. constraints which depend only on positions and not on the velocities, through an extension of the Lagrangian. The Lagrangian of the system without constraints reads

$$L_0(R, \dot{R}) = \int_{t_0}^{t_1} \mathrm{d}t \Big[\sum_i \frac{m_i}{2} \dot{\mathbf{r}}_i^2 - \frac{1}{2} \sum_{i \neq j} U(\mathbf{r}_i - \mathbf{r}_j) \Big]. \tag{8.115}$$

A constraint is introduced as usual through a Lagrange multiplier λ [50]. As the constraint under consideration should hold for all times, λ is a function of t. A simple example of a constraint is the following: particles 1 and 2 have a fixed separation d for all times (this could be the separation of the two atoms in a nitrogen molecule). Such a constraint on the separation is called *bond constraint* – it can formally be written as

$$\sigma[R(t)] = [\mathbf{r}_1(t) - \mathbf{r}_2(t)]^2 - d^2 = 0. \tag{8.116}$$

The Lagrangian for the system with this constraint reads

$$L(R, \dot{R}) = L_0(R, \dot{R}) - \int_{t_0}^{t_1} \mathrm{d}t\, \lambda(t) \{[\mathbf{r}_1(t) - \mathbf{r}_2(t)]^2 - d^2\}. \tag{8.117}$$

The integral over time is needed because the constraint holds for all times between t_0 and t_1. The equations of motion are the Euler–Lagrange equations for this Lagrangian. These equations will depend on the Lagrange parameters, λ, whose values are determined by the requirement that the solution must satisfy the constraint.

A slightly more complicated example is the trimer molecule CS_2 [51]. The linear geometry of this molecule is in principle imposed automatically by the correct bond constraints between the three pairs of atoms. However, the motion of this molecule is described by five positional degrees of freedom: two to define the orientation of the molecule and three for the centre of mass position. The three atoms without constraints have nine degrees of freedom and if three of these are eliminated using the bond constraints, we are left with six degrees of freedom instead of the five required. Therefore one redundant degree of freedom is included in this procedure, which is obviously inefficient. A better procedure is therefore to fix only the distance between the two sulphur atoms:

$$|\mathbf{r}_{S(1)} - \mathbf{r}_{S(2)}|^2 = d^2 \tag{8.118}$$

and to fix the position of the C-atom by a linear vector constraint:

$$(\mathbf{r}_{S(1)} + \mathbf{r}_{S(2)})/2 - \mathbf{r}_C = 0, \tag{8.119}$$

adding up to the four constraints required.

For a molecule, in general a number of atoms forming a 'backbone' set is identified and these are fixed by bond constraints (the two sulphur atoms in our example) and the remaining ones are fixed by linear constraints of the form (8.119). In the case of a planar structure we take three noncollinear atoms as a backbone. These atoms satisfy three bond constraints and the remaining atoms are constrained linearly. In a three-dimensional molecular structure, four backbone atoms are subject to six bond constraints and the remaining ones to a linear vector constraint each. In the constraint procedure, the degrees of freedom of the nonbackbone atoms are eliminated so that the forces they feel are transferred to the backbone. This elimination is always possible for linear constraints such as those obeyed by the nonbackbone atoms.

Let us now return to our CS_2 example. First we write down the equations of motion for all three atoms, following from the extended Lagrangian (the Lagrange parameter for the bond constraint is called λ, that of the linear vector constraint $\boldsymbol{\mu}$):

$$m_S\ddot{\mathbf{r}}_{S(1)} = \mathbf{F}_1 - 2\lambda(\mathbf{r}_{S(1)} - \mathbf{r}_{S(2)}) - \boldsymbol{\mu}/2 \tag{8.120a}$$

$$m_S\ddot{\mathbf{r}}_{S(2)} = \mathbf{F}_2 + 2\lambda(\mathbf{r}_{S(1)} - \mathbf{r}_{S(2)}) - \boldsymbol{\mu}/2 \tag{8.120b}$$

$$m_C\ddot{\mathbf{r}}_C = \mathbf{F}_C + \boldsymbol{\mu}. \tag{8.120c}$$

The linear constraint (8.119) is now differentiated twice with respect to time, and using the equations of motion we obtain

$$\mathbf{F}_C + \boldsymbol{\mu} = \frac{m_C}{2m_S}(\mathbf{F}_1 + \mathbf{F}_2 - \boldsymbol{\mu}). \tag{8.121}$$

We can thus eliminate $\boldsymbol{\mu}$ in the equations of motion for the S-atoms and obtain, with $M = 2m_S + m_C$:

$$m_S\ddot{\mathbf{r}}_{S(1)} = \left(1 - \frac{m_C}{2M}\right)\mathbf{F}_1 - \frac{m_C}{2M}\mathbf{F}_2 + \frac{m_S}{M}\mathbf{F}_C - 2\lambda(\mathbf{r}_{S(1)} - \mathbf{r}_{S(2)}); \tag{8.122a}$$

$$m_S\ddot{\mathbf{r}}_{S(2)} = \left(1 - \frac{m_C}{2M}\right)\mathbf{F}_2 - \frac{m_C}{2M}\mathbf{F}_1 + \frac{m_S}{M}\mathbf{F}_C + 2\lambda(\mathbf{r}_{S(1)} - \mathbf{r}_{S(2)}). \tag{8.122b}$$

These equations define the algorithm for the positions of the S-atoms, and the position of the C-atom is fixed at any time by the linear constraint.

Note that we still have an unknown parameter λ present in the resulting equations: this parameter is fixed by demanding that the bond constraint must hold for $\mathbf{r}_{S(1)}$ and $\mathbf{r}_{S(2)}$ at all times (note that we have not yet used this constraint). It is not easy to eliminate λ from the equations of motion as we have done with $\boldsymbol{\mu}$, as the bond length constraint is quadratic. Instead, we solve for λ at each time step using the constraint equation. We outline this procedure for our example. Suppose we have the positions $\mathbf{r}_{S(1)}$ and $\mathbf{r}_{S(2)}$ at times t and $t-h$ and that for both these time instances the bond constraint is satisfied. According to the equations of motion (8.122) in the

Verlet scheme, predictions for the positions at $t+h$ are given by

$$\mathbf{r}_{S^{(1)}}(t+h) = 2\mathbf{r}_{S^{(1)}}(t) - \mathbf{r}_{S^{(1)}}(t-h) + h^2\left(1 - \frac{m_C}{M}\right)\mathbf{F}_1(t) - h^2\frac{m_C}{M}\mathbf{F}_2 + h^2\frac{m_S}{M}F_C(t) - 2h^2\lambda(t)[\mathbf{r}_{S^{(1)}}(t) - \mathbf{r}_{S^{(2)}}(t)]; \quad (8.123a)$$

$$\mathbf{r}_{S^{(2)}}(t+h) = 2\mathbf{r}_{S^{(2)}}(t) - \mathbf{r}_{S^{(2)}}(t-h) + h^2\left(1 - \frac{m_C}{M}\right)\mathbf{F}_2(t) - h^2\frac{m_C}{M}\mathbf{F}_1 + h^2\frac{m_S}{M}\mathbf{F}_C(t) + 2h^2\lambda(t)[\mathbf{r}_{S^{(1)}}(t) - \mathbf{r}_{S^{(2)}}(t)]. \quad (8.123b)$$

The predictions for the positions at $t+h$ are linear functions of λ and if we substitute them into the bond constraint (8.118), we obtain a quadratic equation for λ which can be solved trivially. Of the two solutions, we keep the smallest value of λ. This means that the bond constraint is now satisfied to computer precision for all times. It should be noted that the value of λ in this procedure will deviate slightly from its value in the exact solution of the continuum case, but the deviation remains within the overall order h^4 error of the integration scheme [52].

We have given the CS_2 example here because it illustrates the general procedure involving linear constraints which are all eliminated from the equations of motion, thereby reducing the degrees of freedom to those of the backbone atoms (the two sulphur atoms in our example). These are confined by quadratic bond constraints. The Lagrange multipliers associated with the latter are kept in the problem and fixed by the bond constraints themselves. After solving for the backbone, the linear constraints fix the positions of the remaining atoms uniquely.

The nitrogen molecule which was discussed in the previous subsection using a direct approach can be treated with the method of constraints. It is a simple problem because there are no linear constraints which have to be used to remove redundant degrees of freedom from the equations of motion, and we are left with the following equations:

$$m_1\ddot{\mathbf{r}}_1 = \mathbf{F}_1 + \lambda(\mathbf{r}_1 - \mathbf{r}_2) \quad (8.124a)$$

$$m_2\ddot{\mathbf{r}}_2 = \mathbf{F}_2 - \lambda(\mathbf{r}_1 - \mathbf{r}_2). \quad (8.124b)$$

The Verlet equations lead again to linear predictions for $\mathbf{r}_1$ and $\mathbf{r}_2$ at the next time step and substituting these into the bond constraint leads to a quadratic equation which fixes the Lagrange multiplier λ. For an implementation, see Problem 8.9.

8.6.3 General procedure: partial constraints

In the previous section we have considered systems consisting of completely rigid molecules. Now we discuss partially rigid molecules, consisting of rigid fragments which can move with respect to each other. The motion of two fragments attached by chemical bonds can be described in terms of stretching, bending and torsion, as described in Section 8.6.2. In general, partial constraints cannot be treated using the methods given previously. Trying to use the constraints to reduce the equations to a smaller set and formulating equations for the rigid fragments in terms of quaternions is quite complicated. Ryckaert *et al.* [51–54] devised a simple and efficient iterative method for treating arbitrary constraints which is now still the most important method for MD with polyatomic molecules. Analogous to the method of constraints for rigid molecules, the rigidity of the fragments can be imposed by constraints, which are all expressed in Cartesian coordinates for simplicity. The Lagrange multipliers are determined after each integration step by substituting the new positions into the constraint equations.

The algorithm, called SHAKE, is formulated within the framework of the Verlet algorithm. The forces experienced by the particles consist of physical and of constraint forces. The constraints are given by $\sigma_k(R) = 0$, where $k = 1, \dots, M$; M is the number of constraints. We denote the physical force on particle i by $\mathbf{F}_i$ and the constraint force is $\sum_{k=1}^{M} \lambda_k \nabla_i \sigma_k(R)$, where λ_k is the Lagrange multiplier which is to be determined. At time step $t = nh$ we have at our disposal the positions at times t and $t - h$. These positions satisfy the constraint equations $\sigma_k(R) = 0$ to numerical precision. The aim is to find the positions at time $t + h$, satisfying the constraint equation. First we calculate the new positions $\tilde{\mathbf{r}}_i(t + h)$ without taking the constraints into account:

$$\tilde{\mathbf{r}}_i(t+h) = 2\mathbf{r}_i(t) - \mathbf{r}_i(t-h) + h^2 \mathbf{F}_i[\mathbf{r}_i(t)]. \tag{8.125}$$

The final positions $\mathbf{r}_i(t + h)$ can be written as

$$\mathbf{r}_i(t+h) = \tilde{\mathbf{r}}_i(t+h) - \sum_{k=1}^{M} \lambda_k \nabla_i \sigma_k[R(t)]. \tag{8.126}$$

The λ_k are found in an iterative procedure. We number the iterations by an index l. In each iteration, a loop over the constraints k is carried out, and in each step of this loop, the Lagrange parameter λ_k and all the particle positions are updated. The positions are updated according to

$$\mathbf{r}_i^{\text{new}} = \mathbf{r}_i^{\text{old}} - h^2 \lambda_k^{(l)} \nabla_i \sigma_k[R(t)]. \tag{8.127}$$

The parameter $\lambda_k^{(l)}$ is found from a first order expansion of $\sigma_k(R)$, requiring that this vanishes:

$$\sigma_k[R^{\text{new}}] \approx \sigma_k[R^{\text{old}}] - h^2 \lambda_k^{(l)} \sum_i \nabla_i \sigma_k[R^{\text{old}}] \nabla_i \sigma_k[R(t)] = 0, \tag{8.128}$$

leading to

$$\lambda_k^{(l)} = \frac{\sigma_k[R^{\text{old}}]}{h^2 \left\{ \sum_i \nabla_i \sigma_k[R^{\text{old}}] \nabla_i \sigma_k[R(t)] \right\}}. \tag{8.129}$$

Each step will therefore shift the positions more closely to the point where they all satisfy the constraint. The iterative process is stopped when all the constraints are smaller (in absolute value) than some small positive number.

The algorithm can be summarised as follows:

```
Calculate R̃(t + h) using (8.125);
Set R^old equal to R̃(t + h);
REPEAT
    Calculate λ_k^(l) from (8.129);
    FOR k = 1 TO M DO
        Set R^old equal to R^new
        Update R^old to R^new using (8.127);
    END FOR
UNTIL Constraints are satisfied.
```

The SHAKE algorithm turns out to be quite efficient: for a system of 48 atoms with 112 constraints, typically 25 iterations are necessary in order to achieve convergence of the constraints within a relative accuracy of 5×10^{-7} [52].

8.7 Long-range interactions

Coulombic and gravitational many-particle systems are of great interest because they describe plasmas, electrolytic solutions, and celestial mechanical systems. The interaction is described by a pair-potential which in three dimensions is proportional to $1/r$ – in two dimensions it is $\ln r$. The long range character of this potential poses problems. First of all, it is not clear whether the potential can be cut off beyond some finite range. One might hope that for a charge-neutral Coulomb system, screening effects could justify this procedure. Unfortunately, for most systems of interest, the screening length exceeds half the linear system size that can be achieved in practice, so we cannot rely on this screening effect to justify cutting off the potential, as this would essentially alter the form of the screening charge cloud. Also, when using the minimum image convention with periodic boundary conditions, equally charged particles tend to occupy opposite ends of a half diagonal of the system unit cell in order to minimise their interaction energy, thus introducing unphysical anisotropies. Therefore, we cannot cut off the potential and all pairs of interacting particles must be taken into account when calculating the forces.

Connected with this is an essential difference in the treatment of periodic or nonperiodic system cells. In the latter case, we simply use the $1/r$ potential (or $\ln r$ in two dimensions), but in the periodic case we must face the problem that in general the sum over the image charges in the periodic replicas does not converge. This can be remedied by subtracting an offset from the potential (note that adding or subtracting a constant to the potential does not alter the forces and hence the dynamics of the system) leading to the following expression for the total configurational energy for a collection of particles with charge (or mass) q_i located at q_i, $i = 1, \ldots, N$:

$$U = \sum_{\mathbf{R}} \sum_{i<j} \frac{q_i q_j}{|\mathbf{r}_i - \mathbf{r}_j + \mathbf{R}|} - \sum_{i<j} q_i q_j \sum\nolimits'_{\mathbf{R}} \frac{1}{\mathbf{R}}. \tag{8.130}$$

Here $\sum_{i<j}$ denotes a sum over i and j running from 1 to N with the restriction $i < j$; furthermore, $\sum_{\mathbf{R}}$ denotes a sum over the locations $\mathbf{R}$ of the system replicas, the prime with the second sum denoting exclusion of $\mathbf{R} = \mathbf{0}$. From now on we shall restrict ourselves to charge-neutral systems with $\sum_i q_i = 0$, for which the second term in (8.130) vanishes. The system then has a dipole moment and the leading term in computing the total energy in PBC is the result of the dipole–dipole interactions between the replicas. Evaluating the sum over the replicas is a difficult problem even for charge-neutral systems and it will be addressed in the next subsection. In Section 8.7.2 we shall then see how the forces can be evaluated more efficiently than in the conventional MD approach where we must sum over all pairs.

8.7.1 The periodic Coulomb interaction

The total configurational energy of the charge-neutral system is given by

$$U = \sum_{\mathbf{R}} \sum_{i<j} \frac{q_i q_j}{|\mathbf{r}_i - \mathbf{r}_j + \mathbf{R}|}; \quad \sum_i q_i = 0. \tag{8.131}$$

It is assumed here that the particles are point particles, that is, their charge distribution is given by a delta-function $\rho_i(\mathbf{r}) = q_i \delta(\mathbf{r} - \mathbf{r}_i)$. In most realistic cases there will be additional short range interactions preventing the particles from approaching each other too closely. We now apply a Fourier transform as defined in Eqs. (4.104)–(4.105) to (8.131). We have

$$\frac{1}{r} = \int \frac{\mathrm{d}^3 k}{(2\pi)^3} \frac{4\pi}{k^2} \mathrm{e}^{i\mathbf{k}\cdot\mathbf{r}}. \tag{8.132}$$

Substituting this into (8.131) and using

$$\sum_{\mathbf{R}} \mathrm{e}^{i\mathbf{k}\cdot\mathbf{R}} = \frac{(2\pi)^3}{V} \sum_{\mathbf{K}} \delta^3(\mathbf{k} - \mathbf{K}) \tag{8.133}$$

where V is the volume of the system and $\mathbf{K}$ are reciprocal lattice vectors, we obtain

$$U = \frac{1}{V} \sum_{\mathbf{K} \neq \mathbf{0}} \sum_{i<j} \frac{\mathrm{e}^{i\mathbf{K}\cdot\mathbf{r}_{ij}}}{K^2} q_i q_j. \tag{8.134}$$

We have not yet made any progress as we have only replaced the infinite sum over $\mathbf{R}$ by another infinite sum over $\mathbf{K}$. It might seem that this sum diverges for small $\mathbf{K}$, but this is not the case for charge-neutral systems: this neutrality is responsible for the exclusion of the $\mathbf{K} = \mathbf{0}$ term, and it ensures convergence of the small-$\mathbf{K}$ terms. Surprisingly, the divergences in the original real-space sum (8.131) were associated with the long range character of the force whereas the divergence in (8.134) is due to the short range (large $\mathbf{K}$) part. In reality, the ions have a finite size, which means that they will repel each other at short distances. This implies that the Coulomb interaction has to be taken into account for ranges beyond some small cut-off r_{core} only, and we can neglect the K-values for $K > 2\pi/r_{\text{core}}$. Of course, this does not yield an exactly spherical cut-off as the reciprocal lattice is cubic, but if the cut-off radius is sufficiently small this will cause no significant errors. Moreover, the core radius can be chosen much smaller than the range of repulsive interaction (which is always present in realistic models) so that this error can be reduced arbitrarily.

In case one insists on having delta-function distributions, or if the cut-off radius is so small that calculating the Fourier sum is still inconveniently demanding, it is possible first to replace the delta-charges by artificial, extended charge distributions with some finite radius and then correcting for this replacement. This is done in the so-called Ewald summation technique. We shall not give a full derivation of the Ewald summation method since it is quite lengthy – it is described elsewhere [55, 56] – but sketch briefly the idea behind this technique. In the Ewald method, the extended charge distribution is taken to be a Gaussian:

$$\rho_i(\mathbf{r}) = q_i \left(\frac{\alpha}{\pi}\right)^{3/2} \exp(-\alpha|\mathbf{r} - \mathbf{r}_i|^2) \tag{8.135}$$

where the normalisation factor is for the three-dimensional case. This charge distribution results in a converging K-sum, and this extension is corrected for by adding the potential resulting from the difference between the point-charge and Gaussian distribution. Since this difference is neutral, it generates a rapidly decaying potential, which can then be treated by the minimum image convention. The total interaction potential for charges q_i located at $\mathbf{r}_i$ is then given as

$$U_{\text{PBC}} = \frac{2\pi}{V} \sum_{\mathbf{K}\neq\mathbf{0}} \left| \sum_i q_i \mathrm{e}^{\mathrm{i}\mathbf{K}\cdot\mathbf{r}_i} \right|^2 \frac{\mathrm{e}^{-K^2/(4\alpha)}}{K^2} + \sum_{i<j} q_i q_j \frac{\operatorname{erfc}(\sqrt{\alpha} r_{ij})}{r_{ij}} - \left(\frac{\alpha}{\pi}\right)^{1/2} \sum_{i=1}^{N} q_i^2 \tag{8.136}$$

where the function erfc is the complementary error function defined in (4.116): erfc $= 1 -$ erf. The first term of the Ewald sum converges rapidly due to the $\exp[-K^2/(4\alpha)]$ term resulting from the Gaussian charge distribution. The second term in the sum is short ranged, so it can be treated in a minimum image convention. The forces can be found by differentiation. The Ewald sum can also be generalised for dipolar interactions (Ewald–Kornfeld method).

In a careful treatment of the Ewald technique, the sum is carried out formally by taking a large volume of some particular shape (e.g. a sphere) containing the system replicas and then this shape is increased. The reason for this is that the sum over the interactions is conditionally convergent, i.e. it depends on the order in which the various contributions are taken into account. This is explained by the fact that the system replicas all have a dipole moment and will hence build up a surface charge at the boundary of the huge volume. The most natural boundary condition (the one which is arrived at in more pedestrian derivations) is consistent with the sphere being embedded in a perfectly conducting medium. For the sphere embedded in a dielectric, a correction must be included [56]. It is important to be aware of this when calculating (say) dielectric properties of a charged system.

8.7.2 Efficient evaluation of forces and potentials

As a result of the long range of the forces, all interacting pairs must be taken into account in the calculation of forces or potentials. The straightforward implementation, considered in the previous sections of this chapter, also called the *particle–particle method* (PP) because all pairs are considered explicitly, would require $\mathcal{O}(N^2)$ steps, but it turns out possible to reduce this to a more favourable scaling. We shall briefly sketch two other methods, and then consider a third one, the *tree method* in greater detail.

In the *particle-mesh* (PM) method, a (usually cubic) grid in space is defined. A mass (or charge) distribution ρ is then defined by assigning part of each particle's mass (or charge) to its four neighbouring grid points according to some suitable scheme. The potential can then be found by solving Poisson's equation on the grid

$$\nabla_{\mathrm{D}}^2 U(\mathbf{r}) = -4\pi\rho(\mathbf{r}) \tag{8.137}$$

where ∇_{D}^2 is the finite-difference version of the Laplace operator. Using fast Fourier transforms (see Appendix A9), this calculation can be carried out in a number of steps proportional to $M \log M$ where M is the number of grid points. Knowing the potential, the force at any position can be found by taking the finite difference gradient of the potential, after a suitable correction for the self-energy resulting from the inclusion of the interaction of a particle with itself in this procedure. This method obviously becomes less accurate for pairs of particles with a small separation, as in that case the Coulomb/gravitation potential is not sufficiently smooth to be represented accurately on the grid. Therefore it is sensible to treat particles within some small range (for example a range comparable to the grid constant) by the PP method. This can be done by splitting the force into a smooth long range (LR) and a short range (SR) part:

$$\mathbf{F} = \mathbf{F}^{\mathrm{LR}} + \mathbf{F}^{\mathrm{SR}}, \tag{8.138}$$

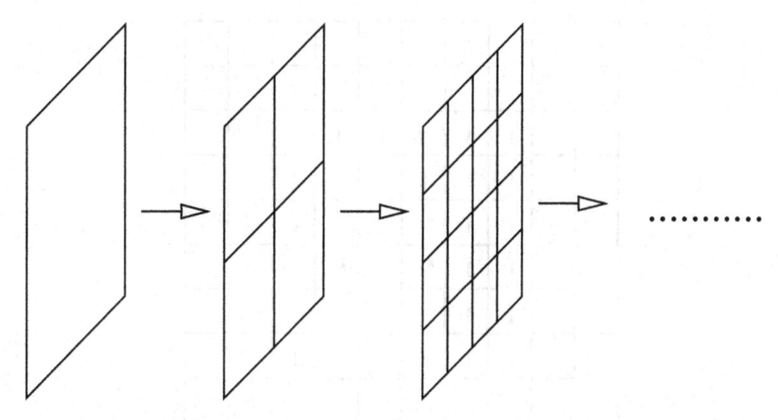

Figure 8.6. Hierarchical subdivision of the full simulation space (a square) into children, grandchildren etc.

such that the short range force vanishes beyond some small range, and the long range force can be calculated accurately on the grid. The splitting can be obtained by considering the long range force as resulting from a particle whose charge (or mass) is distributed over some finite range (homogeneous sphere, Gaussian distribution, ...). The short range force is then the potential resulting from the difference between the point charge and the finite-width distribution (cf. the Ewald method). The long range interactions are treated as in the PM method, and the short range ones can be dealt with using the PP scheme. The resulting method is called the *particle–particle/particle–mesh* (PPPM) or P^3M method. For a detailed description of the PM and PPPM methods, see Ref. [19].

We now describe the *tree-code* algorithm in some detail [57–60]. The amount of computer time involved in the evaluation of the forces in this method is reduced to $\mathcal{O}(N \ln N)$ steps. We describe the Barnes–Hut[57] version in the formulation by van Dommelen and Rundensteiner [61, 62] and restrict ourselves to two-dimensional gravitational (or Coulomb) systems, with an interaction $\ln r$ between two particles of unit mass (or charge) and separation r. The idea of the method is that the force which a mass experiences from a distant cluster of particles can be calculated from a multipole expansion of the cluster. The convergence of the multipole expansion depends on the ratio of the distance from the cluster and its linear size.

The total system volume is hierarchically divided up into blocks. We start with a square shape (level 0) which in a first step is divided into four squares of half the linear size (level 1), and at the next step each of these squares is divided up into four smaller ones etc. We speak of parents and children of squares in this hierarchy – see Figure 8.6. Now consider some square S at level n. It is not justified to apply the multipole expansion to nearest neighbour squares as particles in neighbouring

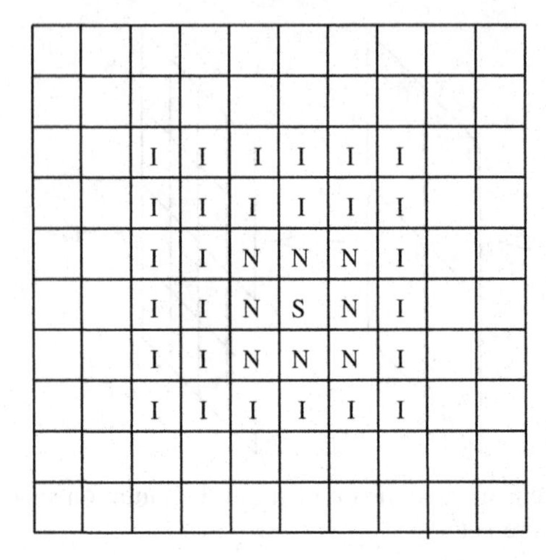

Figure 8.7. Interaction list of a square S at level n. The squares at level n are separated by thin lines, their parents (at level $n-1$) by heavy lines. For the square labelled by S, the squares in the interaction list of a square are denoted by I. The nearest neighbours are labelled by N.

squares might be very close so that the multipole expansion would require far too many terms. These squares will be dealt with at a higher level, so we apply this approximation in each step to at least next nearest squares and skip squares lying in regions that have been treated at previous levels. Therefore, the squares with which the particles in S will interact at the present level are those (1) which are not nearest neighbours of S and (2) whose parent was a nearest neighbour of the parent of S. These squares form the *interaction list* of S. Figure 8.7 shows which squares are in the interaction list of S. It will be clear that all the interactions will be taken into account when proceeding in this way.

More specifically, at level n we carry out two steps.

1. We calculate the multipole moments of each square of the present level.
2. For each particle, we calculate the interactions with the interaction list of the square to which it belongs using the multipole expansion for the particles in the cells.

This process is carried through over $n_{\max} = (\log_2 N)/2$ steps so that for N being an integer power of 4, the squares at the lowest level contain on average one particle. Empty squares are 'pruned' from the tree, that is, they are not divided up any more.

Let us now calculate the number of steps needed in this procedure. We assume that we carry out the multipole expansion up to order M. This number is independent of the number of particles N. At level n, the first step, in which the multipole moments are calculated, requires $N \times M$ steps. The second step requires $N \times M \times K$ steps, where K is the average size of the interaction list, which is at most 27. K and M are fixed numbers, there are $\mathcal{O}(\ln N)$ levels, so the total number of steps scales as $\mathcal{O}(N \ln N)$.

For two dimensions, the multipole moment expansion is very simple if the space is viewed as a complex plane with particles at positions $z = x + iy$. The potential is then given as the real part of $\ln(z)$ and this can easily be expanded in a Taylor expansion around the centre of the cell. For a cluster centred at the origin and consisting of particles of charge q_i at positions z_i, the potential at the point z is given by

$$U(z) = \sum_{i=1}^{N_c} q_i \ln(z - z_i) = a_0 \ln z - \sum_{k=1}^{M} \frac{a_k}{z^k} + \mathcal{O}\left(\frac{R}{z}\right)^{M+1} \tag{8.139}$$

where R is the linear size of the cluster containing N_c particles and the moment expansion coefficients a_k are given by

$$a_0 = \sum_i q_i \quad \text{and} \quad a_k = \sum_{i=1}^{N_c} \frac{q_i z_i^k}{k}, k \geq 1. \tag{8.140}$$

For the field written as a complex number E at the point z we have

$$E(z) = \sum_{k=0}^{M} \frac{a_k}{z^{k+1}} + \mathcal{O}(R/z)^{M+1}. \tag{8.141}$$

From Figure 8.7 it can be seen that a worst case estimate for R/z is 2/3. In practice, fewer than 20 multipole coefficients are necessary to obtain machine accuracy (32 bits).

In fact, it turns out to be possible to reduce the amount of work needed for the force evaluation to $\mathcal{O}(N)$. The resulting method is called the fast multipole method (FMM) – see Refs. [62] and [63].

8.8 Langevin dynamics simulation

Most realistic physical systems are tractable only in a *model*, in which the interesting features of the system are highlighted and in which the less relevant parts are either eliminated or treated in an approximate way. In this spirit we have for example eliminated molecular degrees of freedom in Section 8.6 by considering (parts of) molecules to be rigid. Another example of this approach is the Langevin dynamics technique, the subject of the present section. Consider a solution containing polymers or ions which are much heavier than the solvent molecules. As the kinetic energy is on average divided equally over the degrees of freedom, the ions or polymers will move much more slowly than the solvent molecules. Moreover, because of their large mass, they will change their momenta only after many collisions with the solvent molecules and the picture which emerges is that of the heavy particles forming a system with a much longer time scale than the solvent molecules. This difference in time scale can be employed to eliminate the details of the degrees of freedom of the solvent particles and represent their effect by forces that can be treated in a simple way. This process can be carried out analytically through a projection procedure (see chapter 9 of Ref. [11] and references therein) but here we shall sketch the method in a heuristic way.

How can we model the effect of the solvent particles without taking into account their degrees of freedom explicitly? When a heavy particle is moving through the solvent, it will encounter more solvent particles at the front than at the back. Therefore, the collisions with the solvent particles will *on average* have the effect of a friction force proportional and opposite to the velocity of the heavy particle. This suggests the following equation of motion for the heavy particle:

$$m\frac{\mathrm{d}\mathbf{v}}{\mathrm{d}t}(t) = -\gamma\mathbf{v}(t) + \mathbf{F}(t) \tag{8.142}$$

where γ is the friction coefficient and $\mathbf{F}$ the external or systematic force, due to the other heavy particles, walls, gravitation, etc. It has been noted in Section 7.2.1 that the motion of fluid particles exhibits strong time correlations and therefore the effects of their collisions should show time correlation effects. Time correlations affect the form of the friction term which, in Eq. (8.142), has been taken to be dependent on the *instantaneous* velocity but which in a more careful treatment should include contributions from the velocity at previous times through a memory kernel:

$$m\frac{\mathrm{d}\mathbf{v}}{\mathrm{d}t}(t) = -\int_{-\infty}^{t} \mathrm{d}t'\, \gamma(t-t')\mathbf{v}(t') + \mathbf{F}(t). \tag{8.143}$$

In order to avoid complications we shall proceed with the simpler form (8.142). In the following we shall restrict ourselves to a particle in one dimension; the analysis for more particles in two or three dimensions is similar.

Equation (8.142) has the unrealistic effect that if the external forces are absent, the heavy particle comes to rest, whereas in reality it executes a Brownian motion. To make the model more realistic we must include the rapid variations in the force due to the frequent collisions with solvent particles on top of the coarse-grained friction force. We then arrive at the following equation:

$$m\frac{\mathrm{d}v}{\mathrm{d}t}(t) = -\gamma v(t) + F(t) + R(t) \tag{8.144}$$

where $R(t)$ is a 'random force'. Again, the time correlations present in the fluid should show up in this force, but they are neglected once more and the force is subject to the following conditions.

- As the average effect of the collisions is already absorbed in the friction, the expectation value of the random force should vanish:

$$\langle R(t)\rangle = 0. \tag{8.145}$$

- The values of R are taken to be uncorrelated:

$$\langle R(t)R(t+\tau)\rangle = 0 \quad \text{for } \tau > 0. \tag{8.146}$$

- The values of R are distributed according to a Gaussian:

$$P[R(t)] = (2\pi\langle R^2\rangle)^{-1/2}\exp(-R^2/2\langle R^2\rangle). \tag{8.147}$$

All these assumptions can be summarised in the following prescription for the probability for a set of random forces to occur between t_0 and t_1:

$$P[R_i(t)]_{t_0<t<t_1} \propto \exp\left(-\frac{1}{2q}\int_{t_0}^{t_1} \mathrm{d}t\; R_i^2(t)\right) \tag{8.148}$$

with q a constant to be determined.

Below we consider the numerical integration of the equations of motion for the heavy particles, and in that case it is convenient to assume that the random force is constant over each time step: at step n, the value of the random force is R_n. For this case, the correlation function for the R_n reads

$$\langle R_n R_m\rangle = \frac{\int \mathrm{d}R_n \mathrm{d}R_{n+1}\cdots \mathrm{d}R_m \exp(-1/2q\sum_{l=n}^{m} R_l^2\Delta t)R_n R_m}{\int \mathrm{d}R_n \mathrm{d}R_{n+1}\cdots \mathrm{d}R_m \exp(-1/2q\sum_{l=n}^{m} R_l^2\Delta t)} \tag{8.149}$$

which yields the value 0 for $n \neq m$, in accordance with the previous assumptions. For $n = m$ we find the value $q/\Delta t$, so we arrive at

$$\langle R_n R_m\rangle = \frac{q}{\Delta t}\delta_{nm}. \tag{8.150}$$

For the continuum case $\Delta t \to 0$ (8.150) converges to the δ-distribution function

$$\langle R(t)R(t+\tau)\rangle = q\delta(\tau). \tag{8.151}$$

We now return to the continuum form of the Langevin equation (8.144). This can be solved analytically and the result is

$$v(t) = v(0)\exp(-\gamma t/m) + \frac{1}{m}\int_0^t \exp[-(t-\tau)\gamma/m]R(\tau)\mathrm{d}\tau. \tag{8.152}$$

Because the expectation value of R vanishes we obtain

$$\langle v(t)\rangle = v(0)\exp(-\gamma t/m) \tag{8.153}$$

which is to be expected for a particle subject to a friction force proportional and opposite to the velocity.

The expectation value of v^2 is determined in a similar way. Using (8.151) and (8.144) we find

$$\langle [v(t)]^2 \rangle = v_0^2 \exp(-2\gamma t/m) + \frac{q}{2\gamma m}(1 - \mathrm{e}^{-2\gamma t/m}), \tag{8.154}$$

which for large t reduces to

$$\langle [v(\infty)]^2 \rangle = \frac{q}{2\gamma m}. \tag{8.155}$$

According to (8.152), v depends linearly on the random forces $R(t)$ and as the latter are distributed according to a Gaussian, the same will hold for the velocity. The width is given by (8.155), so we have

$$P[v(t)] = \left(\frac{\gamma m}{\pi q}\right)^{1/2} \exp[-mv(t)^2\gamma/q] \tag{8.156}$$

for large t. This is precisely the Maxwell distribution if we write

$$q = 2k_\mathrm{B}T\gamma, \tag{8.157}$$

so this equation defines the value of q necessary to obtain a system with temperature T. In Chapter 12 we shall discuss Langevin types of equations in a more formal way, using the Fokker–Planck equation.

The velocity autocorrelation function can also be obtained from (8.152):

$$\langle v(0)v(t)\rangle = \langle v(0)^2\rangle \mathrm{e}^{-\gamma t/m}. \tag{8.158}$$

The absence of a long time tail in this correlation function reflects the oversimplifications in the construction of the Langevin equation, in particular the absence of correlations in the random force and the fact that the frictional force does not depend on the 'history' of the system.

The results presented here are easily generalised to more than one dimension. However, including a force acting between the heavy particles causes problems if this force exhibits correlations with the random force, and Eq. (8.157) is no longer valid in that case. Such correlation effects are often neglected and the systematic force is simply added to the friction and the Langevin term.

A further refinement is the inclusion of memory kernels in the forces, similar to the approach in Eq. (8.143). In that case, the random force is no longer uncorrelated – it is constructed with correlations in accordance with the fluctuation-dissipation theorem [64]:

$$\langle R(0)R(t)\rangle = \langle v^2\rangle \gamma(t). \tag{8.159}$$

However, this equation is again no longer valid if external forces are included. The approach with memory kernels has led to a whole industry of so-called generalised Langevin-dynamics simulations [64–67].

The systematic interaction force between the particles in the solvent will affect the friction which these particles are subject to through hydrodynamic effects. This coupling is usually neglected, but a method including these effects has been proposed and implemented [68]. We mention the dissipative particle dynamics (DPD) method which is based on these ideas [69].

An algorithm for simple Langevin dynamics can be formulated starting from the methods given in Section 8.4.1. Suppose the random force is constant during one integration step. Denoting the force during the interval $[0, h]$ by R_+ and that during the interval $[-h, 0]$ by R_-, the random force can directly be included into (8.40):

$$x(h)[1 + \gamma h/2] + x(-h)[1 - \gamma h/2] = 2x(0) + h^2[F(0) + R_+/2 + R_-/2]. \tag{8.160}$$

Therefore, at each step a new value of the random force during the new interval must be drawn from a Gaussian random generator, and this force is to be used together with the random force generated at the previous step in order to predict the new position. This is, however, not always a satisfactory procedure. Normally, the integration time step h is determined by the requirement that the systematic force $\mathbf{F}$ can be assumed to be reasonably constant over a time interval h. This means that the time over which we take the random force to be constant depends on the smoothness of the systematic force. In fact we would prefer to allow for a rapidly varying random force combined with a large time step allowed by the systematic force. This turns out to be possible. Using the statistical properties of the random force, equations of motion can be obtained which are somewhat similar to the ones given here, but with more complicated correlations between the random contributions at subsequent steps – for details see Ref. [70].

It is straightforward to develop a Langevin program for a molecule in a fluid or a gas, using the simple algorithm presented here. For molecules containing chains of at most three chemically bonded atoms, torsion is absent, which reduces the number of forces considerably. Examples are molecules with a tetrahedron conformation, such as CH_4 (methane) and CF_4, and two-dimensional molecules. In Problem 8.9 the construction of a Langevin molecule for methane is considered.

8.9 Dynamical quantities: nonequilibrium molecular dynamics

In the molecular dynamics method, the equations of motion of a classical many-body system are integrated numerically. There is no reason to restrict the applicability of this method to systems in equilibrium. MD is the method of choice for dynamic phenomena in equilibrium or nonequilibrium systems. We speak of nonequilibrium molecular dynamics (NEMD). We consider two examples very briefly here.

There exists a relation between time correlation functions and transport coefficients via the dynamic fluctuation-dissipation theorem [71, 72]. The physical idea behind this theorem is that, in an equilibrium system, particles diffuse and the dynamics of this diffusion tells us something about their ability to transport for example heat or charge. Therefore we can measure transport coefficients by studying the diffusion of the positions or velocities through the system. A disadvantage of measuring transport quantities in this way is that diffusion is often rather slow in equilibrium so that accurate results for transport coefficients are sometimes hard to obtain. Therefore it is useful to apply a field and measure the response to the action of that field directly by keeping track of the motion of the particles (a thermostat must be used in order to prevent the energy from increasing steadily as a result of the interaction with the external field). A complication may arise in connection with periodic boundary conditions, as in that case surface effects may be induced if the applied force is not compatible with the periodicity. Therefore perturbing forces are often chosen sinusoidal with a period compatible with the PBC. An example is provided by the determination of the shear viscosity, caused by fluid layers moving in parallel directions, with different speed, rubbing against each other. The shear viscosity can be measured [73, 74] by applying a force in the x-direction which varies with the coordinate z according to

$$\mathbf{F}(z) = F_0 \cos(kz)\hat{\mathbf{x}} \tag{8.161}$$

where $k = 2\pi/L$, and L is the linear size of the cubic volume. The shear viscosity η can then be measured via the mean velocity in the x-direction of the particles with a given coordinate z:

$$\overline{v_x}(z) = \rho/(k^2\eta)F_0 \cos(kz) \tag{8.162}$$

and this average velocity can easily be determined. In order to improve the estimate one can determine the shear viscosity with various $k_n = 2\pi n/L$ to extrapolate to $k \to 0$.

A second example of NEMD is the transfer of energy between different degrees of freedom. This is of interest in detonation waves. A detonation which traverses a medium of explosive molecules continuously 'recharges' itself by new unstable molecules falling apart, thereby releasing fragments with high velocities. For an unstable molecule to be disrupted it is necessary for the translational energy imparted by a collision with a fast fragment to be transferred to bond length vibrations. For diatomic molecules, the two different degrees of freedom can easily be separated. Holian *et al.* [39, 40] have carried out MD simulations in which the translational and vibrational degrees of freedom were given different temperatures by coupling them to different heat baths which were then turned off or replaced by a single bath (at the higher temperature). In this way it was possible to determine energy transfer rates between the different modes.

Exercises

8.1 [C] For coding the leap-frog method (Eq. (8.7)) two arrays are needed, one containing the velocities at times $t = (n + 1/2)h$, and one for the positions at $t = nh$. The same holds for the velocity-Verlet algorithm.

At first sight it might seem that the Verlet algorithm would need more memory: arrays containing the positions at times $t = nh$, $t = (n - 1)h$ and $t = (n + 1)h$. However, the value $x_i[(n - 1)h]$ can be overwritten by $x_i[(n + 1)h]$. Use this to code the Verlet algorithm such that only two arrays are needed. Test it for a number of particles moving in one dimension and subject to the harmonic oscillator potential.

8.2 The neighbour list proposed by Verlet [8] needs updating every 10–20 integration steps and this update requires of the order of N^2 steps for a system containing N particles. Another bookkeeping device consists of partitioning the system into cubic volumes and keeping track of which particles are to be found in each of these volumes. Consider a two-dimensional $L \times L$ system for convenience. We split this up into $P \times P$ squares of linear size L/P. P is chosen such that the potential can be cut off safely beyond L/P. Suppose we have for each square a list of particles within that volume. These lists will change whenever a particle leaves a square and moves to a neighbouring one. The force evaluation now includes only particle pairs whose members are either in the same or in neighbouring cells.

(a) How many particles are on average to be found in one square?
(b) How many pair forces are on average taken into account in this 'cell method'?
(c) Calculate the gain in speed with respect to the method in which all pair interactions are taken into account, assuming that the particles are distributed more or less homogeneously over the volume.

8.3 The first molecular dynamics simulations were carried out by Alder and Wainwright for hard spheres [14]. The discontinuity in the potential calls for a different approach than that used for smooth potentials. The state of the system is given by the positions $\mathbf{r}_i$ and velocities $\mathbf{v}_i$ (i labels the particles) at some time t_i which is usually the time of the last collision experienced by i. We must calculate the velocity changes for the next pair undergoing a collision.

We consider the elastic collision between two hard spheres, i and j, which are moving with velocities $\mathbf{v}_i$ and $\mathbf{v}_j$. At time t their positions are $\mathbf{r}_i$ and $\mathbf{r}_j$. After the collision, velocities are $\mathbf{v}'_i$ and $\mathbf{v}'_j$ respectively. The sphere diameter is σ.

(a) Show, using energy and momentum conservation, that the changes in velocities of the two particles are given by

$$\Delta \mathbf{v}_i = \mathbf{v}'_i - \mathbf{v}_i = -\Delta \mathbf{v}_j = \frac{\mathbf{r}_{ij}(\mathbf{v}_{ij} \cdot \mathbf{r}_{ij})}{\sigma^2}$$

where $\mathbf{v}_{ij} = \mathbf{v}_i - \mathbf{v}_j$ and $\mathbf{r}_{ij} = \mathbf{r}_i - \mathbf{r}_j$ at the collision.

For each pair of particles we need to know the time at which they will collide (note that because of PBC each pair will indeed collide at some time unless the velocities have very peculiar values). The collision time for pair i, j is found by

$$|\mathbf{r}_{ij} + t\mathbf{v}_{ij}| = \sigma.$$

This is a quadratic equation which yields two solutions for the collision time t. The first time after the current time must be chosen and recorded as the collision time of pair ij.

The simulation is now constructed as follows. At the beginning, the particles are released from a lattice with velocities according to a Boltzmann distribution. For all $N(N-1)/2$ pairs, the collision times are calculated and stored in a sorted list. The first element of this list contains the first collision to take place. For this collision we calculate the new velocities and positions. Then each pair containing at least one of the two collision partners is removed from the list. Their new collision times are calculated and added again to the list in such a way that the latter remains sorted with respect to the collision times.

(b) How does the simulation time scale with the number of particles?

(c) Explain why the kinetic energy of the hard sphere system is rigorously constant.

In order to calculate pressures we must adapt the virial theorem to this system. The virial theorem for smooth forces reads

$$\frac{\beta P}{\rho} = 1 + \frac{1}{3Nk_B T}\left\langle \sum_{i=1}^{N} \mathbf{r}_i \cdot \mathbf{F}_i \right\rangle.$$

The problem is that the force acts over an infinitely small time during which it has an infinite value. Show that for this case the virial theorem reads

$$\frac{\beta P}{\rho} = 1 + \frac{1}{N\langle \mathbf{v}^2 \rangle t} \sum_{\text{collisions}} \mathbf{v}_{ij} \cdot \mathbf{r}_{ij},$$

where the sum is over the collisions taking place within the sampling time t.

8.4 (a) Show that the Verlet algorithm can be written in the form:

$$\begin{pmatrix} p(t+h/2) \\ x(t+h) \end{pmatrix} = \begin{pmatrix} p(t-h/2) + hF[x(t)] \\ x(t) + hp(t-h/2) + h^2 F[x(t)] \end{pmatrix}.$$

(b) Find the Jacobian matrix of this map and show that the Verlet algorithm is symplectic.

8.5 Consider a time-evolution operator acting on vectors in two dimensions, which is described by the symplectic operator $\exp(tA_D)$:

$$z(t) = \exp(tA_D)z(0),$$

$$z = (p, x) = (z_1, z_2).$$

(a) Show that symplecticity implies that

$$\frac{\partial A_1}{\partial p} = -\frac{\partial A_2}{\partial x}.$$

(b) Find a necessary condition to write A_D as $J\nabla_z H_D$. Show that this condition is equivalent to that found in (a).

(c) Show that H_D is a conserved quantity.

8.6 In this problem we consider Andersen's method for keeping the temperature constant during a MD simulation. In particular we want to find the momentum refresh rate R for which the method mimics wall collisions best. The refresh rate is defined such that the average number of velocity updates during a time Δt is equal to $R\Delta t$. Suppose the wall of the system is at temperature T, but the system itself is at a temperature $T + \Delta T$.

(a) Show that the rate at which heat is absorbed by the system is given by

$$\frac{\Delta Q}{\Delta t} \sim \kappa V^{1/3} \Delta T,$$

where κ is the thermal conductivity, defined by $\nabla T = \kappa \mathbf{j}$, where $\mathbf{j}$ is the heat flowing through a unit area per unit time.

(b) Show that the rate at which heat is transferred to a system without walls in Andersen's method is equal to

$$\frac{\Delta Q}{\Delta t} \sim RNk_B \Delta T.$$

(c) Derive from the two equations obtained the optimal rate:

$$R_{opt} \sim \frac{\kappa}{n^{1/3} k_B N^{2/3}}$$

where $n = N/V$.

8.7 [C] In this problem we consider a program for simulating nitrogen molecules in microcanonical MD using the method of constraints. The equations of motion are given in Section 8.6.2 (Eqs. (8.124)). The Lagrange parameters λ occuring in these equations are determined by requiring the constraint to be satisfied by the positions as predicted in the Verlet algorithm. These positions are given in the form

$$\mathbf{r}_i(t + h) = \mathbf{a}_i + \mathbf{b}_i \lambda.$$

The list of particles is grouped into pairs of atoms forming one nitrogen molecule: atoms $2l - 1$ and $2l$ belong to the same molecule. The integration is carried out in a loop over the pairs l – each pair has its own Lagrange parameter λ_l. For reasonable time step sizes the roots λ_l of the constraint equation are real. The smallest of these (in absolute value) is to be chosen. The forces can be calculated as usual, taking only interactions between atoms belonging to different molecules into account. Parameters for the Lennard–Jones interaction are $\varepsilon = 37.3$ K, $\sigma = 3.31$ Å and $d = 0.3296\sigma$.

Periodic boundary conditions are implemented with respect to the centre of mass of the molecules. If a molecule leaves the system cell it is translated back into it (as a whole) according to PBC. Note that determining the momentum from the positions at $t+h$ and $t-h$ after such a translation can cause severe errors: this should be done *before* moving the molecule back into the cell.

(a) Implement this algorithm for liquid nitrogen.

The program can be checked by verifying whether the kinetic energies associated with translational and vibrational degrees of freedom satisfy equipartition. The total kinetic energy K_{tot} can be determined as in the argon case by taking all *atomic* velocities into account. From this, the temperature can be determined as $Nk_{\text{B}}T = 4/5K_{\text{tot}}$ where N is the number of molecules. The translational kinetic energy K_{trans} can be calculated by taking into account the molecular velocities (sums of velocities of the two atoms) and the temperature can be found from this as $Nk_{\text{B}}T = 3/2K_{\text{trans}}$. The average temperatures should be the same for both procedures.

Check whether this requirement is satisfied.

(b) The virial theorem applies as usual: molecular forces should be used and the separation occurring in this theorem is the separation between the centres of mass of the molecules. The correction term is evaluated using $g \equiv 1$ for the correlation function beyond the cut-off distance, where it is assumed that g is independent of distance but also of the angular configuration of the molecular pairs.

(c) Calculate the pressure also using the atomic forces (including the constraint forces), and compare the result with (b).

(d) Calculate the pressure for various temperatures and densities. Cheung and Powles give extensive data on thermodynamic quantities [46]. The table below gives some of the data (in reduced units) obtained by Cheung and Powles.

ρ	T	P	U
0.6964	2.86	8.35	−17.16
0.6964	1.72	1.29	−18.68
0.6220	2.70	2.50	−15.82
0.6220	2.17	0.27	−16.30

8.8 [C] In this problem, we consider the implementation of the Andersen method for simulating a system in the canonical ensemble. Remember that the preferred energy estimator for the Verlet/leap-frog algorithm is

$$E = \sum_i \frac{[\mathbf{p}_i(t+h/2) + \mathbf{p}_i(t-h/2)]^2}{8} + V[R(t)],$$

where R is the combined position coordinate of the system which consists of particles of mass $m = 1$. In view of the form of this estimator, it seems sensible to update the momenta at the same time instances for which we calculate the positions, and it is convenient to define the ith component of the momentum coordinate at time t:

$$\mathbf{p}_i(t) = [\mathbf{p}_i(t+h/2) + \mathbf{p}_i(t-h/2)]/2.$$

(a) Using the leap-frog/Verlet algorithm, show that

$$\mathbf{p}_i(t+h/2) = \mathbf{p}_i(t) + hF_i/2.$$

The refreshed momenta $\mathbf{p}_i(t)$ are drawn from a Maxwell–Boltzmann distribution, and the momenta at time $t + h/2$, which are needed in the Verlet/leap-frog algorithm are then calculated using this last formula.

(b) Implement the Andersen update algorithm for argon and compare the results with the microcanonical program.

(c) Now suppose that the momenta are refreshed at *every* step. Show that in that case we have

$$r_i(t+h) = r_i(t) + h^2F_i/2 + h\zeta_i(t),$$

where $\zeta_i(t)$ is the ith random momentum component generated according to the Maxwell–Boltzmann distribution. This is a kind of Langevin equation. Discuss the difference with the Langevin equation described in Section 8.8.

8.9 [C] In this problem we consider the implementation of the Nosé–Hoover thermostat in the microcanonical MD simulation for Lennard–Jones argon described in Section 8.3. The extension is straightforward – the equations are given in Section 8.5.1. You can verify now that the behaviour of the Nosé–Hoover thermostat is often nonergodic. For $T = 1.5$ and $\rho = 0.8$ the behaviour is as it should be for coupling constant $Q = 1$. You can check that the standard deviation in the temperature is in accordance with Eq. (8.81). For lower temperatures, like $T = 0.85$, $\rho = 1.067$, the temperature exhibits large oscillations. The period of these oscillations depends on Q [26].

8.10 (a) Verify that when we take $g = 3N$ instead of $g = 3N + 1$ in the derivation of the Nosé–Hoover thermostat, the probability density for configurations (P, R) turns out to be:

$$\rho(P,R) = \frac{1}{3N}\left(\frac{2\pi Q}{k_B T}\right)^{1/2} \exp\left[\frac{-\mathcal{H}_0(P,R)(3N+1)}{3Nk_B T}\right].$$

(b) For this choice, verify that quantities sampled in a simulation yield averages as given in Eq. (8.94).

8.11 [C] In this problem, a code for evaluating the potential felt by the particles in a two-dimensional Coulomb (or gravitational) system is developed, using the tree-code method of Section 8.7.2.

Although experienced programmers would be tempted to start building tree structures using pointers and recursive programming for this problem, it can be dealt with using more pedestrian methods. The point is that the squares can be coded by two integers *NX, NY* which are considered as bit-strings. The first of these contains information about the x-coordinate of the square and the second about the y-coordinate. They are ordered linearly: the leftmost column of squares has $NX = 0$, the rightmost column $NX = 2^n - 1$ etc., and a similar coding is adopted for the rows. If squares are neighbours, their respective *NX* and *NY*-codes should differ at most by 1 (and they should not be equal). The codes of the parents can be found simply by shifting the bits of NX and NY one position to the right (least significant direction) and it can therefore easily be checked in the program whether the parents of the squares under consideration are neighbours or not.

The calculation of the multipole moments in each box (Eq. (8.140)) is best done in a loop over the particles, recording its contribution to all the multipole coefficients of the to which square it belongs. Also, the calculation of the interactions (Eq. (8.139)) can be done in a loop over the particles, by executing for each particle a loop over the interaction list of the square to which it belongs.

Proceeding this way, it is not necessary to keep for each square a list of the particles belonging to it. However, at the finest level, the interactions between particles within the same square and between particles in neighbouring boxes must be calculated directly so only for the last step do we need such a list for each square. If you want to economise on memory, you might create a linked list for each square containing the indices of the particles in it, but for a test you may use static arrays.

Compare the results for the tree code with those of a direct calculation, varying the number of terms in the multipole expansion.

8.12 [C] In this problem we consider a simulation of a methane molecule using the Langevin approach. Methane consists of a carbon atom sitting at the centre of a tetrahedron whose vertices are occupied by four H atoms. The C–H bond has a preferred interatomic distance of $2.104a_0$. The stretch-potential associated with the bond length varies as

$$V_{\text{stretch}} = \tfrac{1}{2}\kappa(l - l_0)^2; \quad l_0 = 2.104a_0.$$

The force constant κ has the value $\kappa = 0.30$ (in atomic units). This force acts on both the carbon and the hydrogen atoms and is directed along the C–H bond.

The preferred H–C–H angle is 109° and the potential for this bending angle is

$$V_{\text{bend}} = -\lambda\cos(\varphi - \varphi_0)^2; \quad \varphi_0 = 109°,$$

with a force constant $\lambda = 0.74$. This force lies in the H–C–H plane, and acts on the two H atoms and on the C atom. The forces on the H-atoms are perpendicular to the C–H bonds, and the bending force on the C atom is directed along the bisecting line of the H–C–H angle.

These two 'force fields', bending and stretching, specify the force on each of the atoms. To find the forces, given the position $\mathbf{r}_\text{C}$ of the carbon atom and the four positions $\mathbf{r}_\text{H}$ of the hydrogen atoms, you calculate first the forces on the hydrogen atoms only. The stretch forces can easily be found by calculating the vector $\mathbf{r}_\text{CH} = \mathbf{r}_\text{H} - \mathbf{r}_\text{C}$. The bending force is slightly more difficult. Denoting the two hydrogen atoms of a H–C–H chain as H1 and H2, calulate $\mathbf{r}_\text{CH1}$ and $\mathbf{r}_\text{CH1}$. Then calculate the dot product between these two vectors. From this, the cosine of the bending angle can be found. Moreover, the direction of the force can be found from the cross-product of $\mathbf{r}_\text{CH1}$ and $\mathbf{r}_\text{CH1}$: the bending force on H1 is then perpendicular to this cross product *and* to the vector $\mathbf{r}_\text{CH1}$, and similarly for H2. Knowing the forces on the hydrogen atoms, you can calculate their sum. The force on the carbon atom is then simply the opposite of this, as the sum of all the interparticle forces adds up to zero.

(a) [C] Write routines for calculating the forces on the atoms and use these in an ordinary (microcanonical) MD simulation of the atom. To check the program, you can put the H-atoms on the vertices of a tetrahedron with the C-atom in the centre. If you release the molecule from this conformation with a CH-distance slightly smaller or larger than the equilibrium distance of $2.104a_0$, the molecule should stretch and contract isotropically in an oscillatory fashion.
(b) [C] Keep the temperature of the molecule constant by rescaling the velocities after each time step. Determine the average total energy of the molecule.
(c) [C] Add a Langevin thermostat to the simulation, for example by rescaling the velocities after every time step. Use the algorithm given in the last section for solving the equations of motion with friction. Add a Langevin random force, drawn from a Gaussian distribution with a width

$$\sigma^2 = q/h$$

to the interparticle force. Check that the temperature is given by

$$T = 1/(2\gamma).$$

The temperature is determined from the kinetic energy – we have

$$T = \frac{15}{2} k_B T.$$

Determine the average total energy and compare the result with the program of (b).

References

[1] D. W. Heermann, *Computer Simulations in Statistical Physics*. Heidelberg, Springer, 1986.
[2] M. P. Allen and D. J. Tildesley, *Computer Simulation of Liquids*. Oxford, Oxford University Press, 1989.
[3] W. G. Hoover, *Computational Statistical Mechanics*. Amsterdam, Elsevier, 1991.
[4] D. C. Rapaport, *The Art of Molecular Dynamics Simulation*. New York, Cambridge University Press, 1996.
[5] G. Ciccotti and W. G. Hoover, eds., *Molecular Dynamics Simulation of Statistical-Mechanical Systems Proceedings of the International School of Physics "Enrico Fermi", Varenna 1985*, vol. 97. Amsterdam, North-Holland, 1986.
[6] T. Kihara and S. Koba, 'Crystal structures and intermolecular forces of rare gases,' *J. Phys. Soc. Jpn*, **7** (1952), 348–54.
[7] W. C. Swope, H. C. Andersen, P. H. Berens, and K. R. Wilson, 'A computer simulation method for the calculation of equilibrium constants for the formation of physical clusters of molecules: application to small water clusters,' *J. Chem. Phys.*, **76** (1982), 637–49.
[8] L. Verlet, 'Computer 'experiments' on classical fluids. I. Thermodynamical properties of Lennard–Jones molecules,' *Phys. Rev.*, **159** (1967), 98–103.
[9] S. D. Stoddard and J. Ford, 'Numerical experiments on the stochastic behavior of a Lennard–Jones gas system,' *Phys. Rev. A*, **8** (1973), 1504–12.

[10] J. G. Powles, W. A. B. Evans, and N. Quirke, 'Non-destructive molecular dynamics simulation of the chemical potential of a fluid,' *Mol. Phys.*, **38** (1982), 1347–70.

[11] J. P. Hansen and I. R. McDonald, *Theory of Simple Liquids* 2nd edn. New York, Academic Press, 1986.

[12] L. Verlet, 'Computer 'experiments' on classical fluids. II. Equilibrium correlation functions,' *Phys. Rev.*, **165** (1968), 201–14.

[13] A. Rahman, 'Correlations in the motion of atoms in liquid argon,' *Phys. Rev.*, **136A** (1964), 405–11.

[14] B. J. Alder and T. E. Wainwright, 'Phase transition for a hard sphere system,' *J. Chem. Phys.*, **27** (1957), 1208–9.

[15] S. M. Thompson, 'Use of neighbour lists in molecular dynamics,' *CCP5 Quarterly*, **8** (1983), 20–8.

[16] H. J. C. Berendsen and W. F. van Gunsteren, 'Practical algorithms for molecular dynamics simulations,' in *Molecular Dynamics Simulation of Statistical Mechanical Systems* (G. Ciccotti and W. G. Hoover, eds.), *Proceedings of the International School of Physics "Enrico Fermi", Varenna 1985*, vol. 97. Amsterdam, North-Holland, 1986, pp. 43–65.

[17] J. M. Sanz-Serna, 'Symplectic integrators for Hamiltonian problems: an overview,' *Acta Numerica*, **1** (1992), 243–86.

[18] K. Feng and M.-Z. Qin, 'Hamiltonian algorithms for Hamiltonian systems and a comparative numerical study,' *Comput. Phys. Commun.*, **65** (1991), 173–87.

[19] R. W. Hockney and J. W. Eastwood, *Computer Simulation Using Particles*, 2nd edn. Bristol, Institute of Physics Publishing, 1988.

[20] D. I. Okunbor and R. D. Skeel, 'Explicit canonical methods for Hamiltonian systems,' *Math. Comput.*, **59** (1992), 439–55.

[21] H. Yoshida, 'Construction of higher order symplectic integrators,' *Phys. Lett. A*, **150** (1990), 262–8.

[22] E. Forest, 'Sixth-order Lie group integrators,' *J. Comp. Phys.*, **99** (1992), 209–13.

[23] C. W. Gear, 'The numerical integration of ordinary differential equations of various orders,' report ANL7126, Argonne National Laboratory, 1966.

[24] C. W. Gear, *Numerical Initial Value Problems in Ordinary Differential Equations*. Englewood Cliffs, NJ, Prentice-Hall, 1971.

[25] D. I. Okunbor, 'Energy conserving, Liouville and symplectic integrators,' *J. Comput. Phys.*, **120** (1995), 375–8.

[26] B. L. Holian, A. F. Voter, and R. Ravelo, 'Thermostatted molecular dynamics: How to avoid the Toda demon hidden in Nosé–Hoover dynamics,' *Phys. Rev. E*, **52** (1995), 2338–47.

[27] R. D. Ruth, 'A canonical integration technique,' *IEEE Trans. Nucl. Sci.*, **30** (1983), 2669–71.

[28] K. Feng, 'On difference schemes and symplectic geometry,' in *Beijing Symposium on Differential Geometry and Differential Equations: Computation of Partial Differential Equations* (K. Feng, ed.). Beijing, Science Press, 1985, p. 45.

[29] J. D. Meiss, 'Symplectic maps, variational principles and transport,' *Rev. Mod. Phys.*, **64** (1992), 795–848.

[30] C. Cohen-Tannoudji, B. Diu, and F. Laloë, *Quantum Mechanics*, vols. 1 and 2. New York/Paris, John Wiley/Hermann, 1977.

[31] A. Auerbach and S. P. Friedman, 'Long-time behaviour of numerically computed orbits: small and intermediate timestep analysis of one-dimensional systems,' *J. Comp. Phys.*, **93** (1991), 189–223.

[32] H. C. Andersen, 'Molecular dynamics at constant temperature and/or pressure,' *J. Chem. Phys.*, **72** (1980), 2384–94.

[33] J. M. Haile and S. Gupta, 'Extensions of molecular dynamics simulation method. II. Isothermal systems,' *J. Chem. Phys.*, **79** (1983), 3067–76.

[34] S. Nosé, 'A unified formulation of constant temperature molecular-dynamics methods,' *J. Chem. Phys.*, **81** (1984), 511–19.

[35] W. G. Hoover, A. J. C. Ladd, R. B. Hickman, and B. L. Holian, 'Bulk viscosity via nonequilibrium and equilibrium molecular dynamics,' *Phys. Rev. A*, **21** (1980), 1756–60.

[36] H. J. C. Berendsen, J. P. M. Postma, W. F. van Gunsteren, A. Dinola, and B. Haak, 'Molecular dynamics with coupling to an external bath,' *J. Chem. Phys.*, **81** (1984), 3684–90.

[37] W. G. Hoover, 'Canonical dynamics: equilibrium phase-space distributions,' *Phys. Rev. A*, **31** (1985), 1695–7.

[38] K. Cho and J. D. Joannopoulos, 'Ergodicity and dynamical properties of constant-temperature molecular dynamics,' *Phys. Rev. A*, **45** (1992), 7089–97.

[39] B. L. Holian, 'Simulations of vibrational relaxation in dense molecular fluids,' in *Molecular Dynamics Simulation of Statistical Mechanical systems* (G. Ciccotti and W. G. Hoover, eds.), *Proceedings of the International School of Physics "Enrico Fermi", Varenna 1985*, vol. 97. Amsterdam, North-Holland, 1986, pp. 241–59.

[40] B. L. Holian, 'Simulations of vibrational-relaxation in dense molecular fluids,' *J. Chem. Phys.*, **84** (1986), 3138–46.

[41] G. J. Martyna, M. L. Klein, and M. Tuckerman, 'Nosé–Hoover chains – the canonical ensemble via continuous dynamics,' *J. Chem. Phys.*, **97** (1992), 2635–43.

[42] S. Nosé, 'A molecular dynamics method for simulations in the canonical ensemble,' *Mol. Phys.*, **52** (1984), 255–68.

[43] R. G. Winkler, V. Kraus, and P. Reineker, 'Time reversible and phase-space conserving molecular dynamics at constant temperature,' *J. Chem. Phys.*, **102** (1995), 9018–25.

[44] W. G. Hoover, 'Constant pressure equations of motion,' *Phys. Rev. A*, **34** (1986), 2499–500.

[45] M. Parrinello and A. Rahman, 'Polymorphic transitions in single crystals: a new molecular dynamics method.,' *J. Appl. Phys.*, **52** (1981), 7182–90.

[46] P. S. Y. Cheung and J. G. Powles, 'The properties of liquid nitrogen. IV. A computer simulation,' *Mol. Phys.*, **30** (1975), 921–49.

[47] D. Fincham, 'More on rotational motion of linear molecules,' *CCP5 Quarterly*, **12** (1984), 47–8.

[48] D. J. Evans, 'On the representation of orientation space,' *Mol. Phys.*, **34** (1977), 317–25.

[49] D. J. Evans and S. Murad, 'Singularity-free algorithm for molecular dynamics simulation of rigid polyatomics,' *Mol. Phys.*, **34** (1977), 327–31.

[50] H. Goldstein, *Classical Mechanics*. Reading, Addison-Wesley, 1980.

[51] G. Ciccotti, M. Ferrario, and J. P. Ryckaert, 'Molecular dynamics of rigid systems in cartesian coordinates,' *Mol. Phys.*, **47** (1982), 1253–64.

[52] J. P. Ryckaert, 'The method of constraints in molecular dynamics,' in *Molecular Dynamics Simulation of Statistical Mechanical Systems* (G. Ciccotti and W. G. Hoover, eds.), *Proceedings of the International School of Physics "Enrico Fermi", Varenna 1985*, vol. 97. Amsterdam, North-Holland, 1986, pp. 329–40.

[53] J. P. Ryckaert, G. Ciccotti, and H. J. C. Berendsen, 'Numerical integration of the cartesian equations of motion of a system with constraints: molecular dynamics of n-alkanes,' *J. Comput. Phys.*, **23** (1977), 327–41.

[54] J. P. Ryckaert, 'Special geometrical constraints in the molecular dynamics of chain molecules,' *Mol. Phys.*, **55** (1985), 549–56.

[55] J.-P. Hansen, 'Molecular dynamics simulation of Coulomb systems,' in *Molecular Dynamics Simulation of Statistical Mechanical Systems* (G. Ciccotti and W. G. Hoover, eds.), *Proceedings of the International School of Physics "Enrico Fermi", Varenna 1985*, vol. 97. Amsterdam, North-Holland, 1986, pp. 89–119.

[56] S. W. De Leeuw, J. W. Perram, and E. R. Smith, 'Simulation of electrostatic systems in periodic boundary conditions. I. Lattice sums and dielectric constants,' *Proc. R. Soc. London*, **A373** (1980), 27–56.

[57] J. Barnes and P. Hut, 'A hierarchical $O(N \log N)$ force-calculation algorithm,' *Nature*, **324** (1986), 446–9.

[58] L. Hernquist, 'Performance characteristics of tree codes,' *Astrophysi. J. Suppl.*, **64** (1987), 715–34.

[59] A. W. Appel, 'An efficient program for many-body simulation,' *Siam. J. Sci. Stat. Comput.*, **6** (1985), 85–103.

[60] J. G. Jernigan, 'Direct N-body simulations with a recursive center of mass reduction and regularization,' in *Dynamics of Star Clusters* (J. Goodman and P. Hut, eds.) *IAU Symposium*, vol. 113, Dordrecht, Reidel, 1985, pp. 275–84.

[61] L. van Dommelen and E. A. Rundensteiner, 'Fast, adaptive summation of point forces in the two-dimensional Poisson equation,' *Siam. J. Sci. Stat. Comput.*, **83** (1989), 286–300.

[62] L. Greengard, 'The numerical solution of the N-body problem,' *Comp. Phys.*, **4** (1990), 142–52.

[63] L. Greengard, *The Rapid Evaluation of Potential Fields in Particle Systems*. Cambridge, MIT Press, 1988.

[64] G. Ciccotti and J. P. Ryckaert, 'Computer simulation of the generalized Brownian motion. I. The scalar case,' *Mol. Phys.*, **40** (1980), 141–9.

[65] S. Toxvaerd, 'Solution of the generalised Langevin equation,' *J. Chem. Phys.*, **82** (1985), 5658–62.

[66] F. J. Vesely and H. A. Posch, 'Correlated motion of 2 particles in a fluid. 1. Stochastic equation of motion,' *Mol. Phys.*, **64** (1988), 97–109.

[67] L. G. Nillson and J. A. Prado, 'A time-saving algorithm for generalised Langevin-dynamics simulations with arbitrary memory kernels,' *Mol. Phys.*, **71** (1990), 355–67.

[68] D. L. Ermak and J. A. McCammon, 'Brownian dynamics with hydrodynamic interactions,' *J. Chem. Phys.*, **69** (1978), 1352–60.

[69] P. J. Hoogerbrugge and J. M. V. A. Koelman, 'Simulating microscopic hydrodynamic phenomena with dissipative particle dynamics,' *Europhys. Lett.*, **19** (1992), 155–60.

[70] W. F. van Gunsteren and H. J. C. Berendsen, 'Algorithms for Brownian dynamics,' *Mol. Phys.*, **45** (1982), 637–47.

[71] R. Kubo, H. Ichimura, and T. Usui, *Statistical Mechanics, An Advanced Course*. Amsterdam, North-Holland, 1965.

[72] M. Plischke and H. Bergersen, *Equilibrium Statistical Physics*. Englewood Cliffs, NJ, Prentice-Hall, 1989.

[73] E. M. Gosling, I. R. McDonald, and K. Singer, 'On the calculation by molecular dynamics of the shear viscosity,' *Mol. Phys.*, **26** (1973), 1475–84.

[74] G. Ciccotti and G. Jacucci, 'Direct computation of dynamical response by molecular dynamics: the mobility of a charged Lennard–Jones particle,' *Phys. Rev. Lett.*, **35** (1975), 789–92.

9

Quantum molecular dynamics

9.1 Introduction

In the previous chapter we considered systems of interacting particles. They were treated as classical particles for which the interaction potential is known. We had to solve the classical equations of motion to simulate the behaviour of such a system at some nonzero temperature. Had we added frictional forces, the system would have evolved towards the ground state. In this chapter we discuss methods for simulating interacting atoms and molecules using quantum mechanical calculations. In fact, we consider the nuclei on a classical level but use quantum mechanics for the electronic degrees of freedom. Again, we can use this approach either to simulate a system of interacting particles at a finite temperature, or to find the ground state (minimum energy) configurations of solids and of molecules.

In Chapters 4 to 6 we studied methods for solving the electronic structure of molecular and solid state systems with a static configuration of nuclei (Born–Oppenheimer approximation). Knowledge of the electronic structure includes knowledge of the total energy. Therefore, by varying the positions of the nuclei, we can study the dependence of the total energy on these positions. The energy $E(\mathbf{R}_1, \mathbf{R}_2, \ldots, \mathbf{R}_N)$ as a function of the nuclear positions $\mathbf{R}_i$ is called the *potential surface*. As a simple example, consider the hydrogen molecule. We assume that the molecule is not rotating, so that the nuclear motion is a vibration along the molecular axis. The only relevant parameter describing the relative positions of the two nuclei is their separation X. The force between the nuclei is then given by $F = -\partial E(X)/\partial X$ (see Figure 9.1). These forces are usually parametrised and the parameters are fixed by comparison with quantum mechanical calculations for a few configurations, or by comparison with experimental results. This parametrised form can then be used to calculate the motion of the nuclei on a classical level, for example to find the equilibrium conformation of the molecule, which is the configuration of nuclei that minimises the total energy. This is called the method of *force fields*; it is often used by chemists.

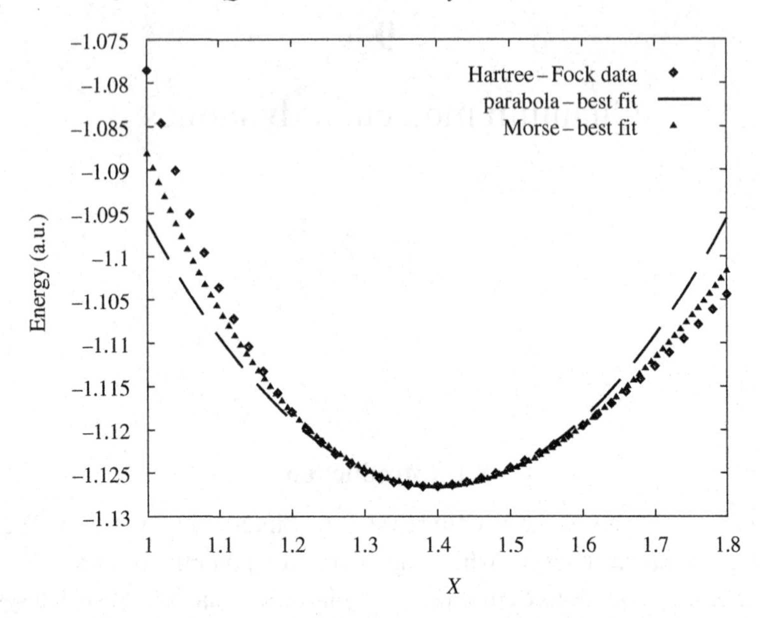

Figure 9.1. The effective potential of the hydrogen nuclei in the hydrogen molecule as a function of the separation X. A harmonic (parabolic) potential and a Morse potential are fitted to the bottom of the well.

The parametrisation of the forces is often carried out for small deviations of the configuration from the equilibrium conformation, so that the potential energy can be approximated quite accurately by harmonic potentials, such as stretching, bending, and torsional potentials, encountered in Section 8.6.1. The motion can then be decomposed into normal modes, by defining new coordinates in terms of which the system can be described as a collection of uncoupled harmonic oscillators. This problem then has an analytic quantum mechanical solution, leading to discretised energy levels which can be compared with experiment. So, although the force field method treats the nuclear motion classically, we can obtain quantum mechanical solutions for the *nuclear* motion from it (within a Born–Oppenheimer approach).

In our example of the hydrogen molecule we can fit the bottom of the potential well shown in Figure 9.1 by a harmonic potential. Since the well is rather asymmetric, a more reliable fit is provided by the Morse potential, for which the spectrum is also known analytically (see Problem A4). For the harmonic approximation $\kappa X^2/2$, the angular frequency $\omega = \sqrt{\kappa/m}$ and the spectral levels are given as $\hbar\omega(n+1/2)$. For the hydrogen molecule, the mass to be used is the reduced mass, which is about half the proton mass (i.e. 918.8 electron masses if we neglect the mass of the two electrons), and we find $\kappa = 0.3850$ (in atomic units) so that the frequency becomes $\nu_{\text{vibr}} = 13.64 \times 10^{13}$ Hz, to be compared with the experimental value $\nu_{\text{vibr}} = 12.48 \times 10^{13}$Hz [1]. [1]

The harmonic approximation works well for low energies. It is used for stretch, bend and torsional energies; see also Section 8.6.1. For higher energies, anharmonic terms can be included in the potentials – see the Morse potential in the hydrogen example. For energies much higher than the spacing between the energy levels, quantum effects do not affect physical properties and a fully classical description is appropriate.

For interacting molecules the force field procedure becomes unfeasible because we would have to calculate energies and variations in energies for all possible relative positions and orientations for pairs or sets of two and more molecules, which becomes exceedingly tedious and (computer) time-consuming. Therefore, in these systems, the intramolecular interactions are modelled by force fields and the intermolecular interactions by atomic pair-interactions as we have seen throughout the previous chapter. Although this approach yields rather accurate results, in particular when the density is not too high, the use of these pair-potentials is not justified for dense systems. Moreover, the energy differences between different atomic conformations are often very small, so that high accuracy is needed for predicting equilibrium structures.

To achieve accuracy in these cases, it is necessary to calculate forces and energies from quantum electronic structure calculations; if this is unfeasible for all possible configurations, we take the more economical approach of calculating forces and energies only for those configurations which actually occur in the simulation. We must therefore perform an electronic structure calculation at every molecular dynamics time step, and derive the *force* on the nuclei from that calculation. The word force is emphasised because the methods described in the first few chapters of this book aimed at calculating the energies and not the forces. Of course, it would be possible to derive the forces from the energies by studying the variations in the latter with nuclear positions but that would require an exceedingly large number of energy calculations. It is better therefore to try to find methods for calculating the forces directly.

The energy of a system of electrons in its ground state ψ_G for a fixed configuration of nuclei $S = (\mathbf{R}_1, \ldots, \mathbf{R}_M)$, where $\mathbf{R}_n$ is the position of the nth nucleus, is given by[2]

$$E = \frac{\langle \psi_G | H(S) | \psi_G \rangle}{\langle \psi_G | \psi_G \rangle}. \tag{9.1}$$

[1] In atomic units, the unit of frequency is $\alpha c/a_0 = 4.13414 \times 10^{14}$ Hz; α is the fine structure constant.

[2] We use the letter S in order to avoid confusion with $R = (\mathbf{r}_1, \ldots, \mathbf{r}_N)$.

The (classical) force on nucleus n is given as the negative gradient ∇_n of the energy with respect to the $\mathbf{R}_n$:

$$\mathbf{F}_n = -\nabla_n E(S) = -\nabla_n \left[\frac{\langle \psi_G | H(S) | \psi_G \rangle}{\langle \psi_G | \psi_G \rangle} \right]. \tag{9.2}$$

It should be noted that there is not only the explicit S-dependence in the Hamiltonian, but the ground state is evaluated for the Hamiltonian with a particular configuration S. Therefore the ground state also depends on S.

The *Hellmann–Feynman theorem* [2, 3], which we discussed for the single-parameter case in Section 5.3, states that we can neglect this dependence: if ψ_G is an eigenstate of the Hamiltonian $H(S)$, we have

$$\begin{aligned}
&(\langle \psi_G | \psi_G \rangle)^2 \nabla_n E \\
&\quad = \left[\langle (\nabla_n \psi_G) | H | \psi_G \rangle + \langle \psi_G | (\nabla_n H) | \psi_G \rangle + \langle \psi_G | H | (\nabla_n \psi_G) \rangle \right] \langle \psi_G | \psi_G \rangle \\
&\quad\quad - \langle \psi_G | H | \psi_G \rangle \left[\langle (\nabla_n \psi_G) | \psi_G \rangle + \langle \psi_G | (\nabla_n \psi_G) \rangle \right],
\end{aligned} \tag{9.3}$$

where we have omitted the S-dependence of the Hamiltonian. Except for the term including $\langle \psi_G | (\nabla_n H) | \psi_G \rangle$, all the terms on the right hand side cancel: this follows directly from the fact that $H\psi_G = E_G \psi_G$ and from the fact that H is Hermitian. So we are left with

$$\nabla_n E = \frac{\langle \psi_G | (\nabla_n H) | \psi_G \rangle}{\langle \psi_G | \psi_G \rangle}. \tag{9.4}$$

In practice we do not know the exact ground state, but we have only a variational approximation to it. Therefore, in actual calculations, the Hellmann–Feynman theorem does not predict the actual forces exactly and the variation of the (approximated) ground state wave function should be taken into account as well. Nevertheless, the Hellmann–Feynman theorem is used quite often for predicting ground state configurations, because the inclusion of other contributions is cumbersome.

9.2 The molecular dynamics method

In principle, all the ingredients for a molecular dynamics simulation using forces calculated from the quantum electronic structure are at our disposal. However, at each step in the MD simulation, a full electronic structure calculation is required, so the method consumes a lot of computer time. In 1985, Car and Parrinello proposed a method in which not only the nuclear positions, but also the electronic states are calculated using MD algorithms. This results in a description of the system in which the electronic structure does not, in general, relax completely to the ground state of the actual configuration of nuclei; however, the calculated electronic structure will follow the exact one rather closely. We start the description of the Car–Parrinello method by recalling the energy functionals of the Hartree–Fock and the density functional theory (see Chapters 4 and 5).

The ground state Hartree–Fock wave function for N electrons can be written as the Slater determinant

$$\Psi_G(R) = \det[\psi_k(\mathbf{x}_i)] = \frac{1}{\sqrt{N!}} \begin{vmatrix} \psi_1(\mathbf{x}_1) & \psi_2(\mathbf{x}_1) & \cdots & \psi_N(\mathbf{x}_1) \\ \psi_1(\mathbf{x}_2) & \psi_2(\mathbf{x}_2) & \cdots & \psi_N(\mathbf{x}_2) \\ \vdots & \vdots & & \vdots \\ \psi_1(\mathbf{x}_N) & \psi_2(\mathbf{x}_N) & \cdots & \psi_N(\mathbf{x}_N) \end{vmatrix}, \tag{9.5}$$

where the ψ_k are one-electron spin-orbitals; $\mathbf{x}_i$ is the combined spin and orbital coordinate of particle i. The spin-orbitals should satisfy the orthonormality requirements

$$\langle \psi_k | \psi_l \rangle = \delta_{kl}. \tag{9.6}$$

The energy is given in terms of the ψ_k as

$$E_{\mathrm{HF}} = \sum_k \langle \psi_k | h | \psi_k \rangle + \frac{1}{2} \sum_{kl} [\langle \psi_k \psi_l | g | \psi_k \psi_l \rangle - \langle \psi_k \psi_l | g | \psi_l \psi_k \rangle]. \tag{9.7}$$

h is the one-electron Hamiltonian and g is the electron–electron Coulomb repulsion (see Chapter 4). Minimisation of this expression with respect to the ψ_k subject to the constraint (9.6) requires the Fock equation to be satisfied:

$$\mathcal{F}\psi_k = \sum_l \Lambda_{kl} \psi_l \tag{9.8}$$

with

$$\mathcal{F}\psi_k = \left[-\frac{1}{2}\nabla^2 - \sum_n \frac{Z_n}{|\mathbf{r} - \mathbf{R}_n|} \right] \psi_k(\mathbf{x}) + \sum_{l=1}^{N} \int \mathrm{d}x' \, |\psi_l(\mathbf{x}')|^2 \frac{1}{|\mathbf{r} - \mathbf{r}'|} \psi_k(\mathbf{x})$$
$$- \sum_{l=1}^{N} \int \mathrm{d}x' \, \psi_l^*(\mathbf{x}') \frac{1}{|\mathbf{r} - \mathbf{r}'|} \psi_k(\mathbf{x}') \psi_l(\mathbf{x}). \tag{9.9}$$

After a unitary transformation of the set $\{\psi_k\}$ (see Section 4.5.2 and Problem 4.7), Eq. (9.8) transforms into

$$\mathcal{F}\psi_k = \varepsilon_k \psi_k. \tag{9.10}$$

Using $\mathcal{F}\psi_k = \delta E_{\mathrm{HF}}/\delta\psi_k$, we can rewrite this as

$$\frac{\delta E_{\mathrm{HF}}}{\delta \psi_k(\mathbf{x})} = \varepsilon_k \psi_k(\mathbf{x}). \tag{9.11}$$

The eigenvalues ε_k are the Fock levels; the energy can be calculated from these.

In density functional theory the energy can be written as a function of the ground state density, which in turn is written in terms of the basis functions as

$$n(\mathbf{r}) = \sum_{k=1}^{N} |\psi_k(\mathbf{r})|^2, \tag{9.12}$$

assuming that the states are ordered according to increasing energy. We have seen already in Chapter 5 that there is no direct expression of the total energy as a function of the density, as the form of the kinetic energy functional of the density is unknown. The energy can however be obtained indirectly via the solutions ψ_k of the Kohn–Sham equations:

$$-\frac{1}{2}\nabla^2\psi_k(\mathbf{r}) + V_{\text{eff}}(\mathbf{r})\psi_k(\mathbf{r}) = \varepsilon_k\psi_k(\mathbf{r}) \tag{9.13}$$

where

$$V_{\text{eff}}(\mathbf{r}) = V_{\text{ion}}(\mathbf{r}) + \int d^3r' \, \frac{n(\mathbf{r}')}{|\mathbf{r}-\mathbf{r}'|} + V_{\text{xc}}[n](\mathbf{r}). \tag{9.14}$$

The exchange correlation potential V_{xc} on the right hand side is the derivative of the exchange correlation energy E_{xc} with respect to $n(\mathbf{r})$.

In terms of the ψ_k, the total DFT energy is written as

$$\begin{aligned} E_{\text{DFT}} = &-\sum_k \frac{1}{2}\langle\psi_k|\nabla^2|\psi_k\rangle + \sum_k \langle\psi_k|V_{\text{ion}}|\psi_k\rangle \\ &+\frac{1}{2}\int d^3r \, d^3r' \, \frac{n(\mathbf{r})n(\mathbf{r}')}{|\mathbf{r}-\mathbf{r}'|} + E_{\text{xc}}[n](\mathbf{r}). \end{aligned} \tag{9.15}$$

The Kohn–Sham equations can be written as

$$\frac{\delta E_{\text{DFT}}}{\delta\psi_k(\mathbf{r})} = \varepsilon_k\psi_k(\mathbf{r}), \tag{9.16}$$

i.e. the same form as (9.11).

Summarising, the total energy, which is the electronic energy (either E_{DFT} or E_{HF}) plus the electrostatic energy of the nuclei, can be written as a functional depending on the orbitals ψ_k and of the nuclear coordinates, collected together in the variable S:

$$E_{\text{tot}} = E_{\text{tot}}(\{\psi_k\}, S), \tag{9.17}$$

where the orbitals ψ_k form an orthonormal set. In both the Hartree–Fock and the density functional theory approach we minimise this energy with respect to the orbitals ψ_k, according to the variational principle. Usually, a finite basis set $\{\chi_r\}$ is used, in terms of which the orbitals are given as

$$\psi_k(\mathbf{r}) = \sum_r C_{rk}\chi_r(\mathbf{r}), \tag{9.18}$$

so that the energy can be written in terms of the C_{rk} and S:

$$E_{\rm tot} = E_{\rm tot}(\{C_{rk}\}, S). \tag{9.19}$$

As the basis functions are often centred on the atomic nuclei, they may contain an explicit S-dependence. Car and Parrinello used the form (9.17) (or (9.19)) with the constraint (9.6) as a starting point for finding the equilibrium conformation (i.e. the minimal energy conformation) by locating the minimum of the total energy as a function of the ψ_k (or the C_{rk}) *and* the nuclear coordinates S. This means that the electronic structure does not have to be calculated exactly for each conformation of nuclei, as both the electronic orbitals and the nuclear positions are varied *simultaneously* in order to locate the minimum.

The minimisation problem of the energy can now be considered as an abstract numerical problem, and any minimisation algorithm can in principle be applied. One possible approach is the simulated annealing method [4], which requires only the energy to be calculated – no force calculations are needed. However, Car and Parrinello assigned, aside from the time dependence of the nuclear coordinates, a *fictitious* time dependence to the electronic wave functions (or, in a linear variational calculation, the C_{rk}), and constructed a dynamical Lagrangian including the electronic wave functions and the nuclear coordinate S with their time derivatives as the variables. This leads to a classical mechanics problem with the energy (9.17) acting as a potential. If a friction term is then added to the equations of motion of this classical system, the degrees of freedom will come to rest after some time, with values corresponding to the minimum of the classical potential, which is the energy of the quantum system at the equilibrium configuration of the nuclei. It is also possible to put the frictional force equal to zero in order to simulate the system at a nonzero temperature.

The Lagrangian of the classical system reads

$$L(\{\psi_k\}, S) = \frac{\mu}{2}\sum_k \dot{\psi}_k^2 + \sum_n \frac{M_n}{2}\dot{\mathbf{R}}_n^2 - E_{\rm tot}(\psi_k, S) + \sum_{kl} \Lambda_{kl}\langle\psi_k|\psi_l\rangle. \tag{9.20}$$

μ is some small mass (see below), and M_n is the actual mass of the nth nucleus, with position $\mathbf{R}_n$. The last term on the right hand side is necessary to ensure orthonormality of the ψ_k; the Λ_{kl} must always be calculated from this requirement. Car and Parrinello suggested that this Lagrangian can be used not only for finding the minimum of the total energy, but also for performing real molecular dynamics simulations at finite temperature. It will be clear that in general, when the nuclei move, the method might not have produced the minimal energy of the electrons before the next nuclear displacement: the calculated electronic structure will 'lag behind' the nuclear motion. Although this retardation effect will occur in reality (the Born–Oppenheimer approximation neglects the fact that the electrons do not have the opportunity to adapt themselves to the changing nuclear configuration at any time), there is no reason to believe that the retardation effect implied by the Car–Parrinello Lagrangian is related to real physical behaviour.

The details of the kinetic energy of the electrons do not matter: what matters is the fact that the mass μ used in this kinetic energy should be small enough to enable the electronic wave function to adapt reasonably well to the changing nuclear configurations. This mass should therefore be much smaller than the nuclear masses. The choice of the mass μ is determined by a trade-off between accuracy and efficiency. If we include friction in the equations of motion, the particular values of neither electronic nor nuclear masses matter, as we shall always end up with zero kinetic energy, at the minimum of the total energy of the system (which is the potential of the Car–Parrinello Lagrangian), although different choices of these masses lead to different rates of convergence towards the energy minimum.

Let us write down the equations of motion for the Car–Parrinello Lagrangian. We must take the orthogonality constraint (9.6) into account using Lagrange parameters $\Lambda_{kl}(t)$. The Euler–Lagrange equations now read

$$\mu\ddot{\psi}_k = -\frac{\partial E_{\text{tot}}}{\partial \psi_k} + 2\sum_l \Lambda_{kl}\psi_l(\mathbf{r}) \tag{9.21}$$

and

$$M_n\ddot{\mathbf{R}}_n = -\frac{\partial E_{\text{tot}}}{\partial \mathbf{R}_n} + \sum_{kl} \Lambda_{kl}\frac{\partial \langle \psi_k|\psi_l\rangle}{\partial \mathbf{R}_n}. \tag{9.22}$$

The last term on the right hand side of the last equation vanishes if the basis functions do not depend on the nuclear positions S. As we know the total energy in both DFT and HF in terms of the orbitals ψ_k and $\mathbf{R}_n$, the energy derivatives occurring in these equations can be evaluated – see the next section.

Instead of assigning a kinetic energy to the orbitals ψ_k, leading to Eq. (9.21), we can assign a kinetic energy to the expansion coefficients C_{rk}. In that case, Eq. (9.21) becomes

$$\mu\ddot{C}_{rk} = -\frac{\partial E_{\text{tot}}}{\partial C_{rk}} + 2\sum_l \Lambda_{kl}\sum_s S_{rs}C_{sl}. \tag{9.23}$$

If μ is allowed to depend on r and k, this equation can be made equivalent to (9.21) but, as argued above, the details of the kinetic energy do not matter that much as long as the electronic degrees of freedom can adapt themselves to the nuclear positions.

If a frictional term is added to the equations of motion, the solution will become stationary after some time, and the left hand side vanishes. Equation (9.21) then becomes an equation similar to the Fock and the Kohn–Sham equations ((9.11) and (9.16)), except for the eigenvalues ε_k being replaced by the matrix elements Λ_{kl}. This is precisely the same difference as we have encountered in the diagonalisation of the Fock matrix (see Section 4.5.2 and above): for $\ddot{\psi}_k = 0$, Eq. (9.21) reduces to an eigenvalue equation after an appropriate unitary transformation of the set $\{\psi_k\}$ and of the Lagrange parameters Λ_{kl}.

The values of the Lagrange parameters Λ_{kl} depend on time: they must be calculated at each MD step such that they guarantee the orthonormality constraint (9.6). This calculational procedure is related to the particular integration scheme used (the Verlet algorithm in our case). In Section 8.6.2 we have encountered this problem already. Car and Parrinello have used the iterative SHAKE algorithm of Ryckaert *et al.* [5] (see Section 8.6.3) to solve for the Λ_{kl}. We return to the problem of calculating the Λ_{kl} in more detail below.

If the nuclear equilibrium configuration is searched for, starting from an initial configuration which might be far off the equilibrium, we are likely to end up in a local energy minimum instead of the global minimum. In this case, we might use the simulated annealing method [4] which allows the system to hop over local energy barriers to arrive at the global minimum.

It is interesting to compare the equations obtained here with the time-dependent Hartree–Fock (TDHF) equations. These are obtained from a variational treatment of the time-dependent Schrödinger equation using Slater determinants constructed from time-dependent spin-orbitals. The time-dependent Schrödinger equation can be derived as the stationarity condition of the functional

$$S = \int \mathrm{d}t \int \mathrm{d}X \; \Psi^*(X,t) \left(\mathrm{i}\hbar \frac{\partial}{\partial t} - H \right) \Psi(X,t) \tag{9.24}$$

with $X = (\mathbf{x}_1, \ldots, \mathbf{x}_N)$. By taking for $\Psi(R,t)$ a Slater determinant with time-dependent orbitals $\psi_k(\mathbf{x},t)$, the stationarity condition leads to the following equation for the spin-orbitals [6]:

$$\mathrm{i}\hbar \frac{\partial}{\partial t} \psi_k(\mathbf{x},t) = \mathcal{F} \psi_k(\mathbf{x},t). \tag{9.25}$$

The TDHF equations lead to a conservation law for the overlap matrix $S_{kl}(t) = \langle \psi_k(t) | \psi_l(t) \rangle$. Hence, if we choose an orthonormal set to start off with at $t = 0$, the set will remain orthonormal in the course of time.

In comparison with the MD equation of motion for the electrons, Eq. (9.21), we see that the second derivative with respect to time is replaced by a first order one, and that there is no Lagrange parameter as a result of the overlap matrix being conserved.

Time-dependent Hartree–Fock is used for studying the quantum dynamics of scattering processes, for example in nuclear physics and in studies of scattering of electrons from atoms.

9.3 An example: quantum molecular dynamics for the hydrogen molecule

In this subsection we work out an application of the Car–Parrinello method to the hydrogen molecule in some detail. Our example is based on the Hartree–Fock calculation of the hydrogen molecule considered in Chapter 4, in particular Problem 4.9. There are two spin-orbitals with opposite spin and the same orbital part. Therefore, the wave function is completely specified by the form of this orbital. We use the GTO basis set of Problem 4.9 with eight basis s-functions χ_r, four on each atom.

The molecular dynamics method can be restricted to the electronic structure part of the total energy, keeping the nuclear positions fixed. We do this first; later we shall extend the method to include nuclear displacements.

9.3.1 The electronic structure

The energy can be written as

$$E_{\text{tot}} = 2\sum_{rs} C_r h_{rs} C_s + \sum_{rstu} C_r C_s C_t C_u \langle rt|g|su\rangle + \frac{1}{X}. \tag{9.26}$$

Note that there is no index k as the two electrons occupy only one orbital. The Fock matrix $\mathbf{F}$ is given by

$$F_{rs} = h_{rs} + \sum_{tu} C_t C_u \langle rt|g|su\rangle \tag{9.27}$$

(all sums over indices r, s, t, u run over the basis states, so in our case from 1 to 8). The normalisation condition for the orbital is

$$\sum_{rs} C_r S_{rs} C_s = 1. \tag{9.28}$$

Therefore, the equation of motion for the C_r (without friction) is given by

$$\begin{aligned}\frac{\mu}{4}\ddot{C}_r &= -\sum_s h_{rs} C_s - \sum_{stu} C_s C_t C_u \langle rt|g|su\rangle - \lambda \sum_s S_{rs} C_s \\ &= -\sum_s (F_{rs} + \lambda S_{rs}) C_s.\end{aligned} \tag{9.29}$$

We shall use the Verlet algorithm for solving the equations of motion. In this form, they read for $\mu = 4$:

$$C_r(t+h) = 2C_r(t) - C_r(t-h) - h^2 \sum_s (F_{rs} + \lambda S_{rs}) C_s(t). \tag{9.30}$$

Suppose we know the $C_r(t)$ and the $C_r(t-h)$. The solution to the equation of motion proceeds in two stages. First we calculate

$$\tilde{C}_r(t+h) = 2C_r(t) - C_r(t-h) - h^2 \sum_s F_{rs} C_s(t). \tag{9.31}$$

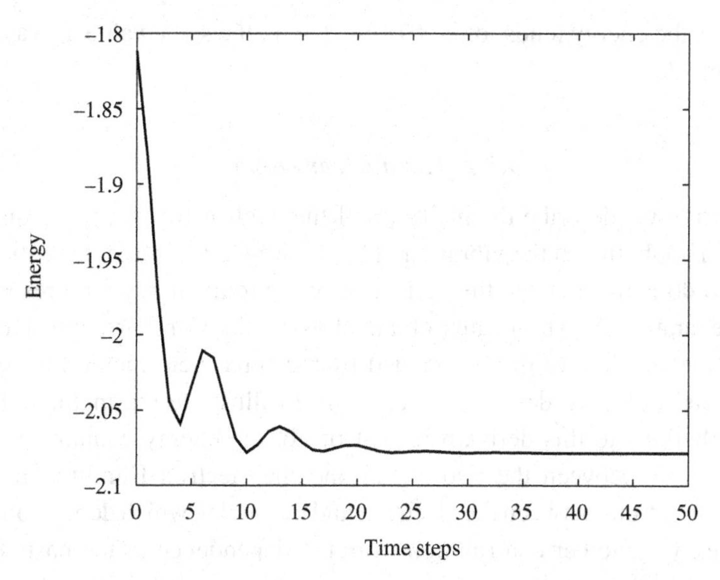

Figure 9.2. Evolution of the energy in a Car–Parrinello simulation of the electronic structure of the hydrogen molecule with separation $X = 1$ between the nuclei, with frictional forces included.

Now we must add an amount $-\lambda S_{rs} C_s(t)$ to this solution, where λ is determined by the requirement that the normalisation condition (9.28) holds:

$$\sum_{rs} \tilde{C}_r(t+h)\tilde{C}_s(t+h) - 2\lambda \sum_{rst} S_{rs}\tilde{C}_r(t+h)S_{st}C_t(t)$$
$$+ \lambda^2 \sum_{rstu} S_{rs}C_s(t)S_{rt}S_{tu}C_u(t) = 1. \tag{9.32}$$

This is a quadratic equation in λ, of which the lowest positive root is needed. The Verlet solution of the equation of motion is now fully defined.

Modifying the HF program of Chapter 4 to calculate the electronic structure is relatively easy, as the Fock matrix and the overlap matrix are calculated already in this program.

PROGRAMMING EXERCISE

Take the program of Problem 4.9 and replace the self-consistency iteration by a molecular dynamics algorithm with friction, using the Verlet algorithm.

A frictional force $-\gamma \dot{C}_r$ is included using the algorithm given in Section 8.4.1 in order to let the system evolve towards the ground state.

Check A reasonable value for the time step is 0.1 (in atomic units) and for the frictional constant γ the value 1 (in atomic units) is chosen. In Figure 9.2, the energy as a function of the 'time' is shown. It is seen that for a nuclear separation of $X = 1$ the energy tends to $-2.078\,547\,6$ a.u., the same value as was found in Problem 4.9.

9.3.2 The nuclear motion

In this section we describe the inclusion of the nuclear forces in the equations of motion and apply this to the vibration of the hydrogen molecule. Essentially, what we have to do is to calculate the derivative of the total energy with respect to the nuclear separation X. The results obtained using the Car–Parrinello HF method are exactly equivalent to those obtained by the force field method as we have a pair potential only; we describe it here only to illustrate the method. There are two contributions to this derivative. First of all, the energy contains a Coulomb interaction $1/X$ between the two nuclei and the electron Hamiltonian contains Coulomb attractions between the electrons and the nuclei, which depend on X. There is, however, yet another contribution from the dependence of the basis functions χ_r on X: remember the basis functions are centred on the nuclei, so varying the positions of the latter changes the matrix elements of the Fock matrix and the overlap matrix. In the following we shall not distinguish explicitly between all these contributions, but it is useful to know that contributions to the forces due to the variation of the basis functions with the nuclear positions are called *Pulay forces* [7]. If the basis functions do not depend on the nuclear coordinates, as is the case with plane wave basis sets, which are often used in conjunction with pseudopotentials, Pulay forces are absent. We shall now calculate the derivatives of the matrix elements of the Fock matrix and the overlap matrix with respect to the nuclear separation in the hydrogen molecule.

Expressions for the various matrix elements were given in Section 4.8. We use notations similar to those used in that section. The overlap matrix was given as

$$S_{\alpha,A;\beta,B} = \langle 1\mathrm{s}, \alpha, A | 1\mathrm{s}, \beta, B\rangle = \left(\frac{\pi}{\alpha+\beta}\right)^{3/2} \exp\left[-\frac{\alpha\beta}{\alpha+\beta}|\mathbf{R}_A - \mathbf{R}_B|^2\right], \quad (9.33)$$

and we see that if both basis functions are centred on the same nucleus ($A = B$), this matrix element does not depend on X. For two basis functions centred on different nuclei, $|\mathbf{R}_A - \mathbf{R}_B| = X$, and we find

$$\frac{\mathrm{d}}{\mathrm{d}X}\langle 1\mathrm{s}, \alpha, A | 1\mathrm{s}, \beta, B\rangle = -2\frac{\alpha\beta}{\alpha+\beta} X S_{\alpha,A;\beta,B}. \quad (9.34)$$

The matrix elements of the kinetic energy operator for two orbitals centred on the same atom are again independent of X, and for the elements between basis functions on different nuclei we have, using $\sigma = \alpha\beta/(\alpha+\beta)$ (see Section 4.8):

$$\left\langle 1\mathrm{s}, \alpha, A \left| -\frac{1}{2}\nabla^2 \right| 1\mathrm{s}, \beta, B\right\rangle = [3\sigma - 2\sigma^2 X^2] S_{\alpha,A;\beta,B}. \quad (9.35)$$

Taking the derivative with respect to X we find

$$\frac{\mathrm{d}}{\mathrm{d}X}\langle 1\mathrm{s},\alpha,A\left|-\frac{1}{2}\nabla^2\right|1\mathrm{s},\beta,B\rangle$$
$$= -4\sigma^2 X S_{\alpha,A;\beta,B} + [3\sigma - 2\sigma^2 X^2]\frac{\mathrm{d}}{\mathrm{d}X}S_{\alpha,A;\beta,B}. \tag{9.36}$$

The Coulomb matrix element is given by

$$\langle 1\mathrm{s},\alpha,A\left|\sum_c \frac{1}{r_c}\right|1\mathrm{s},\beta,B\rangle = \theta\sum_c S_{\alpha,A;\beta,B}F_0(t_c) \tag{9.37}$$

with $\theta = 2\sqrt{(\alpha+\beta)/\pi}$, $t_c = (\alpha+\beta)(PC)^2$ where P is the point

$$\mathbf{R}_P = \frac{\alpha\mathbf{R}_A + \beta\mathbf{R}_B}{\alpha+\beta}, \tag{9.38}$$

$PQ = \mathbf{R}_P - \mathbf{R}_Q$, and C is the position of the nucleus. The sum $\sum_c$ is over the two nuclei. F_0 was given in Section 4.8 – its derivative is given by

$$F_0'(t) = \frac{\mathrm{e}^{-t} - F_0(t)}{2t} \tag{9.39}$$

for $t \neq 0$, and $F_0'(0) = -1/3$. Taking the derivative, we obtain for two basis functions centred on the same nucleus:

$$\frac{\mathrm{d}}{\mathrm{d}X}\langle 1\mathrm{s},\alpha,A\left|\sum_c \frac{1}{r_c}\right|1\mathrm{s},\beta,B\rangle = 2\theta S_{\alpha,A;\beta,B}F_0'(t)X(\alpha+\beta) \tag{9.40}$$

with $t = (\alpha+\beta)X$.

For basis functions centred on different nuclei, we have

$$\frac{\mathrm{d}}{\mathrm{d}X}\langle 1\mathrm{s},\alpha,A\left|\sum_c \frac{1}{r_c}\right|1\mathrm{s},\beta,B\rangle = \theta\frac{\mathrm{d}}{\mathrm{d}X}(S_{\alpha,A;\beta,B})\sum_c[F_0(t_1)+F_0(t_2)]$$
$$+ 2\frac{\theta}{\alpha+\beta}S_{\alpha,A;\beta,B}[F_0'(t_1)\alpha^2 + F_0'(t_2)\beta^2]X. \tag{9.41}$$

where

$$t_1 = \frac{\alpha^2 X^2}{\alpha+\beta}; \tag{9.42a}$$

$$t_2 = \frac{\beta^2 X^2}{\alpha+\beta}. \tag{9.42b}$$

Finally the four-electron matrix element is given by

$$\langle\alpha,A;\gamma,C|g|\beta,B;\delta,D\rangle = \rho S_{\alpha,A;\beta,B}S_{\gamma,C;\delta,D}F_0(t) \tag{9.43}$$

with

$$t = \frac{(\alpha + \beta)(\gamma + \delta)}{\alpha + \beta + \gamma + \delta}(PQ)^2, \tag{9.44}$$

with $\mathbf{R}_P$ as given above and

$$\mathbf{R}_Q = \frac{\gamma \mathbf{R}_C + \delta \mathbf{R}_D}{\gamma + \delta}, \tag{9.45}$$

and

$$\rho = 2\sqrt{\frac{(\alpha + \beta)(\gamma + \delta)}{\pi(\alpha + \beta + \gamma + \delta)}}. \tag{9.46}$$

From this form it follows directly that

$$\begin{aligned} \frac{\mathrm{d}}{\mathrm{d}X}\langle \alpha, A; \gamma, C|g|\beta, B; \delta, D\rangle &= \rho\left(\frac{\mathrm{d}}{\mathrm{d}X} S_{\alpha,A;\beta,B}\right) S_{\gamma,C;\delta,D} F_0(t) \\ &+ \rho S_{\alpha,A;\beta,B}\left(\frac{\mathrm{d}}{\mathrm{d}X} S_{\gamma,C;\delta,D}\right) F_0(t) \\ &+ \rho S_{\alpha,A;\beta,B} S_{\gamma,C;\delta,D} F_0'(t) \frac{(\alpha + \beta)(\gamma + \delta)}{\alpha + \beta + \gamma + \delta}\frac{2(PQ)^2}{X} \end{aligned} \tag{9.47}$$

where we have used the fact that (PQ) is proportional to X in order to obtain the last term on the right hand side.

Using these matrix elements, it is possible to construct the derivatives of the Fock matrix and of the overlap matrix with respect to X, and this gives the force on X which is needed in the Verlet algorithm. Note that the nuclear kinetic energy is given by

$$E_{\text{kin,nucl}} = \frac{M_n}{2}\left[\left(\frac{\dot{X}}{2}\right)^2 + \left(\frac{\dot{X}}{2}\right)^2\right] = \frac{M_n}{4}\dot{X}^2. \tag{9.48}$$

Therefore, in the equation of motion for X, half the proton mass (that is, the reduced mass of the two nuclei) has to be used.

Only the ratio of the masses occurring in the electronic and nuclear kinetic energy is relevant – changing the time step h corresponds to an overall rescaling of the masses. In fact, because the mass occurs in the equation of motion in combination with an acceleration (or, in the kinetic energy, with a velocity squared), rescaling the mass by a factor b and time with a factor $\sqrt{b}$ does not change the calculated motion.

PROGRAMMING EXERCISE

Extend the program of the previous subsection to include the nuclear motion.

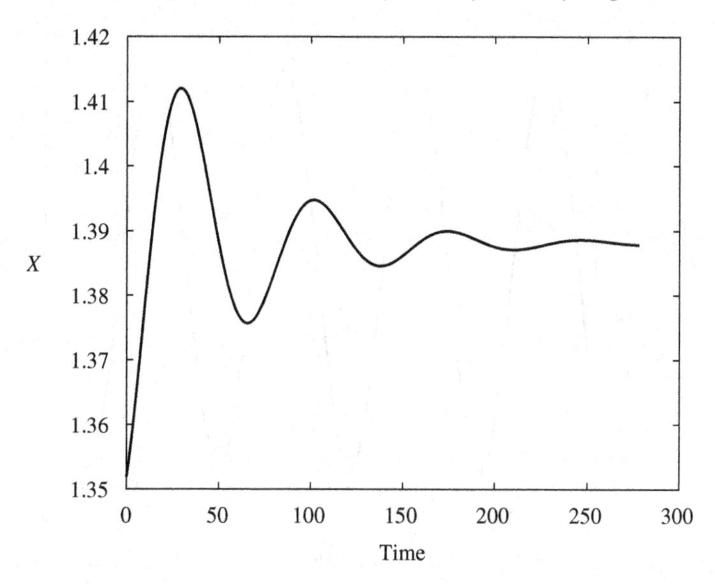

Figure 9.3. The change of the separation X between the nuclei of a hydrogen molecule as a function of time. The number of nuclear integration steps is shown along the X-axis. The nuclear integration step size is 4.3 (in atomic units). The integration step for the electrons was 0.1. Twelve thousand electron integration steps were carried out. The electrons experience a friction with damping constant $\gamma = 1$, and the nuclei are damped with a friction constant of 5.

Check 1 Take the nuclear mass e.g. 1000 times larger than the electron mass. The nuclei will move very slowly in comparison with the electrons because they are so much heavier. If friction is included, the nuclei should end up with zero velocity at their equilibrium spacing, which is at $X = 1.3881a_0$ (within the HF approximation and using exclusively s-basis functions). This is to be compared with the experimental value of $1.401a_0$. The behaviour of X as a function of time is shown in Figure 9.3.

Check 2 If friction is not included, the nuclei will oscillate around their equilibrium separation. Use 1836.15 for the proton mass. The frequency for an initial separation of 1.35 Bohr radii is found to be 13.5×10^{13} Hz, to be compared with the value 13.64×10^{13} Hz obtained above from fitting a parabola to the bottom of the effective potential well in Figure 9.1, and with the experimental value, which is 12.48×10^{13} Hz. The parabola was characterised by a 'spring constant' $\kappa = 0.385$. The behaviour of X as a function of time is shown in Figure 9.4. Check that the results in Figure 9.4 comply with this value (note that the time step in this figure is 4.3 in reduced units).

It is possible and advisable to use fewer integration steps for the nuclear equation of motion than for the electronic one: the nuclei move much more slowly than the

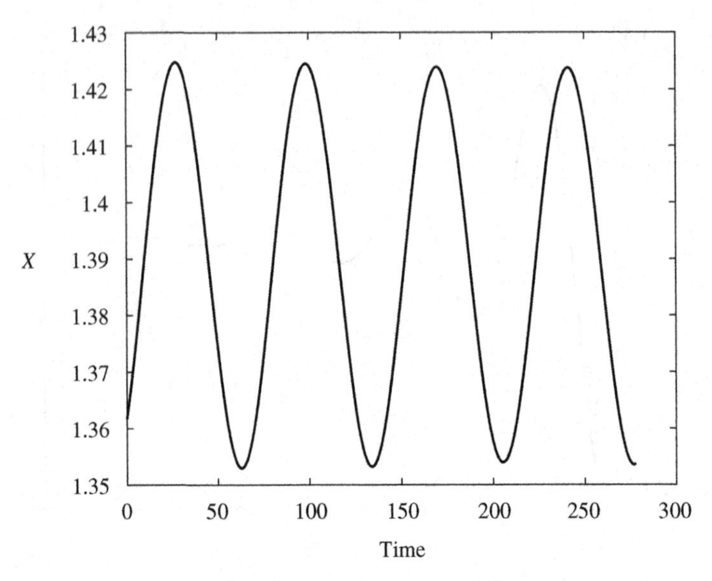

Figure 9.4. The change of the separation X between the nuclei of a hydrogen molecule as a function of time. The number of nuclear integration steps is shown along the X-axis. The nuclear integration step size is 4.3 (in atomic units). The integration step for the electrons was 0.1. Twelve thousand electron integration steps were carried out. The electrons experience a friction with damping constant $\gamma = 1$ during the first 4000 steps; the nuclei experience no friction.

electrons, and a nuclear displacement is computationally expensive because the overlap, Hamilton and Fock matrices have to be calculated again. As the nuclei are moving much more slowly than the electrons this does not affect the overall accuracy significantly, provided the number of electronic integration steps carried out between two nuclear displacements is smaller than $\mathcal{O}(\sqrt{M_n/\mu})$ (see also above).

9.4 Orthonormalisation; conjugate gradient and RM-DIIS techniques

In the previous sections, we have discussed the 'bare-bones' Car–Parrinello method and applied it to a simple system. There is much more to it – quantum molecular dynamics is still a very active field within computational condensed matter research – and the interested reader is referred to the review papers by Payne *et al.* [8], and Marx and Hutter [9] for details. In this section we describe some elements of the Car–Parrinello method in more detail, and briefly describe a variant of it, using conjugate gradients (see Appendix A4) for minimising the electronic energy.

9.4.1 Orthogonalisation of the electronic orbitals

The orthogonalisation of the electronic orbitals is maintained through the Lagrange multipliers Λ_{kl}, whose values therefore vary with time. The procedure to calculate these values depends on the particular integration algorithm used, which is usually the Verlet algorithm. In the previous section you have seen how this is done in the case of one orbital, where only the normalisation matters – for more orbitals, understanding the different procedures is quite subtle.

In the following we shall use the notation

$$\varepsilon = \sum_k \langle \psi_k | H | \psi_k \rangle \tag{9.49}$$

for the total energy for a set of orthonormal orbitals ψ_k. H stands for the Fock matrix in HF, and in DFT it is the Kohn–Sham Hamiltonian. Let us write down the Verlet equations of motion for the electronic orbitals:

$$\psi_k(t+h) = 2\psi_k(t) - \psi_k(t-h) - \frac{2h^2}{\mu}(H\psi_k - \sum_l \Lambda_{kl}\psi_l). \tag{9.50}$$

The – as yet unknown – multipliers Λ_{kl} are symmetric, $\Lambda_{kl} = \Lambda_{lk}$, and therefore represent $N(N+1)/2$ independent values, which are determined by the $N(N+1)/2$ orthonormality conditions. Hence the Lagrange multipliers are uniquely defined. It might therefore be surprising that several different orthogonalisation algorithms exist [8, 10]. The reason is that a unitary transformation of the set of orbitals leaves the set orthonormal: the set $\{\psi_k'\}$ defined by

$$\psi_k' = \sum_l U_{kl}\psi_l \tag{9.51}$$

is orthonormal. Moreover, a unitary transformation leaves the charge density unchanged – remember the DFT energy depends on the density and not on the individual orbitals. Also, the Slater determinants forming the basis functions in the Hartree–Fock theory are invariant under unitary transformations (see Problem 4.7). It should be noted that such a transformation of the set ψ_k is accompanied by a similarity transform of the Lagrange parameters:

$$\Lambda_{kl}' = \sum_{mn} U_{km}^\dagger \Lambda_{mn} U_{nl} \tag{9.52}$$

as can be verified directly from the equation of motion (9.21). Different orthonormalisation algorithms result in sets of orbitals which span the same space of functions but which are slightly rotated with respect to each other.

Such a rotation may have a tremendous effect on the performance of the Verlet algorithm. To see this, consider a permutation of the orbitals (which is a special case of a unitary transformation), carried out between two time steps. This permutation does not affect the density but it may have a disastrous effect on the integration of the equations of motion: the (fictitious) velocities of the permuted orbitals increase suddenly to values of $\mathcal{O}(h^{-1})$, because the permutation disrupts the smooth evolution of the orbitals! However, if the rotation is always close to the unit transformation,

$$U = 1 + h^2 A \tag{9.53}$$

where A is a Hermitian transformation of order one, varying smoothly with time, the Verlet algorithm will still work: apart from the motion governed by the equation of motion, the algorithm might induce some extra forces which cause the orbitals to rotate smoothly in Hilbert space, and this latter motion can be dealt with perfectly by the Verlet algorithm. It is difficult to see whether orthogonalisation algorithms satisfy these requirements and it is therefore easiest to construct the algorithm such that it is equivalent to the unambiguous time evolution resulting from the Verlet algorithm (without extra rotation) to a precision of at least order h^4, which is the overall precision of the Verlet algorithm.

A method which is based on the Verlet algorithm and which solves the Λ_{kl} in (9.50) by the orthogonality requirements is the iterative algorithm called SHAKE by Ryckaert *et al.* [5], which was mentioned in Section 8.6.3. This algorithm was used in the original work of Car and Parrinello [11]. It is straightforward and does not introduce rotations of the set of orbitals. Moreover, it orthogonalises the states to arbitrary precision (depending on the number of iterations performed). For details we refer to the cited literature.

Most other methods first predict the form of the (orthonormal) ψ_k at the next time step with some precision and then perform an additional orthonormalisation of these predicted orbitals by constructing orthonormal linear combinations of them. The idea behind this is that if the prediction is accurate, only a few orthonormalisation iterations are needed. As the Verlet algorithm prescribes an orthonormalisation by mixing in the $\psi_k(t)$ through the Lagrange multipliers (see Eq. (9.50)), and not the $\psi_k(t + h)$, such a final re-orthonormalisation can only be justified if the changes involved are of order h^4 which is the overall accuracy of the Verlet algorithm for a single step. Therefore these algorithms must first predict the new values to order $\mathcal{O}(h^4)$ and the re-orthonormalisation should yield the new states lying close to the old states. Note that after each step orthonormality is of then satisfied to machine precision whereas the error in the integration algorithm is of order h^4.

Let us now consider one such algorithm in detail. Over a time step h, the orbitals ψ_k shift by an amount of order h owing to the fact that they have a velocity. The acceleration, which is caused by the force due to the derivative of the total energy and by the force due to the Lagrange multiplier terms, then gives an additional shift of order h^2. The Hamiltonian force term occurring in the Car–Parrinello equations of motion is given by $-H\psi_k$. First we neglect this force term for simplicity and find Lagrange multipliers which guarantee orthonormality in the absence of forces.

The equation of motion then reads

$$\mu \ddot{\psi}_k = \sum_l \Lambda_{kl} \psi_l \tag{9.54}$$

and the second derivative of the constraint equations gives

$$\langle \ddot{\psi}_k | \psi_l \rangle + \langle \psi_k | \ddot{\psi}_l \rangle + 2 \langle \dot{\psi}_k | \dot{\psi}_l \rangle = 0, \tag{9.55}$$

so that, using the orthonormality of the orbitals, we find

$$\Lambda_{kl} = -\mu \langle \dot{\psi}_k | \dot{\psi}_l \rangle. \tag{9.56}$$

Of course we do not know the exact time derivative of the wave function, but we can use the Taylor expansion

$$\dot{\psi}_k(t) = \frac{\psi_k(t) - \psi_k(t-h)}{h} + \frac{h}{2} \ddot{\psi}_k(t) + \mathcal{O}(h^2), \tag{9.57}$$

and take for the approximate time derivative of the wave function at time t:

$$\dot{\tilde{\psi}}_k(t) = \frac{\psi_k(t) - \psi_k(t-h)}{h}. \tag{9.58}$$

From the time derivative of the constraint equation it follows that

$$\left\langle \frac{\mathrm{d}\psi_k}{\mathrm{d}t} \middle| \psi_l \right\rangle = 0, \quad \text{all } k \text{ and } l \tag{9.59}$$

and using this result, it is easy to show by a Taylor expansion of the $\psi_k(t-h)$ that

$$\langle \dot{\tilde{\psi}}_k(t) | \dot{\tilde{\psi}}_l(t) \rangle = \langle \dot{\psi}_k(t) | \dot{\psi}_l(t) \rangle + \mathcal{O}(h^2). \tag{9.60}$$

As the Lagrange multipliers occur in the Verlet algorithm with a factor h^2 in front of it, the accuracy of order h^2 with which we obtain their values by using this procedure is sufficient (that is, $\mathcal{O}(h^4)$).

Next we include the Hamiltonian force $-H\psi_k$ into the problem. A similar analysis as above (Eqs. (9.54–9.56)) leads to the following equation for the Lagrange multipliers:

$$\Lambda_{kl} = 2\langle \psi_k | H | \psi_l \rangle - \mu \langle \dot{\psi}_k | \dot{\psi}_l \rangle. \tag{9.61}$$

The term $\langle \psi_k | H | \psi_l \rangle$ can easily be calculated in the program. If we use these Lagrange parameters in the Verlet algorithm we obtain a set of orbitals which is accurate to order h^4 and is therefore in particular orthonormal to order h^4. Next we must apply an orthonormalisation algorithm which leaves the orthonormal set close to the near-orthonormal one. One possibility is an iterative algorithm used by Car and Parrinello [12]:

$$\psi_k' = \psi_k - \frac{1}{2} \sum_{l \neq k} \langle \psi_l | \psi_k \rangle \psi_l \tag{9.62}$$

which is repeated until the orbitals do not change any more. By inspection of this algorithm it is seen that if the orbitals are orthonormal up to order h^4, they will change to an extent within that order over a time step. The same holds for the Gram–Schmidt orthonormalisation, which is given by the algorithm

$$\psi'_k = \psi_k - \sum_{l<k} \langle \psi_l | \psi_k \rangle \psi_l. \tag{9.63}$$

In contrast to the previous algorithm, the Gram–Schmidt algorithm depends on the way in which the orbitals are ordered. In particular, the orbital which we take as the first in the Gram–Schmidt process remains unchanged. It is clear from the foregoing analysis that this does not really matter if the states are already orthogonal to order h^4.

Another possible way to orthogonalise the orbitals is to calculate the Lagrange parameters such that the final orbitals will be orthonormal. For example, we first calculate the new orbitals without taking the Lagrange multipliers into account:

$$|\tilde{\psi}_k(t+h)\rangle = 2|\psi_k(t)\rangle - |\psi_k(t-h)\rangle - \frac{2h^2}{\mu} H|\psi_k(t)\rangle, \tag{9.64}$$

and then we calculate the Lagrange parameters X_{ij} such that the orbitals

$$|\psi_k(t+h)\rangle = |\tilde{\psi}_k(t+h)\rangle + \sum_l X_{kl} |\psi_l(t)\rangle \tag{9.65}$$

form an orthonormal set. Obviously, the X_{kl} are related to the Λ_{kl} by

$$X_{kl} = \frac{2h^2}{\mu} \Lambda_{kl}. \tag{9.66}$$

The parameters X_{ij} should satisfy a matrix equation which can conveniently be formulated after introducing the matrices $A_{kl} = \langle \tilde{\psi}_k(t+h) | \tilde{\psi}_l(t+h) \rangle$ and $B_{kl} = \langle \psi_k(t) | \tilde{\psi}_l(t+h) \rangle$ as:

$$\mathbf{X}\mathbf{X}^\dagger + \mathbf{X}\mathbf{B} + \mathbf{B}^\dagger \mathbf{X}^\dagger = \mathbf{I} - \mathbf{A}. \tag{9.67}$$

This can be solved iteratively by the straightforward reformulation

$$\mathbf{X}^{(\mathbf{n+1})} = \tfrac{1}{2}[\mathbf{I} - \mathbf{A} + \mathbf{X}^{(\mathbf{n})}(\mathbf{I} - \mathbf{B}) + (\mathbf{I} - \mathbf{B}^\dagger)\mathbf{X}^{(\mathbf{n})} - \mathbf{X}^{(\mathbf{n})}\mathbf{X}^{(\mathbf{n})\dagger}]. \tag{9.68}$$

As an initial guess, we take $\mathbf{X} = \frac{1}{2}(\mathbf{I} - \mathbf{A})$, which is close to the first guess for Λ_{kl} found above.

Summarising the integration algorithm so far:

We have $|\psi_k(t)\rangle$ and $|\psi_k(t-h)\rangle$.
Find $|\dot{\psi}_k(t)\rangle = (|\psi_k(t)\rangle - |\psi_k(t-h)\rangle)/h$.

Now we may proceed in two ways:

(i) Calculate as a first guess $\Lambda_{kl} = 2\langle\psi_k(t)|H|\psi_l(t)\rangle - \mu\langle\dot{\tilde{\psi}}_k(t)|\dot{\tilde{\psi}}_l(t)\rangle$ and calculate $|\psi(t+h)\rangle$ by the algorithm (9.50). Then refine the solution thus obtained by orthonormalisation.

(ii) Calculate the new solution by the algorithm (9.50) *without* the constraint term. Then calculate the Lagrange parameters X_{ij} by the above algorithm. The constraint term is then added to the solution.

It is possible to turn the Verlet algorithm into a velocity Verlet form. In that case, not only the new orbitals are orthonormalised, but also the constraint $\langle\dot{\psi}_k(t)|\psi_l(t)\rangle$ is satisfied rigorously at all times. The algorithm requires a bit more work and storage. Its advantage is that it can be extended more easily within a Nosé–Hoover thermostat scheme – for details see Refs. [9, 13]. We give the algorithm here for completeness:

$$|\dot{\tilde{\psi}}_k(t+h)\rangle = |\dot{\psi}_k(t)\rangle + \frac{h}{\mu}H|\psi_k(t)\rangle; \tag{9.69a}$$

$$|\tilde{\psi}_k(t+h)\rangle = |\psi_k(t)\rangle + h|\dot{\tilde{\psi}}_k(t+h)\rangle; \tag{9.69b}$$

$$|\psi_k(t+h)\rangle = |\tilde{\psi}_k(t+h)\rangle + \sum_{ij} X_{ij}|\psi_k(t)\rangle; \tag{9.69c}$$

$$|\dot{\psi}'_k(t+h)\rangle = |\dot{\tilde{\psi}}_k(t+h)\rangle + \frac{h}{\mu}H|\psi_k(t+h)\rangle; \tag{9.69d}$$

$$|\dot{\psi}_k(t+h)\rangle = |\dot{\psi}'_k(t+h)\rangle + \sum_{ij} Y_{ij}|\psi_k(t+h)\rangle. \tag{9.69e}$$

The matrix X_{ij} is the same as used in the standard Verlet procedure above; the matrix Y_{ij} is simply calculated in terms of the matrix $C_{kl} = \langle\psi_k(t+h)|\dot{\psi}'_l(t+h)\rangle$ as

$$Y_{kl} = -\frac{C_{kl} + C^{\dagger}_{kl}}{2}. \tag{9.70}$$

If we include friction in the equations of motion, we are allowed more freedom, as the only requirement is that the orbits will become stationary by some damping mechanism. In this case, one can take the Λ_{kl} to be diagonal:

$$\Lambda_{kl} = \varepsilon_k\delta_{kl} \tag{9.71}$$

and the equation of motion for the stationary state leads directly to

$$\varepsilon_k = \langle\psi_k|H|\psi_k\rangle, \tag{9.72}$$

and this form of the Lagrange parameters is then used throughout the simulation, that is, even when if the orbitals move. However, this form of the Lagrange multipliers does not preserve orthonormality if we are not at the energy minimum; therefore, in the simulation, we use these values for the ε_k and perform a Gram–Schmidt orthonormalisation afterwards in order to prevent the orbitals all evolving to the ground state of the Kohn–Sham or Fock equation. The orbitals will then tend to the eigenvalues of the Hamiltonian, as these are the stationary solutions of the equations of motion.

9.4.2 The conjugate gradient method

The Car–Parrinello technique is based upon three ideas. First, the forces are calculated during the simulation, and only for those nuclear configurations which are actually visited. Second, the electronic structure can be determined by minimising the energy using an arbitrary minimisation method, that is, not necessarily by a self-consistency iteration as in Chapters 4 and 5. Car and Parrinello choose the molecular dynamics method for this purpose. Finally, we might pay a price in accuracy by not requiring the electrons to relax to the minimum energy state before each nuclear displacement.

In the conjugate gradients approach, the treatment of the electronic degrees of freedom differs from that in MD simulation methods. The idea is that if we abandon the usual self-consistency iterations which in the conventional HF and DFT approaches lead to the minimum of the electronic energy, we might as well apply any efficient minimisation method to the electronic energy – for example the conjugate gradients method – see Section A4. Using the conjugate gradients method enables us to keep the electronic degrees of freedom much closer to the ground state than in the Car–Parrinello method.

The conjugate gradients technique enables us to calculate a local minimum of an arbitrary smooth function depending on a number of variables. In fact, in our case we must perform the minimisation with a constraint. That is, we must minimise the energy, $E[n]$, as a function of the orbitals, ψ_k, using the gradient of the function

$$E[n] - \sum_{kl} \Lambda_{kl} \langle \psi_k | \psi_l \rangle, \tag{9.73}$$

where Λ_{kl} must be such that orthonormalisation is always ensured. Using the notation of the previous subsection, $\partial E / \partial \psi_k = H | \psi_k \rangle$, we find that the orthonormality condition leads to

$$\Lambda_{kl} = \langle \psi_k | H | \psi_l \rangle; \tag{9.74}$$

see Exercise 9.3. In the conjugate gradients method we need the steepest descent direction, which is the opposite of the gradient of the function (9.73). Note that H depends on the orbitals ψ_k. Neglecting this dependence, we obtain for the steepest descent direction:

$$\zeta_k = -2\Big[H\psi_k - \sum_l \langle\psi_k|H|\psi_l\rangle\psi_k\Big]. \tag{9.75}$$

In a method proposed by Stich *et al.* [14], this approximation is used in the line minimisations. Gillan [15] has devised a conjugate gradients method in which this approximation is not made. Finally, Teter *et al.* [16] have proposed a particularly efficient method in which only one state ψ_k is updated at a time. See Problems 9.3 and 9.5 for examples.

The conjugate gradients method is known to converge slowly for this problem. The reason is that the steepest descent direction, which should be close to the difference between the current orbitals $\{\psi_k\}$ and those that minimise the energy, $\{\tilde{\psi}_k\}$, may not have this property. To see what goes wrong, let us expand $\delta\psi_k = \psi_k - \tilde{\psi}_k$ in the eigenstates ξ_p of the Hamiltonian:

$$\delta\psi_k = \sum_p \alpha_{pk}\xi_p. \tag{9.76}$$

Note that whereas the indices k and l in the previous discussion run only over the occupied electronic states, the index p runs over *all* the eigenstates of the variational Hamiltonian matrix, and their number is equal to the number of states in the variational basis set used. Working out the steepest descent direction for this state, we have

$$\zeta_k = -2\sum_p (\alpha_{pk}\varepsilon_p - \sum_l \Lambda_{kl}\alpha_{pk})\xi_p \tag{9.77}$$

(note that the part involving $\tilde{\psi}_k$ on the right hand side vanishes). The right hand sign will contain important contributions from the high energy levels, and they will spoil the desired proportionality between steepest descent direction and $\delta\psi_k$. This can be remedied by an extension to the method, called *pre-conditioning* [17]. We shall not treat this in detail here – for details see Refs. [8, 16, 17].

The conjugate gradients method can be readily applied to energy minimisation problems. However, if we want to perform a dynamic simulation for the atoms, that is, without friction acting on the nuclei, a rather subtle problem arises. To see this, consider a nucleus which is moving in the positive x-direction with an electronic charge distribution around it, which has converged to the ground state. As the nucleus moves, the charge cloud starts lagging behind and this discrepancy might grow larger and larger with time. But this is not possible in the Verlet algorithm as this symplectic algorithm does not allow for energy drift (see Section 8.4). In the Verlet simulation, the electron charge cloud keeps oscillating around the nucleus so that errors remain bounded, and even cancel out on average. This is an important advantage of the molecular dynamics approach of Car and Parrinello [18–20]. There is, however, the possibility that the electron system absorbs energy from the nuclear system and that the negative charge cloud would therefore oscillate more and more violently around the nuclei. To avoid this, the total simulation time should be kept short, or the electrons and the nuclei should both be coupled to a Nosé–Hoover thermostat (see Section 8.5.1), perhaps at different temperatures [21].

In the conjugate gradients method, the errors are more erratic than oscillatory in nature. Therefore this cancellation effect will not occur in that case, and it is necessary to keep the electronic charge distribution very close to the energy minimum at any time in order to avoid an unstable propagation of the errors spoiling the results. This means that we must perform many conjugate gradient steps, which slows the calculation down considerably. The only way to achieve good performance using the conjugate gradient method is by extrapolating the orbitals at the next time step from the previous ones, so that the conjugate gradients iterations start off from a configuration which is close to the exact energy minimum [22].

Details concerning the conjugate gradients method can be found in Refs. [8, 14, 16, 17].

9.4.3 The RM-DIIS technique

Another widely used technique for finding the optimal orbitals is the RM-DIIS technique. The abbreviation stands for residual minimisation by direct inversion of the iterative subspace. This can be applied to any problem in which a set of orthonormal eigenstates of a large Hamiltonian must be found. We describe the method briefly – for details see Refs. [9, 23, 24].

The eigenfunctions that we seek satisfy

$$(H - \varepsilon_n)|\psi_n\rangle = 0; \quad n = 1, \ldots, N. \tag{9.78}$$

We now want to quantify the deviation of some approximate set of states from this. To this end we reformulate (9.78) as

$$|\psi_n\rangle = |\psi_n\rangle + \frac{1}{H_{nn}}(H - H_{nn})|\psi_n\rangle, \tag{9.79}$$

with

$$H_{nn} = \frac{\langle\psi_n|H|\psi_n\rangle}{\langle\psi_n|\psi_n\rangle}. \tag{9.80}$$

For the eigenstates we obviously have $\varepsilon_n = H_{nn}$. This form may seem a bit arbitrary but it naturally leads to an iterative scheme:

$$|\psi_n^{(j+1)}\rangle = |\psi_n^{(j)}\rangle + \frac{1}{H_{nn}}(H - H_{nn})|\psi_n^{(j)}\rangle. \tag{9.81}$$

The reason the correction term has a prefactor $1/H_{nn}$ is that in this way all eigenvectors have a similar correction, independent of the energy eigenvalue. The main feature of Eq. (9.81) for us is that the correction factor can be used as an error estimate, which we call $|\Delta_j\rangle$:

$$|\Delta_j\rangle = \frac{1}{H_{nn}}(H - H_{nn})|\psi_n^{(j)}\rangle. \tag{9.82}$$

Suppose we have a sequence of estimates $|\psi_n^{(j)}\rangle, j = 1, \ldots, J$ for the optimal states, which are not necessarily constructed according to the recipe of Eq. (9.81). From this sequence we construct a new state as a linear combination of the previous ones

$$|\psi_n^{J+1}\rangle = \sum_{j=1}^{J} d_j |\psi_n^{(j)}\rangle \tag{9.83}$$

such that it has a minimal error. We must, however, beware of the fact that the states can be scaled at will – upon rescaling they still satisfy the same linear equations. However, rescaling affects the norm of the error which we want to minimise. In order to have an unambiguous measure of the error we require that

$$\sum_{j=1}^{J} d_j = 1, \tag{9.84}$$

which turns out to be a convenient choice.

Substituting the linear expansion (9.83) in the expression for the error $|\Delta_{J+1}\rangle$ of the new state $|\psi_n^{J+1}\rangle$, we see that yields a linear combination of the individual errors $|\Delta_j\rangle$ of the $|\psi_n^{(j)}\rangle$:

$$|\Delta_{J+1}\rangle = \sum_{j=1}^{J} d_j |\Delta_j\rangle. \tag{9.85}$$

Now we shall be specific about the minimalisation of the norm of the error. This is given by

$$|\Delta_{J+1}|^2 = \sum_{j,k=1}^{J} d_j d_k \langle \Delta_j | \Delta_k \rangle. \tag{9.86}$$

We abbreviate the matrix elements $\langle \Delta_j | \Delta_k \rangle$ by a_{jk}. Minimising the error norm, respecting the appropriate constraint, which is included through a Lagrange parameter λ, leads to

$$\begin{pmatrix} a_{11} & a_{12} & \ldots & a_{1n} & 1 \\ a_{21} & a_{22} & \ldots & a_{2n} & 1 \\ \vdots & \vdots & \ddots & \vdots & \vdots \\ a_{n1} & a_{n2} & \ldots & a_{nn} & 1 \\ 1 & 1 & \ldots & 1 & 0 \end{pmatrix} \begin{pmatrix} d_1 \\ d_2 \\ \vdots \\ d_n \\ -\lambda \end{pmatrix} = \begin{pmatrix} 0 \\ 0 \\ \vdots \\ 0 \\ 1 \end{pmatrix}. \tag{9.87}$$

The resulting states $|\psi_n^J\rangle$ are not yet orthogonal; an orthogonalisation of the states must be performed after each of the steps described above.

9.4.4 Large systems

Periodic or other type of boundary conditions are not imposed a priori in QMD simulations. Boundary conditions may, however, be introduced by the basis set, which might consist of periodic functions, or functions vanishing at some boundary. There is an increasing interest in nonperiodic systems containing large numbers of atoms; these systems are said to be *mesoscopic*. Examples of mesoscopic systems and phenomena are the scanning tunnelling microscope tip and surface, grain boundaries, quantum dots and wires, biological macromolecules, etc. Often these systems are periodically continued; in the case of grain boundaries, for example, we might consider a system cell with a linear size corresponding to 10 atoms, and containing a grain boundary. Imposing periodic boundary conditions means that we are considering a system containing a periodic array of such grain boundaries.

If the system cell contains large numbers of atoms, it becomes very important to use a method for solving electronic structure and dynamics for which the computer time scales favourably with time. In this subsection we analyse the scaling behaviour for a few methods. Two parameters are important: first, the number $N_{\rm at}$ of atoms in the system, and second, the size $N_{\rm B}$ of the basis set. Of course, these numbers are not independent: they are usually proportional to each other. However, as the number of basis states exceeds the number of atoms sometimes by a factor of 100 or more, it is important to distinguish between the two in the time scaling.

It depends strongly on the type of basis functions used how the time required by a quantum molecular dynamics simulation scales with the number of atoms. In the case of plane wave basis sets, FFT techniques (see Appendix A9) can be used to increase the efficiency of the calculations. The kinetic energy is diagonal in a plane wave basis set; for evaluating the potential energy the FFT transforms the states into real space where the potential energy is diagonal. The plane wave basis leads to a time scaling (complexity) for one integration step in the quantum molecular dynamics method of $N_{\rm at} N_{\rm B} \ln N_{\rm B}$.

Other methods use localised orbitals as basis functions so that the Hamiltonian couples only orbitals on neighbouring atoms – hence the Hamiltonian becomes sparse, and sparse matrix methods can be very efficient, giving essentially a scaling behaviour of $N_{\rm at} N_{\rm B}$. Obviously, this can be used in the calculation of the electronic structure using the conventional self-consistent approach where the Hamilton matrix has to be diagonalised several times. Using localised basis functions, the *recursion method* of Haydock [25–27] is particularly useful as it allows for calculating the density in a number of steps independent of $N_{\rm at}$, and as long as the nuclear positions remain fixed, the method scales as N_B, but with a large pre-factor. The method converges much faster for insulators and semiconductors than for metals [27]. Finally, discretising the degrees of freedom on a cubic lattice can lead to methods with a favourable time scaling [28, 29].

*9.5 Implementation of the Car–Parrinello technique for pseudopotential DFT

In this section we describe how the self-consistent pseudopotential code described in Section 6.7.3 can be extended to include Car–Parrinello dynamics. In the code of Section 6.7.3, we constructed the DFT Hamiltonian, diagonalised it and calculated the density which was then used to construct a new Hamiltonian, and so on until convergence was achieved. We then calculated the total energy from the orbitals, the density and the nuclear positions (which were taken to be fixed).

Now we want to use the derivative of the total energy with respect to the orbital degrees of freedom *and* with respect to the nuclear positions in order to formulate equations of motions which can then be solved using the (velocity) Verlet algorithm. The rather complicated expression for the energy leads to even more complicated expressions for these forces. We start with the orbital forces. They follow directly from the expression for the total energy given in Section 6.7.7. If we calculate the orbital forces from this energy, we obtain by direct differentiation: [9]

$$\frac{\partial E_{\text{total}}}{\partial c_j^*(\mathbf{K})} = \frac{K^2}{2} c_j(\mathbf{K}) + \sum_{\mathbf{K}'} V_{\text{loc}}^*(\mathbf{K}-\mathbf{K}')c_j(\mathbf{K}') + \sum_n \sum_{lm} F_{jlm}^n e^{-i\mathbf{K}\cdot\mathbf{R}_n} Y_{lm}(\hat{\mathbf{K}}) h_{lm}^n p_m^l(K), \tag{9.88}$$

where $V_{\text{loc}}^{\text{all}}$ is the total local potential:

$$V_{\text{loc}}(\mathbf{K}) = \sum_n \Delta V_{\text{loc}}(\mathbf{K}) + V_{\text{xc}}(\mathbf{K}) + 4\pi \frac{n_{\text{tot}}(\mathbf{K})}{K^2}. \tag{9.89}$$

The potentials occurring in this expressions are given in Section 6.7.3.

If we implement this force in a Verlet algorithm in which we reduce the velocities at each time step such as to mimic frictional forces which bring us to the stationary energy minimum, the result should be equal to that obtained in the Kohn–Sham program. For two silicon atoms placed at a distance of 1.05 a.u. in a cell of size 8 a.u., the converged energy is −21.235 347 29 atomic units.

For the dynamics of the nuclei, we need the gradients of the energy with respect to the nuclear coordinates. These can be obtained directly from the expression for the total energy. The nuclear gradients are generated by the local and nonloncal parts of the pseudopotential, and the electrostatic energy. The expressions are:

$$\nabla_{\mathbf{R}_n} E_{\text{local}} = -\Omega \sum_{\mathbf{K}} \mathrm{i}\mathbf{K} V_{\text{local},n}(\mathbf{K}) \mathrm{e}^{-\mathrm{i}\mathbf{K}\cdot\mathbf{R}_n} n^*(\mathbf{K}) \tag{9.90}$$

$$\nabla_{\mathbf{R}_n} E_{\text{nonlocal}} = \sum_j f_j \sum_{l,m\varepsilon n} [(F^n_{jlm})^* h^n_{lm} \nabla_{\mathbf{R}_n} F^n_{jlm} + \nabla_{\mathbf{R}_n} (F^n_{lm})^* h^n_{lm} F^n_{lm}]; \tag{9.91}$$

$$\nabla_{\mathbf{R}_n} E_{\text{ES}} = -\Omega \sum_{\mathbf{K}\neq\mathbf{0}} \mathrm{i}\mathbf{K} \frac{n^*_{\text{tot}}}{K^2} n^n_{\text{core}}(\mathbf{K}) \mathrm{e}^{-\mathrm{i}\mathbf{K}\cdot\mathbf{R}_n} + \nabla_{\mathbf{R}_n} E_{\text{ovrl}}, \tag{9.92}$$

where

$$\nabla_{\mathbf{R}_n} F^n_{lm} = -\frac{1}{\sqrt{\Omega}} \sum_{\mathbf{K}} \mathrm{i}\mathbf{K} \mathrm{e}^{-\mathrm{i}\mathbf{K}\cdot\mathbf{R}_n} c^*_j(\mathbf{K}) Y_{lm}(\hat{\mathbf{K}}) p^l_m(K), \tag{9.93}$$

and

$$\nabla_{\mathbf{R}_n} E_{\text{ovrl}} = \sum\nolimits'_{n'} \sum_{\mathbf{L}} \Bigg\{ \frac{Z_n Z_{n'}}{|\mathbf{R}_n - \mathbf{R}_{n'} - \mathbf{L}|^3} \operatorname{erfc}\left[\frac{|\mathbf{R}_n - \mathbf{R}_{n'} - \mathbf{L}|}{\sqrt{2(\xi_n^2 + \xi_{n'}^2)}}\right]$$
$$+ \frac{2}{\sqrt{\pi}} \frac{1}{\sqrt{\xi_n^2 + \xi_{n'}^2}} \frac{Z_n Z_{n'}}{|\mathbf{R}_n - \mathbf{R}_{n'} - \mathbf{L}|^2} \exp\left[\frac{|\mathbf{R}_n - \mathbf{R}_{n'} - \mathbf{L}|}{\sqrt{2(\xi_n^2 + \xi_{n'}^2)}}\right]\Bigg\}$$
$$\times (\mathbf{R}_n - \mathbf{R}_{n'} - \mathbf{L}). \tag{9.94}$$

The full implementation of the Car–Parrinello is quite cumbersome. For a hydrogen dimer, you should find a result similar to that of Figure 9.4. Note that this frequency depends on the nuclear mass. Similar results should be found for a silicon dimer.

Exercises

9.1 [C] The Car–Parrinello method can be used to find the minimum of any variational energy functional. We use it in this problem for finding the ground state of a particle in a one-dimensional, infinitely deep potential well. This problem was treated in Chapter 3, Section 3.2.1. We use N variational basis functions of the form

$$\chi_r(x) = x^r(x-a)(x+a), \quad r = 0, 1, 2, \ldots, N-1$$

from which a variational state $\psi(x)$ is built as

$$\psi(x) = \sum_r C_r \chi_r(x).$$

The Euler–Lagrange equations for the Lagrangian with potential

$$E[C_r] = \sum_{rs} C_r C_s \langle \chi_r | H | \chi_s \rangle = \sum_{rs} C_r C_s H_{rs}$$

and the normalisation constraint

$$\langle\psi|\psi\rangle = \sum_{rs} C_r C_s \langle\chi_r|\chi_s\rangle = \sum_{rs} C_r C_s S_{rs}$$

are given by

$$\mu \ddot{C}_r = 2\sum_s (H_{rs}C_s - \Lambda S_{rs}C_s).$$

We solve this equation of motion using the Verlet algorithm with friction. Note that Λ is determined by the normalisation condition. Therefore we first perform an integration step with $\Lambda = 0$ and then calculate Λ from (9.32). You are free to choose μ, the frictional constant and the time step h, although they are not independent.

The energy should converge to 2.467 40 as found in Section 3.2.1.

9.2 [C] Extend the program of the previous problem to include more than one state. Each state $\psi_k(x)$ has its own set of coefficients C_{rk}:

$$\psi_k(x) = \sum_r C_{rk}\chi_r(x).$$

We consider K states. There are now $K(K+1)/2$ constraint equations:

$$\langle\psi_k|\psi_l\rangle = \sum_{rs} C_{rk} S_{rs} C_{sl} = \delta_{kl}$$

(interchanging k and l gives the same equation). The Euler–Lagrange equations now become

$$\mu \ddot{C}_{kr} = \sum_l \sum_s (H_{rs} - \Lambda_{kl}) C_{sl}.$$

Because we include friction in the problem, we can take Λ_{kl} diagonal:

$$\Lambda_{kl} = \varepsilon_k \delta_{kl},$$

and at each step we estimate ε_k as

$$\varepsilon_k = \sum_{rs} C_{rk} C_{sk} H_{rs}$$

see Section 9.4.1.

Implement these equations in a computer program and compare your results with those presented in Section 3.2.1.

9.3 [C] Consider again the deep potential well of the previous two problems. We now use the conjugate gradients method for finding the eigenvalues, by using it to minimise the energy functional. It is assumed that you have a conjugate gradient routine available.

Let us first consider the ground state. This can be written as

$$\psi_G(x) = \sum_r C_r \chi_r(x),$$

where χ_r is the basis consisting of polynomials vanishing on the boundaries of the well. The energy functional

$$E[C_r] = \sum_{rs} C_r C_s H_{rs}$$

must be minimised subject to the normalisation constraint

$$\sum_{rs} C_r C_s S_{rs} = 1.$$

The Lagrangian function for this problem is

$$L[C_r] = \sum_{rs} C_r C_s H_{rs} - \lambda \sum_{rs} C_r C_s S_{rs}$$

where λ should be such that the normalisation remains guaranteed when moving in the steepest descent direction. The steepest descent direction ζ is given by

$$\zeta = \sum_r D_r \chi_r(x);$$

$$D_r = -2 \sum_s C_s H_{rs} + 2\lambda \sum_s C_s S_{rs}.$$

(a) Show that

$$\lambda = \langle \psi | H | \psi \rangle.$$

(b) Use this in applying the conjugate gradients method in order to find the ground state. Note that convergence is slow as no preconditioning is applied.

Now we consider the problem of finding more energy eigenstates ψ_k which are expanded in the basis set states as

$$\psi_k(x) = \sum_r C_{rk} \chi_p(x).$$

For N eigenstates we have $N(N+1)/2$ constraints.

The Lagrange function now reads

$$L[C_r] = \sum_k \sum_{rs} C_{rk} C_{sk} H_{rs} - \sum_{kl} \Lambda_{kl} \sum_{rs} C_{rk} C_{sl} S_{rs}.$$

(c) Show that

$$\Lambda_{kl} = \langle \psi_k | H | \psi_l \rangle.$$

(d) Use this form to find the four lowest eigenstates.

9.4 [C] Use the conjugate gradients technique for finding the minimum of the electronic energy of the hydrogen molecule with a fixed configuration of nuclei. This is a straightforward extension of the first program of the previous problem: there is only one normalisation constraint. Note however that the energy functional contains a term which is quartic in the C_r – see Eq. (9.26). Show that the steepest descent direction subject to the constraint is given by

$$D_r = -2 \sum_s F_{rs} C_s + 2\lambda \sum_s S_{rs} C_s$$

with

$$\lambda = \sum_{rs} C_r F_{rs} C_s.$$

Apply the conjugate gradients method to this problem and compare the results with the matrix diagonalisation method of Chapter 4 and the molecular dynamics method of the present chapter.

References

[1] G. M. Barrow, *Introduction to Molecular Spectroscopy*. New York, McGraw-Hill, 1962.
[2] H. Hellmann, *Einführung in die Quantenchemie*. Leipzig, Deuticke, 1937.
[3] R. P. Feynman, 'Forces in molecules,' *Phys. Rev.*, **56** (1939), 340–3.
[4] S. Kirkpatrick, C. D. Gelatt, and M. P. Vecchi, 'Optimization by simulated annealing,' *Science*, **220** (1983), 671–80.
[5] J. P. Ryckaert, G. Ciccotti, and H. J. C. Berendsen, 'Numerical integration of the cartesian equations of motion of a system with constraints: molecular dynamics of n-alkanes,' *J. Comput. Phys.*, **23** (1977), 327–41.
[6] J. W. Negele, 'The mean-field theory of nuclear structure and dynamics,' *Rev. Mod. Phys.*, **54** (1982), 913–1015.
[7] P. Pulay, 'Ab initio calculation of force constants and equilibrium geometries in polyatomic molecules. I. Theory,' *Mol. Phys.*, **17** (1969), 197–204.
[8] M. C. Payne, M. P. Teter, D. C. Allen, T. A. Arias, and J. D. Joannopoulos, 'Iterative minimization techniques for ab initio total-energy calculations: molecular dynamics and conjugate gradients,' *Rev. Mod. Phys.*, **64** (1992), 1045–97.
[9] D. Marx and J. Hutter, 'Ab initio molecular dynamics: theory and implementations,' in *Modern Methods and Algorithms of Quantum Chemistry, NIC series*, vol. 1. Jülich, John von Neumann Institute for Computing, 2000, pp. 301–449.
[10] J. Q. Broughton and F. Khan, 'Accuracy of time-dependent properties in electronic structure calculations using a fictitious Lagrangian,' *Phys. Rev. B*, **40** (1989), 12098–104.
[11] R. Car and M. Parrinello, 'Unified approach for molecular-dynamics and density-functional theory,' *Phys. Rev. Lett.*, **55** (1985), 2471–4.
[12] R. Car and M. Parrinello, 'The unified approach for molecular dynamics and density functional theory,' in *Simple Molecular Systems at Very High Density* (A. Polian, P. Lebouyre, and N. Boccara, eds.), *NATO Advanced Study Institute*, vol. 186, New York, Plenum, 1989, p. 455.
[13] M. E. Tuckerman and M. Parrinello, 'Integrating the Car–Parrinello equations. I. Basic integration techniques,' *J. Chem. Phys.*, **101** (1994), 1302–15.
[14] I. Stich, R. Car, M. Parrinello, and S. Baroni, 'Conjugate-gradient minimisation of the energy functional: a new method for electronic structure calculation,' *Phys. Rev. B*, **39** (1989), 4997–5004.
[15] M. J. Gillan, 'Calculation of the vacancy formation energy in aluminium,' *J. Phys.–Cond Mat*, **1** (1989), 689–711.
[16] M. P. Teter, M. C. Payne, and D. C. Allan, 'Solution of Schrödinger's equation for large systems,' *Phys. Rev. B*, **40** (1989), 12255–63.
[17] P. E. Gill, W. Murray, and M. H. Wright, *Practical Optimization*. London, Academic Press, 1981.
[18] M. C. Payne, 'Error cancellation in the molecular-dynamics method for total energy calculations,' *J. Phys.–Cond. Mat.*, **1** (1989), 2199–210.
[19] R. Car, M. Parrinello, and M. C. Payne, 'Error cancellation in the molecular-dynamics method for total energy calculations: comment,' *J. Phys.–Cond. Mat.*, **3** (1991), 9539–43.
[20] D. K. Remler and P. A. Madden, 'Molecular-dynamics without effective potentials via the Car–Parrinello approach,' *Mol. Phys.*, **70** (1990), 921–66.
[21] P. Bloechl and M. Parrinello, 'Adiabaticity in first-principles molecular dynamics,' *Phys. Rev. B*, **45** (1992), 9413–16.

[22] T. A. Arias, J. D. Joannopoulos, and M. C. Payne, 'Ab initio molecular-dynamics techniques extended to large-length scale systems,' *Phys. Rev. B*, **45** (1992), 1538–49.
[23] P. Pulay, 'Improved SCF convergence acceleration,' *J. Comp. Chem.*, **3** (1982), 556–60.
[24] R. M. Martin, *Electronic Structure*. Cambridge, Cambridge University Press, 2004.
[25] R. Haydock, V. Heine, and M. J. Kelly, 'Electronic structure based on the local atomic environment for tight-binding bands,' *J. Phys. C*, **5** (1972), 2845–58.
[26] R. Haydock, 'The recursive solution of the Schrödinger equation,' in *Solid State Physics* (H. Ehrenreich, F. Seitz, and D. Turnbull, eds.), vol. 35, New York, Academic Press, 1980, pp. 216–94.
[27] A. Gibson, R. Haydock, and J. P. LaFemina, 'Ab initio electronic structure computations with the recursion method,' *Phys. Rev. B*, **47** (1993), 6518–23.
[28] S. Baroni and P. Giannozzi, 'Towards very large-scale electronic structure calculations,' *Europhys. Lett.*, **17** (1992), 547–52.
[29] J. M. Thijssen and J. E. Inglesfield, 'Embedding muffin tins into a finite difference grid,' *Europhys. Lett.*, **27** (1994), 65–70.

10
The Monte Carlo method

10.1 Introduction

In Chapter 8 we saw how a classical many-particle system can be simulated by the MD method, in which the equations of motion are solved for all the particles involved. This enables us to calculate statistical averages of static and dynamic physical quantities. There exists another method, called the Monte Carlo (MC) method, for simulating classical many-particle systems by introducing artificial dynamics based on 'random' numbers.[1] The artificial dynamics used in the MC method prevent us from using it for determining dynamical physical properties in most cases, but for static properties it is very popular.

In fact, every numerical technique in which random numbers play an essential role can be called a 'Monte Carlo' method after the famous Mediterranean casino town, and we shall discuss the method not only as a tool for studying classical many-particle systems, but also as a way of dealing with the more general problem of calculating high-dimensional integrals. In fact, three main types of Monte Carlo simulations can be distinguished:

- *Direct Monte Carlo*, in which random numbers are used to model the effect of complicated processes, the details of which are not crucial. An example is the modelling of traffic where the behaviour of cars is determined in part by random numbers.
- *Monte Carlo integration*, which is a method for calculating integrals using random numbers. This method is efficient when the integration is over high-dimensional volumes (see below).
- *Metropolis Monte Carlo*, in which a sequence of distributions of a system is generated in a so-called Markov chain. This method allows us to study the static properties of classical and quantum many-particle systems. The latter will be discussed in Chapter 12.

[1] As explained in Appendix B, computer-generated random numbers are not truly random, hence the quotes.

Direct Monte Carlo is a powerful method which can be applied to a wide variety of problems inside and outside physics. There is, however, not much to be said about this method as its implementation is as direct as the name suggests. The difficulty usually resides in the modelling aspect: how to represent certain phenomena using random numbers. The implementation of the method is then usually rather straightforward. In the next section we shall briefly discuss MC integration. The Metropolis sampling method will be discussed in the remainder of this chapter. The MC techniques which will be discussed in this chapter are essential for much of the material covered in Chapters 12 and 15 on quantum Monte Carlo methods and lattice field theory simulations.

A general reference on MC techniques is the book by Hammersley and Handscomb [1]. More detailed material concerning Metropolis Monte Carlo methods can be found in the book by Allen and Tildesley [2], two review volumes by Binder [3, 4] and the books by Kalos and Whitlock [5], Binder and Heermann [6] and Barkema and Newman [7].[2]

10.2 Monte Carlo integration

Suppose we want to calculate the integral of a smooth function f on the interval $[a, b]$ on the real axis:

$$I = \int_a^b \mathrm{d}x\, f(x). \tag{10.1}$$

Standard numerical methods for this problem are discussed in Appendix A6, and they usually boil down to calculating the function on a set of equally spaced values x_i (except for Gaussian integration, where the points are not equidistant) and then evaluating the sum

$$I = \frac{(b-a)}{N} \sum_{i=1}^{N} w_i f(x_i), \tag{10.2}$$

where the weights w_i do not depend on f – they determine the accuracy of the method. Usually such methods are based on polynomial approximations of the integrand and their accuracy σ is expressed in terms of a power of the separation h of the integration points: $\sigma \propto h^k \propto N^{-k}$, where k is a positive integer. In Monte Carlo integration we also use Eq. (10.2), with the weights w_i all equal to 1 but the x_i now chosen *randomly*.

It will be clear that if the random coordinates x_i are homogeneously distributed on $[a, b]$, and if N is sufficiently large, the sum (10.2) yields a result close to the

[2] Lecture notes by Frenkel [8] have been helpful in writing part of this chapter.

exact integral. We calculate the variance in the result:

$$\sigma^2 = \left\langle \left(\frac{b-a}{N} \sum_{i=1}^{N} f_i \right)^2 \right\rangle - \left(\left\langle \frac{b-a}{N} \sum_{i=1}^{N} f_i \right\rangle \right)^2. \tag{10.3}$$

The angular brackets denote an average over all possible realisations of the sequence of random coordinates x_i. Carrying out this average for the last term on the right hand side gives the square of the average $\overline{f}$ of the function f on $[a, b]$. Splitting the sums in the first term on the right hand side into a sum with $i = j$ and one with $i \neq j$ leads after some manipulation to

$$\sigma^2 = \frac{(b-a)^2}{N} (\overline{f^2} - \overline{f}^2) \tag{10.4}$$

where the bar denotes an average of the function on $[a, b]$. We see that the error is proportional to the variance of f on $[a, b]$. The fact that $\sigma \propto 1/\sqrt{N}$ is to be expected from the central limit theorem. This scaling is clearly unfavourable compared with standard quadrature methods using equidistant values for the x_i, which yields an error N^{-k} with $k \geq 1$. However, MC integration is more efficient in higher dimensions. To see this, let us consider standard numerical integration, with error $\mathcal{O}(h^k)$, for a d-dimensional integral. For simplicity we assume that the integration volume is a hypercube with side L. This contains $N = (L/h)^d$ points and therefore the error in the result scales as $N^{-k/d}$. The error of the Monte Carlo integration, however, is independent of d; it is still $\mathcal{O}(N^{-1/2})$, since the central limit theorem does not depend on the dimension. Comparing this error with that of the standard method, we see that MC integration is more efficient than an order-k algorithm when $d > 2k$.

This is a rather counterintuitive result: we would expect that using a regular grid for calculating the integral would always be superior to the random distribution of points of the Monte Carlo method. The reason for the superiority of MC integration in higher dimensions is that in a sense the random distribution is more homogeneous than the regular grid. Consider for example a rectangular volume within the integration volume. A homogeneous distribution of the integration points would imply that the number of points within this rectangular volume should be approximately proportional to that volume. If we choose the rectangular volume to have sides parallel to the axes of the point grid used in the standard integration method, it is clear that on increasing the volume size, the number of points it contains increases stepwise whenever a facet of the volume moves through an array of integration points. In this respect random distributions are more homogeneous, since for these distributions such steps in the number of points are extremely unlikely to occur. This heuristic argument can be formalised – see the review by James [9] and references therein.

Several methods have been devised to reduce the error of the Monte Carlo integration method; for a discussion see Ref. [9]. We give a brief overview here. In order to distribute the points more homogeneously over the integration hypercube, it is possible to subdivide the latter into smaller, equally sized subvolumes and to choose an equal number of random points in each subvolume. This is called 'stratified Monte Carlo'.

In many practical cases, the contributions to the integral from different regions in the integration volume vary strongly. The MC method samples the function homogeneously, so if the significant contributions to the integral come primarily from a small region within the integration volume, there will be only a few MC points for sampling the function there, leading to large statistical errors. This effect shows up in (10.4) as a large variance of f over the volume. In order to reduce this contribution to the error, points are concentrated in the regions where f happens to be large (in absolute value). More precisely, let $\rho(x)$ be a function on $[a, b]$ which has more or less the shape of f in the sense that f/ρ is approximately constant over the interval. We furthermore require ρ to be normalised:

$$\int_a^b \mathrm{d}x\ \rho(x) = 1. \tag{10.5}$$

We write the integral over f as follows:

$$\int_a^b \mathrm{d}x\, f(x) = \int_a^b \mathrm{d}x\ \rho(x) \left[\frac{f(x)}{\rho(x)}\right]. \tag{10.6}$$

The function in square brackets is reasonably flat (as ρ is chosen to have more or less the shape of f) and the weight $\rho(x)$ in front of this function can be included in the integral by choosing the random points x_i with distribution $\rho(x)$. The Monte Carlo result for the integral is then still given by (10.2). This reduces the error in the result considerably as we can see by evaluating the variance (10.4) for this case:

$$\sigma^2 = \frac{1}{N}\left\{\int_a^b \left[\frac{f(x)}{\rho(x)}\right]^2 \rho(x)\,\mathrm{d}x - \left(\int_a^b \left[\frac{f(x)}{\rho(x)}\right] \rho(x)\,\mathrm{d}x\right)^2\right\}. \tag{10.7}$$

If we are indeed able to choose ρ such that f/ρ is approximately constant, then as a result of (10.5) the expression in the braces will indeed be much smaller than in the 'crude sampling' case. This method is called *importance sampling Monte Carlo*. *Adaptive Monte Carlo* methods also aim at concentrating the sampling points in those regions where f contributes significantly to the integral, but these methods locate these regions by probing the function at random points and require no a priori knowledge on the function f as in the case of importance sampling.

Note that MC integration is not susceptible to correlations in the random generator. Correlations influence the order in which the points x_i are generated but this does not affect the sum. In fact, it is possible to generate artificial number sequences for which no attempt made to achieve (pseudo-) randomness but which fill a high-dimensional volume very homogeneously so that they are suitable for integration. The resulting method is called 'quasi-Monte Carlo' [9].

10.3 Importance sampling through Markov chains

We now explain the importance sampling method for classical many-particle systems in the canonical or (*NVT*) ensemble; extensions to other ensembles will be discussed later in this chapter. When calculating averages in the (*NVT*) ensemble, the configurations should be weighted according to the Boltzmann factor

$$\rho(X) \propto \exp[-\beta E(X)], \quad \beta = 1/(k_B T). \tag{10.8}$$

This suggests that we apply MC integration with importance sampling, as the phase space over which we must integrate is high-dimensional. The method that first comes to mind for generating phase space points (configurations) with a Boltzmann distribution is the Von Neumann method discussed in Appendix B3. In our case this would mean generating random configurations and accepting them with probability $\exp[-\beta E(X)]$ where the energy scale is assumed to be such that the energy is always positive. However, as the number of configurations with a particular energy increases exponentially with energy, most of the randomly constructed states have very high energy. Hence for finite temperature they will be accepted with a vanishingly small probability and we spend most of our time generating configurations that are then rejected, which is obviously very inefficient.

Another method would be to construct statistically independent configurations with a bias towards lower energies in accordance with the Boltzmann weight. However, recipes for such a construction are difficult and will certainly be very time-consuming. Related to this method is the Metropolis method [10], developed in 1953, which abandons the idea of constructing statistically independent configurations – rather, the configurations are constructed through a so-called Markov chain, in which each new configuration is generated with a probability distribution depending on the previous configuration.

Before considering Markov chains, we consider *truly random*, or *uncorrelated chains*. For an uncorrelated chain, the probability of occurrence of a particular sequence of N objects $X_1, \ldots, X_N$ is statistically uncorrelated:

$$P_N(X_1, X_2, \ldots, X_N) = P_1(X_1)P_1(X_2)\cdots P_1(X_N) \tag{10.9}$$

where $P_1(X)$ is the independent probability of occurrence for the object X (this probability is assumed to be equal for each step). Truly random number sequences are examples of uncorrelated chains (see Appendix B). A Markov chain is different from uncorrelated chains – it is defined in terms of the transition probability $T(X \rightarrow X')$ for having the object X' succeeding object X in the sequence. The probability of having a sequence of objects X_i then becomes

$$P_N(X_1, X_2, \ldots, X_N) = P_1(X_1)T(X_1 \rightarrow X_2)T(X_2 \rightarrow X_3) \cdots T(X_{N-1} \rightarrow X_N). \tag{10.10}$$

The transition probabilities $T(X \rightarrow X')$ are normalised:

$$\sum_{X'} T(X \rightarrow X') = 1. \tag{10.11}$$

As an example, consider a random walk on a two-dimensional square lattice. At every step, the random walker can jump from a point to each of its four nearest neighbours with equal probabilities 1/4. This probability is, however, independent of *how* the walker got there, that is, which path he followed in order to arrive at the present point. An example of a non-Markovian (but correlated) sequence is given by the self-avoiding random walk in which the walker is not allowed to visit a site that has been visited in the past. Therefore, the probability for being at a specific site depends not just on the last position but on the full history of the walk.

We want to generate a Markov chain of system configurations such that they have a distribution proportional to $\exp[-\beta E(X)]$, and this distribution should be independent of the position within the chain and independent of the initial configuration. Under certain conditions, a Markov chain does indeed yield such an invariant distribution, at least for long times, as it needs some time to 'forget' the chosen initial distribution. We shall not go into details, nor give proofs, but summarise these conditions. They are: (i) every configuration which we want to be included in the ensemble should be accessible from every other configuration within a finite number of steps (this property is called *connectedness*, or *irreducibility*) and (ii) there should be no periodicity. Periodicity means that after visiting a particular configuration, it should not be possible to return to the same configuration except after $t = nk$ steps, $n = 1, 2, 3, \ldots$, where k is fixed. A Markov chain that satisfies these criteria is called *ergodic*.

The Metropolis Monte Carlo method consists of generating a Markov chain of configurations with the required invariant distribution, which in our case is the Boltzmann distribution $\exp(-\beta\mathcal{H})$. We must therefore find a transition probability $T(X \rightarrow X')$ which leads to a given stationary distribution $\rho(X)$ (to keep the following analysis general, we use the arbitrary positive function $\rho(X)$ rather than specify this to be the Boltzmann distribution). The number of possible configurations, N, is assumed to be finite as the computer memory in which they are to be stored is finite. The 'matrix' $T(X \rightarrow X')$ therefore has N^2 elements, as opposed to only N elements of the 'vector' $\rho(X)$. This means that there are many different solutions to this problem and we are allowed some freedom in finding one.

Let us introduce the function $\rho(X,t)$ which gives us the probability of occurrence of configuration X at 'time', or Markov step, t (for an ergodic chain, $\rho(x,t)$ becomes independent of t for large t). The change in this function from one step to another is governed by two processes: (i) going from a configuration X at time t to some other configuration X' at $t+1$, leading to a decrease of $\rho(X)$, and (ii) going from some configuration X' at time t to X at time $t+1$, thus causing an increase in $\rho(X)$. These mechanisms can be summarised in the formula

$$\rho(X,t+1) - \rho(X,t) = -\sum_{X'} T(X \to X')\rho(X,t) + \sum_{X'} T(X' \to X)\rho(X',t). \tag{10.12}$$

This equation is called the *master equation*. Remember that we are trying to find the stationary distribution, which is found by requiring $\rho(X,t+1) = \rho(X,t)$, so that we have

$$\sum_{X'} T(X \to X')\rho(X,t) = \sum_{X'} T(X' \to X)\rho(X',t). \tag{10.13}$$

We omit the t-dependence of ρ from now on. It is difficult to find the general solution to this equation, but a particular solution is recognised immediately:

$$T(X \to X')\rho(X) = T(X' \to X)\rho(X') \tag{10.14}$$

for all pairs of configurations X, X'. This solution is called the *detailed balance* solution. Viewing the configurations X as buckets, each containing a certain amount $\rho(X)$ of water, imagine that we make connections with pumps between each pair of buckets. Water is pumped from bucket X to bucket X' with a pumping rate $T(X \to X')\rho(X)$. Condition (10.14) makes sure that the pumping rates between any pair of buckets X, X' are balanced: the flow from X to X' is equal to the flow from X' to X, so that obviously the volumes $\rho(X)$ and $\rho(X')$ do not change. As this holds for any pair of buckets, the whole set of water volumes in the buckets will remain stationary.

Let us now reformulate the detailed balance solution to make it suitable for practical purposes. We write the transtion probability in the form

$$T(X \to X') = \omega_{XX'}A_{XX'}, \tag{10.15}$$

where the matrix $\boldsymbol{\omega}$ is symmetric: $\omega_{XX'} = \omega_{X'X}$. Furthermore, it satisfies $0 \leq \omega_{XX'} \leq 1$, and $\sum_{X'} \omega_{XX'} = 1$. $A_{XX'}$ must lie between 0 and 1 for each pair XX'. Substituting this form of T into the detailed balance equation gives a detailed balance equation for A:

$$\frac{A_{XX'}}{A_{X'X}} = \frac{\rho(X')}{\rho(X)}. \tag{10.16}$$

In order to construct an algorithm, we use $\omega_{XX'}$ as a *trial step probability* and $A_{XX'}$ as an *acceptance probability*. This means that the algorithm proceeds in two stages.

In the first stage, given a state X, we propose a new state X' with a probability given by $\omega_{XX'}$. In the second stage, we compare the weights of the old and the new one, $\rho(X)$ and $\rho(X')$. $A_{XX'}$ is chosen equal to 1 if $\rho(X') > \rho(X)$, and it is chosen equal to $\rho(X')/\rho(X)$ if $\rho(X') < \rho(X)$. Obviously $A_{XX'}$ satisfies condition (10.16). We accept the new state X' with a probability $A_{XX'}$, and we reject it with a probability $1 - A_{XX'}$. If the state X' is accepted, it replaces X; if it is not accepted, the system remains in the state X. Note that if $\rho(X') > \rho(X)$, the state X' is always accepted. We can summarise the Metropolis algorithm as follows:

$$T(X \to X') = \omega_{XX'} A_{XX'}; \tag{10.17a}$$

$$\sum_{X'} \omega_{XX'} = 1; \quad \omega_{XX'} = \omega_{X'X}; \tag{10.17b}$$

$$\omega_{XX'} > 0, \quad \text{for all } X, X'; \tag{10.17c}$$

$$\text{if } \rho(X') < \rho(X): \quad A_{XX'} = \frac{\rho(X')}{\rho(X)}; \tag{10.17d}$$

$$\text{if } \rho(X') \geq \rho(X): \quad A_{XX'} = 1. \tag{10.17e}$$

The question now arises as to how we can accept a state with a probability $A_{XX'} \leq 1$, and reject it with probability $1 - A_{XX'}$. This is done by generating a random number r uniformly between 0 and 1. If $r < A_{XX'}$, the state is accepted, otherwise it is rejected. It is clear that if this procedure is carried out many times with the same probability $A_{XX'}$, the state will be accepted a fraction $A_{XX'}$ of the total number of trials.

Note that because the configurations are generated in a Markov chain, they have correlations inherent to them. The theory of Markov chains guarantees that we arrive at the invariant distribution ρ for long times; however, it may take much longer than the available computer time to reach this distribution.

The total number of statistically independent configurations is given by the total number of steps divided by the correlation 'time', measured in Monte Carlo steps. Note that the number of steps is the *total* number of trials: do not fall into the trap of counting only the *successful* trials as MC steps. As we have generated a sequence of configurations X with a statistical distribution $\exp[-\beta E(X)]$, the ensemble average of a physical quantity A is given by the 'time average'

$$\overline{A} = \frac{1}{n - n_0} \sum_{\nu > n_0}^{n} A_\nu \tag{10.18}$$

where n_0 is the number of steps used for equilibration. Note that the 'time' n is not physical time. This average is exactly the same as for MD simulations discussed near the end of Section 8.2. For the determination of statistical errors in the resulting averages we refer to the discussion in Section 7.4.

In the next subsections we shall work out the canonical ensemble MC method in more detail for two examples of classical many-particle systems, the Ising model and the monatomic fluid.

It should be noted that the detailed balance condition can be fulfilled by algorithms other than the Metropolis method. The *Barker algorithm* [11, 12] reads

$$T(X \to X') = \omega_{XX'} \frac{\rho(X')}{\rho(X) + \rho(X')} \quad \text{for } X \neq X', \tag{10.19a}$$

$$T(X \to X) = 1 - \sum_{X \neq X'} T(X \to X'). \tag{10.19b}$$

where ω satisfies the same criteria as in the Metropolis method. It can easily be shown that this algorithm indeed satisfies the detailed balance condition. The Metropolis solution, however, turns out to be more efficient [2].

We can generalise the Metropolis procedure, in which for some pairs X, X' of configurations, $\omega_{XX'}$ is not equal to $\omega_{X'X}$. In that case, the acceptance criterion for an attempted move $X \to X'$ must be replaced by

$$A_{XX'} = \min(1, q_{XX'}) \tag{10.20}$$

with

$$q_{XX'} = \frac{\omega_{X'X}\rho(X')}{\omega_{XX'}\rho(X)}. \tag{10.21}$$

This is called the *generalised Metropolis method*. In the context of Monte Carlo simulations, the method is also called *smart Monte Carlo*. It can easily be checked that the generalised Metropolis method satisfies detailed balance.

We conclude this section by mentioning another variant of the Metropolis method: the *heat-bath algorithm*. In this algorithm it is assumed that the trial step involves one or a few degrees of freedom, the remaining ones being kept fixed (this is the case in most Metropolis algorithms). The degrees of freedom that may change are denoted x; the remainder of the system, which is kept fixed, is then denoted symbolically as $X - x$. We now assign a new value to x according to

$$P(x) \propto \exp[-\beta \mathcal{H}(x|X - x)], \tag{10.22}$$

where $\mathcal{H}(x|X - x)$ is the Hamiltonian as a function of x, with $X - x$ kept fixed. It is easy to see that this procedure satisfies detailed balance, and that it is equivalent to applying an infinite number of Metropolis steps to x successively, with fixed $X - x$. The practical implementation of this method is often difficult, for reasons to be explained below; however, for lattice models it can be implemented rather straightforwardly.

10.3.1 Monte Carlo for the Ising model

The Ising model was discussed in Chapter 7. Here we consider the Metropolis algorithm for the Ising model in two dimensions (the extension to more than two dimensions is straightforward). In order to formulate the Monte Carlo method, we must make a choice for the matrix $\omega_{XX'}$. For the two-dimensional Ising model on an $L \times L$ square lattice, we take

$$\begin{aligned} \omega_{XX'} &= 1/L^2 \quad \text{if } X \text{ and } X' \text{ differ by one spin;} \\ \omega_{XX'} &= 0 \qquad\quad \text{otherwise.} \end{aligned} \tag{10.23}$$

The realisation of the first stage of the Markov step – generating a trial configuration – is then easy: we select a spin at random, and the trial configuration is the present configuration with the selected spin turned over.

We then calculate the energy difference $\Delta E(X \to X')$ between the old and the trial configuration:

$$\Delta E(X \to X') = E(X') - E(X). \tag{10.24}$$

This is easy as there are only nearest neighbour interactions: the energy difference depends only on the number of neighbours which have the same spin as the selected spin in the old configuration – this is therefore an integer number between 0 and 4. If the energy increases from the old to the new configuration, $\Delta E(X \to X') > 0$, the trial state is accepted with probability $\exp[-\beta \Delta E(X \to X')]$. If, however, the energy decreases, the trial state is always accepted as the new state. Implementation of this algorithm for the two- or three-dimensional Ising model is straightforward; the reader is encouraged to go through this exercise. The average number of steps between two updates of the same spin is equal to L^2. Therefore, the 'time' in a MC simulation is often expressed in units of *Monte Carlo steps per spin* (MCS), 1 MCS being equal to L^2 trials.

PROGRAMMING EXERCISE

Write a program for simulating the nearest neighbour two-dimensional Ising model on a square lattice using the Metropolis Monte Carlo technique.

The following considerations should be borne in mind. It is convenient to have a variable representing the total energy of the system. Usually, βE is recorded rather than the energy E itself, as the behaviour of the system is fully determined by βJ. This must be calculated at the beginning by performing a sweep over the whole lattice. If, however, the initial state is one in which all the spins are the same, the total energy for the $L \times L$ lattice with only nearest neighbour interactions is simply given by $-2\beta J L^2$, where βJ is the coupling constant. Every time a trial configuration is accepted as the new one, we add the energy difference (which is

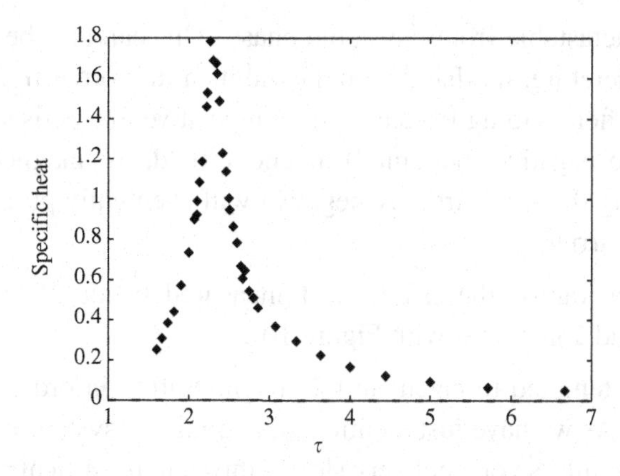

Figure 10.1. Specific heat per site in units of J/k_B of the Ising 20 × 20 square lattice as a function of the coupling constant reduced temperature $\tau = k_B T/J$.

calculated at every step) to the total energy so that we always have the energy of the new state at our disposal. Similarly it is possible to keep track of the magnetisation during the simulation. The approach described here avoids having to calculate these quantities every now and then by summing over the entire lattice.

As the acceptance probability $\exp[-\beta \Delta E(X \to X')]$ can assume only five different values, it is advisable to store these in an array in order to avoid calculating the exponential function over and over again. It is nice to display the lattice after every fixed number of MC steps on the computer screen and give the user the opportunity to change the temperature during the simulation. A magnetic field can also be included. Visual inspection should convince you that the phase transition for zero field takes place somewhere around $\beta J \approx 0.44$, although critical fluctuations make it difficult to locate this transition temperature with satisfactory precision. The specific heat can be determined as the variance of the energy (see Eq. (7.28)) and this should exhibit a peak near the critical temperature (see Figure 10.1).

The initial configuration will be chosen either random (infinite temperature) or as one of the two ground states: all spins either + or −. In the first case, if the temperature at which the system is simulated is lower than the transition temperature, spontaneous magnetisation may not always occur. In fact, several large regions of spin + or of spin − will be formed, separated by boundaries which are relatively smooth in order to minimise their energy. It will now take a very long time before one of the two spin values dominates. This undercooling effect can only be avoided by cooling the system gradually down to the desired temperature, taking care that the cooling rate is particularly slow near T_c. After passing a first order transition in a simulation, it is often impossible to arrive at the equilibrium state within a reasonable time, as the system cannot overcome the free energy barrier separating the metastable from the stable phase. This can also be checked: below the critical temperature, the Ising model exhibits a first order transition triggered by the magnetic field. Going from a state with positive magnetisation and positive magnetic field to negative (but small) magnetic field, the magnetisation will not turn over. Making the field strongly negative will eventually pull the system over the free energy barrier.

Check Produce a graph of the specific heat (measured in units of J/k_B) as a function of $\tau = k_B T/J$ and compare it with Figure 10.1.

You could be tempted to calculate the magnetisation in order to compare this with Figure 7.3. As we have just mentioned, cooling the system down from a high temperature fails unless you cool very slowly through the critical region. However, even if you start with the low temperature phase and heat the system up, you will notice that the magnetisation vanishes for temperatures just below the transition temperature. The point is that in order to let the magnetisation flip from a positive value to a negative one, a domain wall separating the two phases must be built up. Below the transition temperature, domain walls have a positive wall tension, that is, they carry a free energy cost per unit length. Therefore, flipping the magnetisation in the *infinite* system requires an infinite amount of free energy so that this will never happen. In the finite system however, if the wall tension is still finite (and the temperature therefore still below the finite-size critical point), the energy barrier for a magnetisation flip is finite as the domain walls are necessarily finite. Hence the magnetisation will fluctuate around a positive or negative equilibrium value for relatively long periods. But now and then it may switch sign, so that its long time average vanishes. Figure 10.2 shows the typical behaviour of the magnetisation in this phase.

A histogram of the probability of occurrence for various magnetisations shows a double-peaked shape. These problems also show up in the determination of the magnetic susceptibility which can be expressed in terms of the variance of the magnetisation.

Several methods exist for avoiding this problem:

- A restriction to – say – positive magnetisation is built in. This option has the disadvantage that the system is distorted and the consequences of this are not a priori clear. Moreover the average magnetisation is always positive so it becomes hard, if not impossible, to see where it would vanish without this restriction.
- Making a plot of the magnetisation as a function of time and taking averages only on the plateaux where the magnetisation is either positive or negative. This method is, however, difficult to apply close to the transition, as the strong fluctuations there will make it difficult to distinguish these plateaux clearly.

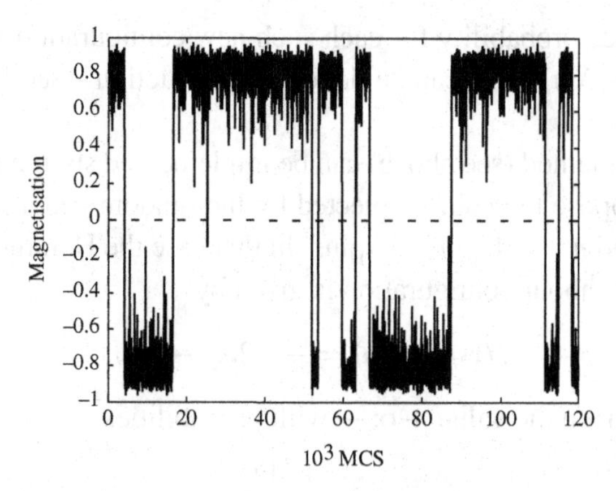

Figure 10.2. Magnetisation of a 20 × 20 Ising lattice as a function of time for an effective coupling constant $J/k_B T = 0.46$. The graph clearly shows plateaux where the magnetisation fluctuates around positive or negative average values.

- Making a histogram of the magnetisation as mentioned above and taking the peak positions as the value of the magnetisation. Again, distinguishing the peak will be difficult when critical fluctuations are strong.

It will be clear that measuring the magnetisation close to the phase transition is a difficult task which should be avoided if possible.

It should be noticed that the Ising model is formulated without a prescription for its dynamical evolution. Consequently, there exists no molecular dynamics method for the Ising model, and Monte Carlo is the only simulation technique for spin systems such as the Ising model. Using the Metropolis method, the Ising model becomes dynamic in the sense that its configurations change with time. The kinetics of this behaviour have been studied, first because it is assumed that the dynamic evolution of Ising-like systems in nature is governed by processes not too dissimilar from Metropolis MC and second because the kinetics are relevant to the reliability of the simulations, in particular near the critical point [13, 14]. Recall that the dynamical behaviour close to the phase transition is expressed in terms of the dynamic critical exponent, z, which describes the divergence of the correlation time (see Section 7.3.2).

Our choice for the matrix $\omega_{XX'}$ is not the only possibility. We may allow for two (or more) spins flipping over at the same step, and these spins might be restricted to be opposite so that the total magnetisation does not change, giving us a constant-magnetisation algorithm. In all cases, instead of selecting the spins randomly, they may be chosen in a regular fashion, for example by scanning through the lattice row by row. In that case, one step in the Markov chain consists of a scan through the whole lattice. $\omega_{X,X'}$ is equal for *any* new configuration X' and it remains to show that the acceptance probability for each such new configuration is proportional to $\exp[-\beta\Delta E(X \to X')]$. This can be done using induction – see Problem 10.7 and Ref. [15].

The heat-bath method (see above) can be implemented straightforwardly for the Ising model. Suppose the spin s_i selected by the random generator is surrounded by n_+ + spins and $n_- = 4 - n_+$ − spins. In that case the Hamiltonian for the spin s_i in the fixed neighbour configuration is given by

$$\mathcal{H}(s_i|S - s_i) = -J(2n_+ - 4)s_i. \tag{10.25}$$

Therefore, s_i is given the value + or − with probabilities

$$P_+ = \frac{e^{(2n_+-4)\beta J}}{e^{(2n_+-4)\beta J} + e^{-(2n_+-4)\beta J}} \quad \text{and} \tag{10.26a}$$

$$P_- = \frac{e^{-(2n_+-4)\beta J}}{e^{(2n_+-4)\beta J} + e^{-(2n_+-4)\beta J}}. \tag{10.26b}$$

Using the heat-bath instead of the Metropolis method results in a substantial decrease of the correlation time. We shall encounter this method again in Chapter 15, where we present results for the correlation times for the scalar lattice field model. We shall return to the Ising model in Section 15.5.1 where we discuss algorithms that are much more efficient near the critical point than the standard MC methods discussed here.

10.3.2 Monte Carlo simulation of a monatomic gas

The Metropolis MC technique enables us to calculate averages of static quantities. Therefore, the momentum degrees of freedom are irrelevant and we integrate these out as in Section 7.2.1, so that the Boltzmann factor depends only on the configurational potential energy:

$$\rho(R) = \exp[-\beta U(R)], \tag{10.27}$$

where R is the combined position coordinate $\mathbf{r}_1, \ldots, \mathbf{r}_N$. The configurational potential energy $U(R)$ is usually written as a sum of pair-potentials as in the previous chapters.

The Monte Carlo procedure for a monatomic gas proceeds as follows. The matrix $\omega_{XX'}$ is chosen such that only one particle may be moved to a new position – it is selected at random and the remaining particles are kept fixed. The new position of the particle is chosen at random with a homogeneous distribution within a cube centred at the old position of the particle.[3] The energy difference is calculated and

[3] The cubic shape is not essential – it is chosen for convenience.

the trial configuration is accepted or rejected as usual, with an acceptance rate $\exp[-\beta \Delta E(X \to X')]$.

The displacement volume remains to be specified. Any cube should lead to the correct behaviour as the requirements of the Markov chain (connectedness and aperiodicity) and the detailed balance condition are fulfilled, but there will obviously be a particular cube size yielding optimum efficiency, where efficiency can be defined as the number of statistically independent configurations that we can generate in a fixed amount of computer time. Obviously, if the cube is chosen very small, particles are allowed to move over small distances only and it will take many MC steps to arrive at a statistically independent configuration. On the other hand, if the cube is chosen too large, the particles will on average make large moves. This changes the configuration of the system to such an extent that the energy will increase strongly in the large majority of cases. The probability that the trial move is accepted will then be vanishingly small and we will spend most of our time generating configurations which are subsequently rejected, so the configuration changes very slowly in that case as well.

A widely adopted rule of thumb is to choose the cube such that the acceptance rate of trial states is on average somewhere between 0.4 and 0.6. Although for hard sphere systems it should perhaps be lower (around 0.1) [11], this rate is generally relied upon as being reasonably efficient [2].

When programming the Metropolis method for the case of argon, much of the MD code (see Section 8.3) can be copied. The particles are again released from face-centred cubic (fcc) lattice positions. Every trial displacement should be performed respecting the periodic boundary conditions. Calling half the linear size of the cube $d_{\max}$, the trial displacement (without correcting for PBC) in the x-direction is given in terms of a random number $0 \le r < 1$ as

$$x_{\text{new}} - x_{\text{old}} = d_{\max}(2r - 1) \tag{10.28}$$

and similarly for y and z. A neighbour list may be kept as in the MD case, but the list must be constructed such that we have at our disposal all the neighbours (at a distance smaller than the neighbour list cut-off $r_{\text{cut-off}}$) for every particle. This means that we need twice as much storage as in the MD method, where only the neighbours with an index higher than the particle under consideration are stored in the list. As in the Ising case, the total potential energy can be updated after every acceptance of a trial configuration. The same can be done with the virial. Physical properties to be calculated are the same as in MD simulations: the pressure, the configurational energy, and the pair correlation function. Any quantity dependent on the positions only can be determined – remember the momenta are not considered in the MC method.

PROGRAMMING EXERCISE

Code the Metropolis method for argon, and compare the results with those of the molecular dynamics program; see Table 8.1.

The heat-bath method turns out impractical in this example. If we were to use the pure heat-bath method, we would have to allow the particle to move to any position in the system cell, and then accept or reject this position using the Von Neumann algorithm (see Appendix B3), but as the vast majority of positions in the cell are unacceptable, this is very inefficient. One could imagine a 'hybrid' heat-bath method, in which we move the particle to a new position in a small sphere or cube centred at the old position with a probability distribution determined by the conditional Hamiltonian $\mathcal{H}(\mathbf{r}_i|R - \mathbf{r}_i)$, using the symbolic notation of Eq. (10.22). This can again be done using the Von Neumann method (see Appendix B3). The difference with the Metropolis method is that the new position is accepted with a probability which is independent of the previous one (except for the sphere or cube being centred on the old position). To apply the Von Neumann method, we should know the minimum of the conditional Hamiltonian, as the acceptance probability in the Von Neumann method may never exceed 1. We might guess a lower bound for this minimum, but this will often be much lower than the actual potential minimum, so that the acceptance rate becomes very small. Because of these difficulties, the heat-bath method is not used for atomic or molecular systems.

Just as in the case of the Ising model, we may sometimes suffer from critical slowing down due to a diverging the correlation time close to a critical point. Methods have been developed which dramatically reduce this effect. A recent breakthrough in this field is the Liu–Luijten algorithm [16].

10.4 Other ensembles

The canonical ensemble is the most natural ensemble for MC simulations. It is, however, possible to simulate other ensembles by the Metropolis MC method. We shall consider the isothermal-isobaric, or (NPT) ensemble, and the grand canonical ensemble. There also exists a microcanonical MC method, [17] but this is seldom used as it is of little practical importance.

10.4.1 The (NPT) ensemble

The (NPT) ensemble is relevant because temperature and pressure are often kept fixed in experiments. A Monte Carlo method for this ensemble was first developed for hard sphere systems by Wood [18, 19] and later extended to smooth potentials by McDonald [20, 21]. We consider the latter case here.

The (NPT) ensemble average of a physical quantity A depending on the positions $R = \mathbf{r}_1, \ldots, \mathbf{r}_N$ is given as

$$\langle A \rangle_{(NPT)} = \frac{\int_0^\infty \mathrm{d}V\, \mathrm{e}^{-\beta PV} \int \mathrm{d}R\, A(R) \mathrm{e}^{-\beta U(R)}}{Q(N,P,T)}. \tag{10.29}$$

Q is the partition function which is related to the Gibbs free energy: $G = -k_\mathrm{B} T \ln Q(N,P,T)$; see Section 7.1. As the volume is allowed to vary, we extend the notion of a configuration to include the volume in addition to the set of particle positions, $R = (\mathbf{r}_1, \mathbf{r}_2, \ldots \mathbf{r}_N)$, with the restriction that the latter should all lie within that volume. A Markov chain must be constructed in which particle moves and volume changes are allowed. It is, however, impossible to change the volume independently from the particle positions, as a decrease of the volume might cause particles close to the wall to fall outside the volume. Therefore, a change in volume must be accompanied by an appropriate rescaling of the particle positions.

To be more specific, let us consider a cubic volume $L \times L \times L$ with the edges along the positive Cartesian axes. We scale the particle positions according to $\mathbf{r}_i = \mathbf{s}_i L$ so that the positions $\mathbf{s}_i$ lie within the unit cube. The average (10.29) can now be written as

$$\langle A \rangle_{(NPT)} = \frac{\int_0^\infty \mathrm{d}V\, \mathrm{e}^{-\beta PV} V^N \int \mathrm{d}S\, A(LS) \mathrm{e}^{-\beta U(LS)}}{Q(N,P,T)} \tag{10.30}$$

where S denotes the combined positions $\mathbf{s}_1, \ldots, \mathbf{s}_N$. When changing the volume, S remains the same; the change in the real positions $\mathbf{r}_i$ is accounted for by a change in the Jacobian V^N and the various factors L in (10.30). The Boltzmann weight of the (NVT) ensemble method is replaced by

$$\rho(V,S) = \mathrm{e}^{-\beta PV} V^N \mathrm{e}^{-\beta U(LS)}. \tag{10.31}$$

A step in the Metropolis Markov chain consists of either a particle move, which is performed exactly like the particle moves in the (NVT) ensemble method, or a volume change. The calculation of the ratio of the weights before and after the volume change consists of calculating the change due to the potential energy,

$$\exp\{-\beta[U(L_\mathrm{new}S) - U(L_\mathrm{old}S)]\} \tag{10.32}$$

and multiplying this by the ratio of the terms involving the volume coordinate:

$$\exp[-\beta P(V_\mathrm{new} - V_\mathrm{old})] \left(\frac{V_\mathrm{new}}{V_\mathrm{old}}\right)^N. \tag{10.33}$$

The product of (10.32) and (10.33) defines the acceptance ratio of the new configuration according to the Metropolis recipe. Eppenga and Frenkel [22] have applied the method with the logarithm of the volume as the extra coordinate rather than the volume itself.

The calculation of the potential energy difference associated with a volume change is rather demanding – as all positions change, we must sum over all pairs of particles. If the potential can be written as a linear combination of powers $(\sigma/r)^k$, the calculation can be performed very fast, because the energy difference in the exponent of (10.32) due to a term $(\sigma/r)^k$ can be evaluated as

$$\sigma^k \sum_{i<j} (L_{\text{after}} \mathbf{s}_{ij})^{-k} = \sigma^k \left[\sum_{i<j} (L_{\text{before}} \mathbf{s}_{ij})^{-k} \right] (L_{\text{before}}/L_{\text{after}})^k; \tag{10.34}$$

that is, this contribution to the potential energy changes simply by an overall scaling factor! If the potential is a linear combination of such powers (e.g. the Lennard–Jones potential), this formula can be used, provided that the contributions to the total potential energy due to each power are stored separately.

A possible problem needs some attention. If the potential is cut off beyond a range $r_{\text{cut-off}}$, we must correct for this in the total energy using Eq. (8.18). The correlation function, g, occurring in this formula is usually replaced by 1, in which case this term does not contribute to the total energy difference in MC steps. Therefore, the cut-off distance is usually kept constant. In the special case of a potential consisting of a sum of powers, where we would like to calculate the potential energy difference using the simple scaling procedure just described, we would like to scale the cut-off with the linear system size, $r_{\text{cut-off}} = L s_{\text{cut-off}}$, so that (10.34) remains valid. It should be noted, however, that in that case the correction to the potential (which depends also on the density N/V, which changes at each rescaling) has to be included in the calculation of the energy difference using Eq. (8.18).

We have the freedom to choose the relative frequency with which particle moves and volume changes are attempted. If the scaling method just described for calculating the potential cannot be used because of invariant intramolecular configurations, or because of more complicated parametrisations of the potential, calculating the potential energy difference due to a volume change becomes quite expensive and volume changes should be attempted at a much lower rate than particle moves. If, however, the method of Eq. (10.34) is applicable, both types of changes can be attempted with equal probability.

10.4.2 The grand canonical ensemble

In 1969, Norman and Filinov [23] developed a Metropolis Monte Carlo method for simulating many-particle systems in the grand canonical ensemble. In this case the temperature, the system volume and the chemical potential are given, and the particle number and pressure vary; their average values can be determined to establish the equation of state. As the particle number does not remain constant, creation and annihilation of particles should be possible. Let us write down the probability distribution of configurations in this ensemble. The grand canonical ensemble average of a configurational physical quantity A is given by

$$\langle A\rangle_{\mu VT} = \frac{\sum_{N=0}^{\infty} 1/N! \mathrm{e}^{\beta\mu N} \Lambda^{-3N} \int \mathrm{d}R_N A(R_N) \mathrm{e}^{-\beta U(R_N)}}{Z_{\mathrm{G}}(\mu, V, T)}; \qquad (10.35)$$

$\Lambda = h/\sqrt{2\pi m k_{\mathrm{B}} T}$ in the case of a monatomic gas – it results from integrating out the momentum degrees of freedom. We have attached a subscript N to the positional coordinate R because the number of particles is not fixed.

It is clear from (10.35) that the configurations are now defined by the number of particles N and by their positions R_N. The weight factor which replaces the Boltzmann factor of the (NVT) ensemble now becomes

$$\rho(N, R_N) = \mathrm{e}^{-\beta U(R_N)} \Lambda^{-3N}/N!\, \mathrm{e}^{\beta\mu N} \qquad (10.36)$$

and the Metropolis algorithm can be applied directly, provided that in addition to the usual particle moves, we allow for particle creations at random positions and annihilations of randomly chosen particles. The algorithm for a Metropolis step then becomes:

- Decide whether the next trial configuration is constructed via a creation, an annihilation, or a particle move according to the probabilities for these processes given by the matrix $\omega_{XX'}$ (note that the trial rates for creation and annihilation should be equal in order to keep $\omega_{XX'}$ symmetric). This choice can be made simply by dividing the interval [0, 1] up into three segments with sizes equal to the respective probabilities, generating a random number uniformly between 0 and 1, and then checking in which segment this number lies.
- If a creation is attempted, a random position in the system is selected and the interactions between a new particle inserted at that position and the remaining ones are added to yield a potential energy difference $\Delta U^+ = U(R_{N+1}) - U(R_N)$. As the probability that the new particle ends up in the volume element $\mathrm{d}^3 r_{N+1}$ is given by $\mathrm{d}^3 r_{N+1}/V$, we accept the creation with a probability

$$\mathrm{e}^{-\beta\Delta U^+} \Lambda^{-3} V/(N+1) \mathrm{e}^{\beta\mu}. \qquad (10.37)$$

- If an annihilation is attempted, one of the existing particles is selected at random, and its interaction with the remaining particles, $\Delta U^- = U(R_{N-1}) - U(R_N)$, is calculated. The annihilation is then accepted with probability

$$\mathrm{e}^{-\beta\Delta U^-} \Lambda^3 N/V \mathrm{e}^{-\beta\mu}. \qquad (10.38)$$

- Particle moves are processed similarly to the canonical case. Only the potential energy difference enters into the acceptance probability.

This form of grand canonical Monte Carlo was presented by Norman and Filinov [23]. Other approaches turn particles into 'ghost' particles instead of annihilating them [24, 25], and these ghosts can be switched on so that they enter the real life of the simulation again. The Norman–Filinov version is, however, more popular.

It is possible to change the relative rates of the creation and annihilation process and correct for this by a suitable change in acceptance rates of the corresponding trial configurations. This was done for example by Saito and Müller-Krumbhaar [26]; see also Problem 10.3. These methods may be useful in situations where the acceptance rate for particle creation in the standard method becomes exceedingly small because of a high value of the chemical potential.

It should be noticed that at high densities, insertion of new particles is likely to fail because the probability of spatial overlap between the new particle and the existing ones becomes very high. In this case, the Boltzmann factor is small, not as a result of a high value of the chemical potential but as a result of the interactions between the new particle and the existing ones. Methods have been devised for locating 'cavities' in such fluids and creating particles preferentially in these regions [27].

More details on grand canonical Monte Carlo methods can be found in Refs. [2, 28].

10.4.3 The Gibbs ensemble

Studying the coexistence of different phases of the same material, and of different species that can transform into each other via chemical reactions, is difficult using the ensembles defined up to now. We know that for coexistence to be possible, the different phases or species must have equal temperature, pressure and chemical potential. We shall use the name 'species' in the context of both chemical mixtures and phase transitions in the following. If we fix pressure and temperature, the chemical potentials of the different species will in general not be equal and one species will grow at the expense of the other, until either one of the two has disappeared or until the chemical potentials are equal. However, it is often very hard to achieve equilibrium in such cases since droplets surrounded by domain walls with a nonvanishing wall tension need a long time to disappear.

Panagiotopoulos [29, 30] has developed a method in which two subsystems are considered with no interface between them so that they are free to exchange particles without having to overcome free energy barriers. The method is called 'Gibbs ensemble method'. The method is quite simple. Consider a volume V which is divided into two subsystems with volumes V_1 and V_2 by a freely movable piston. The volume $V = V_1+V_2$ is fixed, so the total system is described by the (NVT) ensemble. Furthermore, there is a virtual hole in the piston through which particles can move from one subsystem to the other. Most importantly: there are no interactions between the particles of the two subsystems; that is, if a particle moves from V_1 to V_2, the energy difference it feels consists of the interactions with its partners in the new subsystem (V_2) minus the interactions it felt in its previous subsystem (V_1). These moves are actually executed in the Gibbs ensemble MC method, alongside the usual particle moves within the two subsystems and relative changes of the two subsystem volumes. In order to be able to vary the subsystem volumes, we scale the particle positions $R_K = L_K S_K$ where the index $K = 1, 2$ labels the subsystem and $L_K = V_K^{1/3}$. The weight of a configuration with subsystem volumes V_1 and $V_2 = V - V_1$, subsystem particle numbers N_1 and N_2 and subsystem configurations S_1 and S_2 is given by

$$\rho(V_1, N_1, S_1, V_2, N_2, S_2) = \frac{V_1^{N_1}}{N_1!}\frac{V_2^{N_2}}{N_2!}\mathrm{e}^{-\beta U(L_1 S_1)}\mathrm{e}^{-\beta U(L_2 S_2)}. \tag{10.39}$$

This expression follows directly from the weight factors for the ensembles considered in the two previous subsections (the pressure and the chemical potential occur with $V_1 + V_2$ and $N_1 + N_2$ respectively, which are constant).

Equation (10.39) determines the detailed balance transition probabilities in a Metropolis algorithm:

- The matrix $\omega_{XX'}$ is a probability for volume changes, particle transfers from subsystem 1 to 2 or vice versa, and for particle moves in either 1 or 2 respectively. The matrix elements for particle moves are chosen such as to allow for single particle moves only, with equal probability for each particle.
- Particle moves in each subsystem are processed with the probabilities based on the factor
$$\exp\{-\beta[U(LS_K^{\text{new}}) - U(LS_K^{\text{old}})]\}. \tag{10.40}$$
- For subsystem volume changes we find for the acceptance rate (see also Section 10.4.1):
$$\prod_{K=1,2} \exp\{-\beta[U(L_K^{\text{new}} S_K) - U(L_K^{\text{old}} S_K)]\}\left(\frac{V_K^{\text{new}}}{V_K^{\text{old}}}\right)^{N_K}. \tag{10.41}$$
- Acceptance rates for particle transfers from subsytem 1 to subsystem 2 involve the energy difference $\Delta U_K^{\pm}$ for removing ($-$) or adding ($+$) a particle to the subsystem analogous to the grand canonical MC method:
$$\exp[-\beta(\Delta U_2^{+} - \Delta U_1^{-})]\frac{N_1}{N_2 + 1}\frac{V_2}{V_1} \tag{10.42}$$
and similarly for transfers from subsystem 2 to subsystem 1.

These transition rates define the Gibbs ensemble Monte Carlo method. It should be noted that this method is still susceptible to the kind of problems described in connection with the grand canonical Monte Carlo method: moving a particle from one subsystem to another may have a prohibitively low acceptance rate at high densities, because of the large increase of configurational energy in most such attempts.

It is not necessary to separate the two subvolumes by a movable piston in order to arrive at equal pressures in both subsystems: it is also possible [30] to couple both subsystems to a 'pressure bath' with predefined pressure P similar to the (NPT) method described in Section 10.4.1, and imposing no restriction on the total volume $V_1 + V_2$. This method is less suitable for phase coexistence as the coexistence occurs on a line in the (P, T) diagram. Therefore we need to know the exact location of that line, as we have to specify T and P in this method. In the original version, the system will move to the coexistence line by adjusting pressure and chemical potential simultaneously. In the case of chemical equilibrium, however, the coexistence region has a finite width and the constant (P, T) version is useful there. The Gibbs ensemble method has become very popular for studying coexistence equilibria in recent years [29–32].

*10.5 Estimation of free energy and chemical potential

In Section 7.1 we have already mentioned the difficulties involved in calculating free energies. We described briefly the method of thermodynamic integration. Other methods have been proposed and these are easier to understand in the context of MC trials and this is the reason why we discuss them in this chapter. It should, however, be noted that the methods described below are not restricted to MC. Some are applicable within MD, especially in the canonical (NVT) MD method. In view of the equivalence of ensembles, microcanonical MD allows for using these methods too [33].

The reader is referred to Refs. [28, 32] for more details concerning the material in this section.

10.5.1 Free energy calculation

The difficulty in free energy calculation is that it cannot be formulated directly as a 'mechanical average', that is, an ensemble average of functions of the coordinates $\mathbf{r}_i$ (and $\mathbf{p}_i$ in the case of MD). Rather, the free energy must be evaluated as a sum (or integral) over phase space. Clearly, both MD and MC methods sample that part of phase space from which the dominant contribution to the free energy arises; however, this does not provide an estimate of the phase space volume integral, as the frequency with which phase space points are visited is proportional to the Boltzmann weight, with an unknown proportionality factor. Moreover, it is questionable whether the number of points visited in a typical simulation would be sufficient to sample the relevant phase space volume adequately.

It is nevertheless possible to formulate the free energy *difference* between two systems with different interactions as a mechanical ensemble average. The first method for doing this is thermodynamic integration, described in Section 7.1. This method suffers from the fact that many simulations are needed in order to calculate free energy differences. We shall now describe a few alternative procedures for this purpose.

Suppose the two systems have interactions described by the potential functions U_0 and U_1. We show now that the partition function of system 1 can be rewritten as a mechanical average with the Boltzmann factor of system 0. Writing the respective partition functions as Z_0 and Z_1 we have

$$\frac{Z_1}{Z_0} = \frac{\sum_X \mathrm{e}^{-\beta U_1}}{\sum_X \mathrm{e}^{-\beta U_0}} = \frac{\sum_X \mathrm{e}^{-\beta U_0}\mathrm{e}^{-\beta(U_1-U_0)}}{\sum_X \mathrm{e}^{-\beta U_0}} = \langle \mathrm{e}^{-\beta(U_1-U_0)}\rangle_0, \tag{10.43}$$

where $\langle\ldots\rangle_0$ represents an ensemble average in system 0.

Applications of (10.43) are restricted to systems for which the regions they tend to occupy in phase space have appreciable overlap. This can be seen by considering two systems with the same potential function but kept at different temperatures, $U_1 = \alpha U_0$. If $\alpha < 1$, system 1 is at a higher temperature than system 0. If we evaluate the expectation value in (10.43) using MC or MD, the result contains the contributions to Z_1 arising from the equilibrium states of the same system at lower temperature – hence the free energy estimate will be drastically wrong as a result of not taking the overwhelming majority of high energy states contributing to Z_1 into account. On the other hand, if $\alpha > 1$, the system will visit configurations with high energy and rarely assume one of the lower energy configurations which contribute to the partition function of system 1. Sufficient overlap between the phase space volume occupied by the two systems is essential. One way to achieve such overlap is to simulate a *chain* of systems with potentials U_p that interpolate between U_0 and U_1, such that subsequent configurational potential functions U_p, U_{p+1} have sufficient overlap to give reliable results. This means that the amount of computer time needed is comparable to that of thermodynamic integration, which uses a chain of different thermodynamic parameters.

Torrie and Valleau have refined the method by adding an extra weight function $W(X)$ to the average in (10.43), which pushes the system to a different region of phase space such as to reduce the overlap problem. Their method is called 'umbrella sampling' and we refer to their paper for a description of the method and examples. [34, 35].

Bennett [33] writes the free energy difference in another way, by defining the 'Metropolis function' $M(x)$ as $M(x) = \min[1, \exp(-x)]$. Then we have

$$\exp[-\beta(U_0 - U_1)] = M[\beta(U_0 - U_1)]/M[\beta(U_1 - U_0)] \tag{10.44}$$

from which it follows that

$$Z_0 \frac{\sum_X M[\beta(U_1 - U_0)]\mathrm{e}^{-\beta U_0}}{Z_0} = Z_1 \frac{\sum_X M[\beta(U_0 - U_1)]\mathrm{e}^{-\beta U_1}}{Z_1}. \tag{10.45}$$

The quotients in this expression are canonical ensemble averages corresponding to

$$\frac{Z_0}{Z_1} = \frac{\langle M[\beta(U_0 - U_1)]\rangle_1}{\langle M[\beta(U_1 - U_0)]\rangle_0}. \tag{10.46}$$

This equation is now used as follows. We perform two simulations, one with potential U_0, and one with potential U_1. We consider switching the potential in system 0 from U_0 to U_1 as a trial move for which we calculate the acceptance ratio, but which is in fact never carried out. Taking the average of the acceptance ratios of such moves gives the numerator of (10.46). Similarly, the denominator is given as the acceptance ratio of trial switches from U_1 to U_0, evaluated in the system with potential U_1. Again, this method is reliable only for appreciable overlap between the two equilibrium distributions in phase space. Bennett's method can also be extended with a weight function similar to that of Torrie and Valleau, and Bennett calculates the actual form of this function that gives the most accurate results. Bennett furthermore describes an interpolation method to estimate the free energy difference even if the overlap between the distributions is very small. His paper should be consulted for details [33].

10.5.2 Chemical potential determination

Determining the chemical potential is done using relation (7.13), which enables us to extract this thermodynamic quantity as a free energy difference between two systems with N and $N+1$ particles respectively. The exponential of this free energy difference can be written as the fraction of two partition functions:

$$\begin{aligned} \frac{Z_{N+1}}{Z_N} &= \frac{N!\Lambda^{3N}}{(N+1)!\Lambda^{3(N+1)}} \frac{\int \mathrm{d}R_{N+1} \exp[-\beta U(R_{N+1})]}{\int \mathrm{d}R_N \exp[-\beta U(R_N)]} \\ &= \frac{V}{(N+1)\Lambda^3} \left\langle \frac{1}{V} \int \mathrm{d}^3 r_{N+1} \exp(-\beta \Delta U_+) \right\rangle_N = \mathrm{e}^{-\beta\mu} \end{aligned} \tag{10.47}$$

where ΔU_+ is the energy of a particle inserted at $\mathbf{r}_{N+1}$ into the N-particle system. The prefactor $V\Lambda^{-3}/(N+1)$ of the expectation value on the right hand side is $\exp(-\beta\mu_{\text{ideal}})$, where μ_{ideal} is the chemical potential of the ideal gas. The term within angular brackets gives the expectation value of the Boltzmann factor associated with the energy difference for the addition of a particle anywhere in the system. This expectation value is determined via trial additions which are regularly performed but never accepted. After each MC step we generate a position $\mathbf{r}$ at random in the system and calculate the factor $\exp[-\beta\Delta U(\mathbf{r}_{N+1})]$. The factor $1/V$ within the brackets ensures the proper evaluation of the average. These values are then used to calculate the required expectation value by dividing by the number of such trial additions. This method is called 'Widom's particle insertion method'

[36]. The method works well, although problems arise for high densities. In that case the Boltzmann factor is very small for most trials, because the probability that the core of the new particle overlaps with one of the existing particles becomes very high. There have been refinements in which the particle insertions are biased towards the cavities in the fluid [22, 37, 38].

Instead of particle insertions, particle removals could be used to find the chemical potential. In this case we would use the inverse of Eq. (10.47):

$$\frac{Z_{N-1}}{Z_N} = \frac{N!\Lambda^{3N}}{(N-1)!\Lambda^{3(N-1)}} \frac{\int dR_{N-1} \exp[-\beta U(R_{N-1})]}{\int dR_N \exp[-\beta U(R_N)]}$$
$$= NV\Lambda^3 \langle \frac{1}{V} \int d^3 r_N \, \exp(\beta \Delta U_-) \rangle_N = e^{+\beta\mu}. \tag{10.48}$$

However, generalisation of the method proposed above to this case usually fails as the sampling of $\exp(\beta\Delta U_-)$ is very inefficient. The reason is that we are trying to calculate $\langle\exp(\beta\Delta U_-)\rangle$ whereas the Boltzmann weight factor used in the average includes a factor $\exp(-\beta\Delta U_-)$. The latter squeezes the high-ΔU contributions off and these contribute significantly to the average. Shing and Gubbins [37] have formulated an efficient method combining particle insertions and removals.

10.6 Further applications and Monte Carlo methods

10.6.1 Generating ensembles of polymers

An important topic in statistical mechanics is the behaviour of polymers: long and flexible chain-like molecules. These can be studied as a melt (a kind of liquid consisting of polymers) or in solution. A nice review on models, theory and Monte Carlo methods for this problem can be found in Ref. [39]. We focus here on the problem of a dilute polymer solution. In that case, if we can somehow model the effect of the solvent in terms of a simple interaction, the problem reduces to studying ensembles of individual polymers in different conformations. If the solvent is good, then the free energy for a polymer segment which is on all sides embedded in the solvent is lower than that of a polymer segment which is close to another polymer segment, without solvent molecules in between them. This picture boils down to an effective repulsion of the polymer segments.

It is useful to make a model in which the important properties of the polymer are preserved, whereas the details of its structure on the atomic scale have disappeared from the description. This is related to the idea of universality (see Section 7.3.2): only a few major features of the interactions at small length scales influence the behaviour on longer length scales; the details do not matter. We therefore make a 'mesoscopic' model of the polymer. It is a chain consisting of beads: the beads represent segments of the polymer; the segments in turn represent groups of atoms. These atomic groups have a strong short-range interaction, as they are chemically bonded. Remote segments influence each other through a Van der Waals attraction, and they repel when they overlap. However, the solvent effect described above represents another type of repulsion which forces the beads to remain at a minimum distance of at least a few solvent layers. All these characteristics can be summarised in the following polymer model:

- The polymer consists of N beads, which are represented as point particles. Neighbouring beads have a fixed mutual distance. This is the only interaction between them.
- Remote beads feel a repulsion at short distances and a Van der Waals attraction at longer distances. Their interaction can be modelled by a Lennard–Jones shape (see Eq. (7.33)).

Note that if we switch off the Lennard–Jones interaction, the polymer represents a random walk: this is called the *ideal chain* case. If, on the other hand, the beads repel each other, we are dealing with a self-avoiding walk (SAW) as the chain cannot cross itself. Often, polymers are studied on a lattice, which is an even further restriction of the model, but asymptotic (large length scale) behaviour is not sensitive to this. For a review of lattice algorithms, see Ref. [40]. Here we shall discuss algorithms for simulating the off-lattice model. Most of the methods used in this field have a very similar counterpart in the lattice case.

Note that the behaviour of the polymer is independent of temperatures for high temperatures. This is because the Lennard–Jones interaction is then dominated by the repulsive term, which is always noticeable, even when the temperature is very high. The quantities of interest are the end-to-end distance, which is the distance from the first to the last bead, and the radius of gyration (see Problem 1.2). We shall restrict ourselves here to studying the end-to-end distance as a function of N for polymers in two dimensions.

Our aim is to generate (an ensemble of) polymer configurations which are distributed according to the canonical distribution at a given temperature. Then we can calculate physical properties for this temperature as an average over the ensemble. Now let us consider possible steps in a standard Metropolis algorithm. We cannot move a single bead, as we should keep the distance to its neighbours constant. An obvious alternative is to choose a bead at random and change the angle between this bead and its two neighbours by some amount. If the polymer is 'curled up', which turns out to be not unlikely in typical simulations, there are very few such moves that will be accepted. Algorithms based on this type of moves are called 'pivot algorithms': the selected bead acts as a pivot for changing bending or (in three dimensions) dihedral angles.

Another approach is inspired by a particular type of motion of polymers: reptation. This is a snake-type motion which, in the case of the Monte Carlo algorithm, works as follows. A trial step consists of removing the last bead (the 'tail') of the polymer and this is attached to the head. Acceptance of this move proceeds according to the standard Metropolis criterion.

The previous methods are typical Metropolis methods, in the sense that every conformation is strongly correlated with the previous one, and these correlations may result in extremely slow relaxation. Another approach tries to avoid this correlation by generating a new polymer at each step. This must be done quite carefully as simply adding beads will typically result in very unprobable conformations. This is because there will certainly be overlaps or crossings. We should therefore add new segments to the polymer carefully, avoiding high-energy conformations.

A method in which this is done is the *Rosenbluth algorithm* [41]. We start with two beads on positions $(0, 0)$ and $(1, 0)$ (the distance between the beads is taken to be 1). Now we add the third one. This is characterised by the angle made by the three beads. The Metropolis algorithm suggests that it would be wise to add the third one with a distribution $\exp[-E(\theta)/(k_B T)]$, where $E(\theta)$ is the interaction energy of the third bead (of angle θ) with the first two. This is, however, difficult to do if we do not want to use many trial steps. Therefore we discretise the space of possible θ-values to obtain a finite number of, say, six different values of θ, with a random offset and spaced by $2\pi/6$. Now we calculate the weights $w_j^{(l)} = \exp[-E(\theta_j)/(k_B T)]$ for these six θ-values numbered by j (l denotes the bead we are adding), and their sum $W^{(l)}$:

$$W^{(l)} = \sum_j \exp[-E(\theta_j)/(k_B T)]. \tag{10.49}$$

We accept angle j with probability $w_j^{(l)}/W^{(l)}$. This is done by the 'roulette-wheel' algorithm. In fact, what we do is divide up the interval $[0, 1]$ into six segments of size $w_j^{(l)}/W^{(l)}$. Then we calculate a uniform random number between 0 and 1, and we check which segment j it corresponds to. Then we choose the corresponding θ_j. The final Boltzmann weight of the polymer is $\exp[-E_{\text{total}}/(k_B T)]$, where E_{total} is the total energy of the chain. Some reflection may convince you that this is equivalent to the product of the w_j:

$$\exp[-E_{\text{total}}/(k_B T)] = \prod_{l=3}^{N} w_j^{(l)} \tag{10.50}$$

where j denotes the choice which has been made for the angle θ at step l.

We choose as parameters of the Lennard–Jones potential $\epsilon = 0.25$, and $\sigma = 0.8$ (the distance between the polymers is taken to be 1). If the algorithm does what we want it to do, it should generate the polymers distributed according to (10.50).

Calculating the actual probability for a particular configuration to occur, we find

$$P = \prod_{l=3}^{N} \frac{w_j^{(l)}}{W^{(l)}}. \tag{10.51}$$

We see that we are off by a factor $\prod_l W^{(l)}$, so we must correct for this by storing this number into a *weight factor* for calculating properties of the polymers in the generating ensemble. The weighting factor thus compensates for the fact that the actual probability of occurrence does not match (10.50). The method described here is known as 'method A'. If we were to take all θ angles with an equal probability of $1/N_\theta$, where N_θ is the number of angles (six in our example), then we should take for the weights of each polymer $\prod_{l=1}^{N} w_j^{(l)}$ (check this!). This is known as 'method B'.

An interesting quantity to look at is the end-to-end distance as a function of the number of beads N. In fact, the Rosenbluth algorithm is extremely useful for this case as it generates polymer populations for *all* lengths up to the maximum set in the simulation. We have seen that the polymer corresponds to a self-avoiding random walk in two dimensions. It turns out that the end-to-end length scales with N as

$$R \propto N^\nu, \tag{10.52}$$

where ν is an exponent which is 0.75 in two dimensions. The crosses with error bars in Figure 10.3 show the scaling behaviour as determined in the Rosenbluth algorithm. For this figure, 10 000 polymers were grown up to a size of 250 beads. We see that the simulation reproduces the scaling behaviour well for small sizes, but when the sizes exceed 100, the errors become rather large. This seems surprising as the population sizes are equal for all polymer lengths, and the statistics for long polymers should only become better (unless the fluctuations in R increase with polymer length). The reason behind these fluctuations is the fact that the algorithm generally does not suppress high-energy configurations sufficiently; they will be accepted, but with a low weight. This effect becomes more pronounced for long chains – therefore we have a large population with only a few polymers dominating the average, and most members of the population not contributing to better statistics.

A solution to this problem is offered in the 'pruned–enriched Rosenbluth method' (PERM) by Grassberger [42], in which the population evolves towards a more balanced distribution by removing the 'bad' configurations from it and replacing them with copies of the good ones which can then further evolve independently in the simulation. Before going into details, we consider the standard Rosenbluth algorithm formulated in a recursive implementation:

Set PolWeight to 1;
ROUTINE AddBead(Polymer, PolWeight, L)

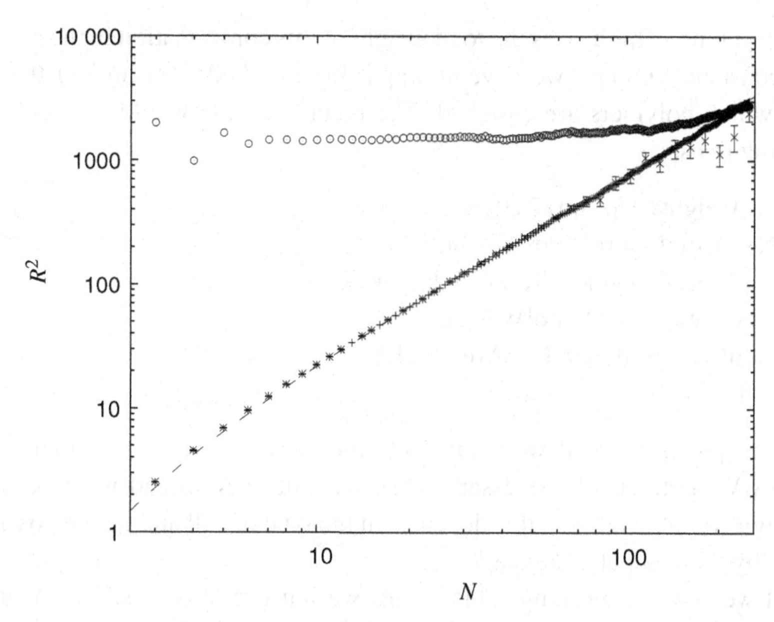

Figure 10.3. Scaling behaviour of the square of the end-to-end length R as a function of N on a log-log scale. Crosses with error bars: Rosenbluth algorithm. Double crosses: PERM algorithm. The circles are the population sizes in the PERM algorithm. The drawn line has the form $a(N-1)^{1.5}$, where a is chosen to fit the data.

```
    Calculate the weights w_j^L and their product W^L;
    Add bead number L;
    PolWeight = PolWeight*W^L;
    IF L<N THEN
        AddBead(Polymer, PolWeight, L+1);
    END IF
END ROUTINE AddBead.
```

A careful read of this pseudocode should convince you that this routine is equivalent to the Rosenbluth algorithm described above. The PERM algorithm cleverly uses this recursive type of procedure in order to perform pruning (i.e. the removal of weak members of the population) and enrichment (the proliferation of the strong members) in a single recursive procedure without having to analyse an entire population. Note that by 'weak' we denote conformations having low weight and vice versa for 'strong'.

Now consider a polymer which has a relatively large weight. This is 'enriched' in the following way. We create two members of this polymer at the next step (i.e. when adding the next bead) and give them each a weight which is half of the weight of their 'parent'. This causes the total weight of this conformation in the population to be constant. Suppose we have an upper limit (called 'UpLim') of the weight above which polymers are enriched. The recursive code would then contain the following lines:

```
IF (PolWeight>UpLim) THEN
    NewWeight = 0.5*PolWeight;
    AddBead(Polymer, NewWeight, L+1);
    NewWeight = 0.5*PolWeight;
    AddBead(Polymer, NewWeight, L+1);
END IF
```

This does precisely what we want. Note the copy we have made of PolWeight into NewWeight. This is necessary when the call to AddBead would change its PolWeight variable on exit; this depends on the type of call and on the possibilities offered by the computer language.

Next we consider pruning. This means we must remove weak members from the population, but we are not allowed to change the distribution. This is realised by removing weak polymers with probability $1/2$, and multiplying the weight of those that happen not to be removed by a factor of 2. The criterion for removal is determined by a lowest weight LowLim. This is done in the following piece of code:

```
IF (PolWeight<LowLim) THEN
    Choose random number R uniformly between 0 and 1;
    IF (R<0.5) THEN
        NewWeight = 2*PolWeight;
        AddBead(Polymer, NewWeight, L+1);
    END IF;
END IF
```

The choice of UpLim and LowLim determine whether the population will grow, shrink or remain stationary. The right choice depends on the average weight 'AvWeight' at step L of the procedure. This average is updated for every new polymer reaching this length. UpLim and LowLim are then taken as multiples of the ratio of this average weight and the weight 'Weight3' at the shortest length (3 beads):

$$\text{UpLim} = \alpha * \text{AvWeight}/\text{Weight3}. \tag{10.53}$$

and similar for LowLim. A good value for α for UpLim is 2, and for LowLim we take 1.2. The values of α may depend on the level L. It is possible to remove this dependence by multiplying the polymer weight by a constant at each addition step.

In this example that constant should be near $1/(0.75N_\theta)$, where N_θ is the number of choices for the angle θ (6 in the example).

PROGRAMMING EXERCISE

You can now try to code the PERM algorithm. If you have done it correctly you should be able to reproduce the double crosses in Figure 10.3. As you can see, they fall on top of the theoretical fit which has $\nu = 3/4$.

It should be mentioned that there exist many more methods than described in this subsection. In particular we mention the configurational bias Monte Carlo (CBMC) method by [43].

10.6.2 Tempering and replica exchange

If we want to simulate a disordered system at low temperatures, we run into the problem that many low-energy states exist. Once the system finds itself in phase space near such a low-energy state, it can escape only with great difficulty: the system is trapped. In order to sample the phase space, we must visit a large set of minima accessible at the relevant temperature, and therefore we should somehow 'push' the system over the barriers separating the minima. The problem which arises is that pushing the system over the barrier is likely to destroy the detailed balance condition. Methods preserving detailed balance and moving the system efficiently through phase space are the *simulated tempering Monte Carlo* [44], the *replica Monte Carlo* [45, 46] and the *replica exchange* [47] method. All these methods are based on the idea that the system can move from one minimum to another by repetitive heating and cooling. The problem is to heat and cool efficiently such that moves to different temperatures have a reasonable acceptance probability and that overall detailed balance is preserved. We briefly describe the replica exchange method in this section.

M replicas of the system under consideration are simulated in parallel, each at a different (inverse) temperature $\beta_1, \beta_2, \ldots, \beta_M$. We take $\beta_1 < \beta_2 < \ldots < \beta_M$. We now alternate a number of ordinary MC steps for each of the replicas by an exchange of the replica configurations, or, equivalently, of the replica temperatures. Exchanging the temperatures just involves assigning different numbers to the temperatures in the two replicas but the temperatures are then no longer ordered. Exchanging the configurations will be more time-consuming.

The partition function describing the collection of replicas is given as

$$Z = \prod_{m=1}^{M} Z(\beta_m) = Z(\beta_1)Z(\beta_2)\ldots Z(\beta_M). \tag{10.54}$$

This can also be written as

$$Z = \sum_{X_m;m=1,\ldots,M} \exp\left(-\sum_{m=1}^{M} \beta_m E_{X_m}\right), \tag{10.55}$$

where X_m denotes a system configuration of replica m. We can interpret this partition function as that of a large system, encompassing all replicas, with a 'Boltzmann factor'

$$P(\beta_1, X_1; \ldots; \beta_M, X_M) = \exp\left(-\sum_{m=1}^{M} \beta_m E_{X_m}\right). \tag{10.56}$$

Note that the β_m are stochastic variables in this formulation. They can exchange their values but are subject to the condition that the set of values taken on by the β_m remains the same. For this ensemble we perform a Metropolis Monte Carlo algorithm. Suppose we let the replicas evolve independently according to the standard Metropolis algorithm. Then obviously detailed balance is satisfied. After a number of steps, however, we exchange the configurations (or the temperatures) pair-wise according to

$$\beta_n, X_n; \beta_m, X_m \rightarrow \beta_n, X_m; \beta_m, X_n. \tag{10.57}$$

This means that a low-temperature replica receives a high(er) temperature configuration. At higher temperatures, in particular above the phase transition temperature, the system moves easily over all the free energy barriers in phase space. If the low-temperature replicas move to such high temperatures and back, their configuration will in general have moved to different (free) energy minima.

We calculate the expectation values of physical quantities A at any of the temperatures by averaging over all replica configurations at that particular temperature (perhaps omitting the first few MCS just after a temperature change). Note that exchanges are performed only between adjacent temperatures (β_m and β_{m+1} if the temperatures are ordered), as the acceptance rate decreases exponentially with temperature difference.

Let us calculate the transition probability for temperature exchange. The ratio between a forward and a backward move is given by

$$\frac{T(\beta_n, X_n; \beta_m, X_m \rightarrow \beta_n, X_m; \beta_m, X_n)}{T(\beta_n, X_m; \beta_m, X_n \rightarrow \beta_n, X_n; \beta_m, X_m)} = \exp[-\beta_m(E_{X_n} - E_{X_m}) - \beta_n(E_{X_n} - E_{X_m})]$$

$$= \exp[-(\beta_m - \beta_n)(E_{X_n} - E_{X_m})]. \tag{10.58}$$

This implies that a trial step in which the temperatures are exchanged is accepted with a probability $\min\{1, \exp[-(\beta_m - \beta_n)(E_{X_n} - E_{X_m})]\}$.

For the method to be successful it is necessary that all Metropolis steps have a reasonable acceptance. It is clear that the acceptance rate is sensitive to the difference between adjacent temperatures. For the acceptance rates to be of order 1, the exponent

$$-(\beta_m - \beta_n)(E_{X_n} - E_{X_m}) \tag{10.59}$$

should be of order 1. Note that before the exchange, E_{X_n} will be close to the equilibrium energy at temperature β_n, and similarly for E_{X_m}. This implies that their difference will be approximately

$$E_{X_n} - E_{X_m} \approx C(\beta_m - \beta_n), \tag{10.60}$$

where C is the (total) specific heat, evaluated near the temperatures β_m and β_n (we assume that $C(\beta)$ does not vary too strongly between both temperatures – note that this is not justified if the temperatures are close to a critical phase transition). We see that for the acceptance rate to be of order 1, we should have

$$\Delta \equiv (\beta_m - \beta_n)(E_{X_n} - E_{X_m}) \approx C(\beta_m - \beta_n)^2 = \mathcal{O}(1). \tag{10.61}$$

From the fact that the total specific heat is an extensive quantity, we see that

$$\beta_m - \beta_n \sim \frac{1}{\sqrt{N}}. \tag{10.62}$$

In practice, the set of β values is determined dynamically, such as to make the acceptance rate approximately constant. This is done as follows. We start with a set of temperatures $\{\beta_n\}$. For these we perform a number of MCS and replica exchanges. The acceptance rates p_m for temperature β_m are stored and then the latter are updated according to the recipe:

$$\beta_1^{\text{new}} = \beta_1;$$

$$\beta_m^{\text{new}} = \beta_{m-1}^{new} + (\beta_m - \beta_{m-1})\frac{p_m}{p_{\text{target}}},$$

where p_{target} is the 'target' acceptance rate which is taken equal for all temperatures. In this way we ensure that the replicas will cycle through all temperatures.

This method has been applied very successfully to spin-glasses and many other examples. For a review, see Ref. [48].

10.6.3 Walking in a rough landscape

In everyday life, we often encounter the problem of finding the optimum solution to a problem among many candidate solutions. An example is the well-known 'travelling salesperson problem' (TSP) which consists of finding the shortest path connecting a set of cities to be visited by a salesperson. This problem is related to design problems in electronic circuits, where the wiring must be done as efficiently as possible. The TSP is an example of so-called combinatorial optimisation problems, as the aim is to find the optimum among all permutations of the cities. Another type of problems is formulated in a continuum phase space. An example is finding the minimum-energy conformation of a polymer. In particular, the problem of protein folding has received much attention. Although the equilibrium conformations of a protein are usually determined by the minimum of the free energy, in many approaches it is a (sometimes phenomenological) potential energy which is to be minimised.

The problems described here have something in common: it is possible to find many good solutions, but there is only one, or at most a few optimal ones. To define the problem, we should first specify what makes a solution the 'best'. This is done by assigning to each possible candidate solution a 'merit function' (or 'fitness function'). In the case of the TSP, the merit function is the length of a path. In the case of the polymer conformation it is the potential energy. The optimum solution is defined as the one with the lowest value of the merit function (if the problem is to find the maximum of some quantity we choose the negative of that quantity to be the merit function).

Now we can define the problems in a more abstract way. It is convenient to consider continuum problems. The candidate solutions (for example the possible conformations) form a phase space, and the merit function has some complicated shape on that space – it contains many valleys and mountains, which can be very steep. The solution we seek corresponds to the lowest valley in the landscape. Note that the landscape is high-dimensional. You may think, naively, that a standard numerical minimum finder can solve this problem for you. However, this is not the case as such an algorithm always needs a starting point, from which it finds the nearest *local* minimum, which is not necessarily the best you can find in the conformation space. The set of points which would go to one particular local minimum when fed into a steepest descent or other minimum-finder (see Appendix A4) is called the *basin of attraction* of that minimum. Once we are in the basin of attraction of the global minimum we can easily find this global minimum; the problem is to find its basin of attraction.

There exist many methods for dealing with this problem. In this section we review a few of these only briefly. One method is to generate configurations at random or on a regular grid in the (high-dimensional) phase space, and then finding the nearest local minima for all these point using a standard function minimisation such as the conjugate gradient method (see Appendix A4). It is however possible to have the simulation let the system probe preferentially those regions where the energy is low. Many of these approaches are based on the ideas presented earlier in this chapter. Note that we simply want to find a (near)-global minimum of the merit function – detailed balance is no longer a concern.

Suppose that you were to find the minimum energy of a polymer studied in the previous subsection. You then could generate a low-temperature ensemble and, for each conformation you encounter, find the nearest local minimum. This has been done by Grassberger for polymer chains consisting of two different types of beads [49]. This method is closely related to *simulated annealing* [50] which is a special version of the Metropolis algorithm. In simulated annealing, the merit function is viewed as an energy. Applying the standard Metropolis ensemble, we would generate configurations distributed according to the Boltzmann factor $\exp[-E/(k_B T)]$. In simulated annealing, we slowly cool the system down. The idea behind this is that at higher temperatures, the system can easily hop over the barriers, thereby probing a large part of the phase space. On cooling the system, it samples only the lower energy domains and finally ends up in one of the deepest valleys. It is obviously efficient to use various samples of the system, so that in fact a population of systems is cooled down and we finally take the one that has reached the lowest energy. Alternatively, one could heat up the system again and then anneal it once more and so on. It is always advisable in these simulations to combine the stochastic algorithm with a deterministic minimum finder such as the conjugate gradient method, in order to efficiently find the deepest point of an attraction basin from each point visited.

The idea behind these method is called 'configurational search' for obvious reasons. Another concept in this field is 'energy sculpting'. This trick tries to overcome the difficulty that a method focusing on the low-energy parts of the phase space will automatically avoid the (sometimes) high (free) energy barriers separating the different energy minima. This is done by replacing the energy (i.e. the merit function) by a modified one. An extreme version of this is the *basin hopping* method by Doye and Wales which was used to find optimal conformations of Lennard–Jones clusters [51]. In this method, the energy of a point is replaced by the energy minimum of the attraction basin the point is in. This implies that all points in an attraction basin have the same (modified) energy. The energy steps up or down when moving from one attraction basin to another – the barriers between the basins are entirely removed.

Finally, another method deserves mention: the genetic algorithm (GA) [52, 53]. This algorithm is inspired by the ideas of evolution theory and employs these to find optimal solutions to combinatorial or continuous problems. These algorithms are based on encoding any point in phase space into a linear chain. This can be a binary chain. For example, let us suppose that our merit function f depends on N variables x_j:

$$f = f(x_1, \ldots, x_N). \tag{10.63}$$

We restrict the range of acceptable values x_j to some finite interval. Within this interval, we allow for a number (256 or 512, say) of different equidistant values. These values can be coded as a bit-string. It is also possible to run the algorithm with the string of reals $x_1, \ldots, x_N$. The algorithm now manipulates a population of such strings. We start with a pool of M individuals. Then we do the following steps.

```
WHILE No acceptable solution found DO
    Calculate fitness (merit function) of all individuals;
    Enrich the fit ones by letting them create identical clones;
    Weed out the individuals with low fitness;
    Mate randomly chosen pairs of the population and
            and have them create offspring;
    Mutate;
END WHILE.
```

The enrichment through cloning and the weeding of low-fitness individuals is carried out along the same lines as the PERM algorithm discussed above. The size of the population should remain more or less constant in this process. Mating is a process in which the two members of a pair of chains are cut into two pieces at some randomly chosen chain position. The left piece cut off from chain 1 is then connected to the right part of chain 2 and vice versa.

Mutation is a process in which one of the bits is chosen at random and then flipped. This is necessary to keep variety in the population. The necessity for this is seen by considering the case in which some segment is the same in all chains. This segment would always remain the same in all steps except when mutation takes place. This interesting method has been used for a large variety of problems [52].

*10.7 The temperature of a finite system

We conclude this chapter with an intriguing aspect of the simulation of finite systems. In molecular dynamics, the microcanonical ensemble is the 'default ensemble', as the solution of the equations of motion leaves the total mechanical energy constant. On the other hand, in Monte Carlo, the Metropolis algorithm naturally leads to the canonical ensemble. We know from equilibrium statistical mechanics that the different ensembles are 'equivalent', which means that physical quantities evaluated with the same values of the thermodynamic quantities in different ensembles are the same up to corrections of order $1/N$. For finite systems, the two therefore cannot be compared to very high accuracy. However, with a careful analysis of the proper definitions of the temperature, we can make comparisons between the two as we shall now show.

Traditionally, the temperature in microcanonical molecular dynamics is calculated from the equipartition theorem. There is however a subtlety in that the number of degrees of freedom for an N-particle system is not $3N$, but $3N - 3$, where the three degrees of freedom of the centre of mass must be subtracted as they are fixed in the molecular dynamics algorithm – only the *internal* momenta contribute to the temperature. This procedure thus leads to the expression

$$k_\mathrm{B}T = 2\frac{\langle K\rangle}{3N-3} \tag{10.64}$$

for the temperature (K is the total kinetic energy). Although this expression is widely used, it is not correct within the microcanonical ensemble if we use the thermodynamic definition of temperature

$$\frac{1}{T} = \left(\frac{\partial S}{\partial E}\right)_{N,V}. \tag{10.65}$$

where $S = k_\mathrm{B} \ln \Omega$, together with the standard expression for Ω:

$$\Omega = k_\mathrm{B} \ln\left[\sum_{\text{allstates}} \delta(E - H_\text{state})\right], \tag{10.66}$$

where H_state is the Hamiltonian of the system.

For a system of N particles in three dimensions, taking into account that the total momentum is conserved, this leads to

$$\Omega = \frac{1}{h^{3N-3}N!}\int \delta[E - H(P,R)]\,\mathrm{d}^{3N-3}P\,\mathrm{d}^{3N}R, \tag{10.67}$$

where h is Planck's constant. Note, however, that other expressions for the entropy can also be used [54]. Working out the derivative of this entropy with respect to the energy as prescribed by (10.65) does not seem very easy, but it can be done when we first use the explicit quadratic expression for the kinetic energy in the expression for the entropy:

$$\Omega = \frac{1}{h^{3N-3}N!}\int \delta\left[E - \frac{p^2}{2m} - V(R)\right]\mathrm{d}^{3N-3}P\,\mathrm{d}^{3N}R. \tag{10.68}$$

Now we write $\mathrm{d}^{3N-3}P = \omega(3N-3)p^{3N-4}\,\mathrm{d}p$ where $p = |P|$ and $\omega(3N-3)$ is the surface of a hypersphere in $3N-3$ dimensions. Furthermore, substituting $K = p^2/(2m)$, the integral becomes

$$\Omega = \frac{1}{h^{3N-3}N!}\int \delta[E - K - V(R)]\omega(3N-3)(2mK)^{(3N-3)/2}\frac{\mathrm{d}K}{2K}\,\mathrm{d}^{3N}R. \tag{10.69}$$

Working out the delta function then leads to

$$\Omega = \frac{\omega(3N-3)}{h^{3N-3}N!}\int [2m(E-V)]^{(3N-5)/2} d^{3N}R. \tag{10.70}$$

Taking the derivative of $S = k_B \ln \Omega$ with respect to energy then leads to the following expression for the temperature:

$$\frac{1}{k_B}\frac{1}{T} = \frac{1}{k_B}\frac{\partial S}{\partial E} = \frac{3N-5}{2}\frac{\int (E-V)^{(3N-7)/2} d^{3N}R}{\int (E-V)^{(3N-5)/2} d^{3N}R}$$
$$= \frac{3N-5}{2}\left\langle \frac{1}{E-V}\right\rangle = \frac{3N-5}{2}\left\langle \frac{1}{K}\right\rangle. \qquad (10.71)$$

The difference between this expression for the temperature and that obtained from the equipartition theorem is of the order of $1/N$, so it is not obvious why people bother about subtracting the 3 from the total number of degrees of freedom when the adopted convention still differs by an order of $1/N$ from the correct value. However, the good reason for sticking to this convention is that it is adopted in most of the MD codes.

We now consider the question how properties obtained in, say, the canonical ensemble, compare with those obtained in the microcanonical ensemble. When we calculate the expectation value of some physical property in one ensemble for particular values of the system parameters, we can measure the expectation values of the variables conjugate to the system parameters and then evaluate the expectation value of the property at hand in any other ensemble. As we have already noted in the beginning of this section, the differences between values obtained in different ensembles will be of the order of $1/N$. To be specific, we may perform a molecular dynamics simulation in the microcanonical ensemble at some energy E^* and calculate some configurational average. If on the other hand we calculate the same configurational average in a canonical Monte Carlo simulation, we should find differences of order $1/N$. We shall now show that it is possible to calculate this difference analytically if the quantity under consideration is the total energy.

The expectation value of the total energy in the canonical ensemble is

$$\langle E\rangle = \frac{\sum_i E_i e^{-\beta E_i}}{\sum_i e^{-\beta E_i}}. \qquad (10.72)$$

The sum can be rewritten when we collect terms with the same energy. According to the definition of entropy we have $\exp[S(E)/k_B]$ states at energy E:

$$\langle E\rangle = \frac{\int E e^{-\beta E + S(E)/k_B}\, dE}{\int e^{-\beta E + S(E)/k_B}\, dE}. \qquad (10.73)$$

We can approximate the term $\rho(E) = \exp[-\beta E + S(E)/k_B]$ as follows:

$$\rho(E) \approx \exp\left[-\beta E^* + S(E^*)/k_B + \frac{\Delta E^2}{2k_B}\frac{\partial^2 S(E^*)}{\partial E^2} + \Delta E\frac{\partial S(E^*)}{\partial E} + \frac{\Delta E^3}{6k_B}\frac{\partial^3 S(E^*)}{\partial E^3}\right]. \qquad (10.74)$$

This is a Taylor expansion of the exponent around its maximum E^* which satisfies

$$\beta = \frac{1}{k_B}\frac{\partial S(E^*)}{\partial E}. \tag{10.75}$$

We now introduce the two parameters

$$\alpha = \frac{1}{k_B}\frac{\partial^2 S(E^*)}{\partial E^2}; \quad \gamma = \frac{1}{k_B}\frac{\partial^3 S(E^*)}{\partial E^3}. \tag{10.76}$$

The expectation value of the energy can be evaluated straightforwardly in terms of these parameters:

$$\langle E\rangle = \frac{\int_{-\infty}^{\infty} e^{-\alpha\Delta E^2}(1+\gamma\Delta E^3)(E^*+\Delta E)\,d\Delta E}{\int_{-\infty}^{\infty} e^{-\alpha\Delta E^2}(1+\gamma\Delta E^3)\,d\Delta E}. \tag{10.77}$$

The leading term is simply E^*. Realising that only even powers of ΔE survive in the Gaussian integrals, we obtain for the correction

$$\langle E\rangle = E^* + \frac{\int_{-\infty}^{\infty} e^{-\alpha\Delta E^2}\gamma\Delta E^4\,d\Delta E}{\int_{-\infty}^{\infty} e^{-\alpha\Delta E^2}\,d\Delta E} = E^* - \frac{3\gamma}{4\alpha^2}. \tag{10.78}$$

The derivatives α and γ can be determined in a way similar to that used to find the temperature calculation above, with the results

$$\alpha = \frac{3N-5}{4}\left[(3N-7)\left\langle\frac{1}{K^2}\right\rangle - (3N-5)\left\langle\frac{1}{K}\right\rangle^2\right], \tag{10.79}$$

and

$$\gamma = \frac{(3N-5)(3N-7)(3N-9)}{8}\left\langle\frac{1}{K^3}\right\rangle - 3\frac{(3N-5)^2(3N-7)}{8}\left\langle\frac{1}{K^2}\right\rangle\left\langle\frac{1}{K}\right\rangle + 1\frac{(3N-5)^3}{4}\left\langle\frac{1}{K}\right\rangle^3. \tag{10.80}$$

From the fact that S and E are both extensive variables, we see that $\alpha \sim 1/N$ (which ensures that the energy fluctuations are of order $1/\sqrt{N}$), whereas $\gamma \sim 1/N^2$. Therefore, the relative correction to the energy $3\gamma/(4\alpha^2)$ is still of order $1/N$. This is significant when N is not too large.

Armed with these expressions it is possible to relate the microcanonical energy E^* to the canonical one. We now give results for a test run involving only eight particles, as for this number the differences are very clear. Accurate simulations for particles with Gaussian repulsion $V(r) = \exp(-4r^2)$ have been performed. We have used this potential because it smooth and does not suffer from the periodicity (it decays rapidly). The kinetic energy K and the expectation values of $1/K$, $1/K^2$

and $1/K^3$ are determined in a molecular dynamics simulation at constant energy in order to calculate α and γ according to (10.79) and (10.80).

Using a target temperature (according to the equipartition theorem) of 1.0, a (measured) temperature of 1.043 (using the correct definition (10.71)), and a total energy of 10.913 ± 0.003 (in reduced units) have been found in our simulation. This is the energy E^*. In the canonical ensemble, we should add the correction $3\gamma^2/(4\alpha^2)$. Adding the correction, we obtain an energy of $11.94 + 0.03$ in natural units. An MC calculation with a temperature of 1.043 gives $E = 11.96$, in good agreement with the prediction. The statistical errors in α and γ are difficult to calculate because the inverse powers of the kinetic energy are correlated quantities. The best method is to calculate the correction from several independent simulations and infer the error from these results, or by data-blocking of the correction over a long run (see Section 7.4). Note that in order to calculate the canonical kinetic energy, $K = 3(N-1)k_B T/2$ should be used rather than $3Nk_B T/2$ in order to relate the energy to the molecular dynamics energy which has three degrees of freedom less (in the canonical ensemble, the equipartition is satisfied even for small particle numbers).

Obviously, the practical value of this calculation is limited – the main point is to show that careful analysis is necessary and possible for systems consisting of small numbers of particles (such as droplets).

Exercises

10.1 Consider a Monte Carlo algorithm for the two-dimensional Ising model in which the sites are scanned in lexicographic order, that is, each row is scanned from left to right, starting with the top row and proceeding towards the bottom row. We want to show that this method satisfies the detailed balance criterion. A sweep through the entire lattice is considered as a single step in the Markov chain.

(a) Explain why this Markov chain is ergodic.
The proof that the transition probabilities satisfy detailed balance is done by recursion. Suppose that the lattice contains N sites, and that for the lattice containing $N-1$ sites the algorithm satisfies detailed balance.

(b) Show that if the Nth spin is flipped according to the usual Metropolis algorithm, the sweep over the lattice with N sites satisfies detailed balance.

10.2 [C] Code the heat-bath algorithm for the Ising model and analyse the correlation time (see Section 7.4), in particular close to the critical temperature. Compare the results with the standard Metropolis algorithm.

10.3 Consider the Norman–Filinov method for a system with a large chemical potential. From Eqs. (10.37) and (10.38) we see that in that case the acceptance rate for creation is much smaller than that for annihilation. Therefore we use a generalised

Metropolis method, in which creations are tried much more often than annihilations. Suppose the trial probabilities for creation and annihilation are P_C and P_A respectively.

Show that the acceptances should be modified as follows:

- The acceptance probability for creation is

$$P_{acc} = \min(1, q_{XX'})$$

with

$$q_{XX'} = e^{-\beta \Delta U^+} \Lambda^{-3} V/(N+1) e^{\beta \mu} \frac{P_A}{P_C}.$$

- Similarly for annihilation:

$$P_{acc} = \min(1, q_{XX'})$$

with

$$q_{XX'} = e^{\beta \Delta U^-} \Lambda^3 N/V e^{-\beta \mu} \frac{P_C}{P_A}.$$

Show that this modification can be implemented by a suitable shift in the chemical potential. Find this shift.

10.4 [C] Consider the methane (CH_4) molecule of Problem 8.12. In that problem we have given the potential energy of the molecule in terms of stretching and bending terms. Write a Monte Carlo simulation for simulating this molecule at a given temperature. Compare the results with those obtained in Problem 8.12.

References

[1] J. M. Hammersley and D. C. Handscomb, *Monte Carlo Methods*. London, Methuen, 1964.

[2] M. P. Allen and D. J. Tildesley, *Computer Simulation of Liquids*. Oxford, Oxford University Press, 1989.

[3] K. Binder, ed., *Applications of the Monte Carlo Method in Statistical Physics, Topics in Current Physics*, vol. 36. Berlin, Springer, 1984.

[4] K. Binder, ed., *Monte Carlo Methods in Statistical Physics*, 2nd edn. *Topics in Current Physics*, vol. 7. Berlin, Springer, 1986.

[5] M. H. Kalos and P. A. Whitlock, *Monte Carlo Methods*. New York, John Wiley, 1986.

[6] K. Binder and D. W. Heermann, eds., *Monte Carlo Simulation in Statistical Physics*. New York, Springer, 1988.

[7] G. T. Barkema and M. E. J. Newman, *Monte Carlo Methods in Statistical Physics*. Oxford, Oxford University Press, 1999.

[8] D. Frenkel, *Monte Carlo Simulations*. Utrecht, Van 't Hoff laboratory, University of Utrecht, The Netherlands, 1988.

[9] F. James, 'Monte Carlo theory and practice,' *Rep. Prog. Phys.*, **43** (1980), 1145–89.

[10] N. Metropolis, A. W. Rosenbluth, M. N. Rosenbluth, A. H. Teller, and E. Teller, 'Equation of state calculations by fast computing machines,' *J. Chem. Phys.*, **21** (1953), 1087–92.

[11] W. W. Wood and J. D. Jacobsen, '*Monte Carlo calculations in statistical mechanics*,' *Proceedings of the Western Joint Computer Conference*, New York, San Francisco Institute of Radio Engineers, 1959, pp. 261–9.

[12] J. A. Barker, 'Monte Carlo calculations of the radial distribution functions for a proton-electron plasma,' *Aust. J. Phys.*, **18** (1965), 119–33.

[13] P. C. Hohenberg and B. I. Halperin, 'Theory of dynamic critical phenomena,' *Rev. Mod. Phys.*, **49** (1977), 435–79.

[14] K. Kawasaki, 'Kinetics of Ising models,' in *Phase Transitions and Critical Phenomena* (C. Domb and M. S. Green, eds.). London, Academic Press, 1972.

[15] W. K. Hastings, 'Monte Carlo methods using Markov chains, and their applications,' *Biometrika*, **57** (1970), 97–109.

[16] J. Liu and E. Luijten, 'Rejection-free geometric cluster algorithm for complex fluids,' *Phys. Rev. Lett.*, **92** (2004), 035504.

[17] M. Creutz, *Quarks, Gluons and Lattices*. Cambridge, Cambridge University Press, 1983.

[18] W. W. Wood, 'Monte Carlo calculations for hard disks in the isothermal-isobaric ensemble,' *J. Chem. Phys.*, **48** (1968), 415–34.

[19] W. W. Wood, 'NpT-ensemble Monte Carlo calculations for the hard disk fluid,' *J. Chem. Phys.*, **52** (1970), 729–41.

[20] I. R. McDonald, 'Monte Carlo calculations for one- and two-component fluids in the isothermal–isobaric ensemble,' *Chem. Phys. Lett*, **3** (1969), 241–3.

[21] I. R. McDonald, 'NpT-ensemble Monte Carlo calculations for binary liquid mixtures,' *Mol. Phys.*, **23** (1972), 41–58.

[22] R. Eppenga and D. Frenkel, 'Monte Carlo study of the isotropic and nematic phases of infinitely thin hard platelets,' *Mol. Phys.*, **52** (1984), 1303–34.

[23] G. E. Norman and V. S. Filinov, 'Investigations of phase transitions by a Monte Carlo method,' *High Temp. (USSR)*, **7** (1969), 216–22.

[24] Z. W. Salsburg, J. D. Jacobson, W. Ficket, and W. W. Wood, 'Application of the Monte Carlo method to the lattice gas model. I. Two dimensional triangular lattice,' *J. Chem. Phys.*, **30** (1959), 65–72.

[25] D. A. Chesnut, 'Monte Carlo calculations for the two-dimensional triangular lattice gas: supercritical region,' *J. Chem. Phys.*, **39** (1963), 2081–4.

[26] Y. Saito and H. Müller-Krumbhaar, '2-Dimensional Coulomb gas: a Monte Carlo study,' *Phys. Rev. B*, **23** (1981), 308–15.

[27] M. Mezei, 'A cavity-based ($TV\mu$) Monte Carlo method for the computer simulation of fluids,' *Mol. Phys.*, **40** (1980), 901–6.

[28] D. Frenkel, 'Free energy computation and first-order phase transitions,' in *Molecular Dynamics Simulation of Statistical Mechanical Systems* (G. Ciccotti and W. G. Hoover, eds.), *Proceedings of the International School of Physics "Enrico Fermi", Varenna 1985*, vol. 97, Amsterdam, North-Holland, 1986, pp. 151–88.

[29] A. Z. Panagiotopoulos, 'Direct determination of phase coexistence properties of fluids by Monte Carlo simulation in a new ensemble,' *Mol. Phys.*, **61** (1987), 813–26.

[30] A. Z. Panagiotopoulos, N. Quirke, and D. J. Tildesley, 'Phase-equilibria by simulation in the Gibbs ensemble: Alternative derivation, generalization and application to mixture and membrane equilibria,' *Mol. Phys.*, **63** (1988), 527–45.

[31] A. Z. Panagiotopoulos, 'Adsorption and capillary condensation of fluids in cylindrical pores by Monte Carlo simulation in the Gibbs ensemble,' *Mol. Phys.*, **62** (1987), 701–19.

[32] D. Frenkel and B. Smit, *Understanding Molecular Simulation*. San Diego, Academic Press, 1996.

[33] C. H. Bennett, 'Efficient estimation of free energy differences from Monte Carlo data,' *J. Comput. Phys.*, **22** (1976), 245–68.

[34] G. M. Torrie and J. P. Valleau, 'Nonphysical sampling distributions in Monte Carlo free energy estimation: umbrella sampling,' *J. Comp. Phys.*, **23** (1977), 187–99.

[35] G. M. Torrie and J. P. Valleau, 'Monte Carlo study of a phase separating liquid mixture by umbrella sampling,' *J. Chem. Phys.*, **66** (1977), 1402–8.

[36] B. Widom, 'Some topics in the theory of fluids,' *J. Chem. Phys.*, **39** (1963), 2808–12.

[37] K. S. Shing and K. E. Gubbins, 'The chemical potential in dense fluids and fluid mixtures via computer simulation,' *Mol. Phys.*, **46** (1982), 1109–28.

[38] M. Fixman, 'Direct simulation of the chemical potential,' *J. Chem. Phys.*, **78** (1983), 4223–6.

[39] J. Baschnagel, J. P. Wittmer, and H. Meyer, 'Monte Carlo simulation of polymers: coarse-grained models,' in *Computational Soft Matter: From Synthetic Polymers to Proteins. Lecture Notes* (N. Attig, K. Binder, H. Grubmüller, and K. Kremer, eds.), NIC Series, vol. 23. Jülich, John von Neumann Institute for Computing, 2004, pp. 83–140.

[40] K. Kremer and K. Binder, 'Monte Carlo simulation of lattice models for macromolecules,' *Comp. Phys. Commum.*, **7** (1988), 259–310.

[41] M. N. Rosenbluth and A. W. Rosenbluth, 'Monte Carlo calculation of the average extension of molecular chains,' *J. Chem. Phys.*, **23** (1955), 356–9.

[42] P. Grassberger, 'Pruned-enriched Rosenbluth method: Simulation of θ polymers of chain length up to 1 000 000,' *Phys. Rev. E*, **56** (1997), 3682–93.

[43] J. I. Siepmann and D. Frenkel, 'Configurational-bias Monte Carlo: a new sampling scheme for flexible chains,' *Mol. Phys.*, **75** (1992), 59–70.

[44] E. Marinari and G. Parisi, 'Simulated tempering: a new Monte Carlo scheme,' *Europhys. Lett.*, **19** (1992), 451–8.

[45] R. H. Swendsen and J.-S. Wang, 'Replica Monte Carlo simulation of spin-glasses,' *Phys. Rev. Lett.*, **57** (1986), 2607–9.

[46] C. Geyer, 'Markov chain Monte Carlo maximum likelihood,' in *Computing Science and Statistics: Proceedings of the 23rd Symposium on the Interface* (E. Keramidas, ed.), Fairfax Station, Interface Foundation of America, 1991, pp. 156–63.

[47] K. Hukushima and K. Nemeto, 'Exchange Monte Carlo method and application to spin glass simulation,' *J. Phys. Soc. Jpn.*, **65** (1996), 1604–8.

[48] D. J. Earl and M. W. Deem, 'Parallel tempering: theory, applications, and new perspectives.' physics/0508111, 2005.

[49] H.-P. Hsu, V. Mehra, and P. Grassberger, 'Structure optimization in an off-lattice protein model,' *Phys. Rev. E*, **68** (2003), 037703.

[50] S. Kirkpatrick, C. D. Gelatt, and M. P. Vecchi, 'Optimization by simulated annealing,' *Science*, **220** (1983), 671–80.

[51] D. J. Wales and J. P. K. Doye, 'Global optimization by basin-hopping and the lowest energy structures of Lennard–Jones clusters containing up to 110 atoms,' *J. Phys. Chem. A*, **101** (1997), 5111–16.

[52] C. R. Reeves and J. E. Rowe, *Genetic Algorithms: Principles and Perspectives*. Dordrecht, Kluwer, 2003.

[53] J. H. Holland, *Adaptation in Natural and Artificial Design*. Ann Arbor, The University of Michigan Press, 1975.

[54] K. Huang, *Statistical Mechanics*, 2nd edn. New York, John Wiley, 1987.

11

Transfer matrix and diagonalisation of spin chains

11.1 Introduction

In Chapters 8 and 10 we studied methods for simulating classical systems consisting of many interacting degrees of freedom. In these methods, a sequence of system configurations is generated, and from this sequence averages of physical quantities, given as functions of the degrees of freedom (positions and momenta, or spins), can be determined. These quantities are called mechanical quantities, and the expressions of their expectation values are called *mechanical averages*.

There exist, however, quantities that cannot be determined straightforwardly using these methods. These quantities include free energies and chemical potentials. The point is that these quantities are not given as a *normalised* average, which for mechanical quantities is replaced by an average over the configurations generated in the simulation. In the previous chapters we have seen that it is not straightforward to find free energies and chemical potentials using MC and MD methods (see Section 10.5).

In this chapter we discuss a method which enables us to find free energies for lattice spin models with very high accuracy; this is the *transfer matrix method.* This method calculates the free energy of a model defined on a strip of finite width and infinite length directly in terms of the largest eigenvalue of a large matrix, the *transfer matrix.* This matrix contains essentially the Boltzmann weights for adding an extra row of spins to the strip. In Chapter 12 we shall see that the transfer matrix is the analogue of the time evolution operator in quantum mechanics. The fact that we can apply matrix methods to both quantum mechanics problems and statistical problems results from the fact that statistical and quantum mechanics are intimately related, as will be shown in the next chapter.

The transfer matrix is an operator acting in a dimension which is one lower than the dimension of the original, classical system. In this lower dimensional space, we must solve a quantum problem. This can be an interesting application by itself, even without reference to a transfer matrix relating the quantum problem to a classical one. We shall discuss one-dimensional quantum problems extensively in this chapter. Not too long ago, a new method was developed for treating this problem very efficiently. It is based on renormalisation and on the density matrix – hence its name, *density matrix renormalisation group*. We shall consider a few applications of this method in the last sections of this chapters.

In the next section we solve the one-dimensional Ising model exactly using the transfer matrix. We shall derive the free energy and the pair correlation function from this matrix. In the following section we shall describe the transfer matrix for two-dimensional spin models and see how transfer matrix techniques allow for numerically exact solutions of two-dimensional models on an infinite strip of finite width. This, in combination with finite-size scaling methods, makes the transfer matrix method very useful for obtaining accurate data for two-dimensional statistical models. For details concerning the numerical transfer matrix method, see Ref. [1].

Later sections are then dedicated to numerically diagonalising quantum chains (Section 11.5), to quantum renormalisation methods (11.6), and a particular form of this, the density matrix renormalisation group method (11.7).

11.2 The one-dimensional Ising model and the transfer matrix

In this section we present the exact solution of the one-dimensional Ising model. To arrive at the solution we must introduce a new concept, the *transfer matrix*, whose largest eigenvalue determines the partition function. The one-dimensional Ising model (see also Section 7.2.2) consists of a chain of spins numbered 1 through L; the spins assume values $s_i = \pm$, and the Hamiltonian is given by

$$\mathcal{H}[\{s_i\}] = \sum_{i=1}^{L}(-Js_is_{i+1} - Hs_i). \tag{11.1}$$

J is the coupling strength between the spins, and the external field H favours spins of one particular sign. Periodic boundary conditions are imposed by identifying the spins at sites 1 and $L+1$. The partition function is

$$Z = \sum_{\{s_i\}} \exp\left[\beta \sum_{i=1}^{L}(Js_is_{i+1} + Hs_i)\right]. \tag{11.2}$$

The sum runs over all possible spin configurations $\{s_i\}$.

The exponent can be written as a product over all nearest neighbour pairs:

$$Z = \sum_{\{s_i\}} \prod_{i=1}^{L} \exp[\beta Js_is_{i+1} + \beta Hs_i]. \tag{11.3}$$

It is therefore convenient to define the *transfer matrix* which contains the contribution of the pair s_i, s_{i+1} to the Boltzmann factor of the full chain:

$$T_{s_i,s_{i+1}} = \langle s_i|T|s_{i+1}\rangle = \exp[\beta(Js_is_{i+1} + Hs_i/2 + Hs_{i+1}/2)]. \tag{11.4}$$

The quantum mechanical Dirac notation is convenient here. The contribution of the magnetic field H has been symmetrically distributed over s_i and s_{i+1} in order to arrive at a symmetric transfer matrix.[1] The transfer matrix has the form:

$$T = \begin{pmatrix} e^{\beta(J+H)} & e^{-\beta J} \\ e^{-\beta J} & e^{\beta(J-H)} \end{pmatrix} \tag{11.5}$$

in a representation with basis vectors $|+\rangle = (1, 0)$ and $|-\rangle = (0, 1)$.

Now we can rewrite the partition function as

$$Z = \sum_{\{s_i\}} \langle s_1|T|s_2\rangle\langle s_2|T|s_3\rangle \cdots \langle s_L|T|s_1\rangle. \tag{11.6}$$

The sum is over all $s_i = \pm 1$, so we may write

$$Z = \sum_{s_1=\pm 1} \langle s_1|T^L|s_1\rangle = \mathrm{Tr}\,(T^L), \tag{11.7}$$

where we have used the completeness property $\sum_{s=\pm 1} |s\rangle\langle s| = I$; I is the 2×2 unit matrix.

The transfer matrix is a 2×2 matrix, which has two eigenvalues, λ_0 and λ_1. We then have

$$Z = \lambda_0^L + \lambda_1^L. \tag{11.8}$$

As the partition function is positive, the eigenvalue with the largest absolute value must be positive. Suppose that this eigenvalue is λ_0, then

$$Z = \lambda_0^L + \lambda_1^L \approx \lambda_0^L. \tag{11.9}$$

The approximation becomes exact for the $L \to \infty$, i.e. the thermodynamic limit. Therefore, using $F = k_B T \ln Z$, we obtain the following result for the free energy per spin:

$$F/L = -k_B T \ln \lambda_0. \tag{11.10}$$

In the case that both eigenvalues are equal, the result remains unaltered as can easily be checked. The transfer matrix method is trivially generalised to models with more than two possible values on each site along the chain – for example when the spins can assume more than two values, or when there is more than one spin per site. It is also possible to include further than nearest neighbour interactions. In these cases the matrix will be larger than 2×2, and numerical diagonalisation will be

[1] Other distributions of these interactions are also possible; the physical properties that we shall calculate are not affected by the choice we make here.

necessary, but for matrix sizes up to $10\,000 \times 10\,000$, diagonalisation of the transfer matrix is a trivial numerical exercise (Appendix A8.2).

Let us now calculate the expectation value of the spin at site m. As a result of translation invariance, the result is independent of m. This expectation value is given by

$$\langle s_m \rangle = \frac{\sum_{\{s_i\}} s_m \prod_{i=1}^{L} \exp[\beta J s_i s_{i+1} + \beta H s_i]}{\sum_{\{s_i\}} \prod_{i=1}^{L} \exp[\beta J s_i s_{i+1} + \beta H s_i]}. \tag{11.11}$$

Introducing again the transfer matrix and using the same argument as above to retain the largest eigenvalue λ_0 with eigenvalue ϕ_0 only, we obtain

$$\langle s_m \rangle = \frac{\sum_{s_1} \langle s_1 | T^{m-1} \hat{s}_m T^{L-m+1} | s_1 \rangle}{\sum_{s_1} \langle s_1 | T^L | s_1 \rangle} = \frac{\langle \phi_0 | \hat{s}_m | \phi_0 \rangle \lambda_0^L}{\lambda_0^L} = \langle \phi_0 | \hat{s} | \phi_0 \rangle. \tag{11.12}$$

In a spinor representation, in which the spin-up and -down states are $(1, 0)$ and $(0, 1)$ respectively, the operator $\hat{s}$ is the Pauli matrix σ_z. There is no subscript m in the rightmost expression as the ground state eigenvector ϕ_0 is independent of m.

We can also calculate the correlation function. This is defined as

$$g(m - n) = \langle s_n s_m \rangle - \langle s_m \rangle \langle s_n \rangle. \tag{11.13}$$

Denoting the eigenvalues of T by λ_α (and higher indices) and the corresponding eigenvectors by ϕ_α (for the one-dimensional Ising model, α assumes only the two values 0 and 1), we have

$$\begin{aligned} g(m - n) = \frac{1}{Z} \sum_{\phi_\alpha \cdots \phi_\epsilon} & \langle \phi_\alpha | T^{m-1} | \phi_\beta \rangle \langle \phi_\beta | \hat{s}_m | \phi_\gamma \rangle \langle \phi_\gamma | T^{m-n} | \phi_\delta \rangle \\ & \times \langle \phi_\delta | \hat{s}_n | \phi_\epsilon \rangle \langle \phi_\epsilon | T^{L-n+1} | \phi_\alpha \rangle - \langle \phi_0 | \hat{s} | \phi_0 \rangle^2. \end{aligned} \tag{11.14}$$

The last eigenvector ϕ_α in the first term is the same as the first eigenvector because of the periodic boundary conditions. One is usually interested in the long-range behaviour of the correlation function:

$$1 \ll |m - n| \ll L. \tag{11.15}$$

This suggests that the major contribution to the correlation function comes from replacing all eigenvalues in (11.14) by the largest one, λ_0. However, inspection shows that the two terms in (11.14) cancel exactly for this choice. The major contribution to the correlation function is therefore obtained by replacing the transfer matrices acting between positions n and m by the largest-but-one eigenvalue, λ_1. In the limit (11.15), all the other choices are much smaller. This results in

$$g(m - n) = |\langle \phi_0 | s | \phi_1 \rangle|^2 \left(\frac{\lambda_1}{\lambda_0} \right)^{|m-n|}. \tag{11.16}$$

We see that the correlation drops off exponentially, unless $\lambda_0 = \lambda_1$, that is, unless the largest eigenvalue is degenerate. This implies that critical behaviour can only occur when the largest eigenvalue is degenerate – recall that critical behaviour is characterised by power-law decay of correlation functions.

In this regard Frobenius' theorem is important. This states that the largest eigenvalue of a positive matrix (i.e. a matrix with all positive elements) of finite size is nondegenerate [2]; that is, our correlation functions will never show critical behaviour. The only way to obtain critical behaviour is by having a matrix of infinite size, which is only possible when the degrees of freedom on a site can assume an infinite number of different values. For a one-dimensional model on a lattice (chain) in which the degrees of freedom on a lattice site can assume a finite number of different values, the correlation function is always exponential.

In Chapter 12 we shall see that the analysis applied here carries over to quantum mechanics, the transfer matrix corresponding to the quantum mechanical time evolution operator $\exp(-itH/\hbar)$, where H is now the quantum Hamiltonian. If this Hamiltonian has eigenvalues E_α, we have $\lambda_\alpha = \exp(-itE_\alpha/\hbar)$. After an analytic continuation into imaginary time $t \to -it$, the eigenvalues become real: $\lambda_\alpha = \exp(-tE_\alpha/\hbar)$ and we see that degeneracy of the largest eigenvalue of the transfer matrix (or time-evolution operator) implies degeneracy of the quantum ground state energy. As an example, consider the one-dimensional Ising model in zero magnetic field. It is easy to see that for this model the quantum Hamiltonian occurring in the exponent of the time-evolution operator can be written as

$$H = -\tilde{J}\sigma_x, \tag{11.17}$$

where

$$\sigma_x = \begin{pmatrix} 0 & 1 \\ 1 & 0 \end{pmatrix} \tag{11.18}$$

is the Pauli matrix, and $\tilde{J}$ is related to J by

$$\tanh(\beta\tilde{J}) = \exp(-2\beta J) \tag{11.19}$$

(see Problem 11.1).

In quantum field theory (see Chapter 15), the ground state is the vacuum (no particles present) and the first excited state is the state with one particle at rest. As the relativistic energy is given by $E = \sqrt{p^2c^2 + m^2c^4}$, the gap between these two states is given by the particle rest energy mc^2. Degeneracy of the ground state implies therefore that the particle mass is zero. A critical point is therefore often identified with a vanishing mass.

11.3 Two-dimensional spin models

We shall now describe the transfer matrix analysis for the two-dimensional Ising model; generalisation to models with more spin degrees of freedom is straightforward. Using the transfer matrix method we can calculate the free energy of an Ising model on an infinite strip of width M with periodic boundary conditions (PBC) connecting the two sides of the strip – it is therefore convenient to imagine the strip to be wrapped around a cylinder. The Hamiltonian of the Ising model in this geometry is given by

$$H = \sum_{i=1}^{M} \sum_{j=-\infty}^{\infty} (-Js_{i,j}s_{i,j+1} - Js_{i,j}s_{i+1,j} - Hs_{i,j}), \tag{11.20}$$

where $s_{1,j} \equiv s_{M+1,j}$ as a result of the PBC.

For this model, the transfer matrix is the contribution to the Boltzmann factor of two adjacent lattice rows (the transfer matrix acts along the axis of the cylinder; the rows are perpendicular to this direction). The possible states of the rows are the indices of the transfer matrix. If the rows contain M sites, the size of the transfer matrix is $2^M \times 2^M$. We represent the configuration of row j by the state $|S_j\rangle$:

$$|S_j\rangle = |s_{1,j}s_{2,j} \ldots s_{M,j}\rangle = |s_{1,j}\rangle \otimes |s_{2,j}\rangle \otimes \ldots \otimes |s_{M,j}\rangle. \tag{11.21}$$

The transfer matrix for rows j and $j+1$ is found as

$$\langle S_j|T|S_{j+1}\rangle = \exp\left[\beta \sum_{i=1}^{M} J\left(\frac{1}{2}s_{i,j}s_{i+1,j} + \frac{1}{2}s_{i,j+1}s_{i+1,j+1} + s_{i,j}s_{i,j+1}\right) + \frac{\beta}{2}H\sum_{i=1}^{M}(s_{i,j} + s_{i,j+1})\right], \tag{11.22}$$

The eigenvalues can now be found from a numerical diagonalisation of this matrix: its largest eigenvalues (in absolute value) determine the free energy and correlation functions of the model on a semi-infinite strip of width M.

Diagonalisation can be performed straightforwardly up to matrices of size $10\,000 \times 10\,000$ (and beyond if one uses powerful machines). However, we need only the largest few eigenvalues and there exist special numerical methods for calculating these which, for sparse matrices, are much more efficient than standard methods. Most convenient is Lanczos' method, which is described in Appendix A8.2. In this method, the matrix enters only via the multiplication with a vector, and this multiplication can be carried out efficiently, in particular for a sparse matrix, provided we take only the nonzero entries into account. Unfortunately, the transfer matrix of the Ising model is not sparse at all – it follows from Eq. (11.22) that it has a nonzero value for each pair of possible row configurations S_j and S_{j+1}.

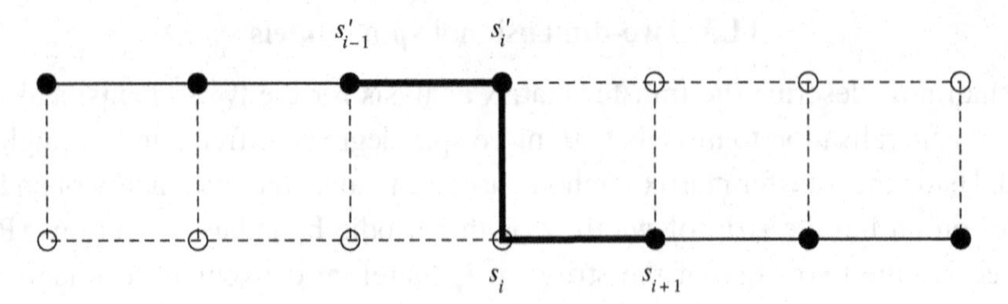

Figure 11.1. A step in the multiplication of a vector with the transfer matrix for the Ising case.

There is, however, a way to perform matrix–vector multiplications efficiently, using the fact that T factorises into a product of sparse matrices:

$$T = T_M T_{M-1} \cdots T_1 \tag{11.23}$$

where the matrix T_i is associated with site number i. To understand the explicit form of the submatrices T_i, it is useful to look at Figure 11.1. Let us call ϕ the state with matrix elements corresponding to the bottom row in that figure. Calling the bottom row configurations $|S\rangle$, and the top row $|S'\rangle$, we want to calculate the elements of the vector $|\psi\rangle = T|\phi\rangle$:

$$\langle S'|\psi\rangle = \langle S'|T|\phi\rangle = \sum_S \langle S'|T|S\rangle\langle S|\phi\rangle \tag{11.24}$$

for each top row configuration S'. The full matrix T contains couplings within the two horizontal rows and on the vertical links between them. The submatrices T_i contain only the couplings on the thick lines in Figure 11.1. We could therefore say that the application of a matrix T_i replaces the lower spin s_i by the upper spin s'_i, leaving the remaining spins unchanged.

We can now give the form of the matrices T_i. It acts between two states which are somehow intermediate between S and S': the state Σ' on the left of T_i represents the spins $s'_1, s'_2, \ldots, s'_i, s_{i+1}, \ldots, s_M$, and the state Σ on the right represents the spins $s'_1, s'_2, \ldots, s'_{i-1}, s_i, \ldots, s_M$. In terms of these states, which have elements σ_j and σ'_j respectively, we have

$$\begin{aligned}\langle\Sigma'|T_i|\Sigma\rangle &= \langle\sigma'_1 \ldots \sigma'_{i-1}\sigma'_i\sigma'_{i+1} \ldots \sigma'_M|T_i|\sigma_1 \ldots \sigma_{i-1}\sigma_i\sigma_{i+1} \ldots \sigma_M\rangle \\ &= \exp\left\{\beta J\left[\frac{1}{2}(\sigma_i\sigma_{i+1} + \sigma'_i\sigma'_{i-1}) + \sigma_i\sigma'_i\right]\right\} \\ &\quad \times \delta_{\sigma_1\sigma'_1}\delta_{\sigma_2\sigma'_2} \cdots \delta_{\sigma_{i-1}\sigma'_{i-1}}\delta_{\sigma_{i+1}\sigma'_{i+1}} \cdots \delta_{\sigma_M\sigma'_M}.\end{aligned} \tag{11.25}$$

The horizontal couplings have a factor $1/2$ because they are taken into account twice: once when the transfer matrix couples the previous row to the current row, and once when the transfer matrix couples the present row to the next one. The form given in (11.25) is not correct for $i = 1$ or $i = M$, because the first new spin, s'_1, does not yet have a left neighbour. Therefore, in the matrix T_1, the term $\beta J\sigma'_1\sigma'_M/2$ is left out and replaced by $\beta J\sigma_1\sigma_M/2$. Similarly, for $i = M$, the coupling $\beta J\sigma_M\sigma_1/2$ is replaced by $\beta J\sigma'_M\sigma'_1/2$.

How do we perform the full matrix–vector multiplication? We introduce a set of intermediate vectors $\psi^{(i)}$, defined by

$$|\psi^{(0)}\rangle \equiv |\phi\rangle \tag{11.26a}$$

$$|\psi^{(i)}\rangle \equiv T_i|\psi^{(i-1)}\rangle, \tag{11.26b}$$

so that $|\psi^{(M)}\rangle = |\psi\rangle$. Suppose we have the vector $|\psi^{(i-1)}\rangle$ at our disposal, that is, we have all its matrix elements $\langle S^{(i-1)}|\psi^{(i-1)}\rangle$ stored in an array (below we shall describe how the different configurations S correspond to the array index). The multiplication of the vector $|\psi^{(i-1)}\rangle$ by the matrix T_i is done as follows ($\Sigma \equiv S^{(i-1)}$; $\Sigma' \equiv S^{(i)}$):

FOR all states $S^{(i)}$ DO
  FOR all $S^{(i-1)}$ equal to $S^{(i)}$ except for spin number i DO
    Calculate $\langle S^{(i)}|\psi^{(i)}\rangle = \langle S^{(i)}|T_i|S^{(i-1)}\rangle\langle S^{(i-1)}|\psi^{(i-1)}\rangle$.
  END FOR
END FOR

To do the evaluation in the inner loop, we need the vector elements $\langle S^{(i-1)}|\psi^{(i-1)}\rangle$ and the matrix elements $\langle S^{(i)}|T_i|S^{(i-1)}\rangle$. The former are stored in an array, and the latter are given by (11.25).

It now remains to find a relation between array and loop indices, and the states S. This is done using binary encoding, using 'bits' b_i which assume values 0 or 1:

$$n = b_0 + b_1 2^1 + b_2 2^2 + \cdots + b_{M-1} 2^{M-1} \tag{11.27}$$

where $b_i = 1$ corresponds to $s_{i+1} = 1$ and $b_i = 0$ to $s_{i+1} = -1$. The matrix element $\langle S^{(i)}|T_i|S^{(i-1)}\rangle$ depends on $s_{i-1}(= s'_{i-1})$, $s_{i+1}(= s'_{i+1})$, and on s_i and s'_i (the latter are not necessarily equal). For an integer n, the ith bit can be found as $b_{i-1} = (n/2^{i-1}) \bmod 2$, where integer division is assumed. Most computer languages have, however, an intrinsic function returning the value of any desired bit of an integer number. It should be noted that in the algorithm given above, the central evaluation for the first and last sites (1 and M) differs slightly from the other ones because the respective transfer matrix elements were defined differently (see above).

Having a routine for evaluating the transfer matrix–vector product, this can be used in the Lanczos algorithm in order to find the lowest eigenvalue and vector.

PROGRAMMING EXERCISE

Write a program for calculating the largest eigenvalue of the transfer matrix of the Ising model. The program involves a considerable amount of bookkeeping and the correctness can be tested only in the very end, when the free energies are actually calculated. When debugging, it is advisable to reserve just sufficient memory for the rows and then compile the code with the range-check option on (in this case it is checked whether indices of arrays addressed in the program lie within the appropriate bounds). Furthermore, an error in the multiplication routine will usually result in the transfer matrix not being Hermitian, which in turn causes the Lanczos routine to converge very slowly, if at all. For strip widths M somewhere between 4 and 12, the Lanczos routine should converge within less than 20 steps.

Check For $M = 6$ and the critical coupling $J/(k_B T) = 0.440\,687$, you should find $\lambda_0 = 276.6004$.

If the program works correctly, we can calculate the free energy. First of all, rather than considering the free energy $F = \ln \lambda_0$ (we leave out the factor $k_B T$), we calculate the free energy *per spin*, given as $f = (\ln \lambda_0)/M$, which should converge to a fixed value for large widths. To analyse the results using finite size scaling, we quote a few results from conformal field theory [3, 4]. In this theory, two-dimensional critical models are labelled by a number, the *central charge*, c, and the possible values of the central charge smaller than one form a discrete set.[2]

The central charge also has a physical interpretation. If we consider the model on a strip of width M, the free energy will scale approximately proportional with M. However, in diagonalising the transfer matrix (or the Hamiltonian), changing M means changing the boundary conditions, and therefore there will be corrections to this scaling behaviour. This mechanism occurs also in electrostatics [5], where it is known as the Casimir effect: two parallel conducting plates will attract each other: the energy per area scales as Const $\times\, M^{-3}$ (zero temperature). It turns out that in our case, the statistical model on a strip, the free energy (per unit length along the strip) scales as

$$F = \ln \lambda_0 = f_{\text{Bulk}} M - \frac{\pi c}{6} \frac{1}{M}. \tag{11.28}$$

f_{Bulk} is the free energy per site of the infinite system. For the Ising model it is known that the central charge is equal to $1/2$. Each value of c defines a set of possible critical exponents of the theory, and indeed, the thermal and magnetic exponent of the Ising model, with values 1 and $1/8$ respectively, are in the set for $c = 1/2$. The determination of the central charge is therefore very useful to classify the model's critical behaviour.

[2] For $c > 1$ there are arguments from supersymmetric field theory that also suggest discrete series, but they are not based on requirements that should hold rigorously.

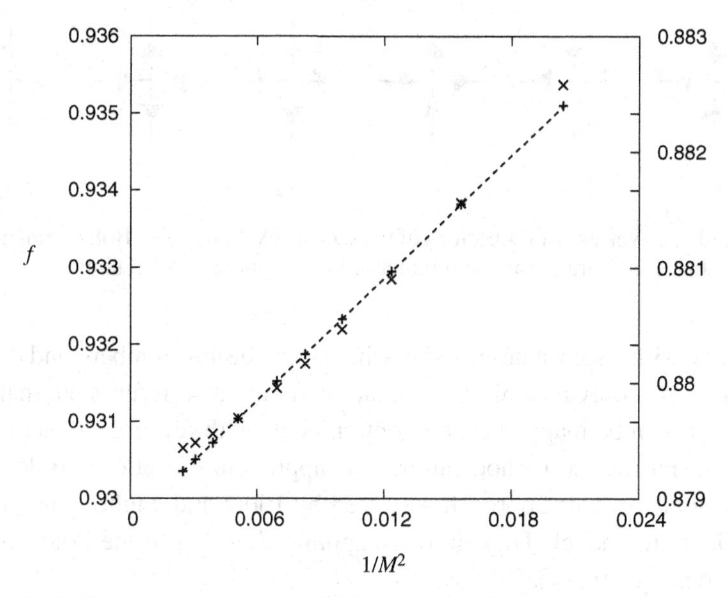

Figure 11.2. Free energy per site of the Ising model on a strip as a function of $1/M^2$. Values for M run from 7 to 20. The pluses correspond to the critical point $J = 0.4407$; the crosses (right axis) are the data for an off-critical point: $J = 0.4$. The straight line through the data has slope $\pi/12$.

Therefore, if we plot our results for the Ising model in the form $(\ln \lambda_0)/M$ vs. $1/M^2$ we should obtain a straight line with slope $\pi/12$. Figure 11.2 shows that this is indeed the case. As this is a finite size correction to the bulk free energy, the eigenvectors λ_0 must be determined with a precision of about six significant digits.

The magnetic exponent can be obtained by comparing the free energies for systems with periodic and antiperiodic boundary conditions, see Ref. [4]. Antiperiodic means the bonds across a seam along the cylinder are antiferromagnetic. The free energy difference then scales with the width M as

$$F_{\mathrm{AP}} - F_{\mathrm{P}} = \frac{2\pi}{M} x, \tag{11.29}$$

where x is the magnetic exponent – indeed, $x = 1/8$ can be found in this way for the Ising model.

11.4 More complicated models

The linear size of the transfer matrix increases exponentially with the strip width. This puts severe limits on the system sizes that can be treated with this method. In particular, this prohibits calculations with reasonable system size for models with larger numbers of degrees of freedom. There exist models, such as the clock or the

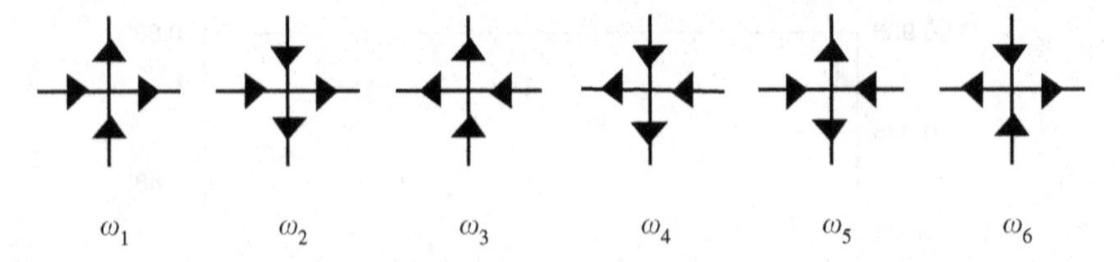

Figure 11.3. Vertex configurations of the six-vertex model. The Boltzmann weight of a lattice configuration is the product of the weights ω_i for all sites.

Potts model, with q spin states per site, where q can be any number,[3] and the transfer matrix method is only feasible for q-values up to about 4. It turns out that many of these models can be mapped exactly onto models with fewer degrees of freedom, and the transfer matrix method can then be applied to the latter in order to obtain critical exponents. This approach was used by Blöte and Nightingale [1, 3, 8] to analyse the Potts model. They used a mapping [7, 9, 10] of the Potts model onto the six-vertex model [11].

In the six-vertex model, the degrees of freedom do not reside on the sites, but on the links of a square lattice. They are two-valued and are represented by arrows on the links. For each site, there are in principle $2^4 = 16$ different configurations of arrows on the four links connected to that site. This number is, however, reduced if we require that the net flow into the sites is zero: the number of incoming and outgoing arrows must be equal for each site. This leaves only six different site configurations or vertices, hence the name of the model. The different vertices are shown in Figure 11.3. Each vertex is assigned a weight, and the Boltzmann weight of a lattice is the product of the vertex weights of all sites. Often the case $\omega_1 = \omega_2 = \omega_3 = \omega_4$; $\omega_5 = \omega_6$ is considered. The model is called the 'F-model' and this particular case has been solved exactly by Lieb [11]. The F-model turns out to be critical, with an infinite order transition (of the Kosterlitz–Thouless type, see Section 15.5.2) for $2\omega_1 = \omega_5$. In Problem 11.2, a transfer matrix implementation for the standard six-vertex is described. In the case of the Potts model, the equivalent six-vertex model is a slightly modified version of the original model, and this has as a consequence that the transfer matrix is no longer Hermitian for q noninteger, which complicates its diagonalisation (for details, see Ref. [1]).

If a transformation onto a simple model is not known, application of the transfer matrix for models in which the degrees of freedom can assume many values might still be possible, using a technique inspired by quantum Monte Carlo methods [12]. This method will be considered in some detail in Section 12.6.

[3] Even noninteger values of q are relevant [6, 7].

11.5 'Exact' diagonalisation of quantum chains

So far in this chapter, we have studied methods for diagonalising the transfer matrices of two-dimensional models. This is equivalent to diagonalising Hamiltonians of one-dimensional quantum systems. In the next few sections we shall consider the analysis of one-dimensional quantum chains more systematically.

You may be sceptical about the use of studying one-dimensional quantum chains. There are, however, good reasons for considering such systems. First, they are equivalent to classical systems in two dimensions, as we have seen extensively in the first half of this chapter. Second, there exist experimental realisations of quasi-one dimensional systems in some particular crystals; see for example Ref. [13]. Last but not least, the fact that one-dimensional chains can be studied successfully using analytical and computational tools makes them useful as testing grounds for new methods which may be successful in higher dimensions too. A nice introduction to the material covered in this and the following sections is Ref. [14].

The quantum chains which are studied most intensively are spin chains and Hubbard-like systems. Spin chains consist of quantum spins located on a one-dimensional lattice. The magnetic spin quantum numbers can assume values $-S$, $-S+1$ up to S, where the maximal spin S is a positive, (half-)integer number. Note that from now on we shall simply call 'spin' the eigenvalue of the z-component of the spin-angular momentum operator (without the factor $\hbar$). A famous example is the Heisenberg (anti-)ferromagnet, which is described by the Hamiltonian:

$$H = J \sum_{i=1}^{L-1} \mathbf{S}_i \mathbf{S}_{i+1}. \tag{11.30}$$

This is a chain with *open ends* as the first and last spins couple only to a single neighbour. If we connect the first and last spin by putting $\mathbf{S}_{L+1} \equiv \mathbf{S}_1$, we have a periodic chain. For positive values of J the chain is antiferromagnetic.

The one-dimensional Hubbard model describes fermions[4] on a chain. The fermions are spin-1/2 particles and because of the Pauli principle there are on each site four possibilities: zero particles, one spin-up or one spin-down particle, or both. The motion of the particles along the chain is represented by a hopping term. There is also a Coulomb interaction, which is assumed to be strongly local: only particles on the same site can feel it. For particles on different sites, the Coulomb interaction is neglected. The Hamiltonian is

$$H = -t \sum_{\sigma;i=1}^{L-1} (c^\dagger_{i,\sigma} c_{i+1,\sigma} + c^\dagger_{i+1,\sigma} c_{i,\sigma}) + U \sum_{i=1}^{L} n_{i,\uparrow} n_{i,\downarrow}. \tag{11.31}$$

[4] Bose–Hubbard models are presently also the subject of research, but the fermion version has been studied most intensively.

The real parameter t represents the hopping rate, and U (which is also real) describes the Coulomb interaction. The fermion creation and annihilation operators $c_{i,\sigma}$ and $c_{i,\sigma}^\dagger$ have a site index i and a spin index $\sigma = \pm 1/2$ (these two values are denoted as $\uparrow$ and $\downarrow$ in the Coulomb term). They satisfy the usual fermion anticommutation relations:

$$\{c_{i,\sigma}^\dagger, c_{j,\sigma'}\} = \delta_{ij}\delta_{\sigma\sigma'}. \tag{11.32}$$

The other anticommutators vanish. Finally, the number-operators $n_{i,\sigma}$ are given by

$$n_{i,\sigma} = c_{i,\sigma}^\dagger c_{i,\sigma}. \tag{11.33}$$

Other models, such as the $t - J$ model, can be related to the Hubbard model, and this model reduces in a particular limit to the Heisenberg chain; we shall not go into details but refer to the literature [15].

In the following, we shall mainly restrict ourselves to the Heisenberg chain, which serves to illustrate the numerical methods suitable for studying quantum chains; we shall briefly mention how these can be adapted to Hubbard-like models where appropriate.

11.5.1 Lanczos diagonalisation of the Heisenberg model

It is quite straightforward to diagonalise the Hamiltonian for the Heisenberg model – this is done using the Lanczos method (see Appendix A8.2). The main problem is writing a procedure for multiplying the Hamiltonian by some given vector $|\psi\rangle$.

The basis vectors of an L-site chain can be chosen as

$$|\Sigma\rangle = |s_1, s_2, \ldots, s_L\rangle, \tag{11.34}$$

where the s_i are the spins. For a spin-1/2 chain these spins assume the values $\pm 1/2$, and for the $S = 1$ chain $1, 0$ and -1. The dimension of the Hilbert space is $(2S+1)^L$. We shall not use this spin representation in our program, but use a mapping of each basis state to an integer just as in the transfer matrix program. First, we change notation and let s_i run from 0 to $2S$ instead of from $-S$ to S. We write

$$|K\rangle = |s_1, s_2, \ldots, s_L\rangle, \tag{11.35}$$

where the integer K is given as

$$K = \sum_{i=1}^{L} s_i M_S^{i-1}; \tag{11.36}$$

$M_S = 2S + 1$. It is instructive to calculate the numbers K for all possible states of a $L = 4$ spin-1/2 chain, for which $s_i = 0, 1$, and the reader is invited to do this.

Before we proceed, we note that the Heisenberg Hamiltonian can be written in the form (see Problem 11.3)

$$H = J\sum_{i=1}^{N-1}\left[\frac{1}{2}(S_i^+S_{i+1}^- + S_i^-S_{i+1}^+) + S_i^zS_{i+1}^z\right]. \qquad (11.37)$$

Furthermore, the Hamiltonian commutes with quite a few symmetry operators. We shall study these in detail in Section 11.5.2.

Multiplication of this Hamiltonian with a given vector $|\psi\rangle$ proceeds in a way analogous to the transfer matrix problem: we 'zip' through the chain and collect all the generated terms into the resulting vector. The heart of the algorithm looks as follows:

```
FOR each site I DO
    FOR N1 = 0 TO (M_S)^(I-1) - 1
        FOR N2 = 0 TO (M_S)^(L-I-1) - 1
            FOR S1 = 0, M_S - 1
                FOR S2 = 0, M_S - 1
                    FOR S1P = 0, M_S - 1
                        FOR S2P = 0, M_S - 1
                            J = N1+N2*(M_S)^(I+1) + S1*(M_S)^(I-1) +
                                          S2*(M_S)^I;
                            JP = N1+N2*(M_S)^(I+1) + S1P*(M_S)^(I-1) +
                                          S2P*(M_S)^I;
                            HPsi(JP) = HPsi(JP)+
                                          HMat(S1P, S2P, S1, S2)*Psi(J);
                        END FOR;
                    END FOR;
                END FOR;
            END FOR;
        END FOR;
    END FOR;
END FOR;
```

It is understood that for i near the boundary, some modification is necessary. The matrix HMat is the part of the Hamiltonian which couples the spins at sites i and $i+1$. This matrix is directly found from the Heisenberg Hamiltonian with $\mathbf{S}_i = \frac{1}{2}(\sigma_x, \sigma_y, \sigma_z)$, and σ_i are the three Pauli matrices.

This procedure can now readily be used in the Lanczos algorithm to find the lowest few eigenvalues.

PROGRAMMING EXERCISE

Program a routine for multiplying the Hamiltonian by a vector. Use this in a Lanczos algorithm to find the ground state and first excited state of a spin-1/2 antiferromagnetic Heisenberg chain. The infinite chain has no energy gap [16]; therefore, the energy gap should decrease with increasing length.

Check For a chain with 12 sites, you should find -5.3873909 for the ground state energy of a periodic chain.

The program can easily be extended to the spin-1 chain. You can then check the value of the Haldane gap [17] between the ground state and the first excited state, which is $0.4107 + 67.9/L^2$ *per site* for an long chain [18]. Note that to extend this program to the Hubbard model, M_S must be four (as there are four possible states for each site) and the Hamiltonian must be adjusted to that model.

11.5.2 Exploiting symmetry

Up to this point, we have hardly mentioned the use of symmetry when diagonalising a Hamiltonian. This is, however, extremely important, as the use of symmetry can result in a huge reduction of the CPU time needed for a calculation. To see this let us assume that we have some symmetry operation, represented by a quantum operator O which commutes with the Hamiltonian. In that case we can organise the eigenstates of the Hamiltonian such that they are also eigenstates of O and vice versa. Now suppose that we can easily find a complete set of eigenstates of O (this often turns out be the case). These states are then organised into subspaces, one for each eigenvalue of O. States within such a (possibly degenerate, and therefore multidimensional) eigenspace will be invariant under action of the Hamiltonian on them (see Problem 11.3). This means that the Hamiltonian will acquire a *block-diagonal* form when formulated with respect to this complete set. This is an extremely useful result. If we have for example a basis on a one-dimensional axis, and a Hamiltonian which is invariant under reflection, we divide up our N basis vectors into a subset of even and one consisting of odd basis functions. If there are equal numbers of each type, the Hamiltonian will have two blocks of size $N/2$ on the diagonal.

What is the use of this? Matrix diagonalisation is an order-N^3 algorithm. Diagonalising two $N/2 \times N/2$ matrices is therefore four times as fast as diagonalising a single $N \times N$ matrix. The invariant subspaces in which we solve our small Hamiltonian matrices are called *sectors*. Studying symmetries and their use in physics is mainly the domain of group theory. The interested reader is advised to consult specialised literature [19]. Here we shall implement the symmetries using pedestrian methods.

For the one-dimensional Heisenberg chain, the implementation of symmetry is very instructive. The symmetry operations are easy to find [20]. The symmetry operations we shall consider are translation symmetry (which holds for the periodic chain), conservation of angular momentum along the z axis (which results from rotational symmetry around that axis), and reflection symmetry (left-right symmetry). In Problem 11.3, these symmetries are considered in some detail.

We can find so many small sectors that it is possible to find *all* eigenvalues (thus not only the lowest ones) for the $L = 12$ Heisenberg chain of the previous section very quickly. We start by considering the rotation symmetry which leads to the conservation of the total angular momentum along the z-axis. This is

$$S_z^{\text{tot}} = \sum_{i=1}^{L} s_i. \tag{11.38}$$

It is clear that the states $|s_1, \ldots, s_L\rangle$ are eigenstates of the S_z^{tot}. We can simply loop over all the states $|K\rangle$ (K given by (11.36)), determine their total spin component along the z axis and reshuffle them into sets of equal S_z. You can try this for the Heisenberg chain and see that it leads to much smaller matrices. These matrices can already be diagonalised on a standard PC. However, an additional major improvement is realised by implementing translation symmetry (which only holds for the periodic case). Eigenstates of the translation operator, which shifts all spins one site to the right, are Bloch states. In fact, for each basis state $|\psi\rangle$ we can generate a Bloch state which is an eigenvector of the translation operator T as follows:

$$|\psi_k\rangle = |\psi\rangle + e^{2\pi ik/L} T|\psi\rangle + e^{4\pi ik/L} T^2|\psi\rangle + \cdots + e^{2\pi i(L-1)k/L} T^{L-1}|\psi\rangle. \tag{11.39}$$

It can easily be checked that the eigenvalue is $e^{-2\pi ik/L}$.

Let us now work out these symmetries for a chain consisting of four spins. The basis states can be grouped into states related to each other by translations.

Sz	0	1	2		3	4
$k=0$	$\|0\rangle$	$\|1\rangle+\|2\rangle+\|4\rangle+\|8\rangle$	$\|3\rangle+\|6\rangle+\|12\rangle+\|9\rangle$	$\|5\rangle+\|10\rangle$	$\|7\rangle+\|14\rangle+\|13\rangle+\|11\rangle$	$\|15\rangle$
$k=1$		$\|1\rangle+i\|2\rangle-\|4\rangle-i\|8\rangle$	$\|3\rangle+i\|6\rangle-\|12\rangle-i\|9\rangle$		$\|7\rangle+i\|14\rangle-\|13\rangle-i\|11\rangle$	
$k=2$		$\|1\rangle-\|2\rangle+\|4\rangle-\|8\rangle$	$\|3\rangle-\|6\rangle+\|12\rangle-\|9\rangle$	$\|5\rangle-\|10\rangle$	$\|7\rangle-\|14\rangle+\|13\rangle-\|11\rangle$	
$k=3$		$\|1\rangle-i\|2\rangle-\|4\rangle+i\|8\rangle$	$\|3\rangle-i\|6\rangle-\|12\rangle+i\|9\rangle$		$\|7\rangle-i\|14\rangle-\|13\rangle+i\|11\rangle$	

All the boxes represent sectors of the Hilbert space: we see that only two sectors are two-dimensional; the others contain only a single state. This means that the diagonalisation can now easily be done by hand.

Note that some boxes are empty: this is because constructing the eigenstate according to the above recipe leads to the zero state. This happens when translating the state over one or two positions turns the state into itself. As the period of these states is smaller than four, the corresponding k-values are spaced by more than one. Note that the prefactors in the different states are given by $\exp(2\pi ikl/L)$, where k is the k-vector, and this prefactor is for the state obtained by translating the first basis state over l sites. It is clear that shifting k by L leaves the states invariant.

We now consider reflection symmetry. This leads to combinations of states with wave vectors k and $-k$. Taking the periodicity $k \rightarrow k+L$ into account, we see that the $k = 3$ line above can be eliminated (it is equivalent to $k = 1$) and that we can take either even or odd combinations of $k, -k$ pairs. In the first case, the prefactor $\exp(2\pi ikl/L)$ is replaced by $\cos(2\pi kl/L)$ and in the second case by $\sin(2\pi kl/L)$.

The program is now set up as follows. A loop over all states is performed. For each new state, the states which can be obtained by translation over 1, 2 and 3 (more generally 1 to $L-1$) positions are considered; if such a state is identical to the first state, we know the periodicity of the state, and hence the nonzero k-vectors. Consider the $|0\rangle$ state as an example. A translation over one site transforms the state into itself – hence, the period in the k-values is $L/1 = L$. This is why we find only one state in the second and rightmost columns of the table above. Similarly, the state $|5\rangle$ has period two, hence we find only two nonzero states in the fifth column. All the other states have period 4, hence period 1 in the acceptable values of k.

Also, the spin S_z^{tot} of each state is determined (note that the translations do not affect this value). We now label the sectors by S_z, k (we omit the superscript 'tot' with S_z). Each time we find a state in this sector we add it to the basis of that sector. Note that we consider only $k = 0, \ldots, L/2$, and construct an even and an odd set (using cosine and sine as indicated above) for each sector S_z, k. When we have gone through all states in this way, we have complete basis sets for each sector. In a following step, we construct the Hamiltonian block matrix $\langle K|H|K'\rangle$ for each sector, where $|K\rangle$ and $|K'\rangle$ are basis states for the sector under consideration. For this calculation we use the procedure written in the previous subsection for calculating $H|K'\rangle$. Then we feed the resulting matrices into a library routine for diagonalising symmetric matrices.

PROGRAMMING EXERCISE

Write a program which implements these symmetries and calculates the full spectrum of the antiferromagnetic spin-1/2 Heisenberg chain.

Check You can check you results by comparing the lowest states with those obtained in the previous section. Furthermore, a useful check is whether the dimensions of all sectors add up to $(M_S)^L$. Also, you can run the program for $L = 4$ and check whether the sectors and basis vectors correspond to those given above, in the Table.

You have now seen group theory 'at work'. The possible improvements in performance are really impressive – for this reason, group theoretical methods are very important in computational physics. A particular example is solid state physics; in Chapter 6 we have already briefly mentioned the use of symmetry, in particular to reduce the Brillouin zone to the so-called 'irreducible wedge'. A nice discussion of the symmetry-issues in quantum systems can be found in Ref. [21].

11.6 Quantum renormalisation in real space

Calculations for systems whose behaviour is characterised by large length scales are usually time-consuming. Often, the systems have short-range interactions, but these generate correlations over large distances. The most obvious example is the Poisson equation for a point charge:

$$\nabla^2 V(\mathbf{r}) = \delta(\mathbf{r}). \tag{11.40}$$

Discretising the Laplace operator on a grid in a first order approximation couples only nearest neighbour lattice sites. On the other hand, the solution has long range character. In Appendix A7.2 we consider the multigrid method for treating this problem. This method uses successive coarsenings and refinements in order to find the solution very efficiently.

The multigrid method is reminiscent of the renormalisation method for critical phenomena. There we perform successive coarse grainings of the system, in order to extract its long-wavelength behaviour which is responsible for the occurrence of critical phenomena. In renormalisation procedures, we usually tend to throw away details relating to the short-wavelength behaviour. This causes the critical point to be found at the wrong value, but critical exponents are still found correctly or to a good approximation.

Often we are interested in the full solution, and not only the long-range behaviour. In these cases we must still treat the short-range behaviour accurately. The multigrid method accomplishes this by the refinements which alternate with the coarsenings. Another possible approach, which is closer to the standard renormalisation procedure, is to take a small sample system and extend the size of this and see whether it is possible to find a self-consistent limit which gives us the solution of the infinite system. Wilson has followed this approach with great success for the Kondo problem [22]. We shall not treat the Kondo problem in detail, but emphasise at the same time that the reason why this numerical renormalisation procedure worked so well is due to the special structure of the Kondo Hamiltonian, which contains couplings that decay exponentially with distance.

The lack of success when applying the renormalisation procedure to other problems has been nagging researchers until White and Noack [18, 23] came up with

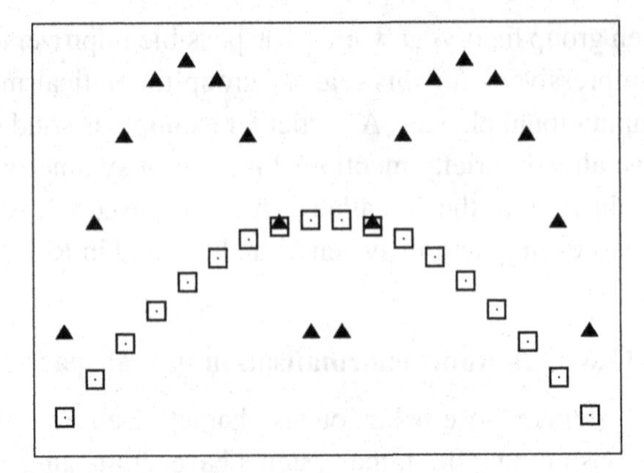

Figure 11.4. Two fixed-boundary solutions for a narrow box (solid triangles) and the ground state (open squares) for a large box. It is clear that the approximation of the ground state will be poor.

a solution for a quantum particle in a potential and for a quantum spin chain. The particle in a potential illustrates where the standard renormalisation procedure fails [18]. Consider a particle in a box with impenetrable walls (infinite potential outside the box). An idea for solving this problem efficiently is first to solve the problem in a box of half the width of the full box and then solve the problem for the full box using the ground state wave functions for the half-width boxes as basis functions in a variational calculation. Figure 11.4 illustrates why this method fails. The Schrödinger equation is solved for fixed boundary conditions, and building the ground state of the full problem from two functions that vanish at the centre is not efficient.

In order to improve the method, White and Noack built the solution of the larger box from solutions obtained for a smaller box *with different kinds of boundary conditions*. Consider two solutions, one of which has fixed zero boundary conditions on both sides, whereas the other has an open boundary condition on one side. The first has value zero, but (in general) a nonzero derivative. The second one has nonzero value at the open boundary, but its derivative is zero. *Any* type of boundary condition can be obtained by linear combination of the two solutions. This notion leads to the following algorithm.

1. Find the solutions of a small box, with fixed and open boundary conditions on both sides (four cases in total). Keep the lowest $M/4$ eigenstates of the Hamiltonian for each case.
2. Construct a Hamiltonian for the large box (twice as large as the small box) with respect to the basis found in step 1, for the same four cases as in step 1. See below for how this is done. The dimension of these Hamiltonians is $2M$.

3. Calculate the eigenstates of these four matrices and keep the lowest M. Now the large box is considered as a small box, and we proceed with step 1.

We shall now fill in the details of the algorithm. When having the small matrices at our disposal, we must construct the large matrices from these. For a large matrix with boundary conditions b and b′ (the b's denote 'fixed' or 'free') on the left and right hand side respectively, the recipe is

$$H^{\mathrm{L}}_{\mathrm{bb}'}(2M) = \begin{pmatrix} H^{\mathrm{S}}_{\mathrm{b,fixed}}(M) & T^{\mathrm{S}}(M) \\ (T^{\mathrm{S}})^{\dagger}(M) & H^{\mathrm{S}}_{\mathrm{fixed,b}'}(M) \end{pmatrix}, \tag{11.41}$$

where we have used the superscripts 'S' and 'L' to denote the matrices for the short and the long chain respectively. The arguments (M) and $(2M)$ specify the sizes of the matrices. We must carry out a similar procedure for the off-diagonal matrix:

$$T^{\mathrm{L}}(2M) = \begin{pmatrix} 0 & 0 \\ T^{\mathrm{S}}(M) & 0 \end{pmatrix}. \tag{11.42}$$

For each of the four possibilities (b, b′) we diagonalise the large matrix, yielding $2M$ eigenvectors for each case. We keep only the $M/4$ corresponding to the lowest energies, yielding a set of M basis states. These states are not necessarily orthogonal, so we use an orthogonalisation procedure such as Gram–Schmidt to transform them into an orthonormal basis. The matrices of the large box are now reduced to size $M \times M$ according to the standard procedure:

$$H^{\mathrm{L}}_{\mathrm{bb}'}(M) = V^{\dagger} H^{\mathrm{L}}_{\mathrm{bb}'}(2M) V, \tag{11.43}$$

where V is an $2M \times M$ matrix whose M columns are the vector representations of the (orthogonal) basis states used.

We start off the procedure with a simple Hamiltonian arising from a real-space discretisation of the Schrödinger equation:

$$H_{\mathrm{fixed,fixed}} = \begin{pmatrix} 2 & -1 \\ -1 & 2 \end{pmatrix}; \quad H_{\mathrm{fixed,free}} = \begin{pmatrix} 2 & -1 \\ -1 & 1 \end{pmatrix} \quad \text{etc.} \tag{11.44}$$

The information concerning the type of boundary conditions is supplied exclusively in these matrices. They are preserved in the subsequent transformations to larger boxes. Note that we start off with $M = 2$ states. Suppose we envisage keeping $M = 8$ states. The 2×2 matrices will lead to a 4×4 matrices of the large box. Of each of these matrices, we keep the lowest two states, in order to arrive at 8×8 matrices at the next step. From then on we continue glueing these matrices together to 16×16 matrices of which we keep only the lowest two eigenstates, etc. It is important to note that after N steps, the box width is given by $2^{N+1} + 1$. This is because initially, with two boxes, we have a box width of 5, and at each step, the small box width (which starts at 2) is doubled.

Table 11.1. *Energies for the particle in a box after* $N = 10$ *steps.*

	Energies
E_0	2.3508×10^{-6}
E_1	9.4032×10^{-6}
E_2	2.1157×10^{-5}
E_3	3.7613×10^{-5}

These are the values $n^2\pi^2/L^2$, with $L = 2049$.

PROGRAMMING EXERCISE

You have now enough information for writing a program for this ‘real-space quantum renormalisation group’.

Check For the starting matrices given, you should after 10 steps arrive at the spectral values given in Table 11.1.

There is an interesting initiative, called ALPS, to construct a C++ library for quantum algorithms [24, 25]. The programs for this and those in the following sections can be found as examples on the ALPS website.

11.7 The density matrix renormalisation group method

In this section we shall describe how the ideas of the previous section can be extended to many-body systems in one dimension. First of all, we note that the renormalisation procedure of the previous section doubled the box size at each step, and thereby the dimension of the Hilbert space. In the case of a quantum chain, adding a single spin to a spin-1/2 chain already doubles the dimension of the Hilbert space, whereas the dimension of the physical space covered increases only by a small fraction. If we were to double the actual (physical) space, the dimension of the Hilbert space would increase by such a large factor that we would make gigantic steps in complexity, which is bound to fail. Therefore we add only a single site (spin) at a time.

To illustrate how the method works, consider a finite chain of which the lower energy states are properly described by a (small) basis set $|m\rangle$ of size M (that is, m runs from 1 to M). The left end of the chain does not couple to a neighbouring spin: we consider an ‘open end’ boundary condition there. When we add a new spin, we have states

$$|m, s\rangle \tag{11.45}$$

We have generated a new basis set of size $M \times M_S$ (remember $M_S = 2S + 1$). The main problem now is to find a procedure for reducing this set to a new one of size M. In order to find such a procedure, we must realise ourselves that the system is always *part of a larger system.* The larger system is called the 'universe', the system under consideration is called the 'system', and the remainder (the universe without the system) is called the 'environment'. Universe, system and environment are denoted by U, S and E respectively.

We want the system U to be in the ground state, and the question is how we can represent the state of our system S. The answer lies in the notion of the *density matrix*. Density matrices are described in many quantum textbooks. We shall briefly recall this concept here. The density operator or matrix ρ can be given as

$$\rho = \sum_i p_i |\psi_i\rangle\langle\psi_i|. \tag{11.46}$$

The states $|\psi_i\rangle$ are accessible to our system, and they occur with probability p_i. This means that the exact state of the system is not known, but we have a set of 'candidate states' $|\psi_i\rangle$ with probabilities p_i. It is easy to see that for an arbitrary Hermitian operator Q, its expectation value is given as

$$\langle Q\rangle = \mathrm{Tr}(\rho Q). \tag{11.47}$$

We distinguish knowing the quantum state of a system, the *pure state* case, from the situation in which we do not have this knowledge, the *mixed state*. The density matrix corresponding to a pure state is $|\psi\rangle$ is obviously given by

$$\rho = |\psi\rangle\langle\psi|. \tag{11.48}$$

Now consider a system U consisting of two parts, S and E. This system is in a pure state $|\psi^{\mathrm{U}}\rangle$. The basis vectors of the system U can be chosen to be of the form $|\psi_\sigma^{\mathrm{S}}\rangle \otimes |\psi_\epsilon^{\mathrm{E}}\rangle$, where the constituents form complete orthonormal basis sets on S and on E. In the following we shall abbreviate these basis states as $|\sigma\epsilon\rangle$. The behaviour of the part S is completely determined by the expectation values of operators Q^{S} acting within the Hilbert space of S. Now let us evaluate this expectation value for the case where the system U is in the (pure) ground state. This is given by

$$\langle Q^{\mathrm{S}}\rangle = \langle\psi^{\mathrm{U}}|Q^{\mathrm{S}}|\psi^{\mathrm{U}}\rangle. \tag{11.49}$$

Now we expand the ground state in our basis set:

$$|\psi^{\mathrm{U}}\rangle = \sum_{\sigma,\epsilon} C_{\sigma,\epsilon}|\sigma\epsilon\rangle. \tag{11.50}$$

The density matrix then becomes

$$\rho = \sum_{\sigma,\epsilon,\sigma',\epsilon'} C_{\sigma\epsilon}C^*_{\sigma'\epsilon'}|\sigma\epsilon\rangle\langle\sigma'\epsilon'|. \tag{11.51}$$

Now we work out Eq. (11.49) using the basis set $|\sigma\epsilon\rangle$. We obtain

$$\langle Q^{\mathrm{S}}\rangle = \sum_{\sigma,\epsilon,\sigma',\epsilon'} C^*_{\sigma\epsilon} C_{\sigma'\epsilon'} \langle\sigma\epsilon|Q^{\mathrm{S}}|\sigma'\epsilon'\rangle. \tag{11.52}$$

We note that Q^{S} only affects the states of S, so the states of E can immediately be contracted. Using orthonormality of the $|\epsilon\rangle$ we obtain

$$\langle Q^{\mathrm{S}}\rangle = \sum_{\sigma,\epsilon,\sigma'} C^*_{\sigma\epsilon} C_{\sigma'\epsilon} \langle\sigma|Q^{\mathrm{S}}|\sigma'\rangle. \tag{11.53}$$

Noting that

$$\rho^{\mathrm{S}} \equiv \mathrm{Tr}_{\mathrm{E}}\rho = \sum_{\epsilon} \langle\epsilon|\rho|\epsilon\rangle = \sum_{\sigma,\epsilon\sigma'} C_{\sigma\epsilon} C^*_{\sigma'\epsilon} |\sigma\rangle\langle\sigma'|, \tag{11.54}$$

we conclude that

$$\langle Q^{\mathrm{S}}\rangle = \mathrm{Tr}_{\mathrm{S}}(\rho^{\mathrm{S}} Q^{\mathrm{S}}). \tag{11.55}$$

What have we just shown? Starting from a *pure* state of the entire (U) system, we have generated a so-called *reduced density matrix* ρ^{S} for S, which in general describes a mixed state, and which is the most complete description of that system. In other words, when a system S is coupled to an environment E, its state must in general be described as a *mixed* state, although system plus environment together (that is, U) are in a *pure* state.

A system whose state is mixed due to coupling with an environment is said to be *entangled* with that environment. The simplest example of entanglement is a system consisting of two degrees of freedom which can both be in two states, $|0\rangle$ and $|1\rangle$. For a state

$$|\psi\rangle = |00\rangle, \tag{11.56}$$

where the first 0 refers to S and the second one to E, the reduced density matrix is $\rho^{\mathrm{S}} = |0\rangle\langle 0|$ which describes a pure state, reflecting the fact that we know that the system S is in state $|0\rangle$. However, the state

$$|\psi\rangle = \frac{1}{\sqrt{2}}(|00\rangle + |11\rangle) \tag{11.57}$$

leads to the reduced density matrix

$$\rho^{\mathrm{S}} = \frac{1}{2}(|0\rangle\langle 0| + |1\rangle\langle 1|) \tag{11.58}$$

which describes a mixed state. The state $|\psi\rangle$ is therefore an example of an entangled state – this particular state is one of the four *Bell states* [26].

Now let us return to the renormalisation procedure. Knowing that U is in the ground state, we should describe S by a mixed state. But which mixed state? After all, we do not know the ground state of U, and therefore we cannot take the trace

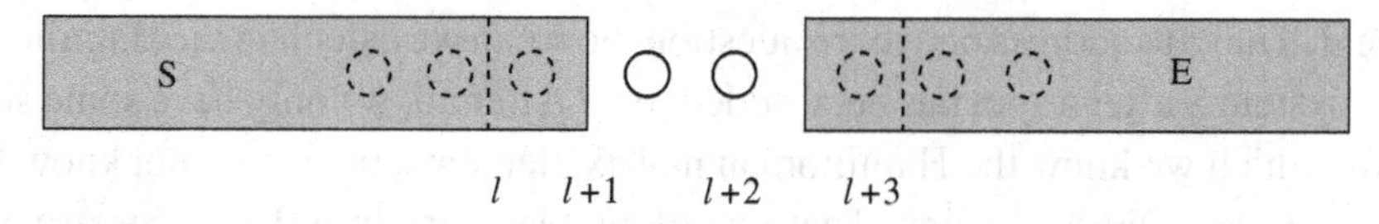

Figure 11.5. Schematic representation of the DMRG procedure. A spin is added to the 'old' system S, which before this addition represented a block of l spins. An environment E is created by reflecting the new S (including the extra spin). Both states are now perfectly known.

over the environment. We can, however, use the information that we have generated on S to create an *artificial* environment E of which all is known. More specifically, we take for E the *reflection* of system S as shown in Figure 11.5. As stated above, we have a basis $|m, s_{l+1}\rangle$ for S. Note that the two indices together belong to S; they can together be viewed as the state $|\sigma\rangle$ of our general discussion of the density matrix above. The reflected system E has a basis $|s_{l+2}, n\rangle$, where we have reversed the indices (and replaced m by n) to represent the structure of E. The combined system U, described by the basis $|m, s_{l+1}, s_{l+2}, n\rangle$ should be in the ground state. If we can find this state, then we can trace out the degrees of freedom of E in order to find the density matrix ρ^{S} describing S. Remember the aim was to find the M most representative basis states of system S, whose Hilbert space has dimension $M \cdot M_S$. We should therefore find the M states which best represent the density matrix ρ^{S}. For this we use the idea behind the singular value decomposition, well known from numerical linear algebra [27]: we simply take the M eigenstates of ρ^{S} with the highest eigenvalues (all eigenvalues of the density matrix are positive). In practice, these eigenvalues rapidly decrease so that the errors made in this procedure are very small. The new density matrix is then given as a truncated expansion

$$\rho^{\mathrm{S}}(M) = \frac{1}{\sum_{m=1}^{M} \lambda_m} \sum_{m=1}^{M} \lambda_m |m\rangle\langle m| \tag{11.59}$$

where $|m\rangle$ are the eigenstates of ρ^{S}. The states $|m\rangle$ are the ones which are carried over to the next stage, analogous to the quantum renormalisation method of the previous section: the Hamiltonian of S is transformed according to

$$H^{\mathrm{S}}(M) = V^{\dagger} H^{\mathrm{S}}(M \cdot M_S) V, \tag{11.60}$$

where V contains the M 'most important' eigenstates of $\rho^{\mathrm{S}}(M \cdot M_S)$. The fact that this procedure is indeed the best we can follow is addressed in Problems 11.4 and 11.5. The fact that the states kept are derived from an estimate of the density matrix of the system S is reflected in the name *density matrix renormalisation group* (DMRG) for this method.

We still have to address one more question: how can we calculate the Hamiltonian of the system S after a spin has been added to it? After all, we only have some states $|m\rangle$ for which we know the Hamiltonian matrix elements, but we do not know how they relate to individual spins. That would be necessary in order to construct the matrix elements of the Hamiltonian in the basis $|m, s_{l+1}\rangle$, as we need the matrix elements of, for example, the operator $S_{+,l}S_{-,l+1}$, where l is the rightmost spin of the 'old' block S, i.e. before adding the new spin s_{l+1} to it. Note, however, that this matrix element can be written as

$$\langle m, s_{l+1}|S_{+,l}S_{-,l+1}|m', s'_{l+1}\rangle = \langle m|S_{+,l}|m'\rangle\langle s_{l+1}|S_{-,l+1}|s'_{l+1}\rangle, \tag{11.61}$$

a product of two matrix elements. So we must keep track not only of the Hamiltonian in the basis at each step, but also of the rightmost spin operators (in a Heisenberg chain these are the operators S_z, S_+ and S_- at site l). If we want to keep track of physical quantities such as correlation functions, we must keep track of the matrix elements of additional operators in a similar way.

The structure of the algorithm is now as follows:

1. Set up an initial chain of length L, as we did above in the straightforward diagonalisation of the Heisenberg chain. This is the (first approximation to the) system U. The dimension of the Hilbert space is called N.
2. Calculate the ground state of this chain using the Lanczos method.
3. Trace out the degrees of freedom of the right half (E) of the chain U to obtain a reduced density matrix ρ^{S}.
4. Diagonalise the reduced transfer matrix of the left half S of U. Fill a matrix V with the M eigenvectors corresponding to the highest eigenvalues of the density matrix ρ^{S} as columns.
5. Reduce the $\sqrt{N} \times \sqrt{N}$ Hamiltonian H of S to a size $M \times M$ using this matrix V. Do the same for the operators S_l acting on the rightmost spin s_l.
6. Add a spin s_{l+1} to S, so that its Hilbert space now has dimension $\sqrt{N} = M \cdot M_S$.
7. Find the ground state of the $N \times N$ Hamiltonian of the system $\mathrm{U} = \mathrm{S} + \mathrm{E}$ which is spanned by the basis $|m, s_{l+1}, s_{l+2}, n\rangle$.
8. Return to step 3.

If we repeat the algorithm until the density matrix no longer, we have reached a fixed point which corresponds to an infinite chain with open ends.

We now specify how to calculate the product of the Hamiltonian with a n arbitrary vector. At each stage we have the Hamiltonian $H^{\mathrm{S,E}}$ of the system S, E without the spin s_{l+1} added. These Hamiltonians are given in the basis $|m\rangle$. We also have the matrix elements of spin operators $S_{\alpha,l}$ and $S_{\beta,l+3}$ with respect to this basis. The total Hamiltonian is then

$$H^{\mathrm{U}}_{m,s_{l+1},s_{l+2},n;m',s'_{l+1},s'_{l+2},n'} = H^{\mathrm{S}}_{mm'} + \sum_{\alpha\beta} J_{\alpha\beta}[S_{\alpha,l}]_{mm'}[S_{\beta,l+1}]_{s_{l+1},s'_{l+1}}$$

$$+ \sum_{\alpha\beta} J_{\alpha\beta}[S_{\alpha,l}]_{s_{l+1},s'_{l+1}}[S_{\beta,l+1}]_{s_{l+2},s'_{l+2}}$$

$$+ \sum_{\alpha\beta} J_{\alpha\beta}[S_{\alpha,l+2}]_{s_{l+2},s'_{l+2}}[S_{\beta,l+3}]_{n,n'} + H^{\mathrm{E}}_{nn'}, \quad (11.62)$$

where it is understood that every term in this expression must be multiplied by a few Kronecker deltas involving all quantum numbers which do not occur in it. The indices $\alpha\beta$ are zz, $+-$ and $-+$ for the Heisenberg Hamiltonian.

Now we consider the complexity of a matrix–vector multiplication. Evaluating the product of the first term in (11.62) with the vector $|\psi\rangle$ is carried out as follows:

$$\langle m, s_{l+1}, s_{l+2}, n|H^{\mathrm{S}}|\psi\rangle = \sum_{m'} H^{\mathrm{S}}_{mm'}\langle m', s_{l+1}, s_{l+2}, n|\psi\rangle, \quad (11.63)$$

and this requires $M^3 M_S^2$ steps, since for each of the $M^2 M_S^2$ elements, a sum over m' must be carried out.

The multiplication for the second term starts by evaluating an intermediate vector $|\phi\rangle$ which is $|\psi\rangle$ multiplied by the term involving $S_{\beta,l+1}$:

$$\langle m, s_{l+1}, s_{l+2}, n|\phi\rangle = \sum_{s'_{l+1}} [S_{\beta,l+1}]_{s_{l+1},s'_{l+1}}\langle m, s'_{l+1}, s_{l+2}, n|\psi\rangle \quad (11.64)$$

which takes $M^2 M_S^3$ steps. Then we multiply the intermediate state $|\phi\rangle$ by the term involving $S_{\alpha,l}$:

$$\langle m, s_{l+1}, s_{l+2}, n|S_{\alpha,l}|\phi\rangle = \sum_{m'} [S_{\alpha,l}]_{m,m'}\langle m', s_{l+1}, s_{l+2}, n|\phi\rangle. \quad (11.65)$$

This is obviously of the same complexity ($M^3 M_S^2$) as the first term.

The remaining terms can now easily be worked out analogous to this procedure.

PROGRAMMING EXERCISE

Write a DMRG program for the one-dimensional Heisenberg chain.

Check You should be able to reproduce the ground state energies for the spin-1/2 and spin-1 chain. The first is analytically known to be $-\ln 2 + 1/2 = -0.443\,147\ldots$ per site, and the second should be found at $-1.40\,148\ldots$ per site. Note that the best way to calculate the energies per site is to subtract the energies found in subsequent steps of the algorithm: this takes into account the energy of the *central* spins and this converges much faster to a fixed value than the total energy divided by the number of sites.

So far, we have considered the simplest DMRG method in fair detail. It is possible to extend the method to periodic chains by not reversing E with respect to S and coupling the right hand side of E to the left hand side of S. Another interesting extension concerns the finite chain instead of the infinite one: sometimes (especially when disorder or incommensurability plays a role) the ground state changes its global structure markedly when varying the chain length, so that studying a fixed length chain is indeed desirable. Another problem in which fixed length chains are very important is finite-size scaling. Although the DMRG method provides a useful way of finding ground states for the infinite chain, finite-size scaling is relevant for finding information about scaling exponents.

The finite-size algorithm works as follows. First the infinite-system algorithm is run for a number of steps, until the universe has the desired length L. Meanwhile, the Hamiltonian and other operators are stored in memory for each system size encountered so far. Then the system S is increased *at the expense of* the environment E, such that U remains constant in size. After E is shrunk to the lowest possible size (one, two or three spins), it is increased again in a stepwise fashion, while reducing the size of S, and this goes on until convergence has been achieved. In more detail, the algorithm for a system of finite size L proceeds along the following steps.

1. Carry out steps of the infinite-size algorithm until the size of the universe (S plus E plus two central spins) is L. During this step, store all the Hamiltonians and other necessary operators (see above) for all sizes in memory.
2. Reduce the matrices (Hamiltonian and necessary spin-operators) of S (which is the left half of the system generated in the previous step) to size M by tracing out E and identifying the most relevant eigenstates of ρ_S, followed by a projection of H_S and the spin-operators onto the space spanned by these most relevant eigenstates. We now have the Hamiltonian H_S for the left half of size $L/2$.
3. Add two spins to H_S, and an environment of length $L/2 - 2$ (the Hamiltonian and other matrices of this has been stored in memory). The system S is therefore increased, but E is *decreased.*
4. The procedure of the last two steps is repeated until the system S has size $L - 3$ (that is, E has size 1, but 2 or 3 is also possible). The matrices for S should still be stored for each size.
5. Now *reduce* the size of S in a similar way to the increase of the size of S. The environment therefore grows. For the environment, the stored matrices are used, whereas for the system, the matrices are updated at each step. Carry on until the system S has reached size 1 (and E has size $L - 3$).
6. Repeat the up and down sweeps of the previous steps until convergence has been achieved.

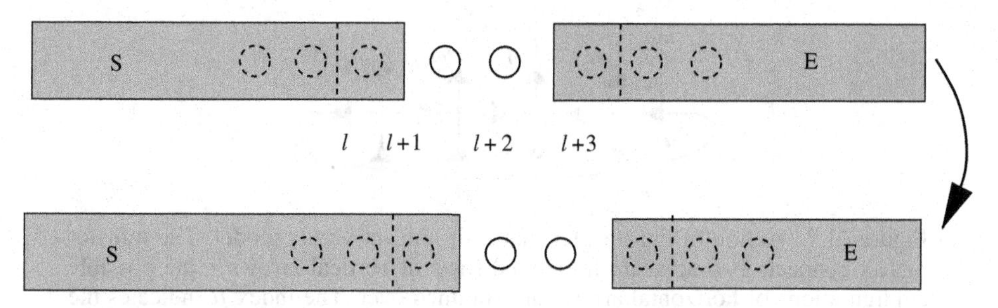

Figure 11.6. The third step in the finite-size DMRG algorithm.

Each time the sweep has reached one of the ends of the chain, the representation of the corresponding end is exact, as it contains only one (or a few) spins. This exact representation is responsible for the improvements in successive sweeps. Step 3 is shown in Figure 11.6. In the particular example of finite-size scaling, the matrices stored for size L can be used for the next size $L - 2$. So part of the work is used later again. Interestingly, the dynamics of polymer chains, which is governed by a Master equation, can also be treated within the DMRG approach, and for reptation and electrophoresis, useful results have been obtained [28, 29].

The major effort in constructing a program for the finite-size chain consists of storing the relevant operator matrices H, S_z etc. for the various sizes in memory. The algorithm can then be implemented straightforwardly. As a check, you can verify whether the ground state energy found corresponds to that found for finite Heisenberg chains in Section 11.5.1.

A major point is whether the DMRG method is applicable to systems in more than one dimension. This turns out to be difficult. Trials have been made in two directions. The first is to consider a two-dimensional system as a one-dimensional chain, 'wrapped up' to fill the plane [30]. The second is to formulate the Hubbard Hamiltonian in k-space, which renders the hopping term diagonal and the Coulomb interaction no longer local. Reviews concerning DMRG can be found in Refs. [31–33].

Exercises

11.1 The Pauli matrix σ_x is given as

$$\sigma_x = \begin{pmatrix} 0 & 1 \\ 1 & 0 \end{pmatrix}.$$

Using $\sigma_x^2 = 1$, show that

$$e^{\beta L \sigma_x} = \cosh(\beta L) + \sigma_x \sinh(\beta L).$$

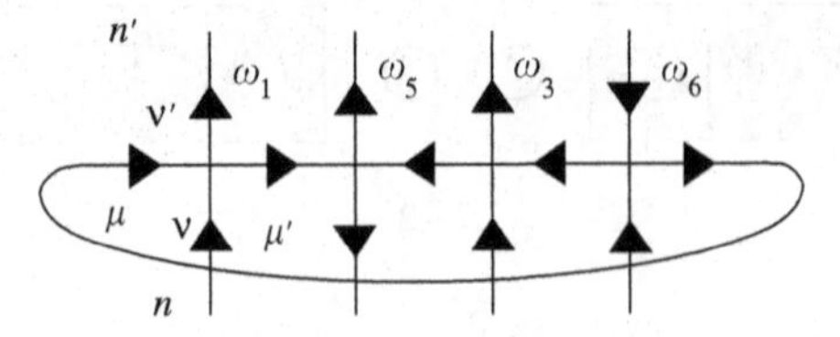

Figure 11.7. Action of the transfer matrix in the six-vertex model. The transfer matrix connects two adjacent horizontal rows of vertical arrows – the possible configurations of horizontal arrows are summed over. The index μ indicates the orientation of the leftmost arrow, and μ' represents the direction of the horizontal arrows in the iterative multiplication process. In this figure, it labels the second horizontal link, as in the first step of the multiplication. Only a single configuration of vertices is shown.

Comparing this with the transfer matrix of the one-dimensional Ising model with zero magnetic field, show that

$$\tanh(\beta L) = \exp(-2\beta J).$$

11.2 In this problem we consider the implementation of the transfer matrix method for the six-vertex model. The vertices and weights are represented in Figure 11.3. The transfer matrix connects a row of vertical arrows to the next one as in Figure 11.7. This implies that for each pair of adjacent rows of vertical arrows a sum must be performed over all possible configurations of horizontal arrows that are compatible with the vertical ones, in the sense that the number of ingoing arrows equals the number of outgoing ones at each site.

The multiplication of the transfer matrix with an arbitrary vector ϕ is again carried out site by site, similar to the procedure followed for the Ising model. Let us first consider the multiplication including the first (leftmost) vertex ($j = 1$). It turns out to be necessary to introduce a vector $\psi(n, \mu, \mu')$ in the multiplication routine. Here, n represents a row configuration of vertical arrows and μ and μ' are indices assuming the values 0 and 1 – they represent the direction of the leftmost horizontal arrow and the horizontal arrow on the link connecting sites j and $j+1$ ($\mu = 0$ denotes a right-pointing arrow, $\mu = 1$ a left-pointing one), which is the second horizontal arrow from the left as we are considering the first (leftmost) vertex.

As the two horizontal arrows of the first site are known, there is a unique relation between the upper and lower vertical arrows. These arrows correspond to the least significant bits ν and ν' which are given by n mod 2 and n' mod 2 respectively ($\nu = 0$ is an upward-pointing arrow, $\nu = 1$ a downward-pointing one). Therefore we have

$$\nu + \mu = \nu' + \mu'. \tag{E11.1}$$

This leads to the following code for the multiplication routine involving the first (leftmost) vertex:

```
FOR n = 0 TO 2^(M-1) - 1 DO {n runs over half the number of
                              row states }
l = 2n;                      {l is a number with least significant
                              bit equal to 0}
    ψ'(l, 0, 0) = ω1 ψ(n);
    ψ'(l, 1, 1) = ω3 ψ(n);
    ψ'(l + 1, 1, 0) = ω6 ψ(n);
    l = l + 1                {l is now a number with least
                              significant bit equal to 1}
    ψ'(l, 0, 0) = ω2 ψ(n);
    ψ'(l, 1, 1) = ω4 ψ(n);
    ψ'(l - 1, 0, 1) = ω6 ψ(n);
END FOR
```

Now the second vertex must be treated. We start from the vector $\psi'(n, \mu, \mu')$ and calculate the next vector $\psi''(n, \mu, \mu')$. The index μ remains the leftmost horizontal arrow (we need this to ensure PBC in the end), and the second corresponds to the link connecting sites 1 and 2. The procedure is analogous to that for the first vertex.

After completing the sequence through the full row, we calculate the resulting vector $\phi(n)$ from the last vector $\psi^{\text{last}}(n, \mu, \mu')$ obtained in the foregoing procedure by taking the trace:

```
FOR N = 0 TO 2^(M-1)
    φ(n) = ψ^last(n, 0, 0) + ψ^last(n, 1, 1);
END FOR
```

The results can be analysed in a similar fashion to the Ising model. For the weights of the F-model, the central charge should be equal to 1.

11.3 The Hamiltonian of the periodic Heisenberg chain can be written as

$$H = J \sum_{l=1}^{L} \mathbf{S}_l \mathbf{S}_{l+1}$$

with $\mathbf{S}_{L+1} \equiv \mathbf{S}_1$.

(a) Show that this can be transformed into

$$H = J \sum_{l=1}^{L} \left[\frac{1}{2}(S_{+,l}S_{-,l+1} + S_{-,l}S_{+,l+1}) + S_{z,l}S_{z,l+1} \right],$$

where $S_{\pm} = S_x \pm iS_y$. Give the matrix forms of $S_{\pm}$ for $s = 1/2$ and $s = 1$.

(b) Show that $S_z = \sum_{l=1}^{L} S_{z,l}$ commutes with H. Also show that

$$S^2 = \left(\sum_{l=1}^{L} \mathbf{S}_l \right)^2$$

commutes with H.

11.4 In the DMRG, we take the set of eigenvectors $|m\rangle$ with the highest eigenvalues of ρ^{S} as representatives of the system S. To show that this is indeed the best procedure, we consider the representation of an arbitrary state $|\psi\rangle$, of the universe U, which can be expanded as

$$|\psi\rangle = \sum_{mn} C_{mn} |mn\rangle,$$

where, as usual, m labels a basis vector of S and n one of E.

We now replace the basis $|m\rangle$ of S by a smaller basis $|u_\alpha\rangle$, $\alpha = 1, \ldots, M$ where we choose the orthonormal basis vectors $|u_\alpha\rangle$ such that the state $|\psi\rangle$ can be optimally represented by them. This means that the norm of the difference between the exact and the expanded state

$$\left| \sum_{mn} C_{mn} |mn\rangle - \sum_{\alpha n} D_{\alpha n} |u_\alpha n\rangle \right|^2$$

is minimal.

We can expand the basis vectors $|u_\alpha\rangle$ in terms of the basis $|m\rangle$:

$$|u_\alpha\rangle = \sum_m u_{\alpha m} |m\rangle.$$

(a) Formulate the error (i.e. the norm of the difference) in terms of the coefficients $D_{\alpha n}$ and $u_{\alpha m}$. Minimise this error with respect to both the $D_{\alpha n}$ and the $u_{\alpha m}$ and show that this leads to the equation

$$-\sum_m C_{nm} u_{\alpha m} + \sum_{m\alpha'} D_{\alpha n} u_{\alpha m} u_{\alpha' m} = 0$$

and

$$-\sum_n C_{mn} D_{\alpha n} + \sum_{n\alpha'} D_{\alpha n} D_{\alpha' n} u_{\alpha' m} = 0.$$

(b) Use the orthonormality of the $|u_\alpha\rangle$ to infer from the first equation:

$$\sum_m C_{mn} u_{\alpha m} = D_{\alpha n}.$$

(c) Substitute this into the second equation to find

$$\sum_{nm'} C_{mn} C_{m'n} u_{\alpha m'} = \sum_{nm'm''} C_{m'n} C_{m''n} u_{\alpha m'} u_{\alpha' m''} u_{\alpha' m}.$$

Using

$$\rho^{\mathrm{S}}_{mm'} = \sum_n C_{mn} C_{m'n},$$

the last equation can be written in the more compact form:

$$\rho^{\mathrm{S}}|u_\alpha\rangle = \sum_{\alpha'} \langle u_\alpha|\rho^{\mathrm{S}}|u_{\alpha'}\rangle |u_{\alpha'}\rangle.$$

Show that the last equation can be satisfied by requiring that $|u_\alpha\rangle$ is a normalised eigenvector of the operator ρ^{S}.

Also show that the above error is minimised by taking the eigenvectors corresponding to the largest eigenvalues of the density operator.

(d) The previous analysis was a bit sloppy as the minimisation is subject to the constraint that $\sum_{\alpha n} D_{\alpha n}|u_\alpha n\rangle$ be normalised, and that the $|u_\alpha\rangle$ are orthonormal. These constraints should be taken into account through Lagrange parameters. Show that this analysis leads to the conclusion that the $|u_\alpha\rangle$ form an invariant set under the action of ρ^{S}, very similar to what we have seen in the derivation of the Hartree–Fock formalism (see Section 4.5.2). Similar to the analysis there, we see that a set of eigenvalues of ρ^{S} can therefore be chosen as a basis.

11.5 We now approach the same problem as in the previous exercise from another point of view. We consider again the Hilbert spaces of the system S, the environment E and the universe U which is the system and environment together. For simplicity, we take all coefficients and basis functions to be real.

Suppose U is in a pure state $|\psi\rangle$. To this state there corresponds a density matrix ρ^{U} which can be reduced to a density matrix ρ^{S} of S or a density matrix ρ^{E} of E. We call λ_α the eigenvalues of ρ^{S} and λ_β those of E. The corresponding eigenstates are denoted $|u_\alpha\rangle$ and $|v_\beta\rangle$ respectively.

Obviously we can write the states $|\psi\rangle$ in terms of the eigenstates $|u_\alpha v_\beta\rangle$:

$$|\psi\rangle = \sum_{\alpha\beta} C_{\alpha\beta}|u_\alpha v_\beta\rangle.$$

(a) Derive the following equations:

$$\lambda_\alpha = \sum_\beta C_{\alpha\beta}^2$$

and

$$\lambda_\beta = \sum_\alpha C_{\alpha\beta}^2.$$

By considering $C_{\alpha\beta}$ as a matrix, show that the set of numbers λ_α must be equal to the set λ_β. In the following, we shall denote both by λ_α.

(b) Show that $|\psi\rangle$ can be expanded as

$$|\psi\rangle = \sum_\alpha \lambda_\alpha |u_\alpha v_\alpha\rangle.$$

and that the eigenvalues of ρ^{U} are $w_\alpha = \lambda_\alpha^2$. Show that the error in the representation of $|\psi\rangle$ on S by using a truncated set of eigenstates as a basis is given as the sum over the discarded weights w_α.

References

[1] H. W. J. Blöte and M. P. Nightingale, 'Critical behaviour of the two-dimensional Potts model with a continuous number of states: a finite size scaling analysis,' *Physica*, **A 112** (1982), 405–65.
[2] F. R. Gantmacher, *The Theory of Matrices*, vol. II. New York, Chelsea, 1959.
[3] H. W. J. Blöte, J. L. Cardy, and M. P. Nightingale, 'Conformal invariance, the central charge, and universal finite size amplitudes at criticality,' *Phys. Rev. Lett.*, **56** (1986), 742–5.
[4] J. Cardy, 'Conformal invariance,' in *Phase Transitions and Critical Phenomena*, vol. 11 (C. Domb and J. Lebowitz, eds.) London, Academic Press, 1987, pp. 55–126.
[5] M. Krech and D. P. Landau, 'Casimir effect in critical systems: a Monte Carlo simulation,' *Phys. Rev. E*, **53** (1996), 4414–23.
[6] R. J. Baxter, *Exactly Solved Models in Statistical Mechanics*. London, Academic Press, 1982.
[7] M. P. M. den Nijs, 'Extended scaling relations for the magnetic critical exponents of the Potts-model,' *Phys. Rev. B*, **27** (1983), 1674–9.
[8] M. P. Nightingale, 'Scaling theory and finite systems,' *Physica*, **A 83** (1976), 561–72.
[9] R. J. Baxter, S. B. Kelland, and F. Y. Wu, 'Equivalence of the Potts model or Whitney polynomial with an ice-type model,' *J. Phys. A*, **9** (1976), 397–406.
[10] B. Nienhuis, 'Coulomb gas formulation of two-dimensional phase transitions,' *Phase Transitions and Critical Phenomena*, vol. 11 (C. Domb and J. Lebowitz, eds.) London, Academic Press, 1987, 1–53.
[11] E. H. Lieb, 'Exact solution of the F model of an antiferroelectric,' *Phys. Rev. Lett.*, **18** (1967), 1046–8.
[12] M. P. Nightingale and H. W. J. Blöte, 'Monte Carlo calculation for the free energy, critical point and surface critical behaviour of three-dimensional Heisenberg ferromagnets,' *Phys. Rev. Lett.*, **60** (1988), 1562–5.
[13] A. N. Vasil'ev, L. A. Ponomarenko, H. Manaka, *et al.* 'Magnetic and resonant properties of quasi-one-dimensional antiferromagnet $LiCuVO_4$,' *Phys. Rev. B*, **64** (2004), 024419.
[14] R. M. Noack and S. R. Manmana, 'Diagonalization- and numerical renormalization-group-based methods for interacting quantum systems,' in *Proceedings of the IX Training Course in the Physics of Correlated Electron Systems and High-Tc Superconductors, 2004, AIP Conf. Proc.* vol. 78. New York, American Institute of Physics, 2005, pp. 93–163.
[15] P. Fulde, *Electron correlations in Molecules and Solids*, 3rd edn. Heidelberg, Springer, 1995.
[16] E. H. Lieb, T. Shultz, and D. Mattis, 'Two soluble models of an antiferromagnetic chain,' *Ann. Phys.*, **16** (1961), 407–66.
[17] F. D. M. Haldane, 'Nonlinear field theory of large-spin Heisenberg antiferromagnets: semi-classically quantized solitons of the one-dimensional easy-axis Néel state,' *Phys. Rev. Lett.*, **50** (1983), 1153–6.
[18] S. R. White, 'Density matrix formulation for quantum renormalization groups,' *Phys. Rev. Lett.*, **69** (1992), 2863–6.
[19] N. G. van Kampen, *Stochastic Processes in Physics and Chemistry*. Amsterdam, North-Holland, 1981.
[20] H. W. Blöte, 'The specific heat of magnetic linear chains,' *Physica*, **79B** (1975), 427–66.
[21] A. L. Malvezzi, 'An introduction to numerical methods in low-dimensional quantum systems,' *Braz. J. Phys.*, **33** (2003), 55–72.
[22] K. G. Wilson, 'The renormalization group: critical phenomena and the Kondo problem,' *Rev. Mod. Phys.*, **47** (1975), 773–840.
[23] S. R. White and R. M. Noack, 'Real-space quantum renormalization groups,' *Phys. Rev. Lett.*, **68** (1992), 3487–90.
[24] ALPS (Algorithms and Libraries for Physics Simulations) homepage. http://alps.comp-phys.org/.

[25] F. Alet *et al.*, 'The ALPS project: open source software for strongly correlated systems.' ar Xiv: cond-mat/0410407, (15 October 2004).

[26] M. A. Nielsen and I. L. Chuang, *Quantum Computing and Quantum Information Theory*. Cambridge, Cambridge University Press, 2000.

[27] W. H. Press, S. A. Teukolsky, W. T. Vetterling, and B. P. Flannery, *Numerical Recipes*, 2nd edn. Cambridge, Cambridge University Press, 1992.

[28] E. Carlon, A. Drzewiński, and J. M. J. van Leeuwen, 'Crossover behavior for long reptating polymers,' *Phys. Rev. E*, **64** (2001), 010801(R).

[29] A. Drzewiński, E. Carlon and J. M. J. van Leeuwen, 'Pulling reptating polymers by one end: magnetophoresis in the Rubinstein–Duke model,' *Phys. Rev. E*, **68** (2003), 061801.

[30] R. M. Noack, S. R. White, and D. J. Scalapino, 'The density matrix renormalization group for Fermion systems, Computer Simulations in Condensed Matter Physics VII'. Berlin, Heidelberg, Springer, 1994, pp. 93–163.

[31] U. Schollwöck, 'The density-matrix renormalization group,' *Rev. Mod. Phys.*, **77** (2003), 259–315.

[32] K. Hallberg, 'Density matrix renormalization: a review of the method and its applications,' in *Theoretical Methods for Strongly Correlated Electrons* (D. Senechal, A-M. Tremblay, and C. Bourbonnais, eds.), CRM Series in Mathematical Physics. Berlin, Heidelberg, Springer, 2003.

[33] I. Peschel, X. Wang, M. Kaulke, and K. Hallberg, eds., *Density Matrix Renormalization: A New Numerical Method in Physics*. Lecture Notes in Physics. Berlin, Heidelberg, Springer, 1999.

12

Quantum Monte Carlo methods

12.1 Introduction

In Chapters 1 to 4 we studied methods for solving the Schrödinger equation for many-electron systems. Many of the techniques described there carry over to other quantum many-particle systems, such as liquid helium, and the protons and neutrons in a nucleus. The techniques which we discussed there were, however, all of a mean-field type and therefore correlation effects could not be taken into account without introducing approximations. In this chapter, we consider more accurate techniques, which are similar to those studied in Chapter 10 and are based on using (pseudo-)random numbers – hence the name 'Monte Carlo' for these methods. In Chapter 10 we applied Monte Carlo techniques to classical many-particle systems; here we use these techniques for studying quantum problems involving many particles. In the next section we shall see how we can apply Monte Carlo techniques to the problem of calculating the quantum mechanical expectation value of the ground state energy. This is used in order to optimise this expectation value by adjusting a trial wave function in a variational type of approach, hence the name *variational Monte Carlo* (VMC).

In the following section we use the similarity between the Schrödinger equation and the diffusion equation in order to calculate the properties of a collection of interacting quantum mechanical particles by simulating a classical particle diffusion process. The resulting method is called *diffusion Monte Carlo* (DMC).

Then we describe the path-integral formalism of quantum mechanics, which is a formulation elaborated by Feynman, based on ideas put forward by Dirac [1], in which a quantum mechanical problem is mapped onto a classical mechanical system (at the expense of increasing the number of degrees of freedom). This classical many-particle system can then be analysed using methods similar to those used in Chapter 10. This is called the *path-integral Monte Carlo* method (PIMC).

The last section of this chapter is dedicated to a stochastic technique, based on diffusion Monte Carlo, for diagonalising the transfer matrix of a lattice spin model on a strip, for cases where the matrix size renders even sparse matrix diagonalisation methods unusable.

Some important applications of quantum Monte Carlo methods are to the electronic structure of molecules [2], to dense helium-4 [3, 4], and to lattice spin-systems [5]. The cited literature also contains detailed accounts of the various methods.

12.2 The variational Monte Carlo method

12.2.1 Description of the method

In Chapter 3 we studied the variational method for finding the ground state and the first few excited states of the quantum Hamiltonian. This was done by parametrising the wave function – in a linear or nonlinear fashion – and then finding the minimum of the expectation value of the energy in the space of parameters occurring in the parametrised (trial) wave function. We described in some detail how this calculation can be carried out if the parametrisation is linear, and we have seen in Chapters 4 to 6 that the choice of basis functions in the linear parametrisation is crucial for the feasibility of the method. Calculating the expectation value of the energy involves integrals over the degrees of freedom of the collection of particles, which can only be carried out if the basis does not include correlations (single-particle picture) and if parts of the integration can be done analytically, for example by using Gaussian basis functions.

In this section we consider the variational method again, but we want to relax some of the above-mentioned restrictions on the trial wave functions and calculate the high-dimensional integrals using Monte Carlo methods, which are very efficient for this purpose as we have seen in Chapter 10. This is called the variational Monte Carlo approach. It should be noted that for some simple atoms, such as hydrogen and helium, the integrations can often be carried out analytically or using direct numerical integration (as opposed to MC integration). However, if there are many more electrons, these methods are no longer applicable.

Let us briefly recall the variational method in the form of an algorithm:

1. Construct the trial many-particle wave function $\psi_{\boldsymbol{\alpha}}(R)$, depending on the S variational parameters $\boldsymbol{\alpha} = (\alpha_1, \ldots, \alpha_S)$. $\psi_{\boldsymbol{\alpha}}$ depends on the combined position coordinate R of all the N particles $R = \mathbf{r}_1, \ldots, \mathbf{r}_N$.
2. Evaluate the expectation value of the energy

$$\langle E \rangle = \frac{\langle \psi_{\boldsymbol{\alpha}} | H | \psi_{\boldsymbol{\alpha}} \rangle}{\langle \psi_{\boldsymbol{\alpha}} | \psi_{\boldsymbol{\alpha}} \rangle}. \tag{12.1}$$

3. Vary the parameters $\boldsymbol{\alpha}$ according to some minimisation algorithm and return to step (1).

The loop stops when the minimum energy is reached according to some criterion. It is the second step in this algorithm that we consider in this section. However, below, we shall describe a variational method in which the parameters $\boldsymbol{\alpha}$ are adjusted according to some numerical scheme within the Monte Carlo simulation.

It turns out that in realistic systems the many-body wave function assumes very small values in large parts of configuration space, so a straightforward procedure using homogeneously distributed random points in configuration space is bound to fail. This suggests that it might be efficient to use a Metropolis algorithm in which a collection of random walkers is pushed towards those regions of configuration space where the wave function assumes appreciable values. Suppose that we can evaluate $H\psi_T$ for any trial function ψ_T, which we shall always assume to be real, and let us define

$$E_L(R) = \frac{H\psi_T(R)}{\psi_T(R)} \tag{12.2}$$

(we omit the $\boldsymbol{\alpha}$-dependence of ψ_T). $E_L(R)$ is called the *local energy*: it is a function that depends on the positions of the particles and it is constant if ψ_T is the exact eigenfunction of the Hamiltonian. The more closely ψ_T approaches the exact wave function (apart from a multiplicative constant), the less strongly will E_L vary with R.

The expectation value of the energy can now be written as

$$\langle E\rangle = \frac{\int \mathrm{d}R\ \psi_T^2(R)E_L(R)}{\int \mathrm{d}R\ \psi_T^2(R)}. \tag{12.3}$$

Let us now construct a Metropolis-walk in the same spirit as in ordinary Monte Carlo calculations, but now with a stationary distribution $\rho(R)$ given by

$$\rho(R) = \frac{\psi_T^2(R)}{\int \mathrm{d}R'\ \psi_T^2(R')}. \tag{12.4}$$

The procedure is now as follows.

Put N walkers at random positions;
REPEAT
  Select next walker;
  Shift that walker to a new position, for example by moving one
    of the particles in the system within a cube with a suitably
    chosen size d;
  Calculate the fraction $p = [\psi_T(R')/\psi_T(R)]^2$, where R' is the new and
    R the old configuration;
  If $p < 1$ the new position is accepted with probability p;
  If $p \geq 1$ the new position is accepted;
UNTIL finished.

The expectation value of the local energy is now calculated as an average over the samples generated in this procedure, excluding a number of steps at the beginning, necessary to reach equilibrium. The decision to stop the simulation is based on the precision achieved and on the available processor time.

The algorithm should work in principle with a single walker. However, chances are that this walker gets stuck in one favourable region surrounded by barriers which are difficult to overcome. Using a large collection of walkers reduces this effect.

12.2.2 Sample programs and results

We demonstrate the VMC approach with some simple programs. Here and in the rest of this chapter, when dealing with many-particle systems, we shall assume units of mass, distance and energy to be such that the kinetic energy operator occurs in the Schrödinger equation as $-\nabla^2/2$.

We start with the harmonic oscillator in one dimension, described by the Hamiltonian (in dimensionless units):

$$H\psi(x) = \left[-\frac{1}{2}\frac{\mathrm{d}^2}{\mathrm{d}x^2} + \frac{1}{2}x^2\right]\psi(x). \tag{12.5}$$

The exact solution for the ground state is given by $\exp(-x^2/2)$ with energy $E_\mathrm{G} = 1/2$; we shall use the trial function $\exp(-\alpha x^2)$. The exact solution lies therefore in the variational subspace. The local energy is given by

$$E_\mathrm{L} = \alpha + x^2\left(\frac{1}{2} - 2\alpha^2\right). \tag{12.6}$$

For $\alpha = 1/2$ the local energy is $1/2$, independent of the position, and we shall certainly find an energy expectation value $1/2$ in that case (this might happen even when the program contains errors!). The crucial test is whether this energy expectation value is a minimum as a function of α. In Table 12.1 we show that this is indeed the case. We also show the variance of the energy. This quantity will be small if E_L is rather flat, and this will be the case when ψ_T is close to the exact ground state: the closer ψ_T is to the ground state wave, the smaller the variance, and this quantity reaches its minimum value at the variational minimum of the energy itself. Again, in this particular case where the trial wave function can become equal to the exact ground state, the variance becomes zero. From the table we see that the variance does indeed decrease to 0 when the ground state is approached. Interestingly, for this simple case, it is possible to calculate the expectation value of the energy as a function of α by integrating the local energy weighted by ψ_T^2. The Gaussian form of the trial wave function makes the integral solvable with the result

$$E_\mathrm{v} = \frac{1}{2}\alpha + \frac{1}{8\alpha}. \tag{12.7}$$

Table 12.1. *Variational Monte Carlo energies.*

Harmonic oscillator					Hydrogen atom			Helium atom		
α	$\langle E\rangle$	$\mathrm{var}(\langle E\rangle)$	E_v	$\mathrm{var}(E)_\mathrm{v}$	α	$\langle E\rangle$	$\mathrm{var}(\langle E\rangle)$	α	$\langle E\rangle$	$\mathrm{var}(\langle E\rangle)$
0.4	0.5124(1)	0.02521(5)	0.5125	0.0253125	0.8	−0.4796(2)	0.0243(6)	0.05	−2.8713(4)	0.1749(2)
0.45	0.50276(4)	0.00556(2)	0.50278	0.00557	0.9	−0.4949(1)	0.0078(2)	0.075	−2.8753(4)	0.1531(2)
1/2	1/2	0	1/2	0	1.0	−1/2	0	0.10	−2.8770(3)	0.1360(2)
0.55	0.50232(6)	0.00454(1)(1)	0.5022727	0.0045558	1.1	−0.4951(2)	0.0121(4)	0.125	−2.8780(4)	0.1223(2)
0.6	0.5084(1)	0.0168(4)	0.508333	0.0168056	1.2	−0.4801(3)	0.058(2)	0.15	−2.8778(3)	0.1114(2)
								0.175	−2.8781(3)	0.1028(2)
								0.20	−2.8767(4)	0.0968(2)
								0.25	−2.8746(10)	0.0883(2)

VMC energies are given for the harmonic oscillator, the hydrogen atom and the helium atom for various values of the variational parameters. In each case, 400 walkers have been used and 30 000 displacements per walker were attempted. The first 4000 of these were removed from the data to ensure equilibrium. The expectation value $\langle E\rangle$ of the ground state energy is given, together with the variance in this quantity, $\mathrm{var}(\langle E\rangle)$. For the harmonic oscillator, the analytical values for the energies and variance are also given (E_v and $\mathrm{var}(E)_\mathrm{v}$).

The same can be done for the variance with the result

$$\mathrm{var}(E)_{\mathrm{v}} = \frac{(1-4\alpha^2)^2}{32\alpha^2}. \tag{12.8}$$

The Monte Carlo results match the analytical values as is clear from the table. Also in Table 12.1 we show results for the hydrogen atom with the Hamiltonian

$$H = -\frac{1}{2}\nabla^2 - \frac{1}{r}. \tag{12.9}$$

The exact ground state with energy $E = -1/2$ is given as e^{-r}; we take variational trial functions of the form $\mathrm{e}^{-\alpha r}$, so that the ground state is again incorporated in the variational subspace. Although we could consider the present problem as a one-dimensional one by using the spherical symmetry of the potential and the ground state wave function, we shall treat it here as a fully three-dimensional problem to illustrate the general approach. For this case, the analytical values of the average local energy and variance can also be calculated. This is left as an exercise for the reader.

The local energy is given by

$$E_{\mathrm{L}}(r) = -\frac{1}{r} - \frac{1}{2}\alpha\left(\alpha - \frac{2}{r}\right). \tag{12.10}$$

It is seen from Table 12.1 that the energy is minimal at the ground state and that its variance vanishes there too.

Finally we consider the helium atom, which we have already studied extensively in Chapters 4 and 5. Constructing good trial functions is a problem on its own – here we shall use the form:

$$\psi(\mathbf{r}_1, \mathbf{r}_2) = \mathrm{e}^{-2r_1}\mathrm{e}^{-2r_2}\mathrm{e}^{r_{12}/[2(1+\alpha r_{12})]} \tag{12.11}$$

where $r_{12} = |\mathbf{r}_1 - \mathbf{r}_2|$. This function consists of a product of two atomic one-electron orbitals and a correlation term. The local energy now has the form:

$$E_{\mathrm{L}}(\mathbf{r}_1, \mathbf{r}_2) = -4 + (\hat{\mathbf{r}}_1 - \hat{\mathbf{r}}_2)\cdot(\mathbf{r}_1 - \mathbf{r}_2)\frac{1}{r_{12}(1+\alpha r_{12})^2} - \frac{1}{r_{12}(1+\alpha r_{12})^3} - \frac{1}{4(1+\alpha r_{12})^4} + \frac{1}{r_{12}} \tag{12.12}$$

With $\hat{\mathbf{r}}$ we denote a unit vector along $\mathbf{r}$, and r_{12} is the distance between the two electrons. Energies and variances are also displayed in Table 12.1. The variance does not have a sharp minimum for the same value of α as the energy. The reason is that most of the variance is due to the trial wave function not being exact, even for the best value of α. The optimum value of the energy, $-2.878\,1 \pm 0.000\,5$, should be compared with the Hartree-Fock value of $-2.861\,7$ a.u. and the DFT value of −2.83 a.u., and with the exact value of −2.903 7 a.u. The VMC value can obviously be improved by including more parameters in the wave function. The wave function is apparently not perfect. One of its deficits can be appreciated by considering the case where one of the electrons is far away from the nucleus and the other electron. Then the trial wave function depends on the position of this particle like the wave function of the helium ion, i.e. it is the asymptotic wave function for an electron in the field of a $Z = 2$ nucleus. In reality, however, the wave function should 'see' a charge $Z = 1$ as the other electron shields off one unit charge.

It is possible to adjust the value of the parameters α in these simulations 'on the fly' [6]. To this end, we need a minimum finder. The most efficient minimum finders use the gradient of the function to be minimised (see Appendix A). This is a problem, as a finite difference calculation of the gradient is bound to fail: the derivatives of stochastic variables are subject to large numerical errors. However, from the analytic derivative of the wave function with respect to α, we can sample this derivative over the population of walkers. From (12.3) we see that

$$\frac{dE}{d\alpha} = 2\left(\left\langle E_L \frac{d \ln \psi_T}{d\alpha}\right\rangle - E\left\langle \frac{d \ln \psi_T}{d\alpha}\right\rangle\right). \tag{12.13}$$

Using a simple damped steepest decent method:

$$\alpha_{\text{new}} = \alpha_{\text{old}} - \gamma \left(\frac{dE}{d\alpha}\right)_{\text{old}}, \tag{12.14}$$

the method then finds the optimal value (and therefore also the energy) for α. This method works remarkably well for the harmonic oscillator, where, starting from $\alpha = 1.2$, the correct value $\alpha = 0.5$ is found in a small fraction of the time needed for accurately evaluating one of the points in Table 12.1. However, the success in this particular case is partly due to the exact solution being in the family of solutions considered. The method is generalised straightforwardly to more parameters. It has been applied successfully to electrons in quantum dots [6].

The reader is invited to write the programs described and check the results with those given in Table 12.1.

12.2.3 Trial functions

The trial wave function for helium, Eq. (12.11), is the two-particle version of the general ground state trial wave function used in quantum Monte Carlo (QMC) calculations of fermionic systems:

$$\psi(\mathbf{x}_1, \ldots, \mathbf{x}_N) = \Psi_{AS}(\mathbf{x}_1, \ldots, \mathbf{x}_N) \exp\left[\frac{1}{2}\sum_{i,j=1}^{N} \phi(r_{ij})\right]. \tag{12.15}$$

Ψ_{AS} is the Slater determinant (see Chapter 4) and ϕ is a function which contains the two-particle correlation effects. For identical bosons, all the minus-signs in the determinant are replaced by pluses. The particular form we chose in the helium case is a simple form of a class called Padé–Jastrow wave functions [7]. Inclusion of three and four point correlations is obviously possible. We shall not go into the problem of finding the best Slater determinants and ϕ-functions but restrict ourselves to a short discussion of the requirements which we can derive for special particle configurations – these are the 'cusp conditions': boundary conditions satisfied at the points where the potential diverges. Near these points the kinetic and potential energy contributions of the Hamiltonian are both very large, and they should cancel out for a large part. This leads to large statistical fluctuations which are avoided by respecting the cusp conditions. In the next section we shall see that these cusp conditions are essential for trial wave functions used in the DMC method. We have already dealt with a similar problem in Chapter 2 of this book, when we found appropriate boundary conditions for the numerical solution of the radial Schrödinger equation with a Lennard–Jones potential, which diverges strongly at $r = 0$. Now we consider singularities in the Coulomb potential.

In the helium atom, the potential diverges when one of the electrons approaches the nucleus, or when the electrons are close to each other. The Schrödinger equation can be solved analytically for these configurations since the Coulomb potential dominates all other terms except the kinetic one. Suppose that one of the electrons, labelled i, is very close to a nucleus (which we take at the origin) with charge Z. In that case the Schrödinger equation becomes approximately

$$\left[-\frac{1}{2}\nabla_i^2 - \frac{Z}{r_i}\right]\psi(\mathbf{r}_1, \ldots, \mathbf{r}_N) = 0. \tag{12.16}$$

Writing out the kinetic energy in spherical coordinates of particle i, we arrive at a radial Schrödinger equation of the form ($r = r_i$)

$$\left[\frac{\mathrm{d}^2}{\mathrm{d}r^2} + \frac{2}{r}\frac{\mathrm{d}}{\mathrm{d}r} + \frac{2Z}{r} - \frac{l(l+1)}{r^2}\right]R(r) = 0. \tag{12.17}$$

If, as is usually the case, the wave function is radially symmetric in $\mathbf{r}_i$ for r_i small, we have exclusively an $l = 0$ contribution, and the two terms containing the factor $1/r$ must cancel (the first term does not contribute for a function which is regular at the origin). For $R(0) \neq 0$ this leads to

$$\frac{1}{R}\frac{\mathrm{d}R}{\mathrm{d}r} = -Z, \qquad r = 0; \tag{12.18}$$

so that $R(r) = \exp(-Zr)$.

For $l > 0$, the radial wave function is written in the form $r^l \rho(r)$ where ρ does not vanish at $r = 0$. Analysing this in a way similar to the $l = 0$ case leads to the cusp condition

$$\frac{1}{\rho(r)} \frac{\mathrm{d}\rho(r)}{\mathrm{d}r} = -\frac{Z}{l+1}. \tag{12.19}$$

Note that this form is the same as (12.18) if we put $l = 0$.

Another cusp condition is found for two electrons approaching each other. Considering the trial wave function of the helium atom, Eq. (12.11), we see that it is the dependence on the separation r between the two electrons which must incorporate the correct behaviour in this limit. The resulting radial equation for the r dependence is the same as for the electron–nucleus cusp except for the $-Z/r$ potential being replaced by $1/r$ (the Coulomb repulsion between the two electrons), and the kinetic term being twice as large (because the reduced mass of the two electrons is half the electron mass):

$$\left[2\frac{\mathrm{d}^2}{\mathrm{d}r^2} + \frac{4}{r}\frac{\mathrm{d}}{\mathrm{d}r} - \frac{2}{r} - \frac{l(l+1)}{r^2}\right] R(r) = 0. \tag{12.20}$$

The cusp condition, written in terms of $\rho(r) = r^{-l}R(r)$, is therefore

$$\frac{1}{\rho(r)} \frac{\mathrm{d}\rho(r)}{\mathrm{d}r} = \frac{1}{2(l+1)}. \tag{12.21}$$

The right hand side reduces to $1/2$ in the usual case of an s-wave function ($l = 0$). For like spins, the value of the wave function must vanish if the particles approach each other; therefore the wave function with lowest energy is a p-state and the right hand side will reduce to 1/4. For a general system, containing more than two electrons, we have this cusp condition for each electron pair ij. It is recommended to have a look at Problem 12.5 to see how cusp conditions are implemented in practice.

12.2.4 Diffusion equations, Green functions and Langevin equations

In the following sections we shall discuss several QMC methods in which the ground state of a quantum Hamiltonian is found by simulating a diffusion process. In the next section for example, we shall use such a simulation to improve on the variational method described above. In this section, we give a brief overview of diffusion and the related equations.

Consider a one-dimensional discrete axis with sites located at na, with integer n. We place a random walker on a site, and this walker jumps from site to site with time intervals h. The walker can only jump from a site to its left or right neighbour. Both jumps have a probability α, and the walker remains at the current position with probability $1 - 2\alpha$. This is clearly a Markov process as described in Section 10.3. We are interested in the probability $\rho(x, t)$ to find the walker at site $x = na$ at time $t = mh$, where n and m are both integer. This probability satisfies the master equation of the Markov process:

$$\rho(x, t + h) - \rho(x, t) = \alpha[\rho(x + a, t) + \rho(x - a, t) - 2\rho(x, t)] \approx \alpha a^2 \frac{\partial^2 \rho(x, t)}{\partial x^2}. \tag{12.22}$$

For small h, the left hand side can be written as $h(\partial\rho/\partial t)$, and defining $\gamma = a^2\alpha/h$, we can write the continuum form of the master equation (for small a) as

$$\frac{\partial \rho(x, t)}{\partial t} = \gamma \frac{\partial^2 \rho(x, t)}{\partial x^2}. \tag{12.23}$$

This equation is called the *diffusion equation*: it describes how the probability distribution of a walker evolves in time. It may equivalently be interpreted as the density distribution for a large collection of independent walkers.

Consider the following function:

$$G(x, y; t) = \frac{1}{\sqrt{4\pi\gamma t}} \mathrm{e}^{-(x-y)^2/(4\gamma t)}. \tag{12.24}$$

This function has the following properties:

- Considered as a function of y and t, keeping x fixed, it is a solution of the diffusion equation for $t > 0$.
- For $t \to 0$, G reduces to a delta-function:

$$G(x, y; t) \to \delta(x - y) \text{ for } t \to 0. \tag{12.25}$$

G is called the *Green's function* of the diffusion equation. This function can be used to write the time evolution of any initial distribution $\rho(x, 0)$ of this equation in integral form:

$$\rho(y, t) = \int \mathrm{d}x\, G(x, y; t)\rho(x, 0), \tag{12.26}$$

which can easily be checked using the properties of G. Inspection of the Green's function shows that it is normalised, that is, $\int \mathrm{d}y\, G(x, y; t) = 1$, independent of x and t.

The Green's function can be interpreted as the probability distribution of a single walker which starts off at position x at $t = 0$. We can use G to construct a new Markov process corresponding to the diffusion equation. We discretise the time in steps Δt. We start with a walker localised at x at $t = 0$. Then we move this walker to a new position y at time Δt with probability distribution $G(x, y; \Delta t)$. From this, we move the walker to a new position z at time $2\Delta t$ with probability distribution $G(y, z; \Delta t)$. We have therefore a Markov process with transition probability given by G:

$$T_{\Delta t}(x \to y) = G(x, y; \Delta t). \tag{12.27}$$

Using the properties of the Green's function it can be shown that the detailed balance condition for the master equation for the Markov process leads to the integral form (12.26), so that the Markov process does indeed model the diffusion process described by (12.23) (check this). The difference between this process and the previous one on the discrete lattice is that we now use the continuum solution of the former version, which should be much more efficient, as a single step in the continuum diffusion process represents a large number of steps in the discrete diffusion process. The Markov process described by (12.27) can be summarised by the equation

$$x(t + \Delta t) = x(t) + \eta\sqrt{\Delta t}, \tag{12.28}$$

where η is a Gaussian random variable with variance 2γ:

$$P(\eta) = \frac{1}{\sqrt{4\pi\gamma}} \mathrm{e}^{-\eta^2/(4\gamma)}. \tag{12.29}$$

This result can be understood by realising that a step in the Markov process (12.27) is distributed according to a Gaussian with width $\sqrt{2\gamma\Delta t}$. In this form, the process is recognised as a Langevin equation for discrete time. Note that a random momentum rather than a random force is added at each step, in contrast to the Langevin equation discussed in Section 8.8.

The general form of the diffusion equation is

$$\frac{\partial \rho}{\partial t} = \mathcal{L}\rho(x, t), \tag{12.30}$$

where $\mathcal{L}$ is a second order differential operator. The formal solution of this equation with a given initial distribution $\rho(x, 0)$ can be written down immediately:

$$\rho(x, t) = \mathrm{e}^{t\mathcal{L}}\rho(x, 0) \tag{12.31}$$

but as this involves the exponential of an operator (which is to be considered as an infinite power series), it is not directly useful. Using Dirac notation, the Green's function can formally be written as

$$G(x, y; t) = \langle x|\mathrm{e}^{t\mathcal{L}}|y\rangle, \tag{12.32}$$

which indeed satisfies the equation (12.31) as a function of y and t, and which reduces to $\delta(x - y)$ for $t = 0$. The diffusion equation can only be used to construct a Markov chain if the Green's function is normalised, in the sense that $\int \mathrm{d}y\ G(x, y; t) = 1$, independent of t. This is not always the case, as we shall now see.

A particular diffusion equation which we shall encounter later in this chapter is

$$\frac{\partial \rho}{\partial \tau} = \frac{1}{2}\frac{\partial^2 \rho(x,\tau)}{\partial x^2} - V(x)\rho(x,\tau). \tag{12.33}$$

This looks very much like the one-dimensional time-dependent Schrödinger equation for a zero-mass particle; in fact, this equation is recovered when we continue the time analytically into imaginary time $\tau = it$ (we use τ for imaginary time). Using (12.31), we can write the solution as

$$\rho(x,\tau) = \mathrm{e}^{\tau(-K-V)}\rho(x,0) \tag{12.34}$$

where K is the kinetic energy operator $K = p^2/2 = -1/2(\partial^2/\partial x^2)$ (p is the momentum operator $p = -\mathrm{i}(\partial/\partial x)$ of quantum mechanics). The exponent cannot be evaluated because the operators K and V do not commute. However, we might neglect Campbell–Baker–Hausdorff (CBH) commutators – this is only justified when τ is small. To emphasise that the following is only valid for small τ, we shall use the notation $\Delta\tau$ instead of τ. We have

$$\mathrm{e}^{-\Delta\tau(K+V)} = \mathrm{e}^{-\Delta\tau K}\mathrm{e}^{-\Delta\tau V} + \mathcal{O}(\Delta\tau^2) \tag{12.35}$$

where the order $\Delta\tau^2$ error term results from the neglect of CBH commutators. To find the Green's function explicitly, we must find the matrix element of the exponential operator on the right hand side. The term involving the potential is not a problem as this is simply a function of x. It remains then to find the matrix elements of the kinetic operator:

$$G_{\mathrm{Kin}}(x,y;\Delta\tau) = \langle x|\mathrm{e}^{-\Delta\tau\hat{p}^2/2}|y\rangle \tag{12.36}$$

where $\hat{p}$ is the momentum operator – we have used the caret ˆ to distinguish the operator from its eigenvalue.

The Green's function can be evaluated explicitly by inserting two resolutions of the unit operator of the form $\int \mathrm{d}p\ |p\rangle\langle p|$ and using the fact that

$$\langle x|p\rangle = \frac{1}{\sqrt{2\pi}}\mathrm{e}^{\mathrm{i}px} \quad (\hbar \equiv 1). \tag{12.37}$$

As the kinetic operator is diagonal in the p-representation, the matrix element is then found simply by performing a Gaussian integral. The result is

$$G_{\mathrm{Kin}}(x,y;\Delta\tau) = \frac{1}{\sqrt{2\pi\,\Delta\tau}}\mathrm{e}^{-(x-y)^2/(2\Delta\tau)}. \tag{12.38}$$

This form is recognised as the Green's function of the simple diffusion equation; indeed our imaginary-time Schrödinger equation reduces to this equation for $V \equiv 0$, and therefore the kinetic part of our Green's function should precisely be equal to the Green's function of the simple diffusion equation. We have derived this form explicitly here, because we need to find the Green's function for a more complicated type of diffusion equation along the same lines below.

The full Green's function for the diffusion equation (12.33) reads:

$$G(x, y; \Delta\tau) = G_{\text{Kin}}(x, y; \Delta\tau)\mathrm{e}^{-\Delta\tau V(y)} + \mathcal{O}(\Delta\tau^2). \tag{12.39}$$

Unfortunately, the term involving the potential destroys the normalisation of the full Green's function, and this prevents us from using it to construct a Markov chain evolution, which is convenient, if not essential, for a successful simulation as we shall see later. We can make the transition rate Markovian by normalising it, which can be done by multiplying the Green's function by a suitable prefactor $\exp(\tau E_{\text{T}})$. Of course we do not know beforehand what the value of this prefactor is, but we shall describe methods for sampling its value in Section 12.3. The new, normalised, Green's function is no longer the proper Green's function for Eq. (12.33), but for a modified form of this equation, in which the potential has been shifted by an amount E_{T}:

$$\frac{\partial \rho}{\partial \tau} = \frac{1}{2}\frac{\partial^2 \rho(x, \tau)}{\partial x^2} - [V(x) - E_{\text{T}}]\rho(x, \tau). \tag{12.40}$$

If we choose E_{T} such that the Green's function is normalised, it describes a Markov process, hence there will be an invariant distribution. This invariant distribution is determined by Eq. (12.40), which for stationary distributions reduces to

$$-\frac{1}{2}\frac{\partial^2 \rho(x)}{\partial x^2} + V(x)\rho(x) = E_{\text{T}}\rho(x), \tag{12.41}$$

which is the stationary Schrödinger equation.

For many problems, it is convenient to construct some Markovian diffusion process which has a predefined distribution as its invariant distribution. This turns out to be possible, and the equation is called the Fokker–Planck (FP) equation. It has the form

$$\frac{\partial \rho(x, t)}{\partial t} = \frac{1}{2}\frac{\partial}{\partial x}\left[\frac{\partial}{\partial x} - F(x)\right]\rho(x, t). \tag{12.42}$$

The 'force' $F(x)$ is related to the invariant distribution $\rho(x)$: the relation is given by

$$F(x) = \frac{1}{\rho(x)}\frac{\mathrm{d}\rho(x)}{\mathrm{d}x}. \tag{12.43}$$

It can easily be checked that $\rho(x)$ satisfies (12.42) when the time derivative occurring in the left hand side of this equation is put equal to zero.

The Green's function can be found along the same lines as that of the kinetic part of the Green's function for the imaginary-time Schrödinger equation. We must work out

$$G(x, y; t) = \langle x|\mathrm{e}^{-\Delta t\hat{p}[\hat{p} - iF(\hat{x})]/2}|y\rangle. \tag{12.44}$$

We again separate the exponent into two terms, one containing $\hat{x}$ and the other $\hat{p}$, at the expense of an $\mathcal{O}(\Delta t^2)$ error. Calculating Gaussian Fourier transforms as before, we obtain the result:

$$G(x, y; \Delta t) = \frac{1}{\sqrt{2\pi \Delta t}} e^{-[y-x-F(x)\Delta t/2]^2/(2\Delta t)}. \tag{12.45}$$

Note that this expression is a first order approximation in Δt of the exact Green's function. This is normalised, and we can therefore use it again for constructing a Markov chain. This is done by moving the random walker first from its old position x to the position $x + F(x)\Delta t/2$ and then adding a random displacement $\eta\sqrt{\Delta t}$, where η is drawn from a Gaussian distribution with a variance 1 (see Eq. (12.29)). In formula, the method reads

$$x(t + \Delta t) = x(t) + \Delta t F[x(t)]/2 + \eta\sqrt{\Delta t}, \tag{12.46}$$

so it is a discrete Langevin equation with 'force' F.

We end this section with a few remarks. First, all results can be extended straightforwardly to higher dimensions. Using a $3N$-dimensional variable R instead of the one-dimensional variable x (R denotes the positions of a set of particles in three dimensions as usual), the Green's function of the simple diffusion equation Eq. (12.23) with $\gamma = 1/2$ is

$$G_{3N}(R, R'; t) = \frac{1}{(2\pi t)^{3N/2}} e^{-(R'-R)^2/(2t)}. \tag{12.47}$$

The Green's function of the Fokker–Planck equation (12.42) becomes

$$G_{3N}(R, R'; \Delta t) = \frac{1}{(2\pi \Delta t)^{3N/2}} e^{-[R'-R-\Delta t\mathbf{F}(R)/2]^2/(2\Delta t)}, \tag{12.48}$$

where $\mathbf{F}(R)$ is a three-dimensional vector, given by

$$\mathbf{F}(R) = \nabla_R \rho(R)/\rho(R). \tag{12.49}$$

You might have been surprised by the way in which the exponential containing noncommuting operators was split in Eq. (12.35). After all, the following splitting

$$e^{-\Delta\tau(V+K)} = e^{-\Delta\tau V/2} e^{-\Delta\tau K} e^{-\Delta\tau V/2} + \mathcal{O}(\Delta\tau^3) \tag{12.50}$$

is more accurate: you can check that the first order CBH commutator vanishes, hence the $\mathcal{O}(\Delta\tau^3)$ error. The reason we use the simpler splitting (12.35) is that diffusion steps are carried out *successively*, hence the rightmost term in the right hand side of (12.50) at one step combines with the leftmost term at the next step, so that the total effect of the more accurate splitting is reduced to a different first and final step. This difference is, however, of the same order of magnitude as the accumulated error of the sequence of steps, and therefore it does not pay to use (12.50).

12.2.5 The Fokker–Planck equation approach to VMC

The VMC method described in Sections 12.2.1 and 12.2.2 has an important disadvantage: typical many-particle wave functions are very small in large parts of configuration space and very large in small parts of configuration space. This means, first, that we might have difficulty in finding the regions where the wave function is large, and second, that attempted moves of walkers from a favourable region (where the wave function is large) will be rejected when they move out of that region. Having a substantial fraction of rejected moves is part of any Metropolis Monte Carlo scheme, and we could live with that if there did not exist a more efficient approach, based on the Fokker–Planck equation described in the previous section.

In this method we try to sample the function $\rho(R) = \psi_T^2(R)$ rather than the trial function $\psi_T(R)$ itself: that is, we use

$$\mathbf{F} = 2\nabla_R \psi_T(R)/\psi_T(R) \tag{12.51}$$

in the FP equation.

The distribution $\rho(R,t)$ can be sampled by simulating a diffusion process. The algorithm is close to that of ordinary VMC. Now we let a collection of walkers diffuse with probabilities given by the Green's function (12.45):

Put N walkers at random positions;
REPEAT
  Select next walker;
  Shift that walker from its current position R to $R + \mathbf{F}(R)\Delta t/2$;
  Displace that walker by an amount $\boldsymbol{\eta}\sqrt{\Delta t}$, where $\boldsymbol{\eta}$ is a
    random vector with a Gaussian distribution (see (12.29) and (12.28));
UNTIL finished.

We see that there is no acceptance/rejection step; this causes the gain in efficiency when using the FP approach.

Note that we have made a time-step error of order $(\Delta t)^2$. It is possible to eliminate this error by combining this Langevin approach with a Metropolis procedure. The point is that we know the form of stationary distribution ρ (it is the square of the trial function ψ_T), and the Langevin process leads to a distribution which is close to but not exactly equal to this distribution. The Metropolis algorithm can give us the desired distribution ρ by acceptance/rejection of the Langevin steps, which themselves are considered as trial moves in the Metropolis algorithm. Referring back to Section 10.3, we call the transition probability of the Langevin equation $\omega_{RR'} = G(R,R';\Delta t)$, where G is given in (12.48). This is not symmetric in R and R' as F depends only on R, and therefore we have to use the generalised Metropolis algorithm, described at the very end of Section 10.3. The Langevin trial move is accepted with probability $\min(1, q_{RR'})$, where

$$q_{RR'} = \frac{\omega_{R'R}\rho(R')}{\omega_{RR'}\rho(R)}. \tag{12.52}$$

Note that the fraction $\omega_{R'R}/\omega_{RR'}$ is in equilibrium approximately equal to the ratio $\rho(R)/\rho(R')$ – if no time step error was made in constructing $\omega_{RR'}$, they would have been exactly equal – so $q_{RR'}$ is always close to 1. The acceptance rate is therefore always high when Δt is taken small, and the method is very efficient. The Metropolis acceptance/rejection step is merely a correction for the time step discretisation error made in the Langevin procedure.

The implementation of the algorithm is straightforward. The resulting energies must be the same as for the standard VMC method, but the error bars are smaller. As an example, an MC simulation for the harmonic oscillator using 300 walkers which perform 3000 steps and $\alpha = 0.4$ yields for the energy expectation in the ordinary VMC program value $E = 0.51 \pm 0.03$, to be compared with $E = 0.515 \pm 0.006$ in the Fokker–Planck program.

Variational Monte Carlo has the advantage that it is simple and straightforward. An important disadvantage is that it relies on the quality of the trial function, hence subtle but important physical effects are sometimes neglected when they are not taken into account when constructing the trial function.

12.3 Diffusion Monte Carlo

12.3.1 Simple diffusion Monte Carlo

The second quantum Monte Carlo method that we consider is the so-called *diffusion* or *projector* Monte Carlo method, abbreviated as DMC. This method does not use variational principles for obtaining ground state properties, but as we shall see, the convergence rate of the practical version of this method relies heavily on the accuracy of the trial functions. The idea of this method has already been sketched in Section 12.2.4. We use the imaginary time form of the time-dependent Schrödinger equation. This is a diffusion equation with a potential. We use the Green's function in the 'normalised' form, i.e. with the normalisation factor $\exp(-\Delta\tau E_\mathrm{T})$ present:

$$G(R, R'; \Delta\tau) = \mathrm{e}^{-\Delta\tau[V(R)-E_\mathrm{T}]}\frac{1}{\sqrt{2\pi\,\Delta\tau}}\mathrm{e}^{-(R-R')^2/\Delta\tau} + \mathcal{O}(\Delta\tau^2). \tag{12.53}$$

This Green's function is a short-time approximation of the imaginary-time operator $\exp[-\tau(H - E_\mathrm{T})]$. If we resolve this operator in its eigenstates $|\phi_n\rangle$, we obtain

$$\mathrm{e}^{-\tau(H-E_\mathrm{T})} = \sum_n |\phi_n\rangle \mathrm{e}^{-\tau(E_n-E_\mathrm{T})}\langle\phi_n|. \tag{12.54}$$

For large τ the ground state energy E_G dominates in the sum by a factor $\exp[-\tau(E_1-E_G)]$; therefore it acts as a projector onto the ground state (for large enough times).

As we have the explicit form of the time-evolution operator at our disposal only in a short-time approximation, we have to perform many short time steps before the distribution will approach the ground state wave function.

In the simulation, a collection of walkers diffuses through configuration space. Every diffusion step consists of two stages: a diffusion step and a *branching step*. In the diffusion step, the walkers are moved to a new position with a transition rate given by the diffusive part of the Green's function, i.e. the part due to the kinetic energy. The term involving the potential is dealt with in the second stage. Suppose we were to assign a weight to each walker, then the effect of the potential term could be taken into account by multiplying this weight for a walker which has arrived at a position R' by a factor $\exp\{-\Delta\tau[V(R') - E_T]\}$.[1] It turns out that this procedure is not very efficient. In the end quite a few walkers might have moved to unfavourable regions and represent small weight, but they require similar computational effort to the more favourable ones. This problem was previously encountered in Section 10.6. It would be more efficient to use computational effort proportional to the significance of the region probed by a particular walker. This is possible by a 'birth and death', or 'pruning and enrichment' (Section 10.6) or *branching* process: poor walkers die, favourable ones give rise to new walkers. More precisely, if a walker moves from a point R to a new point R', we calculate $q = \exp\{-\Delta\tau[V(R') - E_T]\}$. If $q < 1$, the walker survives with a probability q and dies with probability $1 - q$. If $q > 1$, the walker gives birth to either $[q - 1]$ or $[q]$ new ones at R', where $[q]$ represents the integer part (truncation) of q. The probability for having $[q]$ new walkers is given by $q - [q]$, and $[q - 1]$ new walkers will come into existence with the complementary probability $1 + [q] - q$. An efficient way of coding this is to add a uniform random number r between 0 and 1 to q: for $s = q + r$, $[s]$ new walkers are created; if $[s] = 0$ then the walker is deleted.

Finally, we must specify how E_T is found. Remember that this value is ideally chosen such as to normalise the overall transition rate in the process. This is necessary to prevent the population from growing or decreasing steadily. A growing population would cause a steady increase in the computer time per diffusion step, whereas a decrease leads to bad statistics, if not a vanishing population! The energy E_T is in fact determined by keeping track of the change in population and adjusting it at each step in order to keep the population stable. The average value of E_T after many steps will then converge to the ground state energy as we have already seen in Section 12.2.4. Suppose we have a target number of $\tilde{M}$ walkers in our simulation

[1] It is also possible to multiply the weight by $\exp\{-\tau[(V(R') + V(R))/2 - E_T]\}$, which corresponds to the symmetric distribution of the potential terms in the Green's function as in (12.50).

and that after the last branching step their actual number is M, then we adjust E_T as

$$E_{\rm T} = E_0 + \alpha \ln \left(\frac{\tilde{M}}{M} \right) \tag{12.55}$$

where E_0 is close to the ground state energy (our 'best estimate'), and α is some small parameter.

In an algorithmic form, the resulting procedure can be presented as follows:

```
Put the walkers at random positions in configurational space;
REPEAT
    FOR all walkers DO
        Shift walker from its position R to a new position R'
                according to the Gaussian transition probability (12.24);
        Evaluate 1 = exp{−Δτ[V(R') − E_T]};
        Eliminate the walker or create new ones at R',
                depending on s = q + r, where r is random,
                uniform between 0 and 1;
    END FOR;
    Update E_T;
UNTIL finished.
```

The major difference with the variational Monte Carlo method described in the previous section is that the present method does not rely on a trial function and therefore the results have no systematic error due to the trial function being (in general) not exact. There is, however, an error due to the fact that we have split the time-evolution operator into two parts, one depending on the kinetic energy and the other on the potential, by neglecting CBH commutators. By reducing $\Delta\tau$ we can make this error arbitrarily small, but the convergence speed will be reduced accordingly. In Section 12.3.3, we shall describe a Metropolis algorithm to correct for the discretisation error.

The population itself should represent the ground state wave function. For a one-dimensional problem (or a radially symmetric three-dimensional problem) this can be checked by constructing a histogram in which we record the frequencies with which the various positions are occupied. Below we shall give some results of DMC simulations for the harmonic oscillator and the helium atom.

The DMC procedure outlined here might fail in some cases. The distribution of walkers can only represent a density which is positive everywhere. Therefore, it can sample the ground wave function only if the latter is everywhere positive. Fortunately, the ground state of a boson system is indeed everywhere positive. However, for fermions this is no longer the case. Moreover, the Green's function is no longer positive in that case and it is not clear how to perform the diffusion, as

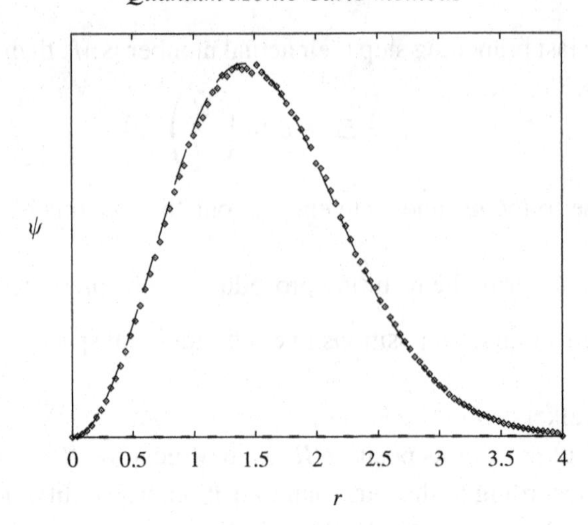

Figure 12.1. Ground state wave function (times r^2) for the three-dimensional harmonic oscillator as resulting from the DMC calculation (dots) compared with the exact form, scaled to match the numerical solution best.

the transition probability should be positive. This is called the *fermion problem*. We shall come back to this later. Another problem arises when the interaction potential assumes strongly negative values. This will be discussed in some detail in the next section and then we shall consider a refinement of the DMC which is not susceptible to this problem.

12.3.2 Applications

We apply the DMC procedure first to the three-dimensional harmonic oscillator. The exact ground state wave function is given by

$$\psi(r) = \frac{1}{(2\pi)^{3/2}} \mathrm{e}^{-r^2/2}; \tag{12.56}$$

the energy is $3/2$ (in dimensionless units). It should be noted that the probability distribution for finding a walker at a distance r from the origin is given by the wave function times r^2, because the volume of a spherical shell of thickness $\mathrm{d}r$ is $2\pi r^2 \mathrm{d}r$. For an average population of 300 walkers executing 4000 steps and a time step $\tau = 0.05$, we find $E_\mathrm{G} = 1.506 \pm 0.015$, to be compared with the exact value $1\frac{1}{2}$. The distribution histogram is shown in Figure 12.1, together with the exact wave function, multiplied by r^2 and scaled in amplitude to fit the DMC results best. Ground state energy and wave function are calculated with reasonable accuracy. Note that these results are obtained without using any knowledge of the exact solution: the diffusion process 'finds' the ground state by itself.

Next we analyse the helium atom using the diffusion Monte Carlo method. This is less successful. The reason is that writing the time-evolution operator as a product of a kinetic and potential energy evolution operator

$$e^{-\Delta\tau(K+V-E_T)} = e^{-\Delta\tau K} e^{-\Delta\tau(V-E_T)} + \mathcal{O}(\Delta\tau^2) \tag{12.57}$$

is not justified when the potential diverges, as is the case with the Coulomb potential at $r = 0$. Formally, this equation is still true, but the prefactor of the $\mathcal{O}(\Delta\tau^2)$ term diverges. However, even if the potential does not diverge but varies strongly, the statistical efficiency of the simulation is low. This is due to the fact that if a walker moves to a very favourable region, it will branch into many copies. But these are all the same, and together they form a rather biased sample of the distribution in that region. It requires some time before they have diffused and branched in order to form a representative ensemble. Frequent occurrence of such strong branching events will degrade the efficiency considerably. Quite generally one can say that the efficiency increases with the flatness of the potential.

There exist, in principle, two ways to solve the divergent potential problem. The first one consists of finding a better alternative to the simple approximation to the time-evolution operator than in (12.57). Such approximations have been devised and we shall consider these in the context of path-integral Monte Carlo (see Section 12.4). The common procedure, however, is to use a *guide function*, which transforms the original Schrödinger equation into a new one with a flatter potential, just as in the case of the Fokker–Planck variational Monte Carlo method. This method will be described in the next section.

12.3.3 Guide function for diffusion Monte Carlo

As we have just seen, the diffusion Monte Carlo method causes problems if the potential is unbounded, and this is the case in almost every many-particle system. Sampling some other function instead of the ground state wave function ψ might cure this problem.

A suitable function is $\rho(R,\tau) = \psi(R,\tau)\Psi_T(R)$ where $\Psi_T(R)$ is some trial function which models the exact wave function in a reasonable way. It turns out that ρ satisfies a Fokker–Planck type of equation:

$$\frac{\partial\rho(R,\tau)}{\partial\tau} = \frac{1}{2}\nabla_R[\nabla_R - \mathbf{F}(R)]\rho(R,\tau) - [E_L(R) - E_T]\rho(R,\tau). \tag{12.58}$$

Here, the 'force' $\mathbf{F}(R)$ is again given as $2\nabla_R\Psi_T(R)/\Psi_T(R)$. This form differs from (12.49) because (12.58) is not a 'pure' Fokker–Planck equation: it contains a 'potential term' $E_L(R) - E_T$. The 'local energy' $E_L(R)$ is given as usual by

$$E_L = \frac{H\Psi_T(R)}{\Psi_T(R)} = \frac{-\nabla^2\Psi_T(R)/2 + V(R)\Psi_T(R)}{\Psi_T(R)}. \tag{12.59}$$

The FP-diffusion term will be used to diffuse the walkers, whereas the 'potential' $E_L(R) - E_T$ is used in a branching process. By writing out all the terms on the left and right hand sides of Eq. (12.58), it can be checked that this equation reduces to the imaginary time-dependent Schrödinger equation (12.33).

The procedure is now a combination of the Fokker–Planck VMC and of the DMC method without guide function: we let the walkers diffuse just as in the Fokker–Planck VMC method, with a transition probability

$$T_{\Delta\tau}(R_n \to R_{n+1}) = \frac{1}{\sqrt{2\pi\Delta\tau}} \exp\{-[R_{n+1} - R_n - \mathbf{F}(R_n)\Delta\tau/2]^2/(2\Delta\tau)\}. \tag{12.60}$$

Then branching is performed, according to the value $q = \exp\{-\Delta\tau[E_L(R) - E_T]\}$. What do we gain by this method? We avoid problems of the kind encountered above with strongly varying potentials. The role of V in standard DMC is now taken over by $E_L(R)$, which is (we hope) rather flat. If $\Psi_T(R)$ were an *exact* eigenstate, then E_L would be independent of R. If Ψ_T is a reasonable approximation to the ground state, then $E_L(R)$ is reasonably flat, and the method will be reliable. It is clear now why the cusp conditions are so important: they guarantee that the trial function converges to the exact solution in those regions where the potential diverges strongly. These are the points that cause problems. The method using trial – or guide – functions was introduced by Kalos [8] and is commonly called *importance sampling Monte Carlo*.

We can again correct for the time step error using a Metropolis procedure, just as we did for VMC in Section 12.2.5. Note that G is not symmetric, so we must use the generalised Metropolis method in order to guarantee detailed balance (see also the variational Fokker–Planck simulation). A trial displacement is accepted with probability

$$\min\left(1, \frac{T_{\Delta\tau}(R' \to R)\rho(R')}{T_{\Delta\tau}(R \to R')\rho(R)}\right) \tag{12.61}$$

and rejected otherwise.

With importance sampling, the algorithm reads:

```
Put the walkers at random positions in configurational space;
REPEAT
    FOR all walkers DO
        Shift walker from its position R to a new position R'
            by first moving it over a distance FΔτ/2 and then
            adding a random displacement according to the
            transition probability (12.24);
        Accept the move with a probability given by (12.61);
        IF Accepted THEN
```

```
        Evaluate q = exp{−Δτ[(E_Local(R') + E_Local(R))/2 − E_T]};
        Eliminate the walker or create new ones at R',
              depending on s = q + r, where r is random,
              uniform between 0 and 1;
      END IF;
    END FOR
    Update E_T using (12.55);
  UNTIL finished.
```

Let us first apply the importance sampling method to the one-dimensional harmonic oscillator. We use the same trial (or guide) function $\Psi_{\mathrm{T}}(x) = \mathrm{e}^{-\alpha x^2}$ as in the VMC simulation. In that case the quantum force is given by

$$F(x) = -4\alpha x, \tag{12.62}$$

and the local energy by Eq. (12.6). Indeed, the local energy is a constant if $\alpha = 1/2$ and it will be slowly varying if α is close to $1/2$. For $\alpha = 0.4$, a target number of 6000 walkers and 4000 steps, we find for the ground state energy $E = 0.5002 \pm 0.0003$ and with $\alpha = 0.6$, $E = 0.4998 \pm 0.0003$.

We can now do the hydrogen and the helium atom problems. For hydrogen we use a guide function $\exp(-\alpha r)$ and a target number of 2000 walkers performing 4000 steps. The local energy is given by (12.10). Obviously, for $\alpha = 1$ we find the exact ground state energy of -0.5 Hartree as the local energy is constant and equal to this value. For $\alpha = 0.9$, we find a ground state energy of $-0.4967(5)$ and for $\alpha = 1.1$ we find $E_{\mathrm{G}} = 0.5035(5)$. Neither of these values agrees with the exact value. The reason is that the guide function should solve the divergence problem at $r = 0$, but it can do this only if the cusp conditions are satisfied. For $\alpha \neq 1$ this is not the case. This shows the importance of the cusp conditions being satisfied for the trial function.

Finally we present results for the helium atom. We use the Padé–Jastrow wave function (12.11). Varying the parameter α gives values above and below the exact energy. If we monitor the variance of the energy, we find a minimum at $\alpha \approx 0.15$ and an energy $E_{\mathrm{G}} = -2.9029(2)$ for 1000 walkers performing 4000 steps. Remember the exact energy is -2.903 and the variational energy for the uncorrelated wave function (the Hartree–Fock energy) is -2.8617 atomic units.

PROGRAMMING EXERCISE

Modify the DMC programs of the previous section to include a guide function and compare the results with those given in this section.

12.3.4 Problems with fermion calculations

We have described how the simulation of a diffusion process can generate an average distribution of random walkers which is proportional to the ground state wave function or (in the case of guide function DMC) to the product of this function and a trial function. But a distribution of walkers can only represent wave functions which are positive everywhere. For bosons, this property is satisfied by the ground state, but the same does not hold in the case of fermions. The difficulties associated with treating fermions in quantum Monte Carlo are generally denoted as 'the fermion problem'. It should be noted that there is no fermion problem in VMC.

The fixed-node method

There are several approaches to the fermion problem. The simplest approximation is the *fixed-node* method, in which the diffusion process is simulated as before, except for steps crossing a node of the trial function being forbidden. The nodes of the trial function divide the configuration space up into simply connected volumes in which the trial wave function has a unique sign. These volumes are separated from each other by nodal surfaces: hypersurfaces on which the wave function vanishes. To understand why the fixed-node method is useful, suppose that we know the nodes of the exact ground state wave function. If we could solve the ground state of the Schrödinger equation in each simply connected region bounded by the nodal surfaces of the ground state wave function with vanishing boundary conditions on these surfaces, this solution would be proportional to the exact ground state of the full Hamiltonian in each region. In the fixed-node solution, we solve the Schrödinger equation in connected regions bounded by the nodal surfaces of the trial function instead of the exact function, and therefore the quality of the solution depends on how close these surfaces are to those of the exact ground state. It can be shown that the resulting energy is a variational upper bound to the exact ground state energy [2]. It should be noted that the fixed-node method often gives a substantial improvement over the variational Monte Carlo method (which does not suffer from the fermion problem).

An additional problem with the fixed-node method is the fact that moves in which two (or any even number of) nodal surfaces are crossed are accepted. This introduces an error as the number of walkers in two regions separated by an even number of node crossings does not necessarily represent the norm of the wave functions on those regions. The degree to which we suffer from this increases with the time step, as a larger time step will result in larger steps to be taken. It introduces an extra time-step bias error which goes by the name *cross–recross error*.

Let us study the nodes more carefully. The requirement that $\psi(\mathbf{x}_1, \ldots, \mathbf{x}_N) = 0$ ($\mathbf{x}_i$ denotes the spin-orbit coordinate of electron i) defines the nodal surfaces. If we assume the spins of the N fermions to be given, then the nodes form $(3N-1)$-dimensional hypersurfaces in the $3N$-dimensional configurational space. The obvious zeroes of ψ whenever $\mathbf{x}_i = \mathbf{x}_j$ for any pair $i \neq j$ define a $(3N-3)$-dimensional scaffolding for the nodal surface structure. This scaffolding does not depend on the particular form of the trial function. A node of a one-electron orbital in the Slater determinant occurring in the wave function should not be confused with a 'fermionic zero', as such an orbital node does not force the many-electron wave function to vanish: one of the electrons, say i, might be at a zero of some orbital, but the wave function also contains contributions with the coordinates of the electrons permuted, and in general the coordinates of the other electrons are different from those of electron i.

Changing the diffusion Monte Carlo method to a fixed-node simulation is easy. Simply add the following step just after having generated a new trial position of a particle, say i. Check whether the trial wave function changes sign for this displacement. If this is the case, the move is not accepted, otherwise proceed as in the boson case. The interested reader can implement the fixed-node extension and test it, for example, for the lithium atom, taking an appropriate Slater determinant for the guide function. More details can be found in Ref. [9].

*The transient estimator method

In view of the variational error present in the fixed-node method it is worthwhile to devise other methods. A method which does not depend on fixed nodal surfaces is the *transient estimator* method. To understand how and why this method works, it is important to realise that the Hamiltonian and hence the time-evolution operator are the same for fermions and for bosons. However, because the time-evolution operator is symmetric with respect to particle permutations, an antisymmetric (fermionic) initial state will remain antisymmetric and a symmetric (bosonic) state remains symmetric.

Let us split an arbitrary fermion wave function ϕ into two parts, ϕ_- and ϕ_+, which contain the negative and positive parts of ϕ respectively (all wave functions depend on all the spin-orbit coordinates $X = (\mathbf{x}_1, \mathbf{x}_2, \ldots, \mathbf{x}_N)$, and on imaginary time τ):

$$\phi_+ = \tfrac{1}{2}(|\phi| + \phi) \tag{12.63a}$$

$$\phi_- = \tfrac{1}{2}(|\phi| - \phi), \tag{12.63b}$$

so that

$$\phi = \phi_+ - \phi_-. \tag{12.64}$$

Now perform two independent DMC calculations, one with ϕ_- and the other with ϕ_+ as a starting distribution, where ϕ is a trial fermion wave function. What will happen? Applying the (exact) imaginary-time evolution operator $T(X \to Y; \tau)$ to ϕ we obtain

$$\begin{aligned}
\phi(Y;\tau) &= \int \mathrm{d}X\; T(X \to Y;\tau)\phi(X;0) \\
&= \int \mathrm{d}X\; T(X \to Y;\tau)\phi_+(X,0) - \int \mathrm{d}X\; T(X \to Y;\tau)\phi_-(X,0) \\
&= \phi_+(Y,t) - \phi_-(Y,t).
\end{aligned} \tag{12.65}$$

This suggests that we can follow the time evolution of ϕ by subtracting $\phi_+(t)$ and $\phi_-(t)$ as produced in the two simulations. As $\phi_-(0)$ and $\phi_+(0)$ are both positive, and as the imaginary time-evolution operator is always positive, the application of the DMC approach causes no problems. In fact, one could also say that if the initial wave function is positive everywhere, it contains no fermion character and hence we have an unambiguous bosonic time evolution for such an initial state. A guide function approach can be used in the two boson simulations.

As the time-evolution operator contains no fermion-like features (see above), both simulations will tend to the bosonic ground state solution for long times. The fermion ground state wave function is an excited state solution of the many-particle Hamiltonian, so the boson ground state contribution to the solution at imaginary time τ will dominate the fermion contribution by a factor $\exp[\tau(E_F - E_B)]$, where E_B and E_F are the fermion and boson ground state energies respectively. Note that this factor grows exponentially with time. The fermion ground state wave function is the *difference* between the two distributions resulting from ϕ_- and ϕ_+, which because of the foregoing analysis are both essentially boson-like. If we are to find a fermion wave function as a small difference of two large, essentially boson wave function distributions we must be prepared for large statistical errors. The analysis given here is represented pictorially in Figure 12.2.

The analysis so far leads to the conclusion that, at the beginning, the difference between the distributions is equal to the trial function ϕ, and for large times it converges to the exact fermion wave function, but it will be buried in the noise of the boson solutions forming the bulk of the two distributions. We might be lucky: if the trial function relaxes to the exact Fermi wave function quickly enough, before the latter is buried in the 'boson noise', then we have an intermediate ('transient') regime in imaginary time during which we might extract useful data from the simulation. The trial energy which is adjusted to keep the respective population sizes stable is no longer a suitable energy estimator as this will converge to the

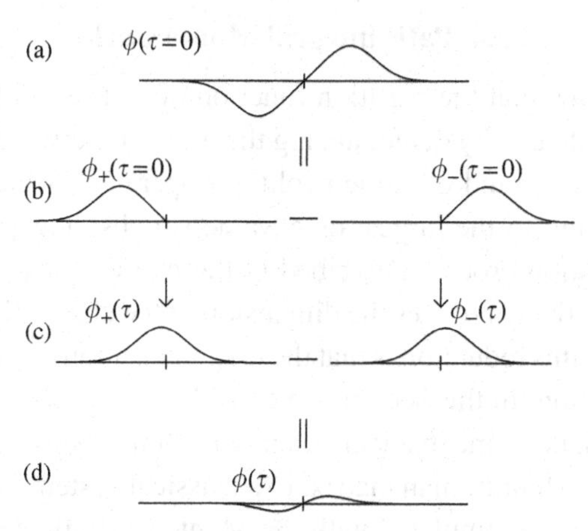

Figure 12.2. Evolution of the distributions in the transient energy estimator method. The wave function $\phi(\tau = 0)$ is shown in (a); it can be written as the difference of the ϕ_+ and ϕ_-. These two functions evolve separately and tend therefore to the same boson ground state solution, as shown in (c). Subtracting the two wave functions in (c) gives the small difference in (d), and this will be soon buried in the noise in the solutions in (c).

boson energy. Therefore we use the 'transient estimator':

$$E_{\mathrm{TE}}(\tau) = \frac{\int \mathrm{d}X\ \phi(\tau)H\phi(\tau = 0)}{\int \mathrm{d}X\ \phi(\tau)\phi(\tau = 0)}$$
$$= \frac{\int \mathrm{d}X\ \phi_-(\tau)H\phi(\tau = 0)}{\int \mathrm{d}X\ [\phi_+(\tau) - \phi_-(\tau)]\phi(\tau = 0)} - \frac{\int \mathrm{d}X\ \phi_+(\tau)H\phi(\tau = 0)}{\int \mathrm{d}X\ [\phi_+(\tau) - \phi_-(\tau)]\phi(\tau = 0)}. \tag{12.66}$$

As the wave function $\phi(\tau)$ converges to the exact fermion ground state, this estimator will indeed relax to the exact fermion energy. As mentioned already, the problem resides in $\phi(\tau)$ to be extracted as the small difference between two large distributions.

The estimator (12.66) is evaluated as follows. At time τ, the walkers occupy points in configuration space which are distributed according to $\phi_\pm(\tau)$. For a walker at the point X in the ϕ_+-simulation we evaluate $H\phi(X, \tau = 0)$ (for the numerator) and $\phi(X, \tau = 0)$ (for the denominator), and sum over walkers. We do the same with the ϕ_- simulation, but now give the contributions a minus sign. The quantity $H\phi(X, \tau = 0)$ can be evaluated because $\phi(X, \tau = 0)$ is a trial function, given in analytic form. The sum is divided by the sum of $\phi(X, \tau = 0)$ over all the walkers.

There exist several extensions to and refinements of the transient estimator method, which are beyond the scope of this book. A common characteristic of these methods is that they are subject to instability in the errors for large τ.

12.4 Path-integral Monte Carlo

In Chapter 11 we saw that the partition function of a classical lattice spin system on a strip can be evaluated by diagonalising the transfer matrix. The transfer matrix can be considered as a kind of 'time-evolution operator', which projects out the eigenvector belonging to the largest eigenvalue (in absolute value). The relation with the time-evolution process described in the previous section is evident. The transfer matrix effectively reduces the dimension of the classical system by one, but the price we pay for this reduction is that the diagonalisation of the transfer matrix is an expensive operation. In this section we consider the reverse transformation: we shall transform a quantum mechanical system in d dimensions, which can be solved by diagonalising the Hamiltonian matrix, to a classical system in $d+1$ dimensions. This system can then be simulated with the Monte Carlo procedures described in Chapter 10. The new formulation enables us to obtain time-dependent properties, or physical quantities of the system at finite temperature. For a very clear discussion of the path-integral concept, see the book by Feynman and Hibbs [10].

12.4.1 Path-integral fundamentals

The path-integral method provides a way to calculate matrix elements and traces of the time-evolution operator of a quantum system in imaginary time:

$$\mathcal{T}(\tau) = \mathrm{e}^{-\tau H} \tag{12.67}$$

which we have encountered in the previous section. If we interpret the imaginary time as an inverse temperature $\tau \leftrightarrow \beta$ and take the trace of the time-evolution operator, we obtain the partition function Z of the quantum system at a finite temperature T:

$$Z(\beta) = \mathrm{Tr}(\mathrm{e}^{-\beta H}) = \int \mathrm{d}R\; \langle R|\mathrm{e}^{-\beta H}|R\rangle. \tag{12.68}$$

R denotes the coordinates of N particles. The path-integral method enables us to sample system configurations with the appropriate Boltzmann factor, so that expectation values for a quantum system at a finite temperature can be evaluated.

The problem with expression (12.68) is that it contains the exponential of the Hamiltonian, which, as mentioned in Section 12.2.4, makes the trace of the time-evolution operator difficult to evaluate. For short times τ (or β), this is not a problem as we can write the Hamiltonian as a sum of several terms (e.g. kinetic and potential energy) which themselves are easily tractable in an exponential – the neglected CBH commutators yield systematic errors of order τ^2. What can we do if τ is not small? In that case, we divide the time τ up into many (say M) small segments $\Delta\tau = \tau/M$ which can be treated in the short-time approximation. For a system consisting of

N spinless particles with coordinates R_i, the partition function can be written as

$$\int \mathrm{d}R_0 \langle R_0|\mathrm{e}^{-\tau H}|R_0\rangle = \int \mathrm{d}R_0\, \mathrm{d}R_1 \ldots \mathrm{d}R_{M-1}$$
$$\langle R_0|\mathrm{e}^{-\Delta\tau H}|R_1\rangle\langle R_1|\mathrm{e}^{-\Delta\tau H}|R_2\rangle \cdots \langle R_{M-1}|\mathrm{e}^{-\Delta\tau H}|R_0\rangle. \tag{12.69}$$

We have inserted $M-1$ unit-operators $\int \mathrm{d}R_i\ |R_i\rangle\langle R_i|$ between the short-time evolution operators. The procedure in which time is divided up into many short segments is called *time-slicing*. The fact that the first and the last state in the product of matrix elements are identical ($|R_0\rangle$) implies that we have periodic boundary conditions in the τ-direction.

We know the matrix elements of the short-time evolution operator: it has been derived in Section 12.2.4:

$$T(R, R'; \Delta\tau) = \langle R|\mathrm{e}^{-\Delta\tau H}|R'\rangle = \frac{1}{(2\pi\,\Delta\tau)^{3N/2}}\mathrm{e}^{-\Delta\tau V(R)}\mathrm{e}^{-(R-R')^2/(2\Delta\tau)}. \tag{12.70}$$

The potential could have been distributed symmetrically over R and R', but we shall see that the final result does not depend on this distribution. The first order CBH commutator can be shown to vanish in this case, so that this short-time approximation is accurate to order $\Delta\tau^2$. Substituting this result into (12.69), we obtain

$$\int \mathrm{d}R_0 \langle R_0|\mathrm{e}^{-\tau H}|R_0\rangle \approx \frac{1}{(2\pi\,\Delta\tau)^{3NM/2}}\int \mathrm{d}R_0\, \mathrm{d}R_1\, \mathrm{d}R_2 \ldots \mathrm{d}R_{M-1}$$
$$\exp\left\{-\Delta\tau \sum_{m=0}^{M-1}\left[\frac{1}{2}\left(\frac{R_{m+1}-R_m}{\Delta\tau}\right)^2 + V(R_m)\right]\right\}. \tag{12.71}$$

In this expression, $R_M = R_0$. The prefactor before the integral seems dangerous in the sense that it explodes when we take the limit $\Delta\tau \to 0$. However, this is balanced by the fact that, of the huge integration volume, only a tiny part gives significant contributions to the integrand – in fact, the smaller we take $\Delta\tau$, the narrower the Gaussian kinetic energy integrands will be and the limit for large M therefore still exists.

You might recognise the summand in the exponent as the Lagrangian (in discrete imaginary time) of the classical many-particle system with coordinates R_i if we take $\Delta\tau \to 0$. The sum is then the *action*, which assumes its minimum for the classical trajectory. The integral is a sum over *all* possible sets of coordinates $R_0, \ldots, R_M$. Such a set denotes a *path* in configuration space. We see that the trace of the time-evolution operator is written as a sum, or rather an integral, over all possible paths. It is important to realise what the classical system represents. The quantum many-particle system we are describing contains N particles, interacting with each other and with an external potential through the potential $V(R)$. We have M copies

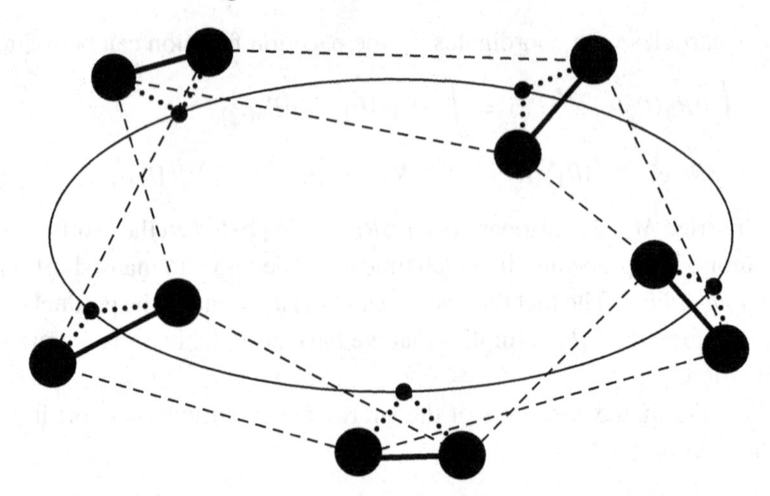

Figure 12.3. Classical system described by the path integral of the two electrons in the helium atom. Periodic boundary conditions are imposed along the quantum imaginary time (the circle). The small full circles denote the helium nuclei, the heavy ones the electrons. The circle is the time axis with periodic boundary conditions. The dashed lines represent harmonic couplings between the electrons of adjacent copies (along the time axis). The heavy solid lines denote the electron–electron interaction, and the heavy dotted lines the electron–nucleus interactions.

of this many-particle system along the quantum imaginary-time direction, so that the classical system consists of NM particles. The first term in the sum in (12.71) derives from the kinetic part of the quantum Hamiltonian, but in the classical system it denotes a harmonic coupling between corresponding particles in adjacent copies: they are connected by springs. Figure 12.3 shows the classical particle system and couplings for the two electrons in helium with $M = 5$.

The quantum partition function for a system of N three-dimensional particles is given as $\mathrm{Tr}\exp(-\beta H)$. The right hand side of Eq. (12.71) can be interpreted as the *classical* partition function of NM particles in three dimensions (without momentum degrees of freedom – these can be thought of as being integrated over), because it is an integral over all the configurations of the coordinates R_i with an appropriate Boltzmann factor. The energy $\mathcal{H}$ of the classical system is identified with the Lagrangian associated with the quantum Hamiltonian H. An unusual feature is the inverse temperature occurring in the denominator of the harmonic interactions of the classical Hamiltonian $\mathcal{H}$ (remember $\Delta\tau = \beta/M$). We see that the path integral maps the partition function of a $3N$-dimensional system onto a $(3N+1)$-dimensional system where the extra dimension can be interpreted either as an imaginary-time or as an inverse-temperature axis – it corresponds to the sub-index i of the R_i.

The path integral provides a very clear insight into the nature of quantum mechanics. Up to now, we have put $\hbar \equiv 1$. Had we kept $\hbar$ in the problem, we would have

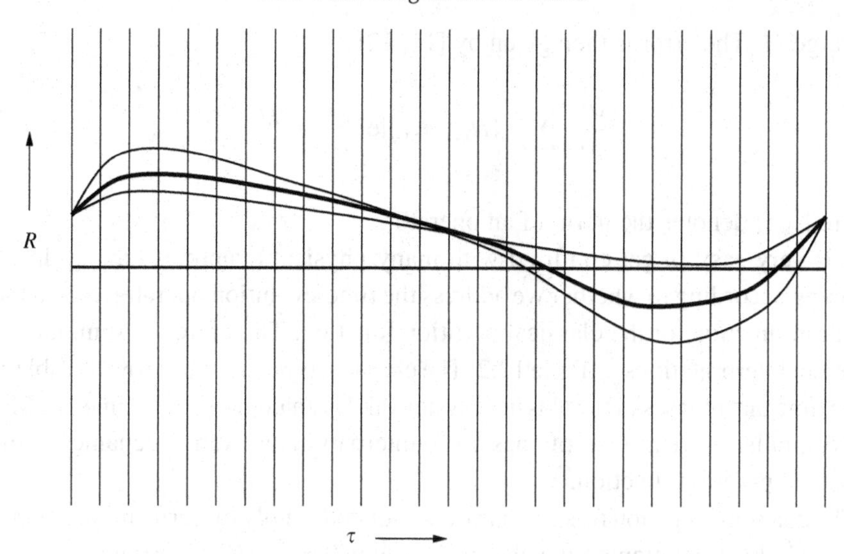

Figure 12.4. The path integral for a one-dimensional system. The vertical axes are R-axes at different times. A path is a set of points given on these axes. The heavy drawn path is the stationary path of the action, which is the solution to the classical equations of motion. The thin lines represent neighbouring paths. For these paths, the action is not stationary, but they are taken into account in the quantum mechanical path integral.

seen that the prefactor in the exponent occurring before the sum was $\Delta\tau/\hbar$ instead of $\Delta\tau$. The classical limit corresponds to $\hbar = 0$, which implies that the path with minimal action dominates all the other paths. This is in fact Hamilton's principle: the classical path corresponds to the minimal action. If we 'switch Planck's constant on', we see a contribution from the nonminimal paths emerging. If we had not identified R_0 with R_M and if we had not integrated over this coordinate, we would have a system with fixed end points, which brings the analogy with classical mechanics even closer. Figure 12.4 gives a pictorial representation of the idea of the path integral.

In this section and in the previous one, we have assumed that the errors in the individual short-time approximations do not add up to significant errors for large times. The justification of this assumption is a theorem, which is usually denoted as the Lie–Trotter–Suzuki formula, which says that for a Hamiltonian H which can be written as the sum of K operators:

$$H = \sum_{k=1}^{K} H_k \tag{12.72}$$

it holds that

$$\mathrm{e}^{-\alpha H} \to (\mathrm{e}^{-\alpha H_1/M}\mathrm{e}^{-\alpha H_2/M}\ldots\mathrm{e}^{-\alpha H_K/M})^M \tag{12.73}$$

for large M. The error is then given by [11, 12]

$$\frac{\alpha^2}{M} \sum_{m>m'} |[H_m, H_{m'}]| e^{-\alpha \sum_m |H_m|}, \tag{12.74}$$

where $|\ldots|$ denotes the norm of an operator.

It is very easy to get confused with many physical quantities having different meaning according to whether we address the time-evolution operator, the quantum partition function, or the classical partition function. Therefore we summarise the different interpretations in Table 12.2. The classical time in the last row of Table 12.2 is the time that elapses in the classical system and is analogous to the time in a Monte Carlo simulation. This quantity has no counterpart in quantum mechanics or in the statistical partition function.

The quantum partition function is now simulated simply by performing a standard Monte Carlo simulation on the classical system. The PIMC algorithm is

```
Put the NM particles at random positions;
REPEAT
    FOR m = 1 TO M DO
        Select a time slice m̃ at random;
        Select one of the N particles at time slice m̃ at random;
        Generate a random displacement of that particle;
        Calculate r = exp[−Δτ(H_new − H_old)];
        Accept the displacement with probability min(1, r);
    END FOR;
UNTIL finished.
```

In this algorithm we have used $\mathcal{H}$ to denote the Hamiltonian of the classical system, which is equal to the Lagrangian occurring in the exponent of the path integral – see Eq. (12.71).

Let us compare the path integral method with the diffusion Monte Carlo approach. In the latter we start with a given distribution and let time elapse. At the end of the simulation the distribution of walkers reflects the wave function at imaginary time τ. Information about the history is lost: physical time increases with simulation time. The longer our simulation runs, the more strongly will the distribution be projected onto the ground state. In the path integral method, we change the positions of the particles along the imaginary-time (inverse-temperature) axis. Letting the simulation run for a longer time does not project the system more strongly onto the ground state – the extent to which the ground state dominates in the distribution is determined by the temperature $\beta = M\Delta\tau$, i.e. for fixed $\Delta\tau$, it is determined by the length of the chain. The PIMC method is not necessarily carried out in imaginary

Table 12.2. *Meaning of several physical quantities in different interpretations of the path integral.*

Quantum mechanics	Quantum statistical mechanics	Classical mechanics	Statistical physics
d-dimensional configuration space	d-dimensional configuration space	d-dimensional subspace of configuration space	d-dimensional configuration space
imaginary time τ	inverse temperature $\beta = 1/k_{\mathrm{B}}T$	1-dimensional axis in configuration space	inverse temperature $\beta = 1/k_{\mathrm{B}}T$
time-evolution operator	Boltzmann operator $\mathrm{e}^{-\beta H}$	–	transfer matrix
kinetic energy	kinetic energy	harmonic interparticle potential	inter-row coupling of transfer matrix
Lagrangian	Lagrangian	Lagrangian	Hamiltonian
path integral	quantum partition function of d-dimensional system	–	partition function of $(d+1)$-dimensional system
classical limit	zero temperature	stationary path	zero temperature

time – there exist versions with real time, which are used to study the dynamics of quantum systems [13–15].

The analysis so far is correct for distinguishable particles. In fact, we have simply denoted a coordinate representation state by $|R\rangle$. For indistinguishable bosons, we should read for this state:

$$|R\rangle = \frac{1}{N!}\sum_{P} |\mathbf{r}_1, \mathbf{r}_2, \ldots, \mathbf{r}_N\rangle, \tag{12.75}$$

where the sum is over all permutations of the positions. The boson character is noticeable when we impose the periodic boundary conditions along the τ-axis, where we should not merely identify $\mathbf{r}_k$ in the last coordinate $|R_M\rangle$ with the corresponding position in $|R_0\rangle$, but also allow for permutations of the individual particle positions in both coordinates to be connected.

This feature introduces a boson entropy contribution, which is particularly noticeable at low temperatures. To see this, let us consider the particles as diffusing from left (R_0) to right (R_M). On the right hand side we must connect the particles to their counterparts on the left hand sides, taking all permutations into account. If the Boltzmann factor forbids large steps when going from left to right, it is unlikely that we can connect the particles on the right hand side to the permuted leftmost positions without introducing a high energy penalty. This is the case when $\tau = \beta$ is small, or equivalently when the temperature is high. This can be seen by noticing that, keeping $\Delta\tau = \beta/N$ fixed, a decrease β must be accompanied by a decrease in the number of segments N. Fewer segments mean less opportunity for the path to wander away from its initial position. On the other hand, we might keep the number of segments constant, but decrease $\Delta\tau$. As the spring constants are inversely proportional to $\Delta\tau$ (see Eq. (12.71)), they do not allow, in that case, for large differences in position on adjacent time slices; hence permutations are quite unlikely. When the temperature is high ($\tau = \beta$ small), large diffusion steps are allowed and there is a lot of entropy to be gained from connecting the particles to their starting positions in a permuted fashion. This entropy effect is responsible for the superfluid transition in ^{4}He [14–16]. Path-integral methods also exist for fermion systems. A review can be found in Ref. [19].

What type of information can we obtain from the path integral? First of all, we can calculate ground state properties by taking β very large (temperature very small). The system will then be in its quantum ground state. The particles will be distributed according to the quantum ground state wave function. This can be seen by considering the expectation value for particle 0 to be at position R_0. This is given by

$$P(R_0) = \frac{1}{Z}\int \mathrm{d}R_1 \mathrm{d}R_2 \ldots \mathrm{d}R_{M-1}$$
$$\langle R_0|\mathrm{e}^{-\Delta\tau H}|R_1\rangle\langle R_1|\mathrm{e}^{-\Delta\tau H}|R_2\rangle \ldots \langle R_{M-1}|\mathrm{e}^{-\Delta\tau H}|R_0\rangle. \tag{12.76}$$

Note that the numerator differs from the path integral (which occurs in the denominator) in the absence of the integration over R_0. Removing all the unit operators we obtain

$$P(R_0) = \frac{\langle R_0|\mathrm{e}^{-\tau H}|R_0\rangle}{\int \mathrm{d}R_0 \langle R_0|\mathrm{e}^{-\tau H}|R_0\rangle}. \tag{12.77}$$

Large τ is equivalent to low temperature. But if τ is large indeed, then the operator $\exp(-\tau H)$ projects out the ground state ϕ_G:

$$\mathrm{e}^{-\tau H} \approx |\phi_G\rangle \mathrm{e}^{-\tau E_G} \langle \phi_G|, \quad \text{large}\,\tau. \tag{12.78}$$

Therefore we have

$$P(R_0) = \frac{1}{Z} \mathrm{e}^{-\tau E_G} |\langle \phi_G|R_0\rangle|^2, \text{large}\,\tau. \tag{12.79}$$

Because of the periodic boundary conditions in the τ direction we obtain the same result for each time slice m. To reduce statistical errors, the ground state can be therefore obtained from the *average* distribution over the time slices via a histogram method.

The expectation value of a physical quantity A for a quantum system at a finite temperature is found as

$$\langle A\rangle_\beta = \frac{\mathrm{Tr}(A\mathrm{e}^{-\beta H})}{\mathrm{Tr}\,\mathrm{e}^{-\beta H}}. \tag{12.80}$$

The denominator is the partition function Z. We can use this function to determine the expectation value of the energy

$$\langle E\rangle_\beta = \frac{\mathrm{Tr}(H\mathrm{e}^{-\beta H})}{Z} = -\frac{\partial}{\partial\beta} \ln Z(\beta). \tag{12.81}$$

If we apply this to the path-integral form of Z, we obtain for the energy per particle (in one dimension):

$$\left\langle \frac{E}{N} \right\rangle_\beta = \frac{M}{2\beta} - \frac{1}{N}(\langle K\rangle - \langle V\rangle). \tag{12.82}$$

with

$$K = \sum_{m=0}^{M-1} \frac{(R_m - R_{m+1})^2}{2\beta^2} \tag{12.83}$$

and V is the potential energy (see also Problem 12.1). The first term in (12.82) derives from the prefactor $1/\sqrt{2\pi\,\Delta\beta}$ of the kinetic Green's function. The angular brackets in the second and third term denote expectation values evaluated in the classical statistical many-particle system. It turns out that this expression for the energy is subject to large statistical errors in a Monte Carlo simulation. The reason is that $1/\beta$ and $\langle K\rangle/(NM)$ are both large, but their difference is small. Herman *et al.* [20] have proposed a different estimator for the energy, given by

$$\left\langle \frac{E}{N} \right\rangle_\beta = \left\langle \frac{1}{M} \sum_{m=0}^{M-1} \left[V(R_m) + \frac{1}{2} R_m \cdot \nabla_{R_m} V(R_m) \right] \right\rangle. \tag{12.84}$$

This is called the *virial energy estimator*, and it will be considered in Problem 12.6.

The virial estimator is not always superior to the direct expression, as was observed by Singer and Smith for Lennard–Jones systems [21]; this is presumably due to the steepness of the Lennard–Jones potential causing large fluctuations in the virial.

12.4.2 Applications

We check the PIMC method for the harmonic oscillator in one dimension. We have only one particle per time slice. The particles all move in a 'background potential', which is the harmonic oscillator potential, and particles in neighbouring slices are coupled by the kinetic, harmonic coupling. The partition function reads

$$Z = \int \mathrm{d}x_0 \ldots \mathrm{d}x_{M-1} \exp\left\{ -\frac{\beta}{M} \sum_{m=0}^{M-1} \left[\frac{(x_m - x_{m+1})^2}{2\Delta\beta^2} + \frac{1}{2} x_m^2 \right] \right\}. \tag{12.85}$$

We have used $\beta = 10$ and $M = 100$. Thirty thousand MCS were performed, of which the first two thousand were deleted to reach equilibrium. The maximum displacement was tuned to yield an acceptance rate of about 0.5. The spacing between the energy levels of the harmonic oscillator is 1; therefore $\beta = 10$ corresponds to large temperature. We find for the energy $E = 0.51 \pm 0.02$, in agreement with the exact ground state energy of $1/2$. The ground state amplitude can also be determined, and it is found to match the exact form $|\psi(x)|^2 = \mathrm{e}^{-x^2}$ very well.

The next application is the hydrogen atom. This turns out to be less successful, just as in the case of the diffusion MC method. The reason is again that writing the time-evolution operator as the product of the exponentials of the kinetic and potential energies is not justified when the electron approaches the nucleus, as the Coulomb potential diverges there – CBH commutators therefore diverge too. The use of guide functions is not possible in PIMC, so we have to think of something else. The solution lies in the fact that the *exact* time-evolution operator over a time slice Δt does not diverge at $r = 0$; we suffer from divergences because we have used the so-called *primitive approximation*

$$T(\mathbf{r} \to \mathbf{r}'; \Delta\tau) = \frac{1}{(2\pi\Delta\tau)^{3/2}} \exp[-(\mathbf{r} - \mathbf{r}')/(2\Delta\tau)] \exp\{-\Delta\tau[V(\mathbf{r}) + V(\mathbf{r}')]/2\} \tag{12.86}$$

to the time-evolution operator. The effect of averaging over all the continuous paths from $(\mathbf{r}, \tau)$ to $(\mathbf{r}', \tau + \Delta\tau)$, as is to be done when calculating the exact time evolution, is that the divergences at $\mathbf{r}, \mathbf{r}' = 0$ are rounded off. So if we could find a better approximation to this exact time evolution than the primitive one, we would not suffer from the divergences any longer. Several such approximations have been developed [22, 23]. They are based either on exact Coulomb potential solutions (hydrogen atom) or on the cumulant expansion. We consider the latter approximation in some detail in Problems 12.2 and 12.3; here we shall simply quote the result:

$$V_{\text{cumulant}}(\mathbf{r}, \mathbf{r}'; \Delta\tau) = \int_0^{\Delta\tau} \mathrm{d}\tau' \frac{\operatorname{erf}[r(\tau')/\sqrt{2}\sigma_{\tau'}]}{r(\tau')}, \tag{12.87a}$$

where

$$\mathbf{r}(\tau') = \mathbf{r} + \frac{\tau'}{\Delta\tau}(\mathbf{r}' - \mathbf{r}) \quad \text{and} \quad \sigma(\tau') = \frac{(\Delta\tau - \tau')\tau'}{\Delta\tau}. \tag{12.87b}$$

The cumulant approximation for V can be calculated and saved in a tabular form, so that we can read it into an array at the beginning of the program, and then obtain the potential for the values needed from this array by interpolation. In fact, for $\Delta\tau$ fixed, V_{cumulant} depends on the norms of the vectors $\mathbf{r}$ and $\mathbf{r}'$ and on the angle between them. Therefore the table is three-dimensional. We discretise r in, say, 50 steps Δr between 0 and some upper limit $r_{\max}$ (which we take equal to 4), and similarly for r'. For values larger than $r_{\max}$ we simply use the primitive approximation, which is sufficiently accurate in that case. For the angle θ in between $\mathbf{r}$ and $\mathbf{r}'$ we store $\cos\theta$, discretised in 20 steps between -1 and 1 in our table. For actual values r, r' and $u = \cos\theta$ we interpolate linearly from the table – see Problem 12.4. Figure 12.5 shows the cumulant potential $V(r = r', \theta = 0; \Delta\tau = 0.2)$, together with the Coulomb potential; the rounding effect of the cumulant approximation is clear. In a path-integral simulation for the hydrogen atom we find a good ground state distribution, shown in Figure 12.6. For the energy, using the virial estimator with the original Coulomb potential (which is of course not entirely correct), we find $E_G = -0.494 \pm 0.014$, using $\Delta\tau = 0.2$, 100 time slices and 60 000 MC steps per particle, of which the first 20 000 were removed for equilibration.

Applying the method to helium is done in the same way. Using 150 000 steps with a chain length of 50 and $\tau = 0.2$, the ground state energy is found as 2.93 ± 0.06 atomic units. Comparing the error with the DMC method, the path-integral method does not seem to be very efficient, but this is due to the straightforward implementation. It is possible to improve the PIMC method considerably as will be described in the next section.

The classical example of a system with interesting behaviour at finite temperature is dense helium-4. In this case the electrons are not taken into account as independent

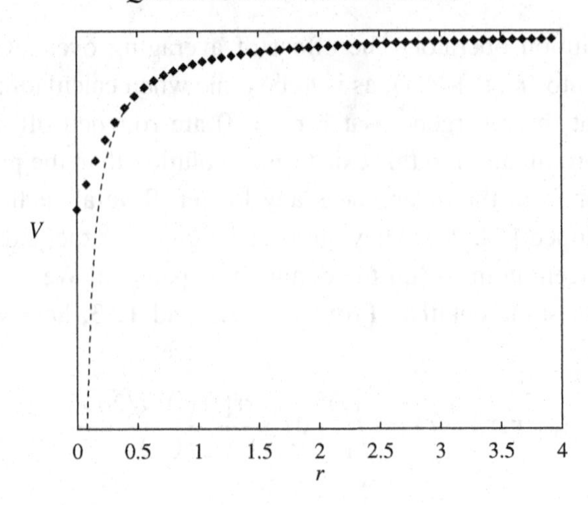

Figure 12.5. The cumulant potential for $\Delta\tau = 0.2$ (diamonds) and the Coulomb potential. It is clearly seen that the cumulant potential is rounded off at $r = 0$.

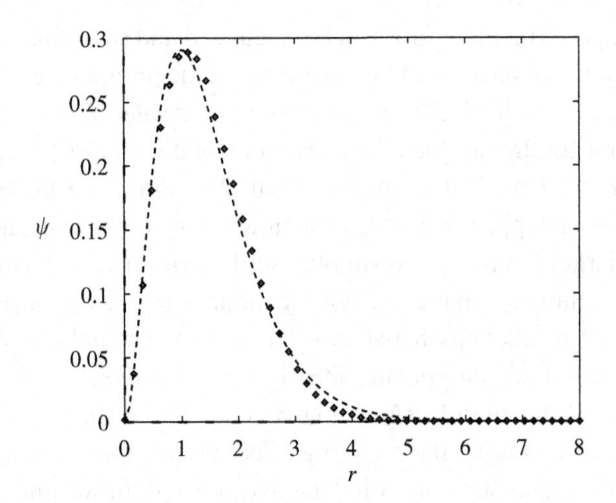

Figure 12.6. PIMC ground state amplitude $|\psi(r)|^2$ (diamonds) and the exact result. Sixty thousand Monte Carlo sweeps with a chain length of 100 and $\tau = 0.2$ were used.

particles; rather, a collection of atoms is considered, interacting through Lennard–Jones potentials. We shall not go into details of implementation and phase diagram, but refer to the work by Ceperley and Pollock [3, 4].

12.4.3 Increasing the efficiency

The local structure of the action enables us to use the heat-bath algorithm instead of the classical sampling rule, in which particles are displaced at random uniformly within a cube (or a sphere). If we update the coordinate R_m, keeping R_{m-1} and R_{m+1} fixed, then in the heat-bath algorithm, the new value R'_m must be generated with distribution

$$\rho(R'_m) = \exp\left[-\Delta\tau\frac{(R'_m - \overline{R}_m)^2}{2\Delta\tau^2} - \Delta\tau V(R'_m)\right] \tag{12.88}$$

where $\overline{R}_m = (R_{m+1} + R_{m-1})/2$. We may sample the new position directly from this distribution by first generating a new position using a Gaussian random generator with width $1/(2\Delta\tau)$ and centred around $\overline{R}_m$, and then accepting or rejecting the new position with a probability proportional to $\exp[-\Delta\tau V(R'_m)]$. This procedure guarantees 100% acceptance for zero potential. If there are hard-core interactions between the particles, the Gaussian distribution might be replaced by a more complicated form to take this into account [4].

A major drawback of the algorithm presented so far is that only one atom is displaced at a time. To obtain a decent acceptance rate the maximal distance over which the atom can be displaced is restricted by the harmonic interaction between successive 'beads' on the imaginary time-chain to $\sim\sqrt{\Delta\tau}$. The presence of the potential V can force us to decrease this step size even further. It will be clear that our local update algorithm will cause the correlation time to be long, as this time is determined by the long-wavelength modes of the chain. As it is estimated that equilibration of the slowest modes takes roughly $\mathcal{O}(M^2)$ Monte Carlo sweeps (see the next chapter), the relaxation time will scale as M^3 single-update steps. This unfavourable time scaling behaviour is well known in computational field theory, and a large part of the next chapter will be dedicated to methods for enhancing the efficiency of Monte Carlo simulations on lattices. An important example of such methods is *normal mode sampling* in which, instead of single particle moves, one changes the configuration via its Fourier modes [24, 25]. If one changes for example the $k = 0$ mode, all particles are shifted over the same distance. The transition probability is calculated either through the Fourier-transformed kinetic (harmonic interaction) term, followed by an acceptance/rejection based on the change in potential, or by using the Fourier transform of the full action. We shall not treat these methods in detail here; in the next chapter, we shall discuss similar methods for field theory.

A method introduced by Ceperley and Pollock divides the time slices up in a hierarchical fashion and alters the values of groups of points in various stages [3, 4]. At each stage the step can be discontinued or continued according to some acceptance criterion. It turns out [4] that with this method it is possible to reduce the relaxation time from M^3 to $M^{1.4}$. The method seems close in spirit to the multigrid Monte Carlo method of Goodman and Sokal, which we shall describe in the next chapter.

It will be clear that for a full boson simulation, moving particles is not sufficient: we must also include permutation moves, in which we swap two springs between particles at subsequent beads, for example. However, the configurations are usually equilibrated for a particular permutation, and changing this permutation can be so drastic a move that permutations are never accepted. In that case it is possible to combine a permutation with particle displacements which adjust the positions to the new permutation [4].

12.5 Quantum Monte Carlo on a lattice

There are several interesting quantum systems which are or can be formulated on a lattice. First of all, we may consider quantum spin systems as generalisations of the classical spin systems mentioned in Chapter 7. An example is the Heisenberg model, with Hamiltonian

$$H_{\text{Heisenberg}} = -J \sum_{\langle ij \rangle} \mathbf{s}_i \cdot \mathbf{s}_j \tag{12.89}$$

where the sum is over nearest neighbour sites $\langle ij \rangle$ of a lattice (in any dimensions), and the spins satisfy the standard angular momentum commutation relations on the same site ($\hbar \equiv 1$):

$$[s^x, s^y] = \frac{\mathrm{i}s^z}{2}. \tag{12.90}$$

Another example is the second quantised form of the Schrödinger equation. This uses the 'occupation number representation' in which we have creation and annihilation operators for particles in a particular state. If the Schrödinger equation is discretised on a grid, the basis states are identified with grid points, and the creation and annihilation operators create and annihilate particles on these grid points. These operators are called $c_i^\dagger$ and c_i respectively, and they satisfy the commutation relations

$$[c_i, c_j] = [c_i^\dagger, c_j^\dagger] = 0; \quad [c_i, c_j^\dagger] = \delta_{ij}. \tag{12.91}$$

In terms of these operators, the Schrödinger equation for a one-dimensional, noninteracting system reads [26]

$$\sum_i -t(c_i^\dagger c_{i+1} + c_{i+1}^\dagger c_i) + \sum_i V_i n_i \tag{12.92}$$

where n_i is the number operator $c_i^\dagger c_i$, and where appropriate boundary conditions are to be chosen.

A major advantage of this formulation over the original version of the Schrödinger equation is that the boson character is automatically taken into account: there is no need to permute particles in the Monte Carlo algorithm. A disadvantage is that the lattice will introduce discretisation errors.

Finally, this model may be formulated for interacting fermions. A famous model of this type is the so-called *Hubbard model*, which models the electrons which are tightly bound to the atoms in a crystalline material. The Coulomb repulsion is restricted to an on-site effect; electrons on different sites do not feel it. The creation and annihilation operators are now called $c^\dagger_{i,\sigma}, c_{i,\sigma}$, where $\sigma = \pm$ labels the spin. They anticommute, except for $[c^\dagger_{i,\sigma}, c_{j,\sigma'}]_+ = \delta_{ij}\delta_{\sigma\sigma'}$. The standard form of the Hubbard model in one dimension reads

$$H = \sum_{i,\sigma} -t[c^\dagger_{i,\sigma}c_{i+1,\sigma} + c^\dagger_{i+1,\sigma}c_{i,\sigma}] + U\sum_i n_{i,\sigma}n_{i,-\sigma} \tag{12.93}$$

where $n_{i,\sigma}$ is the number operator which counts the particles with spin σ at site i: $n_i = c^\dagger_{i,\sigma}c_{i,\sigma}$. The first term describes hopping from atom to atom, and the second one represents the Coulomb interaction between fermions at the same site.

We shall outline the quantum path-integral Monte Carlo analysis for one-dimensional lattice quantum systems, taking the Heisenberg method as the principal example. Extensions to other systems will be considered only very briefly. For a review, see Ref. [5]; see also Ref. [27].

The quantum Heisenberg model is formulated on a chain consisting of N sites, which we shall number by the index i. We have already discussed this model in Section 11.5. The Hilbert space has basis states $|S\rangle = |s_1, s_2, \ldots, s_N\rangle$, where the s_i assume values ± 1; they are the eigenstates of the z-component of the spin operator. The Heisenberg Hamiltonian can be written as the sum of operators containing interactions between *two neighbouring* sites. Let us call H_i the operator $-J\mathbf{s}_i \cdot \mathbf{s}_{i+1}$, coupling spins at sites i and $i+1$. Suppose we have N sites and that N is even. We now partition the Hamiltonian as follows:

$$\begin{aligned} H = H_{\text{odd}} + H_{\text{even}} = (H_1 + H_3 + H_5 + \cdots + H_{N-1}) \\ + (H_2 + H_4 + H_6 + \cdots + H_N). \end{aligned} \tag{12.94}$$

H_i and H_{i+2} commute as the H_i couple only nearest neighbour sites. This makes the two separate Hamiltonians H_{odd} and H_{even} trivial to deal with in the path integral. However, H_{odd} and H_{even} do not commute. It will therefore be necessary to use the short-time approximation.

The time-evolution operator is split up as follows:

$$e^{-\tau H} \approx e^{-\Delta\tau H_{\text{odd}}}e^{-\Delta\tau H_{\text{even}}}e^{-\Delta\tau H_{\text{odd}}}e^{-\Delta\tau H_{\text{even}}}\ldots e^{-\Delta\tau H_{\text{odd}}}e^{-\Delta\tau H_{\text{even}}} \tag{12.95}$$

with a total number of $2M$ exponents in the right hand side; $\Delta\tau = \tau/M$. In calculating the partition function, we insert a unit operator of the form $\sum_S |S\rangle\langle S|$ between

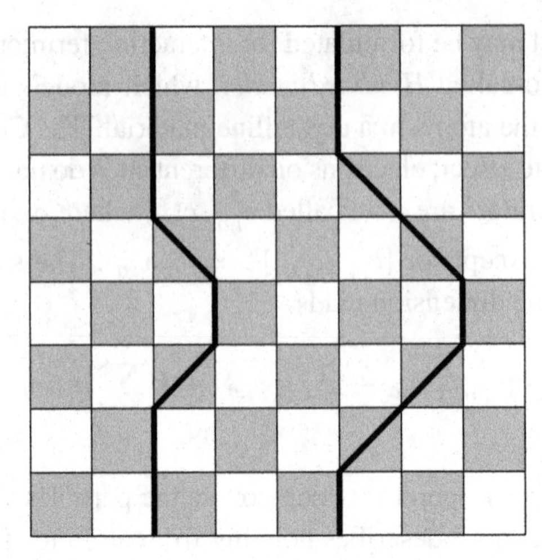

Figure 12.7. The checkerboard decomposition of the space-time lattice. Two world lines are shown.

the exponentials, where $\sum_S$ denotes a sum over all the spins s_i in S:

$$Z = \sum_{S_i,\bar{S}_i} \langle S_0|\mathrm{e}^{-\Delta\tau H_{\mathrm{odd}}}|\bar{S}_0\rangle\langle\bar{S}_0|\mathrm{e}^{-\Delta\tau H_{\mathrm{even}}}|S_1\rangle\langle S_1|\mathrm{e}^{-\Delta\tau H_{\mathrm{odd}}}|\bar{S}_1\rangle$$
$$\times \langle\bar{S}_1|\mathrm{e}^{-\Delta\tau H_{\mathrm{even}}}|S_2\rangle \cdots \langle S_{N/2-1}|\mathrm{e}^{-\Delta\tau H_{\mathrm{odd}}}|\bar{S}_{N/2-1}\rangle$$
$$\times \langle\bar{S}_{N/2-1}|\mathrm{e}^{-\Delta\tau H_{\mathrm{even}}}|S_0\rangle. \tag{12.96}$$

The operators $\exp(\Delta\tau H_{\mathrm{even}})$ and $\exp(\Delta\tau H_{\mathrm{odd}})$ can be expanded as products of terms of the form $\exp(\Delta\tau H_i)$. Each such term couples the spins around a plaquette of the space-time lattice and the resulting picture is that of Figure 12.7, which explains the name 'checkerboard decomposition' for this partitioning of the Hamiltonian. Other decompositions are possible, such as the real-space decomposition [5], but we shall not go into this here.

The simulation of the system seems straightforward: we have a space-time lattice with interactions around the shaded plaquettes in Figure 12.7. At each site of the lattice we have a spin s_{im}, where i denotes the spatial index and m denotes the index along the imaginary-time or inverse-temperature axis. The simulation consists of attempting spin flips, evaluating the Boltzmann weight before and after the change, and then deciding to perform the change or not with a probability determined by the fractions of the Boltzmann weights (before and after). But there is a snake in the grass. The Hamiltonians H_m commute with the total spin operator, $\sum_i s_i^z$; therefore the latter is conserved, i.e.

$$s_{im} + s_{i+1,m} = s_{i,m+1} + s_{i+1,m+1} \tag{12.97}$$

for each plaquette (remember the s_i occurring in this equations are the eigenvalues of the corresponding s_i^z operators). Therefore a single spin flip will never be accepted as it does not respect this requirement. This was already noted in Section 11.5: letting a chain evolve under the Hamiltonian time evolution leaves the system in the 'sector' where it started off. Simple changes in the spin configuration which conserve the total spin from one row to another are spin flips of all the spins at the corners of a nonshaded plaquette.

In the boson and fermion models, where we have particle numbers n_{im} instead of spins, the requirement (12.97) is to be replaced by

$$n_{im} + n_{i+1,m} = n_{i,m+1} + n_{i+1,m+1}. \tag{12.98}$$

In this case the simplest change in the spin configuration consists of an increase (decrease) by one of the numbers at the two left corners of a nonshaded plaquette and a decrease (increase) by one of the numbers at the right hand corners (obviously, the particle numbers must obey $n_{im} \geq 0$ (bosons) or $n_{im} = 0, 1$ (fermions)). Such a step is equivalent to having one particle moving one lattice position to the left (right). The overall particle number along the time direction is conserved in this procedure. The particles can be represented by *world lines*, as depicted in Figure 12.7. The changes presented here preserve particle numbers from row to row, so for a simulation of the full system, one should consider also removals and additions of entire world lines as possible Monte Carlo moves.

Returning to the Heisenberg model, we note that the operator $\exp(-\Delta\tau H_i)$ couples only spins at the bottom of a shaded plaquette to those at the top. This means that we can represent this operator as a 4×4 matrix, where the four possible states $|++\rangle$, $|+-\rangle$, $|-+\rangle$ and $|--\rangle$ label the rows and columns. For the Heisenberg model one finds after some calculation

$$\exp\left[-\Delta\tau\frac{J}{4}\boldsymbol{\sigma}_i\cdot\boldsymbol{\sigma}_{i+1}\right]$$
$$= \begin{pmatrix} e^{-\Delta\tau J/4} & 0 & 0 & 0 \\ 0 & e^{\Delta\tau J/4}\cosh(\Delta\tau J/2) & -e^{\Delta\tau J/4}\sinh(\Delta\tau J/2) & 0 \\ 0 & -e^{\Delta\tau J/4}\sinh(\Delta\tau J/2) & e^{\Delta\tau J/4}\cosh(\Delta\tau J/2) & 0 \\ 0 & 0 & 0 & e^{-\Delta\tau J/4} \end{pmatrix} \tag{12.99}$$

($\boldsymbol{\sigma}$ is the vector of Pauli matrices $(\sigma_x, \sigma_y, \sigma_z)$ – we have $\mathbf{s} = \hbar\boldsymbol{\sigma}/2$; $\hbar \equiv 1$). This matrix can be diagonalised (only a diagonalisation of the inner 2×2 block is necessary) and the model can be solved trivially. Some matrix elements become negative when $J < 0$ (Heisenberg antiferromagnet). This minus-sign problem turns out not to be fundamental, as it can be transformed away by a redefinition of the spins on alternating sites [5, 28].

In the case where, instead of spin-1/2 degrees of freedom, we have (boson) numbers on the sites, the matrix H_1 becomes infinite-dimensional. In that case we must expand $\exp(-\Delta\tau H_i)$ in a Taylor series expansion in $\Delta\tau$. We shall not go into details but refer to the literature [5].

If we have fermions, there is again a minus-sign problem. This turns out to be removable for a one-dimensional chain, but not for two and three dimensions. In these cases one uses fixed-node and transient estimator methods as described above [29].

12.6 The Monte Carlo transfer matrix method

In Chapter 11 we have seen that it is possible to calculate the free energy of a discrete lattice spin model on a strip by solving the largest eigenvalue of the transfer matrix. The size of the transfer matrix increases rapidly with the strip width and the calculation soon becomes unfeasible, in particular for models in which the spins can assume more than two different values. The QMC techniques which have been presented in the previous sections can be used to tackle the problem of finding the largest eigenvalues of the very large matrices arising in such models. Here we discuss such a method. It goes by the name of 'Monte Carlo transfer matrix' (MCTM) method and it was pioneered by Nightingale and Blöte [30].

Let us briefly recall the transfer matrix theory. The elements $T(S', S) = \langle S'|T|S\rangle$ of the transfer matrix T are the Boltzmann weights for adding new spins to a semi-infinite system. For example, the transfer matrix might contain the Boltzmann weights for adding an entire row of spins to a semi-infinite lattice model, or a single spin, in which case we take helical boundary conditions so that the transfer matrix is the same for each spin addition (see Figure 12.8). The free energy is given in terms of the largest eigenvalue λ_0 of the transfer matrix:

$$F = -k_B T \ln(\lambda_0). \tag{12.100}$$

From discussions in Chapter 11 and Section 12.4, it is clear that the transfer matrix of a lattice spin model is the analogue of the time-evolution operator in quantum mechanics.

We now apply a technique analogous to diffusion Monte Carlo to sample the eigenvector corresponding to the largest eigenvalue. In the following we use the terms 'ground state' for this eigenvector, because the transfer matrix can be written in the form $T = \exp(-\tau H)$, so that the ground state of H gives the largest eigenvalue of the transfer matrix. We write the transfer matrix as a product of a normalised transition probability P and a weight factor D. In Dirac notation:

$$\langle S'|T|S\rangle = D(S')\langle S'|P|S\rangle. \tag{12.101}$$

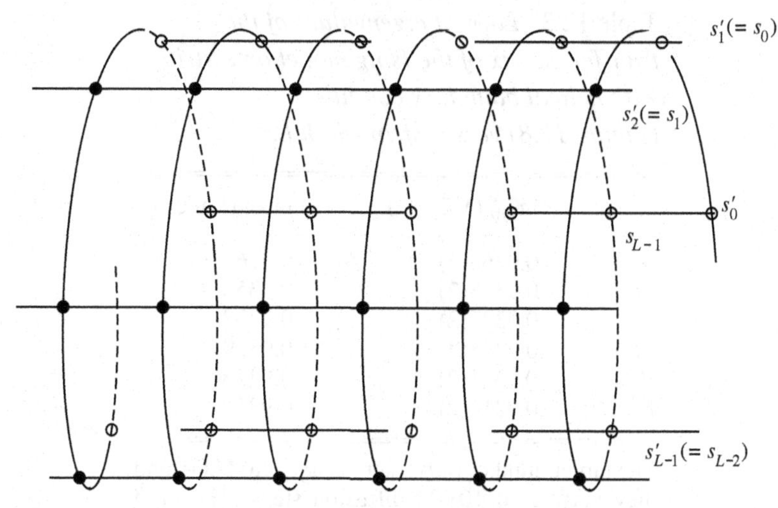

Figure 12.8. Helical boundary conditions for the spin model with nearest neighbour interactions on a strip. A step of the algorithm consists of evolving the 'old' walker S into a new one called S'. This is done by first adding a new 'head' s'_0 of S' according to a probability distribution like (12.104). Then the 'old' components s_{L-2} to s_0 are copied onto s'_{L-1} to s'_1.

The ground state will be represented by a collection of random walkers $\{S_k\}$ which diffuse in configuration space according to the transition probability P. Each diffusion step is followed by a branching step in which the walkers are eliminated or multiplied, i.e. split into a collection of identical walkers, depending on the value of the weight factor $D(S'_k)$.

Let us describe the procedure for a *p-state clock model* with stochastic variables (spins) which assume values

$$\theta = \frac{2\pi n}{p}, n = 0, \ldots, p-1 \tag{12.102}$$

and a nearest neighbour coupling

$$-\frac{\mathcal{H}}{k_B T} = \sum_{\langle ij \rangle} J\cos(\theta_i - \theta_j). \tag{12.103}$$

For $p = 2$ this is equivalent to the Ising model (with zero magnetic field), with J being exactly the same coupling constant as in the standard formulation of this model (Chapter 7). For large p the model is equivalent to the XY model. The XY model will be discussed in Chapter 15 – at this moment it is sufficient to know that this model is critical for all temperatures between 0 and T_{KT}, which corresponds to $\beta J \approx 1.1$ (the subscript KT denotes the Kosterlitz–Thouless phase transition; see Chapter 15). The central charge c (see Section 11.3) is equal to 1 on this critical line.

Table 12.3. *Largest eigenvalues of the transfer matrix of the Ising model on a strip with helical boundary conditions (Figure 12.8) versus strip width L.*

L	$\ln \lambda_0$ (MCTM)	$\ln \lambda_0$ (Lanczos)
6	0.9368(2)	0.9369
7	0.9348(2)	0.9350
8	0.9337(2)	0.9338
9	0.9328(2)	0.9329
10	0.9321(2)	0.9323
11	0.9316(2)	0.9318

The target number of walkers is equal to 5000, and they performed 10 000 diffusion steps. The third column gives the eigenvalues obtained by diagonalising the full transfer matrix using the Lanczos method. These values are determined with high accuracy and are rounded to four significant digits.

The walkers are 'columns' of lattice spins, $(s_0, \ldots, s_{L-1})$, as represented in Figure 12.8. In the diffusion step, a new spin is added to the system, and its value is the s_0-component of the new configuration of the walker. The spin components 1 to $L - 1$ of the new configuration are filled with the components 0 to $L - 2$ of the old walker – the walker is shifted one position over the cylinder. To sample the new s_0'-value, we use the 'shooting method' in which the interval [0, 1] is divided up into p segments corresponding to the conditional probability $P(s_0'|S)$ which is proportional to the Boltzmann factor for adding a spin $s_0' = 0, \ldots, p - 1$ to the existing column S. In our clock model example, we have

$$P(s_0'|S) = \mathrm{e}^{J\cos(s_0'-s_0)+J\cos(s_0'-s_{L-1})}/D(S), \tag{12.104}$$

with normalisation factor

$$D(S) = \sum_{s_0'} \mathrm{e}^{J\cos(s_0'-s_0)+J\cos(s_0'-s_{L-1})}. \tag{12.105}$$

A random number between 0 and 1 is then generated and the new spin value corresponds to the index of the segment in which the random number falls.

The next step is then the assignment of the weight $D(S')$ to the walker with D given in (12.105). Branching is then carried out exactly as in the DMC method. In fact, the weights are also multiplied by a factor $\exp(E_{\text{trial}})$, where E_{trial} is the same for all walkers but varies in time. It is updated as in the DMC method according to

$$E_{\text{trial}} = E_0 - \alpha \ln(N/N_0), \tag{12.106}$$

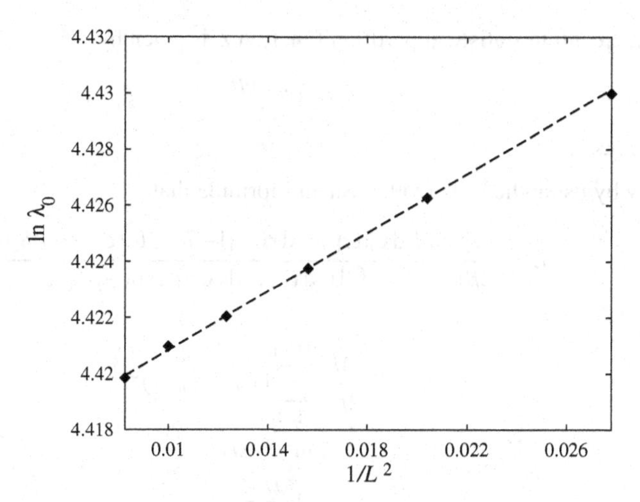

Figure 12.9. The logarithm of the largest eigenvalue of the transfer matrix versus the inverse of the square of the strip width L. The straight line has a slope $\pi/6$ and is adjusted in height to fit the data.

where E_0 is a guess of the trial energy (which should be equal to $-\ln\lambda_0$, λ_0 is the largest eigenvalue), N is the actual number of walkers and N_0 is the target number of walkers. This term aims at stabilising the population size to the target number N_0.

The simplest information we obtain is the largest eigenvalue, which is given as $\exp(E_{\text{trial}})$, where the average value of E_{trial} during the simulation is to be used (with the usual omission of equilibration steps). This can be used to determine central charges. In Table 12.3 we compare the values of this quantity for the Ising model with those obtained by a Lanczos diagonalisation of the transfer matrix. The agreement is seen to be excellent. For the *XY* model, the eigenvalues cannot be found using direct diagonalisation and we can check the MCTM method only by comparing the central charge obtained with the known value: 1 in the low-temperature phase and 0 at high temperatures. In Figure 12.9 we show the results for $\beta J = 1.25$. The points in a graph of the form $\ln\lambda_0$ vs. $1/L^2$ lie on a straight curve with a slope of $\pi/6$ ($c = 1$).

Exercises

12.1 In this problem we consider the virial expression for the energy [20].

In a path-integral QMC simulation for a particle in one dimension in a potential $V(x)$ we want to find the energy E as a function of temperature $T = 1/(k_B\beta)$. We do this by using the thermodynamic relation

$$E = -\frac{\partial \ln Z}{\partial \beta}$$

where the quantum statistical partition function Z is given by

$$Z = \mathrm{Tr} e^{-\beta H}.$$

We take $\hbar \equiv 1$.

(a) Show by using the Lie–Trotter–Suzuki formula that

$$E = \frac{N}{2\beta} + \frac{\int \mathrm{d}x_0 \mathrm{d}x_1 \dots \mathrm{d}x_{M-1}[-T+U] \exp(-\beta S_{\mathrm{cl}})}{\int \mathrm{d}x_0 \mathrm{d}x_1 \dots \mathrm{d}x_{M-1} \exp(-\beta S_{\mathrm{cl}})}$$

with

$$T = \frac{M}{2\beta^2} \sum_{m=0}^{M-1} (x_m - x_{m+1})^2;$$

$$x_0 \equiv x_M;$$

$$U = \frac{1}{M} \sum_{m=0}^{M-1} V(x_m)$$

and

$$S_{\mathrm{cl}} = T + V.$$

(b) Show that

$$\frac{\int \mathrm{d}x_0 \mathrm{d}x_1 \dots \mathrm{d}x_{N-1} \sum_{i=0}^{N-1} x_i (\partial S_{\mathrm{cl}}/\partial x_i)[\exp(-\beta S_{\mathrm{cl}})]}{\int \mathrm{d}x_0 \mathrm{d}x_1 \dots \mathrm{d}x_{N-1} \exp(-\beta S_{cl})} = \frac{N}{\beta}.$$

Hint: use partial integration.

(c) Show that

$$\sum_{m=0}^{M-1} x_m \frac{\partial T}{\partial x_m} = 2T.$$

(d) Show that the energy can also be determined by

$$E = \left\langle \frac{1}{N} \sum_{m=0}^{N-1} \left[V(x_m) + \frac{1}{2} x_m \frac{\partial V}{\partial x_m} \right] \right\rangle.$$

(e) Show that the generalisation to a three-dimensional particle is

$$E = \left\langle \frac{1}{N} \sum_{m=0}^{M-1} \left[V(\mathbf{r}_m) + \frac{1}{2} \mathbf{r}_m \cdot \frac{\partial V}{\partial \mathbf{r}_m} \right] \right\rangle.$$

12.2 A particle moves in three dimensions. It experiences no potential: $V(\mathbf{r}) = 0$. At imaginary time $\tau = 0$ the particle is localised at $\mathbf{r}_1$.

(a) What is the wave function $\psi_0(\mathbf{r}, \tau)$ of the particle for $\tau' > 0$?

(b) We assume that the particle moves from $\mathbf{r}_1$ at time 0 to $\mathbf{r}_2$ at time τ. When we want to evaluate the matrix element

$$\langle \mathbf{r}_1, 0 | \mathbf{r}_2, \tau \rangle,$$

in the path-integral formalism, we should include all paths satisfying these boundary conditions. Using completeness, we can write, with $0 < \tau' < \tau$:

$$\langle \mathbf{r}_1, 0|\mathbf{r}_2, \tau\rangle = \int \mathrm{d}^3r' \langle \mathbf{r}_1, 0|\mathbf{r}', \tau'\rangle\langle \mathbf{r}', \tau'|\mathbf{r}_2, \tau\rangle.$$

Show that the integrand in this equation can be written as

$$\langle \mathbf{r}_1, 0|\mathbf{r}', \tau'\rangle\langle \mathbf{r}', \tau'|\mathbf{r}_2, \tau\rangle = \langle \mathbf{r}_1 0|\mathbf{r}_2, \tau\rangle \frac{1}{(2\pi\sigma_{\tau'})^{3/2}} \mathrm{e}^{-[\mathbf{r}'-\overline{\mathbf{r}}(\tau')]^2/(2\sigma_{\tau'})},$$

with

$$\sigma_{\tau'} = \frac{\tau'(\tau - \tau')}{\tau} \quad \text{and} \quad \overline{\mathbf{r}}(\tau') = \mathbf{r}_1 + \frac{\tau'}{\tau}(\mathbf{r}_2 - \mathbf{r}_1).$$

12.3 In this problem we consider the cumulant expansion analysis for the Coulomb potential [4, 22] using the result of the previous problem.

The cumulant expansion is a well-known expansion in statistical physics [31]. It replaces the Gaussian average of an exponent by the exponent of a sum of averages:

$$\langle \mathrm{e}^{\tau V}\rangle = \mathrm{e}^{\tau\langle V\rangle + \frac{1}{2}(\tau^2\langle V^2\rangle - \langle V\rangle^2)+\cdots}.$$

First we note that the matrix between two positions $\mathbf{r}_1$ and $\mathbf{r}_2$ separated by an imaginary time τ can be written in the following way:

$$\langle \mathbf{r}_1, 0| \exp\left(-\int_0^\tau V(\mathbf{r}'_\tau)\mathrm{d}\tau'\right)|\mathbf{r}_2, \tau\rangle.$$

where the time evolution leading from 0 to τ is that of a free particle and the expression is to be evaluated in a time-ordered fashion.

If we evaluate this in the cumulant expansion approximation retaining only the first term, it is clear that we must calculate

$$\int_0^\tau \mathrm{d}\tau' \int \mathrm{d}^3r' \langle \mathbf{r}_1, 0|\mathbf{r}', \tau'\rangle V(\mathbf{r}')\langle \mathbf{r}', \tau'|\mathbf{r}_2, \tau\rangle.$$

This is done in this problem.

(a) Show that the Fourier transform of the Coulomb potential is $V(\mathbf{k}) = 2\pi/k^2$.

(b) Show that the Fourier transform of the expression derived in Problem 12.2 is given by

$$\mathrm{e}^{-\mathrm{i}\mathbf{k}\cdot\overline{\mathbf{r}}(\tau')-\sigma_{\tau'}k^2/2},$$

with $\sigma_{\tau'}$ and $\overline{\mathbf{r}}(\tau)$ as given in the previous problem.

(c) Show, by transforming back to the $\mathbf{r}$-representation, that the cumulant potential is given by

$$V_{\text{cumulant}}(\mathbf{r}_1, \mathbf{r}_2; \tau) = \int_0^\tau \frac{\operatorname{erf}[\overline{\mathbf{r}}(\tau')/\sqrt{2\sigma_{\tau'}}]}{\overline{\mathbf{r}}(\tau')}\mathrm{d}\tau'.$$

12.4 In the path-integral simulation for the hydrogen atom we use a table in which the cumulant expression for the potential is stored and we want to linearly interpolate this table.

(a) Show that for a two-dimensional table containing values of a function $f(x, y)$ for integer x and y, the value $f(x, y)$ for arbitrary x and y within the boundaries set by the table size is given as

$$f(x, y) = (2 - x - y + [x] + [y])f([x], [y]) \quad (12.107)$$
$$+ (1 + x - [x] - y + [y])f([x] + 1, [y]) \quad (12.108)$$
$$+ (1 + y - [y] - x + [x])f([x], [y] + 1) \quad (12.109)$$
$$+ (x - [x] + y - [y])f([x] + 1, [y] + 1).$$

Here $[x]$ denotes the largest integer smaller than x.

(b) Find analogous expressions for a table with a noninteger (but equidistant) spacing between the table entries and also for a three-dimensional Table.

12.5 [C]

In this problem we consider applying variational Monte Carlo to the hydrogen molecule. There are two complications in comparison with the helium atom. One is the calculation of the local energy which is quite cumbersome, although straightforward. The second one is the cusp condition.

To specify the trial wave function we take the nuclei at positions $\pm s/2$. A one-particle orbital has the form (in atomic units):

$$\phi(\mathbf{r}) = \mathrm{e}^{-|\mathbf{r}-s\hat{\mathbf{x}}/2|/a} + \mathrm{e}^{-|\mathbf{r}+s\hat{\mathbf{x}}/2|/a}$$

where a is some parameter. The two-electron wave function is given as

$$\psi(\mathbf{r}_1, \mathbf{r}_2) = \phi(\mathbf{r}_1)\phi(\mathbf{r}_2)f(r_{12})$$

with f the Jastrow factor

$$f(r) = \exp\left(\frac{r}{\alpha(1 + \beta r_{12})}\right).$$

(a) Show that the Coulomb-cusp condition near the nuclei leads to the relation

$$\frac{1}{1 + \exp(-s/a)} = a.$$

For a given distance s, this equation should be solved numerically to give you the value a.

(b) Show that the electron–electron cusp condition leads to the requirement $\alpha = 2$. This leaves a single parameter β in the wave function.

(c) Now you can implement the hydrogen molecule in VMC. Calculate the energy as a function of the parameters β and s and find the minimum.

(d) You may also evaluate the ground state by applying the method of Harju *et al.* [6] which was described in Section 12.2, in order to update the values of β and s simultaneously.

(e) What would you need in order to calculate the molecular formation energy from this? Note that this is the difference between the energy of the hydrogen molecule and that of two isolated hydrogen atoms. Consider in particular the contribution arising from the nuclear motion.

References

[1] P. A. M. Dirac, *The Principles of Quantum Mechanics*. Oxford, Oxford University Press, 1958.
[2] B. L. Hammond, W. A. Lester Jr, and P. J. Reynolds, *Monte Carlo Methods in Ab Initio Quantum Chemistry*. Singapore, World Scientific, 1994.
[3] D. M. Ceperley and E. L. Pollock, 'Path-integral computation of the low-temperature properties of liquid-He-4,' *Phys. Rev. Lett.*, **56** (1986), 351–4.
[4] D. M. Ceperley, 'Path integrals in the theory of condensed helium,' *Rev. Mod. Phys.*, **67** (1995), 279–355.
[5] H. De Raedt and A. Lagendijk, 'Monte Carlo simulations of quantum statistical lattice models,' *Phys. Rep.*, **127** (1985), 233–307.
[6] A. Harju, S. Siljamäki, and R. M. Nieminen, 'Wigner molecules in quantum dots: a quantum Monte Carlo study,' *Phys. Rev. E*, **65** (2002), 075309.
[7] R. J. Jastrow, 'Many-body problem with strong forces,' *Phys. Rev.*, **98** (1955), 1479–84.
[8] M. H. Kalos, D. Levesque, and L. Verlet, 'Helium at zero temperature with hard-sphere and other forces,' *Phys. Rev. A*, **9** (1974), 2178–95.
[9] R. N. Barnett, P. J. Reynolds, and W. A. Lester Jr, 'Monte Carlo determination of the oscillator strength and excited state lifetime for the Li $2^2S \rightarrow 2^2P$ transition,' *Int. J. Quantum Chem.*, **42** (1992), 837–47.
[10] R. P. Feynman and A. R. Hibbs, *Quantum Mechanics and Path Integrals*. New York, McGraw-Hill, 1965.
[11] M. Suzuki, 'Decomposition formulas of exponential operators and Lie exponents with some applications to quantum-mechanics and statistical physics,' *J. Math. Phys.*, **26** (1985), 601–12.
[12] M. Suzuki, 'Transfer-matrix method and Monte Carlo simulation in quantum spin systems,' *Phys. Rev. B*, **31** (1985), 2957–65.
[13] J. D. Doll, R. D. Coalson, and D. L. Freeman, 'Towards a Monte Carlo theory of quantum dynamics,' *J. Chem. Phys.*, **87** (1987), 1641–7.
[14] V. S. Filinov, 'Calculation of the Feynman integrals by means of the Monte Carlo method,' *Nucl. Phys. B*, **271** (1986), 717–25.
[15] J. Chang and W. H. Miller, 'Monte Carlo path integration in real-time via complex coordinates,' *J. Chem. Phys.*, **87** (1987), 1648–52.
[16] R. P. Feynman, 'The λ-transition in liquid helium,' *Phys. Rev.*, **90** (1953), 1116–17.
[17] R. P. Feynman, 'Atomic theory of the λ-transition in helium,' *Phys. Rev.*, **91** (1953), 1291–301.
[18] R. P. Feynman, 'Atomic theory of liquid helium near absolute zero,' *Phys. Rev.*, **91** (1953), 1301–8.
[19] D. M. Ceperley and E. L. Pollock, 'Path-integral computation techniques for superfluid ^{4}He,' in *Monte Carlo Methods in Theoretical Physics* (S. Caracciolo and A. Fabrocini, eds.). Pisa, Italy, ETS Editrice, 1992, p. 35.
[20] M. F. Herman, E. J. Bruskin, and B. J. Berne, 'On path integral Monte-Carlo simulations,' *J. Chem. Phys.*, **76** (1982), 5150–5.
[21] K. Singer and W. Smith, 'Path integral simulation of condensed phase Lennard–Jones systems,' *Mol. Phys.*, **64** (1988), 1215–31.
[22] D. M. Ceperley, 'The simulation of quantum systems with random walks – a new algorithm for charged systems,' *J. Comp. Phys.*, **51** (1983), 404–22.
[23] E. L. Pollock, 'Properties and computation of the Coulomb pair density matrix,' *Comp. Phys. Comm.*, **52** (1989), 49–60.
[24] M. Takahashi and M. Imada, 'Monte Carlo calculation of quantum-systems,' *J. Phys. Soc. Jpn*, **53** (1984), 963–74.
[25] J. D. Doll, R. D. Coalson, and D. L. Freeman, 'Solid-fluid phase transition of quantum hard-spheres at finite temperatures,' *Phys. Rev. Lett.*, **55** (1985), 1–4.

[26] M. Plischke and H. Bergersen, *Equilibrium Statistical Physics*. Englewood Cliffs, NJ, Prentice-Hall, 1989.

[27] J. E. Hirsch, 'Discrete Hubbard–Stratonovich transformation for fermion lattice models,' *Phys. Rev. B*, **28** (1983), 4059–61.

[28] J. W. Negele and H. Orland, *Quantum Many-Particle Systems*. Redwood City, Addison-Wesley, 1988.

[29] D. B. F. ten Haaf, H. J. M. van Bemmel, J. M. J. van Leeuwen, W. van Saarloos, and D. M. Ceperley, 'Proof for an upper bound in fixed-node Monte Carlo for lattice fermions,' *Phys. Rev. B*, **51** (1995), 13039–45.

[30] M. P. Nightingale and H. W. J. Blöte, 'Monte Carlo calculation for the free energy, critical point and surface critical behaviour of three-dimensional Heisenberg ferromagnets,' *Phys. Rev. Lett.*, **60** (1988), 1562–5.

[31] N. G. van Kampen, *Stochastic Processes in Physics and Chemistry*. Amsterdam, North-Holland, 1981.

13

The finite element method for partial differential equations

13.1 Introduction

When we consider a partial differential equation, such as the ubiquitous Laplace equation

$$\nabla^2 \phi(\mathbf{r}) = 0, \tag{13.1}$$

together with some boundary condition(s), the obvious way of solving it that comes to mind is to discretise this equation on a regular grid, hoping that this grid can match the boundary in some way. Then we solve the discretised problem using, for example, iterative methods such as the Gauss–Seidel or conjugate gradients method (see Appendix A7.2). For many problems, this approach is adequate, but if the problem is difficult in the sense that it has a lot of structure on small scales in some region of the domain, or if the boundary has a complicated shape which is difficult to match with a regular grid, it might be useful to apply methods that allow for flexibility of the grid on which the solution is formulated. In this chapter we discuss such a method, the *finite element method*.

One way of looking at the finite element method (FEM) is by realising that many partial differential equations can be viewed as solution methods for variational problems. In the case of the Laplace equation with zero boundary condition, for example, finding the stationary solution of the functional

$$\int_D [\nabla \phi(\mathbf{r})]^2 \, \mathrm{d}^d r, \tag{13.2}$$

where the integral is over the d-dimensional domain D and where we confine ourselves to functions $\phi(\mathbf{r})$ which vanish on the domain boundary, yields the same solution as that of the Laplace equation – in fact, the Laplace equation is the Euler equation for this functional (see the next section).

The integral can be discretised by dividing up the domain D into *elements* of – in principle – arbitrary shape and size, and assuming a particular form of the solution within each element, a linear function for example, together with continuity conditions on the element boundaries. It turns out that finding the solution boils down to solving a sparse matrix problem, which can be treated by conjugate gradient methods, see (see Appendix A7.2).

In this chapter we discuss the finite element method, error estimation, and principles of local grid refinement. This will be done for two different problems: the Poisson/Laplace equation, and the equations for elastic deformation of a solid. Both problems will be considered in two dimensions only. The aim is to explain the ideas behind the finite element methods and adaptive refinement without going into too much detail. For a more rigorous and complete treatment, the reader is referred to the specialised literature [1–5].

Some special topics will be covered in the remaining sections: local adaptive grid refinement, dynamics, and, finally, the coupling of two descriptions, finite element and molecular dynamics, in order to describe phenomena at very different length scales occurring in one system.

Most of the sections describe implementation of FEM for standard problems. The reader is invited to try the implementation by him- or herself.

13.2 The Poisson equation

As mentioned in the previous section, the Laplace equation can easily be formulated in a variational way. The same holds for the Poisson equation:

$$\nabla^2\phi(\mathbf{r}) = f(\mathbf{r}), \tag{13.3}$$

with appropriate boundary conditions. We assume Dirichlet boundary conditions on the edge of the domain, which we take as a simple square of size $L \times L$. The functional whose stationary solution satisfies this equation is

$$J[\phi(\mathbf{r})] = \int_D \{[\nabla\phi(\mathbf{r})]^2 + f(\mathbf{r})\phi(\mathbf{r})\}\, \mathrm{d}^d r, \tag{13.4}$$

as is easily verified using Green's first identity [6] together with the fact that ϕ vanishes on the boundary. From now on, we shall use $\mathrm{d}\Omega$ to denote a volume element occurring in integrals.

We now divide up the square into triangular elements, and assume that the solution $\phi(\mathbf{r})$ is linear within each element:

$$\phi(x, y) = a_i + b_i x + c_i y \tag{13.5}$$

within element i. Now consider a grid point. This will in general be a vertex of more than one triangle. Naturally, we want to assign a single value of the solution

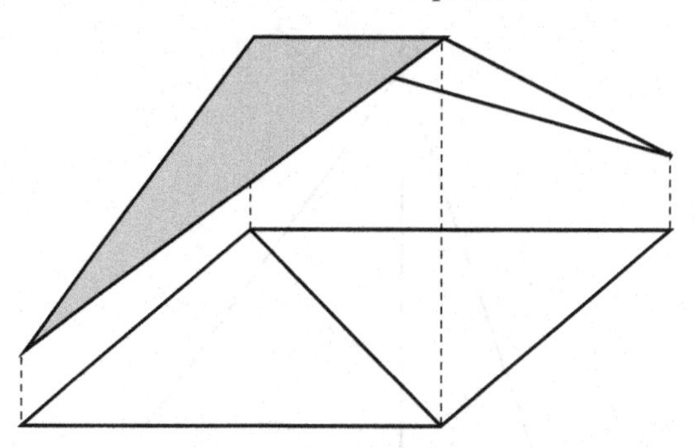

Figure 13.1. Two adjacent triangles on a square (ground plane) with a linear function $\phi(\mathbf{r})$ shown as the height (vertical) coordinate on both triangles. As ϕ is linear for each triangle, the requirement that the values of the two triangles are the same at their two shared vertices ensures continuity along their edges.

to that point, so we require the solution within each triangle sharing the same vertex to have the same value at that vertex. Linearity of the solution within the triangles then makes the solution continuous over each triangle edge (Figure 13.1). We see that for each triangle, the solution is characterised by three constants, a_i, b_i and c_i. They can be fixed by the values of the solution at the three vertices of the triangle. It is also possible to use rectangles as elements. In that case, we must allow for one more degree of freedom of the solution (as there are now four vertices), and the form may then be

$$\phi(x, y) = a_i + b_i x + c_i y + d_i xy. \tag{13.6}$$

It is also possible to use quadratic functions on the triangles:

$$\phi(x, y) = a_i + b_i x + c_i y + d_i xy + e_i x^2 + f_i y^2, \tag{13.7}$$

requiring six conditions. In that case, we use the midpoints of the edges of the triangles as additional points where the solution must have a particular value. We shall restrict ourselves in this book to linear elements. In three dimensions, the linear solution requires four parameters to be fixed, and this can be done by using tetrahedra as elements (a tetrahedron has four vertices). The triangle and the tetrahedron are the elements with nonzero volume which are bounded by the *smallest possible* number of sides in two and three dimensions respectively. Such elements are called *simplices*. In one dimension, an element with this property is the line segment.

Now that we have a discrete representation of our solution by considering just its values on the vertices of the grid, we must find the expression for the integral within the approximations made (i.e. linear behaviour of the solution within the elements). To do this we digress a bit to introduce *natural coordinates*. For a triangle these are linear coordinates which have a value 1 at one of the vertices and zero at the two

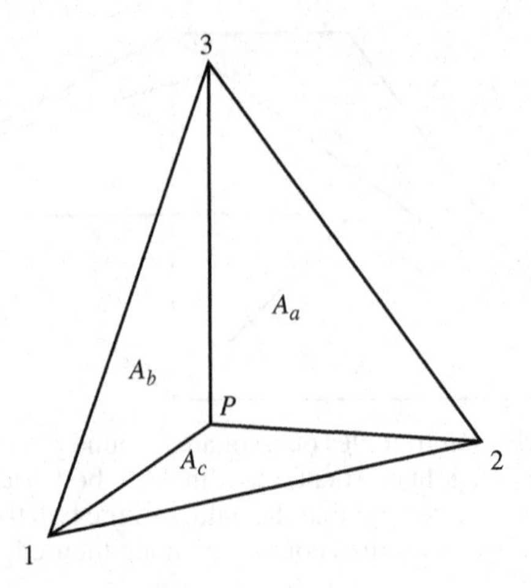

Figure 13.2. The areas A_a, A_b and A_c for any point P within the triangle. The A_i are used to define the natural coordinates ξ_i of the point P. A is the total surface area.

others. Any point P within the triangle can be defined by specifying any two out of three natural coordinates, ξ_a, ξ_b or ξ_c. These are defined by

$$\xi_i = \frac{A_i}{A}, \quad i = a, b, c, \tag{13.8}$$

where A_i, A are the surface areas shown in Figure 13.2. The natural coordinates satisfy the requirement

$$\xi_a + \xi_b + \xi_c = 1. \tag{13.9}$$

The x- and y-coordinates of a point can be obtained from the natural coordinates by the linear transformation

$$\begin{pmatrix} 1 \\ x \\ y \end{pmatrix} = \begin{pmatrix} 1 & 1 & 1 \\ x_a & x_b & x_c \\ y_a & y_b & y_c \end{pmatrix} \begin{pmatrix} \xi_a \\ \xi_b \\ \xi_c \end{pmatrix} \tag{13.10}$$

where (x_a, y_a) are the Cartesian coordinates of vertex a etc.

The reverse transformation

$$\begin{pmatrix} \xi_a \\ \xi_b \\ \xi_c \end{pmatrix} = \frac{1}{2A} \begin{pmatrix} x_b y_c - x_c y_b & y_{bc} & x_{cb} \\ x_c y_b - x_b y_c & y_{ca} & x_{ac} \\ x_a y_b - x_b y_a & y_{ab} & x_{ba} \end{pmatrix} \begin{pmatrix} 1 \\ x \\ y \end{pmatrix}, \tag{13.11}$$

with $2A = \det(A) = x_{ba}y_{ca} - x_{ca}y_{ba}$ and $x_{ab} = x_a - x_b$ etc., translates the x, y coordinates into natural coordinates. All these relations can easily be checked.

Having natural coordinates, we can construct a piecewise linear approximation to the solution from the values of the solution at the vertices of the triangles. Calling

ϕ_a, ϕ_b and ϕ_c these values at the corresponding vertices, the solution inside the triangle is given by

$$\tilde{\phi}(\mathbf{r}) = \phi_a \xi_a + \phi_b \xi_b + \phi_c \xi_c. \tag{13.12}$$

In order to evaluate integrals over the triangular element, we use the following formula:

$$\int_A \xi_a^k \xi_b^l \xi_c^m \, d\Omega = 2A \frac{k!\, l!\, m!}{(2 + k + l + m)!} \tag{13.13}$$

for non-negative integers k, l and m, and for a, b and c assuming values 1, 2 and 3. Remember, $d\Omega$ is the volume element.

We now have all the ingredients for solving the Laplace equation using triangular finite elements. First, note that the integral (13.4) now becomes a quadratic expression in the values ϕ_i at the grid points. This quadratic expression can be written in the form

$$J[\boldsymbol{\phi}] = -\boldsymbol{\phi}^{\mathrm{T}} \mathbf{K} \boldsymbol{\phi} - \mathbf{r}^{\mathrm{T}} \boldsymbol{\phi} \tag{13.14}$$

(we work out the specific form of the expressions below). Here, $\boldsymbol{\phi}$ is the vector whose elements are the values of the solutions at the grid points, $\mathbf{K}$ is a symmetric matrix, and $\mathbf{r}$ is a vector. Minimising this expression leads to the matrix equation

$$\mathbf{K}\boldsymbol{\phi} = \mathbf{r}. \tag{13.15}$$

The matrix–vector product on the left hand side can be evaluated as a sum over all triangles. Within a triangle, we deal with the values on its vertices.

To be more specific, let us calculate

$$\int_{\text{elem}} (\nabla\phi)^2 \, d\Omega. \tag{13.16}$$

Using (13.11) we have

$$\nabla \xi_a = \frac{1}{2A} (y_{bc}, x_{cb}) \tag{13.17}$$

and similar expressions for the other two natural coordinates. From these we have, with the parametrisation (13.12),

$$\nabla\tilde{\phi} = \frac{1}{2A} [\phi_a (y_{bc}, x_{cb}) + \phi_b (y_{ca}, x_{ac}) + \phi_c (y_{ab}, x_{ba})]. \tag{13.18}$$

Note that on the left and right hand side, we have two-dimensional vectors, which are given in row form on the right hand side. Also note that the components of the vector are constant over the triangle, which is natural as the solution is assumed to be linear within the triangle. The integral is the norm of this constant vector squared times the surface area of the triangle. Obviously this yields a quadratic expression in ϕ_a, ϕ_b and ϕ_c of a form similar to (13.14), but now formulated for a single triangle.

This equation is defined by a matrix **k** which is called the *local stiffness matrix* for the triangle under consideration. Introducing the vectors

$$\mathbf{b} = \begin{pmatrix} y_{bc} \\ y_{ca} \\ y_{ab} \end{pmatrix} \tag{13.19}$$

and

$$\mathbf{c} = \begin{pmatrix} x_{cb} \\ x_{ac} \\ x_{ba} \end{pmatrix} \tag{13.20}$$

we see, after some calculation, that the stiffness matrix **k** can be evaluated as

$$\mathbf{k} = \mathbf{b}\mathbf{b}^{\mathrm{T}} + \mathbf{c}\mathbf{c}^{\mathrm{T}}, \tag{13.21}$$

which leads to the result

$$[\mathbf{k}] = \frac{1}{4A} \begin{pmatrix} y_{bc}^2 + x_{cb}^2 & y_{bc}y_{ca} + x_{cb}x_{ac} & y_{bc}y_{ab} + x_{cb}x_{ba} \\ y_{bc}y_{ca} + x_{cb}x_{ac} & y_{ca}^2 + x_{ac}^2 & y_{ca}y_{ab} + x_{ac}x_{ba} \\ y_{bc}y_{ab} + x_{cb}x_{ba} & y_{ca}y_{ab} + x_{ac}x_{ba} & y_{ab}^2 + x_{ba}^2 \end{pmatrix}. \tag{13.22}$$

We not only need the matrix representing the Laplacian operator, but we must also evaluate the integral containing the source term $f(\mathbf{r})$ in Eq. (13.4). The continuous function $f(\mathbf{r})$ is approximated by a piecewise linear function $\tilde{f}$ on the triangles – just like the solution $\phi(\mathbf{r})$. For a particular triangle with vertices a, b, and c, we have

$$\tilde{f}(\mathbf{r}) = f_a\xi_a + f_b\xi_b + f_c\xi_c. \tag{13.23}$$

We must multiply this by the linear approximation for ϕ, Eq. (13.12), and then integrate over the element, using (13.13). The result must then be differentiated with respect to ϕ_a, ϕ_b, ϕ_c, which results in a vector element

$$r_a = 2A\left(\frac{f_a}{12} + \frac{f_b}{24} + \frac{f_c}{24}\right), \tag{13.24}$$

and similar for r_b and r_c.

The matrix–vector multiplication can be carried out as a loop over all triangular elements where for each element the stiffness matrix is applied to the three vertices of that triangle. Note that the stiffness matrix should always act on the *old* vector containing the field values ϕ_a and that the result should be added to the new vector (which initially is set to zero; see the next section). If we have a matrix–vector multiplication and a right hand side of the form (13.15), we can apply the conjugate gradients method to solve the matrix equation.

We have overlooked one aspect of the problem: if a triangle contains vertices on the boundary where the value of the solution is given (Dirichlet boundary condition), the corresponding values of $\phi(\mathbf{r})$ are known and therefore not included in the vector.

In that case we apply the stiffness matrix only to those points which are in the interior of the system. That is, we update only the interior points – the values at the boundary points remain unchanged.

13.2.1 Construction of a finite element program

The program should contain an array in which the equilibrium positions of the vertices are stored. Furthermore, we need a vector containing the displacements of each vertex. Note that locations and displacements are both two-dimensional in our case. Furthermore there is an array containing the relevant stiffness matrices for the triangles. For each triangle, we must know the indices of its three vertices. From this we can calculate

- The stiffness matrix for the triangle;
- The force vector of the triangle, which is the right hand side of the matrix equation to be solved.

The heart of the program is the multiplication of the field vector by the stiffness matrixes of the triangle. This can be done as follows.

```
Set the new global field vector to zero;
FOR each triangle DO
    Store the three old values of the field at the vertices
            in a local 3-vector;
    Multiply this vector by the stiffness matrix;
    Add the result to the appropriate entries of the new global
            field vector;
END FOR
```

You should now be able to write such a program. If you study the the problem of a point charge (delta function) on a 40×40 square grid, which is divided up into 3200 rectangular triangles with two 45° angles, you need 118 conjugate gradient iterations to achieve convergence of the residue (the L_2 norm of the vector $[K][\phi] - [r]$) within 10^{-10}. Obviously, this is the error in the solution of the matrix equation. The numerical error introduced by the discretisation of the grid may be (and will be) substantially larger.

13.3 Linear elasticity

13.3.1 The basic equations of linear elasticity

For many materials, deformations due to applied forces can to a good approximation be calculated using the equations of linear elasticity. These equations are valid in

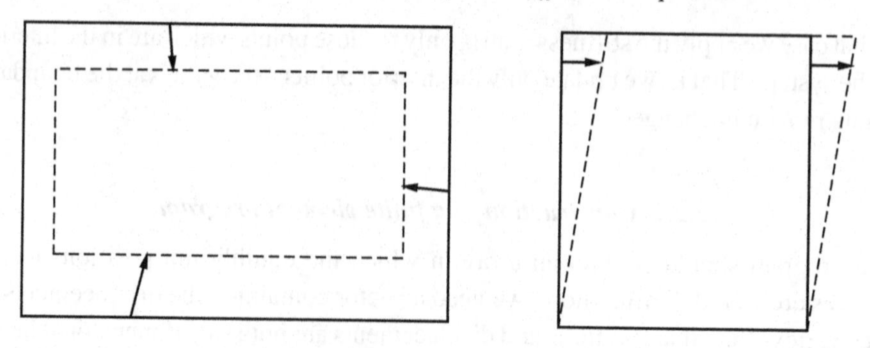

Figure 13.3. Two types of deformation, compression (left) and shear (right).

particular when the deformation is relatively small so that the total energy of the deformed system can be well approximated by a second order Taylor expansion.

There are two types of deformation. The first is compression or expansion of the system, and the second is shear. These effects are shown in Figure 13.3. We restrict ourselves to homogeneous isotropic systems in two dimensions. Then the resistance of a material to the two types of deformation is characterised in both cases by an elastic constant – in the literature either the Lamé constants λ and μ are used, or the Young modulus E and Poisson ratio ν. They are related by

$$E = \frac{\mu(3\lambda + 2\mu)}{\lambda + \mu} \tag{13.25a}$$

$$\nu = \frac{\lambda}{2(\lambda + \mu)}. \tag{13.25b}$$

To formulate the equations of deformation, consider the displacement field $\mathbf{u}(\mathbf{r})$. This vector field is defined as the displacement of the point $\mathbf{r}$ as a result of external forces acting on the system. These forces may either be acting throughout the system (gravity is an example) or on its boundary, like pushing with a finger on the solid object. In the equilibrium situation, the forces balance each other inside the material. So, if we identify a small line (a planar facet in three dimensions) with a certain orientation somewhere inside the object, the forces acting on both sides of this line should cancel each other. These forces vary with the orientation of the line or facet, as can be seen by realising that in an isotropic medium and in the absence of external forces, the force is always normal to the line (it is due to the internal, isotropic pressure). Another way to see this is by considering gravity. This acts on a horizontal facet from above but not on a vertical facet. Therefore it is useful to define the *stress tensor* σ_{ij} which gives the jth component of the force acting on a small facet with a normal along the ith Cartesian axis.

The stress plus the body forces results in the displacement. It is important that the actual value of the displacement matters less than its derivative: if we displace two points connected by a spring over a certain distance, the forces acting between the two points do not change. What matters is the difference in displacement of neighbouring points. Information concerning this is contained in the *strain* ε_{ij}. It is defined as

$$\varepsilon_{ij} = \frac{\mathrm{d}u_i}{\mathrm{d}x_j}. \tag{13.26}$$

For an isotropic, homogeneous material in two dimensions, only three components of stress and strain are important:

$$\sigma_{xx} \quad \text{and} \quad \varepsilon_{xx}; \tag{13.27a}$$

$$\sigma_{yy} \quad \text{and} \quad \varepsilon_{yy}; \tag{13.27b}$$

$$\sigma_{xy} \quad \text{and} \quad 2\varepsilon_{xy}. \tag{13.27c}$$

Stress and strain are related by Hooke's law:

$$\boldsymbol{\sigma} = \mathbf{C}\boldsymbol{\varepsilon}, \tag{13.28}$$

where $\boldsymbol{\sigma}$ is the vector $(\sigma_{xx}, \sigma_{yy}, \sigma_{xy})^{\mathrm{T}}$ and similarly for $\boldsymbol{\varepsilon}$. $\mathbf{C}$ is the elastic matrix:

$$\mathbf{C} = \frac{E}{1-\nu^2}\begin{pmatrix} 1 & \nu & 0 \\ \nu & 1 & 0 \\ 0 & 0 & \frac{1}{2}(1-\nu^2) \end{pmatrix}. \tag{13.29}$$

The body is at rest in a state where all forces are in balance. The force balance equation reads

$$\mathbf{D}^{\mathrm{T}}\boldsymbol{\sigma} + \mathbf{f} = 0 \tag{13.30}$$

with

$$\mathbf{D} = \begin{pmatrix} \frac{\partial}{\partial x} & 0 \\ 0 & \frac{\partial}{\partial y} \\ \frac{\partial}{\partial y} & \frac{\partial}{\partial x} \end{pmatrix} \tag{13.31}$$

This matrix can also be used to relate $\boldsymbol{\varepsilon}$ and $\mathbf{u}$:

$$\boldsymbol{\varepsilon} = \mathbf{D}\mathbf{u}. \tag{13.32}$$

There are two types of boundary conditions: parts of the boundary may be free to move in space, and other parts may be fixed. You may think of a beam attached to a wall at one end. In the example which we will work out below, we only include gravity as a (constant) force acting on each volume element of the system.

Just as in the case of the Laplace equation, we must find an integral formulation of the problem, and approximate the various relevant functions by some special form on the elements. As before, we will choose piecewise linear functions on the elements. Note that in this case we approximate each of the two components of the displacement field by these functions.

13.3.2 Finite element formulation

The finite element formulation can be derived from the continuum equations if we can formulate the latter as a variational problem for a functional expression which is an integral formulation of the problem.

To find this formulation in terms of integrals, we introduce the so-called 'weak formulation', for the force balance equation, which has the form:

$$\int_\Omega (\delta \mathbf{u})^{\mathrm{T}} (\mathbf{D}^{\mathrm{T}} \boldsymbol{\sigma} + \mathbf{f}) \, \mathrm{d}\Omega = 0. \tag{13.33}$$

Here, $\delta \mathbf{u}$ is an *arbitrary* displacement field satisfying the appropriate boundary conditions. Using (13.32) this integral equation is cast into the form

$$\int_\Omega (\delta \boldsymbol{\varepsilon})^{\mathrm{T}} \boldsymbol{\sigma} \, \mathrm{d}\Omega = -\int_\Omega (\delta \mathbf{u})^{\mathrm{T}} \mathbf{f} \, \mathrm{d}\Omega \tag{13.34}$$

We then can divide up the space Ω into N elements (triangles for two dimensions) and write

$$\sum_{e=1}^{N} \int_{\Omega_e} (\delta \boldsymbol{\varepsilon}_e)^{\mathrm{T}} \boldsymbol{\sigma}_e \, \mathrm{d}\Omega_e = -\sum_{e=1}^{N} \int_{\Omega_e} (\delta \mathbf{u}_e)^{\mathrm{T}} \mathbf{f}_e \, \mathrm{d}\Omega_e. \tag{13.35}$$

From this we can derive the form of the stiffness matrix for the elastic problem. First note that the variables of the problem are the deformations $\mathbf{v}_n$ on the vertices n. This means that for each triangle we have six variables (two values at each of the three vertices). Therefore, the stiffness matrix is 6×6. The deformations do not enter as such into the problem but only through the strain tensor. We have

$$\boldsymbol{\varepsilon}_e = \begin{pmatrix} \varepsilon_{xx} \\ \varepsilon_{yy} \\ 2\varepsilon_{xy} \end{pmatrix} = \mathbf{D}\mathbf{u}_e = \begin{pmatrix} \frac{\partial}{\partial x} & 0 \\ 0 & \frac{\partial}{\partial y} \\ \frac{\partial}{\partial y} & \frac{\partial}{\partial x} \end{pmatrix} \begin{pmatrix} u_x \\ u_y \end{pmatrix}. \tag{13.36}$$

This tensor, however, is linearly related to the $\mathbf{v}_i$. We write the displacement field as

$$\mathbf{u} = \mathbf{v}_a \xi_a + \mathbf{v}_b \xi_b + \mathbf{v}_c \xi_c. \tag{13.37}$$

The ξ_i depend on x and y – the relation is given in Eq. (13.11). From this, and from (13.17), we find

$$\boldsymbol{\varepsilon}_e = \begin{pmatrix} y_b - y_c & 0 & y_c - y_a & 0 & y_a - y_b & 0 \\ 0 & y_c - y_b & 0 & x_a - x_c & 0 & x_b - x_a \\ x_c - x_b & y_b - y_c & x_a - x_c & y_c - y_a & x_b - x_a & y_a - y_b \end{pmatrix} \begin{pmatrix} v_{ax} \\ v_{ay} \\ v_{bx} \\ v_{by} \\ v_{cx} \\ v_{cy} \end{pmatrix}. \tag{13.38}$$

We call the 3×6 matrix on the right hand side B. Using the relation

$$\boldsymbol{\sigma} = \mathbf{C}\boldsymbol{\varepsilon}, \tag{13.39}$$

we can rewrite the element integral of the left hand side of Eq. (13.35) as

$$\int_{\Omega_e} (\delta\mathbf{v})^{\mathrm{T}}\mathbf{B}^{\mathrm{T}}\mathbf{C}\mathbf{B}\mathbf{v}\; \mathrm{d}\Omega_e, \tag{13.40}$$

where $\mathbf{v}$ is a six-dimensional vector, $\mathbf{B}$ is a 3×6 matrix and $\mathbf{C}$ a 3×3 matrix.

Note that there is no dependence on the coordinates x and y in this expression. This can be traced back to the fact that we can express the integrand in terms of the strain, which contains derivatives of the deformation $\mathbf{u}$ which in turn is a *linear* function within the element. The integral is obtained by multiplying the constant integrand by the surface area A of the integrand. The stiffness matrix $\mathbf{k}$ is therefore given by

$$\mathbf{k} = A\mathbf{B}^{\mathrm{T}}\mathbf{C}\mathbf{B}. \tag{13.41}$$

This is a 6×6 matrix which connects the six-dimensional vectors $\mathbf{v}$.

The right hand side of Eq. (13.35) also involves an integral expression. This contains the external force. Taking this to be gravity, it is constant. We must evaluate the integral

$$\mathbf{f}_e \cdot \int_{\Omega_e} [(\delta\mathbf{v})_a\xi_a + (\delta\mathbf{v})_b\xi_b + (\delta\mathbf{v})_c\xi_c]\, \mathrm{d}\Omega_e. \tag{13.42}$$

This can be written in the form

$$\mathbf{f}_e \cdot \mathbf{G}\delta\mathbf{v}, \tag{13.43}$$

where $\mathbf{G}$ is the 2×6 matrix:

$$\frac{A}{3}\begin{pmatrix} 1 & 0 & 1 & 0 & 1 & 0 \\ 0 & 1 & 0 & 1 & 0 & 1 \end{pmatrix}. \tag{13.44}$$

We have now reworked (13.35) to the form

$$\delta\mathbf{v}^{\mathrm{T}} \cdot \mathbf{K}\mathbf{v} = \delta\mathbf{v}^{\mathrm{T}} \cdot \mathbf{G}^{\mathrm{T}}\mathbf{f}, \tag{13.45}$$

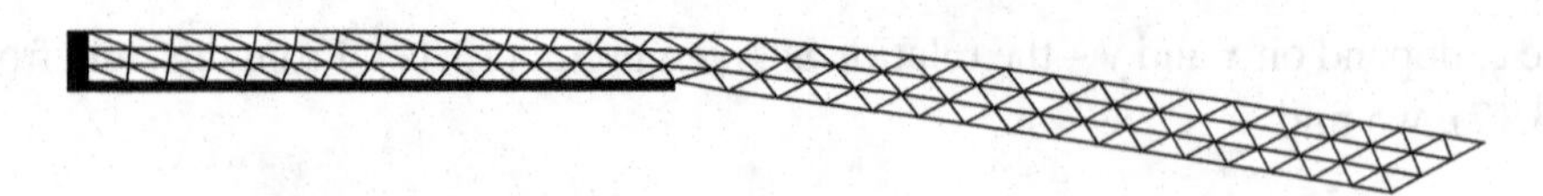

Figure 13.4. Deformation of a beam attached to a vertical wall, calculated with the finite element method. The beam is supported on half of its base.

where $\mathbf{v}$ now represents the vector of *all* displacements (that is, for the whole grid), $\mathbf{K}$ is the *full* stiffness matrix, which can be evaluated as a careful sum over the stiffness matrices for all triangles in the same spirit as described for the Laplace equation in Section 13.2.1, and the right hand side is a vector defined on the full grid. The dimension of the matrix problem is $2N$, where N is the number of vertices. If points are subject to Dirichlet boundary conditions, they are excluded from the vectors and matrices, so that for actual problems the dimension is less than $2N$. The matrix equation found must hold for all $\delta\mathbf{v}$, which can only be true when

$$\mathbf{Kv} = \mathbf{Gf} \tag{13.46}$$

and this can be solved for using the conjugate gradients method. In Figure 13.4, the result of a deformation calculation is shown for a beam with the left end attached to a wall.

13.4 Error estimators

Like every numerical method, the finite element method is subject to errors of several kinds. Apart from modelling errors and errors due to finite arithmetic precision in the processor, the discretisation errors are important, and we will focus on these. Obviously the discretisation error can be made small by reducing the grid constant homogeneously over the lattice, but this can only be done at the cost of increasing the computer time needed to arrive at a stable solution. It might be that the error is due to only a small part of the system under consideration, and reducing the mesh size in those regions which are already treated accurately with a coarse mesh is unnecessary and expensive overkill.

It is therefore very useful to have available a *local estimator* of the error which tells us for a particular region or element in space what its contribution to the overall error is. In that case, we can refine the mesh only in those regions where it is useful. In this section, we first address the problem of formulating such a local error estimator and then describe a particular refinement strategy for triangular meshes.

One type of local error estimator is based on the notion that, unlike the displacement field, the *stress* usually is not continuous over the element boundaries. If a number of triangles meet at a particular mesh point, they will all have slightly different values of their stress components (recall that the stress is defined in terms of first derivatives of the displacement fields). A more accurate solution would lead to continuous stresses and this can be achieved by some suitable averaging of the stress components at the mesh points. To be specific, the nodal stress at a mesh point p would be given by

$$\sigma_p = \frac{1}{\sum_{\text{elems}} w_{\text{elem}}} \sum_{\text{elems}} w_{\text{elem}} \sigma_{\text{elem}} \tag{13.47}$$

where the stresses on the right hand side (in the sum) are the result of the finite element calculation; the weights w_{elem} may be taken equal or related to the surface area of the elements sharing the vertex p. The error is then the difference between the 'old' stresses resulting from the calculation and the improved values based on the recipe above. We shall refer to the 'old' stress, resulting from the FEM calculation, as the FEM stress.

The question arises how the weights w_{elem} can be chosen optimally. One answer to this question is provided by the *projection method* [7–10]. In this method we seek a *continuous*, piecewise linear stress field, which deviates to a minimal extent from the FEM stress. The deviation can be defined as the L_2-norm of the difference between the FEM stress and the continuous stress σ_{C} which is a piecewise linear FEM-type expansion, based on the values $\boldsymbol{\sigma}_p$:

$$\Delta = \int_\Omega (\sigma_{\text{C}} - \sigma_{\text{FEM}})^{\text{T}} (\sigma_{\text{C}} - \sigma_{\text{FEM}})\, \mathrm{d}\Omega. \tag{13.48}$$

We write the continuous stress within a particular triangle (a, b, c), as usual, in the form

$$\sigma_{\text{C}} = \sigma_a \xi_a + \sigma_b \xi_b + \sigma_c \xi_c, \tag{13.49}$$

where σ_a etc. are the values of the stresses at the three vertices (as the stress is continuous, it must be single-valued at the mesh points). The optimal approximation of the actual stress is defined by those values of σ_{C} at the vertices for which the deviation Δ is minimal. This directly leads to the condition

$$\frac{\partial \Delta[\sigma_{\text{C}}]}{\partial \sigma_p} = 2 \int_\Omega \left(\frac{\partial \sigma_{\text{C}}}{\partial \sigma_p} \right)^{\text{T}} (\sigma_{\text{C}} - \sigma_{\text{FEM}})\, \mathrm{d}\Omega = 0. \tag{13.50}$$

As the continuous stress field is a linear function of the values at the mesh points, we immediately obtain

$$\sum_q \int_\Omega \xi_p \xi_q \sigma_q \, \mathrm{d}\Omega - \int_\Omega \xi_p \sigma_{\text{FEM},p} \, \mathrm{d}\Omega = 0. \tag{13.51}$$

This expression needs some explanation. For the point p, the points q run over p and all its neighbours. The functions ξ_p and ξ_q are defined within the same triangle;

$\sigma_{\text{FEM},p}$ is the (constant) stress in that triangle. Therefore we can evaluate the product of the matrix

$$\int_{\Omega} \xi_p \xi_q \, \mathrm{d}\Omega \tag{13.52}$$

with the vector σ_p again as a sum over all triangular elements.

We can evaluate the resulting matrix equation in exactly the same way as the full matrix equation which we have solved in order to find the displacement. However, the present problem is usually solved within about 10% of the time needed for the full elasticity problem.

It is interesting to calculate the local error. For the problem we are focusing on, a beam attached to a wall, the corners where the beam is attached to the wall are the points where the error is maximal.

There exist other methods for calculating the local error. Superconvergent patch recovery (SPR) is based on the notion that the error oscillates throughout the elements – hence, there exist points where the error vanishes. Even if those points cannot be found, some points can be identified where the error is an order of magnitude better than average. These points usually are somewhere near the centre of the elements – the vertices are the worst possible points. Using the values at the superconvergent points, a much more accurate stress field can be constructed, and the difference between this field and the FEM field is used as the local error. For details see Refs. [8, 11, 12].

13.5 Local refinement

The local error can be used to decide which elements should be refined. Local refinement of triangles is a subtle problem mainly for two reasons. The first is that when a triangle is refined by dividing it up into two triangles as in Figure 13.5(a), the resulting triangles might have an awkward shape. The point is that narrow, long triangles are not suitable for FEM calculations because they give rise to large errors. Therefore, it is good practice to construct the new triangles by bisecting the longest edge of a triangle.

The second problem is that if we perform such a bisection, another triangle, sharing the same long edge, should be partitioned as well, as it would be impossible to have continuity of the solution otherwise (see Fig. 13.5(b)). Rivara therefore devised the following refinement procedure [13]:

- If a triangle needs refinement, we bisect its longest edge;
- If this edge is also the largest edge of the neighbouring triangle, this triangle should also be divided via bisection of the same edge;
- If the edge is not the longest edge of the neighbouring triangle, this triangle should be refined by bisecting its longest edge.

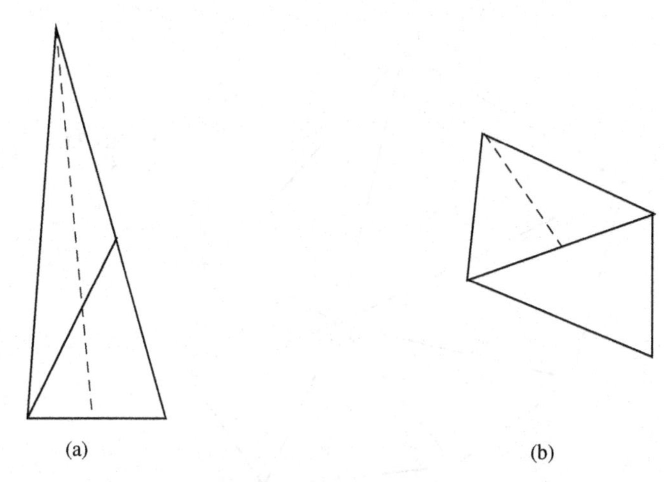

Figure 13.5. Refinement of triangular grid. (a) Two ways of partitioning a triangle – partitioning according to the dashed line is undesired. (b) It is not allowed to partition a triangle by bisecting an edge without partitioning its neighbour along the same edge.

This procedure is recursive in nature. It boils down to the following algorithm, starting from the triangle T which is to be refined:

```
ROUTINE RefineTriangle(T)
    Find the longest edge E of T;
    IF E is not the longest edge of the neighbouring triangle T′ THEN
        RefineTriangle(T′);
    END IF;
    Create a new mesh point by bisection of E;
END ROUTINE RefineTriangle.
```

Note that the routine does not generate triangles, but vertices. It is important to store the information concerning which vertices are neighbours. The new triangles can then be constructed from these data. In order to do this, we must make sure that, for each vertex, we have an array containing the neighbours of that vertex, ordered anticlockwise. If the vertex is a boundary point, the list starts with the leftmost neighbour and proceeds until the rightmost neighbour is reached. For vertices in the bulk, there is obviously no natural ‘first’ and ‘last’ neighbour: the first point is the right neighbour of the last point, and obviously the last point is then the left neighbour of the first.

For each vertex, all neighbouring triangles can be found by taking the vertex itself together with two subsequent neighbours. In this way, however, a triangle in the bulk would be counted three times. A way to define the triangle uniquely is by

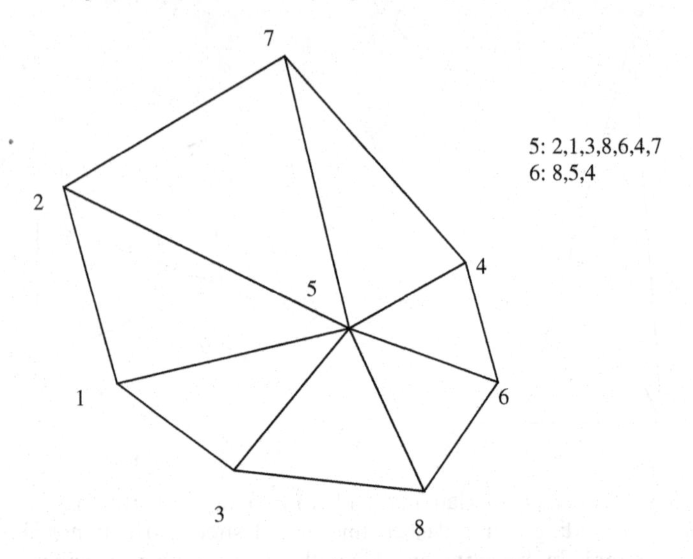

Figure 13.6. The data structure proposed by Rivara [13]. The points of the mesh are numbered in some way, and for each point, the neighbours are kept in a list. The list for a central point (number 5) is cyclic: the first point is connected to the last one, whereas a boundary point has a noncyclic neighbour list. Each triangle is counted only from the vertex with the lowest index possible.

requiring that the vertex we start from has a lower index than the two neighbours with which it forms the triangle.

The data structure is clarified in Figure 13.6.

Once we have a list of vertices with a list of neighbours for each vertex according to the rules specified above, the triangles can be generated straightforwardly:

```
FOR each vertex DO
    FOR each triangle spanned by the vertex
                    and two of its subsequent neighbours DO
        IF the central vertex has a lower index than the two neighbours THEN
            Add triangle to the list of triangles;
        END IF;
    END FOR;
END FOR
```

The line ‘FOR each triangle spanned by the vertex and two of its subsequent neighbours DO’ is different for edge points, where we only look at subsequent neighbour pairs *between the first and the last* neighbouring vertex, than for interior points, where we also include the pair formed by the first and the last neighbouring vertex.

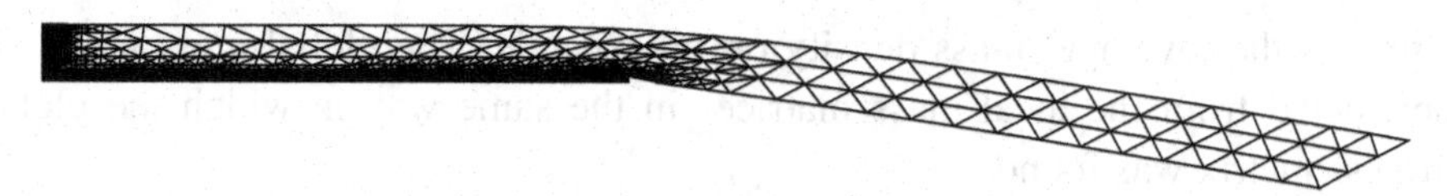

Figure 13.7. Deformed elastic beam which is attached to a vertical wall and supported over half its length. The difference from Figure 13.4, which shows the same beam, is the local refinement of the elements.

When the refinement procedure is carried out, we simply add the new vertices to the list of vertices. After the mesh has been refined, we construct the new list of triangles using the above algorithm.

The question is what the best measure of the error would be. We could take the L_2 norm of the difference between σ and σ_{FEM}. There are many other possibilities, and a very common one is the 'energy norm', defined as

$$e_{\text{E}} = \int_{\Omega} (\sigma - \sigma_{\text{FEM}})^{\text{T}} \mathbf{C} (\sigma - \sigma_{\text{FEM}}) \, \text{d}^3 r. \tag{13.53}$$

Figure 13.7 shows the deformation of a beam which is attached to a wall and to a horizontal line over part of its lower edge. As is to be expected, the mesh is strongly refined near the sharp edge where the horizontal fixed line ends.

The use of adaptive refinement may give tremendous acceleration when a highly accurate solution is wanted for a heterogeneous problem.

13.6 Dynamical finite element method

In the previous sections we have assumed that dissipative forces remove all the kinetic energy so that an elastic object subject to forces will end up in a shape in which its potential energy is minimal. We may, however, also consider nondissipative dynamics in the elastic limit. We treat this case by formulating the total energy as a sum of the elastic energy, the work done by external forces and the kinetic energy:

$$H = \frac{1}{2} \int_{\Omega} \boldsymbol{\varepsilon}^{\text{T}}(\mathbf{r}) \mathbf{C} \boldsymbol{\varepsilon}(\mathbf{r}) \, \text{d}^3 r + \int_{\Omega} \mathbf{f}(\mathbf{r}) \cdot \mathbf{u}(\mathbf{r}) \, \text{d}^3 r + \frac{1}{2} \int_{\Omega} \rho(\mathbf{r}) \dot{\mathbf{u}}^2(\mathbf{r}) \, \text{d}^3 r. \tag{13.54}$$

We can perform the integrals as above, taking the mass density constant over a triangle, leading to

$$\mathbf{M}\ddot{\mathbf{v}} = -\mathbf{K}\mathbf{v} + \mathbf{G}\mathbf{f}. \tag{13.55}$$

The matrix $\mathbf{M}$ is the *mass matrix*. Putting the expressions for the natural coordinates in the integral containing the mass density, we find for the mass matrix $\mathbf{m}$ of a single triangle

$$m_{pq} = \frac{\rho A}{12} (1 + \delta_{pq}). \tag{13.56}$$

Here ρ is the (average) mass density on the triangle. The global mass matrix is constructed from the local mass matrices in the same way in which the global stiffness matrix was found.

Adding dynamics to the program is a relatively small addition to the static program which was described in the previous sections. The solution of the equations of motion, however, is a bit more involved. This equation is not diagonal in the mass as is the case in the many-body dynamics of molecular dynamics simulations. Formulating the discrete solution using the midpoint rule

$$\mathbf{M}[\mathbf{u}(t+h)+\mathbf{u}(t-h)-2\mathbf{u}(t)] = h^2(-\mathbf{K}\mathbf{v}+\mathbf{G}\mathbf{f}) \tag{13.57}$$

shows that, knowing the solution $\mathbf{u}$ at the times t and $t-h$, we can predict its value at $t+h$ by solving an implicit equation. We can again use the conjugate gradient method for this purpose. This algorithm should be applied at each time step. As the solution to be found is close to the solution we had at the last time step, the conjugate gradient method will converge in general much faster than for a stationary state problem for which the initial solution is still far away from the final one (in the first case we speak of a *transient problem*). The difference between the two problems is the same as that between solving the diffusion equation (transient) and the Poisson/Laplace equation. It is also possible to add friction to the dynamics. A damping matrix is then introduced which has a shape similar to the mass matrix, but this is multiplied by the first time derivative of $\mathbf{u}$ rather than the second derivative. Obviously, the eigenvalues of the damping matrix must be negative (otherwise, there would be no damping).

A dynamical simulation shows an object wobbling as a result of external forces or of being released from a nonequilibrium state. In general, we see elastic waves propagating through the material.

13.7 Concurrent coupling of length scales: FEM and MD

If we exert strong forces on an object, there will be deviations from elastic behaviour due to the fact that a second order approximation of the potential energy in terms of the strain breaks down. New phenomena may then occur: in the first place, we see a change in speed of the elastic waves; moreover they start interacting, even in the bulk.[1] The most spectacular deviation from elastic behaviour occurs when we break the material. The elastic description fails completely in that case. In fact, when an object is broken or cut, the bonds between rows of atom pairs are broken and an accurate description should therefore include atomic details, preferably at the quantum level. The problem is that, although such a description is adequate for

[1] Elastic waves can also interact at the boundary of an object by coupling between the transverse and longitudinal components.

processes taking place near the fissure and far away from it, it becomes unfeasible when we want to include substantially large (parts of) objects. You may ask why we would bother about the processes far from a fissure, since the deviations of the atoms from their equilibrium positions are very small there. However, the energy released by breaking a bond will generate elastic waves into the bulk, which, when the bulk is small, will bounce back at the boundary and reinject energy to the fissure region. It is possible to couple an atomic description to an elastic medium which then carries the energy sufficiently far away . This is done by *concurrent coupling of length scales* [14, 15]. In this technique a quantum mechanical tight-binding description is applied to the region where the most essential physics is taking place: in our example this is the breaking of atomic bonds. The surrounding region is described by classical MD. Farther away, this description is then replaced by an elastic one, which is treated by finite elements. We shall not describe the full problem here – for this we refer to the papers by Broughton, Rudd and others [14, 15]. We shall, however, show that elastic FEM can be coupled to MD in a sensible way.

From the chapter on MD, it is clear that we would like to have dynamics described by a Hamiltonian. The dynamic FEM method has this property, and this is also the case for the MD method. We must ensure that this requirement is satisfied by the coupling regime. The coupling between FEM and MD is called *handshaking*. To show how this coupling is realised and to check that it gives sensible behaviour, we consider a 2D rectangular strip through which an elastic wave is travelling. The left hand side of the strip is treated using the FEM, the right hand side by MD. In order to realise the handshaking protocol, the finite element grid should approach atomic resolution near the boundary – grid points next to the boundary should coincide with equilibrium atomic positions of the MD system.

Within the MD, we use a Lennard–Jones potential as in Chapter 8. The equilibrium configuration with this potential in two dimensions is a triangular lattice. The situation is shown in Figure 13.8. The vertical dashed line in Figure 13.8 separates the FEM from the MD region. The Hamiltonian is composed of three parts: a FEM Hamiltonian for the points inside the FEM region, a MD Hamiltonian for the points in the MD region, and a handshaking Hamiltonian which contains the forces that the MD particles exert on the FEM points and vice versa. In order to have a smooth transition from one region to the other, this handshake Hamiltonian interpolates between a FEM and a MD Hamiltonian. It is built up as follows.

- The FEM triangles in the shaded region carry half of the FEM Hamiltonian; that is, we formulate the usual FEM Hamiltonian in this region, but multiply it by $1/2$.
- The points of the shaded region lying right of the dashed vertical line couple via a MD Hamiltonian to the points on the left, but this Hamiltonian is also multiplied by 1/2. Note that such couplings involve in general more than nearest neighbour points on the triangular grid – we neglect those here.

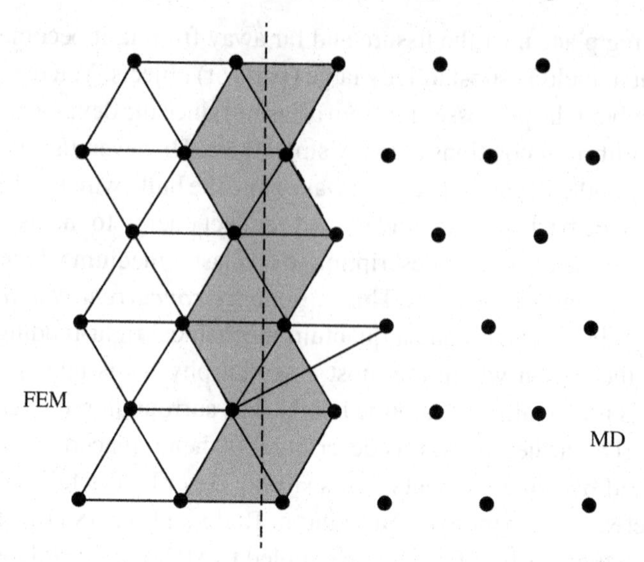

Figure 13.8. Strip modelled partly by finite elements and partly by molecular dynamics.

Three remarks are in place. In the original formulation [14], the MD region is three-dimensional, whereas the FEM region is only two-dimensional. The transition is made by averaging the MD points over the z-direction which is taken perpendicular to the FEM grid. Here we shall consider the strictly two-dimensional case for simplicity. The second remark concerns the treatment of the FEM masses. As we have seen above, the mass matrix couples the kinetic degrees of freedom at the vertices of the FEM triangles. However, in the handshake region, we strictly want to assign the mass of a real atom to the point. For this reason, we use the *lumped mass* approximation in the finite element description. In this formulation, we assign one-third of the mass of each triangle to each of its vertices. This means that the mass matrix has become diagonal, so that the numerical integration of the equations of motion has become much simpler as the solution of an implicit equation at each time step is avoided. The FEM mass is derived from the MD equilibrium by requiring that the same amount of mass is present per unit area in both descriptions.

The final remark is that the boundaries of the system in the MD and FEM description do not fit onto each other. In the FEM description, the triangles are taken uniform, but a MD system with a boundary will have a slightly smaller distance between the outermost layers than in the bulk, as a result of the fact that the next nearest-neighbour interactions pull the outermost particles slightly more towards the interior of the system. This deviation is minor so we do not correct for it.

Obviously, we could take periodic boundary conditions in the transverse direction, which implies a cylindrical description (this could be useful for describing a carbon nanotube). However, this is not compatible with longitudinal waves, as the Poisson ratio causes the system to expand where it is longitudinally compressed and vice versa. A periodic boundary condition would not allow for this to happen and cause unphysical, strong stresses to build up.

Once a FEM and a MD program are working, it is not so much work to couple them along the lines decribed above. We use the velocity Verlet algorithm which naturally splits into two steps, separated by a force calculation.
First step:

$$\mathbf{p}_i(t+h) = \mathbf{p}_i(t) + \frac{h}{2}\mathbf{F}_i[\mathbf{r}(t)]; \tag{13.58a}$$

$$\mathbf{r}_i(t+h) = \mathbf{r}_i(t) + h\mathbf{p}_i(t); \tag{13.58b}$$

Calculate $\mathbf{F}_i[\mathbf{r}(t)]$;

Second step:

$$\mathbf{p}_i(t+h) = \mathbf{p}_i(t+h) + \frac{h}{2}\mathbf{F}_i[\mathbf{r}(t+h)]$$

These steps must be kept in mind when setting up the algorithm for the full system. This algorithm looks as follows:

```
Calculate MD forces;
Calculate FEM forces;
Copy locations of leftmost MD points to a shadow array
          in the FEM region;
Copy locations of rightmost FEM points to a shadow array
          in the MD region;
FOR TimeStep = 1, MaxStep DO
  Set Initial values of boundary points;
  Do first integration step (see Eq. (13.58a));
  Copy locations of leftmost MD points to a shadow array
          in the FEM region;
  Copy locations of rightmost FEM points to a shadow array
          in the MD region;
  Calculate forces in FEM region, including those on the MD particles;
  Calculate forces in MD region, including those on the FEM particles;
  Add FEM forces acting on MD particles to MD forces;
  Add MD forces acting on FEM particles to FEM forces;
  Do second integration step [see Eq. (13.58b)];
END FOR.
```

Obviously, we should investigate which elasticity matrix should be used in the FEM domain. This is fully determined by the MD interaction, for which we take the pairwise Lennard-Jones interaction. We can evaluate the elastic constants by allowing the MD unit cell to deform, as is done in the Parrinello–Rahman method [16]. Another method is to measure the stretch resulting from a force applied to the left- and rightmost particles for a strip of atoms, fully described by MD. The lateral shrink as a result of the end-to-end stretch then gives us the Poisson ratio. For simplicity, we shall consider here the $T = 0$ limit, for which we can calculate the elasticity matrix analytically from the pair potential. The idea is that we can Taylor-expand the total energy per unit area with respect to the strain to second order, which corresponds precisely to how the elasticity matrix is defined: the change in energy per unit area resulting from a strain field $\boldsymbol{\varepsilon}$ is given by

$$\delta V = \frac{1}{2\Omega} \int_{\Omega} \boldsymbol{\varepsilon}^{\mathrm{T}} \mathbf{C} \boldsymbol{\varepsilon} \, \mathrm{d}^2 r. \tag{13.59}$$

In our case, we have for the total energy per unit area at small deviations, in the bulk:

$$\delta V = \frac{1}{2\Omega} \sum_{i \neq j} \left[\frac{\partial V(\mathbf{R}_0)}{\partial r_{ij}^{\alpha}} (\delta r_i^{\alpha} - \delta r_j^{\alpha}) + \frac{1}{2} \frac{\partial^2 V(\mathbf{R}_0)}{\partial r_{ij}^{\alpha} \partial r_{ij}^{\beta}} (\delta r_i^{\beta} - \delta r_j^{\beta})(\delta r_i^{\alpha} - \delta r_j^{\alpha}) \right]. \tag{13.60}$$

Greek indices α and β denote the Cartesian coordinates – they are summed over according to the Einstein summation convention. Equation (13.60) is nothing but a Taylor expansion to second order for the potential in terms of the coordinates. In equilibrium, the second term vanishes as the total force on each particle vanishes. We may write $\delta r_i^{\alpha} = u_i^{\alpha}$, where u_i has precisely the same meaning as in the formulation of the finite element method: it is the deviation from equilibrium of coordinate α of particle i. Now we write $\mathbf{u}_{ij} = \mathbf{r}_{ij} - \mathbf{a}_{ij}$, where $\mathbf{a}_{ij}$ is the relative coordinate of particles i and j in equilibrium. We therefore have $u_{ij}^{\alpha} = a_{ij}^{\beta} \varepsilon_{\alpha\beta}$, so that we obtain

$$\delta V = \frac{1}{4\Omega} \sum_{i \neq j} a_{ij}^{\alpha} \varepsilon_{\alpha\beta} \frac{\partial^2 V(\mathbf{R}_0)}{\partial r_{ij}^{\beta} \partial r_{ij}^{\gamma}} \varepsilon_{\gamma\delta} a_{ij}^{\delta}, \tag{13.61}$$

and we can write

$$\tilde{C}_{\alpha\beta\gamma\delta} = \frac{1}{2\Omega} \sum_{i \neq j} a_{ij}^{\alpha} \frac{\partial^2 V(\mathbf{R}_0)}{\partial r_{ij}^{\beta} \partial r_{ij}^{\gamma}} a_{ij}^{\delta}. \tag{13.62}$$

For a pair potential, this can be worked out further to yield

$$\tilde{C}_{\alpha\beta\gamma\delta} = \frac{1}{2\Omega}\sum_{i\neq j}\left\{\frac{1}{r_{ij}^2}\left[V''(r_{ij}) - \frac{1}{r_{ij}}V'(r_{ij})\right]a_{ij}^{\alpha}a_{ij}^{\beta}a_{ij}^{\gamma}a_{ij}^{\delta}\right\}. \tag{13.63}$$

We have used the tilde ($\sim$) for the elasticity matrix because it is given in terms of the xy components of the strain. The relation with the $\mathbf{C}$ matrix given above for two dimensions, which used $(\partial u_x/\partial y + \partial u_y/\partial x)/2$ as the third component, is given by

$$C_{11} = \tilde{C}_{xxxx}, \qquad C_{22} = \tilde{C}_{yyyy} \tag{13.64a}$$

$$C_{12} = \tilde{C}_{xxyy}, \qquad C_{21} = \tilde{C}_{yyxx} \tag{13.64b}$$

$$C_{13} = \frac{1}{2}(\tilde{C}_{xxxy} + \tilde{C}_{xxyx}), \qquad C_{23} = \frac{1}{2}(\tilde{C}_{yyxy} + \tilde{C}_{yyyx}) \tag{13.64c}$$

$$C_{33} = \frac{1}{4}(\tilde{C}_{xyxy} + \tilde{C}_{xyyx} + \tilde{C}_{yxxy} + \tilde{C}_{yxyx}) \tag{13.64d}$$

For a Lennard–Jones potential we find, in reduced units:

$$C = \begin{pmatrix} 76.8 & 25.5 & 0 \\ 25.6 & 76.8 & 0 \\ 0 & 0 & 25.6 \end{pmatrix}. \tag{13.65}$$

From this we find, for the case of plane stress: $\nu = 1/3$ and $E = 68$. The fact that $\nu = 1/3$ shows an important shortcoming of a pair potential: irrespective of the specific form of the potential, a pair potential always leads to $\nu = 1/3$.

Exercises

13.1 In this problem, we study the natural coordinates for triangles. We consider an 'archetypical' triangle as shown in Figure 13.9. Now consider a mapping of this triangle to some other triangle, also shown in Figure 13.9. This can be obtained from the archetypical one by a translation over the vector $\mathbf{r}_{aa'}$, followed by a linear transformation. The matrix $\mathbf{U}$ of this linear transformation can be found as

$$\mathbf{U} = \begin{pmatrix} x'_b - x'_a & x'_c - x'_a \\ y'_b - y'_a & y'_c - y'_a \end{pmatrix}, \tag{13.66}$$

where (x'_a, y'_a) are the Cartesian coordinates of the vector $\mathbf{r}_{a'}$ etc. We have

$$\begin{pmatrix} x' \\ y' \end{pmatrix} = \mathbf{U}\begin{pmatrix} x \\ y \end{pmatrix} + \begin{pmatrix} x_a \\ y_a \end{pmatrix} \tag{13.67}$$

Now we take for the natural coordinates in the archetypical triangle x, y and $1-(x+y)$. It is clear that these coordinates assume the value 1 on a, b and c respectively and vanish at the other points. We want the linear transformation of these coordinates to have the same property. We therefore consider the function

$$g(x', y') = f(x, y)$$

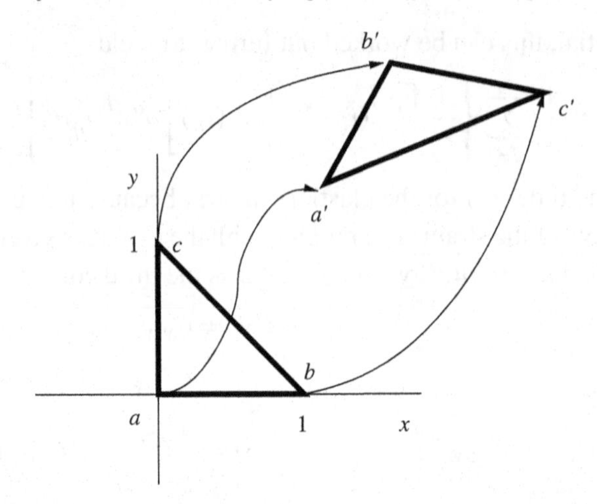

Figure 13.9. Archetypical triangle with two angles of 45° and sides 1, oriented along the x- and y-axes. Another triangle is shown, which can be obtained from the archetypical one through a linear transformation.

where $f(x,y) = x$, say, and where x', y' are the images of x, y under the transformation $\mathbf{U}$. It now is straightforward to verify that the expressions for the natural coordinates (13.11) are correct.

References

[1] O. C. Zienkiewicz and R. L. Taylor, *The Finite Element Method: Its Basis and Fundamentals*, 6th edn. Oxford/Burlington (MA), Elsevier Butterworth-Heinemann, 2005.

[2] O. C. Zienkiewicz and R. L. Taylor, *The Finite Element Method for Solid and Structural Mechanics*, 6th edn. Oxford/Burlington (MA), Elsevier Butterworth-Heinemann, 2005.

[3] O. C. Zienkiewicz and R. L. Taylor, *The Finite Element Method for Fluid Dynamics*, 6th edn. Oxford/Burlington (MA), Elsevier Butterworth-Heinemann, 2005.

[4] K. J. Bathe, *Finite Element Procedures*. Upper Saddle River, NJ, Prentice Hall, 1996.

[5] J. Mackerle, *A Primer for Finite Elements in Elastic Structures*. New York, Wiley, 1999.

[6] W. Kaplan, *Advanced Calculus*, 4th edn. Reading (MA), Addison–Wesley, 1991.

[7] J. T. Oden and H. J. Brauchli, 'On the calculation of consistent stress distributions in finite element calculations,' *Int. J. Numer. Meth. Eng.*, **3** (1971), 317–25.

[8] E. Hinton and J. S. Campbell, 'Local and global smoothing of discontinuous finite element functions using a least squares method,' *Int. J. Numer. Meth. Eng.*, **8** (1974), 461–80.

[9] O. C. Zienkiewicz and J. Z. Zhu, 'A simple error estimator and adaptive procedures for practical engineering analysis,' *Int. J. Numer. Meth. Eng.*, **24** (1987), 337–57.

[10] O. C. Zienkiewicz and R. L. Taylor, *The Finite Element Method*, vol. I. London, McGraw-Hill, 1988.

[11] J. Barlow, 'Optimal stress locations in finite element models,' *Int. J. Numer. Meth. Eng.*, **10** (1976), 243–51.

[12] J. Barlow, 'More on optimal stress points, reduced integration, element distortions and error estimation,' *Int. J. Numer. Meth. Eng.*, **28** (1989), 1487–504.

[13] M.-C. Rivara, 'Design and data structure of fully adaptive, multigrid, finite element software,' *ACM Trans. Math. Software*, **10** (1984), 242–64.

[14] J. Q. Broughton, F. F. Abraham, N. Bernstein, and E. Kaxiras, 'Concurrent coupling of length scales: methodology and application,' *Phys. Rev. B*, **60** (1999), 2391–403.

[15] R. E. Rudd and J. Q. Broughton, 'Concurrent coupling of length scales in solid state systems,' *Phys. Stat. Sol. (b)*, **217** (2000), 251–91.

[16] M. Parrinello and A. Rahman, 'Polymorphic transitions in single crystals: a new molecular dynamics method,' *J. Appl. Phys.*, **52** (1981), 7182–90.

14

The lattice Boltzmann method for fluid dynamics

14.1 Introduction

Flow problems are widely studied in engineering because of their relevance to industrial processes and environmental problems. Such problems belong to the realm of macroscopic phenomena which are formulated in terms of one or more, possibly nonlinear, partial differential equations. If there is no possibility of exploiting symmetry, allowing for separation of variables, these equations are usually solved using finite element or finite difference methods.

The standard problem is the flow of a one-component, isotropic nonpolar liquid, which is described by the Navier–Stokes equations. These equations are based on mass and momentum conservation, and on the assumption of isotropic relaxation towards equilibrium. Finite element methods have been described in Chapter 13 for elasticity; the methods described there may be extended and adapted to develop codes for computational fluid dynamics (CFD), which are widely used by engineers. Such an extension is beyond the scope of this book.

A finite element solution of the Navier–Stokes equations may sometimes become cumbersome when the boundary conditions become exceptionally complicated, as is the case with flow through porous media where the pore diameter becomes very small (and the number of pores very large). Other cases where finite element methods run into problems are multiphase or binary systems, where two different substances or phases exist in separate regions of space. These regions change their shape and size in the course of time. Usually, the finite element points are taken on the system boundaries, but that implies complicated bookkeeping, in particular when the topology of the regions changes, as is the case in coalescence of droplets.

A possible solution to these problems characterized by difficult topologies is to use regular, structured grids, and let the interfaces move over the grid. In that case, we need special couplings for nearest neighbour grid points that are separated by the interface (such as a phase boundary). This 'immersed interface' method has been pioneered by LeVeque and Li [1]. Another method is to replace the macroscopic, finite element approach by a microscopic particle approach, such as molecular dynamics, but this usually requires so many particles that it is no longer feasible to do useful simulations. Since the Navier–Stokes equations are rather universal in that they only include two viscosities and the mass density as essential parameters, it seems that the details of the interactions between the particles do not all matter: only few features of these interactions will survive in the macroscopic description. This is in some sense similar to the macroscopic description of elasticity, where the elasticity tensor, which is based on two elastic or Lamé constants, is the only fingerprint surviving from the microscopic details of the interactions.

This suggests that an alternative approach may be to use a 'mock fluid', consisting of 'mock particles' with very simple microscopic properties which are tuned to yield the correct hydrodynamic behaviour. This scheme is adopted in the *lattice Boltzmann* approach. The particles are put on a lattice and can move only to neighbouring lattice sites. Interactions between the particles and relaxation effects are included, and the resulting system yields hydrodynamic behaviour on large scales. This method has been applied sucessfully to binary flows, complex geometries, objects moving in a flow, etc.

We start this chapter by deriving the Navier–Stokes equations from the Boltzmann equation in the continuum (Section 14.2). Then we formulate the Boltzmann approach on a discrete lattice (Section 14.3) and consider an example. In Section 14.4, we apply the method to binary systems. Finally, in Section 14.5, we show that the lattice Boltzmann model leads in some limit to the Navier–Stokes equations for fluids in the incompressible limit. For more information about the method and its applications, the interested reader may consult Refs. [2–4].

14.2 Derivation of the Navier–Stokes equations

In this section we present a derivation of the Navier–Stokes equations from an approximate Boltzmann equation through a Chapman–Enskog procedure [5]. This works as follows. We start by defining the particle distribution function $n(\mathbf{r}, \mathbf{v}, t)$ which gives us the number density of particles with velocity $\mathbf{v}$ inside a small cell located at $\mathbf{r}$, at time t. This distribution for $\mathbf{r}$ and $\mathbf{v}$ will change in time because particles have a velocity and therefore move to a different position, and because the particles collide, which results in exchange of momentum and energy. The evolution of the distribution function is described by the well-known *Boltzmann equation* which describes a dilute system. The picture behind the Boltzmann equation is that of particles moving undisturbed through phase space most of the time, but experiencing every now and then a collision with other particles, and these collisions are considered to be instantaneous. The Boltzmann equation works very well, even in cases where the substance is not so dilute, such as a fluid. The Boltzmann equation reads:

$$\frac{\partial n}{\partial t} + \mathbf{v} \cdot \nabla_{\mathbf{r}} n = \left(\frac{\mathrm{d}n}{\mathrm{d}t} \right)_{\text{collisions}}. \tag{14.1}$$

The second term describes the change due to particle flow, and the right hand side is the change due to collisions.

If the particles were simply to flow according to their initial velocity, without interaction, equilibrium would never be reached: the role of the collisions is to establish *local equilibrium*, that is, a distribution which is in equilibrium in a small cell with fixed volume, constant temperature, density and average velocity $\mathbf{u}$. We know this equilibrium distribution; it is given as

$$n^{\text{eq}}(\mathbf{r}, \mathbf{v}) = \left(\frac{m\pi}{2k_B T} \right)^{3/2} n(\mathbf{r}) \exp\{-m[\mathbf{v} - \mathbf{u}(\mathbf{r})]^2/(2k_B T)\}, \tag{14.2}$$

which holds for cells small enough to justify a constant potential. Once the liquid is in (local) equilibrium, the collisions will not push it away from equilibrium. It can be shown that the collisions have the effect of increasing the entropy – hence they generate heat.

Before we continue, we note that the mass must *always* be conserved, whether there are collisions or not. The mass density is found as

$$\rho(\mathbf{r}, t) = \int mn(\mathbf{r}, \mathbf{v}, t) \, \mathrm{d}^3 v = mn(\mathbf{r}). \tag{14.3}$$

Its evolution can be calculated by integrating equation (14.1), multiplied by the single particle mass m, over the velocity:

$$\frac{\partial \rho(\mathbf{r}, t)}{\partial t} + \int m\mathbf{v} \cdot \nabla_{\mathbf{r}} n(\mathbf{r}, \mathbf{v}, t) \, \mathrm{d}^3 v = \int \left(m \frac{\mathrm{d}n}{\mathrm{d}t} \right)_{\text{collisions}} \mathrm{d}^3 v. \tag{14.4}$$

The second term of this equation can be written as $\nabla \cdot \mathbf{j}(\mathbf{r}, t)$ where $\mathbf{j}$ denotes the mass flux, or momentum density, of the fluid:

$$\mathbf{j}(\mathbf{r}, t) = \int m\mathbf{v} n(\mathbf{r}, \mathbf{v}, t) \mathrm{d}^3 v = \rho \mathbf{u}, \tag{14.5}$$

where $\mathbf{u}$ is the average local velocity. This equation can be checked using (14.2). The collisions change the velocity distribution, but not the mass density of the particles. Hence the right hand side of (14.4) vanishes and we obtain the familiar continuity equation:

$$\frac{\partial \rho(\mathbf{r}, t)}{\partial t} + \nabla \cdot \mathbf{j}(\mathbf{r}, t) = 0. \tag{14.6}$$

Another interesting equation describes the conservation of momentum. We would like to know how $\mathbf{j}(\mathbf{r}, t)$ changes with time. This is again evaluated straightforwardly by multiplying Eq. (14.1) by $\mathbf{v}$ and integrating over the velocity. Using the indices α and β for the Cartesian coordinates, we obtain

$$\frac{\partial j_\alpha}{\partial t} + \int m v_\alpha \sum_\beta v_\beta \partial_\beta n(\mathbf{r}, \mathbf{v}, t) \mathrm{d}^3 v = \int m v_\alpha \left(\frac{\mathrm{d}n}{\mathrm{d}t} \right)_{\text{collisions}} \mathrm{d}^3 v, \tag{14.7}$$

where ∂_β denotes a derivative with respect to the coordinate r_β. For the right hand side, a similar statement can be made as for the equivalent term in the mass equation: although individual particles involved in a collision change their momenta, the *total* momentum is conserved on collision. After thus putting the right hand side to zero, we write (14.7) in shorthand notation as

$$\frac{\partial j_\alpha}{\partial t} + \partial_\beta \Pi_{\alpha\beta} = 0, \tag{14.8}$$

where we have introduced the momentum flow tensor

$$\Pi_{\alpha\beta} = \int m v_\alpha v_\beta n(\mathbf{r}, \mathbf{v}, t) \mathrm{d}^3 v, \tag{14.9}$$

and where we have used the Einstein summation convention in which repeated indices (in this case β) are summed over.

Assuming that we are in equilibrium, we can evaluate the momentum tensor by substituting for $n(\mathbf{r}, \mathbf{v}, t)$ the form (14.2):

$$\begin{aligned} \Pi^{\text{eq}}_{\alpha\beta} &= \int v_\alpha v_\beta \rho(\mathbf{r}) \exp[-m(\mathbf{v} - \mathbf{u})^2/(2k_{\mathrm{B}}T)] \mathrm{d}^3 v \\ &= \rho(\mathbf{r})(k_{\mathrm{B}} T \delta_{\alpha\beta} + u_\alpha u_\beta). \end{aligned} \tag{14.10}$$

This result can be derived by separately considering $\alpha = \beta$ and $\alpha \neq \beta$, and working out the appropriate Gaussian integrals (using the substitution $\mathbf{w} = \mathbf{v} - \mathbf{u}$). Noting that $\rho k_{\mathrm{B}} T$ equals the pressure P,[1] we arrive at the following two equations:

$$\frac{\partial \rho(\mathbf{r}, t)}{\partial t} + \nabla \cdot \mathbf{j}(\mathbf{r}, t) = 0 \text{ (mass conservation)}; \tag{14.11a}$$

$$\frac{\partial (\rho \mathbf{u})}{\partial t} + \nabla_{\mathbf{r}} \cdot (P\mathbf{I} + \rho \mathbf{u}\mathbf{u}) = 0 \text{ (momentum conservation)}. \tag{14.11b}$$

Using the first equation, we can rewrite the second as

$$\frac{\partial \mathbf{u}(\mathbf{r}, t)}{\partial t} + [\mathbf{u}(\mathbf{r}, t) \cdot \nabla_{\mathbf{r}}] \mathbf{u}(\mathbf{r}, t) = -\frac{1}{\rho(\mathbf{r}, t)} \nabla_{\mathbf{r}} P(\mathbf{r}, t). \tag{14.12}$$

The equations (14.11a) and (14.11b) or (14.12) are the *Euler equations* for a fluid in equilibrium. These equations neglect dissipative effects.

When the fluid is not everywhere in local equilibrium, the collisions will drive the system towards equilibrium – hence their effect can no longer be neglected.

[1] Here, we consider the fluid as an ideal gas; a realistic equation of state may be used instead.

As mentioned above, the additional currents which arise on top of the equilibrium currents increase the entropy and are therefore called *dissipative*. Hence these terms describe the viscous effects in the fluid.

We now split the distribution function into an equilibrium and a nonequilibrium part:

$$n(\mathbf{r}, \mathbf{v}, t) = n^{\mathrm{eq}}(\mathbf{r}, \mathbf{v}) + n^{\mathrm{noneq}}(\mathbf{r}, \mathbf{v}, t). \tag{14.13}$$

The equilibrium term satisfies (14.2).

How can we represent the effect of the collision term? There is an approach due to Maxwell, which is based on the assumption that all relaxation processes have the same, or are dominated by a single, relaxation time τ. In that case:

$$\left(\frac{\mathrm{d}n(\mathbf{r}, \mathbf{v}, t)}{\mathrm{d}t}\right)_{\text{collisions}} = -\frac{n(\mathbf{r}, \mathbf{v}, t) - n^{\mathrm{eq}}(\mathbf{r}, \mathbf{v})}{\tau} = -\frac{n^{\mathrm{noneq}}}{\tau}. \tag{14.14}$$

As mentioned above, the collisions do not change the mass conservation equation, which should always be valid. The equation for the flux will, however, acquire a contribution from the nonequilibrium part of the distribution function, as we shall see. The mass flux can still be written as $\rho\mathbf{u}$. Moreover, the collisions leave the total momentum unchanged.

The flux $\mathbf{j}$ occurring in the mass conservation equation also occurs in the momentum conservation equation. In this second equation, the momentum flux $\Pi_{\alpha\beta}$ occurs, which we have calculated above *assuming equilibrium*. If we consider the evolution of this flux using the Boltzmann equation, we see that the collision effects enter explicitly in this momentum flux.

To find the lowest-order contribution to a systematic expansion of the density, we replace n on the left hand side of the Boltzmann equation by its equilibrium version:

$$\frac{\partial n^{\mathrm{eq}}(\mathbf{r}, \mathbf{v})}{\partial t} + \mathbf{v} \cdot \nabla_{\mathbf{r}} n^{\mathrm{eq}} = -\frac{n^{\mathrm{noneq}}(\mathbf{r}, \mathbf{v}, t)}{\tau}. \tag{14.15}$$

This is an *explicit* equation for the nonequilibrium term. It can be shown that this is an expansion in the parameter ℓ/L, where ℓ is the mean free path, and L is the typical length scale over which the hydrodynamic quantities vary [6]. Note that if we integrate this equation over the velocity, the right hand side vanishes as the collisions do not affect the mass density.

The momentum flux is defined in (14.9). This is calculated from the density $n(\mathbf{r}, \mathbf{v}, t)$ and it can therefore be split into an equilibrium and a nonequilibrium part. The equilibrium part was calculated in Eq. (14.10), and the nonequilibrium part will now be calculated using (14.15):

$$\Pi_{\alpha\beta}^{\text{noneq}} = \int m v_\alpha v_\beta n^{\text{noneq}}\, \mathrm{d}^3 v$$
$$= -\tau \left[\int m v_\alpha v_\beta \frac{\partial n^{\text{eq}}}{\partial t}\, \mathrm{d}^3 v + \int m v_\alpha v_\beta \mathbf{v} \cdot \nabla_{\mathbf{r}} n^{\text{eq}} \mathrm{d}^3 v \right]. \tag{14.16}$$

Before we proceed to work out (14.16) further, we note that the tensor $\Pi_{\alpha\beta}^{\text{noneq}}$ has an important property: its trace vanishes. This can be seen by writing out this trace:

$$\sum_\alpha \Pi_{\alpha\alpha}^{\text{noneq}} = \int v^2 n^{\text{noneq}}(\mathbf{r}, \mathbf{v}, t) \mathrm{d}^3 v. \tag{14.17}$$

Realizing that this expression represents the change in the average kinetic energy due to the collisions, we immediately see that it vanishes as the (instantaneous) collisions leave the total energy invariant:

$$\mathrm{Tr}\, \Pi^{\text{noneq}} = 0. \tag{14.18}$$

For the calculation of the nonequilibrium stress tensor, Eq. (14.16), we use the following equations, which can easily be seen to hold for the equilibrium distribution:

$$\int m n^{\text{eq}}(\mathbf{r}, \mathbf{v}) \mathrm{d}^3 v = \rho(\mathbf{r}); \tag{14.19a}$$

$$\int m(v_\alpha - u_\alpha)(v_\beta - u_\beta) n^{\text{eq}}(\mathbf{r}, \mathbf{v}) \mathrm{d}^3 v = \rho \frac{k_\mathrm{B} T}{m} \delta_{\alpha\beta} = P \delta_{\alpha\beta}; \tag{14.19b}$$

$$\dot{u}_\alpha = -\sum_\beta u_\beta \partial_\beta u_\alpha - \frac{1}{\rho}(\partial_\alpha P); \tag{14.19c}$$

where in the last equation it is understood that the velocities are those evaluated for the equilibrium distribution: this equation is the Euler equation, (14.12).

We first work out the first term in the square brackets on the right hand side in (14.16), which can also be written as $\partial_t \Pi_{\alpha\beta}^{\text{eq}}$ (we use ∂_t to denote a derivative with respect to t). After some manipulation, using Eqs. (14.10), (14.11a) and (14.11b) (or (14.19c)), this can be written as

$$\partial_t \Pi_{\alpha\beta}^{\text{eq}} = \partial_t (P \delta_{\alpha\beta} + \rho u_\alpha u_\beta)$$
$$= \dot{P} \delta_{\alpha\beta} - \sum_\gamma [\partial_\gamma(\rho u_\gamma) u_\alpha u_\beta + \rho u_\alpha u_\gamma (\partial_\gamma u_\beta) + \rho u_\beta u_\gamma (\partial_\gamma u_\alpha)]$$
$$- u_\beta \partial_\alpha P - u_\alpha \partial_\beta P. \tag{14.20}$$

The second term in the square brackets of (14.16) can be written, using the quantity $w_\alpha = v_\alpha - u_\alpha$ (see also Eq. (14.10) and (14.19b)), in the form

$$\int (u_\alpha + w_\alpha)(u_\beta + w_\beta)(u_\gamma + w_\gamma)\partial_\gamma n^{\mathrm{eq}}(\mathbf{r}, \mathbf{v})\, \mathrm{d}^3 v$$

$$= \partial_\gamma(\rho u_\alpha u_\beta u_\gamma + u_\alpha P\delta_{\beta\gamma} + u_\beta P\delta_{\alpha\gamma} + u_\gamma P\delta_{\alpha\beta}). \tag{14.21}$$

This term can now be worked out to yield

$$\sum_\gamma [u_\alpha u_\beta u_\gamma \partial_\gamma \rho + \rho u_\beta u_\gamma(\partial_\gamma u_\alpha) + \rho u_\alpha u_\gamma(\partial_\gamma u_\beta) + \rho u_\alpha u_\beta(\partial_\gamma u_\gamma)$$

$$+ \partial_\gamma(Pu_\gamma)\delta_{\alpha\beta} + \partial_\gamma(Pu_\alpha)\delta_{\beta\gamma} + \partial_\gamma(Pu_\beta)\delta_{\alpha\gamma}]. \tag{14.22}$$

Adding the two terms of Eq. (14.16), many terms occurring in (14.20) and (14.22) cancel. The ones that remain are

$$P(\partial_\beta u_\alpha + \partial_\alpha u_\beta) + \delta_{\alpha\beta}\left\{\dot{P} + \sum_\gamma [u_\gamma(\partial_\gamma P) + P\partial_\gamma u_\gamma]\right\}. \tag{14.23}$$

The terms

$$\dot{P} + \sum_\gamma u_\gamma(\partial_\gamma P) \tag{14.24}$$

can be calculated using (14.19b) and the equilibrium distribution. When we write this out, we obtain, again with $w_\alpha = v_\alpha - u_\alpha$:

$$\frac{\partial}{\partial t}\int mw^2 n\, \mathrm{d}^3 w + \sum_\gamma u_\gamma \partial_\gamma \int mw^2 n\, \mathrm{d}^3 w$$

$$= \int mw^2\left(\frac{\partial n}{\partial t} + \sum_\gamma u_\gamma \partial_\gamma n\right)\mathrm{d}^3 w = -\frac{1}{\tau}\int mw^2 n^{\mathrm{noneq}}\, \mathrm{d}^3 w. \tag{14.25}$$

This is the trace of the tensor

$$-\frac{1}{\tau}\int mw_\alpha w_\beta n^{\mathrm{noneq}}\, \mathrm{d}^3 v. \tag{14.26}$$

Now we use the fact that $\mathrm{Tr}\Pi^{\mathrm{noneq}}$ vanishes. This can only happen when the trace occurring in the last equation cancels the trace of the remaining terms in the expression for Π^{noneq}. This tensor must therefore be

$$\Pi^{\mathrm{noneq}} = -P\tau\left(\partial_\alpha u_\beta + \partial_\beta u_\alpha - \tfrac{2}{3}\delta_{\alpha\beta}\partial_\gamma u_\gamma\right). \tag{14.27}$$

Using this, we can formulate the momentum conservation equation, with $\nu = \tau k_{\mathrm{B}} T/m$, as

$$\frac{\partial \mathbf{u}}{\partial t} + \mathbf{u}\cdot\nabla\mathbf{u} = -\frac{1}{\rho}\nabla P + \nu\nabla^2\mathbf{u} + \frac{1}{3}\nu\nabla(\nabla\cdot\mathbf{u}). \tag{14.28}$$

The mass conservation equation and the momentum conservation equation together are insufficient to give us the four unknown field: ρ, u_x, u_y and P. We therefore need an additional equation, which may be ρ = constant for an incompressible fluid, or $P \propto \rho$ for the isothermal case (which we have analysed here). Note that the case where ρ = constant also implies $\nabla \cdot \mathbf{u} = 0$ from the continuity equation, which in turn causes the last term in the last equation to become negligible.

14.3 The lattice Boltzmann model

The idea of the lattice Boltzmann method is to use a 'toy model' for the liquid, which combines the properties of mass, momentum and energy conservation (only the first two are used in the derivation of the fluid equations) and isotropic relaxation of the stress. In the simplest case, the Maxwell *Ansatz* of a single dominant relaxation time is used. Some time ago, the *lattice gas* model was studied intensively [7–9]. In this model, the fluid consists of particles which can occupy lattice sites. In two dimensions, the lattice can be square (with links to nearest and to next nearest, diagonal neighbours) or hexagonal. The time is also taken discrete, and at subsequent time steps, particles are allowed to move to a neighbouring position, according to rules which guarantee mass and momentum conservation. Because of the discrete nature of the model, the fluctuations in this method are substantial, and an alternative formulation was developed: the lattice Boltzmann model. This is also formulated on a lattice, and the time is discrete, but the particles are replaced by *densities* on the lattice sites.

Let us specify the formulation on a hexagonal lattice. This is shown on the left hand side of Figure 14.1. The arrows indicate possible velocities of particles at each site: the particles with one of these velocities travel in one time step to the neighbour at the other end of the link. On the hexagonal lattice, the velocities are all equal in size: one lattice constant per time step. Particles can also stand still. In a lattice gas cellular automaton, individual particles are distinguished; in the lattice Boltzmann method, however, we only consider *densities* of particles with a particular velocity: n_i is the density of particles with velocity $\mathbf{v}_i$, which may be directed along a lattice link or may be 0.

To be more specific, we have seven possible velocities at each link (including the zero-velocity particles), and therefore we have seven densities, one for each velocity. At each time step, two processes take place, which can be directly related to the Boltzmann equation:

- At each site, all particles with a nonzero speed move to the neighbouring site to which their velocity points (see the description above);
- At each site, the distribution is relaxed to the equilibrium distribution.

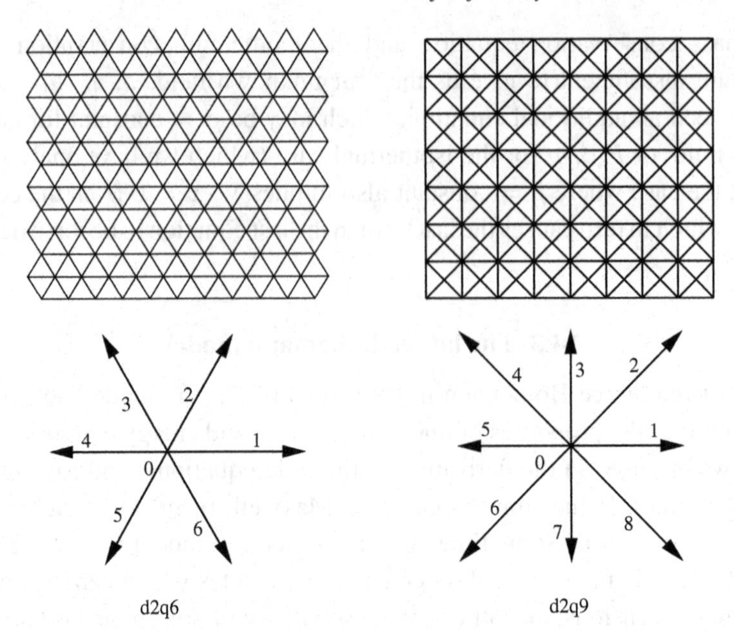

Figure 14.1. The two grids used in two-dimensional lattice Boltzmann simulations. On the left hand side is the hexagonal grid (d2q6; see text) with the possible particle velocities. On the right hand side, the d2q9 grid is shown. A square grid is not suitable, as on such a grid all momenta along each horizontal and each vertical grid line would be conserved.

The second step represents the effect of the collisions.

This algorithm is in fact a realisation of the following equation which governs the evolution of the distribution in time:

$$n_i(\mathbf{r} + c\Delta t\mathbf{e}_i, t + \Delta t) = n_i(\mathbf{r}, t) - \frac{1}{\tau}[n_i(\mathbf{r}, t) - n_i^{\text{eq}}(\mathbf{r}, t)]. \tag{14.29}$$

Here, i labels the seven possible directions $\mathbf{e}_i$ from each site ($i = 0$ represents the rest particles), and n^{eq} is the equilibrium distribution. Furthermore, c is related to the velocity at which the particles with $i \neq 0$ move; in practice we take c and Δt both equal to 1 in the end, but for derivations it is useful to keep them as parameters in the problem.

The first step can easily be carried out: for each site, we loop over the six nonzero speeds, and move the particles to the neighbouring site. Care must be taken to distinguish those which have not yet been moved from the 'newcomers', to avoid the latter being moved twice. The relaxation is then straightforward, provided we know the equilibrium distribution. We can construct this equilibrium distribution as a second order polynomial in terms of the velocity components u_α:

$$n_i^{\text{eq}}(\mathbf{u}) = n_i^{\text{eq}}(0)(1 + Au_\alpha e_{i\alpha} + Bu_\alpha u_\alpha + Cu_\alpha u_\beta e_{i\alpha} e_{i\beta}), \tag{14.30}$$

where α and β are Cartesian coordinates x and y; the dependence on the positions is left out for convenience. All terms in this expression have been chosen to be isotropic, i.e. there is no preferred direction.

An additional condition is Gallilean invariance: all moments

$$M^j(\mathbf{u}) = \sum_{i=0}^{6} (\mathbf{u} - \mathbf{e}_i)^j n_i^{\text{eq}} \tag{14.31}$$

should satisfy $M^j(\mathbf{u}) = M(0)$, for $j = 0, 1, 2, \ldots$ After some calculation, this condition leaves [10, 11]

$$n_i^{\text{eq}}(\mathbf{u}) = w_i \frac{\rho}{m} \left(1 + \frac{4}{c^2} u_\alpha e_{i\alpha} - \frac{2}{c^2} u_\alpha u_\alpha + \frac{8}{c^4} u_\alpha u_\beta e_{i\alpha} e_{i\beta} \right) \tag{14.32}$$

as the only possibility for the equilibrium distribution. In (14.32), an implicit summation over repeating Cartesian indices (α, β) is understood. Furthermore,

$$w_0 = 1/2; \tag{14.33a}$$

$$w_i = 1/12, \quad i = 1, \ldots, 6. \tag{14.33b}$$

For a square lattice, we should allow the particles to visit the four neighbours along the two Cartesian directions, and the four next-nearest neighbours as shown in Figure 14.1. This is called the 'd2q9' model because it is two-dimensional and at each site there are nine possible velocities. For this model, the equilibrium distribution can be derived similarly – the result is:

$$n_i^{eq} = w_i \frac{\rho}{m} \left(1 + \frac{3}{c^2} e_{i\alpha} u_\alpha + \frac{9}{2c^4} e_{i\alpha} e_{i\beta} u_\alpha u_\beta - \frac{3}{2} \frac{u_\alpha u_\alpha}{c^2} \right). \tag{14.34}$$

Here,

$$w_i = \begin{cases} 4/9 & \text{for } i = 0; \\ 1/9 & \text{for } i = 1, 3, 5, 7; \\ 1/36 & \text{for } i = 2, 4, 6, 8. \end{cases} \tag{14.35}$$

We now have all the ingredients to build the lattice Boltzmann program, if we know how to handle the boundary conditions. In fluid dynamics, the boundary condition at a wall is usually defined to correspond to zero velocity (this is called a 'stick' boundary condition as opposed to 'slip' boundary conditions). The simplest scheme to realise this is to use the 'bounce back' boundary condition. This assumes that boundaries lie in between neighbouring grid points. For a point lying just inside the system, we move its particles pointing to a neighbouring point outside the system to that neighbour and reverse its velocity there. On average, this boils down to having zero velocity in between the two particles. More accurate implementations for the boundary conditions have been developed [12, 13].

In Section 14.5 we show that the lattice Boltzmann model leads to the Navier–Stokes equations for an incompressible fluid in the limit of small velocities. There, it will be shown that the only parameter of the algorithm, which is the relaxation time τ, is related to the viscosity ν by

$$\nu = \frac{2\tau - 1}{6}\frac{\Delta x^2}{\Delta t}, \tag{14.36}$$

where Δx, Δt are the lattice constant and the time steps (which are usually taken equal to 1).

PROGRAMMING EXERCISE

Construct a lattice Boltzmann code for the flow through a two-dimensional pipe which we imagine to be horizontal. This is a rectangle, where on the left hand side we supply fluid, which is drained on the right hand side. Use the d2q9 lattice. On the lateral boundaries, the bounce back rule is used to ensure stick boundary conditions. The easiest way to realise the flow is by imposing a pressure gradient over the system from left to right. This means that on each segment of the fluid (corresponding to a point of the Boltzmann lattice), a constant force is acting. This has the effect of increasing all velocities along the direction of the flow at each time step by the same (small) amount.

The algorithm is set up as follows.

Move the density n_i to the appropriate neighbour;
Reverse the new velocities on points beyond the system boundaries;
Calculate velocities at each point;
Add a small velocity along the direction of the pressure gradient;
Calculate equilibrium distribution at each point;
Relax the densities at the points inside the system according to

$$n_i^{\text{new}} = (1 - 1/\tau)n_i^{\text{old}} + n_i^{\text{eq}}/\tau.$$

Check If your program works correctly, you should obtain a parabolic flow profile throughout the pipe. Note that the simulation is only reliable for small velocities. The parabola should have curvature $\nabla P/(\rho\nu)$. The viscosity ν is related to the relaxation time τ as $\nu = (2\tau - 1)/6$ (see the next section). See also Problem 14.1.

14.4 Additional remarks

The lattice Boltzmann method works on a hexagonal lattice and on a square lattice provided the stresses can relax isotropically. This requirement forces us to include the next nearest neighbours into the possible moves on the square lattice. In three dimensions, a so-called FCHC lattice lattice must be used [7, 8], which contains moves to the neighbours along Cartesian directions, as well as moves to neighbours at relative positions $(1, 1, 0)$, $(1, 0, -1)$ etc. We shall not go further into this; an extensive literature exists on the three-dimensional version of the model (see most references in this chapter).

An interesting aspect of the lattice Boltzmann method is that it can easily be extended to problems that are usually difficult to treat using other methods. These include binary or multiphase systems and objects moving in the flow. Here we concentrate on binary systems. Two methods are predominant in this field: the first was proposed by Shan and Chen [14] and the second by Swift *et al.* [15]. We shall adopt the first method here for its simplicity. Shan and Chen start by simulating two fluids in parallel. These fluids can be conveniently be denoted by a 'colour': say, red and blue particles. In principle, the fluids may have different viscosities (thus different values for the relaxation times τ) – we shall take them to be equal in our description. The two fluids interact through some potential which has the form

$$V(\mathbf{r}, \mathbf{r}') = G_{cc'}(\mathbf{r} - \mathbf{r}')\rho_c(\mathbf{r})\rho_{c'}(\mathbf{r}'). \tag{14.37}$$

The indices c and c' denote the colours and $\rho_c = \sum_i m_c n_{c,i}$. The kernel $G_{cc'}$ is zero for equal colours $c = c'$ (which assume the values r and b for red and blue respectively). Furthermore, $G_{\text{rb}}(\mathbf{r}, \mathbf{r}')$ is only nonzero when $\mathbf{r}$ and $\mathbf{r}'$ are lattice neighbours.

The average velocities of the fluids are calculated for the two fluids together:

$$\mathbf{u} = \frac{m_\text{r} \sum_i \mathbf{e}_i n_{\text{r},i} + m_\text{b} \sum_i \mathbf{e}_i n_{\text{b},i}}{\sum_i (m_\text{r} n_{\text{r},i} + m_\text{b} n_{\text{b},i})}. \tag{14.38}$$

The relaxation of the distributions at each site is with respect to the equilibrium distribution based on this average velocity.

From this potential we can derive a force by taking the (discrete) gradient. This directly leads to an extra force on a particle with colour c located at $\mathbf{r}$ of the form

$$\frac{\text{d}p_c(\mathbf{r})}{\text{d}t} = -\rho_c(\mathbf{r}) \sum_i G_{cc',i} \rho_{c'}(\mathbf{r} + \mathbf{e}_i) \tag{14.39}$$

where $c \neq c'$. The interaction $G_{cc',i}$ assumes different values for nearest (nn) and next nearest neighbours (nnn) respectively. They must be tuned to make the force isotropic – in our case this means

$$G_{nnn} = \sqrt{2} G_{nn}. \tag{14.40}$$

The extra force is included in the lattice Boltzmann equation which now reads

$$n_{c,i}(\mathbf{r}+\mathbf{e}_i,t+1) = n_{c,i}(\mathbf{r},t) - \frac{1}{\tau}[n_{c,i}(\mathbf{r},t) - n_i^{\mathrm{eq}}(\mathbf{r},t)] + \frac{\mathrm{d}p_c}{\mathrm{d}t}(\mathbf{r},t). \tag{14.41}$$

In summary, the heart of the algorithm reads (the change in velocity related to the force $\mathrm{d}p_c/\mathrm{d}t$ is called Δv):

Calculate average velocities from (14.38);
FOR each site $\mathbf{r}_1$ DO
  Calculate local densities $\rho_1(c)$
    for each colour c;
  FOR each neighbour $\mathbf{r}_2$ of $\mathbf{r}_1$ DO
    Calculate density $\rho_2(c')$ of that neighbour;
    Subtract $\Delta v \rho_2(c')$ from the velocity of colour c at $\mathbf{r}_1$;
  END FOR;
END FOR.

You must first put the two colours on each site with a fixed concentration ratio of, for example, 2 to 1. It turns out that for high values of Δv, the simulation becomes unstable. This is due to the fact that when the different colours repel each other too strongly, we get excessively high velocities, and this results in negative densities. For $\tau = 1$ the divergence sets in for $\Delta v > 0.11$ on the d2q9 grid (for a concentration ratio of 2:1). The instability is due to the speed exceeding the sound speed.

Now the simulation is fully defined and can be implemented (see Problem 14.3, which also addresses the analysis of the data). The result shows the formation of bubbles which grow by coalescence. The bubbles should satisfy Laplace's law, which states that the jump in pressure when going from inside the bubble to the outside should be proportional to the inverse of the radius of the bubble [16].

Figure 14.2 shows the pressure drop for different bubbles as a function of inverse radius. Clearly, our simulation satisfies Laplace's law.

*14.5 Derivation of the Navier–Stokes equation from the lattice Boltzmann model

In this section, we shall show how the Navier–Stokes equations can be recovered in the incompressible limit from the lattice Boltzmann model. The derivation is based on two major ingredients [15, 17, 18]:

- A systematic expansion of $n_i(\mathbf{r},t)$ in the time step Δt is made;
- Terms of the form $(u/c_s)^j$ are neglected beyond a certain order j. The quantity u/c_s, where c_s is the sound speed, is known as the *Mach number*, M.

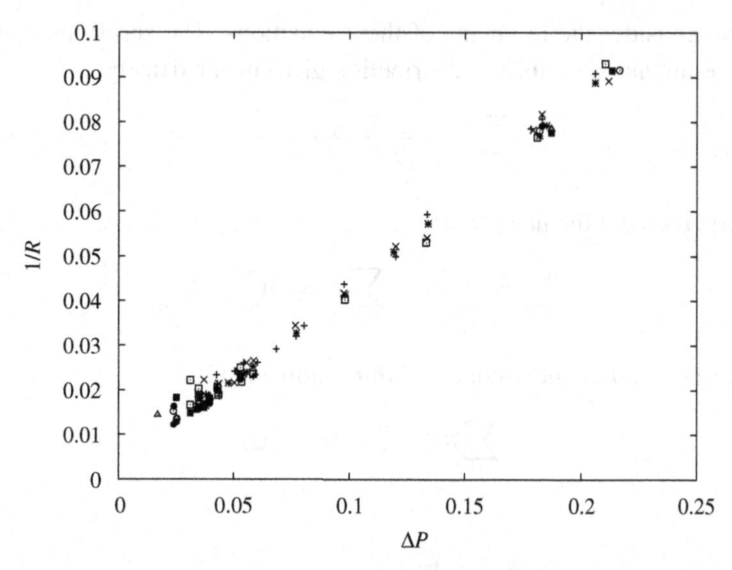

Figure 14.2. Pressure drop across droplet boundary. Data are taken from a long run: different symbols correspond to different times. The points are seen to fall on a straight line. In the simulation we have taken $\tau = 1$, and $\Delta v = 0.085$. The times vary between 6000 and 14 000 steps.

We start with the systematic expansion of n in powers of Δt. Taylor-expanding Eq. (14.29), we have

$$-\frac{n_i(\mathbf{r},t) - n_i^{\text{eq}}(\mathbf{r},t)}{\tau} = \sum_{k=0}^{\infty} \frac{(\Delta t)^k}{k!} D_i^k n_i(\mathbf{r},t) \tag{14.42}$$

where we have introduced the total differential operator

$$D_i = \frac{\partial}{\partial t} + c e_{i\alpha} \partial_\alpha. \tag{14.43}$$

We call $n_i^{(k)}(\mathbf{r},t)$ an approximation of $n_i(\mathbf{r},t)$ which is correct to order k in Δt. We then have:

$$n_i^{(0)}(\mathbf{r},t) = n_i^{\text{eq}}(\mathbf{r},t) \tag{14.44a}$$

$$n_i^{(1)}(\mathbf{r},t) = n_i^{\text{eq}}(\mathbf{r},t) - \tau \Delta t D_i n_i^{\text{eq}}(\mathbf{r},t) \tag{14.44b}$$

$$n_i^{(2)}(\mathbf{r},t) = n_i^{\text{eq}}(\mathbf{r},t) - \tau \Delta t D_i n_i^{(1)}(\mathbf{r},t) - \tau \frac{\Delta t^2}{2} D_i^2 n_i^{\text{eq}}(\mathbf{r},t). \tag{14.44c}$$

Substituting the second equation for $n_i^{(1)}$ into the third, we obtain

$$\begin{aligned} n_i^{(2)}(\mathbf{r},t) = {} & n_i^{\text{eq}}(\mathbf{r},t) - \tau \Delta t (\partial_t + c e_{i\alpha} \partial_\alpha) n_i^{\text{eq}}(\mathbf{r},t) \\ & - \tau \Delta t^2 \left(\tfrac{1}{2} - \tau\right)(\partial_t + c e_{i\alpha} \partial_\alpha)^2 n_i^{\text{eq}}(\mathbf{r},t). \end{aligned} \tag{14.45}$$

Now we consider the moments of these equations. The zeroth moment of the nonequilibrium and equilibrium distribution gives us the density:

$$\sum_i mn_i = \sum_i mn_i^{\text{eq}} = \rho \tag{14.46}$$

and the first moment the mass flux:

$$\sum_i mce_{i\alpha}n_i = \sum_i mce_{i\alpha}n_i^{\text{eq}} = j_\alpha \tag{14.47}$$

We furthermore define the second and third moments:

$$\sum_i mc^2 e_{i\alpha}e_{i\beta}n_i \equiv \Pi_{\alpha\beta} \tag{14.48}$$

and

$$\sum_i mc^3 e_{i\alpha}e_{i\beta}e_{i\gamma}n_i \equiv \Gamma_{\alpha\beta\gamma} \tag{14.49}$$

We now take the zeroth moment of Eq. (14.45). Realising that this moment is identical for n_i^{eq} and $n_i^{(2)}$, we obtain

$$\partial_t\rho + \partial_\alpha j_\alpha = \left(\tau - \tfrac{1}{2}\right)\Delta t(\partial_t^2\rho + 2\partial_t\partial_\alpha j_\alpha + \partial_\alpha\partial_\beta\Pi_{\alpha\beta}^{(0)}) + \mathcal{O}(\Delta t^2). \tag{14.50}$$

For the first moment, we obtain

$$\partial_t j_\alpha + \partial_\beta\Pi_{\alpha\beta}^{(0)} = \left(\tau - \tfrac{1}{2}\right)\Delta t(\partial_t^2 j_\alpha + 2\partial_t\partial_\beta\Pi_{\alpha\beta}^{(0)} + \partial_\beta\partial_\gamma\Gamma_{\alpha\beta\gamma}^{(0)}) + \mathcal{O}(\Delta t^2). \tag{14.51}$$

Now we note that the $\mathcal{O}(\Delta t)$ term on the right hand side of the Eq. (14.50) can be written as

$$\partial_t^2\rho + 2\partial_t\partial_\alpha j_\alpha + \partial_\alpha\partial_\beta\Pi_{\alpha\beta}^{(0)} = \partial_t(\partial_t\rho + \partial_\alpha j_\alpha) + \partial_\alpha(\partial_t j_\alpha + \partial_\beta\Pi_{\alpha\beta}^{(0)}). \tag{14.52}$$

From Eqs. (14.50) and (14.51), the two terms on the right hand side of this equation can easily be seen to be of order Δt, so we see that

$$\partial_t\rho + \partial_\alpha j_\alpha = 0 + \mathcal{O}(\Delta t^2). \tag{14.53}$$

We have now recovered the continuity equation to order Δt^2.

For the other moment equation, (14.51), we can argue along the same lines that the first half of the $\mathcal{O}(\Delta t)$ term is close to zero, and we are left with

$$\partial_t j_\alpha + \partial_\beta\Pi_{\alpha\beta}^{(0)} = \left(\tau - \tfrac{1}{2}\right)\Delta t(\partial_t\partial_\beta\Pi_{\alpha\beta}^{(0)} + \partial_\beta\partial_\gamma\Gamma_{\alpha\beta\gamma}^{(0)}) + \mathcal{O}(\Delta t^2). \tag{14.54}$$

To proceed, we must work out the tensors $\Pi^{(0)}$ and $\Gamma^{(0)}$. In order to do this, we note that, on the hexagonal grid, the following relations hold:

$$\sum_i e_{i\alpha} e_{i\beta} = 3\delta_{\alpha\beta}; \tag{14.55a}$$

$$\sum_i e_{i\alpha} e_{i\beta} e_{i\gamma} e_{i\eta} = \frac{3}{4}(\delta_{\alpha\beta}\delta_{\gamma\eta} + \delta_{\alpha\gamma}\delta_{\beta\eta} + \delta_{\alpha\eta}\delta_{\beta\gamma}), \tag{14.55b}$$

whereas similar moments containing an odd number of $e_{i\alpha}$ vanish. Together with (14.32), this yields

$$\Pi^{(0)}_{\alpha\beta} = \rho u_{i\alpha} u_{i\beta} + P\delta_{\alpha\beta}, \tag{14.56}$$

where P is found using the ideal gas law relation with the kinetic energy as in Section 14.2. Furthermore

$$\Gamma^{(0)}_{\alpha\beta\gamma} = \rho \frac{c^2}{3}(u_\alpha \delta_{\beta\gamma} + u_\beta \delta_{\alpha\gamma} + u_\gamma \delta_{\alpha\beta}). \tag{14.57}$$

Substituting these equations into (14.54), we obtain

$$\partial_t j_\alpha + \partial_\beta \Pi^{(0)}_{\alpha\beta} = \left(\tau - \frac{1}{2}\right)\Delta t \left[\partial_t \partial_\beta (\rho u_\alpha u_\beta + P\delta_{\alpha\beta}) + \rho \frac{c^2}{3}(\partial_\beta \partial_\beta u_\alpha + 2\partial_\alpha \partial_\beta u_\beta)\right] + \mathcal{O}(\Delta t^2). \tag{14.58}$$

In vector notation, this reads

$$\partial_t \mathbf{j} + \nabla \cdot \Pi^{(0)} = \left(\tau - \frac{1}{2}\right)\Delta t \left[\nabla\left(\frac{\partial P}{\partial t}\right) + \rho\nabla \cdot \left(\frac{\partial \mathbf{uu}}{\partial t}\right) + \rho\frac{c^2}{3}\nabla^2 \mathbf{u}\right] + \mathcal{O}(\Delta t^2). \tag{14.59}$$

To obtain the last form, we have neglected the term $\nabla(\nabla \cdot \mathbf{u})$, which is small in the incompressible limit as can be seen from the continuity equation. Furthermore, in this limit, the time derivatives on the right hand side are of the order of M^2 – hence we are left with

$$\partial_t \mathbf{j} + \rho\nabla(\mathbf{uu}) = \nabla p + \rho\nu\nabla^2 \mathbf{u}, \tag{14.60}$$

where we have neglected corrections beyond Δt^2 and $\Delta t M^2$ and where we have used (14.36).

This is the Navier–Stokes equation in the incompressible limit for which $\nabla \cdot \mathbf{u} = 0$. This derivation does not depend on the fact that we have used the hexagonal lattice – it yields the same result for the d2q9 model.

Exercises

14.1 Derive from the Navier–Stokes equation that the flow in a two-dimensional pipe forms a parabolic velocity profile. Check that the curvature (i.e. the second derivative) of the parabola is given by $\nabla P/(\nu\rho)$. Check also that the velocity Δv which is added at each step is related to the pressure by $\rho\Delta v = \nabla P/c$. In the simulation, we usually take the lattice constant Δx and the time step Δt equal to 1. Check that the velocity profile of your simulation matches the result for $\nu = (2\tau - 1)/6$.

14.2 [C] Extend the simulation of the parabolic flow by putting an object in the flow. This can be a fixed block. Check your results by inspecting the flow in the channels on both sides of the block to see whether the total flow is the same through the rectangular pipe.

14.3 [C] Write a program for droplet formation in a binary system along the lines indicated in Section 14.4.

14.4 [C] In this problem we check Laplace's law. In order to do this, you need to identify the droplets and find the pressure drop across their boundaries. To this end you step over all lattice sites and check whether there is a large predominance of one colour. If this is the case, you use either the back-track or the Hoshen–Kopelman algorithm which are both described in Section 15.5.1, in order to identify all sites. Furthermore, you store the largest pressure value in the current droplet.

References

[1] R. J. LeVeque and Z. Li, 'The immersed interface method for elliptic equations with discontinuous coefficients and singular sources,' *Siam J. Numer. Anal.*, **31** (1994), 1019–44.

[2] S. Succi, *The Lattice Boltzmann Equation for Fluid Dynamics and Beyond*. Oxford, Oxford University Press, 2001.

[3] R. Benzi, S. Succi, and M. Vergassola, 'The lattice Boltzmann equation: theory and application,' *Phys. Rep.*, **222** (1992), 145–97.

[4] D. H. Rothman and S. Zaleski, *Lattice-Gas Cellular Automata: Simple Models of Complex Hydrodynamics*. Cambridge, Cambridge University Press, 2004.

[5] S. Chapman and T. G. Gowling, *The Mathematical Theory of Non-Uniform Gases*. Cambridge, Cambridge University Press, 1970.

[6] K. Huang, *Statistical Mechanics*, 2nd edn. New York, John Wiley, 1987.

[7] U. Frisch, B. Hasslacher, and Y. Pomeau, 'Lattice-gas automata for the Navier–Stokes equation,' *Phys. Rev. Lett.*, **56** (1986), 1505–8.

[8] U. Frisch, D. d' Humières, B. Hasslacher, P. Lallemand, and Y. Pomeau, 'Lattice gas hydrodynamics in two and three dimensions,' *Complex Systems*, **1** (1987), 649–707.

[9] S. Wolfram, 'Cellular automaton fluids 1: basic theory,' *J. Stat. Phys.*, **45** (1986), 471–526.

[10] R. D. Kingdon, P. Schofield, and L. White, 'A lattice Boltzmann model for the simulation of fluid flow,' *J. Phys. A*, **25** (1992), 3559–66.

[11] H. Chen, S. Chen, and H. Matthaeus, 'Recovery of the Navier Stokes equation using a lattice Boltzmann method,' *Phys. Rev. A*, **45** (1992), 5339–442.

[12] D. P. Ziegler, 'Boundary conditions for lattice Boltzmann simulations,' *J. Stat. Phys.*, **71** (1992), 1171–7.

[13] R. S. Maier, R. S. Bernard, and D. W. Grunau, 'Boundary conditions for the lattice Boltzmann method,' *Phys. Fluids*, **8** (1996), 1788–801.

[14] X. Shan and H. Chen, 'Lattice Boltzmann model for simulating flows with multiple phases and components,' *Phys. Rev. B*, **47** (1993), 1815–19.

[15] M. R. Swift, E. Orlandini, W. R. Osborn, and J. M. Yeomans, 'Lattice Boltzmann simulations of liquid–gas and binary fluid systems,' *Phys. Rev. E*, **54** (1996), 5041–52.
[16] C. Isenberg, *The Science of Soap Films and Soap Bubbles*. New York, Dover, 1992.
[17] X. He and L.-S. Luo, 'Lattice Boltzmann model for the incompressible Navier–Stokes equation,' *J. Stat. Phys.*, **88** (1997), 927–44.
[18] Z. Guo and N. Wang, 'Lattice BGK model for incompressible Navier–Stokes Equation,' *J. Comp. Phys.*, **165** (2000), 288–306.

15

Computational methods for lattice field theories

15.1 Introduction

Classical field theory enables us to calculate the behaviour of fields within the framework of classical mechanics. Examples of fields are elastic strings and sheets, and the electromagnetic field. Quantum field theory is an extension of ordinary quantum mechanics which not only describes extended media such as string and sheets, but which is also supposed to describe elementary particles. Furthermore, ordinary quantum many-particle systems in the grand canonical ensemble can be formulated as quantum field theories. Finally, classical statistical mechanics can be considered as a field theory, in particular when the classical statistical model is formulated on a lattice, such as the Ising model on a square lattice, discussed in Chapter 7.

In this chapter we shall describe various computational techniques that are used to extract numerical data from field theories. *Renormalisation* is a procedure without which field theories cannot be formulated consistently in continuous space-time. In computational physics, we formulate field theories usually on a lattice, thereby avoiding the problems inherent to a continuum formulation. Nevertheless, understanding the renormalisation concept is essential in lattice field theories in order to make the link to the real world. In particular, we want to make predictions about physical quantities (particle masses, interaction constants) which are independent of the lattice structure, and this is precisely where we need the renormalisation concept.

Quantum field theory is difficult. It does not belong to the standard repertoire of every physicist. We try to explain the main concepts and ideas before entering into computational details, but unfortunately we cannot give proofs and derivations, as a thorough introduction to the field would require a book on its own. For details, the reader is referred to Refs. [1–5]. In the next section we shall briefly describe what quantum field theory is and present several examples. In the following section, the procedure of numerical quantum field theory will be described in the context of renormalisation theory. Then we shall describe several algorithms for simulating field theory, and in particular methods for reducing critical slowing down, a major problem in numerical field theory computations. Finally, we shall consider some applications in quantum electrodynamics (QED) and quantum chromodynamics (QCD).

15.2 Quantum field theory

To understand quantum field theory, it is essential to be accustomed to the path-integral formalism (Section 12.4), so let us recall this concept briefly.

Consider a single particle in one dimension. The particle can move along the x-axis and its trajectory can be visualised in $(1+1)$-dimensional space-time. Fixing initial and final positions and time to (t_i, x_i), (t_f, x_f) respectively, there is (in general) one particular curve in the (t, x)-plane, the *classical trajectory*, for which the action S is stationary. The action is given by

$$S(x_\text{i}, x_\text{f}; t_\text{i}, t_\text{f}) = \int_{t_\text{i}}^{t_\text{f}} \mathrm{d}t \; L(x, \dot{x}, t) \tag{15.1}$$

where $L(x, \dot{x}, t)$ is the Lagrangian. In quantum mechanics, nonstationary paths are allowed too, and the probability of going from an initial position x_i to a final position x_f is given by

$$\int [\mathcal{D}x(t)] \, \mathrm{e}^{-\mathrm{i}S/\hbar} = \langle x_\text{f} | \, \mathrm{e}^{-\mathrm{i}(t_\text{i} - t_\text{f})H/\hbar} | x_\text{i} \rangle, \tag{15.2}$$

where H is the Hamiltonian of the system. The integral $\int [\mathcal{D}x(t)]$ is over all possible paths with fixed initial and final values x_i and x_f respectively. If we send Planck's constant $\hbar$ to zero, the significant contributions to the path integral will be more and more concentrated near the stationary paths, and the stationary path with the lowest action is the only one that survives when $\hbar = 0$.

Now consider a field. The simplest example of a field is a one-dimensional string, which we shall consider as a chain of particles with mass m, connected by springs such that in equilibrium the chain is equidistant with spacing a. The particles can move along the chain, and the displacement of particle n with respect to the equilibrium position is called ϕ_n. Fixed, free or periodic boundary conditions can be imposed. The chain is described by the action

$$S = \frac{1}{2} \int_{t_\text{i}}^{t_\text{f}} \left\{ \sum_n \frac{1}{2} m \dot{\phi}_n^2(t) - A \left[\frac{\phi_{n+1}(t) - \phi_n(t)}{a} \right]^2 \right\} \mathrm{d}t. \tag{15.3}$$

A is a constant, and some special conditions are needed for the boundaries. In a quantum mechanical description we again use the path integral, which gives us the probability density for the chain to go from an initial configuration $\Phi^{\mathrm{i}} = \{\phi_n^{(\mathrm{i})}\}$ at time t_{i} to a final configuration $\Phi^{\mathrm{f}} = \{\phi_n^{(\mathrm{f})}\}$ at time t_{f} (note that $\Phi^{(\mathrm{i,f})}$ denotes a complete chain configuration):

$$\int [\mathcal{D}\Phi(t)]\, \mathrm{e}^{-\mathrm{i}S/\hbar} \tag{15.4}$$

where the path integral is over all possible configurations of the field Φ in the course of time, with fixed initial and final configurations.

We now want to formulate this problem in continuum space. To this end we replace the discrete index n by a continuous index x_1, and we replace the interaction term occurring in the summand by the continuous derivative:

$$S = \frac{1}{2}\int_{t_{\mathrm{i}}}^{t_{\mathrm{f}}} \mathrm{d}t \int \mathrm{d}x_1 \left\{ m\dot{\phi}(t, x_1)^2 - A\left[\frac{\partial\phi(t, x_1)}{\partial x_1}\right]^2 \right\}. \tag{15.5}$$

The field $\phi(t, x_1)$ can be thought of as a sheet whose shape is given as a height $\phi(t, x_1)$ above the $(1+1)$ dimensional space-time plane. In the path integral, we must sum over all possible shapes of the sheet, weighted by the factor $\mathrm{e}^{\mathrm{i}}S/\hbar$. The field can be rescaled at will, as it is integrated over in the path integral (this rescaling results in an overall prefactor), and the time and space units can be defined such as to give the time derivative a prefactor $1/c$ with respect to the spatial derivative (c is the speed of light), and we obtain

$$S = \int \mathrm{d}^2x\, \frac{1}{2}\partial_\mu\phi(x)\partial^\mu\phi(x), \tag{15.6}$$

where we have used x to denote the combined space-time coordinate $x \equiv (t, x_1) \equiv (x_0, x_1)$. From now on, we put $c \equiv \hbar \equiv 1$ and we use the Einstein summation convention according to which repeated indices are summed over. The partial space-time derivatives $\partial^\mu, \partial_\mu$ are denoted by:

$$\partial_\mu = \frac{\partial}{\partial x^\mu}, \quad \partial^\mu = \frac{\partial}{\partial x_\mu}. \tag{15.7}$$

Furthermore we use the Minkowski metric:

$$a_\mu a^\mu = a_0^2 - \mathbf{a}^2. \tag{15.8}$$

The fact that we choose $c \equiv \hbar \equiv 1$ leaves only one dimension for distances in space-time, and masses and energies. The dimension of inverse distance is equal to the energy dimension, which is in turn equal to the mass dimension.

Using partial integration, we can reformulate the action as

$$S = -\int \mathrm{d}^2x\, \frac{1}{2}\phi(x)\partial_\mu\partial^\mu\phi(x), \tag{15.9}$$

where we have assumed periodic boundary conditions in space and time (or vanishing fields at the integral boundaries, which are located at infinity) to neglect integrated terms.

If, apart from a coupling to its neighbours, each particle had also been coupled to an external harmonic potential $m^2\phi^2/2$, we would have obtained

$$S = -\int d^2x\, \frac{1}{2}\phi(x)(\partial_\mu\partial^\mu + m^2)\phi(x). \tag{15.10}$$

Note that the Euler–Lagrange equations for the field are

$$(\partial^\mu\partial_\mu + m^2)\phi(x) = 0; \tag{15.11}$$

which is recognised as the Klein–Gordon equation, the straightforward covariant generalisation of the Schrödinger equation.[1]

Quantum field theory is often used as a theory for describing particles. The derivation above started from a chain of particles, but these particles are merely used to formulate our quantum field theory, and they should not be confused with the physical particles which are described by the theory. The difference between the two can be understood as follows. Condensed matter physicists treat wave-like excitations of the chain (i.e. a one-dimensional 'crystal') as particles – they are called *phonons*. Note that the 'real' particles are the atoms of the crystal. In field theory, the only 'real' particles are the excitations of the field.

In fact, we can imagine that a wave-like excitation pervades our sheet, for example

$$\phi(t, x_1) = e^{ipx_1 - i\omega t} \tag{15.12}$$

(here x_1 denotes the spatial coordinate). This excitation carries a momentum p and an energy $\hbar\omega$, and it is considered as a particle. We might have various waves as a superposition running over the sheet: these correspond to as many particles.

Let us try to find the Hamiltonian corresponding to the field theory presented above (the following analysis is taken up in some detail in Problems 15.2 and 15.3). We do this by returning to the discretised version of the field theory. Let us first consider the ordinary quantum description of a single particle of mass 1, moving in one dimension in a potential $mx^2/2$. The Hamiltonian of this particle is given by

$$H = \frac{p^2}{2} + \frac{m}{2}x^2 \tag{15.13}$$

with $[p, x] = -i$. In the example of the chain we have a large number of such particles, but each particle can still be considered as moving in a harmonic potential, and after some calculation we find for the Hamiltonian:

$$H = \sum_n [\hat{\pi}_n^2 + (\hat{\phi}_n - \hat{\phi}_{n-1})^2 + m^2\hat{\phi}_n^2], \tag{15.14}$$

[1] The Klein–Gordon equation leads to important problems in ordinary quantum mechanics, such as a nonconserved probability density and an energy spectrum which is not bounded from below.

with

$$[\hat{\pi}_n, \hat{\phi}_l] = -\mathrm{i}\delta_{nl}. \tag{15.15}$$

The hats are put above ϕ and π to emphasise that they are now operators. The Hamiltonian can be diagonalised by first Fourier transforming and then applying operator methods familiar from ordinary quantum mechanics to it. The result is [2,6]

$$H = \frac{1}{2}\int_{-\pi}^{\pi} \mathrm{d}k\ \omega_k \hat{a}_k^{\dagger}\hat{a}_k \tag{15.16}$$

where $\hat{a}_k^{\dagger}$ is a creation operator: it creates a Fourier mode

$$\phi_n = \mathrm{e}^{ikn} \tag{15.17}$$

and $\hat{a}_k$ is the corresponding destruction or annihilation operator. In the ground state (the 'vacuum') there are no modes present and the annihilation operator acting on the ground state gives zero:

$$a_k|0\rangle = 0. \tag{15.18}$$

The Fourier modes represent energy quanta of energy $\hbar\omega$; the *number operator* $n_k = a_k^{\dagger}a_k$ acting on a state $|\psi\rangle$ counts the number of modes (quanta) with wave vector k, present in that state. The Hamiltonian (15.16) operator then adds all the energy quanta which are present in the state.

In fact, $\hat{a}_k$ is given in terms of the Fourier transforms of the $\hat{\phi}$ and $\hat{\pi}$ operators:

$$\hat{a}_k = \frac{1}{\sqrt{4\pi\omega_k}}[\omega_k\hat{\phi}_k + \mathrm{i}\hat{\pi}_k], \tag{15.19}$$

analogous to the definition of creation and annihilation operators for the harmonic oscillator. The frequency ω is related to k by

$$\omega_k = \omega_{-k} = \sqrt{m^2 + 2(1 - \cos k)}. \tag{15.20}$$

For small k we find the continuum limit:

$$\omega_k = \sqrt{m^2 + k^2} \tag{15.21}$$

which is the correct dispersion relation for a relativistic particle with mass m (in units where $c = \hbar = 1$).

We see that the Hamiltonian (15.16) of a particular configuration is simply given as the sum of the energies of a set of one-particle Hamiltonians (remember these particles are nothing but Fourier-mode excitations of the field): the particles do not interact. Therefore, the field theory considered so far is called *free field theory*. The eigenstates of the free field theory are

$$|k_1, \ldots, k_M\rangle = \hat{a}_{k_1}^{\dagger} \ldots \hat{a}_{k_M}^{\dagger}|0\rangle \tag{15.22}$$

for arbitrary M, which denotes the number of particles present. It is possible to have the same k_i occurring more than once in this state (with an appropriate normalisation factor): this means that there is more than one particle with the same momentum. The state $|0\rangle$ is the vacuum state: it corresponds to having no particles. The lowest energy above the vacuum energy is that corresponding to a single particle at rest ($k = 0$): the energy is equal to the mass. In Section 11.2 we have seen that for a statistical field theory the inverse of the lowest excitation energy is equal to the correlation length:

$$m \approx 1/(\xi a). \tag{15.23}$$

However, this holds for a statistical field theory where we do not have complex weight factors; these can be made real by an analytical continuation of the physics into imaginary time: $t \to it$ (see also Section 12.2.4). In that case the (continuous) action in d space-time dimensions reads

$$S = \int \mathrm{d}^d x \, \frac{1}{2}[\partial_\mu \phi \partial^\mu \phi + m^2 \phi^2] \tag{15.24}$$

where now

$$\partial_\mu \phi \partial^\mu \phi = (\nabla \phi)^2 + \left(\frac{\partial}{\partial t} \phi \right)^2 \tag{15.25}$$

i.e. the Minkowski metric has been replaced by the Euclidean metric. The matrix elements of the time-evolution operator now read $\exp(-S)$ instead of $\exp(-S/\mathrm{i})$ (for $\hbar \equiv 1$). We have now a means to determine the particle mass: simply by measuring the correlation length. In the free field theory, the inverse correlation length is equal to the mass parameter m in the Lagrangian, but if we add extra terms to the Lagrangian (see below) then the inverse correlation length (or the physical mass) is no longer equal to m.

It might seem that we have been a bit light-hearted in switching to the Euclidean field theory. Obviously, expectation values of physical quantities can be related for the Minkowski and Euclidean versions by an analytic continuation. In the numerical simulations we use the Euclidean metric to extract information concerning the Hamiltonian. This operator is the same in both metrics – only the time evolution and hence the Lagrangian change when going from one metric to the other. Euclidean field theory can therefore be considered merely as a trick to study the spectrum of a quantum Hamiltonian of a field theory which in reality lives in Minkowski space.

If we add another term to the Lagrangian:

$$S = \int \mathrm{d}^d x \left\{ \frac{1}{2} \phi(x)(-\partial_\mu \partial^\mu + m^2)\phi(x) + V[\phi(x)] \right\}, \tag{15.26}$$

where V is not quadratic (in that case it would simply contribute to the mass term), then interactions between the particles are introduced. Usually one considers

$$V = g\phi^4(x)/2 \tag{15.27}$$

and the Lagrangian describes the simplest interesting field theory for interacting particles, the *scalar ϕ^4 theory*. The name 'scalar' denotes that $\phi(x)$ is not a vector. Vector theories do exist, we shall encounter examples later on. When a potential is present, the energy is no longer a sum of one-particle energies: the particles interact.

We have mentioned the probability of going from a particular initial state to another (final) state as an example of the problems studied in field theory. Our experimental knowledge on particles is based on scattering experiments. This is a particular example of such a problem: given two particles with certain initial states, what are the probabilities for different resulting reaction products? That is, which particles do we have in the end and what are their momenta? In scalar field theory we have only one type of particle present. As we have seen in the first chapter of this book, experimental information on scattering processes is usually given in terms of scattering cross sections. These scattering cross sections can be calculated – they are related to an object called the *S-matrix*, which is defined as

$$S_{\rm fi} = \lim_{\substack{t_{\rm i}\to-\infty \\ t_{\rm f}\to\infty}} \langle\psi_{\rm f}|U(t_{\rm i},t_{\rm f})|\psi_{\rm i}\rangle. \tag{15.28}$$

Our initial state is one with a particular set of initial momenta as in (15.22), and similarly for the final state; $U(t_{\rm i},t_{\rm f})$ is the time-evolution operator,[2] and the states $\psi_{\rm i,f}$ usually contain a well-defined number of free particles with well-defined momenta (or positions, depending on the representation).

Scattering cross sections are expressed directly in terms of the S-matrix, and the latter is related to the *Green's functions* of the theory by the so-called Lehmann–Symanzik–Zimmermann relation, which can be found for example in Ref. [2]. These Green's functions depend on a set of positions $x_1,\ldots,x_n$ and are given by

$$\mathcal{G}(x_1,\ldots,x_n) = \int [\mathcal{D}\phi]\phi(x_1)\cdots\phi(x_n)\,{\rm e}^{-S/\hbar} \Big/ \int [\mathcal{D}\phi]\,{\rm e}^{-S/\hbar}. \tag{15.29}$$

Note that x_i is a space-time vector, the subscripts do not denote space-time components. The scattering cross sections are evaluated in the Euclidean metric; the Minkowskian quantities are obtained by analytical continuation: $t \to {\rm i}t$.

As the initial and final states in a scattering experiment are usually given by the particle momenta, we need the Fourier transform of the Green's function, defined as

$$\begin{aligned}
&\mathcal{G}(k_1,\ldots,k_n)\delta(k_1+\cdots+k_n)(2\pi)^d \\
&\quad = \int {\rm d}^d x_1 \ldots {\rm d}^d x_n\, {\rm e}^{{\rm i}k_1\cdot x_1+\cdots+{\rm i}k_n\cdot x_n}\mathcal{G}(x_1,\ldots,x_n).
\end{aligned} \tag{15.30}$$

[2] More precisely, the time-evolution operator is that for a theory with an interaction switched on at a time much later than $t_{\rm i}$ and switched off again at a time long before $t_{\rm f}$.

The d-dimensional delta-function reflects the energy–momentum conservation of the scattering process, which is related to the space-time translation invariance of the Green's function.

For the free field theory it is found that

$$\mathcal{G}(k, -k) = \frac{1}{k^2 + m^2} \tag{15.31}$$

which leads to the real-space form:

$$\mathcal{G}(x - x') = \frac{\mathrm{e}^{-|x-x'|m}}{|x - x'|^{\eta}}; \quad \text{large } |x - x'|, \tag{15.32}$$

with $\eta = (d - 1)/2$. We see that the Green's function has a finite correlation length $\xi = 1/m$. Higher-order correlation functions for the free field theory can be calculated using *Wick's theorem*: correlation functions with an odd number of ϕ-fields vanish, but if they contain an even number of fields, they can be written as a symmetric sum over products of pair-correlation functions, for example

$$\begin{aligned}\mathcal{G}(x_1, x_2, x_3, x_4) &= \langle \phi(x_1)\phi(x_2)\phi(x_3)\phi(x_4)\rangle \\ &= \langle \phi(x_1)\phi(x_2)\rangle\langle \phi(x_3)\phi(x_4)\rangle + \langle \phi(x_1)\phi(x_3)\rangle\langle \phi(x_2)\phi(x_4)\rangle \\ &\quad + \langle \phi(x_1)\phi(x_4)\rangle\langle \phi(x_2)\phi(x_3)\rangle. \end{aligned} \tag{15.33}$$

In fact, it is well known that for stochastic variables with a Gaussian distribution, all higher moments can be formulated similarly in terms of the second moment. Wick's theorem is a generalisation of this result.

15.3 Interacting fields and renormalisation

The free field theory can be solved analytically: all the Green's functions can be given in closed form. This is no longer the case when we are dealing with interacting fields. If we add, for example, a term $g\phi^4$ to the free field Lagrangian, the only way to proceed analytically is by performing a perturbative analysis in the coupling constant g. It turns out that this gives rise to rather difficult problems. The terms in the perturbation series involve integrals over some momentum coordinates, and these integrals diverge! Obviously our predictions for physical quantities must be finite numbers, so we seem to be in serious trouble. Since this occurs in most quantum field theories as soon as we introduce interactions, it is a fundamental problem which needs to be faced.

To get a handle on the divergences, one starts by controlling them in some suitable fashion. One way to do this is by cutting off the momentum integrations at some large but finite value Λ. This renders all the integrals occurring in the perturbation series finite, but physical quantities depend on the (arbitrary) cut-off Λ which is still unacceptable. Another way to remove the divergences is by formulating the theory on a discrete lattice. This is of course similar to cutting off momentum integrations, and the lattice constant a used is related to the momentum cut-off by

$$a \sim 1/\Lambda. \tag{15.34}$$

Such cut-off procedures are called *regularisations* of the field theory.

We must remove the unphysical cut-off dependence from the theory. The way to do this is to allow the coupling constant and mass constants of the theory to be dependent on cut-off and then require that the cut-off dependency of the Green's functions disappears.[3] There are infinitely many different Green's functions and it is not obvious that these can all be made independent of cut-off by adjusting only the three quantities m, g and ϕ. Theories for which this *is* possible are called *renormalisable*. The requirement that all terms in the perturbative series are merely *finite*, without a prescription for the actual values, leaves some arbitrariness in the values of field scaling, coupling constant and mass. We use experimental data to fix these quantities.

To be more specific, suppose we carry out the perturbation theory to some order. It turns out that the resulting two-point Green's function $\mathcal{G}(k, -k)$ assumes the form of the free-field correlation function (15.31) with a finite mass parameter plus some cut-off dependent terms. Removing the latter by choosing the various constants of the theory (m, g, scale of the field) in a suitable way, we are left with

$$\mathcal{G}(k, -k) = 1/(k^2 + m_{\mathrm{R}}^2) \tag{15.35}$$

where m_{R} is called the 'renormalised mass' – this is the physical mass which is accessible to experiment. This is not the mass which enters in the Lagrangian and which we have made cut-off dependent: the latter is called the 'bare mass', which we shall now denote by m_{B}. The value of the renormalised mass m_{R} is not fixed by the theory, as the cut-off removal is prescribed up to a constant. We use the experimental mass to 'calibrate' our theory by fixing m_{R}. In a similar fashion, we use the experimental coupling constant, which is related to the four-point Green's function, to fix a renormalised coupling constant g_{R} (see the next section).

The renormalisation procedure sounds rather weird, but it is certainly not some arbitrary *ad hoc* scheme. The aim is to find bare coupling constants and masses, such that the theory yields cut-off independent physical (renormalised) masses and couplings. Different regularisation schemes all lead to the same physics. We need as many experimental data as we have parameters of the theory to adjust, and having fixed these parameters we can predict an enormous amount of new data (in particular, all higher order Green's functions). Moreover, the requirement that the

[3] In addition to mass and coupling constant, the field is rescaled by some factor.

theory is renormalisable is quite restrictive. For example, only the ϕ^4 potential has this property; changing the ϕ^4 into a ϕ^6 destroys the renormalisability of the theory.

In computational physics we usually formulate the theory on a lattice. We then choose values for the bare mass and coupling constant and calculate various physical quantities in units of the lattice constant a (or its inverse). Comparison with experiment then tells us what the actual value of the lattice constant is. Therefore the procedure is somehow the reverse of that followed in ordinary renormalisation, although both are intimately related. In ordinary renormalisation theory we find the bare coupling constant and mass as a function of the cut-off from a comparison with experiment. In computational field theory we find the lattice constant as a function of the bare coupling constant from comparison with experimental data.

Let us consider an example. We take the Euclidean ϕ^4 action in dimension $d = 4$:

$$S = \frac{1}{2} \int d^4x \{[\partial_\mu \phi(x)][\partial^\mu \phi(x)] + m^2\phi^2(x) + g\phi^4(x)\} \tag{15.36}$$

and discretise this on the lattice, with a uniform lattice constant a. Lattice points are denoted by the four-index $n = (n_0, n_1, n_2, n_3)$. A lattice point n corresponds to the physical point $x = (an_0, an_1, an_2, an_3) = an$. The discretised lattice action reads

$$S_{\text{Lattice}} = \frac{1}{2} \sum_n a^4 \left\{ \sum_{\mu=0}^{3} \left[\frac{\phi(n+e_\mu) - \phi(n)}{a} \right]^2 + m^2\phi_n^2 + g\phi^4(n) \right\}. \tag{15.37}$$

We rescale the ϕ-field, the mass and the coupling constant according to

$$\phi(n) \to \phi(n)/a; \quad m \to m/a \quad \text{and} \quad g \to g, \tag{15.38}$$

to make the lattice action independent of the lattice constant a:

$$S_{\text{Lattice}} = \frac{1}{2} \sum_n \left\{ \sum_{\mu=0}^{3} [\phi(n+e_\mu) - \phi(n)]^2 + m^2\phi_n^2 + g\phi^4(n) \right\}. \tag{15.39}$$

Now we perform a Monte Carlo or another type of simulation for particular values of m and g. We can then 'measure' the correlation length in the simulation. This should be the inverse of the experimental mass, measured in units of the lattice constant a. Suppose we know this mass from experiment, then we can infer what the lattice constant is in real physical units.

Life is, however, not as simple as the procedure we have sketched suggests. The problem is that in the generic case, the correlation length is quite small in units of lattice constants. However, a lattice discretisation is only allowed if the lattice constant is *much smaller* than the typical length scale of the physical problem. Therefore, the correlation length should be an order of magnitude larger than the lattice constant. Only close to a critical point does the correlation length assume values much larger than the lattice constant. This means that our parameters m and g should be chosen close to a critical point. The ϕ^4 theory in $d = 4$ dimensions has one critical line [7], passing through the point $m = g = 0$, the massless free field case. Therefore, m and g should be chosen very close to this critical line in order for the lattice representation to be justifiable.

As the experimental mass of a particle is a fixed number, varying the lattice constant a forces us to vary g and m in such a way that the correlation length remains finite. Unfortunately this renders the use of finite size scaling techniques impossible: the system size L must always be larger than the correlation length: $a \ll \xi < L$.[4]

The fact that the lattice field theory is always close to a critical point implies that we will suffer from *critical slowing down*. Consider a Monte Carlo (MC) simulation of the field theory. We change the field at one lattice point at a time. At very high temperature, the field values at neighbouring sites are more or less independent, so after having performed as many MC attempts as there are lattice sites (one MCS), we have obtained a configuration which is more or less statistically independent from the previous one. If the temperature is close to the critical temperature, however, fields at neighbouring sites are strongly correlated, and if we attempt to change the field at a particular site, the coupling to its neighbours will hardly allow a significant change with respect to its previous value at that site. However, in order to arrive at a statistically independent configuration, we need to change the field over a volume of linear size equal to the correlation length. If that length is large, it will obviously take a very long time to change the whole region, so this problem gets worse when approaching the critical point. Critical slowing down is described by a dynamic critical exponent z which describes the divergence of the decay time τ of the dynamic correlation function (see Chapter 7, Eq. (7.73)):

$$\tau = \xi^z, \tag{15.40}$$

where ξ is the correlation length of the system.

In recent years, much research has aimed at finding simulation methods for reducing the critical time relaxation exponent. In the following section we shall describe a few straightforward methods developed for simulating quantum field theories, using the ϕ^4 scalar field theory in two dimensions as a testing model. In Section 15.5 we shall focus on methods aiming at reducing critical slowing down. We shall then also discuss methods devised for the Ising model and for a two-dimensional model with continuous degrees of freedom.

[4] In the case where physical particles are massless, so that the correlation length diverges, finite size scaling can be applied. Finite size scaling applications in massive particle field theories have, however, been proposed; see Ref. [8].

In the final sections of this chapter, simulation methods for gauge field theories (QED, QCD) will be discussed.

15.4 Algorithms for lattice field theories

We start by reviewing the scalar Euclidean ϕ^4 field theory in d dimensions in more detail. The continuum action is

$$S_\mathrm{E} = \frac{1}{2}\int \mathrm{d}^d x[\partial_\mu\phi(x)\partial^\mu\phi(x) + m^2\phi^2(x) + g\phi^4(x)] \tag{15.41}$$

(the subscript E stands for Euclidean). For $g = 0$, we have the free field theory, describing noninteracting spinless bosons. Performing a partial integration using Green's first identity, and assuming vanishing fields at infinity, we can rewrite the action as

$$S_\mathrm{E} = \frac{1}{2}\int \mathrm{d}^d x[-\phi(x)\partial_\mu\partial^\mu\phi(x) + m^2\phi^2(x) + g\phi^4(x)]. \tag{15.42}$$

The scalar field theory can be formulated on a lattice by replacing derivatives by finite differences. We can eliminate the dependence of the lattice action on the lattice constant by rescaling the field, mass and coupling constant according to

$$\hat{\phi}_n = a^{d/2-1}\phi(an); \quad \hat{m} = am; \quad \hat{g} = a^{4-d}g. \tag{15.43}$$

For the four-dimensional case, $d = 4$, we have already given these relations in the previous section. Later we shall concentrate on the two-dimensional case, for which the field ϕ is dimensionless. In terms of the rescaled quantities, the lattice action reads:

$$S_\mathrm{E}^\mathrm{Lattice} = \frac{1}{2}\sum_n\left[-\sum_\mu \hat{\phi}_n\hat{\phi}_{n+\mu} + (2d + \hat{m}^2)\hat{\phi}_n^2 + \hat{g}\hat{\phi}_n^4\right]. \tag{15.44}$$

The arguments n are vectors in d dimensions with integer coefficients and the sum over μ is over all positive and negative Cartesian directions. The action (15.44) is the form which we shall use throughout this section and it will henceforth be denoted as S. From now on we shall omit the carets from the quantities occurring in the action (15.44). As we shall simulate the field theory in the computer, we must make the lattice finite – the linear size is L.

We now describe the analytical solution of this lattice field theory for the case $g = 0$ (free field theory). The free field theory action is quadratic and can be written in the form

$$S_\mathrm{E} = \frac{1}{2}\sum_{nl}\phi_n K_{nl}\phi_l, \tag{15.45}$$

where

$$K_{nl} = (2d + m^2)\delta_{nl} - \sum_{\mu} \delta_{n,l+\mu}. \tag{15.46}$$

Defining Fourier-transformed fields as usual:

$$\phi_k = \sum_n \phi_n \, \mathrm{e}^{ik\cdot n}; \tag{15.47a}$$

$$\phi_n = \frac{1}{L^d} \sum_k \phi_k \, \mathrm{e}^{-ik\cdot n}, \tag{15.47b}$$

where n and l run from 0 to $L - 1$, periodic boundary conditions are assumed, and the components of k assume the values $2m\pi/L$, $m = 0, \ldots, L-1$. Then we can rewrite the free-field action as

$$S_{\mathrm{E}} = \frac{1}{2L^{2d}} \sum_k \phi_k K_{k,-k} \phi_{-k}, \tag{15.48}$$

as $K_{k,-k}$ are the only nonzero elements of the Fourier transform $K_{k,k'}$:

$$\begin{aligned} K_{k,k'} &= L^d \left[-\sum_{\mu} 2\cos(k_\mu) + (2d + m^2) \right] \delta_{k,-k'} \\ &= L^d \left[4 \sum_{\mu} \sin^2 \frac{k_\mu}{2} + m^2 \right] \delta_{k,-k'} \end{aligned} \tag{15.49}$$

where the sum is now only over the positive μ directions; k_μ is the μ-component of the Fourier wave vector k.

The partition function

$$\begin{aligned} Z &= \int [\mathcal{D}\phi_k] \exp\left(-\frac{1}{2L^{2d}} \sum_k \phi_k K_{k,-k} \phi_{-k} \right) \\ &= \int [\mathcal{D}\phi_k] \exp\left(-\frac{1}{2L^{2d}} \sum_k |\phi_k|^2 K_{k,-k} \right) \end{aligned} \tag{15.50}$$

(up to a normalisation factor) is now a product of simple Gaussian integrals, with the result ($N = L^d$):

$$Z = (2\pi N^2)^{N/2} / \prod_k \sqrt{K_{k,-k}} = (2\pi N^2)^{N/2} / \sqrt{\det K} = (2\pi N^2)^{N/2} \sqrt{\det(K^{-1})}. \tag{15.51}$$

The partition function appears as usual in the denominator of expressions for expectation values. We can calculate for example the two-point correlation or Green's function $\langle\phi_n\phi_l\rangle$. The Fourier transform of this correlation function can be found quite easily:

$$\langle\phi_n\phi_l\rangle = \frac{1}{L^{2d}}\sum_{k,k'}\langle\phi_k\phi_{k'}\rangle \mathrm{e}^{ik\cdot n}\mathrm{e}^{ik'\cdot l}; \tag{15.52a}$$

$$\langle\phi_k\phi_{k'}\rangle = \frac{L^{2d}}{K_{k,-k}}\delta_{k,-k'}. \tag{15.52b}$$

Taking the small-k limit in (15.49) and (15.52) leads to the form (15.31), as it should be. Taking $k = 0$, we find

$$\langle\phi_{k=0}^2\rangle = \left\langle\left(\sum_n \phi_n\right)^2\right\rangle = L^d\zeta/m_\mathrm{R}^2, \tag{15.53}$$

where the factor ζ on the right hand side represents the square of the scaling factor of the field – from (15.43), $\zeta = a^{d-2}$. This equation enables us therefore to determine ζ/m_R in a simulation simply by calculating the average value of $\langle\Phi^2\rangle$, $\Phi = \sum_n \phi_n$.

We have seen that according to Wick's theorem, the correlation functions to arbitrary order for free fields can always be written as sums of products of two-point correlation functions. If we switch on the ϕ^4 interaction, we will note deviations from this Gaussian behaviour to all higher orders. Renormalisation ideas suggest that it should be possible to express all higher order correlation functions in terms of second and fourth order correlation functions, if the arguments of the Green's function are not too close (that is, much farther apart than the cut-off a). The second order Green's functions are still described by the free field form (15.52), but with m in the kernel $K_{k,k'}$ being replaced by a *renormalised mass*, m_R. The deviations from the Gaussian behaviour manifest themselves in fourth and higher order correlation functions. Therefore a natural definition of the renormalised coupling constant g_R is

$$g_\mathrm{R} = \frac{\langle\Phi^4\rangle - 3\langle\Phi^2\rangle^2}{\langle\Phi^2\rangle^2} \tag{15.54}$$

where $\Phi = \sum_n \phi_n$.[5] Equations (15.53) and (15.54) are used below to measure the renormalised mass and coupling constant in a simulation.

15.4.1 Monte Carlo methods

The problem of calculating expectation values for the interacting scalar field theory is exactly equivalent to the problem of finding expectation values of a statistical field theory. Therefore we can apply the standard Monte Carlo algorithms of Chapter 10

[5] This renormalisation scheme corresponds to defining the renormalised coupling constant as the four-point one-particle irreducible (OPI) Green's function in tree approximation at momentum zero [2, 4, 5].

straightforwardly in order to sample field configurations with Boltzmann weight $\exp(-S[\phi])$. Starting point is the action (15.44). An obvious method is the Metropolis MC algorithm, in which lattice sites are visited at random or in lexicographic order, and at the selected site a change in the field is attempted by some random amount. The change in the field is taken from a random number generator either uniformly within some interval or according to a Gaussian distribution (with a suitable width). Then we calculate the change in the action due to this change in the field. The trial value of the field is then accepted as the field value in the next step according to the probability

$$P_{\text{Accept}} = \mathrm{e}^{-S[\phi_{\text{new}}]+S[\phi_{\text{old}}]} \tag{15.55}$$

where the exponent on the right hand side is the difference between the action of the new and old field at the selected site, keeping the field at the remaining sites fixed. If $P_{\text{Accept}} > 1$, then the new configuration is accepted.

In Chapter 10 we have already encountered another method which is more efficient as it reduces correlations between subsequent configurations: the heat-bath algorithm. In this algorithm, the trial value of the field is chosen independently of the previous value. Let us call $W_\phi[\phi_n]$ the Boltzmann factor $\mathrm{e}^{-S[\phi]}$ for a field which is fixed everywhere except at the site n. We generate a new field value at site n according to the probability distribution $W_\phi[\phi_n]$. This is equivalent to performing infinitely many Metropolis steps at the same site n successively. The new value of ϕ_n can be chosen in two ways: we can generate a trial value according to some distribution $\rho(\phi_n)$ and accept this value with probability proportional to $W_\phi[\phi_n]/\rho(\phi_n)$, or we can directly generate the new value with the required probability $W_\phi[\phi_n]$. The Gaussian free field model will serve to illustrate the last method.

Consider the action (15.44). If we vary ϕ_n, and keep all the remaining field values fixed, we see that the minimum of the action occurs for $\bar{\phi}_n = \sum_\mu \phi_{n+\mu}/(2d+m^2)$, where the sum is over *all* neighbouring points, i.e. for both positive and negative directions. The Boltzmann factor $W_\phi[\phi_n]$ as a function of ϕ_n for all remaining field values fixed is then a Gaussian centred around $\bar{\phi}_n$ and with a width $1/\sqrt{2d+m^2}$. Therefore, we generate a Gaussian random number r with a variance 1, and then we set the new field value according to

$$\phi_n = \bar{\phi}_n + r/\sqrt{2d+m^2}. \tag{15.56}$$

An advantage of this method is that no trial steps have to be rejected, which obviously improves the efficiency.

Unfortunately, this method is not feasible when a ϕ^4 interaction is present as we cannot generate random numbers with an $\exp(-x^4)$ distribution. Therefore we treat this term with an acceptance/rejection step as described above. This is done as follows. First we generate a 'provisional' value of the field ϕ_n according with a Gaussian distribution $\rho(\phi_n)$, according to the procedure just described. Then we accept this new field value with a probability $\exp(-g\phi_n^4/2)$. If g is not too large (and this will be the case in most of the examples given below), then the acceptance rate will still be reasonably close to 1 and not too many trial steps are rejected. If g is large, then a different procedure for generating the trial field value should be followed [9].

There is an intimate relation between the heat-bath method described here and the Gauss–Seidel method for finding the solution of the Poisson equation (see Appendix A7.2). In the Gauss–Seidel method, the sites are visited in lexicographic order (the same can be done in the heat-bath method), and ϕ_n is set equal to $\bar{\phi}_n$ without adding a Gaussian random number to it. In Appendix A7.2 the problem of slow convergence of the numerical solution of the Poisson problem will be addressed: it turns out that the relaxation time, measured in sweeps over the entire lattice, scales as the square of the linear lattice size. The amount of computer time involved in one lattice sweep also scales linearly with the lattice volume, so the total time needed to obtain results within a certain level of accuracy scales with the volume squared. Because of this power-law scaling behaviour of the standard Poisson solvers, one might call this problem 'critical': the relaxation time scales with the system size in a way similar to a system subject to critical fluctuations. The relation between Poisson solvers and free field theory leads us to apply clever methods for solving Poisson's equation to the problem of generating configurations with a probability density $\exp(-S[\phi])$. In Appendix A, successive over-relaxation (SOR), the use of fast Fourier transforms (FFT), and the multigrid method are mentioned, and we shall see that all of these methods have their counterpart in Monte Carlo.

Successive over-relaxation is a method for increasing the efficiency of the Gauss–Seidel method. The idea behind this method is that if we update the sites in lexicographic order, half of the neighbours of the site being updated have already been updated and the other half are still to be treated. In SOR, a compensation is built in for the fact that half of the neighbouring sites have not yet been updated. Site n is being updated according to

$$\phi_n^{\text{new}} = \phi_n^{\text{old}} + \omega(\bar{\phi}_n - \phi_n^{\text{old}}). \tag{15.57}$$

It can be shown that the optimal value for ω is close to 2: in that case the relaxation time, which scales as L^2 (measured in lattice sweeps) in the Gauss–Seidel method is reduced to L (see Appendix A7.2 and Ref. [10]). Adler has shown that the relaxation time for a Monte Carlo algorithm where a Gaussian random number is added to ϕ^{new} [11]:

$$\phi_n^{\text{new}} \to \phi_n^{\text{new}} + \sqrt{\omega(2-\omega)}r/\sqrt{2d+m^2}, \tag{15.58}$$

is equal to that of the corresponding Poisson solver algorithm, that is, the relaxation time will now scale as L. We should obviously check that the SOR method still satisfies detailed balance. This is left as an exercise (Problem 15.5).

The SOR method works well for models with quadratic interactions. Including a ϕ^4 term renders the method less suitable (see however Ref. [12]). Fortunately, the physically more interesting gauge theories which will be discussed later in this chapter are quadratic. A problem with this method is that the optimal value of the over-relaxation parameter ω, which is 2 in the case of the scalar free field theory, is not known in general and has to be determined empirically.

We have encountered the most straightforward methods for simulating the scalar field theory. Most of these methods can easily be generalised to more complicated field theories. Before discussing different methods, we shall analyse the behaviour of the methods presented so far.

15.4.2 The MC algorithms: implementation and results

The implementation of the algorithms presented in the previous sections is straightforward. The reader is encouraged to try coding a few and to check the results given below.

To obtain the renormalised mass and coupling constant, Eqs. (15.53) and (15.54) can be used. However, it is nice to measure the full two-point correlation function. This can be found by sampling this function for pairs of points which lie in the same column or in the same row. To obtain better statistics, nonhorizontal and nonvertical pairs can be taken into account as well. To this end we construct a histogram, corresponding to equidistant intervals of the pair separation. We keep two arrays in the program, one for the value of the correlation function, and the other for the average distance r corresponding to each histogram column. At regular time intervals we perform a loop over all pairs of lattice sites. For each pair we calculate the closest distance within the periodic boundary conditions according to the minimum image convention. Suppose this distance is r_{ij}. We calculate to which column this value corresponds, and add the product of the field values at the two sites $\phi_i\phi_j$ to the correlation function array. Furthermore we add r_{ij} to the average distance array. After completing the loop over the pairs, we divide the values in the correlation function array and in the average distance array by the number of pairs that contributed to these values. The final histogram must contain the time averages of the correlation function values thus evaluated, and this should be written to a file.

We can now check whether the scalar ϕ^4 theory is renormalisable. This means that if we discretise the continuum field theory using finer and finer grids, the resulting physics should remain unchanged. Equation (15.43) tells us how we should change the various parameters of the theory when changing the grid constant. We now

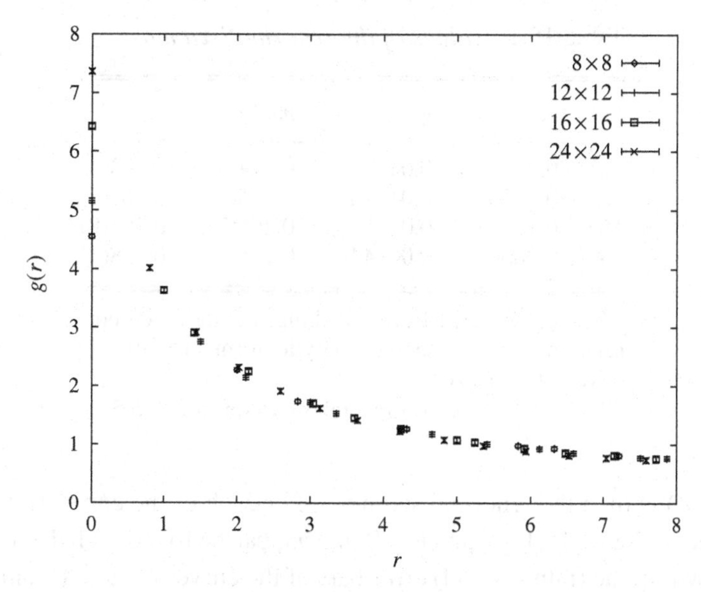

Figure 15.1. The correlation function of the interacting scalar field theory for various lattice sizes. The mass and coupling parameters for the different lattice sizes have been scaled such as to keep the physical lattice size constant. The x-axis has been scaled accordingly. The values have been determined using the histogram method described in the text.

present results for a field theory which on an 8×8 lattice is fixed by the parameter values $m = 0.2$ and $g = 0.04$. Note that both m and g should be close to the critical line (which passes through $m = 0, g = 0$) to obtain long correlation lengths justifying the discretisation. According to (15.43) we use $m = 0.1$ and $g = 0.01$ on a 16×16 lattice, etc. The results are obtained using a heat-bath algorithm using 30 000 steps (8×8) to 100 000 steps (24×24). Figure 15.1 shows the correlation functions for various lattice sizes, obtained using the heat-bath algorithm. The horizontal axis is scaled proportional to the lattice constant (which is obviously twice as large for an 8×8 lattice as for a 16×16 lattice). The vertical axis is scaled for each lattice size in order to obtain the best collapse of the various curves. It is seen that for length scales beyond the lattice constant, scaling is satisfied very well. Only on very small length scales do differences show up, as is to be expected.

The correlation functions obtained from the simulations can be compared with the analytic form, which can be obtained by Fourier transforming

$$\mathcal{G}_{k,-k} = \frac{Z}{4\sum_\mu \sin(k_\mu/2) + m_\mathrm{R}^2} \tag{15.59}$$

(see Section 15.3 and Eq. (15.53)). The parameter m_R that gives the best match to the correlation function obtained in the simulation (with an optimal value of the

Table 15.1. *Values of the renormalised mass.*

L	m	g	m_R [a]	m_R [b]
8	0.2	0.04	0.374(5)	0.363(4)
12	0.1333	0.01778	0.265(5)	0.265(7)
16	0.1	0.01	0.205(7)	0.204(8)
24	0.06667	0.004444	0.138(4)	0.138(4)

[a] Values obtained from matching the measured correlation function to the analytic form (15.59), for different grid sizes.
[b] Values obtained from matching to formula (15.53).

parameter Z) is then the renormalised mass. For each of the correlation functions represented in Figure 15.1, parameters Z and m_R can be found such that the analytic form lies within the (rather small) error bars of the curves obtained from the simulation. In Table 15.1, the values of the renormalised mass as determined using this procedure are compared with those obtained using (15.53). Excellent agreement is found. It is seen that for the larger lattices, the renormalised mass is more or less inversely proportional to the linear lattice size. The physical mass, however, should be independent of the lattice size. This is because masses are expressed in units of the inverse lattice constant, and the lattice constant is obviously inversely proportional to the linear lattice size L if the lattice represents the same physical volume for different sizes.

The determination of the renormalised coupling constant is difficult. We use Eq. (15.54), but this is subject to large statistical errors. The reason is that the result is the difference of two nearly equal quantities, and this difference is subject to the (absolute) error of these two quantities – hence the *relative* error of the difference becomes very large. The renormalised coupling constant should not depend on the lattice size for large sizes, as it is dimensionless. Table 15.2 shows the results. The errors are large and it is difficult to check whether the renormalised coupling constant remains the same, although the data are compatible with a coupling constant settling at a size-independent value of $g \approx 0.11$ for large lattices.

15.4.3 Molecular dynamics

How can we use molecular dynamics for a field theory formulated on a lattice, which has no intrinsic dynamics?[6] The point is that we assign a *fictitious* momentum degree

[6] The dynamics are here defined in terms of the evolution of the field configuration and not in terms of the time axis of the lattice.

Table 15.2. *Values of the renormalised mass and coupling constant.*

L	m	g	m_R	m_R
8	0.05	0.1	0.456(3)	0.20(4)
12	0.03333	0.04444	0.332(3)	0.18(7)
16	0.025	0.025	0.260(3)	0.13(5)
24	0.016667	0.01111	0.184(2)	0.12(6)
32	0.0125	0.00625	0.1466(7)	0.10(4)

Values obtained from (15.53) and (15.54), for different lattice sizes. Various methods (see later sections) are used.

of freedom to the field at each site (the Car–Parrinello method is based on a similar trick; see Chapter 9). As we have seen in Chapters 7 and 8, for a dynamical system the probability distribution of the coordinate part can be obtained by integrating out the momentum degrees of freedom, and this should be the desired distribution $e^{-S[\phi]}$. Therefore, we simply add a kinetic energy to the action in order to obtain a classical Hamiltonian (which should not be confused with the field theory's quantum Hamiltonian):

$$\mathcal{H}_{\text{class}} = \sum_n \frac{p_n^2}{2} + S[\phi]. \tag{15.60}$$

Integrating out the momentum degrees of freedom of the classical partition function, we obtain the Boltzmann factor of the action back again (up to a constant):

$$\int [\mathcal{D}p_n]\, e^{-\mathcal{H}_{\text{class}}[p_n,\phi_n]} = \text{Const} \cdot e^{-S[\phi]}. \tag{15.61}$$

The Andersen method

The classical Hamiltonian gives rise to classical equations of motion which can be solved numerically. These equations yield trajectories with constant energy (up to numerical errors). But we want trajectories representing the canonical ensemble, and in Chapter 8 we studied various methods for obtaining these. In the Andersen refreshed molecular dynamics method, the momenta are refreshed every now and then by replacing them with a new value drawn from a random generator with a Maxwell distribution. In field theories, one often replaces all momenta at the same time with regular intervals between these updates (the method is usually denoted as the *hybrid method*). That is, first the equations of motion are solved for a number of time steps, and then *all* momenta are replaced by new values from the Maxwell random generator. Then the equations of motion are solved again for a number of steps and so on [13–16]. The exact dynamical trajectory plus the momentum update can be considered as one step in a Markov chain whose invariant distribution is the canonical one. We do not obtain the exact dynamical trajectory but a numerical approximation to it, and the errors made can be corrected for in a procedure which will be discussed in the next section. In Chapter 8 we mentioned that the Andersen method does indeed lead to the canonical distribution of the coordinate part. We shall prove this statement now.

First, it is useful to consider 'symmetric' Markov steps: these consist of an integration of the equations of motion over a time $\Delta t/2$, then a momentum refreshing, and then again an integration over a time $\Delta t/2$. Such a step can schematically be represented as follows:

$$\Phi_\text{i}, P_\text{i} \stackrel{\Delta t/2}{\longrightarrow} \Phi_\text{m}, P_\text{m} \quad \text{Refresh} : \Phi_\text{m}, P_\text{rm} \stackrel{\Delta t/2}{\longrightarrow} \Phi_\text{f}, P_\text{f}.$$

Energy conservation during the microcanonical trajectories implies

$$H(\Phi_\text{i}, P_\text{i}) = H(\Phi_\text{m}, P_\text{m}); \tag{15.62a}$$

$$H(\Phi_\text{m}, P_\text{rm}) = H(\Phi_\text{f}, P_\text{f}). \tag{15.62b}$$

The steps occur with a probability

$$T(\Phi_\text{i}, P_\text{i} \to \Phi_\text{f}, P_\text{f}) = \delta(\Phi_\text{f} - \Phi_\text{microcanonical}) \exp(-P_\text{rm}^2/2), \tag{15.63}$$

where the delta-function indicates that Φ_f is uniquely determined by the microcanonical trajectory, which depends of course on the initial configuration $\Phi_\text{i}, P_\text{i}$, on the refreshed momentum P_rm, and on the integration time (which is fixed).

The trial steps are ergodic, and the master equation of the Markov chain

$$\sum_{\Phi',P'} \rho(\Phi, P) T(\Phi, P \to \Phi', P') = \sum_{\Phi',P'} \rho(\Phi', P') T(\Phi', P' \to \Phi, P) \tag{15.64}$$

will have an invariant solution. However, the detailed balance condition for this chain is slightly modified. The reason is that we need to use the time-reversibility of the microcanonical trajectories, but this reversibility can only be used when we reverse the momenta. Therefore we have

$$\frac{\rho(\Phi', P')}{\rho(\Phi, P)} = \frac{T(\Phi, P \to \Phi', P')}{T(\Phi', -P' \to \Phi, -P)} \tag{15.65}$$

(note that $\rho(\Phi, P) = \exp[-P^2/2 - S(\Phi)]$ is symmetric with respect to $P \leftrightarrow -P$). The transition probability in the denominator of the right hand side corresponds to the step in the numerator traversed backward in time (see the above diagram of a symmetric trial step). The fraction on the right hand side is clearly equal to $\exp[(P_\text{mr}^2 - P_\text{m}^2)/2]$. Using Eq. (15.62), it then follows that the invariant distribution is given as $\rho(\Phi, P) = \exp[-P^2/2 - S(\Phi)]$.

The procedure can be implemented straightforwardly. The equations of motion in the leap-frog form read

$$p_n(t+h/2) = p_n(t-h/2) + hF_n(t); \tag{15.66a}$$

$$\phi_n(t+h) = \phi_n(t) + hp_n(t+h/2), \tag{15.66b}$$

where the force $F_n(t)$ is given by

$$F_n(t) = \sum_\mu [\phi_{n+\mu}(t)] - (2d+m^2)\phi_n(t) - 2g\phi_n^3(t), \tag{15.67}$$

where $\sum_\mu$ denotes a sum over *all* neighbours. Refreshing the momenta should be carried out with some care. We refresh the momenta at the time steps t for which the field values ϕ_n are evaluated in the leap-frog algorithm. However, we need the momenta in the leap-frog algorithm at times precisely halfway between these times. Therefore, after the momentum update, we must propagate the momenta over half a time step h:

$$p_n(t+h/2) = p_n(t) + hF_n(t)/2, \tag{15.68}$$

and then the integration can proceed again.

This method contains a tunable parameter: the refresh rate. It turns out that the efficiency has a broad optimum as a function of the refresh rate [16]. Having around 50 steps between the all-momenta updates with a time step $h = 0.1$ is quite efficient. If we refresh after every time step, the system will essentially carry out a random walk in phase space as the small steps made between two refreshings are nearly linear, and the direction taken after each refreshment step is approximately random. If we let the system follow its microcanonical trajectory for a longer time, it will first go to a state which is relatively uncorrelated with respect to the previous one. The momentum refreshings then ensure that the canonical distribution is satisfied; however, the fact that the energy is not conserved, but may change by an amount (on average) of $\mathcal{O}(h^2)$ during the MD trajectory, causes deviations from the canonical distribution of the same order of magnitude.

This method is obviously more efficient than refreshing after each step, as the distance covered by a random walker increases only as the square root of the number of steps made. If we wait too long between two refreshings, the simulation samples only a few different energy surfaces which is not representative for the canonical ensemble. The optimum refresh rate is therefore approximately equal to the correlation time of the microcanonical system.

The Metropolis-improved MD method

The leap-frog algorithm introduces systematic errors into the numerical simulation, and the distribution will therefore not sample to the exact one. That is not necessarily a bad thing: we can always write the distribution which is sampled by the MD trajectory as $\exp(-S_D[\phi])$, where the action S_D differs by some power of h from the continuum action [17]:

$$S_D[\phi] = S[\phi] + \mathcal{O}(h^k) \tag{15.69}$$

for some positive k. The discrete action may renormalise to a continuum limit with slightly different parameters, but as the behaviour of the model is calibrated in the end by matching calculated physical quantities to the experimental values, our model with discrete time step might still describe the correct continuum limit. Indeed, Batrouni *et al.* [17] show that the discrete time action in the Langevin limit (i.e. the case in which the momenta are refreshed at every time step, see below) is a viable one at least to first order in h. A problem is that the difference between the discrete and the continuum actions makes it difficult to compare the results of the MD simulation with an MC simulation of the same system with the same values of the parameters.

The discretisation error can be corrected for in exactly the same way as in the variational and diffusion quantum Monte Carlo method (see, for example, the discussion near the end of Section 12.2.5). The idea is to consider the leap-frog MD trajectories as a trial step in a Monte Carlo simulation. The energy before and after this trial step is calculated, and the trial step is accepted with probability $\exp(-\mathcal{H}_{\text{class}}^{\text{new}} + \mathcal{H}_{\text{class}}^{\text{old}})$ (note that $\mathcal{H}_{\text{class}}$ is a classical 'energy' which includes kinetic and potential energy). If it is rejected, the momenta are refreshed once more and the MD sequence starts again. This method combines the Andersen refreshment steps with microcanonical trajectory acceptance/rejection steps. In the previous subsection we saw that the refreshment step satisfies a modified detailed balance condition which ensures the correct (canonical) distribution. Now we show that the microcanonical trajectories plus the acceptance/rejection step, satisfy a similar detailed balance with a canonical invariant distribution.

We write the transition probability in the form of a trial step probability $\omega_{\Phi,P;\Phi',P'}$ and a Metropolis acceptance/rejection probability $A_{\Phi,P;\Phi',P'}$. The trial step probability is determined by the numerical leap-frog trajectory and hence is nonzero only for initial and final values compatible with the leap-frog trajectory. Time-reversibility of the leap-frog algorithm implies that

$$\omega_{\Phi,P;\Phi',P'} = \omega_{\Phi',-P';\Phi,-P}. \tag{15.70}$$

The acceptance probability is given as usual by

$$A_{\Phi,P;\Phi',P'} = \min\{1, \exp[\mathcal{H}_{\text{class}}(\Phi, P) - \mathcal{H}_{\text{class}}(\Phi', P')]\}. \tag{15.71}$$

The acceptance step is invariant under $P \leftrightarrow -P$ as the momenta occur only with an even power in the Hamiltonian.

From this, it follows immediately that the modified detailed balance condition holds:

$$\frac{\rho(\Phi', P')}{\rho(\Phi, P)} = \frac{\omega_{\Phi,P;\Phi',P'} A_{\Phi,P;\Phi',P'}}{\omega_{\Phi',-P';\Phi,-P} A_{\Phi',P';\Phi,P}} = \exp[\mathcal{H}_{\text{class}}(\Phi, P) - \mathcal{H}_{\text{class}}(\Phi', P')]. \tag{15.72}$$

We see that without momentum refreshings, the canonical distribution is a stationary distribution of the Markov process. However, for small time steps in the leap-frog algorithm, the changes in the classical Hamiltonian are very small, and convergence will be extremely slow. That is the reason why these steps are combined with momentum refreshings, which are compatible with a canonical invariant distribution too, but which cause more drastic changes in the energy. This method is usually called the *hybrid Monte Carlo method* [18].

The important advantage of this Metropolis-improved MD method is that the time step of the leap-frog algorithm can be stretched considerably before the acceptance rate of the Metropolis step drops too low. This causes the correlation time for the 'microcanonical' part, measured in time steps, to be reduced considerably. We have put the quotes around 'microcanonical' because the energy is not conserved very well with a large time step. If the time step is taken too large, the Verlet method becomes unstable (see Appendix A7.1). In practice one often chooses the time step such that the acceptance rate becomes about 80%, which is on the safe side, but still not too far from this instability limit.

It should be noted that the acceptance rate depends on the difference in the total energy of the system before and after the trial step. The total energy is an extensive quantity: it scales linearly with the volume. This implies that discrete time step errors will increase with volume. To see how strong this increase is [19], we note that the error in coordinates and momenta after many steps in the leap-frog/Verlet algorithm is of order h^2 per degree of freedom (see Problem A3). This is then the deviation in the energy over the microcanonical trajectory, and we shall denote this deviation ΔH_{MD}. The energy differences obtained including the acceptance/rejection step are called ΔH_{MC}; that is, if the trajectory is accepted, ΔH_{MC} is equal to ΔH_{MD}, but if the step is rejected, $\Delta H_{\text{MC}} = 0$. If ΔH_{MD} averaged over all possible initial configurations was to vanish, the acceptance rate would always be larger than 0.5, as we would have as many positive as negative energy differences (assuming that the positive differences are on average not much smaller or larger than the negative ones), and all steps with negative and some of the steps with positive energy difference would be accepted. However, the net effect of the acceptance/rejection step is to *lower* the energy, and since the energies measured with this step included remain on average stationary, $\langle \Delta H_{\text{MD}} \rangle$ must be positive. The fact that the energy remains stationary implies that $\langle \Delta H_{\text{MC}} \rangle = 0$ and this leads to an equation for

$\langle \Delta H_{\text{MD}} \rangle$:

$$\langle \Delta H_{\text{MC}} \rangle = 0 = \frac{\sum_{\{\Phi,P\}} P_{\text{acc}}(\Delta H_{\text{MD}}) \Delta H_{\text{MD}}}{\sum_{\{\Phi,P\}}}. \tag{15.73}$$

Using $P_{\text{acc}} = \min[1, \exp(-\Delta H_{\text{MD}})]$, and expanding the exponent, we find

$$0 = \langle \Delta H_{\text{MD}} \rangle - \langle \theta(\Delta H_{\text{MD}})(\Delta H_{\text{MD}})^2 \rangle \tag{15.74}$$

where the theta-function restricts ΔH_{MD} to be positive: $\theta(x) = 0$ for $x < 0$ and 1 for $x > 0$. We see that $\langle \Delta H_{\text{MD}} \rangle$ is indeed positive and we furthermore conclude that $\langle \Delta H_{\text{MD}} \rangle = \mathcal{O}(h^4 V)$ for of the order of V degrees of freedom. For the average acceptance value we then find

$$\langle P_{\text{acc}} \rangle = \langle \min(1, \mathrm{e}^{-\Delta H_{\text{MD}}}) \rangle \approx \mathrm{e}^{-\langle \Delta H_{\text{MD}} \rangle} = \mathrm{e}^{-\alpha h^4 V} \tag{15.75}$$

where α is of order one. Therefore, in order to keep the acceptance rate constant when increasing the volume, we must decrease h according to $V^{-1/4}$, which implies a very favourable scaling.

The Langevin method

Refreshing the momenta after every MD step leads to a Langevin-type algorithm. Langevin algorithms have been discussed in Sections 8.8 and 12.2.4. In Section 8.8 we applied a Gaussian random force at each time step. In the present case we assign Gaussian random values to the momenta at each time step as in Section 12.2.4. In that case the two steps of the leap-frog algorithm can be merged into one, leading to the algorithm:

$$\phi_n(t+h) = \phi_n(t) + \frac{h^2}{2} F_n(t) + h R_n(t). \tag{15.76}$$

The random numbers R_n are drawn from a Gaussian distribution with a width of 1 – it is a Gaussian momentum, not a force (hence the pre-factor h instead of h^2). Comparing the present approach with the Fokker–Planck equation discussed in Section 12.2.4, we see that when we take ρ of the Fokker–Planck equation (12.42) equal to $\exp(-S[\phi])$, Eq. (12.46) reduces to (15.76) if we put $\Delta t = h^2$. This then shows immediately that the Langevin algorithm guarantees sampling of the configurations weighted according to the Boltzmann distribution.

An advantage of this algorithm is the memory saving resulting from the momenta not being required in this algorithm but, as explained in the previous section, the method is not very efficient because the system performs a random walk through phase space. The reason we treat this method as a separate one here is that there exists an improved version of it which is quite efficient [17]. We shall discuss this algorithm in Section 15.5.5.

Implementation

All the MD algorithms described can be implemented without difficulty. The details of the leap-frog and Langevin algorithm can be found in Chapter 8. Moreover, calculation of the correlation function is described in Section 15.4.2. The programs can all be tested using the results presented in that section.

15.5 Reducing critical slowing down

As we have already seen in Section 7.3.2, systems close to the critical point suffer from *critical slowing down*: the correlation time τ diverges as a power of the correlation length. This renders the calculation of the critical properties very difficult, which is unfortunate as these properties are usually of great interest: we have seen in this chapter that lattice field theories must be close to a critical point in order to give a good description of the continuum theory. In statistical mechanics, critical properties are very often studied to identify the critical exponents for various universality classes.

For most systems and methods, the critical exponent z, defined by

$$\tau = \xi^z, \tag{15.77}$$

is close to 2. For Gaussian models, the value $z = 2$ of the critical exponent is related to the convergence time of the simple Poisson solvers, which can indeed be shown to be equal to 2 (see Appendix A7.2). The value of 2 is related to the fact that the vast majority of algorithms used for simulating field theories are *local*, in the sense that only a small number of degrees of freedom (mostly one) is changed at a step. For systems characterised by domain walls (e.g. the Ising model), the exponent 2 can be guessed by a crude heuristic argument. The major changes in the system configuration take place at the domain walls, as it takes less energy to move a wall than to create new domains. In one sweep, the sites neighbouring a domain wall have on average been selected once. The domain wall will therefore move over a distance 1. But its motion has a random walk nature. To change the configuration substantially, the domain wall must move over a distance ξ, and for a random walk this will take of the order of ξ^2 steps.

Over the past ten years or so, several methods have been developed for reducing the correlation time exponent z. Some of these methods are tailored for specific classes of models, such as the Ising and other discrete spin models. All methods are variations of either the Metropolis method, or of one of the MD methods discussed in the previous section. In this section we shall analyse the different methods in some detail. Some methods are more relevant to statistical mechanics, such as those which are suitable exclusively for Potts models, of which the Ising model is a special case, but we treat them in this chapter because the ideas behind the methods developed for field theories and statistical mechanics are very similar.

As the local character of the standard algorithms seems to be responsible for the critical slowing down present in the standard methods, the idea common to the methods to be discussed is to update the stochastic variables *globally*, that is, all in one step. How this is done can vary strongly from one method to the other, but the underlying principle is the same for all of them.

15.5.1 The Swendsen–Wang method

We start with the cluster method of Swendsen and Wang (SW) [20], and explain their method for the Ising model in d dimensions, discussed already in Section 7.2 and 10.3.1. The SW method is a Monte Carlo method in which the links, rather than the sites, of the Ising lattice are scanned in lexicographic order. For each link there are two possibilities:

1. The two spins connected by this link are opposite. In that case the interaction between these spins is deleted.
2. The two spins connected by the link are equal. In that case we either delete the bond or 'freeze' it, which means that the interaction is made infinitely strong. Deletion occurs with probability $p_{\rm d} = {\rm e}^{-2\beta J}$ and freezing with probability $p_{\rm f} = 1 - p_{\rm d}$.

This process continues until we have visited every link. In the end we are left with a model in which all bonds are either deleted or 'frozen', that is, their interaction strength is either 0 or ∞. This means that the lattice is split up in a set of disjoint clusters and within each cluster the spins are all equal. This model is simulated trivially by assigning at random a new spin value $+$ or $-$ to each cluster. Then the original Ising bonds are restored and the process starts again, and so on.

Of course we must show that the method does indeed satisfy the detailed balance condition. Before doing so, we note that the method does indeed lead to a reduction of the dynamic critical exponent z of the two-dimensional Ising model to the value 0.35 presented by Swendsen and Wang,[7] which is obviously an important improvement with respect to the value $z = 2.125$ for the standard MC algorithm. The reason the method works is that flipping blocks involves flipping many spins in one step. In fact, the Ising (or more generally, the Potts model) can be mapped on a cluster model, where the distribution of clusters is the same as for the SW clusters [22]. The average linear cluster size is proportional to the correlation length, and this will diverge at the phase transition. Therefore, the closer we are to the critical point, the larger the clusters are and the efficiency will increase accordingly.

[7] From a careful analysis, Wolff has found exponents $z = 0.2$ and $z = 0.27$ for the 2D Ising model, depending on the physical quantity considered [21].

In the SW method, any configuration can be reached from any other configuration, because there is a finite probability that the lattice is partitioned into L^d single-spin clusters which are then given values $+$ and $-$ at random. Furthermore it is clear that the method does not generate periodicities in time and it remains to be shown that the SW method satisfies detailed balance. We do this by induction. We show that the freezing/deleting process for some arbitrary bond does not destroy detailed balance, so carrying out this process for every bond in succession does not do so either.

Every time we delete or freeze a particular bond ij we change the Hamiltonian of the system:

$$H \to H_0 + V_{ij}. \tag{15.78}$$

H is the Hamiltonian in which the bond is purely Ising-like. H_0 is the Hamiltonian without the interaction of the bond ij, and V_{ij} represents an interaction between the spins at i and j which is either ∞ (in the case of freezing) or 0 (if the bond has been deleted); the remaining bonds do not change. We write the detailed balance condition for two arbitrary configurations S and S' for the system with Hamiltonian H as follows:

$$\frac{T(S \to S')}{T(S' \to S)} = \mathrm{e}^{-\beta[H(S')-H(S)]} = \frac{T_0(S \to S')}{T_0(S' \to S)} \mathrm{e}^{-\beta J(s'_i s'_j - s_i s_j)}, \tag{15.79}$$

where T_0 is the transition probability for the Hamiltonian H_0 and we have explicitly split off the contribution from the bond ij. In the last equality we have used the detailed balance condition for the system with Hamiltonian H_0.

In the SW algorithm, we must decide for a bond ij whether we delete or freeze this bond. The transition probability of the system after this step can be written as

$$T(S \to S') = T_\mathrm{f}(S \to S')P_\mathrm{f}(S) + T_\mathrm{d}(S \to S')P_\mathrm{d}(S). \tag{15.80}$$

Here, $P_{\mathrm{d,f}}(S)$ is the probability that we delete (d) or freeze (f) the bond ij in spin configuration S. $T_\mathrm{d}(S \to S')$ is the transition probability with a deleted bond, and therefore $T_\mathrm{d} = T_0$, and T_f is the transition probability when the bond is frozen. The latter is equal to T_0 in the case that the spins s_i, s_j are equal in both S and S' and it is zero in the case that they are unequal in S' (they must be equal in S, otherwise they could not have been frozen).

Let us consider the detailed balance condition for the transition probability in (15.80):

$$\frac{T(S \to S')}{T(S' \to S)} = \frac{T_\mathrm{f}(S \to S')P_\mathrm{f}(S) + T_\mathrm{d}(S \to S')P_\mathrm{d}(S)}{T_\mathrm{f}(S' \to S)P_\mathrm{f}(S') + T_\mathrm{d}(S' \to S)P_\mathrm{d}(S')} = \mathrm{e}^{-\beta[H(S')-H(S)]}. \tag{15.81}$$

We show that this condition is indeed satisfied, using (15.79). Let us assume that s_i and s_j are equal in both S and S'. In that case $P_\mathrm{f}(S) = 1 - \exp(-2\beta J)$ and

$P_{\rm d}(S) = \exp(-2\beta J)$ respectively and we have

$$\frac{T(S \to S')}{T(S' \to S)} = \frac{T_{\rm f}(S \to S')[1 - \exp(-2\beta J)] + T_{\rm 0}(S \to S')\exp(-2\beta J)}{T_{\rm f}(S' \to S)[1 - \exp(-2\beta J)] + T_{\rm 0}(S' \to S)\exp(-2\beta J)}. \tag{15.82}$$

Since the pair $s_i s_j$ is equal in both S and S', the transition probability $T_{\rm f} = T_{\rm 0}$ and we see that (15.79) does hold for the transition probability after the SW step.

If before and after the step the spins s_i and s_j are unequal, $T_{\rm f}$ vanishes in both numerator and denominator, and it is clear that (15.79) holds in this case too. Suppose $s_i = s_j$ in S and that the corresponding pair s'_i, s'_j in S' is unequal. In that case we have

$$\frac{T(S \to S')}{T(S' \to S)} = \frac{T_{\rm f}(S \to S')[1 - \exp(-2\beta J)] + T_{\rm 0}(S \to S')\exp(-2\beta J)}{T_{\rm 0}(S' \to S)}. \tag{15.83}$$

The denominator in the right hand side contains only the term with a deleted bond because starting from the configuration S' in which s'_i and s'_j are unequal, we can only delete the bond. The transition probability $T_{\rm f}(S \to S')$ occurring in the numerator obviously vanishes, and we see that for this case detailed balance, Eq. (15.79), is again satisfied.

It is instructive to code the SW method. First a sweep through the lattice is performed in which all the bonds are either frozen or deleted. This poses no difficulties. Then the clusters must be identified. This can be done using 'back-tracking' and is most conveniently coded recursively. It works as follows. A routine BackTrack(x, y) is written, which scans the cluster containing the site given by the Cartesian (integer) components (x, y). Start at site (x, y) and check whether this site has already been visited. If this is not the case, leave a flag there as a mark that the cluster site has now been visted, and scan the neighbouring sites in a similar way by recursive calls. The resulting routine looks more or less as follows (for $d = 2$):

```
ROUTINE BackTrack(x, y)
    IF NOT Visited (x,y) THEN
        Mark (x, y) as being visited;
        IF (Frozen(x, y, x + 1, y)) THEN
            BackTrack(x + 1, y);
        IF (Frozen(x, y, x, y + 1)) THEN
            BackTrack(x, y + 1);
        IF (Frozen(x, y, x − 1, y)) THEN
            BackTrack(x − 1, y);
```

```
        IF (Frozen(x, y, x, y − 1)) THEN
            BackTrack(x, y − 1);
    END IF
END BackTrack.
```

Frozen($x1, y1, x2, y2$) is a Boolean function which returns TRUE if the nearest neighbour bond between ($x1, y1$) and ($x2, y2$) is frozen and FALSE otherwise. Periodic boundaries should be implemented using a modulo operator or function, and it is convenient to decide before scanning the cluster whether it is going to be flipped and, if yes, to do so during the recursive scanning (alongside putting the Visited flag). On exit, the cluster is scanned and all its sites marked as visited. In a sweep through all values i and j, all clusters will be found in this way; note that the computer time needed to scan a cluster in the back-track algorithm scales linearly with the cluster volume (area).

Another algorithm for detecting all the clusters in the system is that of Hoshen and Kopelman. This algorithm does not use recursion. It scales linearly with the lattice size and it is more efficient than back-tracking (30–50%) but it is somewhat more difficult to code. Details can be found in the literature [23].

The time scaling exponent z can be determined from the simulations. Note that the time correlation of the magnetisation is useless for this purpose as the clusters are set to arbitrary spin values after each sweep, so that the magnetisation correlation time is always of order 1. Therefore, we consider the time correlation function of the (unsubtracted) susceptibility per site. This is defined as

$$\chi = \frac{1}{L^{2d}}\left\langle\left(\sum_i s_i\right)^2\right\rangle. \tag{15.84}$$

Its time correlation function is

$$C_\chi(k) = \frac{\sum_{n=1}^{N} \chi_{n+k}\chi_n}{\sum_{n=1}^{N} \chi_n^2} \tag{15.85}$$

where the indices n and k are 'time' indices, measured in MC steps per spin.

The susceptibility can be determined directly from the lattice configuration after each step using (15.84), but it is possible to obtain a better estimate by realising that when the system is divided up into clusters c of area N_c, the average value of χ is given by

$$\chi = \frac{1}{L^{2d}}\left\langle\left(\sum_c N_c s_c\right)^2\right\rangle \tag{15.86}$$

where s_c is the spin value of cluster c. We can write this as

$$\chi = \frac{1}{L^{2d}} \left\langle \sum_c N_c s_c \sum_{c'} N_{c'} s_{c'} \right\rangle, \tag{15.87}$$

and by summing over $s_c = \pm 1$ for all the clusters we obtain the average of this value for all possible cluster-spin configurations. Then only the terms $c = c'$ survive and we are left with

$$\chi = \frac{1}{L^{2d}} \left\langle \sum_c N_c^2 \right\rangle. \tag{15.88}$$

This is the so-called 'improved estimator' for the unsubtracted susceptibility [24, 25]. This estimator gives better results because the average over all possible cluster-spin configurations is built into it.

The correlation time can be determined from the values of χ at the subsequent MC steps in the usual way (see Section 7.4). For a detailed analysis of the dynamic exponent for various cases, see Ref. [21].

PROGRAMMING EXERCISE

Code the SW algorithm for the two-dimensional Ising model. Determine the time relaxation exponent and compare this with the value found for the single-spin flip algorithm.

Wolff has carried out the cluster algorithm in the microcanonical ensemble [24], using a microcanical MC method proposed by Creutz [26]. He fixed the number of unequal bonds to half the number of total bonds and found considerable improvement in the efficiency.

15.5.2 Wolff's single cluster algorithm

Wolff has proposed a different cluster method for eliminating critical slowing down for Potts spin systems, and an extension of this method and the SW method to a special class of continuous spin models [27]. We start with Wolff's modification of the SW method for the Ising model. In Wolff's method, at each step a *single* cluster is generated, as opposed to the SW model in which the entire lattice is partitioned into clusters. The single cluster is constructed according to the same rules as the SW clusters. We start with a randomly chosen spin and consider its neighbours. Only equal neighbours can be linked to the cluster by freezing the bonds between them – this happens with probability $1 - e^{-2\beta J}$. The cluster is extended in this way until no more spins are added to it. Then all the spins in the cluster are flipped.

It will be clear that the cluster generated in this way is a SW cluster and therefore the method satisfies detailed balance. The difference between the two methods is that the selection of the cluster to be grown can be viewed as throwing a dart at the lattice [21] with equal probability of hitting any of the sites – the probability of hitting a SW cluster (SWC) of size N_{SWC} is N_{SWC}/L^d (for d dimensions), thereby favouring the growth of large clusters. Because of this preference for large clusters it is expected that the single cluster version changes the configuration on average more drastically in the same amount of time and that statistically independent configurations are generated in fewer steps. This turns out to be the case in the 3D Ising model, where the single cluster algorithm yields time correlation exponents 0.28 or 0.14 (depending on the correlation function studied) as opposed to 0.5 for the SW algorithm. For the 2D Ising model only a small increase in efficiency has been measured [21].

It is convenient to generate the clusters in a recursive way. Each MC step consists of selecting a random site (*Location*) on the lattice. *ClusterSpin* is minus the spin at this location (the spins are flipped when added to the cluster). The algorithm is then as follows:

```
ROUTINE GrowCluster(Location, ClusterSpin):
  Flip Spin at Location;
  Mark Spin as being added to Cluster;
  IF right-hand neighbour not yet added THEN
    TryAdd(RightNeighbour, ClusterSpin);
  ...Similar for other neighbours...
END GrowCluster.

ROUTINE TryAdd (Location, ClusterSpin):
  Determine Spin at Location;
  IF Spin opposite to ClusterSpin THEN
    IF Random number < 1 − e^{−2J} THEN
      GrowCluster(Location, ClusterSpin);
    END IF;
  END IF;
END TryAdd.
```

Measuring correlation times requires some care, as a step in the Wolff algorithm consists of flipping one cluster instead of (on average) half of the total number of spins in the lattice as in the SW algorithm. The correlation time $\bar{\tau}_{\mathrm{W}}$ expressed in numbers of single cluster flips must therefore be translated into the single cluster correlation time τ_W expressed in SW time steps:

$$\tau_W = \bar{\tau}_W \frac{\langle N_{1C} \rangle}{L^d}. \tag{15.89}$$

The average single cluster size $\langle N_{1C} \rangle$ occurring on the right hand side is the improved estimator for the (unsubtracted) susceptibility per site:

$$\langle N_{1C} \rangle = \left\langle \frac{N_{SWC}}{L^d} N_{SWC} \right\rangle = \chi. \tag{15.90}$$

This formula can be understood by realising that the probability of generating a SW cluster of size N_{SWC} in the single cluster algorithm is equal to N_{SWC}/L^d. To evaluate the average cluster size we must multiply this probability with N_{SWC} and take the expectation value of the result.

PROGRAMMING EXERCISE

Implement Wolff's single cluster algorithm and compare the results with the SW algorithm – see also Ref. [21].

In many statistical spin systems and lattice field theories, the spins are not discrete but they assume continuous values. Wolff's algorithm was formulated for a particular class of such models, the O(N) models. These models consist of spins, which are N-dimensional unit vectors, on a lattice. Neighbouring spins $\mathbf{s}_i, \mathbf{s}_j$ interact – the interaction is proportional to the scalar product $\mathbf{s}_i \cdot \mathbf{s}_j$. An example which is relevant to many experimental systems (superfluid and superconducting materials, arrays of coupled Josephson junctions ...) is the O(2) or XY model, in which the spins are unit vectors $\mathbf{s}_i$ lying in a plane, so that they can be characterised by their angle θ_i with the x-axis, $0 \leq \theta_i < 2\pi$.

For simulations, it is important that relevant excitations in $O(N)$ models are smooth variations of the spin orientation over the lattice (except near isolated points – see below). This implies that changing the value of a single angle θ_i somewhere in the lattice by an amount of order 1 is likely to lead to an improbable configuration; hence the acceptance rate for such a trial change is on average very small. The only way of achieving reasonable acceptance rates for changing a single spin is by considerably restricting the variation in the orientation of the spin allowed in a trial step. This, however, will reduce the efficiency because many MC steps are then needed to arrive at statistically independent configurations. A straightforward generalisation of the SW or single cluster algorithm in which all spins in some cluster are reversed is bound to fail for the same reason, as this destroys the smoothness of the variation of the spins at the cluster boundary.

Wolff has proposed a method in which the spins in a cluster are modified to an extent depending on their orientation [27]. It turns out that his method can be

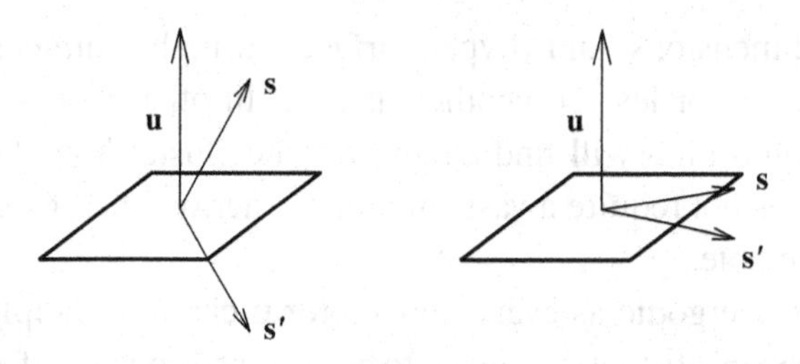

Figure 15.2. Spin flips in the Wolff algorithm for the $O(3)$ model.

formulated as an embedding of an Ising model into an $O(N)$ model [28]. First a random unit vector **u** is chosen. Every spin $\mathbf{s}_i$ is then split into two components: the component along **u** and that perpendicular to **u**:

$$\mathbf{s}_i^{\parallel} = (\mathbf{s}_i \cdot \mathbf{u})\mathbf{u} \tag{15.91a}$$

$$\mathbf{s}_i^{\perp} = \mathbf{s}_i - \mathbf{s}_i^{\parallel}. \tag{15.91b}$$

We keep $\mathbf{s}_i^{\perp}$ and $|\mathbf{s}_i^{\parallel}|$ fixed. The only freedom left for the $O(N)$ spins is to flip their parallel component:

$$\mathbf{s}_i = \mathbf{s}_i^{\perp} + \epsilon_i |\mathbf{s}_i^{\parallel}|\mathbf{u}, \quad \epsilon_i = \pm 1. \tag{15.92}$$

A flip in the sign ϵ_i corresponds to a reflection with respect to the hyperplane perpendicular to **u** (see Figure 15.2). The interaction of the model with the restriction on the fluctuations that only flips of the parallel components are allowed, can now be described entirely in terms of the ϵ_i:

$$\mathcal{H}[\epsilon_i] = \sum_{\langle ij \rangle} J_{ij}\epsilon_i\epsilon_j \tag{15.93a}$$

$$J_{ij} = J|\mathbf{s}_i^{\parallel}|\,|\mathbf{s}_j^{\parallel}|. \tag{15.93b}$$

This Ising Hamiltonian is now simulated using the single cluster or the SW algorithm. After choosing the unit vector **u**, we calculate the ϵ_i for the actual orientations of the spins and then we allow for reflections of the $\mathbf{s}_i$ (that is, for spin flips in the ϵ_i system).

This method is more efficient than the standard single spin-update method because large clusters of spins are flipped at the same time. But why is the acceptance rate for such a cluster update not exceedingly small? The point is that the amount by which a spin changes, depends on its orientation (see Figure 15.2): for a spin more or less perpendicular to **u**, the change in orientation is small. This translates itself into the coupling J_{ij} being small for spins $\mathbf{s}_i$, $\mathbf{s}_j$ nearly perpendicular to **u**. For spins parallel to **u**, the coupling constant J_{ij} is large and these spins will almost certainly be frozen to the same cluster. The cluster boundaries will be the curves (in two dimensions, and (hyper)surfaces in higher dimensions) on which the spins $\mathbf{s}_i$ are more or less perpendicular to $\mathbf{u}$. In other words, if we provide a direction $\mathbf{u}$, the algorithm will find an appropriate cluster boundary such that the spin reflection does not require a vast amount of energy. Therefore, the acceptance rate is still appreciable.

The procedure is ergodic as every unit vector $\mathbf{u}$ can in principle be chosen and there is a finite probability that the cluster to be swapped consists of a single spin. The isolated spin-update method is therefore included in the new algorithm. Detailed balance is satisfied because the new Ising Hamiltonian (15.93) is exactly equivalent to the original $O(N)$ Hamiltonian under the restriction that only the reflection steps described are allowed in the latter.

The implementation of the method for the two-dimensional *XY* model proceeds along the same lines as described above for the Ising model. Apart from selecting a random location from which the cluster will be grown, a unit vector $\mathbf{u}$ must be chosen, simply by specifying its angle with the X-axis. Each spin is flipped when added to the cluster. If we try to add a new spin $\mathbf{s}_i$ to the cluster (in routine 'TryAdd'), we need the spin value of its neighbour $\mathbf{s}_j$ in the cluster. The freezing probability P_f is then calculated as

$$P_\text{f} = 1 - \exp[-2J(\mathbf{s}_i \cdot \mathbf{u})(\mathbf{s}_j \cdot \mathbf{u})] \tag{15.94}$$

(note that the cluster spin $\mathbf{s}_i$ has already been flipped, in contrast to $\mathbf{s}_j$). The spin $\mathbf{s}_j$ is then added to the cluster with this freezing probability. Instead of considering continuous angles between 0 and 2π, it is possible to consider an *n-state clock model*, which is an *XY* model with the restriction that the angles allowed for the spins assume the values $2j\pi/n$, $j = 0, \ldots, n-1$ [29]; see also Section 12.6. At normal accuracies, the discretisation of the angles will not be noticed for n greater than about 20. The cosines and sines needed in the program will then assume n different values only and these can be calculated in the beginning of the program and stored in an array.

PROGRAMMING EXERCISE

Write a Monte Carlo simulation program for the *XY* model, using Wolff's cluster algorithm. If the program works correctly, it should be possible to detect the occurrence of the so-called Kosterlitz-Thouless phase transition (note that this occurs only in two dimensions). This is a transition which has been observed experimentally in helium-4 films [30] and Josephson junction arrays [31]. We shall briefly describe the behaviour of the *XY* model.

Apart from excitations which are smooth throughout the lattice – *spin-waves* – the *XY* model exhibits so-called *vortex excitations*. A pair of vortices is shown in

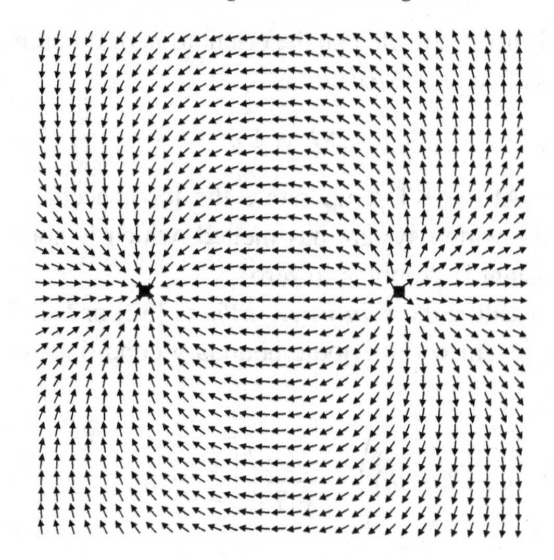

Figure 15.3. A pair of vortices, one with positive and one with negative vorticity.

Figure 15.3. The vortices can be assigned a *vorticity* which is roughly the 'winding' number of the spins along a closed path around the vortex – the vorticity assumes values 2π, -2π (for higher temperatures 4π etc. can also occur). An isolated vortex requires an amount of energy which scales logarithmically with the lattice size and is thus impossible in a large lattice at finite temperature. However, vortex pairs of opposite vorticity are possible; their energy depends on the distance between the vortices and is equal to the Coulomb interaction (which is proportional to $\ln R$ for two dimensions) for separations R larger than the lattice constant. The system can only contain equal numbers of positive and negative vortices. At low temperatures the vortices occur in bound pairs of opposite vorticity (to be compared with electrical dipoles), and the spin-waves dominate the behaviour of the model in this phase. It turns out that the correlations are long-ranged:

$$\langle \theta_i - \theta_j \rangle \propto \frac{1}{|\mathbf{r}_i - \mathbf{r}_j|^{x_{\mathrm{T}}}}, \tag{15.95}$$

for large separation $|\mathbf{r}_i - \mathbf{r}_j|$, with a critical exponent x_{T} which varies with temperature. At the KT transition, the dipole pairs unbind and beyond the transition temperature T_{KT} we have a fluid of freely moving vortices (to be compared with a plasma).

Imagine you have an *XY* lattice with fixed boundary conditions: the spins have orientation $\theta = 0$ on the left hand side of the lattice and you have a handle which enables you to set the fixed value δ of the spin orientation of the rightmost column of *XY*-spins. Turning the handle from $\delta = 0$ at low temperatures, you will feel a resistance as if it is attached to a spring. This is due to a nonvanishing amount of free energy which is needed to change the orientation of the spins on the right hand column. This excess free energy scales as

$$\Delta F \propto \Gamma \delta^2 \tag{15.96}$$

for small angles δ. At the KT temperature the force needed to pull the handle drops to zero, as the vortex system has melted, which is noticeable through the proportionality constant Γ dropping to zero.

The quantity Γ is called the *spin-wave stiffness* [32] or *helicity modulus*. It can be calculated in a system with periodic boundary conditions using the following formula [32]:

$$\Gamma = \frac{J}{2L^2}\left\{\left\langle \sum_{\langle ij\rangle} \cos(\theta_i - \theta_j)\right\rangle - \frac{J}{k_B T}\left\langle\left[\sum_i \sin(\theta_i - \theta_{i+\hat{e}_x})\right]^2\right\rangle \right.$$
$$\left. - \frac{J}{k_B T}\left\langle\left[\sum_i \sin(\theta_i - \theta_{i+\hat{e}_y})\right]^2\right\rangle\right\}. \tag{15.97}$$

From the Kosterlitz-Thouless theory [29,33,34] it follows that the helicity modulus has a universal value $\Gamma = 2k_B T_{KT}/\pi$ at the KT transition. The drop to zero is smooth for finite lattices but it becomes steeper and steeper with increasing lattice size. Figure 15.4 shows Γ/J as a function of $k_B T/J$. The line $\Gamma/J = 2(k_B T/J)/\pi$ is also shown and it is seen that the intersection of the helicity modulus curve with this line gives the value from which the helicity modulus drops to zero. You can check your program by reproducing this graph.

Edwards and Sokal have found that for the *XY* model the dynamic critical exponent in the low-temperature phase is zero or almost zero [28].

15.5.3 Geometric cluster algorithms

The cluster algorithms described so far flips or rotates spins on a lattice. In fact, Wolff's version of the algorithm for the *XY* model boils down to flipping an Ising spin. Cluster algorithms strongly rely on a reflection symmetry of the Hamiltonian: flipping all spins does not affect the Hamiltonian. This is the reason that a simple distinction can be made for pairs which may be frozen and those which certainly will not. The same holds for the q-state Potts model: there we have a symmetry under permutation of all spin values. Another way of looking at this is that flipping a large cluster in an Ising model with a magnetic field, yields an energy loss or gain proportional to the volume (surface in two dimensions), which leads to very low acceptance rates in phases with a majority spin. Hence cluster algorithms will be less efficient for such systems.

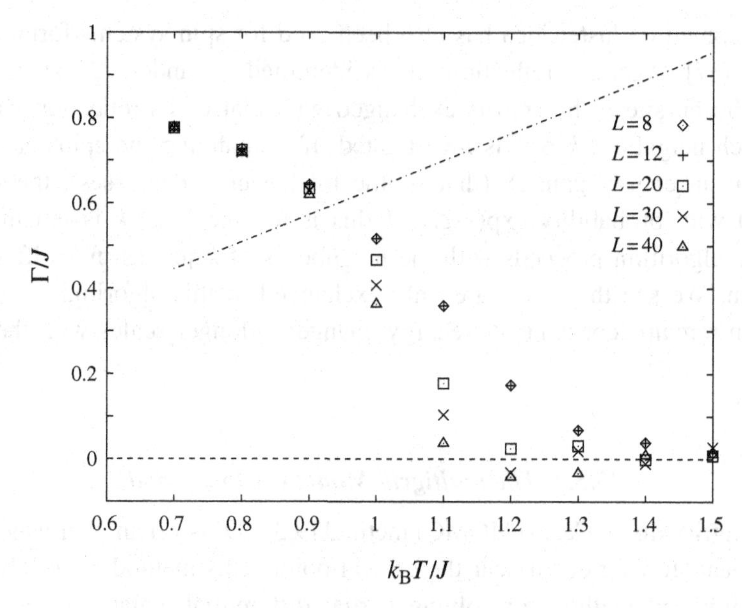

Figure 15.4. The helicity modulus in units of the coupling constant J of the XY model vs. the inverse coupling constant in units of $k_B T$. The intersection of the helicity modulus curves with the straight line gives the value from which the helicity modulus starts dropping to zero.

If we switch on a magnetic field, the spin-flip symmetry is broken and the cluster algorithm can no longer be used. Another problem is that it is not immediately clear whether and how cluster ideas may be generalised to systems that are not formulated on a lattice. A step towards a solution was made by Dress and Krauth [35], who used *geometric* symmetries to formulate a cluster algorithm for particles moving in a continuum. Usually, a reflection of the particles with respect to a point chosen randomly in the system is used. The interaction between these particles is considered to be a hard-core interaction, but long(er)-ranged interaction may also be present. The problem with the algorithm is that the decision to displace a particle is made based on the hard-core part. Other interactions are included in the acceptance criterion, and this leads to many rejections. This problem was solved by Liu and Luijten who take all interactions into account [36]. They start by identifying a random reflection point and then choose an initial particle at random. This and other particles having nonnegligible interaction with the first particles are then candidates to be reflected. This is done one by one, taking all interactions into account, and each time a reflection of a particle would result in a decrease Δ of the energy, the particle is reflected with probability $\exp(-|\Delta|)$. If the energy increases, the particle is not reflected. This algorithm promises to be valuable for the analysis of dense liquids.

The geometric cluster idea has also been used for spin systems formulated on a lattice [37]. Again, a reflection site is identified at random. Next, for a randomly chosen site i, the spin is exchanged with that of its reflection partner i'. Then each neighbour k of i is investigated. If exchanging the spins at k and k' results in an energy gain Δ (that is, the total energy decreases), the move is accepted with probability $\exp(-\Delta)$; if this is not the case, k is left unaltered. Then the algorithm proceeds with the neighbours of k just as in Wolff's cluster algorithm. We see that spins are only exchanged in this algorithm, so that the total spin remains constant: the energy change no longer scales with the cluster volume.

15.5.4 The multigrid Monte Carlo method

The multigrid Monte Carlo (MGMC) method [9,38,39] is yet another way of reducing critical slowing down near the critical point. This method is closely related to the multigrid method for solving partial differential equations described in Appendix A7.2, and readers not familiar with this method should go through that section first; see also Problem A7.

Multigrid ideas can be used to devise a new Monte Carlo algorithm which reduces critical slowing down by moving to coarser and coarser grids and updating these in an MC procedure with a restricted form of the Hamiltonian.

To be specific, let us start from a grid at level l; a field configuration on this grid is called ψ. The Hamiltonian on this grid is called $H_l[\psi]$. The coarse grid is the grid at level $l-1$, and configurations on this coarse grid are denoted by ϕ. Now consider the prolongation operation $P_{l,l-1}$ described in Appendix A7.2, which maps a configuration ϕ on the coarse grid to a configuration ψ on the fine grid by copying the value of ϕ on the coarse grid to its four nearest neighbours on the fine grid:

$$P_{l,l-1} : \phi \to \psi; \tag{15.98a}$$

$$\psi(2i+\mu, 2j+\nu) = \phi(i,j), \tag{15.98b}$$

where μ and ν are ± 1. We now consider a restricted Hamiltonian $H_{l-1}[\delta\phi]$, which is a function of the coarse grid configuration $\delta\phi$, depending on the fine grid configuration ψ which is kept fixed:

$$H_{l-1}[\delta\phi] = H_l[\psi + P_{l,l-1}(\delta\phi)]. \tag{15.99}$$

We perform a few MC iterations on this restricted Hamiltonian and then we go to the coarser grid at level $l-2$. This process is continued until the lattice consists of a single site, and then we go back by copying the fields on the coarser grid sites to the neighbouring sites of the finer grids, after which we again perform a few MC steps, and so on.

The algorithm reads, in recursive form:

ROUTINE MultiGridMC(l, ψ H_l)
  Perform a few MC sweeps: $\psi \rightarrow \psi'$;
  IF (l >0) THEN
    Calculate the form of the Hamiltonian
      on the coarse grid: $H_{l-1}[\delta\phi] = H_l[\psi' + P_{l,l-1}(\delta\phi)]$;
    Set $\delta\phi$ equal zero;
    MultiGridMC($l - 1, \delta\phi, H_{l-1}$);
  ENDIF;
  $\psi'' = \psi' + P_{l,l-1}\delta\phi$;
  Perform a few MC sweeps: $\psi'' \rightarrow \psi'''$;
END MultiGridMC.

The close relation to the multigrid algorithm for solving Poisson's equation, given in Appendix A7.2, is obvious.

The MC sweeps consist of a few Metropolis or heat-bath iterations on the fine grid field ψ. This step is ergodic as the heat bath and Metropolis update is ergodic. Note that the coarse grid update in itself is not ergodic because of the restriction imposed on fine grid changes (equal changes for groups of four spins) – the Metropolis or heat-bath updates are essential for this property.

We should also check that the algorithm satisfies detailed balance. Again, the Metropolis or heat-bath sweeps respect detailed balance. The detailed balance requirement for the coarse grid update is checked recursively. A full MCMG step satisfies detailed balance if the coarse grid update satisfies detailed balance. But the coarse grid update satisfies detailed balance if the coarser grid update satisfies detailed balance. This argument is repeated until we reach the coarsest level ($l = 1$). But at this level we perform only a few MC sweeps, which certainly satisfy detailed balance. Therefore, the full algorithm satisfies detailed balance.

There is one step which needs to be worked out for each particular field theory: constructing the coarse Hamiltonian H_{l-1} from the fine one, H_l. We do not know a priori whether new interactions, not present in the fine Hamiltonian, will be generated when constructing the coarse one. This often turns out to be the case. As an example, consider the scalar interacting ϕ^4 field theory. The terms ϕ^2 and ϕ^4 generate linear and third powers in ϕ when going to the coarser grid. Moreover, the Gaussian coupling $(\phi_n - \phi_{n+\mu})^2$ generates a term $\phi_n - \phi_{n+\mu}$. Therefore, the

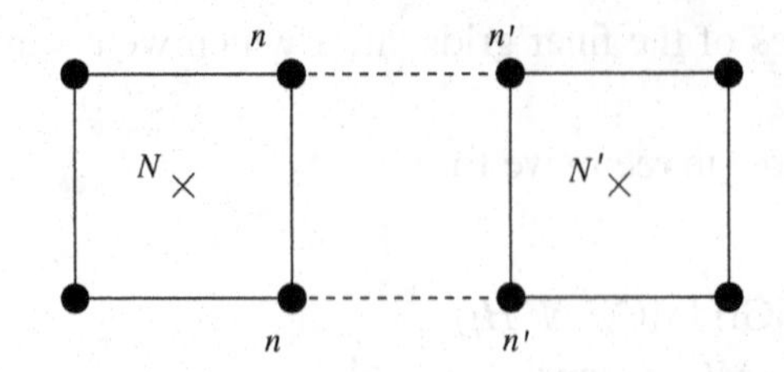

Figure 15.5. Two neighbouring blocks on a fine lattice with coarse lattice sites N and N'.

Hamiltonians which we must consider have the form:

$$\mathcal{H}[\psi] = \frac{1}{2}\left\{\sum_{\langle nn'\rangle}\left[J_{n,n'}(\psi_n - \psi_{n'})^2 + \sum_{\mu} K_{n,n'}(\psi_n - \psi_{n'})\right]\right.$$
$$\left. + \sum_n \left[L_n\psi_n + M_n\psi^2 + T_n\psi_n^3 + G_n\psi_n^4\right]\right\}. \tag{15.100}$$

Restricting this Hamiltonian to a coarser grid leads to new values for the coupling constants.

We denote the sites of the new grid by N, N'. Furthermore, $\sum_{nn'|NN'}$ denotes a sum over sets n, n' of neighbouring points, which belong to different neighbouring blocks of four sites belonging to N and N' respectively as in Figure 15.5. Finally, $\sum_{n|N}$ denotes a sum over the fine grid sites n belonging to the block N. With this notation, the new coupling constants on the coarse grid can be written in terms of those on the fine grid:

$$J_{NN'} = \sum_{nn'|NN'} J_{nn'}; \quad K_{NN'} = \sum_{nn'|NN'} [K_{nn'} + 2J_{nn'}(\psi_n - \psi_{n'})];$$
$$L_N = \sum_{n|N}(L_n + 2M_n\psi_n + 3T_n\psi_n^2 + 4G_n\psi_n^3); \tag{15.101}$$
$$M_N = \sum_{n|N}(M_n + 3T_n\psi_n + 6G_n\psi_n^2);$$
$$T_N = \sum_{n|N}(T_n + 4G_n\psi_n); \quad G_N = \sum_{n|N} G_n.$$

With this transformation, the MCMG method can be implemented straightforwardly. It can be shown that critical slowing down is completely eliminated for Gaussian type actions, so it will work very well for the ϕ^4 theory close to the Gaussian fixed point. However, the ϕ^4 theory has more than one critical point in two dimensions. One of these points has Ising character: for this point, the coefficient of the quadratic term is negative, whereas the coefficient g of ϕ^4 is positive. This means that the field has two opposite minima. For this model, the MCMG method does not perform very well. This can be explained using a heuristic argument. Suppose the field assumes values very close to $+1$ or -1. Consider a block of four spins on the fine lattice which belong to the same coarse lattice site. Adding a nonzero amount, ϕ_N, to these four spins will only be accepted if they are either all equal to -1, so that an amount of 2 can be added, or if they are all equal to $+1$ so that we can subtract 2 from each of them. The probability that all spins in a block have equal values becomes smaller and smaller when coarsening the lattice more and more, so the efficiency of the MCMG method is degraded severely for this case. However, it still turns out to be more efficient by a factor of about 10 than the standard heat-bath method.

15.5.5 The Fourier-accelerated Langevin method

We have encountered the Langevin method for field theories in Section 15.4.3. This method suffered from slow convergence as a result of small, essentially random, steps being taken, causing the system to perform a random walk in phase space. In 1985, Batrouni *et al.* proposed a more efficient version of the Langevin method in which the fields are updated globally [17]. This is done by Fourier-transforming the field, and then applying the Langevin method to the Fourier modes, rather than to the local fields. That this is a valid approach can be seen as follows. We have seen that MD methods can be applied to fields after assigning fictitious momenta to the field variables. In the MD method we have assigned a momentum p_n to each field variable ϕ_n. It is, however, perfectly possible to assign the momenta not to each individual field variable, but to linear combinations of the field variables. After integrating out the momenta we shall again find a Boltzmann distribution of the field variables.

In addition we have the freedom to assign a different time step to each linear combination of field variables. As we have seen in Section 9.3.2, this is equivalent to changing the mass associated with that variable, but we shall take the masses all equal to 1, and vary the time step.

In the Fourier-accelerated Langevin method, we assign momenta p_k to each Fourier component ϕ_k of the field. Furthermore, we choose a time step h_k for each k individually. To be specific, we write the action S in terms of Fourier transformed fields, and construct the following classical Hamiltonian expressed in terms of Fourier modes:

$$\mathcal{H}_{\text{Class}} = \sum_k \left\{ \frac{p_k^2}{2} + S[\phi_k] \right\}. \qquad (15.102)$$

By integrating out the momenta it is clear that an MD simulation at constant temperature for this Hamiltonian leads to the correct Boltzmann distribution of the field. In the Langevin leap-frog form, the equation of motion reads

$$\phi_k(t + h_k) = \phi_k(t) - \frac{h_k^2}{2} \frac{\partial S[\phi_k(t)]}{\partial \phi_k} + h_k R_k, \tag{15.103}$$

where R_k is the Fourier transform of a Gaussian random number with a variance of 1 (see below). Fourier transforms are obviously carried out using the fast Fourier transform (see Appendix A9).

For a free field model, the dynamical system described by the Hamiltonian (15.102) can be solved trivially, as the Hamiltonian does not contain couplings between the different ks. In that case the action can be written as

$$S[\phi_k] = \frac{1}{2L^{2d}} \phi_k K_{k,-k} \phi_{-k}. \tag{15.104}$$

$K_{k,k'}$ is the free field propagator given in (15.49). The Hamiltonian describes a set of uncoupled harmonic oscillators with periods $T_k = 2\pi/\sqrt{K_{k,-k}}$. The algorithm will be unstable when one of the time steps h_k becomes too large with respect to T_k (see Appendix A7.1). The most efficient choice for the time steps is therefore

$$h_k = \alpha T_k = \alpha \frac{2\pi}{\sqrt{4 \sum_\mu \sin^2(k_\mu/2) + m^2}}, \tag{15.105}$$

where α is some given, small fraction, e.g. $\alpha = 0.2$. If we take all the h_k smaller than the smallest period, then the slower modes would evolve at a much smaller rate than the fast modes. By adopting convention (15.105), the slow modes evolve at exactly the same rate as the fast ones. Therefore, critical slowing down will be completely eliminated for the free field model. For the interacting field with a ϕ^4 term present, the time steps are taken according to (15.105), but with the renormalised mass occuring in the denominator [17].

A remark is in place here. As the method uses finite time steps, it is not the continuum action which is simulated, but the discrete version which deviates to some order of the time steps from the continuum one. Therefore, comparisons with MC or hybrid algorithms are not straightforward. The time steps chosen here are such that the time step error is divided homogeneously over the different modes.

The algorithm for a step in the Fourier-accelerated Langevin method is as follows:

ROUTINE LangStep(ψ)
  Calculate forces F_n in real space;
  FFT: $F_n \rightarrow F_k$;
  FFT: $\phi_n \rightarrow \phi_k$;
  Generate random forces R_k;

Update ϕ_k using (15.103) with time steps (15.105);
$\overline{\text{FFT}}$: $\phi_k \rightarrow \phi_n$;
END LangStep.

We have used 'FFT' for the forward transform (from real space to reciprocal space) and '$\overline{\text{FFT}}$' for the backward transform. The random forces R_k can be generated in two ways. The simplest way is to generate a set of random forces R_n on the real space grid, and then Fourier-transform this set to the reciprocal grid. A more efficient way is to generate the forces directly on the Fourier grid. The forces R_k satisfy the following requirements: $R_k = R^*_{-k}$, as a result of the R_n being real; and the variance satisfies $\langle |R_k|^2 \rangle = \langle R_n^2 \rangle = 1$. Thus, for $k \neq -k$ (modulo $2\pi/L$), the real and imaginary part of the random force R_k both have width $1/\sqrt{2}$. If $k = -k$ (modulo $2\pi/L$) then the random force has a real part only, which should be drawn from a distribution with width 1.

15.6 Comparison of algorithms for scalar field theory

In the previous sections we have described seven different methods for simulating the scalar field theory on a lattice. We now present a comparison of the performance of the different methods (Table 15.3). We have taken $m = 0.1$ and $g = 0.01$ as the bare parameters on a 16×16 lattice. The simulations were carried out on a standard workstation. The results should not be taken too seriously because different platforms and different, more efficient codings could give different results. Moreover, some methods can be parallelised more efficiently than others, which is important when doing large-scale calculations (see Chapter 16). Finally, no real effort has been put into optimising the programs (except for standard optimisation at compile time), so the results should be interpreted as trends rather than as rigorous comparisons.

We give the CPU time needed for one simulation step and the correlation time, measured in simulation steps. The error in the run time is typically a few per cent, and that in the correlation time is typically between 5 and 10 per cent. For each method we include results for an 8×8 and a 16×16 lattice to show how the CPU time per step and the correlation time scale with the lattice size. All programs give the correct results for the renormalised mass and coupling constant, which have been presented before. The number of MC or MD steps in these simulations varied from 30 000 to 100 000, depending on the method used. In the Andersen method, we used 100 steps between momentum refreshing for $h = 0.05$ and 50 steps for $h = 0.1$. The time steps used in the hybrid algorithm were chosen such as to stabilise the acceptance rate at 70%. In this algorithm, 10 steps were used between the updates. The time step $h = 0.2$ given for the Fourier-accelerated Langevin method is in fact

Table 15.3. *Comparison between different methods for simulating the scalar quantum field theory on a lattice.*

Method	Described in section	Lattice size	Time constant h	Correlation time	CPU time	Overall efficiency
Metropolis	15.4.1	8		28	110	0.32
Metropolis	15.4.1	16		97	416	0.10
Heat-bath	15.4.1	8		6.5	103	1.49
Heat-bath	15.4.1	16		24	392	0.43
Andersen	15.4.3	8	0.05	170	32	0.063
Andersen	15.4.3	8	0.1	85	31	0.38
Andersen	15.4.3	16	0.1	110	124	0.29
Hybrid	15.4.3	8	0.365	15	45	1.48
Hybrid	15.4.3	16	0.22	60	175	0.38
Langevin	15.4.3	8	0.1	560	82	0.022
Langevin	15.4.3	16	0.1	2200	322	0.0056
Multigrid	15.5.4	8		1.5	1440	0.46
Multigrid	15.5.4	16		1.5	5750	0.46
Four/Lang	15.5.5	8	0.2	10	118	0.85
Four/Lang	15.5.5	16		10	523	0.76

The time units are only relative: no absolute run times should be deduced from them. The correlation time is measured in simulation steps (MD steps or MCS). For the methods with momentum refreshment, the correlation time is measured in MD steps. The overall efficiency in the last column is the inverse of (CPU time $\times$ correlation time). For 16×16 lattices this number has been multiplied by four.

the proportionality constant between h_k and the inverse propagator:

$$h_k = \frac{0.2}{\sqrt{K_{k,-k}}}. \tag{15.106}$$

From the table it is seen that for small lattices the heat-bath and the hybrid methods are most efficient. For larger lattices, the multigrid and Fourier-accelerated Langevin methods take over, where the latter seems to be more efficient. However, its efficiency decreases logarithmically with size (that of the multigrid remains constant) and comparisons of the values obtained with this method and MC simulations are always a bit hazardous, although these results may be very useful in themselves.

15.7 Gauge field theories

15.7.1 The electromagnetic Lagrangian

The scalar field theory is useful for some applications in particle physics and statistical mechanics. However, the fundamental theories describing elementary particles have a more complicated structure. They include several kinds of particles, some of which are fermions. Intermediate particles act as 'messengers' through which other particles interact. It turns out that the action has a special kind of local symmetry, the so-called 'gauge symmetry'.

Global symmetries are very common in physics: rotational and translational symmetries play an important role in the solution of classical and quantum mechanical problems. Such symmetries are associated with a transformation (rotation, translation) of the full space, which leaves the action invariant. Local symmetries are operations which vary in space-time, and which leave the action invariant. You have probably met such a local symmetry: electrodynamics is the standard example of a system exhibiting a local gauge symmetry. The behaviour of electromagnetic fields in vacuum is described by an action defined in terms of the four-vector potential $A_\mu(x)$ (x is the space-time coordinate) [38]:

$$S_{\text{EM}} = \frac{1}{4}\int d^4x F_{\mu\nu}F^{\mu\nu} \equiv \int d^4x \, \mathcal{L}_{\text{EM}}(\partial_\mu A^\nu) \tag{15.107a}$$

with

$$F_{\mu\nu} = \partial_\mu A_\nu - \partial_\nu A_\mu. \tag{15.107b}$$

$\mathcal{L}_{\text{EM}} = \frac{1}{4}F_{\mu\nu}F^{\mu\nu}$ is the electromagnetic Lagrangian. The gauge symmetry of electrodynamics is a symmetry with respect to a particular class of space-time-dependent shifts of the four-vector potential $A_\mu(x)$:

$$A_\mu(x) \to A_\mu(x) + \partial_\mu \chi(x), \tag{15.108}$$

where $\chi(x)$ is some scalar function. It is easy to check that the action (15.107a) is indeed invariant under the gauge transformation (15.108). If sources j_μ are present [$j = (\rho, \mathbf{j})$ where ρ is the charge density, and $\mathbf{j}$ the current density], the action reads

$$S_{\text{EM}} = \frac{1}{4}\int d^4x (F_{\mu\nu}F^{\mu\nu} + j_\mu A^\mu). \tag{15.109}$$

The Maxwell equations are found as the Euler–Lagrange equations for this action. The action is gauge-invariant if the current is conserved, according to

$$\partial_\mu j^\mu(x) = 0. \tag{15.110}$$

A quantum theory for the electromagnetic field (in the absence of sources) is constructed proceeding in the standard way, by using the action (15.107a) in the path integral. If we fix the gauge, for example by setting $\partial_\mu A^\mu = 0$ (Lorentz gauge), the transition probability for going from an initial field configuration A_i at t_i to A_f at t_f for imaginary times (we use Euclidean metric throughout this section) is given by

$$\langle A_\text{f}; t_\text{f} | A_\text{i}; t_\text{i} \rangle = \int [\mathcal{D}A_\mu] \exp\left[-\frac{1}{\hbar}\int_{t_\text{i}}^{t_\text{f}} dt \, L_{\text{EM}}(\partial_\mu A^\nu)\right] \tag{15.111}$$

where the path integral is over all vector potential fields that are compatible with the Lorentz gauge and with the initial and final vector potential fields at times $t_{\rm i}$ and $t_{\rm f}$ respectively. If we do not fix the gauge, this integral diverges badly, whereas for a particular choice of gauge, the integral converges.

Just as in the case of scalar fields, the excitations of the vector potential field are considered as particles. These particles are massless: they are the well-known *photons*. The electromagnetic field theory is exactly solvable: the photons do not interact, so we have a situation similar to the free field theory. The theory becomes more interesting when electrons and positrons are coupled to the field. These particles are described by vector fields $\psi(x)$ with $D = 2^{[d/2]}$ components for d-dimensional space-time ($[x]$ denotes the integer part of x), so $D = 4$ in four-dimensional space-time ($d = 4$). The first two components of the four-vector correspond to the spin-up and spin-down states of the fermion (e.g. the electron) and the third and fourth components to the spin-up and -down components of the anti-fermion (positron). The Euler–Lagrange equation for a fermion system interacting with an electromagnetic field is the famous *Dirac equation*:

$$[\gamma^\mu(\partial_\mu - {\rm i}eA_\mu) + m]\psi(x) = 0. \tag{15.112}$$

The objects γ_μ are Hermitian $D \times D$ matrices obeying the anti-commutation relations:

$$[\gamma_\mu, \gamma_\nu]_+ = \gamma_\mu\gamma_\nu + \gamma_\nu\gamma_\mu = 2\delta_{\mu\nu} \tag{15.113}$$

(in Minkowski metric, $\delta_{\mu\nu}$ is to be replaced by $g_{\mu\nu}$). The Dirac equation is invariant under the gauge transformation (15.108) if it is accompanied by the following transformation of the ψ:

$$\psi(x) \to {\rm e}^{{\rm i}e\chi(x)}\psi(x). \tag{15.114}$$

The action from which the Dirac equation can be derived as the Euler–Lagrange equation is the famous quantum electrodynamics (QED) action:

$$S_{\rm QED} = \int {\rm d}^4x \Big[-\bar{\psi}(x)(\gamma^\mu\partial_\mu + m)\psi(x) \\ + {\rm i}eA_\mu(x)\bar{\psi}(x)\gamma^\mu\psi(x) - \frac{1}{4}F_{\mu\nu}(x)F^{\mu\nu}(x)\Big]. \tag{15.115}$$

Here, $\psi(x)$ and $\bar{\psi}(x)$ are independent fields. The Dirac equation corresponds to the Euler–Lagrange equation of this action with

$$\bar{\psi}(x) = \psi^\dagger(x)\gamma^0. \tag{15.116}$$

The QED action itself does not show the fermionic character of the ψ-field, which should, however, not disappear in the Lagrangian formulation. The point is that the ψ field is not an ordinary c-number field, but a so-called *Grassmann field*.

Grassmann variables are anticommuting numbers – Grassmann numbers a and b have the properties:

$$ab + ba = 0. \tag{15.117}$$

In particular, taking $a = b$, we see that $a^2 = 0$. We do not go into details concerning Grassmann algebra [4, 5, 39] but mention only the result of a Gaussian integration over Grassman variables. For a Gaussian integral over a vector $\boldsymbol{\psi}$ we have the following result for the components of $\boldsymbol{\psi}$ being ordinary commuting, or Grassmann anticommuting numbers:

$$\int \mathrm{d}\psi_1 \ldots \mathrm{d}\psi_N \exp(-\boldsymbol{\psi}^{\mathrm{T}} A \boldsymbol{\psi}) = \begin{cases} \sqrt{\dfrac{(2\pi)^N}{\det(A)}} & \text{commuting;} \\ \sqrt{\det(A)} & \text{anticommuting.} \end{cases} \tag{15.118}$$

The matrix A is symmetric. In quantum field theories such as QED, we need a Gaussian integral over complex commuting and noncommuting variables, with the result:

$$\int \mathrm{d}\psi_1 \, \mathrm{d}\psi_1^* \ldots \mathrm{d}\psi_N \, \mathrm{d}\psi_N^* \exp(-\boldsymbol{\psi}^{\dagger} A \boldsymbol{\psi}) = \begin{cases} (2\pi)^N / \det(A) & \text{commuting;} \\ \det(A) & \text{anticommuting} \end{cases} \tag{15.119}$$

for a Hermitian matrix A. Fortunately the Lagrangian depends only quadratically on the fermionic fields, so only Gaussian integrals over Grassmann variables occur in the path integral.

15.7.2 Electromagnetism on a lattice: quenched compact QED

Physical quantities involving interactions between photons and electrons, such as scattering amplitudes, masses and effective interactions, can be derived from the QED Lagrangian in a perturbative analysis. This leads to divergences similar to those mentioned in connection with scalar fields, and these divergences should be renormalised properly by choosing values for the bare coupling constant e and mass m occurring in the Lagrangian such that physical mass and coupling constant become finite; more precisely, they become equal to the experimental electron mass and the charge which occurs in the large-distance Coulomb law in three spatial dimensions:

$$V(r) = \frac{e^2}{4\pi\epsilon_0 r} \tag{15.120}$$

(for short distances, this formula is no longer valid as a result of quantum corrections).

Instead of following the perturbative route, we consider the discretisation of electrodynamics on a lattice (the Euclidean metric is most convenient for lattice

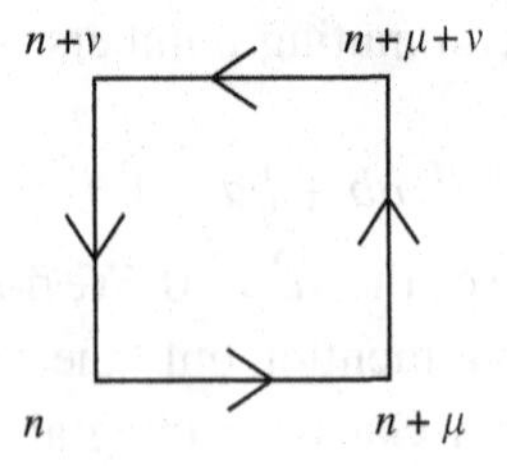

Figure 15.6. A lattice plaquette at site n with sides μ and ν used in (15.122).

calculations, so it is assumed throughout this section). This is less straightforward than in the scalar field case as a result of the greater complexity of the QED theory. We work in a space-time dimension $d = 4$. We first consider the discretisation of the photon gauge field and describe the inclusion of fermions below. An important requirement is that the gauge invariance should remain intact. Historically, Wegner's Ising lattice gauge theory [40] showed the way to the discretisation of continuum gauge theories. We now describe the lattice formulation for QED which was first given by Wilson [41], and then show that the continuum limit for strong coupling is the conventional electromagnetic gauge theory.

We introduce the following objects, living on the *links* μ of a square lattice with sites denoted by n:

$$U_\mu(n) = \exp[ieaA_\mu(n)] = \exp[i\theta_\mu(n)] \tag{15.121}$$

where we have defined the dimensionless scalar variables $\theta_\mu = eaA_\mu$. The action on the lattice is then written as a sum over all plaquettes, where each plaquette carries an action (see Figure 15.6):

$$\begin{aligned} S_{\text{plaquette}}(n; \mu\nu) = \text{Re}[1 - U_\mu(n)U_\nu(n+\mu) \\ \times U_\mu^*(n+\mu+\nu)U_\nu^*(n+\nu)]. \end{aligned} \tag{15.122}$$

Note that the effect of complex conjugation is a sign-reversal of the variable θ_μ. The Us are orientation-dependent: $U_\mu(n) = U_{-\mu}^\dagger(n)$. The constant 1 has been included in (15.122) to ensure that the total weight of a configuration with all θ-values being equal to zero vanishes. Note that the integration over θ is over a range 2π, so it does not diverge, in contrast to the original formulation, where the gauge must be fixed in order to prevent the path integral on a finite lattice from becoming infinite (see also the remark after Eq. (15.111)). The plaquette action can also be written as

$$S_{\text{plaquette}}(n; \mu\nu) = \sum_{\mu\nu} [1 - \cos(\theta_{\mu\nu}(n))] \tag{15.123}$$

where the argument of the cosine is the sum over the θ-variables around the plaquette as in Figure 15.6:

$$\theta_{\mu\nu}(n) = \theta_\mu(n) + \theta_\nu(n+\mu) - \theta_\mu(n+\mu+\nu) - \theta_\mu(n+\nu). \tag{15.124}$$

The total action

$$S_{\text{LQED}} = \sum_{n;\mu\nu} S_{\text{plaquette}}(n;\mu\nu) \tag{15.125}$$

occurs in the exponent of the time-evolution operator or of the Boltzmann factor (for field theory in imaginary time). The partition function of the Euclidean field theory is

$$Z_{\text{LQED}}(\beta) = \int_0^{2\pi} \prod_{n,\mu} \mathrm{d}\theta_\mu(n) \exp(-\beta S_{\text{LQED}}), \tag{15.126}$$

where the product $\prod_{n,\mu}$ is over all the links of the lattice. For low temperature (large β), only values of θ close to 0 (mod 2π) will contribute significantly to the integral. Expanding the cosine for small angles, we can extend the integrals to the entire real axis and obtain

$$S_{\text{LQED}}(\beta \text{ large}) = \sum_{n,\mu,\nu} \frac{1}{2} \left(\sum_{\mu\nu} \theta(n) \right)^2. \tag{15.127}$$

Using $\theta_\mu = eaA_\mu$ and the fact that the lattice constant a is small, we see that the action can be rewritten as

$$\beta S_{\text{LQED}} \approx \frac{\beta}{4} \int \frac{\mathrm{d}^4 x}{a^4} [a^4 \, \mathrm{e}^2 F_{\mu\nu}(x) F^{\mu\nu}(x)], \tag{15.128}$$

where now the summation is over *all* $\mu\nu$, whereas in the sums above (over the plaquettes), μ and ν were restricted to positive directions. The $F_{\mu\nu}$ are defined in Eq. (15.107b). Taking $\beta = 1/e^2$ we recover the Maxwell Lagrangian:

$$\beta S_{\text{LQED}} \approx \frac{1}{4} \int \mathrm{d}^4 x F_{\mu\nu} F^{\mu\nu} \tag{15.129}$$

in the continuum limit.

What are interesting objects to study? Physical quantities are gauge-invariant, so we search for gauge-invariant correlation functions. Gauge invariance can be formulated in the lattice model as an invariance under a transformation defined by a lattice function $\chi(n)$ which induces a shift in the $\theta_\mu(n)$:

$$\theta_\mu(n) \to \theta_\mu(n) + \frac{\chi(n+\mu) - \chi(n)}{a}. \tag{15.130}$$

This suggests that gauge-invariant correlation functions are defined in terms of a sum over θ_μ over a closed path: in that case a gauge transformation does not induce a change in the correlation function since the sum over the finite differences of the gauge function $\chi(n)$ over the path will always vanish. Furthermore, as correlation functions usually contain products of variables at different sites, we consider the

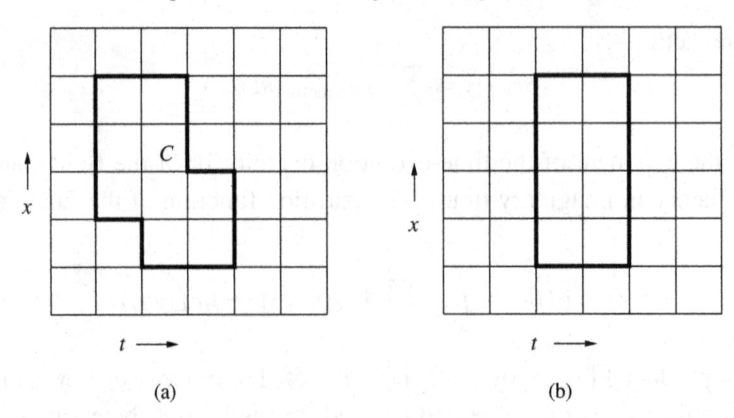

Figure 15.7. The Wilson loop on a two-dimensional square lattice. (a) A general Wilson loop. (b) A two-fermion loop in a gauge field theory with infinite-mass fermions.

so-called *Wilson loop* correlation function:

$$W(C) = \left\langle \prod_{n,\mu \in C} e^{i\theta_\mu(n)} \right\rangle, \tag{15.131}$$

where the product is over all links n, μ between site n and its neighbour $n+\mu$ lying on the closed loop C (see Figure 15.7) [41].

The Wilson loop correlation function has a physical interpretation. Suppose we create at some time t_i a fermion–antifermion pair, which remains in existence at fixed positions up to some time t_f, at which the pair is annihilated again. Without derivation we identify the partition function of the gauge field in the presence of the fermion–antifermion pair with the Wilson loop correlation function in Figure 15.7b times the vacuum partition function – for a detailed derivation see Refs. [6,42,43]. Now let us stretch the loop in the time direction, $T = t_\text{f} - t_\text{i} \to \infty$. The effective interaction between two electrons at a distance R is given by the difference between the ground state energy in the presence of the fermion-antifermion pair (which we denote by 2f) and the ground state energy of the vacuum:

$$V(R) = \left\langle \psi_\text{G}^{(2\text{f})} |H| \psi_\text{G}^{(2\text{f})} \right\rangle_{2\text{f}} - \left\langle \psi_\text{G}^{(\text{vac})} |H| \psi_\text{G}^{(\text{vac})} \right\rangle_{\text{vac}}. \tag{15.132}$$

This expression can, however, be evaluated straightforwardly in the Lagrangian picture. We have

$$e^{-TV(R)} = \frac{Z(C)}{Z} = W(C) \tag{15.133}$$

where C is the rectangular contour of size T (time direction) and R (space direction); $Z(C)$ is the partition function evaluated in the presence of the Wilson loop; and

Z is the vacuum partition function. Note that the fact that T is taken large guarantees that the ground state of the Hamiltonian is projected out in the simulation.

By varying the coupling β, different results for the value of the Wilson loop correlation function are found. For large loops we have either the 'area law':

$$W(C) = \exp(\text{Const.} \times \text{Area within loop}), \tag{15.134}$$

or the 'perimeter law' :

$$W(C) = \exp(\text{Const.} \times \text{Perimeter of loop}), \tag{15.135}$$

with additional short-range corrections. Let us consider the area law. In that case we find $V(R) \propto R$, which means that the two particles cannot be separated: pulling them infinitely far apart requires an infinite amount of energy. We say that the particles are *confined*. On the other hand, the perimeter law says that $V(R)$ is a constant (it is dominated by the vertical parts of length T), up to corrections decaying to zero for large R (for the confined case, these corrections can be neglected). For $d = 4$ one finds after working out the dominant correction term $V(R) \propto -(e^2/R)$, i.e. Coulomb's law [42]. We see that the lattice gauge theory incorporates two different kinds of gauge interactions: confined particles and electrodynamics. The analysis in which the fermions are kept at fixed positions corresponds to the fermions having an infinite mass. It is also possible to allow for motion of the fermions by allowing the loops of arbitrary shape, introducing gamma-matrices in the resulting action. The procedure in which the fermions are kept fixed is called 'quenched QED' – in quenched QED, vacuum polarisation effects (caused by the fact that photons can create electron–positron pairs) are not included.

We know that electrodynamics does not confine electrons: the lattice gauge theory in four dimensions has two phases, a low-temperature phase in which the interactions are those of electrodynamics, and a high-temperature phase in which the particles are confined [44] ('temperature' is inversely proportional to the coupling constant β). The continuum limit of electrodynamics is described by the low-temperature phase of the theory. Why have people been interested in putting QED on a grid? After all, perturbation theory works very well for QED, and the lattice theory gives us an extra phase which does not correspond to reality (for QED). The motivation for studying lattice gauge theories was precisely this latter phase: we know that quarks, the particles that are believed to be the constituents of mesons and hadrons, are confined: an isolated quark has never been observed. Lattice gauge theory provides a mechanism for confinement! Does this mean that quarks are part of the same gauge theory as QED, but corresponding to the high-temperature phase? No: there are reasons to assume that a quark theory has a more complex structure than QED, and moreover, experiment has shown that the interaction between quarks vanishes when they come close together, in sharp contrast with the confining phase of electrodynamics in which the interaction energy increases linearly with distance. The high-temperature phase of the gauge theory considered so far is always confining, so this does not include this short-distance decay of the interaction, commonly called 'asymptotic freedom' (G. 't Hooft, unpublished remarks at the Marseille conference on gauge theories, 1972; see also Refs. [6,42,43,45]). We shall study the more complex gauge theory which is believed to describe quarks later; this theory is called 'quantum chromodynamics' (QCD).

The lattice version of quantum electrodynamics using variables θ_μ ranging from 0 to 2π is often called $U(1)$ lattice gauge theory because the angle θ_μ parametrises the unit circle, which in group theory is called $U(1)$. Another name for this field theory is 'compact QED' because the values assumed by the variable θ_μ form a compact set, as opposed to the noncompact A_μ field of continuum QED. Compact QED can be formulated in any dimension, and in the next section we discuss an example in 1 space + 1 time dimension.

15.7.3 A lattice QED simulation

We describe a QED lattice simulation for determining the inter-fermion potential. We do this by determining the Wilson loop correlation function described in the previous section. Only the gauge field is included in the theory – the fermions have a fixed position, and the photons exchanged between the two cannot generate fermion–antifermion pairs (vacuum polarisation). This is equivalent to assigning an infinite mass to the fermions. We use a square lattice with periodic boundary conditions.

We consider the (1 + 1)-dimensional case. This is not a very interesting theory by itself – it describes confined fermions, as the Coulomb potential in one spatial dimension is confining:

$$V(x) \propto |x|, \tag{15.136}$$

but we treat it here because it is simple and useful for illustrating the method. The theory can be solved exactly (see Problem 15.6) [42]: the result is that the Wilson loop correlation function satisfies the area law

$$W = \exp(-\alpha A) \tag{15.137}$$

(A is the area enclosed within the loop) with the proportionality constant α given in terms of the modified Bessel functions I_n:

$$\alpha = -\ln\left[\frac{I_1(\beta)}{I_0(\beta)}\right]; \tag{15.138}$$

β is the coupling constant (inverse temperature). In fact the area law holds only for loops much smaller than the system size; deviations from this law occur when the linear size of the loop approaches half the system size.

The system can be simulated straightforwardly using the Metropolis algorithm, but we shall use the heat-bath algorithm because of its greater efficiency. We want the coefficient α to be not too large, as large values of α cause $W(C)$ to decay very rapidly with size, so that it cannot be distinguished from the simulation noise for loops of a few lattice constants. From (15.138) we see that β must be large in that case – we shall use $\beta = 10$. This causes the probability distribution $P(\theta_\mu)$ for some θ_μ, embedded in a particular, fixed configuration of θ_μ on neighbouring links, to be sharply peaked. Therefore it is not recommended to take θ_μ random between 0 and 2π and then accept with probability $P(\theta_\mu)$ and retry otherwise, as in this approach most trial values would end up being rejected. We shall therefore first generate a trial value for θ_μ according to a Gaussian probability distribution.

The distribution $P(\theta)$ has the form

$$P(\theta) = \exp\{-\beta[\cos(\theta - \theta_1) + \cos(\theta - \theta_2)]\} \tag{15.139}$$

where θ_1 and θ_2 are fixed; they depend on the θ-values on the remaining links of the plaquettes containing θ. The sum of the two cosines can be rewritten as

$$\cos(\theta - \theta_1) + \cos(\theta - \theta_2) = 2\cos\left(\frac{\theta_1 - \theta_2}{2}\right)\cos\left(\theta - \frac{\theta_1 + \theta_2}{2}\right). \tag{15.140}$$

We define

$$\tilde{\beta} = 2\beta\cos\left(\frac{\theta_1 - \theta_2}{2}\right) \tag{15.141a}$$

and

$$\phi = \theta - \frac{\theta_1 + \theta_2}{2} \tag{15.141b}$$

so that our task is now to generate an angle ϕ with a distribution $\exp(-\tilde{\beta}\cos\phi)$. We distinguish between two cases: (i) $\tilde{\beta} > 0$. In that case the maximum of the distribution occurs at $\phi = \pi$. A Gaussian distribution centred at π and with width $\sigma = \pi/(2\sqrt{\tilde{\beta}})$ and amplitude $\exp(\tilde{\beta})$ is always close to the desired distribution. The Gaussian random numbers must be restricted to the interval $[-\pi, \pi]$. Therefore the algorithm becomes:

REPEAT
    REPEAT
        Generate a Gaussian random variable $-r$ with width 1;
        $\phi = \sigma r$;
    UNTIL $-\pi \le \phi \le \pi$;
    $\phi \to \phi + \pi$;

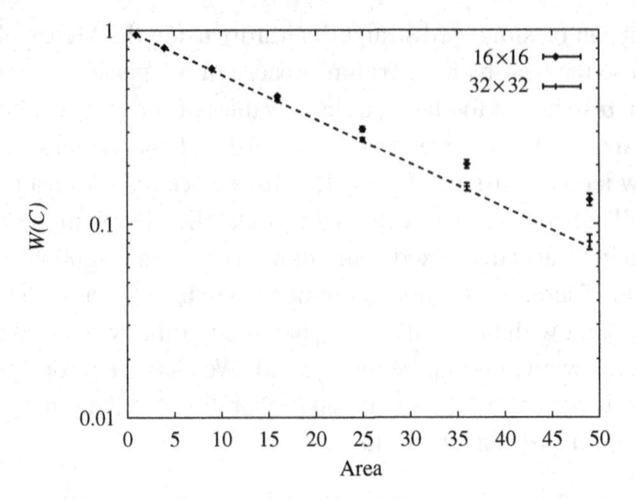

Figure 15.8. The Wilson loop correlation function as a function of the enclosed area for (1+1)-dimensional lattice QED. Note that the vertical scale is logarithmic, so that the straight line is compatible with the area law. The values were determined in a heat-bath simulation using 40 000 updates (first 2000 rejected).

Accept this trial value with probability

$$\exp\{-\tilde{\beta}[\cos\phi + 1 - (\phi - \pi)^2/(2\sigma^2)]\};$$

UNTIL Accepted;

$$\theta = \phi + \frac{\theta_1 + \theta_2}{2}.$$

(ii) If $\tilde{\beta} < 0$ then the distribution is centred around $\phi = 0$. In that case, we do not shift the Gaussian random variable over π. The reader is invited to work out the analogue of the algorithm for case (i).

In the simulation we calculate the averages of square Wilson loops, given in Eq. (15.131) (it should be emphasised that for the area law it is not required to have $T \gg R$). This is done by performing a loop over all lattice sites and calculating the sum of the θ_μ over the square loop with lower left corner at the current site. The expectation values for different square sizes can be calculated in a single simulation. Figure 15.8 shows the average value of the Wilson loop correlation functions as a function of the area enclosed by the loop for a 16×16 and a 32×32 lattice. The straight line is the theoretical curve with slope α as in (15.138). From this figure it is seen that the area law is satisfied well for loops which are small with respect to half the lattice size. By implementing free boundary conditions, the theoretical curve can be matched exactly, but this requires a little more bookkeeping.

15.7.4 Including dynamical fermions

In real problems studied by particle physicists, fermions are to be included into the lattice action, and not in a quenched fashion as done in the previous section. In this section we focus on dynamical fermions, which cause two problems. First, a straightforward discretisation of the fermion action leads to 2^d species of uncoupled fermions to be included into the problem (in d space-time dimensions) instead of the desired number of fermion species ('flavours'). Second, we have not yet discussed the problem of including the fermion character in a path integral simulation. We first consider the 'fermion doubling problem' and then sketch how simulations can actually be carried out for theories including fermions.

Fermions on a lattice

When calculating the path integral for free fermions for which the Lagrangian is quadratic in the fermion fields, the following Gaussian Grassmann integral must be evaluated:

$$\int [\mathcal{D}\psi \mathcal{D}\bar{\psi}] e^{-\bar{\psi} M \psi}, \tag{15.142}$$

where the kernel M is given as

$$M = m + \gamma^\mu \partial_\mu. \tag{15.143}$$

The expression in the exponent is shorthand for an integral over the space-time coordinates. Discretising the theory on the lattice and Fourier-transforming the fields and M we find that the latter becomes diagonal:

$$M_{k,k'} = \left(m + \frac{\mathrm{i}}{a} \sum_\mu \gamma_\mu \sin(k_\mu a) \right) \delta(k + k'). \tag{15.144}$$

The lattice version of M is therefore the Fourier transform of this function.

There is a problem with this propagator. The continuum limit singles out only the minima of the sine as a result of the factor $1/a$ in front of it. These are found not only near $k = 0$ but also near $ka = (\pm\pi, 0, 0, 0)$ (in four dimensions) etc., because of the sine function having zeroes at 0 and π. This causes the occurrence of two different species of fermions per dimension, adding up to 16 species for four-dimensional space-time. It turns out that this degeneracy can be lifted only at the expense of breaking the so-called 'chiral symmetry' for massless fermions [46]. Chiral symmetry is a particular symmetry which is present in the Dirac equation (and action) for massless particles. Suppose chiral symmetry is present in the lattice version of the action. This symmetry forbids a mass term to be present, and the renormalised theory should therefore have $m = 0$. A lattice action which violates chiral symmetry might generate massive fermions under renormalisation.

One could ignore the doubling problem and live with the fact that the theory now contains 2^d different species of fermions. However, it is also possible to lift up the unwanted parts of the propagator by adding a term proportional to $1 - \cos(ak_\mu)$ to it, which for k near 0 does not affect the original propagator to lowest order, but which lifts the parts for $k_\mu a = \pm\pi$ such that they are no longer picked up in the continuum. The method is referred to as the *Wilson fermion method.* The resulting propagator is

$$M_k = m + \frac{\mathrm{i}}{a}\sum_\mu \gamma_\mu \sin(ak_\mu) + \frac{r}{a}\sum_\mu [1 - \cos(ak_\mu)]. \tag{15.145}$$

This form is very convenient because it requires only a minor adaptation of a program with the original version of the propagator. In real space, and taking the lattice constant a equal to 1, the Wilson propagator reads in d space-time dimensions:

$$M_{nl} = (m + 4r)\delta_{nl} - \frac{1}{2}\sum_\mu [(r + \gamma_\mu)\delta_{l,n+\mu} + (r - \gamma_\mu)\delta_{l,n-\mu}]. \tag{15.146}$$

The disadvantage of this solution is that the extra terms destroy any chiral symmetry, which is perhaps a bit too drastic.

In a more complicated method half of the unwanted states are removed by doubling the period, so that the Brillouin zone of the lattice is cut-off at $\pi/(2a)$ instead of π/a. This is done by putting different species of fermions on alternating sites of the lattice. Although this removes the unwanted fermions, it introduces new fermions which live on alternate sites of the lattice. The resulting method is called the *staggered fermion method.* The staggered fermion method respects the chiral symmetry discussed above and is therefore a better option than the Wilson fermion method. It is, however, more complicated than the Wilson fermion method and we refrain from a discussion here, but refer to the original literature [47, 48] and later reviews [6, 43, 49].

Algorithms for dynamical fermions

If we want to include dynamical rather than quenched fermions into our lattice field theory, we must generate configurations of anticommuting fermion fields. As it is not clear how to do this directly and as this may cause negative probabilities, various alternatives using results for Gaussian integrals over Grassmann variables (see Section 15.7.1) have been developed. We shall explain a few algorithms for an action consisting of a bosonic part, S_{Boson}, defined in terms of the boson field, $A(x)$, coupled to the fermion field, $\psi, \bar{\psi}$, via the fermion kernel, $M(A)$:

$$S = S_{\mathrm{Boson}}(A) + \int \mathrm{d}^d x\, \bar{\psi}(x) M(A) \psi(x). \tag{15.147}$$

The QED Lagrangian in (15.115) has this form.

Integrating out the fermion part of the path integral using (15.119) leads to a path integral defined entirely in terms of bosons:

$$\int [\mathcal{D}A][\mathcal{D}\bar{\psi}][\mathcal{D}\psi]\, \mathrm{e}^{-[S_{\mathrm{Boson}}(A)+\bar{\psi}M(A)\psi]} = \int [\mathcal{D}A] \det[M(A)]\mathrm{e}^{-\mathcal{L}_{\mathrm{Boson}}(A)}$$
$$= \int [\mathcal{D}A]\, \mathrm{e}^{-\mathcal{L}_{\mathrm{Boson}}(A)+\ln[\det(M(A))]} \tag{15.148}$$

(the inverse temperature β is included in the action). Although the determinant of $M(A)$ is real and usually positive, $M(A)$ is not necessarily a positive definite Hermitian matrix (a positive definite matrix has real and positive eigenvalues). It is therefore sometimes useful to consider the matrix

$$W(A) = M^{\dagger}(A)M(A), \tag{15.149}$$

in terms of which the path integral can be written as

$$\int [\mathcal{D}A]\mathrm{e}^{-\mathcal{L}_{\mathrm{Boson}}(A)+\frac{1}{2}\ln[\det(W(A))]}. \tag{15.150}$$

Now suppose that we want to perform a Metropolis update of the A-field. The acceptance probability for a trial change $A \to A'$ is

$$P_{\mathrm{Accept}}(A \to A') = \mathrm{e}^{-S_{\mathrm{Boson}}(A')+S_{\mathrm{Boson}}(A)} \frac{\det[M(A')]}{\det[M(A)]}$$
$$= \mathrm{e}^{-S_{\mathrm{Boson}}(A')+S_{\mathrm{Boson}}(A)} \sqrt{\frac{\det[W(A')]}{\det[W(A)]}}. \tag{15.151}$$

This is very expensive to evaluate since we must calculate the determinant of the very large matrix $M(A)$ [or $W(A)$] at every step. A clever alternative follows from the observation that if the field A changes on one site only (as is usually the case), very few elements of the matrix $M(A)$ change, which allows us to perform the calculation more efficiently [50]. Another interesting suggestion is that of Bhanot *et al.* who propose to evaluate the fraction of the determinants as follows [51]:

$$\frac{\det[W(A')]}{\det[W(A)]} = \frac{\int[\mathcal{D}\phi][\mathcal{D}\phi^*]\,\mathrm{e}^{-\phi^{\dagger}W(A)\phi}}{\int[\mathcal{D}\phi][\mathcal{D}\phi^*]\,\mathrm{e}^{-\phi^{\dagger}W(A')\phi}}, \tag{15.152}$$

where ϕ is a boson field for which we can use the algorithms given earlier in this chapter. Defining $\Delta W = W(A') - W(A)$, we can express the ratio in terms of an expectation value:

$$\frac{\det[W(A')]}{\det[W(A)]} = \langle \exp(\phi^{\dagger}\Delta W\phi)\rangle_{A'} = 1/\langle \exp(-\phi^{\dagger}\Delta W\phi)\rangle_{A}. \tag{15.153}$$

It is now possible to calculate this average by updating the field ϕ in a heat-bath algorithm. As the matrix $M(A)$ is local (it couples only nearest neighbours), $W(A)$ is local as well (it couples up to next nearest neighbours). Therefore the heat-bath algorithm can be carried out efficiently (it should be possible to apply the SOR method to this method). Each time we change the field A, the matrix $W(A)$ changes and a few heat-bath sweeps for the field ϕ have to be carried out. The value of the fraction of the determinants is determined as the geometrical average of the two estimators given in Eq. (15.153).

The most efficient algorithms for dynamical fermions combine a molecular dynamics method for the boson fields with a Monte Carlo approach for the fermionic part of the action. We describe two of these here. The first one is a Langevin approach, proposed by Batrouni *et al.* [17], and suitable for Fourier acceleration. It is based on two observations: first, $\det(M)$ can be written as $\exp[\mathrm{Tr}\ln(M)]$, and second, if ξ_n is a complex Gaussian random field on the lattice, so that

$$\langle \xi_l^\dagger \xi_n \rangle = \delta_{nl} \tag{15.154}$$

(the brackets $\langle\rangle$ denote an average over the realisations of the Gaussian random generator), then the trace of any matrix K can be written as

$$\mathrm{Tr}(K) = \sum_{nl} \langle \xi_n^\dagger K_{nl} \xi_l \rangle. \tag{15.155}$$

In the Langevin approach, the force is given by the derivative of the action with respect to the boson field. In the presence of fermions, the action reads (see Eq. (15.148)):

$$S = S_{\mathrm{Boson}} - \mathrm{Tr}\ln[M(A)]. \tag{15.156}$$

Therefore the derivative has the form

$$\frac{\partial S(A)}{\partial A_n} = \frac{\partial S_{\mathrm{Boson}}(A)}{\partial A_n} - \mathrm{Tr}\left[M^{-1}(A)\frac{\partial M(A)}{\partial A_n}\right]. \tag{15.157}$$

To evaluate the trace, we make use of the auxiliary field ξ:

$$\mathrm{Tr}\left[M^{-1}(A)\frac{\partial M(A)}{\partial A_n}\right] = \left\langle \xi^\dagger M^{-1}(A)\frac{\partial M(A)}{\partial A_n}\xi \right\rangle \tag{15.158}$$

$$= \sum_{ijl} \left\langle \xi_i^* M_{ij}^{-1}(A) \left[\frac{\partial M(A)}{\partial A_n}\right]_{jl} \xi_l \right\rangle. \tag{15.159}$$

In the Langevin equation we do not calculate the average over the ξ by generating many random fields for each step, but instead we generate a single random Gaussian vector ξ at every MD step, and evaluate the terms in angular brackets in (15.158) only for this configuration. Below we justify this simplification. The MD step reads therefore (see Eq. (15.103)):

$$A_n(t+h) = A_n(t) + \frac{h^2}{2}\left[-\frac{\partial S_{\text{Boson}}(A)}{\partial A_n} + \xi^\dagger M^{-1}(A)\frac{\partial M(A)}{\partial A_n}\xi\right] + h\eta_n. \tag{15.160}$$

The A-fields occurring between the square brackets are evaluated at time t. To evaluate the second term in the square brackets we must find the vector ψ satisfying

$$M(A)\psi = \xi, \tag{15.161}$$

so that the algorithm reads

$$A_n(t+h) = A(t) + \frac{h^2}{2}\left[-\frac{\partial S_{\text{Boson}}(A)}{\partial A_n} + \psi^\dagger \frac{\partial M(A)}{\partial A_n}\xi\right] + h\eta_n. \tag{15.162}$$

Finding the vector ψ is time-consuming. Use is made of the sparseness of the matrix $M(A)$ in order to speed up the calculation.[8] Note that this calculation is done only once per time step in which the full boson field is updated.

In the Langevin equation we generate a set of configurations which occur with a probability distribution given by the action (or rather an approximation to it because of time discretisation). If we evaluate the average distribution with respect to the random noise fields η and ξ, the average over the ξ-field gives us back the trace via equation (15.155), therefore we were justified in replacing the average over the noise field by the value for the actual noise field. It must be noted that Fourier acceleration is implemented straightforwardly in this fermion method: after the force is evaluated with the noise field ξ, it is Fourier-transformed, and the leap-frog integration proceeds as described in Section 15.5.5.

Finally we describe a combination of MD and MC methods [52] which can be formulated within the hybrid method of Duane *et al.* [18]; see also Section 15.4.3.

A first idea is to replace the determinant by a path integral over an auxiliary boson field:

$$\det[M(A)] = \int [\mathcal{D}\phi][\mathcal{D}\phi^*]\, \mathrm{e}^{-\phi^\dagger M^{-1}(A)\phi}. \tag{15.163}$$

We want to generate samples of the auxiliary field ϕ with the appropriate weight. Equation (15.163) is, however, a somewhat problematic expression as it involves the inverse of a matrix which moreover is not Hermitian. If we have an even number of fermion flavours, we can group the fermion fields into pairs, and each pair yields

[8] The conjugate gradient method (Appendix A8.1) is applied to this matrix problem.

a factor $\det[M(A)]^2$ which can be written as

$$\det[M(A)]^2 = \int [\mathcal{D}\phi][\mathcal{D}\phi^*]\, \mathrm{e}^{-\phi^\dagger [M(A)^\dagger M(A)]^{-1}\phi} = \int [\mathcal{D}\phi][\mathcal{D}\phi^*]\, \mathrm{e}^{-\phi^\dagger [W(A)]^{-1}\phi} \tag{15.164}$$

with $W(A)$ defined in (15.149). Note that we need an even number of fermion flavours here, because we cannot simply replace the matrix $W(A)$ by its square root in the following algorithm (see also the beginning of this subsection). This partition function is much more convenient than (15.163) for generating MC configurations of the field ϕ. This is done by an exact heat-bath algorithm, in which a Gaussian random field ξ_n is generated, and the field ϕ is then found as

$$M(A)\xi = \phi. \tag{15.165}$$

For staggered fermions (see previous subsection) it turns out that $W(A) = M(A)^\dagger M(A)$ couples only even sites with even sites, or odd sites with odd sites [53, 54]. Therefore the matrix $W(A)$ factorises into an even–even (ee) and an odd–odd one (oo), so that we can write

$$\det[W(A)] = \det[W(A)_{\mathrm{ee}}]\det[W(A)_{\mathrm{oo}}]. \tag{15.166}$$

The matrices $W(A)_{\mathrm{ee}}$ and $W(A)_{\mathrm{oo}}$ are identical – therefore we have:

$$\det[M(A)] = \int [\mathcal{D}\phi][\mathcal{D}\phi^*]\, \mathrm{e}^{-\phi^\dagger [W_{\mathrm{ee}}(A)]^{-1}\phi}. \tag{15.167}$$

The matrix W_{ee} is Hermitian and positive; it can be written as $W_{\mathrm{ee}} = M_{\mathrm{eo}}^\dagger(A)M_{\mathrm{oe}}(A)$, where the two partial matrices M_{eo} and M_{oe} are again identical, so we can use the heat-bath algorithm as described, with $M(A)$ replaced by $M_{\mathrm{eo}}(A)$.

The full path integral contains only integrations over boson fields:

$$Z = \int [\mathcal{D}A][\mathcal{D}\phi][\mathcal{D}\phi^*]\, \mathrm{e}^{-S_{\mathrm{Boson}}(A) - \phi^\dagger W^{-1}(A)\phi} \tag{15.168}$$

where subscripts ee for W should be read in the case of staggered fermions. We want to formulate a molecular dynamics algorithm for the boson field A, but generate the auxiliary field configurations ϕ with an MC technique. This procedure is justified because the MD trajectory between an acceptance/rejection decision is reversible, and the acceptance/rejection step ensures detailed balance.

We assign momenta to the boson field A only:

$$Z = \int [\mathcal{D}A][\mathcal{D}\phi][\mathcal{D}\phi^*][\mathcal{D}P]\, \mathrm{e}^{-1/2\sum_n P_n^2(x) - S_{\mathrm{Boson}}(A) - \phi^\dagger W^{-1}(A)\phi}. \tag{15.169}$$

The equations of motion for the field A and its conjugate momentum P are then given by

$$\dot{A}_n = P_n; \tag{15.170a}$$

$$\dot{P}_n = -\frac{\partial S_{\text{Boson}}}{\partial A_n} - \sum_{lm} \phi_l^\dagger \frac{\partial [W_{lm}^{-1}(A)]}{\partial A_n} \phi_m. \tag{15.170b}$$

The difficult part is the second equation which involves the derivative of the inverse of $W(A)$. The key observation is now that

$$\frac{\partial W^{-1}(A)}{\partial A_n} = W^{-1}(A) \frac{\partial W(A)}{\partial A_n} W^{-1}(A), \tag{15.171}$$

so that we need the vector η with

$$W(A)\eta = \phi. \tag{15.172}$$

This can be found using a suitable sparse matrix algorithm. Using this η-field, the equation of motion for P simply reads:

$$\dot{P}_n = -\frac{\partial S_{\text{Boson}}}{\partial A_n} - \eta^\dagger \frac{\partial (M^\dagger M)}{\partial A_n} \eta. \tag{15.173}$$

Summarising, a molecular dynamics update consists of the following steps:

ROUTINE MDStep
  Generate a Gaussian random configuration ξ;
  Calculate $\phi = M(A)\xi$;
  Calculate η from $(M^\dagger M)\eta = \phi$;
  Update the boson field A and its conjugate momentum field P using

$$P(t + h/2) = P(t - h/2) - h\left[\frac{\partial S_{\text{Boson}}}{\partial A_n} + \eta \frac{\partial (M^{\text{T}} M)}{\partial A_n} \eta\right] \text{ and}$$

$$A(t + h) = A(t) + hP(t + h/2).$$

END MDStep

We see that in both the Langevin and the hybrid method, the most time-consuming step is the calculation of a (sparse) matrix equation at each field update step (in the above algorithm this is the step in the third line).

15.7.5 Non-abelian gauge fields; quantum chromodynamics

QED is the theory for charged fermions interacting through photons, which are described by a real-valued vector gauge field A_μ. Weak and strong interactions are described by similar but more complicated theories. A difference between these theories and QED is that the commuting complex phase factors $U_\mu(n)$ of QED are replaced by noncommuting matrices, members of the group SU(2) (for the weak interaction) or SU(3) (strong interaction). Furthermore, in quantum chromodynamics (QCD), the SU(3) gauge theory for strong interactions, more than one fermion flavour must be included. In this section we focus on QCD, where the fermions are the *quarks*, the building blocks of mesons and hadrons, held together by the gauge particles, called *gluons*. The latter are the QCD analogue of photons in QED.

Quarks occur in different species, or 'flavours' ('up', 'down' 'strange'...); for each species we need a fermion field. In addition to the flavour quantum number, each quark carries an additional colour degree of freedom: red, green or blue. Quarks form triplets of the three colours (hadrons, such as protons and neutrons), or doublets consisting of colour–anticolour (mesons): they are always observed in colourless combinations. Quarks can change colour through the so-called *strong interactions*. The gluons are the intermediary particles of these interactions. They are described by a gauge field of the SU(3) group (see below). The gluons are massless, just as the photons in QED.

The U(1) variables of QED were parametrised by a single compact variable θ ($U = \exp(\mathrm{i}\theta)$). In QCD these variables are replaced by SU(3) matrices. These matrices are parametrised by eight numbers corresponding to eight gluon fields A_μ^a, $a = 1, \ldots, 8$ (gluons are insensitive to flavour). The gluons are massless, just like the photons, because inclusion of a mass term $m^2 A_\mu A^\mu$, analogous to that of the scalar field, destroys the required gauge invariance.

Experimentally, quarks are found to have almost no interaction at short separation, but when the quarks are pulled apart their interaction energy becomes linear with the separation, so that it is impossible to isolate one quark. The colour interaction carried by the gluon fields is held responsible for this behaviour. There exists furthermore an intermediate regime, where the interaction is Coulomb-like.

The fact that the interaction vanishes at short distances is called 'asymptotic freedom'. It is possible to analyse the behaviour of quarks and gluons in the short-distance/small coupling limit by perturbation theory, which does indeed predict asymptotic freedom (G. 't Hooft, unpublished remarks, 1972; and Refs. [55,56]). The renormalised coupling constant increases with increasing distance, and it is this coupling constant which is used as the perturbative parameter. At length scales of about 1 fm the coupling constant becomes too large, and the perturbative expansion breaks down. This is the scale of hadron physics. The breakdown of perturbation theory is the reason that people want to study SU(3) gauge field theory on a computer, as this allows for a nonperturbative treatment of the quantum field theory. The lattice formulation has an additional advantage. If we want to study the time evolution of a hadron, we should specify the hadron state as the initial state. But the hadron state is very complicated! If we take the lattice size in the time direction large enough, the system will find the hadron state 'by itself' because that is the ground state, so that this problem does not occur.

The QCD action has the following form ($i = 1, 2, 3$ denotes the colour degree of freedom of the quarks, f the flavour):

$$S_{\text{QCD}} = \int \mathrm{d}^4x \left\{ \frac{1}{4} F^a_{\mu\nu} F^{a\mu\nu} + \sum_f \sum_{ij} \bar{\psi}^i_f \gamma^\mu \left(\delta_{ij} \partial_\mu + \mathrm{i}g \frac{A^a_\mu}{2} \lambda^a_{ij} \right) \psi^j_f \right.$$
$$\left. + \sum_f \sum_i m_f \bar{\psi}^i_f \psi^i_f \right\}. \tag{15.174}$$

The matrices λ^a are the eight generators of the group SU(3) (they are the Gell–Mann matrices, the analogue for SU(3) of the Pauli matrices for SU(2)), satisfying

$$\mathrm{Tr}(\lambda_a \lambda_b) = \delta_{a,b}. \tag{15.175}$$

The m_f are the quark masses, and $F^a_{\mu\nu}$ is more complicated than its QED counterpart:

$$F^a_{\mu\nu} = \partial_\mu A^a_\nu - \partial^\nu A^a_\mu - g f^{abc} A^b_\mu A^c_\nu; \tag{15.176}$$

the constants f^{abc} are the structure constants of $SU(3)$, defined by

$$[\lambda^a, \lambda^b] = 2\mathrm{i} \sum_c f^{abc} \lambda^c. \tag{15.177}$$

The parameter g is the coupling constant of the theory; it plays the role of the charge in QED. A new feature of this action is that the f^{abc}-term in (15.176) introduces interactions between the gluons, in striking contrast with QED where the photons do not interact. This opens the possibility of having massive gluon bound states, the so-called 'glueballs'.

When we regularised QED on the lattice, we replaced the gauge field A_μ by variables $U_\mu(n) = e^{\mathrm{i}eA_\mu(n)}$ living on a link from site n along the direction given by μ. For QCD we follow a similar procedure: we put SU(3) matrices $U_\mu(n)$ on the links. They are defined as

$$U_\mu(n) = \exp\left(\mathrm{i}g \sum_a A^a_\mu \lambda^a / 2 \right). \tag{15.178}$$

The lattice action is now constructed in terms of these objects. The gauge part of the action becomes

$$S_{\text{Gauge}} = \frac{1}{4} F^a_{\mu\nu} F^{a\mu\nu} \rightarrow -\frac{1}{g^2} \mathrm{Tr}[U_\mu(n) U_\nu(n+\mu) U^\dagger_\mu(n+\nu) U^\dagger_\nu(n)$$
$$+ \text{Hermitian conjugate}]. \tag{15.179}$$

The quark part of the action, which includes the coupling with the gluons, reads in the case of Wilson fermions (see above):

$$S_{\text{Fermions}} = \sum_n (m+4r)\bar{\psi}(n)\psi(n) - \sum_{n,\mu} [\bar{\psi}(n)(r-\gamma^\mu)U_\mu(n)\psi(n+\mu) + \bar{\psi}(n+\mu)(r+\gamma^\mu)U_\mu^\dagger(n)\psi(n)]. \quad (15.180)$$

An extensive discussion of this regularisation, including a demonstration that its continuum limit reduces to the continuum action (15.174), can be found, for example, in Rothe's book [43]. The lattice QCD action

$$S_{\text{LQCD}} = S_{\text{Gauge}} + S_{\text{Fermions}} \quad (15.181)$$

can now be simulated straightforwardly on the computer, although it is certainly complicated. We shall not describe the procedure in detail. In the previous sections of this chapter we have described all the necessary elements, except for updating the gauge field, which is now a bit different because we are dealing with matrices as stochastic variables as opposed to numbers. Below we shall return to this point.

Simulating QCD on a four-dimensional lattice requires a lot of computer time and memory. A problem is that the lattice must be rather large. To see this, let us return to the simpler problem of quenched QCD, where the quarks have infinite mass so that they do not move; furthermore there is no vacuum polarisation in that case. The Wilson loop correlation function is now defined as

$$W(C) = \text{Tr} \prod_{(n,\mu)\epsilon C} U_\mu(n), \quad (15.182)$$

where the product is to be evaluated in a *path-ordered* fashion, i.e. the matrices must be multiplied in the order in which they are encountered when running along the loop. This is different from QED and reflects the fact that the Us are noncommuting matrices rather than complex numbers. This correlation function gives us the quenched inter-quark potential in the same way as in QED. In this approximation, perturbative renormalisation theory can be used to find an expansion for the potential at short distances in the coupling constant, g, with the result:

$$V(R,g,a) = \frac{C}{4\pi R}\left[g^2 + \frac{22}{16\pi^2}g^4 \ln\frac{R}{a} + \mathcal{O}(g^6)\right]. \quad (15.183)$$

Here C is a constant. We see that the coefficient of the second term increases for large R, rendering the perturbative expansion suspect, as mentioned before. The general form of this expression is

$$V(R,g,a) = \alpha(R)/R, \quad (15.184)$$

in other words, a 'screened Coulomb' interaction. Equation (15.183) can be combined with the requirement that the potential should be independent of the renormalisation cut-off a

$$a\frac{\mathrm{d}V(r,g,a)}{\mathrm{d}a} = 0 \qquad (15.185)$$

to find a relation between the coupling constant g and the lattice constant a. To see how this is done, see Refs. [6,43,45]. This relation reads

$$a = \Lambda_0^{-1}(g^2\gamma_0)^{\gamma_1/(2\gamma_0^2)} \exp[-1/(2\gamma_0 g^2)][1 + \mathcal{O}(g^2)]. \qquad (15.186)$$

This implies that g decreases with decreasing a, in other words, for small distances the coupling constant becomes small. From (15.183) we then see that the potential is screened to zero at small distances. This is just the opposite of ordinary screening, where the potential decays rapidly for large distances. Therefore, the name 'anti-screening' has been used for this phenomenon, which is in fact the asymptotic freedom property of quarks. The constants γ_0 and γ_1 are given by $\gamma_0 = (11 - 2n_f/3)/(16\pi^2)$ and $\gamma_1 = (102 - 22n_f/3)/(16\pi^2)^2$ repectively (n_f is the number of flavours), and Λ_0 is an integration constant in this derivation, which must be fixed by experiment. Any mass is given in units of a^{-1}, which in turn is related to g through the mass constant Λ_0. The important result is that if we do not include quark masses in the theory, only a single number must be determined from experiment, and this number sets the scale for all the masses, such as the masses of glueballs, or those of massive states composed of zero-mass quarks. Therefore, after having determined Λ_0 from comparison with a single mass, all other masses and coupling constants can be determined from the theory, that is, from the simulation.

Nice as this result may be, it tells us that if we simulate QCD on a lattice, and if we want the lattice constant a to be small enough to describe the continuum limit properly, we need a large lattice. The reason is that the phase diagram for the SU(3) lattice theory is simpler than that of compact QED in four dimensions. In the latter case, we have seen that there exist a Coulomb phase and a confined phase, separated by a phase transition. In lattice QCD there is only one phase, but a secret length scale is set by the lattice parameter for which (15.186) begins to hold. The lattice theory will approach the continuum theory if this equation holds, that is, if the lattice constant is sufficiently small. If we want to include a hadron in the lattice, we need a certain physical dimension to be represented by the lattice (at least a 'hadron diameter'). The small values allowed for the lattice constant and the fixed size required by the physical problem we want to describe cause the lattice to contain a very large number of sites. Whether it is allowable to take the lattice constant larger than the range where (15.186) applies is an open question, but this cannot be relied upon.

In addition to the requirement that the lattice size exceeds the hadronic scale, it must be large enough to accomodate small quark masses. The reason is that there exist excitations ('Goldstone bosons') on the scale of the quark mass. The quark masses that can currently be included are still too high too predict the instability of the ρ-meson, for example.

At the time of writing, many interesting results on lattice QCD have been obtained and much is still to be expected. A very important breakthrough is the formulation of improved staggered fermion (ISF) actions, which approximate the continuum action to higher order in the latice constant than the straightforward lattice formulations discussed so far [57–59]. This makes it possible to obtain results for heavy quark, and even for lighter ones, important properties have been or are calculated [59], such as decay constants for excited hadron states.

An interesting state of matter is the *quark–gluon plasma*, which is the QCD analog of the Kosterlitz-Thouless phase transition: the hadrons can be viewed as bound pairs or triplets of quarks, but for high densities and high temperatures, the 'dielectric' system may 'melt' into a 'conducting', dense system of quarks and gluons. This seems to have been observed very recently after some ambiguous indications. It turns out that this state of matter resembles a liquid. Lattice gauge theorists try to match these results in their large-scale QCD calculations. For a recent review, see Ref. [60].

To conclude, we describe how to update gauge fields in a simulation. In a Metropolis approach we want to change the matrices $U_\mu(n)$ and then accept or reject these changes. A way to do this is to fill a list with 'random SU(3)' matrices, which are concentrated near the unit matrix. We multiply our link matrix $U_\mu(n)$ by a matrix taken randomly from the list. For this step to be reversible, the list must contain the inverse of each of its elements. The list must be biased towards the unit matrix because otherwise the changes in the matrices become too important and the acceptance rate becomes too small. Creutz has developed a clever heat bath algorithm for SU(2) [6, 61]. Cabibbo and Marinari have devised an SU(3) variant of this method in which the heat bath is successively applied to SU(2) subgroups of SU(3) [62].

Exercises

15.1 Consider the Gaussian integral

$$I_1 = \int_{-\infty}^{\infty} \mathrm{d}x_1 \ldots \mathrm{d}x_N \; \mathrm{e}^{-\mathbf{x}A\mathbf{x}}$$

where $\mathbf{x} = (x_1, \ldots, x_N)$ is a real vector and A is a Hermitian and positive $N \times N$ matrix (positive means that all the eigenvalues λ_i of A are positive).

(a) By diagonalising A, show that the integral is equal to

$$I_1 = \sqrt{\frac{(2\pi)^N}{\prod_{i=1}^{N} \lambda_i}} = \sqrt{\frac{(2\pi)^N}{\det(A)}}.$$

(b) Now consider the integral

$$I_2 = \int \mathrm{d}x_1\, \mathrm{d}x_1^* \ldots \mathrm{d}x_N\, \mathrm{d}x_N^*\, \mathrm{e}^{-\mathbf{x}^\dagger A \mathbf{x}}$$

where $\mathbf{x}$ is now a complex vector. Show that

$$I_2 = \frac{(2\pi)^N}{\det(A)}.$$

15.2 In this problem and the next we take a closer look at the free field theory. Consider the one-dimensional, periodic chain of particles with harmonic coupling between nearest neighbours, and moving in a harmonic potential with coupling constant m^2. The Lagrangian is given by

$$\mathcal{L} = \frac{1}{2} \sum_{n=-\infty}^{\infty} [\dot{\phi}_n^2 - (\phi_n - \phi_{n+1})^2 - m^2 \phi_n^2].$$

We want to find the Hamiltonian $\mathcal{H}$ such that

$$\int [\mathcal{D}\phi_n]\, \mathrm{e}^{-S} = \langle \Phi_\mathrm{i} | \mathrm{e}^{-(t_\mathrm{f}-t_\mathrm{i})\mathcal{H}} | \Phi_\mathrm{f} \rangle$$

where

$$S = \int_{t_\mathrm{i}}^{t_\mathrm{f}} \mathcal{L}[\phi_n(t)]\, \mathrm{d}t$$

and the path integral $\int [\mathcal{D}\phi_n]$ is over all field configurations $\{\phi_n\}$ compatible with Φ_i at t_i and Φ_f at t_f.

We use the Fourier transforms

$$\phi_k = \sum_n \phi_n\, \mathrm{e}^{ikn}; \quad \phi_n = \int_0^{2\pi} \frac{\mathrm{d}k}{2\pi} \phi_k\, \mathrm{e}^{-ikn}.$$

(a) Show that from the fact that ϕ_n is real, it follows that $\phi_k = \phi_{-k}^*$, and that the Lagrangian can be written as

$$\mathcal{L} = \frac{1}{2} \int_0^{2\pi} \frac{\mathrm{d}k}{2\pi} \{|\dot{\phi}_k|^2 - \phi_{-k}[m^2 + 2(1 - \cos k)]\phi_k\}.$$

This can be viewed as a set of uncoupled harmonic oscillators with coupling constant $\omega_k^2 = m^2 + 2(1 - \cos k)$.

(b) In Section 12.4 we have evaluated the Hamiltonian for a harmonic oscillator. Use the result obtained there to find

$$\mathcal{H} = \frac{1}{2} \int_0^{2\pi} \frac{\mathrm{d}k}{2\pi} \{\hat{\pi}(k)\hat{\pi}(-k) + \hat{\phi}(-k)[m^2 + 2(1 - \cos k)]\hat{\phi}(k)\},$$

where the hats denote operators; $\hat{\pi}(k)$ is the momentum operator conjugate to $\hat{\phi}(k)$ – they satisfy the commutation relation

$$[\hat{\pi}(k), \hat{\phi}(-k')] = \mathrm{i} \sum_n \mathrm{e}^{\mathrm{i}k(k-k')n} = 2\pi\delta(k - k'),$$

where the argument of the delta-function should be taken modulo 2π.

(c) To diagonalise the Hamiltonian we introduce the operators

$$\hat{a}_k = \frac{1}{\sqrt{4\pi\omega_k}}[\omega_k\hat{\phi}(k) + \mathrm{i}\hat{\pi}(k)];$$

$$\hat{a}_k^\dagger = \frac{1}{\sqrt{4\pi\omega_k}}[\omega_k\hat{\phi}^\dagger(k) - \mathrm{i}\hat{\pi}^\dagger(k)].$$

Show that

$$[a_k, a_{k'}] = [a_k, a_{-k'}^\dagger] = \delta(k - k').$$

(d) Show that $\mathcal{H}$ can be written in the form

$$\mathcal{H} = \frac{1}{2}\int_0^{2\pi} \mathrm{d}k\; \omega_k(a_k^\dagger a_k + a_k a_k^\dagger) = \int_0^{2\pi} \mathrm{d}k\; \omega_k\left(a_k^\dagger a_k + \frac{1}{2}\right).$$

15.3 Consider the path integral for the harmonic chain of the previous problem. We have seen that the Lagrangian could be written as a k-integral over uncoupled harmonic-oscillator Lagrangians:

$$\mathcal{L} = \int_0^{2\pi} \mathrm{d}k\mathcal{L}(k) = \frac{1}{2}\int_0^{2\pi} \mathrm{d}k\left[|\dot{\phi}(k)|^2 - \omega_k^2|\phi(k)|^2\right].$$

We discretise the time with time step 1 so that

$$\dot{\phi}(k,t) \to \phi(k,t+1) - \phi(k,t).$$

(a) Show that the Lagrangian can now be written as a two-dimensional Fourier integral of the form:

$$\mathcal{L} = -\frac{1}{2}\int \frac{\mathrm{d}^2 q}{(2\pi)^2}\tilde{\omega}_q^2|\phi(q)|^2$$

with

$$\tilde{\omega}_q^2 = m^2 + 2(1 - \cos q_0) + 2(1 - \cos q_1);$$

q_0 corresponds to the time component and q_1 to the space component.

(b) Show that in the continuum limit (small q), the two-point Green's function in q-space reads

$$\langle\phi_q\phi_{q'}\rangle = \frac{1}{m^2 + q^2}\delta_{q,-q'}.$$

15.4 [**C**] The multigrid Monte Carlo program for the ϕ^4 field theory can be extended straightforwardly to the *XY* model. It is necessary to work out the coarsening of the Hamiltonian. The Hamiltonian of the *XY* model reads

$$\mathcal{H} = -\sum_{\langle n,n'\rangle} J\cos(\phi_n - \phi_{n'}).$$

In the coarsening procedure, the new coupling constant will vary from bond to bond, and apart from the cosines, sine interactions will be generated. The general form which must be considered is therefore

$$\mathcal{H} = -\sum_{\langle nn'\rangle}[J_{nn'}\cos(\phi_n - \phi_{n'}) + K_{nn'}\sin(\phi_n - \phi_{n'})].$$

The relation between the coarse coupling constants $J_{NN'}, K_{NN'}$ and the fine ones is

$$J_{NN'} = \sum_{nn'|NN'} [J_{nn'} \cos(\phi_n - \phi_{n'}) + K_{nn'} \sin(\phi_n - \phi_{n'})];$$

$$K_{NN'} = \sum_{nn'|NN'} [K_{nn'} \cos(\phi_n - \phi_{n'}) - J_{nn'} \sin(\phi_n - \phi_{n'})];$$

see Figure 15.5.

(a) Verify this.
(b) [**C**] Write a multigrid Monte Carlo program for the *XY* model. Calculate the helicity modulus using (15.97) and and check the results by comparison with Figure 15.4.

15.5 In this problem we verify that the SOR method for the free field theory satisfies detailed balance.

(a) Consider a site n, chosen at random in the SOR method. The probability distribution according to which we select a new value for the field ϕ_n in the heat bath method is

$$\rho(\phi_n) = \exp[-a(\phi_n - \bar{\phi}_n)^2/2],$$

where $\bar{\phi}_n$ is the average value of the field at the neighbouring sites. In the SOR method we choose for the new value ϕ'_n at site n:

$$\phi'_n = \tilde{\phi}_n + r\sqrt{\omega(2-\omega)/a},$$

where

$$\tilde{\phi}_n = \omega\bar{\phi}_n + (1-\omega)\phi_n$$

and where r is a Gaussian random number with standard deviation 1. Show that this algorithm corresponds to a transition probability

$$T(\phi_n \to \phi'_n) \propto \exp\left[-\frac{a}{\omega(2-\omega)}(\phi'_n - \tilde{\phi}_n)^2\right].$$

(b) Show that this transition probability satisfies the detailed balance condition:

$$\frac{T(\phi_n \to \phi'_n)}{T(\phi'_n \to \phi_n)} = \frac{\exp[-a(\phi'_n - \bar{\phi}_n)^2/2]}{\exp[-a(\phi_n - \bar{\phi}_n)^2/2]}.$$

15.6 The Wilson loop correlation function for compact QED in $(1+1)$ dimensions can be solved exactly. Links in the time direction have index $\mu = 0$, and the spatial links have $\mu = 1$. We must fix the gauge in order to keep the integrals finite. The so-called *temporal gauge* turns out convenient: in this gauge, the angles θ_0 living on the time-like bonds are zero, so that the partition sum is a sum over angles θ_1 on spatial links only. Therefore there is only a contribution from the two space-like sides of the rectangular Wilson loop. The Wilson loop correlation function is defined as

$$W(C) = \frac{\int_0^{2\pi} \prod_{n,\mu} d\theta_\mu(n)\, e^{\beta \cos\left[\sum_{n;\mu\nu} \theta_{\mu\nu}(n)\right]}\, e^{i \sum_{(n,\mu)\epsilon C} \theta_\mu(n)}}{\int_0^{2\pi} \prod_{n,\mu} d\theta_\mu(n)\, e^{\beta \cos\left[\sum_{n;\mu\nu} \theta_{\mu\nu}(n)\right]}}.$$

A plaquette sum over the θ angles for a plaquette with lower-left corner at n reduces in the temporal gauge to:

$$\sum_{n;\mu\nu} \theta_{\mu\nu}(n) = \theta_1(n_0, n_1) - \theta_1(n_0 + 1, n_1).$$

(a) Show that in the temporal gauge the Wilson loop sum can be written as

$$\sum_{(n,\mu)\epsilon C} \theta_\mu(n) = \sum_{(n;\mu\nu)\epsilon A} \sum_{n;\mu\nu} \theta_{\mu\nu}(n)$$

where A is the area covered by the plaquettes enclosed by the Wilson loop.

(b) Use this to show that the Wilson loop correlation function factorises into a product of plaquette-terms. Defining

$$\theta_P(n) = \sum_{n;\mu\nu} \theta_{\mu\nu}(n),$$

where P denotes the plaquettes, we can write:

$$W(C) = \frac{\int \prod_P d\theta_P \exp[\beta \cos\theta_P + i\theta_P]}{\int \prod_P d\theta_P \exp[\beta \cos\theta_P]}.$$

(c) Show that this leads to the final result:

$$W(C) = \left[\frac{I_1(\beta)}{I_0(\beta)}\right]^A$$

where $I_n(x)$ is the modified Bessel function and A is the area enclosed by the Wilson loop.

References

[1] R. Balian and J. Zinn-Justin, eds., *Méthodes en théorie des champs / Methods in Field Theory, Les Houches Summer School Proceedings*, vol. XXVIII. Amsterdam, North-Holland, 1975.

[2] C. Itzykson and J.-B. Zuber, *Quantum Field Theory*. New York, McGraw-Hill, 1980.

[3] S. Weinberg, *The Quantum Theory of Fields*, vols. 1 and 2. Cambridge, Cambridge University Press, 1995.

[4] D. Bailin and A. Love, *Introduction to Gauge Field Theory*. Bristol, Adam Hilger, 1986.

[5] J. Zinn-Justin, *Quantum Field Theory and Critical Phenomena*, 3rd edn. New York, Oxford University Press, 1996.

[6] M. Creutz, *Quarks, Gluons and Lattices*. Cambridge, Cambridge University Press, 1983.

[7] M. Lüscher and P. Weisz, 'Scaling laws and triviality bounds in the lattice-ϕ^4 theory. 1. One-component model in the symmetric phase,' *Nucl. Phys. B*, **290** (1987), 25–60.

[8] M. Lüscher, 'Volume dependence of the energy spectrum in massive quantum field theories (I). Stable particle states,' *Commun. Math. Phys.*, **104** (1986), 177–206.

[9] J. Goodman and A. D. Sokal, 'Multigrid Monte Carlo method for lattice field theories,' *Phys. Rev. Lett.*, **56** (1986), 1015–18.

[10] W. H. Press, S. A. Teukolsky, W. T. Vetterling, and B. P. Flannery, *Numerical Recipes*, 2nd edn. Cambridge, Cambridge University Press, 1992.

[11] S. L. Adler, 'Over-relaxation method for the Monte Carlo evaluation of the partition function for multiquadratic actions,' *Phys. Rev. Lett.*, **23** (1981), 2901–4.

[12] C. Whitmer, 'Over-relaxation methods for Monte Carlo simulations of quadratic and multiquadratic actions,' *Phys. Rev. D*, **29** (1984), 306–11.

[13] S. Duane and J. B. Kogut, 'Hybrid stochastic differential-equations applied to quantum chromodynamics,' *Phys. Rev. Lett.*, **55** (1985), 2774–7.

[14] S. Duane, 'Stochastic quantization versus the microcanonical ensemble – getting the best of both worlds,' *Nucl. Phys. B*, **275** (1985), 398–420.

[15] E. Dagotto and J. B. Kogut, 'Numerical analysis of accelerated stochastic algorithms near a critical temperature,' *Phys. Rev. Lett.*, **58** (1987), 299–302.

[16] E. Dagotto and J. B. Kogut, 'Testing accelerated stochastic algorithms in 2 dimensions – the SU(3) $\times$ SU(3) spin model,' *Nucl. Phys. B*, **290** (1987), 415–68.

[17] G. G. Batrouni, G. R. Katz, A. S. Kronfeld, *et al.*, 'Langevin simulation of lattice field theories,' *Phys. Rev. D*, **32** (1985), 2736–47.

[18] S. Duane, A. D. Kennedy, B. J. Pendleton, and D. Roweth, 'Hybrid Monte Carlo,' *Phys. Lett. B*, **195** (1987), 216–22.

[19] M. Creutz, 'Global Monte Carlo algorithms for many-fermion systems,' *Phys. Rev. D*, **38** (1988), 1228–38.

[20] R. H. Swendsen and J.-S. Wang, 'Nonuniversal critical dynamics in Monte Carlo simulations,' *Phys. Rev. Lett.*, **58** (1987), 86–8.

[21] U. Wolff, 'Comparison between cluster Monte Carlo algorithms in the Ising models,' *Phys. Lett. B*, **228** (1989), 379–82.

[22] C. M. Fortuin and P. W. Kasteleyn, 'On the random cluster model. I. Introduction and relation to other models,' *Physica*, **57** (1972), 536–64.

[23] J. Hoshen and R. Kopelman, 'Percolation and cluster distribution. I. Cluster multiple labeling technique and critical concentration algorithm,' *Phys. Rev. B*, **14** (1976), 3438–45.

[24] U. Wolff, 'Monte Carlo simulation of a lattice field theory as correlated percolation,' *Nucl. Phys. B*, **FS300** (1988), 501–16.

[25] M. Sweeny, 'Monte Carlo study of weighted percolation clusters relevant to the Potts models,' *Phys. Rev. B*, **27** (1983), 4445–55.

[26] M. Creutz, 'Microcanonical Monte Carlo simulation,' *Phys. Rev. Lett.*, **50** (1983), 1411–14.

[27] U. Wolff, 'Collective Monte Carlo updating for spin systems,' *Phys. Rev. Lett.*, **69** (1989), 361–4.

[28] R. G. Edwards and A. D. Sokal, 'Dynamic critical behaviour of Wolff's collective-mode Monte Carlo algorithm for the two-dimensional O(n) nonlinear σ model,' *Phys. Rev. D*, **40** (1989), 1374–7.

[29] B. Nienhuis, 'Coulomb gas formulation of two-dimensional phase transitions,' *Phase Transitions and Critical Phenomena*, vol. 11 (C. Domb and J. L. Lebowitz, eds.). London, Academic Press, 1987.

[30] D. J. Bishop and J. D. Reppy, 'Study of the superfluid transition in two-dimensional ^{4}He films,' *Phys. Rev. Lett.*, **40** (1978), 1727–30.

[31] C. J. Lobb, 'Phase transitions in arrays of Josephson junctions,' *Physica B*, **126** (1984), 319–25.

[32] T. Ohta and D. Jasnow, '*XY* model and the superfluid density in two dimensions,' *Phys. Rev. B*, **20** (1979), 139–46.

[33] M. Kosterlitz and D. J. Thouless, 'Ordering, metastability and phase transitions in two-dimensional systems,' *J. Phys. C*, **6** (1973), 1181–203.

[34] M. Plischke and H. Bergersen, *Equilibrium Statistical Physics*. Englewood Cliffs, NJ, Prentice-Hall, 1989.

[35] C. Dress and W. Krauth, 'Cluster algorithm for hard spheres and related systems,' *J. Phys. A*, **28** (1995), L597–601.
[36] J. Liu and E. Luijten, 'Rejection-free geometric cluster algorithm for complex fluids,' *Phys. Rev. Lett.*, **92** (2004), 035504.
[37] J. R. Heringa and H. W. J. Blöte, 'Geometric cluster Monte Carlo simulation,' *Phys. Rev. E*, **57** (1998), 4976–8.
[38] J. D. Jackson, *Classical Electrodynamics*, 2nd edn. New York, John Wiley, 1974.
[39] J. W. Negele and H. Orland, *Quantum Many-particle Systems*. Frontiers in Physics, Redwood City, Addison-Wesley, 1988.
[40] F. Wegner, 'Duality in generalized Ising models and phase transitions without local order parameters,' *J. Math. Phys.*, **12** (1971), 2259–72.
[41] K. G. Wilson, 'Confinement of quarks,' *Phys. Rev. D*, **10** (1974), 2445–59.
[42] J. B. Kogut, 'An introduction to lattice gauge theory and spin systems,' *Rev. Mod. Phys.*, **51** (1979), 659–713.
[43] H. J. Rothe, *Lattice Gauge Theories: An Introduction*. Singapore, World Scientific, 1992.
[44] A. H. Guth, 'Existence proof of a nonconfining phase in four-dimensional U(1) lattice gauge theory,' *Phys. Rev. D*, **21** (1980), 2291–307.
[45] J. B. Kogut, 'The lattice gauge theory approach to quantum chromodynamics,' *Rev. Mod. Phys.*, **55** (1983), 775–836.
[46] H. B. Nielsen and M. Ninomiya, 'Absence of neutrinos on a lattice. 1. Proof by homotopy theory,' *Nucl. Phys. B*, **185** (1981), 20–40.
[47] J. B. Kogut and L. Susskind, 'Hamiltonian form of Wilson's lattice gauge theories,' *Phys. Rev. D*, **11** (1975), 395–408.
[48] T. J. Banks, R. Myerson, and J. B. Kogut, 'Phase transitions in Abelian lattice gauge theories,' *Nucl. Phys. B*, **129** (1977), 493–510.
[49] J. Kuti, 'Lattice field theories and dynamical fermions,' in *Computational Physics. Proceedings of the 32nd Scottish University Summer School in Physics* (R. D. Kenway and G. S. Pawley, eds.). Nato ASI, 1987, pp. 311–78.
[50] F. Fucito and G. Marinari, 'A stochastic approach to simulations of fermionic systems,' *Nucl. Phys. B*, **190** (1981), 266–78.
[51] G. Bahnot, U. M. Heller, and I. O. Stamatescu, 'A new method for fermion Monte Carlo,' *Phys. Lett. B*, **129** (1983), 440–4.
[52] J. Polonyi and H. W. Wyld, 'Microcanonical simulation of fermion systems,' *Phys. Rev. Lett.*, **51** (1983), 2257–60.
[53] O. Martin and S. Otto, 'Reducing the number of flavors in the microcanonical method,' *Phys. Rev. D*, **31** (1985), 435–7.
[54] S. Gottlieb, W. Liu, D. Toussaint, R. L. Renken, and R. L. Sugar, 'Hybrid molecular dynamics algorithm for the numerical simulation of quantum chromodynamics,' *Phys. Rev. D*, **35** (1988), 2531–42.
[55] D. J. Gross and F. Wilczek, 'Ultra-violet behavior of non-abelian gauge theories,' *Phys. Rev. Lett.*, **30** (1973), 1343–6.
[56] H. D. Politzer, 'Reliable perturbation results for strong interactions,' *Phys. Rev. Lett.*, **30** (1973), 1346–9.
[57] S. Naik, 'On-shell improved action for QCD with Susskind fermions and the asymptotic freedom scale,' *Nucl. Phys. B*, **316** (1989), 238–68.
[58] G. P. Lepage, 'Flavor-symmetry restoration and Szymanzik improvement for staggered quarks,' *Phys. Rev. D*, **59** (1999), 074502.
[59] C. T. H. Davies and G. P. Lepage, 'Lattice QCD meets experiment in hadron physics.' hep-lat/0311041, 2003.

[60] O. Philipsen, 'The QCD phase diagram at zero and small baryon density.' hep-lat/0510077, 2005.

[61] M. Creutz, 'Monte Carlo study of quantized SU(2) gauge theory,' *Phys. Rev. D*, **21** (1980), 2308–15.

[62] N. Cabibbo and E. Marinari, 'A new method for updating $SU(N)$ matrices in computer-simulations of gauge-theories,' *Phys. Lett. B*, **119** (1982), 387–90.

16

High performance computing and parallelism

16.1 Introduction

It is not necessary to recall the dramatic increase in computer speed and the drop in cost of hardware over the last two decades. Today, anyone can buy a computer with which all of the programs in this book can be executed within a reasonable time – typically a few seconds to a few hours.

On the other hand, if there is one conclusion to be drawn from the enormous amount of research in computational physics, it should be that for most physical problems, a realistic treatment, one without severe approximations, is still not within reach. Quantum many-particle problems, for example, can only be treated if the correlations are treated in an approximate way (this does not hold for quantum Monte Carlo techniques, but there we suffer from minus-sign problems when treating fermions; see Chapter 12). It is easy to extend this list of examples.

Therefore the physical community always follows the developments in hardware and software with great interest. Developments in this area are so fast that if a particular type of machine were presented here as being today's state of the art, this statement would be outdated by the time the book is on the shelf. We therefore restrict ourselves here to a short account of some general principles of computer architecture and implications for software technology. The two main principles are pipelining and parallelism. Both concepts were developed a few decades ago, but pipelining became widespread in supercomputers from around 1980 and has found its way into most workstations, whereas parallelism has remained more restricted to the research community and to more expensive machines. The reason for this is that it is easier to modify algorithms to make them suitable for pipelining than it is for parallelism. Recently, the dual-core processor has started to find its way into consumer PCs.

Conventional computers are built according to the *Von Neumann* architecture, shown in the figure below:

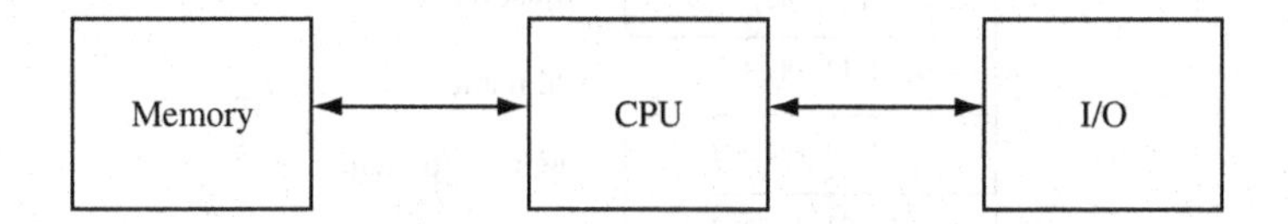

There is one processor, the CPU (central processing unit), which communicates with the internal memory and with the I/O devices such as keyboards, screens, printers, tape streamers and disks. Therefore every piece of data to be manipulated must pass through the processor which can perform elementary operations at a fixed rate of one operation per *clock cycle*. The clock cycle time determines the maximal speed at which the computer can operate. This limit on the performance speed with this type of architecture is called the *Von Neumann bottleneck*.

In the next two sections the two methods for overcoming the Von Neumann bottleneck, pipelining and parallelism, are discussed, and in the last section we present a parallel molecular dynamics algorithm in some detail.

16.2 Pipelining

16.2.1 Architectural aspects

Pipelining or *vector processing* is closely related to a pipeline arrangement in a production line in a factory. Consider the addition of two floating point numbers, 0.92×10^4, and 0.93×10^3, in a computer. We shall assume that the numbers in our computer are represented according to the decimal (base 10) system – in reality mostly a base 2 (binary) or 16 (hexadecimal) system is used. The exponents (powers of 10) are always chosen such that the mantissas (0.92 and 0.93) lie between 0.1 and 1.0.

The computer will first change the representation of one of the two numbers so that it has the same exponent as the other, in our case $0.93 \times 10^3 \rightarrow 0.093 \times 10^4$, then the two mantissas (0.92 and 0.093) are added to give the result 1.013×10^4 and then the representation of this number will be changed into one in which the mantissa has a value between 0.1 and 1.0: $1.013 \times 10^4 \rightarrow 0.1013 \times 10^5$. All in all, a number of steps must be carried out in the processor in order to perform this floating point operation:

Load the two numbers to be added from memory;
Compare exponents;
Line exponents up;
Add mantissas;

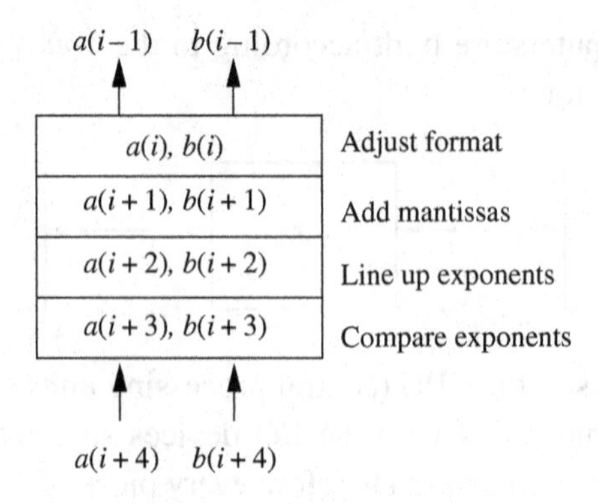

Figure 16.1. A pipeline for adding vectors.

Shift exponent such that the mantissa lies between 0.1 and 1.0;
Write result to memory.

Disregarding the load from and write to memory, we still have four steps to carry out for the addition. Each of these steps requires at least one clock cycle, and a conventional processor has to wait until the last step has been completed before it can accept a new command.

A pipeline processor, however, can perform the different operations needed to add two floating point numbers at the same time (in parallel). This is of no use when only two numbers are added, as this calculation must be completed before starting execution of the next statement. However, if we have a sequence of similar operations to be carried out, like in the addition of two vectors:

FOR $i = 1$ TO N DO
  $c[i] = a[i] + b[i]$;
END FOR

then it is possible to have the processor comparing the exponents of $a[i+3]$ and $b[i+3]$, lining up the exponents of $a[i+2]$ and $b[i+2]$, adding the mantissas of $a[i+1]$ and $b[i+1]$ and putting $a[i]$ and $b[i]$ into the right format *simultaneously*. Of course this process acts at full speed only after $a[4]$ and $b[4]$ have been loaded into the processor and only until $a[N]$ and $b[N]$ have entered it. Starting up and emptying the pipeline therefore represent a small overhead. Figure 16.1 shows how the process works and also renders the analogy with the pipeline obvious.

In the course of this pipeline process, one addition is carried out at each clock cycle. We call an addition or multiplication of real numbers a *floating point operation* (FLOP). We see that the pipeline arrangement makes it possible to perform one FLOP (FLOP) per clock cycle. Multiplication and division require many more than four clock cycles in a conventional processor, and if each of the steps involved in the multiplication or division can be executed concurrently in a pipeline processor, the speed-up can be much higher than for addition. Often, pipeline processors are able to run a pipeline for addition and multiplication simultaneously, and the maximum obtainable speed, measured in floating point operations per second, is then increased by a factor of two. A pipeline processor with a clock time of 5 ns can therefore achieve a peak performance of $2/(5 \times 10^{-9}) = 400$ MFLOPs per second (1 MFLOP = 10^6 FLOPS).

In practice, a pipeline processor will not reach its peak performance for various reasons. First, the overheads at the beginning and the end of the pipeline slightly decrease the speed, as we have seen. More importantly, a typical program does not exclusively contain simultaneous additions and multiplications of pairs of large vectors. Finally, for various operations, pipeline machines suffer from slow memory access. This is a result of the fact that memory chips work at a substantially slower rate than the processor, simply because faster memory is too expensive to include in the computer. The typical difference in speed of the two is a factor of four (or more). This means that the pipeline cannot feed itself with data at the same rate at which it operates, nor can it get rid of its data as fast as they are produced. In so-called *vector processors*, the data are read into a fast processor memory, the so-called *vector registers*, from which they can be fed into the pipelines at a processor clock-cycle rate. This still leaves the problem of feeding the vector registers at a fast enough rate, but if the same vectors are used more than once, a significant increase increase in performance is realised.

Because of this problem, which may cause the performance for many operations to be reduced by a factor of four, various solutions have been developed. First of all, the memory is often organised into four or more *memory banks*. Suppose that we have four banks, and that each bank can be accessed only once in four processor clock cycles (i.e. the memory access time is four times as slow as the processor clock cycle), but after accessing bank number one, bank number two (or three or four) can be accessed immediately at the next clock cycle. Therefore, the banks are cycled through in succession (see Figure 16.2). For this to be possible, vectors are distributed over the memory banks such that for four banks, the first bank contains the elements $1, 5, 9, 13$ etc., the second one the elements $2, 6, 10, 14 \ldots$ and so on. Cycling through the memory banks enables the processor to fetch a vector from memory at a rate of one element per clock cycle. In the case of the vector-addition code above, this is not enough as two numbers are required per clock cycle. Also, if we want to carry out operations on the vector elements $1, 5, 9 \ldots$ of a vector, the memory access time increases by a factor of four. Such problems are called *memory bank conflicts*.

Another device for solving the slow memory problem is based on the observation that a program will often manipulate a restricted amount of data more than once. Therefore a relatively small but fast memory, the so-called *cache memory*, is included. This memory acts as a buffer between the main memory and the processor, and it contains the data which have been accessed most recently by the processor.

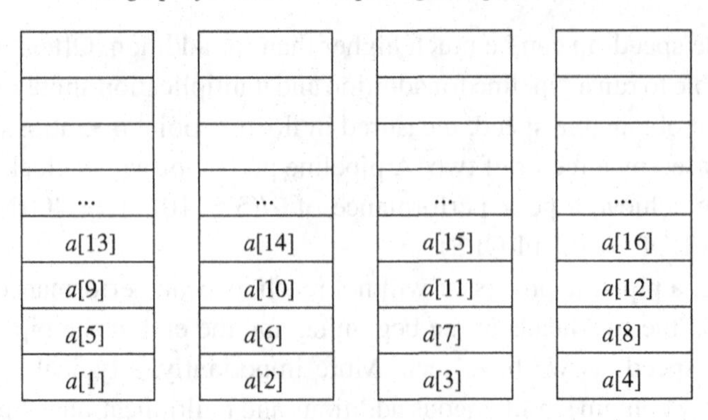

Figure 16.2. Distribution of one-dimensional array over four memory banks.

Therefore, if a first pipeline using vectors a, b and c is executed, it will cause a rather large overhead because these arrays have to be loaded from main memory into the cache. However, a subsequent pipeline using (a subset of) a, b and c will run much faster because these vectors are still stored in the cache. Obviously the cache paradigm is based on statistical considerations, and for some programs it may not improve the performance at all because most data are directly fetched from main memory (*cache trashing*). Sometimes, the cache divided into several levels of increasing speed and cost, but decreasing size. Level-1 cache is built into the processor; level-2 cache is either part of the processor, or external.

16.2.2 Implications for programming

For a pipeline process to be possible, it should be allowed to run without interruption. If the processor receives commands to check whether elements of the vector being processed are zero, or to check whether the loop must be interrupted, it cannot continue the pipeline and performance drops dramatically. Pipelines usually work in vector processing mode, in which the processor receives a command such as 'calculate the scalar product' or 'add two vectors' with as operands the memory locations (addresses) of the first elements of the vectors, the length of the pipeline and the address of the output value (scalar or vector). How can we tell the processor to start a pipeline rather than operate in conventional mode? This is usually done by the compiler which recognises the parts of our program which can be pipelined. In the language Fortran 90 the statement

```
c = a + b,
```

for one-dimensional arrays `a`, `b` and `c` is equivalent to the vector addition considered above. This can easily be recognised by a compiler as a statement which can be executed in pipeline mode.

If we want our program to be pipelined efficiently, there are a few dos and don'ts which should be kept in mind.

- Avoid conditional statements ('if-statements') within loops. Also avoid subroutine or function calls within loops.
- In the case of nested loops whose order can be interchanged, the longest loop should always be the interior loop.
- Use standard, preferably machine-optimised software to perform standard tasks.
- If possible, use indexed loops ('do-loops') rather than conditional loops ('while-loops').

A few remarks are in order. In the third item, use of standard software for standard tasks is recommended. There exists a standard definition, called LAPACK, of linear algebra routines. Often, vendors provide machine-optimised versions of these libraries; these should always be used if possible.

If a function call or IF-statement cannot be avoided, consider splitting the loop into two loops: one in which the function calls or if-statements occur, and another in which the vectorisable work is done. Some computers contain a *mask-register* which contains some of the bits which are relevant to a certain condition (e.g. the sign bit for the condition $a[i] > 0$). This mask register can be used by the processor to include the condition in the pipeline process. Finally, compiling with the appropiate optimisation options will, for some or perhaps most of the above recommendations, automatically implement the necessary improvements in the executable code.

16.3 Parallelism

16.3.1 Parallel architectures

Parallel computers achieve increase in performance by using multiple processors, which are connected together in some kind of network. There is a large variety of possible arrangements of processors and of memory segments over the computer system. Several architecture classifications exist. The most famous classification is that of Flynn who distinguishes the following four types of architectures [1,2]:

- SISD: Single Instruction Single Data stream computers;
- MISD: Multiple Instruction Single Data stream computers;
- SIMD: Single Instruction Multiple Data stream computers;
- MIMD: Multiple Instruction Multiple Data stream computers.

The single/multiple instruction refers to the number of instructions that can be performed concurrently on the machine, and the single/multiple data refers to whether the machine can process one or more data streams at the same time.

We have already described the SISD type: this is the Von Neumann architecture, in which there is one processor which can process a single data stream: the sequence of data which is fetched from or written to memory by the processor. The second type, MISD is not used in practice but it features in the list for the sake of completeness. The two last types are important for parallel machines. SIMD machines consist of arrays of processor elements which have functional units controlled by a single control unit: this unit sends a message to all processors which then all carry out the same operation on the data stream accessed by them. MIMD machines consist of an array of processors which can run entire programs independently, and these programs may be different for each processor. In very large-scale scientific problems, most of the work often consists of repeating the same operation over and over. SIMD architectures are suitable for such problems, as the processor elements can be made faster than the full processors in a MIMD machine at the same cost. The latter obviously offer greater flexibility and can be used as multi-user machines, which is impossible for SIMD computers.

Another classification of parallel machines is based on the structure of the memory. We have *distributed memory* and *shared memory* architectures. In the latter, all processors can all access the same memory area using some communication network (sometimes they can also communicate the contents of vector registers or cache to the other processors). In distributed systems on the other hand, each processor has its own local memory, and processors can interact with each other (for example to exchange data from their local memories) via the communication network. Some machines can operate in both modes. Shared memory architectures are easier to program than distributed memory computers as there is only a single address space. Modifying a conventional program for a shared memory machine is rather easy. However, memory access is often a bottleneck with these machines. This will be clear when you realise that memory bank conflicts are much more likely to occur when 10 processors are trying to access the memory instead of a single one. Distributed memory systems are more difficult to program, but they offer more potential if they are programmed adequately. The shared memory model is adopted in several supercomputers and parallel workstations containing of the order of 10 powerful vector processors. However, the most powerful machines at present and for the future seem to be of the distributed memory type, or a mixture of both.

In parallel machines, the *nodes*, consisting of a processor, perhaps with local (distributed) memory, must be connected such that data communication can proceed fast, but keeping an eye on the overall machine cost. We shall mention some of these configurations very briefly. The most versatile option would be to connect each processor to each of the others, but this is far too expensive for realistic designs. Therefore, alternatives have been developed which are either tailored to particular problems, or offer flexibility through the possibility of making different connections

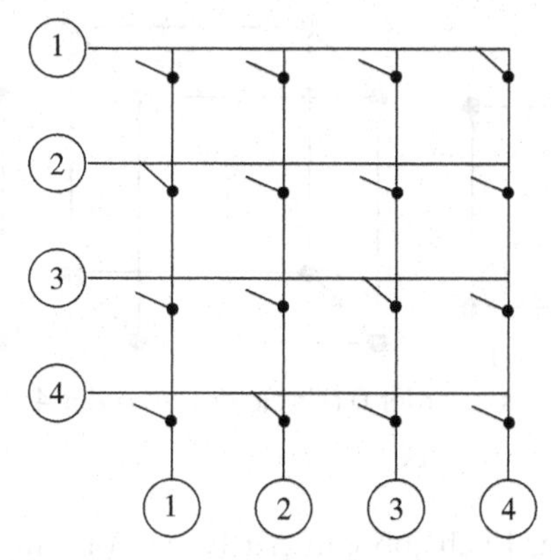

Figure 16.3. A crossbar switch connecting four processors. Each processor has two I/O ports and these are connected together via the crossbar switch. Denoting by (i,j) a connection from i to j, the switch configuration shown is (1,4), (2,1), (3,3), (4,2).

using *switches*. An example of an architecture with switches is a *crossbar switch*, shown in Figure 16.3. On each row and on each column, only a single switch may be connected – in this way binary connections are possible for any pair-partitioning of the set of processors.

Another way to interconnect processors is to link them together in a grid or chain, where each processor can communicate with its nearest neighbours on the grid or chain. Usually, the grids and chains are periodic, so that they have the topology of a ring or a torus. Rings and two- and three-dimensional tori have the advantage that they are *scalable*: this means that with increasing budget, more processors can be purchased and the machine performance increased accordingly. Furthermore, some problems in physics and engineering map naturally onto these topologies, such as a two- or three-dimensional lattice field theory with periodic boundary conditions which maps naturally onto a torus of the same dimension.

The problem of sending data from one processor to the other in the most efficient way is called *routing*. In older machines, this data-traffic, called *message passing*, was often a bottleneck for overall performance, but today this is less severe (although still a major concern).

Another type of network is the *binary hypercube*. This consists of 2^d nodes, which are labelled sequentially. If the labels of two nodes differ by only one bit, they are connected. The hypercube is shown for $d = 1$ to 4 in Figure 16.4. This network has many more links than the ones which have essentially a two-dimensional layout, and the substructures of the hypercube include multi-dimensional grids and trees.

In connection with parallel processors, *Amdahl's law* is often mentioned. This law imposes an upper limit on the increase in performance of a program by running

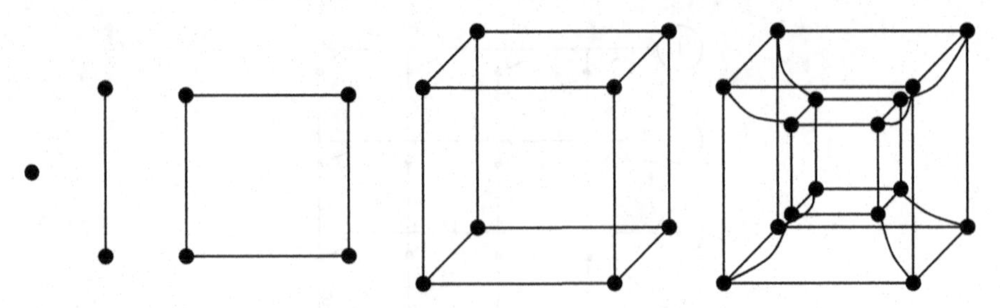

Figure 16.4. Hypercubes of order 0 (single dot) to order 4 (two cubes, one inside the other).

it on a parallel machine. Each job can partly be done in parallel, but there will always be some parts that have to be carried out sequentially. Suppose that on a sequential computer, a fraction p_{seq} of the total effort consists of jobs which are intrinsically sequential and that a fraction $p_{\text{par}} = 1 - p_{\text{seq}}$ can in principle be carried out in parallel. If we ran the program on a parallel machine and distributed the parallel parts of the code over P processors, the total run time would decrease by a factor

$$\frac{t_{\text{seq}}}{t_{\text{par}}} = \frac{p_{\text{seq}} + p_{\text{par}}}{p_{\text{seq}} + p_{\text{par}}/P} = \frac{1}{p_{\text{seq}} - (1 - p_{\text{seq}})/P} \tag{16.1}$$

(t_{seq} and t_{par} are the total run times on the sequential and parallel machine respectively). The *processor efficiency* is defined by

$$\text{P.E.} = \frac{1}{P p_{\text{seq}} + (1 - p_{\text{seq}})}. \tag{16.2}$$

The processor efficiency is equal to 1 if all the work can be evenly distributed over the P processors.

Even if we could make P arbitrarily large, the maximum decrease in runtime can never exceed $1/p_{\text{seq}}$ – this is Amdahl's law – and the processor efficiency will decrease as $1/P$. We see that for larger P, increase of performance decays to zero. Amdahl's law should not discourage us too much, however, since if we have a larger machine, we usually tackle larger problems, and as a rule, the parallelisable parts scale with a higher power of the problem size than the sequential parts. Therefore, for really challenging problems, a processor efficiency of more than 90% can often be achieved.

16.3.2 Programming implications

Writing programs which exploit the potential of parallel computers is different from vectorising, as there are many more operations and procedures that can be parallelised than can be vectorised. Moreover, we might want to tailor our program to the particular network topology of the machine we are working on. Ideally, compilers would take over this job from us, but unfortunately we are still far from this utopia. There exist various programming paradigms which are in use for parallel architectures. First of all, we have *data-parallel* programming versus *message-passing* programming. The first is the natural option for shared memory systems, although in principle it is not restricted to this type of architecture.

In data-parallel programming, all the data are declared in a single program, and at run-time these data must be allocated either in the shared memory or suitably distributed over the local memories, in which case the compiler should organise this process. For example, if we want to manipulate a vector $a[N]$, we declare the full vector in our program, and this is either allocated as a vector in the shared memory or it is chopped into segments which are allocated to the local memory of the processors involved. The message-passing model, however, is suitable only for distributed memory machines. Each processor runs a program (they are either all the same or all different, according to whether we are dealing with a SIMD or MIMD machine) and the data are allocated locally by that program. This means that if the program starts with a declaration of a real variable a, this variable is allocated at each node – together these variables may form a vector.

As an example we consider the problem in which we declare the vector $a[N]$ and initialise this to $a[i] = i$, $i = 1, \ldots, N$. Then we calculate:

FOR $i = 1$ TO N DO
  $a[i] = a[i] + a[((i-1) \text{ MOD N}) + 1]$;
END DO

In Fortran 90, in the data-parallel model, this would read:

```
1  INTEGER, PARAMETER :: N=100            ! Declaration of
                                          ! array size
2  INTEGER, DIMENSION(N)  :: A, ARight    ! Declare arrays

3  DO I=1, N                              ! Initialise A

4    A(I)=I

5  END DO
6  ARight=CSHIFT(A, SHIFT=-1, DIM = 1)    ! Circular shift
                                          ! of A
7  A = A + ARight                         ! Add A and
                                          ! ARight
                                          ! result stored
                                          ! in B
```

In this program, the full vectors `A` and `ARight` are declared – if we are dealing with a distributed memory machine, it is up to the compiler to distribute these over the nodes. The command `CSHIFT` returns the vector A, left-shifted in a cyclical fashion over one position. In the last statement, the result is calculated in vector notation (Fortran 90 allows for such vector commands). This program would run equally well on a conventional SISD machine – the compiler must find out how it distributes the data over the memory and how the `CSHIFT` is carried out.

In a message-passing model, the program looks quite different:

```
1  INTEGER, PARAMETER :: N =100              ! Declaration of
                                             ! array size
2  INTEGER :: VecElem, RVecElem              ! Elements of
                                             ! the vector
3  INTEGER :: MyNode, RightNode, &           ! Variables to
              LeftNode                       ! contain node
                                             ! addresses
4  MyNode = whoami()                         ! Determine node
                                             ! address
5  VecElem = MyNode                          ! Initialise
                                             ! vector element
6  RightNode = MOD(MyNode+1, N)+1            ! Address of
                                             ! right neighbour
7  LeftNode = MOD(Mynode-1+N, N)+1           ! Address of left
                                             ! neighbour
8  CALL send&get(VecElem, LeftNode, &        ! Circular
                 RVecElem, RightNode)        ! left-shift
9  VecElem = VecElem + RVecElem              ! Calculate sum
```

This program is more difficult to understand. In the program it is assumed that there are at least 100 free processors. They are numbered 1 through 100. The function `whoami()` returns the number of the processor – at each processor, `MyNode` will have a different value. The variable `VecElem` is an `INTEGER` which is allocated locally on each processor: the same name, `VecElem` is used to denote a collection of different numbers. Referring to this variable on processor 11 gives in general a different result than on processor 37. The right and left neighbours are calculated in statements 6 and 7. Statement 8 contains the actual message passing: `VecElem`, the element stored at the present processor, is sent to its left neighbour and the element of the right hand neighbour, `RVecElem`, is received at the present processor. In statement 9 this is then added to `VecElem`. A popular system for parallel communication in the message-passing paradigm is the *message-passing system* MPI-2, which is highly portable and offers extensive functionality.

Another classification of programming paradigms is connected with the organisation of the program. A simple but efficient way of parallelisation is the *master–slave*

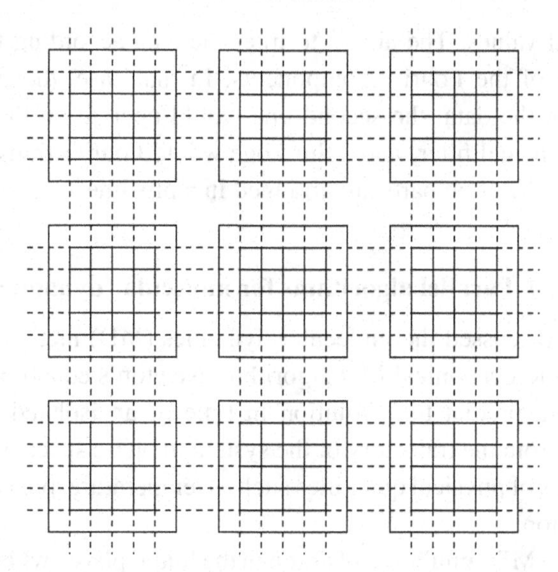

Figure 16.5. The Ising model partitioned over the nodes of a two-dimensional torus. The nodes are indicated as the heavy squares and on each node part of the lattice is shown. The dashed lines represent the couplings between spins residing on neighbouring nodes.

or *farmer–worker* model. In this model, one processor is the farmer (master) who tells the workers (slaves) to do a particular job. The jobs are carried out by the workers independently, and they report the results to the farmer, not to each other. An example is the solution of the Schrödinger equation for a periodic solid. As we have seen in Chapter 6, Bloch's theorem can be used, which tells us that the Schrödinger equation for each particular Bloch vector $\mathbf{k}$ in the Brillouin zone can be carried out independently. So, if the effective one-electron potential is known, the farmer will tell the workers to diagonalise the Hamiltonian for a set of $\mathbf{k}$-vectors allocated previously to the worker nodes. If the workers have finished the diagonalisation, they send the result back to the farmer who will calculate the charge density for example. Parts of the latter calculation could also be distributed over the workers.

Another model is the peer-to-peer model, in which the nodes are equivalent; they exchange data without one processor telling the others what to do. An example is an MC program for the Ising lattice which is partitioned over a two-dimensional torus. Each node performs the Metropolis or heat-bath algorithm for the part of the lattice which has been allocated to it, but it needs the neighbouring spins residing on other nodes. The idea is represented in Figure 16.5.

Finally, in the macro-pipelining model, each node performs a subtask of a large job. It yields processed data which are then fed into the next node for further processing. The nodes thus act in a pipelined mode, hence the name macro-pipelining, as opposed to micro-pipelining which takes place within a (pipeline-) processor. An example is the processing of a huge amount of time-series consisting of a set of N real values. The aim is to filter the data according to some scheme defined in terms of the Fourier components. In that case, the first node would Fourier-transform the data, the second one would operate on the Fourier coefficients (this is the actual filter) and a third one would Fourier-transform the results back to real time; the three parts are arranged in a pipeline.

16.4 Parallel algorithms for molecular dynamics

In Chapter 8 we discussed the molecular dynamics (MD) method in some detail. In the standard microcanonical MD algorithm, Newton's equations of motion are solved in order to predict the evolution in time of an isolated classical many-particle system. From the trajectory of the system in phase space we can determine expectation values of physical quantities such as temperature, pressure, and the pair correlation function.

We consider the MD simulation of argon in the liquid phase, which was discussed rather extensively in Chapter 8. The total interaction energy is a sum of the Lennard–Jones potentials between all particle pairs. The force on a particular particle is therefore the sum of the Lennard–Jones forces between this and all the other particles in the system. The evaluation of these forces is the most time-consuming step of the simulation, as this scales as the square of the number N of particles in the system. It therefore seems useful to parallelise this part of the simulation.

There exist essentially two methods for doing this. The first method is based on a spatial decomposition of the system volume: the system is divided up into cells, and the particles in each cell are allocated to one node. The nodes must have the same topology as the system cells, in the sense that neighbouring cells should be allocated to neighbouring nodes whenever possible. The particles interact with the other particles in their own cell, and with the particles in neighbouring cells. Interactions between particles in non-neighbouring cells are neglected. The smallest diameter of a cell must therefore be large enough to justify this approximation: in other words, the Lennard–Jones interaction must be small enough for this distance and beyond. A problem is that particles may tend to cluster in some regions in phase space, in particular when there is phase separation or when the system is close to a second order phase transition. Then some nodes may contain many more particles than others. In that case the nodes with fewer particles remain inactive for a large part of the time. This is a general problem in parallelism, called *load balancing*. Moreover, particles move and may leave the cell they are in, and therefore the allocation of the particles to the cells must be updated from time to time (similar to the procedure in Verlet's neighbour list: see Section 8.2).

The simplest version of this algorithm is realised by dividing up the system along one dimension. In this way, one obtains slices of the system, the thickness of which

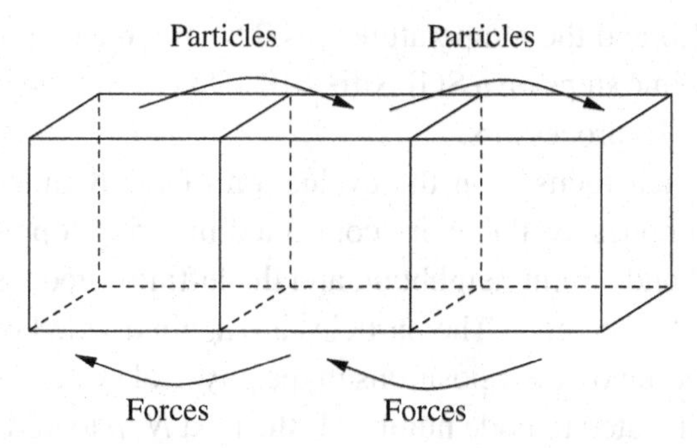

Figure 16.6. The parallel molecular dynamics procedure. At each time step, the particle positions are sent to the right neighbours. There, the forces on the local particles, and on the particles just received from the left box, are calculated. These forces are then sent back to the left slice so that they can be added to the total force calculated there. Now the particles can be moved according to the Verlet algorithm.

should exceed the cut-off of the Lennard–Jones force. The force evaluation is then implemented according to the following scheme.

```
Send particles in slice to the right neighbour;
Receive guest particles from left neighbour;
FOR all particle pairs in present slice DO
    Calculate forces for particles in present slice due to
            particles in this slice and store their values;
END FOR;
FOR all particles in present slice DO
    FOR all particles received from left neighbour DO
        Calculate forces between 'resident' and 'guest' particle;
        Store force on resident particle in resident force array;
        Store force on 'guest' particle in a sending array
    END FOR;
END FOR;
Send forces on guest particles to left neighbour;
Receive forces on resident particles from right neighbour;
Add these last forces to total forces on resident particles;
Move particles according to Verlet algorithm.
```

The procedure is represented in Figure 16.6.

To evaluate this method, we give results for the performance of an MD simulation for a rectangular box of size $NL \times L \times L$. This was divided up into N cubic $L \times L \times L$ boxes, each of which was allocated to a different processor. In total $N \times 2048$ particles were present in the system, where N is the number of processors.

The density was 1.0 and the temperature was 0.8 (in reduced units). Running this program for 1000 time steps on a SGI Altix MIMD system took 20 ± 1 seconds per step using 4, 8 and 16 processors.

A second approach focuses on the evaluation of the double sum in the force calculation. The p nodes available are connected in a ring topology: each node is connected to a left and a right neighbour, and the leftmost node is connected to the rightmost node and vice versa. The particles are again divided over the nodes, but not in a way depending on their positions in the physical system: of the N particles, the first N/p are allocated to node number 1, the next N/p to node number 2 and so on (we suppose that N/p is an integer for simplicity). Each node therefore contains the positions and momenta of the particles allocated to it (we shall use the leap-frog algorithm for integrating the equations of motion) – these are called the *local particles*. In addition to these data, each node contains memory to store the forces acting on its local particles, and sufficient memory to contain positions and momenta of 'travelling particles', which are to be sent by the other nodes. The number of travelling particles at a node equals the number of local particles, and therefore we need to keep free memory available for $6N/p$ real numbers in order to be able to receive these travelling particle data.

In the first step of the force calculation, all the interactions of local particles with other local particles are calculated, as described in Chapter 8. Then each node copies its local positions and momenta to those of the travelling particles. Each node then sends the positions and momenta of its travelling particles to the left, and it receives the data sent by its right neighbour into its travelling particle memory area. At this point, node 1, which still contains the momenta and positions of its local particles in its local particle memory area, contains the positions and momenta of the local particles of node 2 in its travelling memory area. The travelling particles of node 1 have been sent to node p.

Now the contributions to the forces acting on the local particles of each node resulting from the interactions with the travelling particles are calculated and added to the forces. Then the travelling particles are again sent to the left neighbour and received from the right neighbour (modulo PBC), and the calculation proceeds in the same fashion. When each segment of travelling particles has visited all the nodes in this way, all the forces have been calculated, and the new local positions and momenta are calculated according to the leap-frog scheme in each processor. Then the procedure starts over again, and so on. The motion of the particles over the nodes in the force calculation is depicted in Figure 16.7. This algorithm is called the *systolic algorithm* [3], because of the similarity between the intermittent data-traffic and the heart beat.

We now give the systolic algorithm in schematic from, using a data-parallel programming mode. This means that at each node, memory is allocated for the local particles, forces and travelling particles, but the data contained in this memory are

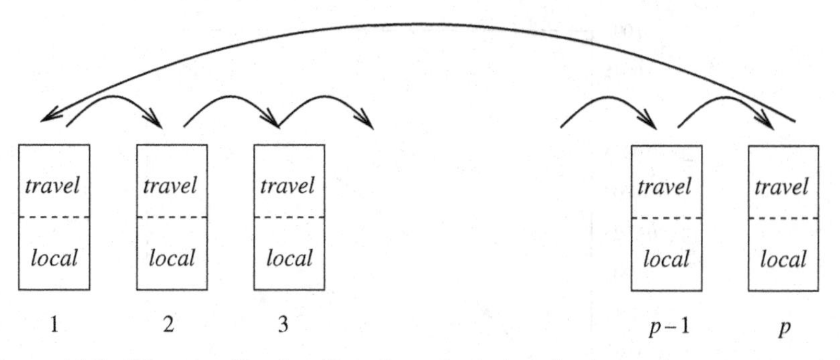

Figure 16.7. The systolic algorithm for calculating the forces in a many-particle system with pair-wise interactions.

different for each node. It is assumed that the local particles are initialised correctly at the beginning – in practice this is done by a so-called *host processor*, which is either one of the nodes, or a so-called front-end computer, which usually executes the sequential parts of a program.

```
ROUTINE NodeStep
    REPEAT
        Calculate forces on local particles;
        Perform leap-frog step for local particles;
    UNTIL Program stops;
END ROUTINE Node Step;
ROUTINE Calculate forces on local particles
    Calculate contributions from local particles;
    Copy local particles to travelling particles;
    DO K = 1, P − 1
        Send travelling particles to the left (modulo PBC);
        Receive data from right neighbour and load into
                memory used by travelling particles;
        Calculate contributions to local forces from travelling particles;
    END DO
END ROUTINE Calculate forces on local particles;
```

The systolic algorithm does not suffer from load-balance problems, as the work to be executed at each node is the same. Only if the number of particles N is not equal to an integer times the number of nodes does the last processor contain fewer particles than the remaining ones, so that a slight load-imbalance arises.

In the form described above, the systolic algorithm does not exploit the fact that for every pair ij, the force exerted by particle j on particle i is opposite to the force exerted by i on j, so that these two force contributions need only one calculation. However, a modification exists which removes this disadvantage [4]. A drawback

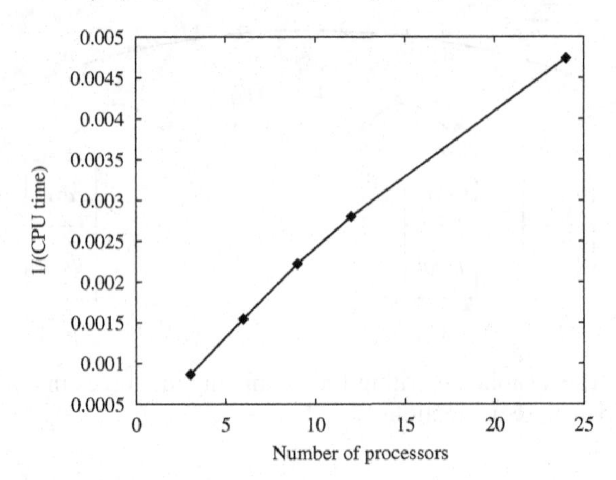

Figure 16.8. Execution speed (given as the inverse of the execution time in seconds) of the systolic MD algorithm for 864 particles executing 750 steps on a Transtech Paramid parallel computer.

of the systolic algorithm is the fact that the interactions between *all* particle pairs are taken into account in the force calculation, whereas neglecting contributions from pairs with a large separation, in the case of rapidly decaying forces, can reduce the number of calculations considerably without affecting the accuracy noticeably. Of course we can leave out these calculations in the systolic algorithm as well, but this does not reduce the total number of steps (although some steps will be carried out faster) and it requires the use of IF-statements within loops, which may degrade performance of nodes with pipelining facilities (see Section 16.2.2). Therefore, the efficiency gain by parallelising the force calculation (and the integration of the equation of motion) may not be dramatic for rapidly decaying interactions.

In Figure 16.8, the execution speed is shown as a function of the number of processors for an 864 Lennard–Jones system, integrating 750 time steps. A parallel program is called *scalable* if the execution speed is proportional to the number of processors used. We see that for larger numbers of processors, the scalability gets worse as a result of the increasing amount of interprocessor communication. This is noticeable here because the runs were performed on a rather old machine: a Transtech Paramid. On modern machines, the interprocessor communication is so fast that this algorithm would be seen to be almost perfectly scalable.

References

[1] M. J. Flynn, 'Very high speed computing systems,' *Proc. IEEE*, **54** (1966), 1901–9.
[2] M. J. Flynn, 'Some computer architectures and their effectiveness,' *IEEE Trans. Comp.*, **C-21** (1972), 948–60.
[3] W. D. Hillis and J. Barnes, 'Programming a highly parallel computer,' *Nature*, **326** (1987), 27–30.
[4] A. J. van der Steen, ed., *Aspects of Computational Science: A Textbook on High-Performance Computing*. Den Haag, National Computing Facilities Foundation, 1995.

Appendix A

Numerical methods

A1 About numerical methods

In computational physics, many mathematical operations have to be carried out numerically. Techniques for doing this are studied in numerical analysis. For a large variety of numerical problems, commercial or public domain software packages are available, and on the whole these are recommended in preference to developing all the numerical software oneself, not only because this saves time but also because available routines are often coded by professionals and hence are hard to surpass in generality and efficiency. To avoid eventual problems with using existing numerical software, it is useful to have some background in numerical analysis. Moreover, knowledge of this field enables you to write codes for special problems for which no routine is available. This chapter reviews some numerical techniques which are often used in computational physics.

There are several ways of obtaining 'canned' routines. Commercially available libraries (such as NAG, IMSL) are of very high quality, but often the source code is not available, which might prevent your software from being portable. However, several libraries, such as the ones quoted, are available at many institutes and companies for various types of computers ('platforms'), so that in practice this restriction is not so severe.

Via the internet, it is possible to obtain a wide variety of public domain routines; a particularly useful site is `/www.netlib.org/`. Most often these are provided in source code. Another cheap way of obtaining routines is by purchasing a book on numerical algorithms containing listings of source codes and, preferably, a CD (or an internet address) with these sources. A well-known book is *Numerical Recipes* by Press *et al.* [1]. This is extremely useful: it contains very readable introductions to various subjects in numerical mathematics and describes many algorithms in detail. It explains the pros and cons of the methods and, most importantly, it explains what can go wrong and why. Source codes are provided on diskette or CD in C, Fortran and Pascal, although in some cases people have found it better to use routines from NAG or Netlib. Apart from *Numerical Recipes* there are many excellent books in the field of numerical mathematics [2–6] and the interested reader is advised to consult these.

This appendix serves as a refresher to those readers who have some knowledge of the subject. Novice readers may catch an idea of the methods and can look up the details in a specialised book. The choice of problems discussed in this appendix is somewhat biased: although several methods described here are not used in the rest of the book, the emphasis is on those that are.

A2 Iterative procedures for special functions

Physics abounds with special functions: functions which satisfy classes of differential equations or given by some other prescription, and which are usually more complicated than simple sines, cosines or exponentials. Often we have an iterative prescription for determining these functions. Such is the case for the solutions to the radial Schrödinger equation for a free particle:

$$\left[-\frac{1}{2}\frac{\mathrm{d}^2}{\mathrm{d}r^2}+\frac{l(l+1)}{2r^2}\right][rR_l(r)] = E[rR_l(r)], \tag{A.1}$$

where the units are chosen such that $\hbar^2/m \equiv 1$ (it is always useful to choose such natural units to avoid cumbersome exponents). The solutions R_l of (A.1) are known as the *spherical Bessel functions* $j_l(kr)$ and $n_l(kr)$, $k = \sqrt{2E}$. The j_l are regular for $r = 0$ and the n_l are irregular (singular). For $l = 0, 1$, the spherical Bessel functions are given by:

$$\begin{aligned} j_0(x) &= \frac{\sin(x)}{x}; & n_0(x) &= -\frac{\cos(x)}{x}; \\ j_1(x) &= \frac{\sin(x)}{x^2} - \frac{\cos(x)}{x}; & n_1(x) &= -\frac{\cos(x)}{x^2} - \frac{\sin(x)}{x}. \end{aligned} \tag{A.2}$$

For higher l, we can find the functions by:

$$s_{l+1}(x) + s_{l-1}(x) = \frac{2l+1}{x} s_l(x) \tag{A.3}$$

where s_l is either j_l or n_l. Equation (A.3) gives us a procedure for determining these functions numerically. Knowing for example j_0 and j_1, Eq. (A.3) determines j_2 and so on. A three-point recursion relation has two independent solutions, and one of these may grow strongly with l. If the solution we are after damps out and the other one grows with l, the solution is unstable with respect to errors that will always sneak into the solution owing to the fact that numbers are represented with finite precision in the computer. It turns out that j_l damps out rapidly with increasing l, so that it is sensitive to errors of this kind, especially when l is significantly larger than x. One can avoid such inaccuracies by performing the recursion downwards instead of upwards. This is done by starting at a value l_{top} lying significantly higher than the one we want the Bessel function for. We put $s_{l_{\text{top}}+1}$ equal to zero and $s_{l_{\text{top}}}$ equal to some small value δ. Then, the recursion procedure is carried out downward until one arrives at the desired l. Notice that the normalisation of the solution is arbitrary (since it is determined by the arbitrary small number δ). To normalise the solution, we continue recursion down to $l = 0$ and then use the fact that $(j_0 - xj_1)\cos x + xj_0 \sin x = 1$ which determines the normalisation constant.

A problem is of course with how large an l_{top} one should start. In many cases this can be read off from the asymptotic expression for the functions to be determined.

A3 Finding the root of a function

An important problem in numerical analysis is that of finding the root of a one-dimensional, real function f:

$$f(x) = 0. \tag{A.4}$$

If this problem cannot be solved analytically, numerical alternatives have to be used, and here we shall describe three of these, the regula falsi, the secant method and the Newton–Raphson method. Of course, the function f may have more than one or perhaps no root in the interval under consideration; some knowledge of f is therefore necessary for the algorithms to arrive at the desired root. If a continuous function has opposite signs at two points x_1 and x_2, it must have at least one root between these points. If we have two points x_1 and x_2 where our function has equal signs, we can extend the interval beyond the point where the absolute value of f is smallest until the resulting interval brackets a root. Also, if the number of roots between x_1 and x_2 is even, the function has the same sign on both points; if we suspect roots to lie within this interval, we split the interval up into a number of smaller subintervals and check all of them for sign changes. We suppose in the following that the user roughly knows the location of the root (it might be bracketed, for example), but that its precise location is required.

The efficiency of a root-finding method is usually expressed in terms of a convergence exponent y for the error in the location of the root. Since root-finding methods are iterative, the expected error, δ_n, at some step in the iteration can be expressed in terms of the error at the previous step, δ_{n-1}:

$$\delta_n = \text{Const.} \times |\delta_{n-1}|^y. \tag{A.5}$$

The *regula falsi*, or *false position* method, starts with evaluating the function at two points, x_1 and x_2, lying on either side of a single root, with $x_1 < x_2$. Therefore the signs of $f(x_1) \equiv f_1$ and $f(x_2) \equiv f_2$ are opposite. A first guess for the root is found by interpolating f linearly between x_1 and x_2, resulting in a location

$$x_3 = \frac{x_2 f_1 - x_1 f_2}{f_1 - f_2}. \tag{A.6}$$

If $f_3 = f(x_3)$ and f_1 have opposite signs, we can be sure that the root lies between x_1 and x_3 and the procedure is repeated for the interval $[x_1, x_3]$, else the root lies in the interval $[x_3, x_2]$ and this is used in the next step. The procedure stops when either the size of the interval lies below a prescribed value, or the absolute value of the function is small enough.

Instead of checking which of the intervals, $[x_1, x_3]$ or $[x_3, x_2]$ contains the root, we might simply use the two x-values calculated most recently to predict the next one. In that case we use an equation similar to (A.6):

$$x_{n+1} = \frac{x_n f_{n-1} - x_{n-1} f_n}{f_{n-1} - f_n}. \tag{A.7}$$

Here, x_n is the value of x obtained at the nth step, so the points x_n and x_{n-1} do not necessarily enclose the root. This method, called the *secant method*, is slightly more efficient than the regula falsi; *regula falsi method* it has a convergence exponent equal to the golden mean 1.618 .... If the function has a very irregular behaviour, these methods become less efficient and it is more favourable in that case to take at each step the middle of the current bracket interval as the new point, and then check whether the root lies in the left or in the right half of the current interval. This method is called the bisection method. As the size of the intervals is halved at each step, the exponent y takes on the value 1. For more elaborate and fault-proof methods, see Ref. [1].

For the Newton–Raphson method, the derivative of the function must be calculated, which is not always possible. Again, a prediction based on a linear approximation of the function f is made. Starting with a point x_0 lying close to the root, the values of $f(x_0) \equiv f_0$ and of its derivative $f'(x_0) \equiv f_0'$ suffice to make a linear interpolation for f, resulting in a guess x_1 for the location of the root:

$$x_1 = x_0 - \frac{f_0}{f_0'}. \tag{A.8}$$

Just as in the secant method or the regula falsi, this procedure is repeated until the convergence criterion is satisfied. Notice that for a successful guess of the root, the sign of the derivative of the function f must not change in the region where the points x_i are located. The convergence of this method is quadratic, that is, the scaling exponent y is equal to 2.

A4 Finding the optimum of a function

Suppose we want to locate the point where a function f assumes its minimum (for a maximum, we consider the function $-f$ and apply a minimisation procedure).

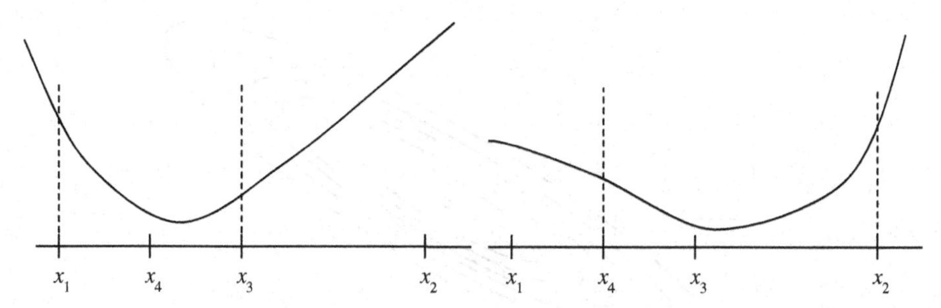

Figure A.1. Bracketing the minimum using the bisection method. The point x_4 is chosen midway between x_1 and x_3. The two graphs show two possibilities for the interval at the next step of the iteration. These intervals are indicated by the dotted lines.

We take f to depend on a single variable at first. A sufficient (but not a necessary) condition to have a local minimum in the interval $[x_1, x_2]$ is to have a point x_3 between x_1 and x_2 where f assumes a value smaller than $f(x_1)$ and $f(x_2)$:

$$x_1 < x_3 < x_2; \tag{A.9a}$$

$$f(x_3) < f(x_1) \quad \text{and} \quad f(x_3) < f(x_2). \tag{A.9b}$$

There must always exist such a triad of points closely around the local minimum. We use this condition to find an interval in which f has a minimum; we say that the interval $[x_1, x_2]$ 'brackets' the minimum. Then we can apply an algorithm to narrow this interval down to some required precision. In order to find the first bracketing interval, consider two initial points x_1 and x_2 with $f(x_2) < f(x_1)$. Then we choose a point x_3 beyond x_2 and check whether the value of f in x_3 is still smaller, or whether we are going 'uphill' again. If that is the case we have bracketed the minimum, otherwise we continue in the same manner with the points x_1 and x_3.

We consider now an algorithm for narrowing down the bracketing interval for the case that we know the derivative of the function f. Then we can use the methods of the previous section, applied to the derivative of f rather than to f itself. If the derivative is not known, we can apply a variant of the bisection method described in the previous section. The procedure is represented in Figure A.1. Suppose we have points x_1, x_2, x_3 satisfying (A.9). We take a new point x_4 either halfway in between x_1 and x_3, or between x_3 and x_2. Suppose we take the first option. Then either x_4 is the lowest point of (x_1, x_4, x_3) or x_3 is the lowest point of (x_4, x_3, x_2). In the first case, the new interval becomes $[x_1, x_3]$, with x_4 as the point in between, in the second it becomes $[x_4, x_2]$, with x_3 as the point in between, where f assumes a value lower than at the boundary points of the interval. There exist also methods for minimising a function if we do not know the derivative. Usually, more elaborate but more efficient procedures are used, which go by the name of *Brent's method*; for a description see Refs. [1, 7].

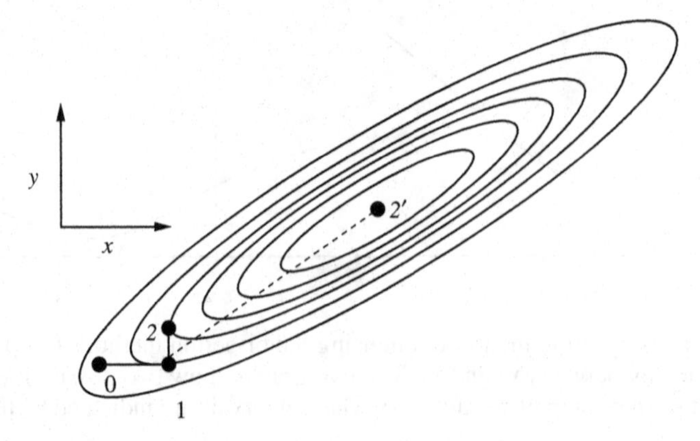

Figure A.2. Finding the minimum of a quadratic function f depending on two variables. The ellipses shown are the contour lines of the function f. The solid lines represent the steps followed in the steepest decsent method. In the conjugate gradient method, the minimum is found in two steps, the second step being the dotted line.

Next we turn to the minimisation of a function depending on more than one variable. We assume that the gradient of the function is known. To get a feeling for this problem and the solution methods, imagine that you try to reach the nearest lowest point in a mountainous landscape. We start off at some point $\mathbf{x}_0$ and we restrict ourselves to piecewise linear paths. An obvious approach is the *method of steepest descent.* Starting at some point $\mathbf{x}_0$, we walk on a straight line through x_0 along the negative gradient $\mathbf{g}_0$ at $\mathbf{x}_0$:

$$\mathbf{g}_0 = -\nabla f(\mathbf{x}_0). \tag{A.10}$$

The points $\mathbf{x}$ on this line can be parametrised as

$$\mathbf{x} = \mathbf{x}_0 + \lambda \mathbf{g}_0, \quad \lambda > 0. \tag{A.11}$$

We minimise f as a function of λ using a one-dimensional minimisation method as described at the beginning of this section. At the minimum, which we shall call $\mathbf{x}_1$, we have

$$\frac{\mathrm{d}}{\mathrm{d}\lambda} f(\mathbf{x}_0 + \lambda \mathbf{g}_0) = \mathbf{g}_0 \cdot \nabla f(\mathbf{x}_1) = 0. \tag{A.12}$$

From this point, we proceed along the negative gradient $\mathbf{g}_1$ at the point $\mathbf{x}_1$: $\mathbf{g}_1 = -\nabla f(\mathbf{x}_1)$. Note that the condition (A.12) implies that $\mathbf{g}_0$ and $\mathbf{g}_1$ are perpendicular: at each step we change direction by an angle of 90°.

Although the steepest descent method seems rather efficient, it is not the best possible minimisation method. To see what can go wrong, consider the situation shown in Figure A.2, for a function depending on two variables, x and y. We start off from the point $\mathbf{x}_0$ along the negative gradient, which in Figure A.2 lies along the x-axis. At the point $\mathbf{x}_1$, the gradient is along the y-axis:

$$\frac{\partial f}{\partial x} = 0. \tag{A.13}$$

It now remains to set the derivative along y to zero as well. To this end, we move up the y-axis, but unfortunately the gradient in the x-direction will start deviating from 0 again, so that, when we arrive at the point $\mathbf{x}_2$ where the y-gradient vanishes, we must move again in the x-direction, whereby the y-gradient starts deviating from zero and so on. Obviously, it may take a large number of steps to reduce both derivatives to zero – only in the case where the contour lines are circular will the minimum be found in two steps. The more eccentric the elliptic contours are, the more steps are needed in order to find the minimum, and the minimum is approached along a zigzag path.

To see how we can do a better job, we first realise that close to a minimum a Taylor expansion of f has the form

$$f(\mathbf{x}) = f_0 + \tfrac{1}{2}\mathbf{x}\cdot\mathbf{H}\mathbf{x} + \cdots, \tag{A.14}$$

where we have assumed that the minimum is located at the origin and that f assumes the value f_0 there (for the general case, a similar analysis applies). $\mathbf{H}$ is the *Hessian matrix*, defined as

$$H_{ij} = \frac{\partial^2 f(\mathbf{x}_0)}{\partial x_i \partial x_j}. \tag{A.15}$$

Note that $\mathbf{H}$ is symmetric if the partial derivatives of f are continuous; moreover $\mathbf{H}$ is a positive definite matrix (positive definite means that all the diagonal elements of $\mathbf{H}$ are positive), otherwise we would not be dealing with a minimum but with a saddle point or a maximum. For the moment, we restrict ourselves to quadratic functions of the form (A.14) – below we shall see how the same method can be applied to general functions. The method we describe is called the *conjugate gradient method.*

If we move from a point $\mathbf{x}_i$ along a direction $\mathbf{h}_i$, we find the point $\mathbf{x}_{i+1}$ by a line minimisation. For this point $\mathbf{x}_{i+1}$ we have

$$\mathbf{h}_i \cdot \nabla f(\mathbf{x}_{i+1}) = 0. \tag{A.16}$$

We want to move in a new direction $\mathbf{h}_{i+1}$ along which the gradient along $\mathbf{h}_i$ remains zero; therefore the gradient is allowed to change only in a direction perpendicular to $\mathbf{h}_i$. If we move from the point $\mathbf{x}_i$ over a vector $\lambda\mathbf{h}_{i+1}$, the change in the negative gradient is given by

$$\Delta\mathbf{g} = -\lambda\mathbf{H}\mathbf{h}_{i+1} \tag{A.17}$$

and we want this change to be orthogonal to the direction $\mathbf{h}_i$ of the previous line minimisation, and this implies that

$$\mathbf{h}_i \cdot \mathbf{H}\mathbf{h}_{i+1} = 0. \tag{A.18}$$

Vectors $\mathbf{h}_i$ and $\mathbf{h}_{i+1}$ satisfying this requirement are called *conjugate*. Since we have guaranteed the gradient along $\mathbf{h}_i$ to remain zero along the new path, and as the second minimisation will annihilate the gradient along $\mathbf{h}_{i+1}$ (which is independent of $\mathbf{h}_i$), at the next point the gradient is zero along two independent directions, and in the two-dimensional case we have arrived at the minimum of f: in two dimensions, only two steps are needed to arrive at the minimum of a quadratic function. For n dimensions, we need n steps to arrive at the minimum. This is less obvious than it seems. Remember that the second step (along $\mathbf{h}_{i+1}$) was designed such that it does not spoil the gradient along the direction $\mathbf{h}_i$. The third step will be designed such that it does not spoil the gradient along the direction $\mathbf{h}_{i+1}$. But does it leave the gradient along $\mathbf{h}_i$ unchanged as well? For this to be true at all stages of the iteration, *all* the directions $\mathbf{h}_i$ should be conjugate:

$$\mathbf{h}_i \cdot \mathbf{H}\mathbf{h}_j = 0, \text{all} i \neq j. \tag{A.19}$$

If $\mathbf{H}$ is the unit matrix, this requirement is equivalent to the set $\{\mathbf{h}_i\}$ being orthogonal. For a symmetric and positive Hessian matrix, the set $\{\mathbf{H}^{1/2}\mathbf{h}_i\}$ is orthogonal for a conjugate set $\{\mathbf{h}_i\}$.

To see perhaps more clearly where the magic of the conjugate gradients arises from, we define new vectors $\mathbf{y}_i = \sqrt{\mathbf{H}}\mathbf{h}_i$. We can always take the square root of $\mathbf{H}$ as this matrix has positive, real eigenvalues (see also Section 3.3). In terms of these new vectors we have the trivial problem of minimising:

$$\mathbf{y} \cdot \mathbf{y}. \tag{A.20}$$

Starting at some arbitrary point, we can simply follow a sequence of orthogonal directions along each of which we minimise this expression. Now it is trivial that these minimisations do not spoil each other. But the fact that the vectors in the set y_i are othogonal implies

$$\mathbf{h}_i \cdot \mathbf{H}\mathbf{h}_j = 0 \tag{A.21}$$

for $i \neq j$, that is, the directions $\mathbf{h}_i$ form a conjugate set. Each sequence of conjugate search directions therefore corresponds to a sequence of orthogonal search directions $\mathbf{y}_i$.

How can we construct a sequence of conjugate directions? It turns out that the following recipe works. We start with the negative gradient $\mathbf{g}_0$ at $\mathbf{x}_0$ and we take $\mathbf{h}_0 = \mathbf{g}_0$. The following algorithm constructs a new set of vectors $\mathbf{x}_{i+1}$, $\mathbf{g}_{i+1}$, $\mathbf{h}_{i+1}$ from the previous one ($\mathbf{x}_i$, $\mathbf{g}_i$, $\mathbf{h}_i$):

$$\mathbf{g}_{i+1} = \mathbf{g}_i - \lambda_i \mathbf{H}\mathbf{h}_i; \tag{A.22a}$$

$$\mathbf{h}_{i+1} = \mathbf{g}_{i+1} + \gamma_i \mathbf{h}_i; \tag{A.22b}$$

$$\mathbf{x}_{i+1} = \mathbf{x}_i + \lambda_i \mathbf{h}_i \tag{A.22c}$$

with

$$\lambda_i = \frac{\mathbf{g}_i \cdot \mathbf{g}_i}{\mathbf{g}_i \cdot \mathbf{H}\mathbf{h}_i}; \tag{A.23a}$$

$$\gamma_i = -\frac{\mathbf{g}_{i+1} \cdot \mathbf{H}\mathbf{h}_i}{\mathbf{h}_i \cdot \mathbf{H}\mathbf{h}_i}. \tag{A.23b}$$

For this algorithm, the following properties can be verified:

(i) $\mathbf{g}_i \cdot \mathbf{g}_{i+1} = 0$ for all i.
(ii) $\mathbf{h}_{i+1} \cdot \mathbf{H}\mathbf{h}_i = 0$, for all i.
(iii) $\mathbf{g}_{i+1} \cdot \mathbf{h}_i = 0$ for all i (this is equivalent to (A.16)).
(iv) $\mathbf{g}_i \cdot \mathbf{g}_j = 0$ for all $i \neq j$.
(v) $\mathbf{h}_i \cdot \mathbf{H}\mathbf{h}_j = 0$ for all $i \neq j$.

The proof of these statements will be considered in Problem A.2. It is now easy to derive the following alternative formulas for λ_i and γ_i:

$$\lambda_i = \frac{\mathbf{g}_i \cdot \mathbf{h}_i}{\mathbf{h}_i \cdot \mathbf{H}\mathbf{h}_i}; \tag{A.24a}$$

$$\gamma_i = \frac{\mathbf{g}_{i+1} \cdot \mathbf{g}_{i+1}}{\mathbf{g}_i \cdot \mathbf{g}_i}. \tag{A.24b}$$

For an arbitrary (i.e. nonquadratic) function, the conjugate gradient method will probably be less efficient. However, it is expected to perform significantly better than the steepest descent method, in particular close to the minimum where f can be approximated well by a quadratic form. A problem is that the above prescription cannot be used directly as we do not know the matrix $\mathbf{H}$. However, in this case λ_i can be found from a line minimisation: x_{i+1} should be such that it is a minimum of f along the line $\mathbf{x}_i + \lambda_i \mathbf{h}_i$, in other words $\mathbf{h}_i \cdot \nabla f(\mathbf{x}_i + \lambda_i \mathbf{h}_i) = 0$. For a quadratic function as considered above, this condition does indeed reduce to (A.24a); for a general function we use the bisection minimisation or Brent's method. We know then the value of $\mathbf{x}_{i+1}$. The gradients $\mathbf{g}_{i+1}$ can be calculated directly as $\mathbf{g}_{i+1} = -\nabla f(\mathbf{x}_{i+1})$. We use $\mathbf{g}_{i+1}$ and $\mathbf{g}_i$, together with equation (A.24b) to find γ_{i+1}, and use this in (A.22b) to find $\mathbf{h}_{i+1}$. We see that the function values and gradients of f suffice to construct a sequence of directions which in the case of a quadratic function are conjugate.

A5 Discretisation

In the equations of physics, most quantities are functions of continuous variables. Obviously, it is impossible to represent such functions numerically in a computer since real numbers are always stored using a finite number of bits (typically 32 or 64) and therefore only a finite number of different real numbers is available. But even storing a function for all arguments allowed by the computer resolution is impossible since the memory required for this would exceed any reasonable bounds. Instead of representing the functions with the maximum precision possible, they are usually defined on a much coarser grid, specified by the user and independent of the representation of real numbers in the computer. Most problems can be solved with sufficient accuracy using discretised variables. One should, however, in all cases be careful in choosing the discretisation step; a balance must be found between the number of values and the level of precision .

Suppose we want to solve Newton's equations for the motion of a satellite orbiting around the Earth. Its path is quite smooth, the velocity changes relatively slowly so that a relatively large and constant time interval yields accurate results. If we consider a rocket launched from the Earth which should orbit around the Moon, most of the path between Earth and Moon will be smooth so that a large time step is possible, but when the rocket comes close to the Moon, its velocity may change strongly because of the Moon's attraction, and a smaller time step will then be necessary to keep the representation accurate.

Very many numerical methods are based on discretisation. If a function is represented on a discrete grid, it is possible to reconstruct the function by interpolation. Interpolation often consists of constructing a polynomial that assumes the same value as the discretised function on the grid points. The larger the number of points taken into account, the higher the order of the interpolation polynomial. A higher order implies a more accurate interpolation, but too high an order often results in strongly oscillatory behaviour of the polynomial between the grid points, so that it deviates strongly from the original function there. Interpolations are often used to derive numerical methods with a high order of accuracy. Examples can be found in the next two sections. When discretising a function, it should always be kept in mind that the interval must be chosen such that the main features of the function are preserved in the discretised representation.

A6 Numerical quadratures

Numerical integration (or *quadrature*) of a continuous and bounded function on the interval $[a, b]$ can be done straightforwardly. One defines an equidistant grid on the interval $[a, b]$: the grid points x_n are given by

$$x_n = a + nh \tag{A.25}$$

with $h = (b - a)/N$ and the index n running from 0 to N. It is clear that a (crude) approximation to the integral is given by the sum of the function values on the grid points:

$$\int_a^b f(x)\mathrm{d}x \approx h \sum_{n=0}^{N-1} f(x_n). \tag{A.26}$$

Such an approximation is useless without an estimate of the error in the result. If f is continuous and bounded on the interval $[x_n, x_n + h]$, its value does not deviate from $f(x_n)$ more than Mh, with M some finite constant, depending on the function f and the values of a and b. Therefore, the relative error in this result is $\mathcal{O}(h)$: doubling the number of integration points reduces the error by a factor of about 2.

A way of approximating the integral of a differentiable function f to second order in h using the grid points given in Eq. (A.25) is the following. Approximate f on the interval $[x_n, x_{n+1}]$ to first order in h by

$$f(x_n) + \frac{(x - x_n)}{h}[f(x_{n+1}) - f(x_n)] + \mathcal{O}(h^2), \; n = 0, \ldots, N-1. \tag{A.27}$$

Integrating this form analytically on the interval, we obtain

$$\int_a^b f(x)\mathrm{d}x = h\left[\frac{1}{2}f(x_0) + f(x_1) + f(x_2) + \cdots + f(x_{N-1}) + \frac{1}{2}f(x_N)\right] + \mathcal{O}(h^2). \tag{A.28}$$

This is called the trapezoidal rule.

If f is twice differentiable, we can approximate it by piecewise quadratic functions. For a single interval using three equidistant points, we have

$$\int_a^b f(x)\mathrm{d}x = \frac{b-a}{6}\left[f(a) + 4f\left(\frac{a+b}{2}\right) + f(b)\right] + \mathcal{O}[(b-a)^5], \tag{A.29}$$

which is one order of magnitude better than expected, because there is a cancellation of errors at the left and right hand boundaries, resulting from the symmetry of this formula. If the interval $[a, b]$ is split up into $N/2$ subintervals, each of size $2h$, we obtain the following expression:

$$\int_a^b f(x)\mathrm{d}x = h\frac{1}{3}[f(x_0) + 4f(x_1) + 2f(x_3) + 4f(x_4) + \cdots \\ + 2f(x_{N-2}) + 4f(x_{N-1}) + f(x_N)] + \mathcal{O}(h^4). \tag{A.30}$$

This is an example of *Simpson's rule*.

It would be easy to extend this list of algorithms. They all boil down to making some polynomial approximation of the integrand and then integrating the approximate formula analytically, which is trivial for a piecewise polynomial. Also, many tricks exist for integrating functions containing singularities, but these are beyond the scope of this discussion. In all cases, the order of the accuracy is equal to the number of points used in fitting the polynomial, since for n points one can determine a polynomial of order $n - 1$ having the same value as the function to be integrated on all n points.

A very efficient method consists of repeating the trapezoidal rule and performing it for successive values of h, each having half the size of the previous one. This yields a sequence of approximants to the integral for various values of h. These approximants can be fitted to a polynomial and the value for this polynomial at $h = 0$ is a very accurate approximation to the exact value. This is called the *Romberg method.*

The famous Gauss integration method works essentially the same way as the polynomial methods for equidistant points described above, but for the grid points x_n the zeroes of the Legendre polynomials are taken, and on the interval $[a, b]$ the function f is approximated by Legendre polynomials. Legendre polynomials P_l have the property of being orthonormal on the interval $[-1, 1]$:

$$\int_{-1}^{1} P_l(x)P_{l'}(x)\mathrm{d}x = \delta_{ll'}. \tag{A.31}$$

The advantage of the Gauss–Legendre method is that its accuracy is much better than that of other methods using the same number of integration points: the accuracy of an N-point Gauss–Legendre method is equivalent to that of an equidistant-point method using $2N$ points!

We give no derivations but just present the resulting algorithm on the interval $[-1, 1]$:

$$\int_{-1}^{1} \mathrm{d}x f(x) = \sum_{n=1}^{N} w_n f(x_n) + \mathcal{O}(h^{2N}). \tag{A.32}$$

Here, x_n are the zeroes of the Legendre polynomial P_N, h is $2/N$, x_n and w_n can be found in many books. Moreover, there exist programs to generate them [1].

There exist other Gauss integration methods for nonbounded intervals [1].

A7 Differential equations

Many physical theories boil down to one or more differential equations: for example Newton's second law, the Schrödinger equation or Maxwell's equations. It is therefore of primary importance to the computational physicist to have available reliable and efficient methods for solving such equations. As we have seen in Chapter 3 of this book, we can often determine the stationary functions of a functional instead of solving the differential equations directly (these equations can even be derived using such a stationarity condition), but here we restrict ourselves to the direct solution of the differential equations. In the field of numerical analysis, many methods have been developed to solve differential equations. Here we shall not treat these methods in great detail, but review the most common ones and their properties.

What makes a method 'good' depends on the problem at hand, and a number of criteria can be distinguished:

- **Precision and speed**. Higher precision will generally cost more time, higher speed will yield less precise results.

- **Stability**. In some methods, errors in the starting values or errors due to the discrete numerical representation tend to grow during integration, so that the solutions obtained deviate sometimes strongly from the exact ones. If the errors tend to grow during integration, the method is called *unstable*. It is essentially the same phenomenon as we encountered in the discussion of recursion in Appendix A2.
- **Implementation**. Very complicated algorithms are sometimes less favourable because of the time needed to implement them correctly. This criterion is of course irrelevant when using existing routines or programs (e.g. from numerical libraries).
- **Flexibility**. In all methods, the coordinates are discretised: some methods demand a fixed discretisation interval. These are less useful for problems with solutions having strongly varying behaviour (somewhere very smooth and elsewhere oscillating rapidly).
- **Symmetry**. For particular types of differential equations we would like the numerical method to share symmetry properties of the original equation; an example is time reversibility which might be present in the equation of motion of a particle. In Chapter 8, symplectic symmetry properties of Hamiltonian equations and particular integration schemes are discussed.

There are other criteria, such as analyticity of the functions occurring in the differential equation, which make some methods more suitable than others.

A7.1 Ordinary differential equations

We now describe a number of numerical algorithms for the solution of this type of equation. In one dimension, a first order differential equation looks like

$$\dot{x}(t) = f[x(t), t]. \tag{A.33}$$

We call the variables x and t 'space' and 'time' respectively, although in various problems they will represent completely different quantities. In the following we integrate always from $t = 0$. In practice, one integrates from arbitrary values of t, but the methods described here are trivially generalised.

By writing $y(t) = \dot{x}(t)$, a second order equation

$$\ddot{x}(t) = f[x(t), t] \tag{A.34}$$

can be transformed into two differential equations of the form (A.33):

$$\dot{x}(t) = y(t) \tag{A.35a}$$

$$\dot{y}(t) = f[x(t), t] \tag{A.35b}$$

and the methods that will be discussed are easily generalised to this two-dimensional case.

Stability

As noted above, some differential equations are susceptible to instabilities in the numerical solution. In first order, one-dimensional equations, taking a small time step usually guarantees stability, but in higher dimensions or for higher orders it is not so easy to avoid instabilities. The reason is that such equations have in general more than one independent solution; the initial values determine how these are combined into the actual solution. The situation often occurs that the solution we are after is a damped one (in the direction in which we integrate), but there exists a growing solution to the equation too which will mix into our numerical solution and spoil the latter as it grows, just as in the recursion problems discussed in Appendix A2. If possible, one should integrate in the other direction with the appropriate initial conditions, so that the required solution is the growing one, as instabilities are absent in that case.

Sometimes, however, the problem cannot be solved so easily, and this is particularly the case when the two solutions have a very different time scale. There might, for example, be a combination of a fast and a slow oscillation, where the time scale of the fast one determines the time step for which the integration is stable. More tricky is the existence of a very strongly damped term in the solution, which is easily overlooked (because it damps out so quickly) but might spoil the numerical solution if the time step used is not adapted to the strongly damped term. Equations that suffer from this type of problem are called 'stiff equations' [4]. For more details about stiff equations see Ref. [4].

The Runge–Kutta method

This is a frequently used method for the solution of ordinary differential equations. Like most methods for solving differential equations, the Runge–Kutta method can be considered as a step-wise improvement of Euler's forward method, the latter predicting the solution to the differential equation (A.33), stating that, given $x(0)$, $x(t)$ for $t = h$ is given by

$$x(h) = x(0) + hf[x(0), 0] + \mathcal{O}(h^2). \tag{A.36}$$

It is clear that if we use Euler's rule halfway across the interval $(0, h)$ to estimate $x(h/2)$, and evaluate $f(x, t)$ for this value of x and $t = h/2$, we obtain a better estimate for $x(h)$:

$$\begin{aligned} k_1 &= hf[x(0), 0] \\ k_2 &= hf\left[x(0) + \frac{1}{2}k_1, h/2\right] \\ x(h) &= x(0) + k_2 + \mathcal{O}(h^3). \end{aligned} \tag{A.37}$$

This is called the *midpoint rule*. The fact that the error is of order h^3 can be verified by expanding $x(t)$ in a power series in t to second order. One can systematically derive similar rules for higher orders. The most popular one is the fourth order rule:

$$\begin{aligned}
k_1 &= hf(x, 0)\\
k_2 &= hf\left(x + \frac{1}{2}k_1, h/2\right)\\
k_3 &= hf\left(x + \frac{1}{2}k_2, h/2\right)\\
k_4 &= hf(x + k_3, h)\\
x(h) &= x(0) + \frac{1}{6}(k_1 + 2k_2 + 2k_3 + k_4) + \mathcal{O}(h^5).
\end{aligned} \tag{A.38}$$

We shall not present a derivation of this rule, because it involves some tedious algebra. Although the name 'Runge–Kutta method' is used for the method for arbitrary order, it is usually reserved for the fourth order version.

The Runge–Kutta method owes its popularity mostly to the possibility of using a variable discretisation step: at every step, one can decide to use a different time interval without difficulty. A severe disadvantage for some applications is the fact that the function $f(x, t)$ has to be evaluated four times per integration step. Especially when integrating Newton's equations for many-particle systems, as is done in molecular dynamics, the force evaluations demand a major part of the computer time, and the Runge–Kutta method with its large number of force evaluations becomes therefore very inefficient.[1]

It is possible to implement the Runge–Kutta method with a prescribed maximum error, δ, to the resulting solution, by adapting the step size automatically. In this method the fact that the Runge–Kutta rule has an order h^5 error is used to predict for which time step the desired accuracy will be achieved. For each time step t, the value of x at time $t+h$ is calculated twice, first using a time interval h and then using two intervals, each of length $h/2$. By comparing the two results, it is possible to estimate the time step needed to determine the resulting value with the prescribed precision δ. If the absolute value of the difference between the two results is equal to Δ, the new time step h' is given by $h' = (15/16)h(\delta/\Delta)^{1/5}$, as can easily be verified. It is recommended to use the Runge–Kutta rule only in this implementation, because when using a constant time step the user gets no warning whatsoever about this step being too large (for example when the function f is varying strongly) and the solution may deviate appreciably from the exact one without this being noticed.

[1] The fact that Runge–Kutta methods are not symplectic is another reason why they are not suitable for this application – see Chapter 8.

Verlet algorithm

The Verlet algorithm is a simple method for integrating second order differential equations of the form

$$\ddot{x}(t) = F[x(t), t]. \tag{A.39}$$

It is used very often in molecular dynamics simulations because of its simplicity and stability (see Chapter 8). It had been mentioned already at the end of the eighteenth century [8, 9]. It has a fixed time discretisation interval and it needs only one evaluation of the 'force' F per step. The Verlet algorithm is easily derived by adding the Taylor expansions for the coordinate x at $t = \pm h$ about $t = 0$:

$$x(h) = x(0) + h\dot{x}(0) + \frac{h^2}{2}F[x(0), 0] + \frac{h^3}{6}\dddot{x}(0) + \mathcal{O}(h^4) \tag{A.40}$$

$$x(-h) = x(0) - h\dot{x}(0) + \frac{h^2}{2}F[x(0), 0] - \frac{h^3}{6}\dddot{x}(0) + \mathcal{O}(h^4) \tag{A.41}$$

leading to

$$x(h) = 2x(0) - x(-h) + h^2 F[x(0), 0] + \mathcal{O}(h^4). \tag{A.42}$$

Knowing the values of x at time 0 and $-h$, this algorithm predicts the value of $x(h)$. We always need the last two values of x to produce the next one. If we only have the initial position, $x(0)$, and velocity, $v(0)$, at our disposal, we approximate $x(h)$ by

$$x(h) = x(0) + v(0)h + \frac{h^2}{2}F[x(0), 0] + \mathcal{O}(h^3). \tag{A.43}$$

The algorithm is invariant under time reversal (as is the original differential equation). This means that if, after having integrated the equations for some time, time is reversed by swapping the two most recent values of x, exactly the same path in phase space should be followed backward in time. In practice, there will always be small differences as a result of finite numerical precision of the computer.

As we shall see in Problem A.3, the accumulated error in the position after a large number of integration steps is of order $\mathcal{O}(h^2)$. As there is no point in calculating the velocities with higher precision than that of the positions, we can evaluate the velocities using the simple formula

$$v(0) = \frac{x(h) - x(-h)}{2h} \tag{A.44}$$

which gives the velocity with the required $\mathcal{O}(h^2)$ error.

Using the velocities at half-integer time steps:

$$v[(n + 1/2)h] = \{x[(n + 1)h] - x(nh)\}/h + \mathcal{O}(h^2), \tag{A.45}$$

we may reformulate the Verlet algorithm in the *leap-frog* form:

$$v(h/2) = v(-h/2) + hF[x(0), 0] \tag{A.46a}$$

$$x(h) = x(0) + hv(h/2). \tag{A.46b}$$

Note that this form is exactly equivalent to the Verlet form, so the error in the positions is still $\mathcal{O}(h^4)$. This form is less sensitive to errors resulting from finite precision arithmetic in the computer than the Verlet form (A.42).

Note that the time step is not easily adapted in the Verlet/leap-frog algorithm. This is, however, possible, either by changing the coefficients of the various terms in the equation, or by reconstructing the starting values with the new time step by interpolation.

Consider the Verlet solution for the harmonic oscillator $\ddot{x} = -Cx$:

$$x(t + h) = 2x(t) - x(t - h) - h^2 Cx(t). \tag{A.47}$$

The analytic solution to this algorithm (not the exact solution to the continuum differential equation) can be written in the form $x(t) = \exp(i\omega t)$, with ω satisfying

$$2 - 2\cos(\omega h) = h^2 C. \tag{A.48}$$

If $h^2 C > 4$, ω becomes imaginary, and the analytical solution becomes unstable. Of course we would not take h that large in actual applications, as we always take it substantially smaller than the period of the continuum solution. It is, however, useful to be aware of this instability, especially when integrating systems of second order differential equations, which reduce to a set of coupled harmonic equations close to the stationary solutions. High-frequency degrees of variables are then often easily overlooked.

Numerov's method

This is an efficient method for solving equations of the type

$$\ddot{x}(t) = f(t)x(t) \tag{A.49}$$

of which the stationary Schrödinger equation is an example [10]. Numerov's method makes use of the special structure of this equation in order to have the fourth order contribution to $x(h)$ cancel, leading to a form similar to the Verlet algorithm, but accurate to order h^6 (only even orders of h occur because of time-reversal symmetry). The Verlet algorithm (see above) was derived by expanding $x(t)$ up to second order around $t = 0$ and adding the resulting expression for $t = h$ and for $t = -h$. If we do the same for Eq. (A.49) but now expand $x(t)$ to order six in t, we obtain

$$x(h) + x(-h) - 2x(0) = h^2 f(0)x(0) + \frac{h^4}{12} x^{(4)}(0) + \frac{h^6}{360} x^{(6)}(0) + \mathcal{O}(h^8) \tag{A.50}$$

with $x^{(4)}$ being the fourth and $x^{(6)}$ the sixth derivative of x with respect to t. As these derivatives are not known, this formula is not useful as such. However, we can evaluate the fourth derivative as the discrete second derivative of the second derivative of $x(t)$ taken at $t = 0$. Using the differential equation (A.49) we then obtain

$$x^{(4)}(0) = \frac{x(h)f(h) + x(-h)f(-h) - 2x(0)f(0)}{h^2}, \tag{A.51}$$

and then, after switching to another variable $w(t) = [1 - (h^2/12)f(t)]x(t)$, Eq. (A.50) becomes

$$w(h) + w(-h) - 2w(0) = h^2 f(0)x(0) + \mathcal{O}(h^6). \tag{A.52}$$

Now we see that, using a second order integration scheme for w (i.e. using only two values of the solution to predict the next one), $x(h)$ is known to order h^6. Whenever $x(t)$ is required, it can be calculated as $x(t) = [1 - (h^2/12)f(t)]^{-1}w(t)$. Note that the integration error over a fixed interval scales as h^4, that is, two orders less than the single-step error – see the analysis in Problem A.3.

The implementation of the initial conditions is not straightforward: when not dealt with carefully enough, we might obtain errors which at the end of the integration interval scale worse than $\mathcal{O}(h^4)$. Usually the initial value $x(0)$ and the derivative $\dot{x}(0)$ are known. From these, we need to predict the value at $x(h)$ with an accuracy of at least $\mathcal{O}(h^6)$. This can be done by first subtracting the Taylor expansions for $x(h)$ and $x(-h)$ rather than adding them as was done in the derivation of the Numerov algorithm. The result is

$$2h\dot{x}(0) = [1 - h^2 f(h)/6]x(h) - [1 - h^2 f(-h)/6]x(-h) + \mathcal{O}(h^5). \tag{A.53}$$

The derivative $\dot{x}(0)$ is known: together with the Numerov algorithm, we have two equations for $x(h)$ and $x(-h)$. Eliminating the latter, we are left with

$$x(h) = \frac{[2 + 5h^2 f(0)/6][1 - h^2 f(-h)/12]x(0) + 2h\dot{x}(0)[1 - h^2 f(-h)/6]}{[1 - h^2 f(h)/12][1 - h^2 f(-h)/6] + [1 - h^2 f(-h)/12][1 - h^2 f(h)/6]}. \tag{A.54}$$

Starting with $x(0)$ and $x(h)$, the Numerov algorithm can be applied straightforwardly.

More details about this method and its applications can be found in Ref. [11].

Finite difference methods

The Verlet algorithm is a special example of this class of methods. Finite difference algorithms exist for various numerical problems, like interpolation, numerical differentiation and solving ordinary differential equations [12]. Here we consider the solution of a differential equation of the form

$$\dot{x}(t) = f[x(t), t], \tag{A.55}$$

and in passing we touch upon the interpolation method.

For any function x depending on the coordinate t, one can build the following table (the entries in this table will be explained below):

$$\begin{array}{cccccccc}
-3 & x_{-3} & & & & & & \\
 & & \delta x_{-5/2} & & & & & \\
-2 & x_{-2} & & \delta^2 x_{-2} & & & & \\
 & & \delta x_{-3/2} & & \delta^3 x_{-3/2} & & & \\
-1 & x_{-1} & & \delta^2 x_{-1} & & \delta^4 x_{-1} & & \\
 & & \delta x_{-1/2} & & \delta^3 x_{-1/2} & & \delta^5 x_{-1/2} & \\
0 & x_0 & & \delta^2 x_0 & & \delta^4 x_0 & & \delta^6 x_0 \\
 & & \delta x_{1/2} & & \delta^3 x_{1/2} & & \delta^5 x_{1/2} & \\
1 & x_1 & & \delta^2 x_1 & & \delta^4 x_1 & & \\
 & & \delta x_{3/2} & & \delta^3 x_{3/2} & & & \\
2 & x_2 & & \delta^2 x_2 & & & & \\
 & & \delta x_{5/2} & & & & & \\
3 & x_3 & & & & & &
\end{array}$$

The first column contains equidistant time steps at separation h, and the second one the values x_t for the times t of the first column. We measure the time in units of h. The third column contains the differences between two subsequent values of the second one; therefore they are written just halfway between these two values. The fourth column contains differences of these differences and so on. Formally, except for the first column, the entries are defined by

$$\begin{aligned}
\delta^j x_{i-1/2} &= \delta^{j-1} x_i - \delta^{j-1} x_{i-1} \\
\delta^j x_i &= \delta^{j-1} x_{i+1/2} - \delta^{j-1} x_{i-1/2} \quad \text{and} \\
\delta^0 x_i &\equiv x_i.
\end{aligned} \tag{A.56}$$

It is possible to build the entire table from the second column only, or from the values $\delta^{2j} x_0$ and $\delta^{2j+1} x_{1/2}$ (that is, the elements on the middle horizontal line and the line just below), because the table is highly redundant.

Such a table can be used to construct an interpolation polynomial for x_t. First we note that

$$\begin{aligned}
x_{\pm 1} &= x_0 \pm \delta x_{\pm 1/2} \\
x_{\pm 2} &= x_0 \pm 2\delta x_{\pm 1/2} + \delta^2 x_{\pm 1} \\
x_{\pm 3} &= x_0 \pm 3\delta x_{\pm 1/2} + 3\delta^2 x_{\pm 1} \pm \delta^3 x_{\pm 3/2}
\end{aligned} \tag{A.57}$$

and so on. In these expressions the binomial coefficients are recognised. From these equations, it can be directly seen that

$$x_t = x_0 + t\delta x_{1/2} + \frac{t(t-1)}{2}\delta^2 x_1 + \frac{t(t-1)(t-2)}{3!}\delta^3 x_{3/2} + \cdots \tag{A.58}$$

is a polynomial which coincides with x_t for $t = 1, 2, 3$. One can also build higher order polynomials in the same fashion. For negative t-values, this formula reads

$$x_t = x_0 + t\delta x_{-1/2} + \frac{t(t+1)}{2}\delta^2 x_{-1} + \frac{t(t+1)(t+2)}{3!}\delta^3 x_{-3/2} + \cdots \tag{A.59}$$

These formulas are called *Newton interpolation formulas.*

We now use this table to integrate differential equations of type (A.55). Therefore we consider a difference table, not for x, but for $\dot{x}$:

$$\begin{array}{rllll}
-2 & \dot{x}_{-2} & & \delta^2\dot{x}_{-2} & \\
 & & \delta\dot{x}_{-3/2} & & \delta^3\dot{x}_{-3/2} \\
-1 & \dot{x}_{-1} & & \delta^2\dot{x}_{-1} & \\
 & & \delta\dot{x}_{-1/2} & & \\
0 & \dot{x}_0 & & & \\
 & & & & \\
1 & & & &
\end{array}$$

For simplicity we did not make this table too 'deep' (up to third differences). Suppose the equation has been integrated up to $t = 1$, so x_1 is known, and we would like to calculate x_2. This is possible by adding an extra row to the lower end of the table. This can be done because from (A.55), $\dot{x}_1 = f(x_1, 1)$ and the table can be extended as follows:

$$\begin{array}{rllll}
-2 & \dot{x}_{-2} & & \delta^2\dot{x}_{-2} & \\
 & & \delta\dot{x}_{-3/2} & & \delta^3\dot{x}_{-3/2} \\
-1 & \dot{x}_{-1} & & \delta^2\dot{x}_{-1} & \\
 & & \delta\dot{x}_{-1/2} & & \delta^3\dot{x}_{-1/2} \\
0 & \dot{x}_0 & & \delta^2\dot{x}_0 & \\
 & & \delta\dot{x}_{1/2} & & \\
1 & \dot{x}_1 & & &
\end{array}$$

Using this table, an interpolation polynomial for $\dot{x}_t$ can be constructed. The Newton interpolation polynomial (A.59) for $\dot{x}$ reads

$$\dot{x}_t = \dot{x}_1 + (t-1)\delta\dot{x}_{1/2} + \frac{(t-1)t}{2!}\delta^2\dot{x}_0 + \frac{(t-1)(t+1)t}{3!}\delta^3\dot{x}_{-1/2} + \cdots \quad \text{(A.60)}$$

We can now use this as an extrapolation polynomial to calculate x_2 (we assumed x_1 to be known). Using

$$x_2 = x_1 + \int_1^2 \dot{x}_t \mathrm{d}t, \quad \text{(A.61)}$$

one finds, using (A.60):

$$x_2 = x_1 + \dot{x}_1 + \frac{1}{2}\delta\dot{x}_{1/2} + \frac{5}{12}\delta^2\dot{x}_0 + \frac{3}{8}\delta^3\dot{x}_{-1/2}. \quad \text{(A.62)}$$

This value can then be used to add another row to the table, and so on.

Up to order 6, the formula reads

$$x_2 = x_1 + \dot{x}_1 + \frac{1}{2}\delta\dot{x}_{1/2} + \frac{5}{12}\delta^2\dot{x}_0 + \frac{3}{8}\delta^3\dot{x}_{-1/2} + \frac{251}{720}\delta^4\dot{x}_{-1} + \frac{95}{288}\delta^5\dot{x}_{-3/2} + \frac{19087}{60480}\delta^6\dot{x}_{-2} + \cdots \quad \text{(A.63)}$$

This equation is known as Adams' formula. Note that during the integration it is needed to renew the lower diagonal of the table only and therefore we need not store the entire table in memory.

To start the algorithm, we need the solution to the differential equation at the first few time steps to be able to set up the table. This solution can be generated using a Runge–Kutta starter, for example. The starting points should, however, be of the same order of accuracy as Adams' method itself. If this is not the case, we can use a special procedure which improves the accuracy of the table iteratively [13].

Finite difference methods are often very efficient as they can be of quite high order. As soon as the integration becomes inaccurate because of too large a time step, the deepest (i.e. the rightmost) differences tend to diverge and will lead to an overflow. This means that absence of such trouble implies high accuracy.

One can also construct so-called *predictor-corrector* methods in this way. In these methods, after every integration step the last value of x_t (in our case this is x_2) is calculated again, now using also $\dot{x}_t$ which is determined by the value of x_t and the differential equation (A.55). This procedure is repeated until the new value for x_t is close enough to the previous one. Predictor-corrector methods are seldom used nowadays, since using a sufficiently small integration step makes the time-consuming corrector cycle superfluous.

Bulirsch–Stoer method

This method is similar to the Romberg method for numerical integration. Suppose the value of x_t is wanted, x_0 being known. Over the interval $[0, t]$, the equation can be integrated using a simple method with a few steps. Then the method is repeated with a larger number of steps and so on. The resulting predictions for x_t are stored as a function of the inverse of the number of steps used. Then these values are used to build an interpolation polynomial which is extrapolated to an infinite number of steps. The efficiency of this method is another reason why predictor-corrector methods are seldom used.

A7.2 Partial differential equations

In mathematics, one usually classifies partial differential equations (PDE) according to their characteristics, leading to three different types: parabolic, elliptic and hyperbolic. In numerical analysis, this classification is less relevant, and only two different types of equations are distinguished: initial value and boundary value problems. We study examples of both categories. In the next two sections, we describe finite difference methods for these two types and then we discuss several other methods for partial differential equations very briefly.

Initial value problems

An important example of this class of problems is the diffusion equation or the time-dependent Schrödinger equation (see Section 12.2.4 for a discussion of the relation between these two equations) which is mathematically the same as the diffusion equation, the only difference being a factor i before the time derivative. In the following, we shall consider the Schrödinger equation, but the analysis is the same for the diffusion equation.

Consider the one-dimensional time-dependent Schrödinger equation (we take $\hbar = 2m = 1$):

$$\mathrm{i}\frac{\partial \psi(x,t)}{\partial t} = -\frac{\partial^2 \psi(x,t)}{\partial x^2} + V(x)\psi(x,t), \tag{A.64}$$

or, in a more compact notation:

$$\mathrm{i}\frac{\partial \psi(x,t)}{\partial t} = H\psi(x,t) \tag{A.65}$$

with H standing for the Hamilton operator. First we discretise H at L equidistant positions on the real axis, with separation Δx. The value of ψ at the point $x_j = j\Delta x$, $j = 1, \ldots, L$, is denoted by ψ_j. We can then approximate the second order derivative with respect to x on the jth position as follows:

$$\frac{\partial^2 \psi(x_j, t)}{\partial x^2} = \frac{1}{\Delta x^2}[\psi_{j-1}(t) + \psi_{j+1}(t) - 2\psi_j(t)] + \mathcal{O}(\Delta x^2). \tag{A.66}$$

We denote the discretised Hamiltonian by H_D, so that the Schrödinger equation can be written as

$$\mathrm{i}\frac{d\psi_j(t)}{dt} = H_D \psi_j(t). \tag{A.67}$$

As a next step, we discretise time using intervals Δt. Indicating the nth time step with an upper index n to the wave function ψ, the time derivative of ψ_j^n at this time step can be approximated by $(\psi_j^{n+1} - \psi_j^n)/\Delta t$ (with an error of order Δt), so that the discretised form of Eq. (A.64) is given by

$$\mathrm{i}\frac{1}{\Delta t}(\psi_j^{n+1} - \psi_j^n) = \frac{1}{\Delta x^2}[-\psi_{j-1}^n(t) - \psi_{j+1}^n(t) + 2\psi_j^n(t)] + V_j \psi_j^n \tag{A.68}$$

or, in shorthand notation:

$$\psi_j^{n+1} = (1 - \mathrm{i}\Delta t H_D)\psi_j^n. \tag{A.69}$$

The stability of this method can be investigated using the so-called Von Neumann analysis. In this analysis, one considers the analytical solution to Eq. (A.69) for V_j constant:

$$\psi_j^n = \xi^n \exp(\mathrm{i}kj\Delta x). \tag{A.70}$$

The wave vector k depends on the boundary conditions, and ξ and k are related – the relation can be found by substituting this solution in Eq. (A.68). If there exists such a solution with $|\xi| > 1$, it follows that a small perturbation in the solution tends to grow exponentially, because an expansion of a generic small perturbation in terms of the solutions (A.70) will almost certainly contain the $|\xi| > 1$ modes. For our algorithm, we find

$$\xi = 1 - \mathrm{i}\Delta t\left[\frac{4}{\Delta^2 x}\sin^2(k\Delta x/2) + V_j\right] \tag{A.71}$$

which means that in all cases $|\xi| > 1$, indicating that this method is always unstable. Therefore, we shall consider a modification of the method.

First, the wave function occurring in the right hand side of (A.69) is calculated at time $t = (n+1)\Delta t$. We then arrive at a form which reads in shorthand:

$$\psi_j^{n+1} = \psi_j^n - \mathrm{i}\Delta t H_D \psi_j^{n+1}. \tag{A.72}$$

This algorithm seems impractical, however, because to determine ψ^{n+1} on the left hand side of (A.72) it is needed on the right hand side: it is an *implicit* form. We can make this explicit by an inversion:

$$\psi_j^{n+1} = \sum_{j'} (1 + \mathrm{i}\Delta t H_\mathrm{D})^{-1}_{jj'} \psi_{j'}^n. \tag{A.73}$$

As the matrix in brackets is tridiagonal, the inversion can be done efficiently – it requires only $\mathcal{O}(L)$ steps, i.e. of the same order as the other steps of the algorithm. The inversion method will be discussed in Appendix A8.1.

Performing the Von Neumann analysis for this method, we find

$$\xi = \frac{1}{1 + i\Delta t[(4/\Delta^2 x)\sin^2(k\Delta x/2) + V_j]}. \tag{A.74}$$

This looks more promising than the first method: for *every* choice of Δt and Δx one gets $|\xi| < 1$, so that the method is stable. It is important to take Δt small as the time-derivative of ψ_i has a first order accuracy in Δt only. The accuracy of the discretised second derivative with respect to x is still of order Δx^2.

We have neglected an important point: we would like ψ^n to be normalised at every time step. This is unfortunately not the case for the methods discussed up to now. Writing the exact solution to the continuous differential equation (A.64) as

$$\psi(x,t) = \exp(-\mathrm{i}tH/\hbar)\psi(x,0), \tag{A.75}$$

we notice that for all times t, the norm of ψ is equal to that at $t = 0$ since $\exp(-\mathrm{i}tH/\hbar)$ is a unitary operator:

$$\langle\psi(t)|\psi(t)\rangle = \langle\exp(-\mathrm{i}tH/\hbar)\psi(0)|\exp(-\mathrm{i}tH/\hbar)\psi(0)\rangle = \langle\psi(0)|\psi(0)\rangle. \tag{A.76}$$

The operator $1 \pm \mathrm{i}\Delta t H_\mathrm{D}$ is in general not unitary. If we are able to find a unitary operator which gives us at the same time a stable method, we have the best method we can think of (apart from finite accuracy inherent to discretisation). It turns out that the operator

$$\frac{1 - \mathrm{i}\Delta t H_\mathrm{D}/2}{1 + \mathrm{i}\Delta t H_\mathrm{D}/2} \tag{A.77}$$

does the job, and surpasses expectations: not only is it unitary and stable, it is even correct to *second* order in Δt. This can be seen by realising that using this matrix is equivalent to taking the average of H acting on ψ^n and ψ^{n+1}, or by noticing that the matrix in (A.77) is a second order approximation to $\exp(-\mathrm{i}\Delta t H/\hbar)$ in Δt.

This method is recommended for solving the time-dependent Schrödinger equation using finite difference methods. It is known as the *Crank–Nicholson* method. Note that the presence of the denominator in the operator (A.77) makes this an implicit method: a matrix inversion must be carried out at every time step, requiring $\mathcal{O}(L)$ calculations, which is the overall time scaling of the algorithm.

In the Crank–Nicholson and in the implicit scheme we need to solve a tridiagonal matrix equation. This is rather straightforward using back-substitution, and you may want to use a library routine for this purpose. However, when periodic boundaries are used, the matrix is no longer tridiagonal, but has elements in the upper right and lower left corner:

$$H = \begin{pmatrix} a_1 & b & 0 & \dots & 0 & b \\ b & a_2 & b & \dots & 0 & 0 \\ \vdots & \vdots & \vdots & \ddots & \vdots & \vdots \\ 0 & 0 & 0 & \dots & a_{N-1} & b \\ b & 0 & 0 & \dots & b & a_N \end{pmatrix}, \tag{A.78}$$

where a_k and b are complex numbers. In order to solve this equation we use the *Sherman–Morrison* formula. This is a formula which gives the inverse of a matrix of the form (in Dirac notation):

$$\mathbf{A} + |\mathbf{u}\rangle\langle\mathbf{v}|. \tag{A.79}$$

Applied to our problem, the Sherman–Morrison formula leads to a prescription for finding the solution of:

$$\mathbf{H}|\psi\rangle = (\mathbf{A} + |\mathbf{u}\rangle\langle\mathbf{v}|)|\psi\rangle = |\mathbf{b}\rangle. \tag{A.80}$$

In fact we should find solutions to the two following problems:

$$\mathbf{A}|\phi\rangle = |\mathbf{b}\rangle; \tag{A.81a}$$

$$\mathbf{A}|\chi\rangle = |\mathbf{u}\rangle. \tag{A.81b}$$

It is then easily shown that

$$|\psi\rangle = |\phi\rangle - \left(\frac{\langle\mathbf{v}|\phi\rangle}{1 + \langle\mathbf{v}|\chi\rangle}\right)|\chi\rangle \tag{A.82}$$

gives the correct solution.

In order to apply this recipe to our problem, we take

$$|\mathbf{u}\rangle = \begin{pmatrix} -a_1 \\ 0 \\ \vdots \\ 0 \\ b \end{pmatrix}; \quad |\mathbf{v}\rangle = \begin{pmatrix} 1 \\ 0 \\ \vdots \\ 0 \\ (-b/a_1)^* \end{pmatrix}. \tag{A.83}$$

The complex conjugate in the last element of $|\mathbf{v}\rangle$ should not be forgotten! The matrix $\mathbf{A}$ is identical to the matrix $\mathbf{H}$ except for the upper left and lower right diagonal elements. The first should be replaced by $a_1' = 2a_1$ and the last by $a_N' = \alpha_N + b^2/a_N$. It can easily be checked that these values yield the correct matrix to appear in the matrix equation. We need to solve two tridiagonal matrix equations for $|\phi\rangle$ and $|\chi\rangle$ to obtain the solution to the problem with periodic boundaries.

Finally, we describe a fourth method which can be applied to initial value problems: the *split operator* method. The idea of this method is based on principles which have been used extensively in Chapter 12. It consists of splitting the Hamiltonian into the kinetic and potential operator, and using representations in which these operators are diagonal. To be specific, the operator $V(x)$ is diagonal in the x-representation, whereas the kinetic energy term, $T = p^2/(2m)$ is diagonal in the p-representation. The two representations are connected through a Fourier transform:

$$\langle x|p\rangle = \frac{1}{\sqrt{2\pi}} \mathrm{e}^{\mathrm{i}px/\hbar}. \tag{A.84}$$

The time-evolution operator is split into three terms:

$$\mathrm{e}^{-\mathrm{i}\Delta t(V+T)/\hbar} = \mathrm{e}^{-\mathrm{i}\Delta tV/(2\hbar)}\mathrm{e}^{-\mathrm{i}\Delta tT/\hbar}\mathrm{e}^{-\mathrm{i}\Delta tV/(2\hbar)} + \mathcal{O}(\Delta t^3) \tag{A.85}$$

Suppose we know the initial state in the x-representation. Then we act on it with the front term (exponent of the potential) which is diagonal. Then we Fourier-transform the result and multiply it by the second term (exponent of the kinetic energy) and then, after Fourier-transforming back again, we multiply the result by the last factor (potential). This method has a similar efficiency to the Crank–Nicholson method.

Boundary value problems

We consider the Poisson equation for the potential of a charge distribution as a typical example of this category:

$$\nabla^2\psi(\mathbf{r}) = -\rho(\mathbf{r}). \tag{A.86}$$

∇^2 is the Laplace operator and ψ is the potential resulting from the charge distribution ρ. Furthermore, there are boundary conditions, which are generally of the Von Neumann or of the Dirichlet type (according to whether the derivative or the value of the potential is given at the boundary respectively). In this section we shall discuss iterative methods for solving this type of equation. For more details the reader is referred to standard books on the subject [14, 15].

We restrict ourselves to two dimensions, and we can discretise the Laplace equation on a square lattice, analogous to the way in which this was done for the Schrödinger equation in the previous subsection:

$$\nabla^2\psi(\mathbf{r}) = \left(\frac{\partial^2}{\partial x^2} + \frac{\partial^2}{\partial y^2}\right)\psi(\mathbf{r}); \tag{A.87a}$$

$$\begin{aligned}\nabla_{\mathrm{D}}^2\psi_{i,j} &= \frac{1}{\Delta x^2}(\psi_{i+1,j} + \psi_{i-1,j} + \psi_{i,j+1} + \psi_{i,j-1} - 4\psi_{ij}) \\ &= \nabla^2\psi(\mathbf{r}) + \mathcal{O}(\Delta x^2).\end{aligned} \tag{A.87b}$$

This leads to the discretised form of (A.86):

$$\nabla^2_{\mathrm{D}}\psi_{ij} = -\rho_{ij}. \tag{A.88}$$

The grid on which the Laplace operator is discretised is called a *finite difference grid*. An obvious approach to solving (A.88) is to use a relaxation method. We can interpret (A.88) as a self-consistency equation for ψ: on every grid point (i,j), the value ψ_{ij} must be equal to the average of its four neighbours, plus $\Delta x^2 \rho_{ij}/4$. So, if we start with an arbitrary trial function ψ^0, we can generate a sequence of potentials ψ^n according to

$$\psi^{n+1}_{i,j} = \frac{1}{4}(\psi^n_{i+1,j} + \psi^n_{i-1,j} + \psi^n_{i,j+1} + \psi^n_{i,j-1}) + \frac{\Delta x^2}{4}\rho_{ij}. \tag{A.89}$$

If we interpret the upper index n as the time, we recognise in this equation an initial value problem. Indeed, the stationary solutions of the initial value problem,

$$\frac{\partial \psi}{\partial t}(\mathbf{r}, t) = \nabla^2\psi(\mathbf{r}, t) + \rho(\mathbf{r}, t), \tag{A.90}$$

is a solution to (A.86). Equation (A.89) is the discretised version of (A.90) with $\Delta x^2 = 4\Delta t$. In fact, we have turned the boundary value problem into an initial value problem, and we are now after the stationary solution of the latter. The method given in (A.89) is called the *Jacobi method.*

Unfortunately, the relaxation to the stationary solution is very slow. For an $L \times L$ lattice with periodic boundary conditions, and taking $\rho \equiv 0$, we can find solutions similar to the Von Neumann modes of the previous subsection. These are given as:

$$\psi(j, l; t) = \mathrm{e}^{-\alpha t}\mathrm{e}^{\pm \mathrm{i}k_x j}\mathrm{e}^{\pm \mathrm{i}k_y l}; \tag{A.91}$$

$$k_x = \frac{2n}{L}\pi; \quad k_y = \frac{2m}{L}\pi; \quad n, m = 0, 1, 2, \ldots$$

$$\mathrm{e}^{-\alpha} = [\cos(k_x) + \cos(k_y)]/2.$$

Obviously, the mode with $k_x = k_y = 0$ remains constant in time, as it is a solution of the stationary equation. From the last equation we see that when k_x and k_y are not both zero, α is positive, so that stability is guaranteed. We see that the nontrivial modes with slowest relaxation (smallest α) occur for $(n, m) = (1, 0)$, $(0, 1)$, $(L-1, 0)$, or $(0, L-1)$. For these modes, we have $\alpha = \mathcal{O}(1/L^2)$. This shows that if we discretise our problem on a finer grid in order to achieve a more accurate representation of the continuum solution, we pay a price not only via an increase in the time per iteration but also in the convergence rate, which forces us to perform more iterations before the solutions converge satisfactorily. Doubling of L causes the relaxation time to be increased by a factor of four.

We can change the convergence rate by another discretisation of the continuum initial value problem (A.90), in which instead of ψ_{ij}^{n+1} in (A.89), a linear combination of this function and the 'old one', ψ_{ij}^{n} is taken:

$$\psi_{ij}^{n+1} = (1-\omega)\psi_{ij}^{n} + \omega\left[\frac{1}{4}(\psi_{i+1,j}^{n} + \psi_{i-1,j}^{n} + \psi_{i,j+1}^{n} + \psi_{i,j-1}^{n})\right] + \frac{\Delta x^2}{4}\rho_{ij}. \tag{A.92}$$

Another way of looking at this method is to consider it as a discretisation of (A.86) with $\omega \Delta x^2 = 4\Delta t$. The solutions to this equation are again modes of the form (A.91), but now with

$$\mathrm{e}^{-\alpha} = 1 - \omega[1 - (\cos k_n + \cos k_y)/2]. \tag{A.93}$$

We see that the relaxation time of the slowest modes is still of order L^2. This method is called the *damped Jacobi method.*

Yet another type of iteration is the one in which we scan the rows of the grid in lexicographic order and overwrite the old values of the solution, ψ_{ij}^{n}, with the new ones, ψ_{ij}^{n+1}, during the scan so that these new values are then used instead of the old ones in the evaluation of the right hand side of the iteration Eq. (A.89). This method is called the *Gauss–Seidel method.* In formula this leads to the rule:

$$\psi_{i,j}^{n+1} = \frac{1}{4}(\psi_{i+1,j}^{n} + \psi_{i-1,j}^{n+1} + \psi_{i,j+1}^{n} + \psi_{i,j-1}^{n+1}) + \frac{\Delta x^2}{4}\rho_{ij}. \tag{A.94}$$

In this method, the slowest modes still relax with $\mathrm{e}^{-\alpha} \approx 1 - \mathcal{O}(1/L^2)$ although the prefactor turns out to be half that of the Jacobi method. An improvement can be achieved by considering an extension similar to the damped Jacobi method, by combining the old and new solutions according to

$$\psi_{ij}^{n+1}(\omega) = (1-\omega)\psi_{ij}^{n} + \frac{\omega}{4}[(\psi_{i+1,j}^{n} + \psi_{i-1,j}^{n+1} + \psi_{i,j+1}^{n} + \psi_{i,j-1}^{n+1})]. \tag{A.95}$$

The method can be shown to converge for $0 \leq \omega < 2$ and there exists some optimal value for ω for which the relaxation rate is given by $\mathrm{e}^{-\alpha} \approx 1 - \mathcal{O}(1/L)$, substantially better than all the methods described up to now. This optimal value turns out to be close to 2. This can be understood by a heuristic argument. In a Gauss–Seidel update of a particular site, half of its neighbours have their old, and the other half the new values. We would obviously like the new value at the present site to be equal to the average of the new values of all its neighbours. We therefore compensate for the fact that half of the neighbours still have the old value by multiplying the change in the potential ψ at each site by a factor of two. The precise optimal value for ω is related to the parameter α for the slowest mode in the simple Jacobi method, which is in general unknown (we have found this parameter above from the exact solution which is in general unknown). In practice, a theoretical approximation of this optimal value is used, or it is determined empirically using the Jacobi method. This method is called *successive over-relaxation.*

Summarising, we can say that the iterative methods discussed in this section are subject to slow mode relaxation problems; some more than others. The reason for the slow relaxation is that the update of the value of the solution at the grid points takes only nearest neighbour points into account and acts hence on a short range. It will take many iterations in order to update a long-wavelength mode. In the next subsections we shall consider ways to get round this problem.

The conjugate gradient method for elliptic differential equations

For ease of notation and for generality we denote the PDE problem as

$$\mathbf{Ax} = \mathbf{b} \tag{A.96}$$

where x_{ij} is the solution; $\mathbf{x}$ and $\mathbf{b}$ are considered as *vectors* whose indices are the grid points (i, j) (in two dimensions), and $\mathbf{A}$ is a *matrix* with similar indices. When solving Poisson's equation, $\mathbf{A}$ is the matrix corresponding to the discretised Laplace operator, as defined in (A.87). Using this notation, the Jacobi relaxation method can be represented as

$$x_{ij}^{n+1} - x_{ij}^{n} = \sum_{kl} \mathbf{A}_{ij,kl} x_{kl}^{n} - b_{ij}. \tag{A.97}$$

The result on the left hand side can be viewed as the extent to which the solution $\mathbf{x}^n$ fails to satisfy the equation (A.96). The left hand side is called the *residual* $\mathbf{r}$ of the trial solution $\mathbf{x}^n$.

We would like to find the value $\mathbf{x}$ for which the residual vanishes. It turns out that this is possible using the conjugate gradients method. To this end, we consider the function

$$f(\mathbf{x}) = \tfrac{1}{2}\mathbf{x} \cdot \mathbf{Ax} - \mathbf{b} \cdot \mathbf{x}. \tag{A.98}$$

The gradient of this function is

$$\nabla f(\mathbf{x}) = \mathbf{Ax} - \mathbf{b}. \tag{A.99}$$

Therefore, when we can minimise this equation, we have solved the matrix equation (A.96). But we have already encountered a method which is very efficient at doing this: the conjugate gradient method (see Appendix A4). We now give this algorithm in pseudocode for the present problem.

$\mathbf{h} = \mathbf{b}$; ($\mathbf{b}$ is the right hand side of (A.96)
$\mathbf{g} = \mathbf{b}$;
$r_2 = \mathbf{g} \cdot \mathbf{g}$;
WHILE r_2 is not small enough DO

$$\mathbf{k} = \mathbf{A}\mathbf{h};$$
$$\lambda = r_2/(\mathbf{h} \cdot \mathbf{k});$$
$$\mathbf{g} = \mathbf{g} - \lambda\mathbf{k};$$
$$r_2' = \mathbf{g} \cdot \mathbf{g};$$
$$\gamma = r_2'/r_2;$$
$$\mathbf{x} = \mathbf{x} + \lambda\mathbf{h};$$
$$\mathbf{h} = \mathbf{g} + \gamma\mathbf{h};$$
$$r_2 = r_2';$$

END WHILE.

The algorithm can be directly related to the discussion of the conjugate gradient method in Appendix A4.

The convergence of the algorithm is fast: $\mathcal{O}(L)$ steps are needed, each of which takes L^d floating point operations (as a step involves a sparse matrix–vector multiplication). We see that we only need a matrix–vector multiplication routine. The algorithm is so simple to implement and so efficient that it should always be preferred over the Gauss–Seidel method.

Fast Fourier transform methods

The first method uses the fast Fourier transform (FFT), which enables us to calculate the Fourier transform of a function defined on a one-dimensional grid of N points in a number of steps which scales as $N \log N$, in contrast with the direct method of evaluating all the Fourier sums, which takes $\mathcal{O}(N^2)$ steps. The FFT method will be described in Appendix A9. We now explain how the FFT method can be used in order to solve PDEs of the type discussed in the previous section. On a two-dimensional square grid of size $L \times L$, the FFT requires $\mathcal{O}(L^2 \log L)$ steps. In the following we assume periodic boundary conditions.

The idea behind using Fourier transforms for these equations is that the Laplace operator is diagonal in Fourier space. For our model problem of the previous section, we have

$$\sum_{jm} (\nabla_{\mathrm{D}}^2 \psi_{jm}) \mathrm{e}^{\mathrm{i}(k_x j + k_y m)} = (2\cos k_x + 2\cos k_y - 4)\hat{\psi}_{k_x k_y}, \tag{A.100}$$

where $\hat{\psi}$ is the Fourier transform of ψ. The wave numbers k_x, k_y assume the values $2\pi n/L$. We see that acting with the Laplace operator on ψ corresponds to multiplying by the factor $2\cos k_x - 2\cos k_y - 4$ in k-space. If we are to determine the solution to the PDE $\nabla_{\mathrm{D}}^2 \psi = -\rho$, we perform a Fourier transform on the left and right sides of this equation, and we arrive at

$$(4 - 2\cos k_x - 2\cos k_y)\hat{\psi}_{k_x k_y} = \hat{\rho}_{k_x k_y}. \tag{A.101}$$

This equation is easily solved because it is diagonal: we have simply L^2 independent trivial equations. The solution in direct space is found by Fourier transforming $\hat{\psi}$ back to real space.

The computational cost is that of performing the necessary Fourier transforms, and this requires $\mathcal{O}(L^2 \log L)$ floating point operations in two dimensions. The method is therefore very efficient with respect to relaxation methods, which require L^2 floating point operations per scan over the lattice. Remember this is to be multiplied by the number of iterations, which is at least $\mathcal{O}(L)$ for satisfactory accuracy (in the SOR method), so that the total time needed for these methods scales as $\mathcal{O}(L^3)$.

Multigrid methods

Multigrid methods are based on iterative ideas (see 'Boundary value problems' above) and aim – crudely speaking – at increasing the relaxation rate by updating the solution on blocks of grid points alongside the short-range update on the grid points themselves. The method is based on residual minimisation; see 'The conjugate gradient method' above. For some trial solution ψ_{ij} with residual r_{ij}, we consider the difference between ψ and the exact solution χ

$$\mathbf{A}(\psi - \chi) = r, \tag{A.102}$$

hence, if we would have a solution δ of the equation

$$\mathbf{A}\delta = r, \tag{A.103}$$

the exact solution χ would be given by

$$\chi = \psi - \delta. \tag{A.104}$$

If we have performed a few iterations, e.g. Gauss–Seidel iterations, the residual will contain on average an important contribution from the long-wavelength components and only weak short-wavelength modes, since iteration methods tend to eliminate the short-wavelength errors in the solution faster than the long-wavelength ones, because a Gauss–Seidel update involves only nearest neighbours. In the multigrid method, the remaining smooth components are dealt with in a coarser grid. We shall roughly describe the idea of the multigrid method; details will be given below. The coarse grid is a grid with half as many grid points along each cartesian direction, so in our two-dimensional case the number of points on the coarse grid is one-fourth of the number of fine grid points. A function defined on the fine grid is *restricted* to the coarser grid by a suitable coarsening procedure. We apply this coarsening procedure to the residual and perform a few Gauss–Seidel iterations on the result. The idea is that the wavelength of the long-wavelength modes is effectively twice as small on the coarse grid, so hence the long-wavelength modes can be dealt with more efficiently. The solution on the coarse grid must somehow be

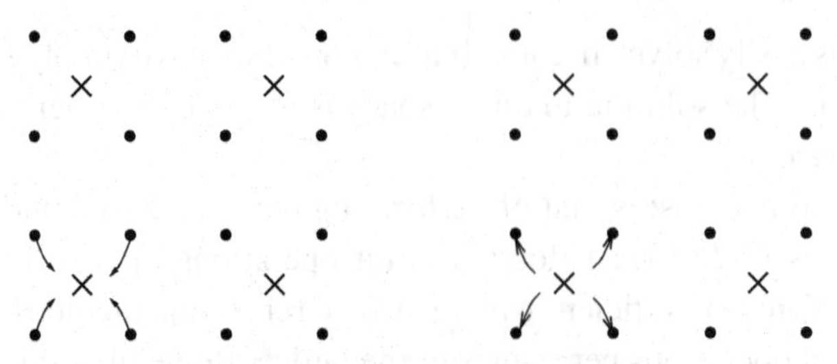

Figure A.3. Restriction and prolongation. The coarse grid points are marked by × and the fine grid points by •. The arrows in the left hand figure represent the restriction, and those in the right hand figure, the prolongation map.

transferred to the fine grid, and the mapping used for this purpose is called *prolongation*. On the fine grid, finally, a few Gauss–Seidel iterations are then carried out to smooth out errors resulting from the somewhat crude representation of the solution at the higher level. Schematically we can represent the method described by the following diagram:

$$\begin{array}{llcl} \text{Coarse:} & & \nearrow \;\; \delta = \mathbf{A}^{-1} r \to \delta \;\; \searrow & \\ \text{Fine:} & \text{Gauss–Seidel} \to r & & \text{Gauss–Seidel} \end{array}$$

The obvious extension then is to play the same game at the coarse level, go to a 'doubly coarse' grid etc., and this is done in the multigrid method.

We now fill in some details. First of all, we must specifiy a *restriction mapping*, which maps a function defined on the fine grid onto the coarse grid. We consider here a coarse grid with sites being the centres of half of the plaquettes of the fine grid as in Figure A.3. The value of a function on the coarse grid point is then given as the average value of the four function values on the corners of the plaquette. Using ψ for a function on the fine grid and ϕ for a function on the coarse grid, we can write the restriction operator as

$$\phi_{ij} = (P_{l-1,l}\psi)_{ij} = \tfrac{1}{4}(\psi_{2i,2j} + \psi_{2i+1,2j} + \psi_{2i,2j+1} + \psi_{2i+1,2j+1}). \qquad \text{(A.105)}$$

ϕ_{ij} can be thought of as centred on the plaquette of the four ψ functions at the right hand side. We also need a prolongation mapping from the coarse grid to the fine grid. This consists of copying the value of the function on the coarse grid to the four neighbouring fine grid points as in the right hand side of Figure A.3. The formula for the prolongation mapping $(P_{l,l-1})$ is

$$\psi_{ij} = (P_{l,l-1}\phi)_{ij} = \phi_{i/2,j/2} \qquad \text{(A.106)}$$

where $i/2$ and $j/2$ denote (truncated) integer divisions of these indices.

We need the matrix $\mathbf{A}$ on the coarse grid. However, we know its form only on the fine grid. We use the most natural option of taking for the coarse form also a

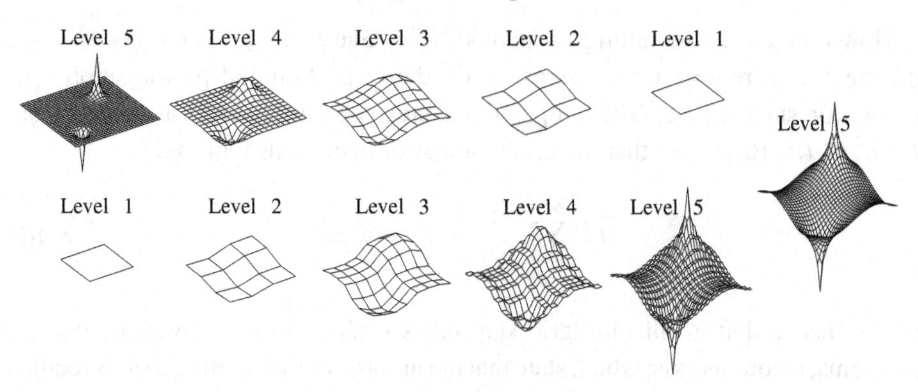

Figure A.4. The evolution of the trial solution to Poisson's equation in two dimensions with a positive and a negative point charge. The upper row shows the solution during coarsening, and the lower row during refinement. The rightmost picture is the rightmost configuration of the lower row with some extra Gauss–Seidel iterations performed.

Laplace operator coupling neighbouring sites:

$$\nabla_D^2 \phi = \frac{1}{4\Delta x^2}(\phi_{i+1,j} + \phi_{i-1,j} + \phi_{i,j+1} + \phi_{i,j-1} - 4\phi_{i,j}), \tag{A.107}$$

where Δx is taken twice as large as on the fine grid. We now have all the ingredients for the multigrid method at our disposal and we can write down the general algorithm in a recursive form:

ROUTINE MultiGrid(*Level*, ψ, *Residual*)
  Perform a few Gauss–Seidel iterations: $\psi \to \psi'$;
  IF (*Level*>0) THEN
    Calculate *Residual*;
    Restrict *Residual* to coarser grid: $\to$ *Residual'*;
    Set ϕ equal zero;
    Multigrid(*Level-1*, ϕ, *Residual'*);
  END IF;
  Prolongate $\phi :\to \phi'$;
  $\psi'' = \psi' - \phi'$;
  Perform a few Gauss–Seidel iterations: $\psi'' \to \psi'''$;
END MultiGrid.

The number of Gauss–Seidel iterations before and after the coarsening procedure must be chosen – typical values are two to five iterations. This algorithm can be coded directly (see Problem A.7). Figure A.4 shows how the method works for a two-dimensional Poisson problem with a positive and a negative charge.

How efficient is the multigrid method? The Gauss–Seidel iterations on a grid of size $L \times L$ require $\mathcal{O}(L^2)$ steps, as do the restriction and prolongation operations for such a grid. These operations must be carried out on grids of size $L^2, L^2/4, L^2/16, \ldots$, so that the total number of operations required is

$$L^2 \sum_{n=0}^{\log_2 L} \frac{1}{2^{2n}} = L^2 \frac{1}{1 - 1/4} \tag{A.108}$$

which means that a full multigrid step takes $\mathcal{O}(L^2)$ steps. For various types of problems, theorems exist which state that the number of full multigrid steps required to achieve convergence for Poisson's equation is independent of L, so that the method has the best possible time scaling, that is time $\propto L^2$.

Obviously, the definition of the coarse grid and the restriction and prolongation operators are not unique; in setting up the multigrid method there are quite a few options at each stage. We have only given one example in order to expose the method. For more details, the specialised literature should be consulted [16–19].

A8 Linear algebra problems

A8.1 Systems of linear equations

This section is devoted to an overview of matrix calculations, a topic that we have touched on several times in previous sections. A first example is the solution of a set of N linear equations of N unknowns. In a matrix formulation, the equations can be formulated as

$$\begin{pmatrix} a_{11} & a_{12} & \cdots & a_{1N} \\ a_{21} & a_{22} & \cdots & a_{2N} \\ \vdots & \vdots & & \vdots \\ a_{N1} & a_{N2} & \cdots & a_{NN} \end{pmatrix} \begin{pmatrix} x_1 \\ x_2 \\ \vdots \\ x_N \end{pmatrix} = \begin{pmatrix} b_1 \\ b_2 \\ \vdots \\ b_N \end{pmatrix}, \tag{A.109}$$

which can be written in a more compact way as

$$\mathbf{Ax} = \mathbf{b}. \tag{A.110}$$

Here, $\mathbf{A}$ is a nonsingular $N \times N$ matrix with elements a_{ij}, $\mathbf{x}$ is an N-dimensional vector containing the unknowns and $\mathbf{b}$ is a known N-dimensional vector.

Dense matrices

A *dense* matrix is one whose elements are mostly nonzero, so all its elements have to be taken into account in a solution method involving such a matrix, such as the solution of a system of linear equations.

The method used for this problem is straightforward. We use the fact that the rows of $\mathbf{A}$ can be interchanged if the corresponding elements of $\mathbf{b}$ are also interchanged, and it is also allowed to replace a row of $\mathbf{A}$ by a linear combination of that row and another, provided similar transformations are performed to the elements of $\mathbf{b}$. These properties can now be used to transform $\mathbf{A}$ into an upper triangular matrix. This is done proceeding step by step from the left column of $\mathbf{A}$ to the right one, making the elements below the diagonal for each column equal to zero. Suppose we have zeroed the lower parts of columns 1 through to $m-1$. The equation now looks as follows:

$$\begin{pmatrix} a'_{11} & a'_{12} & \cdots & a'_{1,m-1} & a'_{1m} & \cdots & a'_{1N} \\ 0 & a'_{22} & \cdots & a'_{2,m-1} & a'_{2m} & \cdots & a'_{2N} \\ \vdots & \ddots & \ddots & \vdots & \vdots & & \vdots \\ 0 & & \ddots & a'_{m-1,m-1} & a'_{m-1,m} & \cdots & a'_{mN} \\ 0 & \cdots & \cdots & 0 & a'_{mm} & \cdots & a'_{mN} \\ 0 & \cdots & \cdots & 0 & a'_{m+1,m} & \cdots & a'_{m+1,N} \\ \vdots & & & \vdots & \vdots & & \vdots \\ 0 & \cdots & \cdots & 0 & a'_{Nm} & \cdots & a'_{NN} \end{pmatrix} \begin{pmatrix} x_1 \\ x_2 \\ \vdots \\ \vdots \\ \vdots \\ \vdots \\ \vdots \\ x_N \end{pmatrix} = \begin{pmatrix} b'_1 \\ b'_2 \\ \vdots \\ \vdots \\ \vdots \\ \vdots \\ \vdots \\ b'_N \end{pmatrix} \tag{A.111}$$

Now we must treat column number m. To do this, we take rows $i = m+1$ to N of $\mathbf{A}'$, a'_{mi}/a'_{mm} times the mth row of $\mathbf{A}'$ and subtract, which causes all elements a'_{im}, that is, the elements of the mth column below the diagonal, to vanish.

Obviously, there is a problem when $a'_{m,m}$ is equal to zero. In that case we have to search from this diagonal element downward until we find an element $a'_{i,m}$, $i > m$, which is nonzero. We then interchange rows m and i and we proceed in the same way as described above. The calculation takes a number of m^2 steps for column m. Summing over m, we obtain a total number of $\mathcal{O}(N^3)$ steps. Unfortunately, for a generic matrix which does not contain many zeroes, the N^3 scaling cannot be altered by using other methods.

Finally, we are left with an upper triangular matrix $\mathbf{A}''$ (and a right hand side $\mathbf{b}''$) with elements a''_{ij} (b''_i):

$$\begin{pmatrix} a''_{11} & a''_{12} & \cdots & a''_{1,N-1} & a''_{1N} \\ 0 & a''_{22} & \cdots & a''_{2,N-1} & a''_{2N} \\ \vdots & \vdots & \ddots & \vdots & \vdots \\ 0 & 0 & \cdots & a''_{N-1,N-1} & a''_{N-1,N} \\ 0 & 0 & \cdots & 0 & a''_{NN} \end{pmatrix} \begin{pmatrix} x_1 \\ x_2 \\ \vdots \\ \vdots \\ x_N \end{pmatrix} = \begin{pmatrix} b''_1 \\ b''_2 \\ \vdots \\ \vdots \\ b''_N \end{pmatrix} \tag{A.112}$$

We can now solve for the x_i straightforwardly: first, we see that its last element x_N is equal to b''_N / a''_{NN}. Then the remaining x_i are found as

$$x_i = \left(b''_i - \sum_{j=i+1}^{N} a''_{ij} x_j \right) / a''_{ii}. \tag{A.113}$$

The latter procedure is called *back-substitution*. It requires only $\mathcal{O}(N^2)$ steps since finding x_i requires $\mathcal{O}(N - i)$ steps.

For large matrices, the procedure described above is often unstable. This is because the element a'_{mm}, which is used to zero the elements below it, may become small. As the factor by which the lower rows are multiplied contains $1/a'_{mm}$, it is clear that the elements of these lower rows may become very large. Combinations of small and large numbers often cause numerical instabilities in the computer. To avoid such problems in matrix calculations it is advised to use *pivoting*. Pivoting means that the largest element (in absolute value) a'_{im}, $i = m, \ldots, N$ is looked for. If this largest element is not a'_{mm}, then rows i and m are interchanged so that the largest possible number now occurs in the denominator of the multiplication factor. This largest element is called the *pivot*.

Up to now, we have described a stable and efficient method for solving the fundamental problem of linear algebra, Eq. (A.110), finding $\mathbf{x}$ for *one* given vector $\mathbf{b}$. However, we may often have to face the problem of solving for several, or even many, vectors $\mathbf{b}_i$. For the case of several $\mathbf{b}$s (by several we mean less than the dimension of the matrix $\mathbf{A}$), we can perform the row changes needed to diagonalise $\mathbf{A}$ on the corresponding elements of all the $\mathbf{b}_i$ simultaneously. It is even possible, by initially putting the $\mathbf{b}_i$ equal to the N unit vectors, to calculate the inverse of $\mathbf{A}$. However, instead of following the procedure described above, it is preferable to zero not only the part below but also above the diagonal of $\mathbf{A}$ (meanwhile performing the same changes to the elements of the $\mathbf{b}_i$s on the right hand side) and finally to normalise the remaining diagonal elements of $\mathbf{A}$ to 1, so that it is now in unit matrix form. The $\mathbf{b}_i$ are then the columns of the inverse of $\mathbf{A}$, and no back-substitution is necessary.

For many $\mathbf{b}_i$ (i.e. more than the dimension of $\mathbf{A}$) it is tempting to use the inverse $\mathbf{A}^{-1}$ of $\mathbf{A}$ to solve for the solutions $\mathbf{x}$ by

$$\mathbf{x} = \mathbf{A}^{-1}\mathbf{b} \tag{A.114}$$

but this procedure is quite susceptible to round-off errors and therefore not recommended. A stable and efficient method for solving Eq. (A.110) for many vectors $\mathbf{b}_i$ is the *LU* decomposition. In this method, one decomposes $\mathbf{A}$ into two matrices $\mathbf{L}$ and $\mathbf{U}$ with $\mathbf{A} = \mathbf{L} \cdot \mathbf{U}$. The matrix $\mathbf{L}$ is lower, and $\mathbf{U}$ upper triangular. Because $\mathbf{L}$ and $\mathbf{U}$ contain together $N(N+1)$ nonzero elements and $\mathbf{A}$ only N^2, the problem is redundant and we are allowed to put the diagonal elements of **L** equal to 1. In that case we have

$$\begin{pmatrix} a_{11} & a_{12} & \cdots & a_{1N} \\ a_{21} & a_{22} & \cdots & a_{2N} \\ \vdots & \vdots & & \vdots \\ a_{N1} & a_{N2} & \cdots & a_{NN} \end{pmatrix}$$

$$= \begin{pmatrix} 1 & 0 & \cdots & 0 & 0 \\ l_{21} & 1 & \cdots & 0 & 0 \\ l_{31} & l_{32} & \cdots & 0 & 0 \\ \vdots & \vdots & & \vdots & \vdots \\ l_{N-1,1} & l_{N-1,2} & \cdots & 1 & 0 \\ l_{N1} & l_{N2} & \cdots & l_{N,N-1} & 1 \end{pmatrix} \begin{pmatrix} u_{11} & u_{12} & \cdots & u_{1,N-1} & u_{1N} \\ 0 & u_{22} & \cdots & u_{1,N-1} & u_{2N} \\ \vdots & 0 & & \vdots & \vdots \\ \vdots & 0 & & \vdots & \vdots \\ 0 & 0 & \cdots & u_{N-1,N-1} & u_{N-1,N} \\ 0 & 0 & \cdots & 0 & u_{NN} \end{pmatrix}, \tag{A.115}$$

l_{ij} and u_{ij} being the matrix elements of **L** and **U** respectively. The decomposition matrices **L** and **U** are now easily determined: multiplying the first row of **L** with the columns of **U** and putting the result equal to a_{1i} it is found that the first row of **U** is equal to the first row of **A**. We then proceed, determining l_{21} by multiplying the second row of **L** with the first column of **U**, and then finding the elements of the second row of **U** and so on.

Having the *LU* decomposition for **A**, it is easy to solve for an unknown vector **x** in the linear equation (A.110). First we search for the vector **y** obeying

$$\mathbf{L}\mathbf{y} = \mathbf{b} \tag{A.116}$$

and then for a vector **x** which satisfies

$$\mathbf{U}\mathbf{x} = \mathbf{y}. \tag{A.117}$$

Using $\mathbf{A} = \mathbf{LU}$, it then immediately follows that $\mathbf{Ax} = \mathbf{b}$. Both problems can be solved using back-substitution, and thus require only $\mathcal{O}(N^2)$ calculations [finding **L** and **U** requires $\mathcal{O}(N^3)$ steps].

Sparse matrices

Differential operators, which frequently occur in physics, can be discretised on a finite difference grid, leading to the differential operator being represented by a *sparse* matrix: the overwhelming majority of the matrix elements are equal to zero. A typical example is the discrete Laplacian mentioned previously in the context of partial differential equations. There, space was discretised on a grid and the discrete Laplacian coupled nearest neighbour points on the grid only, so that the matrix representation of this operator contains only nonzero elements for indices corresponding to neighbouring or identical grid points. In two dimensions with a grid constant h we obtain

$$\nabla^2\psi(\mathbf{r}) \to \frac{1}{h^2}(\psi_{i+1,j} + \psi_{i-1,j} + \psi_{i,j+1} + \psi_{i,j-1} - 4\psi_{i,j}). \tag{A.118}$$

In numerical problems involving sparse matrices, it usually pays off to exploit the sparseness: this enables us in many cases to reduce the effort involved in solving matrix problems from $\mathcal{O}(N^3)$ to $\mathcal{O}(N)$.

As an example, let us consider a tridiagonal matrix. This arises when discretising the one-dimensional Laplace operator on a grid with fixed boundary conditions. The tridiagonal matrix has the form

$$\mathbf{A} = \begin{pmatrix} a_1 & b_1 & 0 & \cdots & 0 & 0 & 0 \\ c_2 & a_2 & b_2 & \cdots & 0 & 0 & 0 \\ \vdots & \vdots & \vdots & \vdots & \vdots & \vdots & \vdots \\ 0 & 0 & 0 & \cdots & c_{N-1} & a_{N-1} & b_{N-1} \\ 0 & 0 & 0 & \cdots & 0 & c_N & a_N \end{pmatrix}, \tag{A.119}$$

and we want to solve the equation $\mathbf{Ax} = \mathbf{d}$. This is done as in the previous subsection, by first zeroing the lower diagonal followed by backward substitution. Implementation is straightforward. Only subdiagonal elements are to be eliminated and there is only one candidate row to be subtracted off the row containing that subdiagonal element: the row just above it. All steps amount to order N.

Dedicated algorithms exist for matrices with different patterns (e.g. band-diagonal, scattered, striped) of nonzero elements. There exist also methods which rely exclusively on simple matrix operations such as a matrix–vector multiplication. You only have to supply an efficient routine for multiplying the matrix under consideration with an arbitrary vector to such a sparse matrix routine. An example is the conjugate gradient method for solving the problem $\mathbf{Ax} = \mathbf{b}$. We have encountered the conjugate gradient method for minimising an arbitrary smooth function in Appendix A4, and as a sparse-matrix method in Appendix A7.2.

A8.2 Matrix diagonalisation

We now turn to another important problem in linear algebra: that of finding the eigenvalues and eigenvectors of a matrix $\mathbf{A}$. This problem is commonly referred to as the *eigenvalue problem* for matrices. Expressing the matrix with respect to the basis formed by its eigenvectors, it assumes a diagonal form. In physics, solving the eigenvalue problem for general operators occurs frequently (solving the stationary Schrödinger equation is an eigenvalue problem for example) and in computational physics, it is usually transformed into matrix form in order to solve it efficiently in a computer.

Dense matrices: the Householder method

The majority of the matrices to be diagonalised in computational physics are Hermitian – the elements h_{ij} of an Hermitian matrix $\mathbf{H}$ satisfy $h_{ij} = h^*_{ji}$. If a physical problem leads to a non-Hermitian matrix to be diagonalised, it is always advisable to try casting the problem in a form such that the matrix becomes Hermitian, as the diagonalisation is generally more efficient for this case.

Here we restrict ourselves to the subclass of real symmetric matrices $\mathbf{S}$ because this procedure is completely analogous to that for the complex Hermitian case. Diagonalising the matrix involves finding an orthogonal matrix $\mathbf{O}$ which brings $\mathbf{S}$ to diagonal form $\mathbf{s}$:

$$\mathbf{s} = \mathbf{O}^{\mathbf{T}}\mathbf{S}\mathbf{O}. \tag{A.120}$$

The columns of $\mathbf{O}$ are the eigenvectors, which form an orthonormal set. We first construct a sequence of matrices $\mathbf{O}_i$ which, when applied one after another in a fashion as indicated by (A.120), transform $\mathbf{S}$ to a tridiagonal form. In the same spirit as the treatment of the elimination procedure described in the previous section, we assume that we have brought the matrix to tridiagonal form except for the upper left $i \times i$ block and construct the matrix $\mathbf{O}_i$ which tridiagonalises $\mathbf{S}$ one step further.

$$\mathbf{S}' = \begin{pmatrix} s_{jk} & & \mathbf{c} & & & & \\ \mathbf{c}^T & & s_{ii} & b_i & & & \\ & & b_i & d_{i+1} & b_{i+1} & & \\ & & & b_{i+1} & \ddots & \ddots & \\ & & & & \ddots & \ddots & b_{N-1} \\ & & & & & b_{N-1} & d_N \end{pmatrix} \tag{A.121}$$

The matrix $\mathbf{O}_i$ which does the job is given by

$$\mathbf{O}_i = \mathbf{I} - 2\mathbf{u}\mathbf{u}^T/(|\mathbf{u}|^2), \tag{A.122}$$

where $\mathbf{I}$ is the $N \times N$ unit matrix and $\mathbf{u}$ is the vector of size $i - 1$ given in terms of the vector $\mathbf{c}$ occurring at the right end and the bottom of the nontridiagonalised part of $\mathbf{S}'$ with zeroes at positions $i \dots N$:

$$\mathbf{u} = \mathbf{c} - |\mathbf{c}|\mathbf{e}_{i-1}. \tag{A.123}$$

So, $\mathbf{u}\mathbf{u}^T$ is an $N \times N$ matrix, which can be written as

$$\begin{pmatrix} u_1\mathbf{u}^T & 0 & \cdots & 0 \\ u_2\mathbf{u}^T & 0 & \cdots & 0 \\ \vdots & 0 & \cdots & 0 \\ u_{i-1}\mathbf{u}^T & 0 & \cdots & 0 \\ 0\cdots 0 & 0 & \cdots & 0 \\ \vdots & \vdots & & \vdots \\ 0\cdots 0 & 0 & \cdots & 0 \end{pmatrix} \tag{A.124}$$

from which it is readily seen that multiplication of $\mathbf{S}'$ with the matrix $\mathbf{O}_i^T$ from the left, zeroes the upper $i-2$ elements of the right column $\mathbf{c}$ of the not yet diagonalised block in $\mathbf{S}'$. Similarly, multiplying the resulting matrix from the right with $\mathbf{O}_i$ zeroes the $i-2$ leftmost elements of the ith row, and we arrive at the form (A.121) but now tridiagonalised one step further.

We still have to solve the eigenvalue problem for the tridiagonalised matrix obtained in this way. This is rather difficult and we give only an outline of the procedure. The method consists again of constructing a sequence of matrices which are successive orthogonal transformations of the original matrix. After some time, the latter takes on a lower triangular form (or upper triangular, depending on the transformation) with the eigenvalues appearing on the diagonal. This algorithm is called the *QL* or *QR* algorithm according to whether one arrives at a lower or upper tridiagonal form. For tridiagonal matrices, this procedure scales as $\mathcal{O}(N^2)$.

It is clear that the tridiagonalisation procedure preceding the *QR* (or *QL*) step takes $\mathcal{O}(m^2)$ steps for stage m of the process. Therefore, the procedure as a whole takes $\mathcal{O}(N^3)$ steps. The $\mathcal{O}(N^3)$ time complexity makes the diagonalisation of large matrices very time consuming.

If the matrix is not Hermitian, the Householder method does not work. The algorithm to be used in that case is the Hessenberg method, which we shall not discuss here [1].

Sparse matrices

For diagonalisation problems, we can exploit sparseness of the matrices involved, just as in the case of linear systems of equations, discussed in Appendix A8.1. An example of a sparse diagonalisation problem is when we discretise a quantum Hamiltonian on a cubic grid, leading to a sparse matrix, as we have seen in Appendix A7.2.

A multiplication of such a matrix with a vector takes only $\mathcal{O}(N)$ steps. For sparse matrices like this, there exist special algorithms for solving the eigenvalue problem, requiring fewer steps than the Householder method, which does not exploit the sparseness. In most cases we need only the lowest part of the eigenvalue spectrum of the discretised operator, as it is only for these eigenvalues that the discrete eigenfunctions represent the continuum solutions adequately. We can then use the *Lanczos* or *recursion* method. This method yields the lowest few eigenvalues and eigenvectors in $\mathcal{O}(N)$ steps, thus gaining two orders of efficiency with respect to Householder's method [20–22].

The Lanczos method works for Hermitian matrices, of which the Hamiltonian matrix for a quantum system is an example. We denote the matrix to be diagonalised by $\mathbf{A}$. We start with an arbitrary vector ψ_0, normalised to one and construct a second vector, ψ_1, in the following way:

$$\mathbf{A}\psi_0 = a_0\psi_0 + b_0\psi_1. \tag{A.125}$$

We require ψ_1 to be normalised and perpendicular to ψ_0; ψ_1, a_0 and b_0 are then defined uniquely:

$$\begin{aligned} a_0 &= \langle\psi_0|\mathbf{A}|\psi_0\rangle \\ b_0\psi_1 &= \mathbf{A}\psi_0 - a_0\psi_0 \\ ||\psi_1|| &= 1. \end{aligned} \tag{A.126}$$

Next we construct a normalised vector ψ_2 by letting $\mathbf{A}$ act on ψ_1:

$$\mathbf{A}\psi_1 = c_1\psi_0 + a_1\psi_1 + b_1\psi_2. \tag{A.127}$$

The vector ψ_2 is taken perpendicular to ψ_0 and ψ_1 and normalised, so ψ_2, a_1, b_1 and c_1 are determined uniquely too. We proceed in the same way, and at step p we have

$$\mathbf{A}\psi_p = \sum_{q=0}^{p-1} c_p^{(q)}\psi_q + a_p\psi_p + b_p\psi_{p+1} \tag{A.128}$$

where ψ_{p+1} is taken orthogonal to all its predecessors ψ_q, $q \leq p$. Using the fact that $\mathbf{A}$ is Hermitian we find for $q < p - 1$:

$$c_p^{(q)} = \langle\psi_q|\mathbf{A}\psi_p\rangle = \langle\mathbf{A}\psi_q|\psi_p\rangle = 0. \tag{A.129}$$

The last inequality follows from the fact that we have encountered q already in the iteration procedure, so ψ_p is perpendicular to $\mathbf{A}\psi_q$. In the same way we find for $c_p^{(p-1)}$:

$$c_p^{(p-1)} = \langle\psi_{p-1}|\mathbf{A}\psi_p\rangle = \langle\mathbf{A}\psi_{p-1}|\psi_p\rangle = b_{p-1}. \tag{A.130}$$

So after p steps we obtain

$$\mathbf{A}\psi_p = b_{p-1}\psi_{p-1} + a_p\psi_p + b_p\psi_{p+1}, \tag{A.131}$$

where ψ_{p+1} is normalised and orthogonal to ψ_p and ψ_{p-1}. By the foregoing analysis, it is then orthogonal to all the previous ψ_q $(q < p - 1)$. We see that $\mathbf{A}$, expressed with respect to the basis ψ_p, takes on a tridiagonal form; the eigenvalues of the $p \times p$ tridiagonal matrix obtained after p Lanczos-iterations will converge to the eigenvalues of the original matrix. It can be shown that the lowest and highest eigenvalues converge most rapidly in this process, so that the method can be used successfully for these parts of the spectrum. The number of steps needed to achieve sufficient accuracy depends strongly on the spectrum; for example, if a set of small eigenvalues is separated from the rest by a relatively large gap, convergence to these small eigenvalues is fast.

A9 The fast Fourier transform

A9.1 General considerations

Fourier transforms occur very often in physics. They can be used to diagonalise operators, for example when an equation contains the operator ∇^2 acting on a function ψ. After Fourier-transforming the equation, this differential operator becomes a multiplicative factor $-k^2$ in front of the Fourier transform of ψ, where $\mathbf{k}$ is the wave vector. In quantum mechanics the stationary solutions for a free particle are found in this way: these are plane waves with energy $E = \hbar^2 k^2/2m$.

In data analysis, the Fourier transform is often used as a tool to reduce the data: in music notation one uses the fact that specifying tones by their pitch (which is nothing but a frequency) requires much less data than specifying the real-time oscillatory signal. A further application of Fourier transform is filtering out high-frequency noise present in experimental data.

The fast Fourier transform, or FFT, is a method to perform the Fourier transform much more efficiently than the straightforward calculation. It is based on the Danielson–Lanczos lemma (see the next subsection) and uses the fact that in determining the Fourier transform of a discrete periodic function, the factor e^{ikx} assumes the same value for different combinations of x and k. We shall explain the method in the next subsection, emphasising the main idea and why it works, but avoid the details involved in coding it up for the computer. Here we recall the definition and some generalities concerning the Fourier transform.

First, we define the Fourier transform for the one-dimensional case. Generalisation to more dimensions is obvious. We consider first the Fourier transform of a periodic function $f(x)$ with period L, defined on a discrete set of N equidistant points with spacing $h = L/N$:

$$x_j = jL/N, \quad j = 0, \dots, N-1. \tag{A.132}$$

Note that in contrast to our discussion of partial differential equations, L does not denote the number of grid points (N), but a distance. In that case, the Fourier transform $\hat{f}$ is also a function defined on discrete points, k_n, defined by

$$k_n = \frac{2n\pi}{L}, \quad n = 0, \ldots, N-1. \tag{A.133}$$

The Fourier transform reads

$$\hat{f}(k_n) = \sum_{j=0}^{N-1} f(x_j) e^{2\pi i n j/N}. \tag{A.134}$$

We can also define the backward transform:

$$f(x_j) = \frac{1}{L} \sum_j \hat{f}(k_n) e^{-2\pi i n j/N}. \tag{A.135}$$

If the points are not discrete, the Fourier transform is defined for the infinite set of k-values $2\pi n/L$ with $n = 0, \ldots \infty$, and the sum over the index j is replaced by an integral:

$$\hat{f}(k_n) = \int_0^L \mathrm{d}x f(x) e^{ik_n x}. \tag{A.136}$$

If the function is not periodic, L goes to infinity, and the Fourier transform is defined for every real k as

$$\hat{f}(k) = \int_{-\infty}^{\infty} \mathrm{d}x f(x) e^{ikx}, \tag{A.137}$$

with backward transform

$$f(x) = \frac{1}{2\pi} \int_{-\infty}^{\infty} \mathrm{d}k \hat{f}(k) e^{-ikx}. \tag{A.138}$$

When Fourier-transforming a nonperiodic function on a computer, it is usually restricted to a finite interval using discrete points within that interval. Furthermore, the function is considered to be continued periodically outside this interval. This restriction to a discrete set combined with periodic continuation should always be handled carefully. For example, after Fourier-transforming and back again, we must use the result only on the original interval: extending the solution outside this interval leads to periodic continuation of our function there. The usual problems involved in discretisation should be taken care of – see Appendix A5.

A9.2 How does the FFT work?

To determine the discrete Fourier transform directly, the sum over j is carried out for each k_n, that is, N times a sum of N terms has to be calculated, which amounts to a total number of multiplications/additions of N^2. In the fast Fourier transform, it is possible to reduce the work to $\mathcal{O}(N \log_2 N)$ operations. This is an important difference: for large N, $\log_2 N$ is much smaller than N. For example, $\log_2 10^6 \approx 20$!

From now on, we shall use the notation $f(x_j) = f_j$ and $\hat{f}(x_n) = \hat{f}_n$. We assume that the number of points N is equal to a power of 2: $N = 2^m$. If this is not the case, we put the function for j-values from N upward to the nearest power of 2 to zero and do the transform on this larger interval. The Danielson–Lanczos lemma is simple. The points x_j fall into two subsets: one with j even and one with j odd. The Fourier transforms of these subsets are called $\hat{f}_{2|0}$ and $\hat{f}_{2|1}$ This notation indicates that we take the indices j modulo 2 and check if the result is equal to 0 (j even) or 1 (j odd). We obtain

$$\begin{aligned}\hat{f}_n &= \sum_{j=0}^{N-1} f_j e^{2\pi inj/N} \\ &= \sum_{j=0}^{N/2-1} f_{2j} e^{4\pi inj/N} + e^{2\pi in/N} \sum_{j=0}^{N/2-1} f_{2j+1} e^{4\pi inj/N} \\ &= \hat{f}_{2|0} + e^{2\pi n/N} \hat{f}_{2|1}.\end{aligned} \tag{A.139}$$

The Fourier transform on the full set of N points is thus split into two transforms on sets containing half the number of points. If we were to calculate these sub-transforms using the direct method, each would take $(N/2)^2$ steps. Adding them as in the last line of (A.139) takes another N steps, so in total we have $N^2/2 + N$ as opposed to N^2 operations if we had applied the direct method to the original series. Where does the gain in speed come from? Compare the transforms for k_0 and $k_{N/2}$ respectively. For any point x_j with j even, we have the exponential $\exp(2\pi inj/N)$ which are equal for these two k-values, but in the direct method, these exponentials are calculated twice! The same holds for odd j-values and for other pairs of k_n-values spaced by $k_{N/2}$. Indeed, the two sub-transforms $\hat{f}_{2|0}$ and $\hat{f}_{2|1}$ are periodic with period $N/2$ instead of N – therefore their evaluation for all $j = 0, \ldots, N-1$ takes only $(N/2)^2$ steps, and the total work involved in finding the Fourier transform is reduced approximately by a factor of two. That is not yet the $N \log_2 N$ promised above. But this can be found straightforwardly by applying the transform of (A.139) to the sub-transforms of length $N/2$. One then splits the sum into four sub-transforms of length $N/4$. The reader can verify that the new form becomes

$$\hat{f}_n = \hat{f}_{4|0} + e^{4\pi ik_n/N} \hat{f}_{4|2} + e^{2\pi ik_n/N} \hat{f}_{4|1} + e^{6\pi ik_n/N} \hat{f}_{4|3}. \tag{A.140}$$

The same transformation can be performed recursively m times (remember $N = 2^m$) to arrive at $N \log_2 N$ sub-transforms of length 1, which are trivial.

It will be clear that the FFT can be coded most easily using recursive programming. When recursion is avoided for reasons of efficiency, a bit more bookkeeping is required: see Refs. [1, 23, 24].

Exercises

A.1 [C] Write routines for generating the spherical Bessel functions j_l and n_l according to the method described in Appendix A2. For j_l, downward recursion is necessary to obtain accurate results.

The upper value $L_{\max}$ from which the downward recursion must start must be sufficiently large. In this problem, we take the following value for N:

$$L_{\max} = \max\left\{\left[\frac{3[x]}{2}\right] + 20, l + 20\right\}$$

where $[x]$ is the largest integer smaller than x and l is the value we want the spherical Bessel function for.

Write a function that yields $j_l(x)$. Check the result using the following values:

$$j_5(1.5) = 6.696\,205\,96 \cdot 10^{-4}$$

$$n_5(1.5) = -94.236\,110\,1.$$

A.2 Check the statements (i)–(v) given on page 565. In all cases, proofs by induction can be given. The following hints may be useful:

(i) Use (A.22a) and (A.23a).
(ii) Use (A.22a), (A.22b) and (A.23b).
(iii) Use (A.22a) and (A.22b).

The last two items are proven together. Equations (A.22a) and (A.22b), and statement (i) are used in the proof.

A.3 [C] Consider the Verlet algorithm:

$$x_{n+1} = 2x_n - x_{n-1} + h^2 F[x_n, n] + \mathcal{O}(h^4)$$

where $x_n = x(t + nh)$. The $\mathcal{O}(h^4)$ means that the deviation of the exact solution from the numerical one is smaller than $\alpha_n h^4$, where α_n are finite numbers that are nonvanishing for $h \to 0$. We write the exact solution of the differential equation at times nh as $\tilde{x}_n$, and the numerical solution x_n deviates from the latter by an amount δ_n:

$$x_n = \tilde{x}_n + \delta_n.$$

(a) Write down an equation for δ_n and show that integration over an interval of finite width using a number of steps of size h, yields an error in the final result which, in the absence of divergences in the force (or the solution), is at most $\mathcal{O}(h^2)$.
(b) Carry out a similar analysis for the Numerov algorithm, showing that the final accuracy is at most $\mathcal{O}(h^4)$.
(c) Discuss when the 'worst case' error ($\mathcal{O}(h^2)$ for Verlet and $\mathcal{O}(h^4)$ for Numerov) is found.
(d) [C] Check the results of (a), (b) and (c) by writing routines for the two algorithms.

A.4 [C] Consider the Schrödinger equation in one dimension for a particle moving in the *Morse potential*:

$$\left[-\frac{\mathrm{d}^2}{\mathrm{d}x^2} + V_0(\mathrm{e}^{-2x} - 2\mathrm{e}^{-x})\right]\psi(x) = E\psi(x).$$

We shall take $\hbar^2/2m = 1$ in the following. This equation can be solved analytically (it can be mapped onto the Laplace equation which plays an important role in the solution of the hydrogen atom [25]), the bound state energies are given by

$$E_n = -V_0 \left(1 - \frac{n + \frac{1}{2}}{\sqrt{V_0}}\right), \qquad n = 0, 1, 2, \ldots$$

In this problem it is assumed that you have a routine available for integrating the one-dimensional Schrödinger equation: you can write a Numerov routine or use a library routine. We want to determine the bound state spectrum numerically. This is done in the following way. First, a range $x_{\rm max}$ is defined, beyond which we can safely approximate the potential by $V(x) = 0$; $x_{\rm max} \approx 10$ is a good value for this range. For some energy E, the Numerov routine can be used to integrate the Schrödinger equation numerically up to the range $x_{\rm max}$ yielding a solution $u(x)$, and beyond $x_{\rm max}$, the solution, which we denote as $v(x)$, is known analytically:

$$v(x) = A\mathrm{e}^{-qx} + B\mathrm{e}^{qx},$$
$$q = \sqrt{-E}.$$

Consider the *Wronskian* W:

$$W = u'(x_{\rm max})v(x_{\rm max}) - u(x_{\rm max})v'(x_{\rm max})$$

(the prime $'$ denotes a derivative). For a bound state, the coefficient B in the solution v should vanish and we have for the Wronskian:

$$W_{B=0} = [u'(x_{\rm max}) + qu(x_{\rm max})]\mathrm{e}^{-qx_{\rm max}}.$$

The matching condition between the analytical solution v and the numerical one (u), is equivalent to the Wronskian becoming equal to zero, and we thus have to find the energies for which the Wronskian with $B = 0$ vanishes. These energies can be found using, for example, the secant method.

(a) [C] Write a function which uses the Numerov procedure for solving the Schrödinger equation and returns the Wronskian with $B = 0$ as a function of the energy E.
(b) [C] Write a code for the secant method of Appendix A3. Test this with some simple function you know the roots of, for example $\sin x$.
(c) [C] Use the secant code for finding the energies for which the Wronskian with $B = 0$ vanishes. Note that the energies lie between 0 and $-V_0$. The search starts from $E = -V_0$, and after increasing the energy by a small step (which is predefined in the program or by the user), it is checked if the Wronskian changes sign. If this is the case, the secant method is executed for the last energy interval. When the root is found, the energy is increased again until the Wronskian changes sign. The procedure is repeated until the energy becomes positive.

   Check if the energies match the exact values given above.

A.5 Starting from Newton's interpolation formula, derive the Gauss interpolation formula to order three:

$$x_{i+t} = x_i + t\delta x_{i+1/2} + \frac{t(t-1)}{2!}\delta^2 x_i + \frac{t(t^2-1)}{3!}\delta^3 x_{i+1/2} + \cdots$$

A.6 (a) Show explicitly that equating the time derivative $(\psi_i^{n+1} - \psi_i^n)/\Delta t$ to the average value of the Hamiltonian acting on ψ_i^{n+1} and on ψ_i^n (see Eqs. (A.67), (A.68)), yields the Crank–Nicholson algorithm.

(i) Show that the operator

$$\frac{1 - \frac{1}{2}\mathrm{i}\Delta t H}{1 + \frac{1}{2}\mathrm{i}\Delta t H}$$

is a second order approximation in Δt to $\exp(-\mathrm{i}\Delta t H)$.

A.7 [C] Consider Poisson's equation for two opposite point charges on a square of linear size Lh (see Figure A.4). We take L to be a multiple of 4. The charges are placed at positions $\mathbf{r}_{1/4,1/4} = Lh(1/4, 1/4)$ and $\mathbf{r}_{3/4,3/4} = Lh(3/4, 3/4)$, so that the charge distribution is given as

$$\rho(\mathbf{r}) = \delta(\mathbf{r} - \mathbf{r}_{1/4,1/4}) - \delta(\mathbf{r} - \mathbf{r}_{3/4,3/4}).$$

We discretise Poisson's equation on an $L \times L$ grid with periodic boundary conditions, and with grid constant h. The Laplace operator is given in discretised form in Appendix A7.2 ('Boundary value problems') and the discretised charge distribution is given in terms of Kronecker deltas:

$$\rho(i,j) = (\delta_{i,L/4}\delta_{j,L/4} - \delta_{i,3L/4}\delta_{j,3L/4})/h^2.$$

As the Laplace operator and the charge distribution both contain a pre-factor $1/h^2$, this drops out of the equation.

(a) [C] Solve Poisson's equation (A.88) for this charge distribution using the Gauss–Seidel iteration method.

(b) [C] Apply the multigrid method using the prolongation and discretisation mappings described in Appendix A7.2 ('Multigrid methods') to solve the same problem.

(c) [C] Compare the performance (measured as a time to arrive at the solution within some accuracy) for the methods in (a) and (b). Check in particular that the number of multigrid steps needed to obtain convergence is more or less independent of L.

References

[1] W. H. Press, S. A. Teukolsky, W. T. Vetterling, and B. P. Flannery, *Numerical Recipes*, 2nd edn. Cambridge, Cambridge University Press, 1992.

[2] R. W. Hamming, *Numerical Methods for Scientists and Engineers*. International Series in Pure and Applied Mathematics, New York, McGraw-Hill, 1973.

[3] G. E. Forsythe, M. A. Malcolm, and C. B. Moler, *Computer Methods for Mathematical Computations*. Englewood Cliffs, NJ, Prentice-Hall, 1977.

[4] J. Stoer and R. Bulirsch, *Introduction to Numerical Analysis*. New York, Springer Verlag, 1977.
[5] D. Kincaid and W. Cheney, *Numerical Analysis*. Belmont, CA, Wadsworth, 1991.
[6] D. Greenspan and V. Casulli, *Numerical Analysis for Applied Mathematics, Science and Engineering*. Redwood City, CA, Addison-Wesley, 1988.
[7] R. P. Brent, *Algorithms for Minimization without Derivatives*. Englewood Cliffs, NJ, Prentice-Hall, 1973.
[8] D. Levesque and L. Verlet, 'Molecular dynamics and time reversibility,' *J. Stat. Phys.*, **72** (1993), 519–37.
[9] J. Delambre, *Mem. Acad. Turin*, **5** (1790), 1411.
[10] B. Numerov, *Publ. l'Observ. Astrophys. Central Russie*, **2** (1933), 188.
[11] D. R. Hartree, *Numerical Analysis*, 2nd edn. London, Oxford University Press, 1958.
[12] L. M. Milne-Thomson, *The Calculus of Finite Differences*. London, Macmillan, 1960.
[13] S. Herrick, *Astrodynamics*, vol. 2. New York, Van Nostrand Reinhold, 1972.
[14] R. S. Varga, *Matrix Iterative Analysis*. Englewood Cliffs, NJ, Prentice-Hall, 1962.
[15] D. M. Young, *Iterative Solution of Large Linear Systems*. New York, Academic Press, 1971.
[16] W. Hackbusch and U. Trottenberg, eds., *Multigrid Methods*, Berlin, Springer, 1982.
[17] W. Hackbusch, *Multigrid Methods and Applications*. Berlin, Springer, 1984.
[18] P. Wesseling, *An Introduction to Multigrid Methods*. Chichester, John Wiley, 1992.
[19] W. L. Briggs, *A Multigrid Tutorial*. Philadelphia, SIAM, 1987.
[20] H. H. Roomany, H. W. Wyld, and L. E. Holloway, 'New method for the Hamiltonian formulation for lattice spin systems,' *Phys. Rev. D*, **21** (1980), 1557–63.
[21] R. Haydock, V. Heine, and M. J. Kelly, 'Electronic structure based on the local atomic environment for tight-binding bands,' *J. Phys. C*, **5** (1972), 2845–58.
[22] C. G. Paige,, 'Computational variants of the Lanczos method for the eigenproblem,' *J. Inst. Math. Appl.*, **10** (1972), 373–81.
[23] J. W. Cooley and J. W. Tukey, 'An algorithm for the machine calculation of complex Fourier series,' *Math. Comp.*, **19** (1965), 297–301.
[24] R. J. Higgins, 'Fast Fourier transform: an introduction with some minicomputer experiments,' *Am. J. Phys.*, **44** (1976), 766–73.
[25] A. Messiah, *Quantum Mechanics*, vols. 1 and 2. Amsterdam, North-Holland, 1961.

Appendix B
Random number generators

B1 Random numbers and pseudo-random numbers

Random numbers are used in many simulations, not only of gambling tables but also of particle accelerators, fluids and gases, surface phenomena, traffic and so forth. In all these simulations some part of the system responsible for the behaviour under investigation is replaced by events generated by a random number generator, such as particles being injected into the system, whereas the source itself is not considered. Here we discuss various methods used for generating random numbers and study the properties of these numbers.

Random numbers are characterised by the fact that their value cannot be predicted. More precisely, if we construct a sequence of random numbers, the probability distribution for a new number is independent of all the numbers generated so far. As an example, one may think of throwing a die: the probability of throwing a 3 is independent of the results obtained before. Pure random numbers may occur in experiments: for a radioactive nucleus having a certain probability of decay, it is not possible to predict *when* it will decay. There is an internet service, `http://www.fourmilab.ch/hotbits/` which creates random numbers in this way and sends them over the internet (a careful correction has been carried out to remove any bias resulting in a majority of either 1 or 0). These numbers are *truly* random [1, 2].

On the other hand, random numbers as generated by a computer are not truly random. In all computer generators the new numbers are generated from the previous ones by a mathematical formula. This means that the new value is fully determined by the previous ones! However, the numbers generated with these algorithms often have properties making them very suitable for simulations. Therefore they are called *pseudo-random* numbers. In fact, a 'good' random number generator yields sequences of numbers that are difficult to distinguish from sequences of pure random numbers, although it is not possible to generate pseudo-random numbers that are completely indistinguishable from pure ones. In the following, we drop the prefix 'pseudo' when we are dealing with random numbers generated in a computer.

Random numbers from standard generators are uniformly distributed over the interval [0, 1]. This means that each number between 0 and 1 has the same probability of occurrence, although in reality only values on a dense discrete set are possible because of the finite number of bits used in the representation of the numbers. It is also possible to make nonuniform random number generators: in that case, some numbers have a higher probability of occurrence than others (nonuniform generators will be considered in Appendix B3). We define a distribution function P such that $P(x)\mathrm{d}x$ gives us the probability of finding a random number in the interval $(x, x + \mathrm{d}x)$. For the uniform distribution we have

$$P(x) = \begin{cases} 1 & \text{for } 0 \leq x \leq 1 \\ 0 & \text{else.} \end{cases}$$

A more precise criterion for the quality of random number generators involves the absence of correlations. The absence of correlations can be expressed in terms of more complicated distribution functions. As an example, we would like the distribution function $P(x_i, x_{i+1})$, giving the probability for the successive occurrence of two random numbers x_i and x_{i+1}, to satisfy

$$P(x_i, x_{i+1}) = P(x_i)P(x_{i+1}) = 1. \tag{B.1}$$

Similar conditions are required for $P(x_i, x_j)$ with $j - i > 1$ and also for distribution functions of more than two variables.

B2 Random number generators and properties of pseudo-random numbers

In a computer, (pseudo-)random numbers are always represented by a finite number of bits. This means that they can conveniently be interpreted as integers. If these integer numbers are distributed uniformly, we can obtain real numbers by dividing them by the largest integer that can be generated.

As a first example, we consider the most commonly used generator, the *linear congruent* or *modulo* generator [3]. Given fixed integers a, c and m, we have the following prescription for generating the next random integer x_i from the previous one (x_{i-1}):

$$x_i = (a \cdot x_{i-1} + c) \bmod m, i > 0. \tag{B.2}$$

The initial number x_0, which must be chosen before starting the process of generating numbers, is called the *seed* of the generator. A real number between 0 and 1 is obtained by dividing x_i by m. In most cases c is taken to be equal to 0. We now see a problem arising when $x_0 = 0$: in that case, all subsequent numbers remain equal to 0, which can hardly be taken for a random sequence. The choice for the seed must therefore be made carefully. Since $x = 0$ is ruled out, the maximum number of different random numbers is $m-1$. It turns out that there exist special combinations of a and m allowing for $m-1$ different random numbers to be generated. If the first number of the sequence reappears, the sequence will repeat itself and the random character is obviously lost.

To show that the choice of a and m is indeed a subtle one, consider $a = 12$ and $m = 143$. In that case:

$$\begin{aligned} x_{i+1} &= 12x_i \bmod 143 \\ x_{i+2} &= 12^2 x_i \bmod 143 \\ &= 144x_i \bmod 143 = x_i \end{aligned} \tag{B.3}$$

so the sequence has a length of only 2! It is not too difficult to see that the following conditions are necessary and sufficient for obtaining a maximum length of the random number sequence:

- x_0 is relatively prime to m, that is, x_0 and m have no common factors;
- a is a primitive element modulo m, that is, it is the integer with the largest *order modulo m* possible. The order modulo m, denoted by λ, for a number a is the smallest integer number λ for which it holds that $a^\lambda \mod m = 1$.

The maximum length of the random number sequence is λ for the primitive element a. If m is prime, this length is equal to $m-1$. For $m = 2^r$, the maximum length is 2^{r-1}.

If we use r bits to represent the random numbers of our generator, there are in principle 2^r different numbers possible. The choice $m = 2^r$ is a convenient one since calculations involving computer words are carried out modulo 2^r after cutting off beyond the r least significant bits. In particular, for 32-bit integers with the first bit acting as a sign-bit, r can be chosen as 31. Primitive elements for this value of m are all integers a with $a \mod 8 = 3$ or 5. A random generator which has been used frequently is IBM's `randu` which used $m = 2^{31}$ and $a = 65\,539$. This turns out to have poor statistical properties (see below), and moreover it is not really portable as it depends on the computer's word length and on a specific handling of the overflow in integer multiplication. A better choice is $m = 2^{31} - 1$, which is a prime, leading to a maximal sequence length of $2^{31} - 2 \approx 2 \times 10^9$. A primitive element for this m is $a = 16\,807$ [4]. Schrage has given a method for calculating $x_i \cdot a \mod m$ without causing overflow, even if $x_i \cdot a$ is larger than the computer word size [5, 6]; see also Problem B.1. In fact, 16 807 is not the only primitive element modulo $2^{31} - 1$: there are more than half a million of them! Extensive research has been carried out to find the 'best' ones among them and those to which Schrage's method is applicable [7]. It turns out that 16 807 is not the very best, but it is not bad at all. Moreover, people feel safe using it, because it has been tested more extensively than any other multiplier and has not exhibited really dangerous behaviour in any test so far [4]. The sequence lengths in all these examples might seem large, but in practice they are not sufficient for large-scale simulations, so that sometimes one has to look for generators with larger periods.

Before treating another type of random number generator, we discuss statistical deficiencies intrinsic to all types of generators, although some suffer more from them than others. It turns out that when pairs of subsequent random numbers from a sequence are considered, fairly small correlations are found between them. However, if triples or higher multiples of subsequent random numbers are taken, the correlations become stronger. This can be shown by taking triples of random numbers, for example, and considering these as the indices of points in three-dimensional space. For pure random numbers, these points should fill the unit cube homogeneously, but for pseudo-random sequences from a modulo generator, the points fall near a set of parallel planes. A theoretical upper bound to the number of such planes in dimension d has been found [8] – it is given by $(d!m)^{1/d}$. In general one can say that the better the random number generator, the more planes fill the d-dimensional hypercube. It is not clear to what extent such deviations from pure random sequences influence the outcome of simulations using random number generators: this depends on the type of simulation.

In the case of the modulo random number generator, the importance of correlations varies with the multiplier a. A small value of the multiplier a for example results in a small random number x_i to be followed by a few more relatively small numbers, and this is of course highly correlated. To obtain a good multiplier, extensive tests have to be performed [4, 7].

It should be stressed that bad random number generators abound in currently available software, and it is recommended to check the results of any simulation with different random number generators. Surprisingly, random generators may be better on paper than others but have less favourable properties in particular simulations. It has been noted, for example, that the simple modulo generator has better properties in cluster Monte Carlo simulations (see Section 15.5.1) than several others that perform better in general statistical tests [9], because the formation of the clusters induces a higher sensitivity to long-range correlations.

A second example of a random number generator, which suffers less from correlations than the modulo generator, is a shift-register generator. This works as follows. The random numbers are strings of, say, r bits. Each bit of the next number in the sequence is determined by the corresponding bits of the previous n numbers. If we denote the kth bit of the ith number in the sequence by $b_i^{(k)}$, we can write down the production rule for the new bits:

$$b_i^{(k)} = (c_1 b_{i-1}^{(k)} + c_2 b_{i-2}^{(k)} + \cdots + c_n b_{i-n}^{(k)}) \bmod 2 \tag{B.4}$$

where the numbers c_i are all equal to 0 or 1. We see that the shift-register generator is a generalisation of the modulo generator. The maximum cycle length is $2^n - 1$. This maximum length is obtained for special combinations of the numbers c_i. Of course, the algorithm presented can also be used to generate rows of random bits [10].

A simple form of this generator is the one for which only two c_i are nonzero:

$$x_i^{(k)} = \left(x_{i-q}^{(k)} + x_{i-p}^{(k)}\right) \bmod 2. \tag{B.5}$$

The sum combined with the mod 2 condition is precisely the XOR operation which can be executed very fast in a computer. 'Magical' (p, q) pairs yielding an optimal cycle length are: (98, 27), (521, 32) and finally (250, 103), which is frequently used.

Before the generator (B.4) can be started, the first n random numbers have to be known already. These can be generated with a modulo generator. Since only a limited number of starting values is needed, correlations between these are not yet noticeable [11].

B3 Nonuniform random number generators

Random numbers with a nonuniform distribution are usually constructed starting from a uniform generator. In this section, we show how this can be done. As a first example we consider a generator with a Gaussian distribution:

$$P(x) = \mathrm{e}^{-(x-x_0)^2/2\sigma^2}. \tag{B.6}$$

From the central limit theorem, which states that the sum of many uncorrelated random numbers is distributed according to a Gaussian with a width proportional to $1/\sqrt{N}$, it follows directly that we can obtain a Gaussian distribution just by adding n uniform random numbers; the higher n the more accurate this distribution will match the Gaussian form. If we want to have a Gaussian with a width σ and a centre $\bar{x}$, we transform the sum S of n uniform random numbers according to

$$x = \bar{x} - 2\sigma(S/n - 1/2)\sqrt{3} \tag{B.7}$$

This method for achieving a Gaussian distribution of the random numbers is very inefficient as we have to generate n uniform random numbers to obtain a single Gaussian one. We discuss more efficient methods below.

More generally, one can make a nonuniform random number generator using a real function f, and for each number x generated by a uniform generator, taking $f(x)$ as the new nonuniform random number, where f is a function designed such as to arrive at the prescribed distribution P. As the number of random numbers lying between x and $x + \mathrm{d}x$ is proportional to $\mathrm{d}x$ and the same number of nonuniform random numbers $y = f(x)$ will lie between y and $y + \mathrm{d}y$ with $\mathrm{d}y/\mathrm{d}x = f'(x)$, the density of the numbers $y = f(x)$ is given by $1/f'(x)$, so this should be equal to the prescribed distribution $P(y)$:

$$1/f'(x) = P(y) \text{ with} \tag{B.8a}$$

$$y = f(x). \tag{B.8b}$$

We must construct a function f that yields the prescribed distribution P, i.e. one that satisfies (B.8b). To this end, we use the following relation between f and its inverse f^{-1}:

$$(f^{-1})'(y)f'(x) = 1 \tag{B.9}$$

from which it follows that

$$P(y) = (f^{-1})'(y). \tag{B.10}$$

There is a restricted number of distributions for which such a function f can be found, because it is not always possible to find an invertible primitive function to the distribution P. A good example for which this *is* possible is the Maxwell distribution for the velocities in two dimensions. Taking the Boltzmann factor $1/(k_B T)$ equal to $1/2$ for simplicity, the velocities are distributed according to

$$P(v_x, v_y)\mathrm{d}v_x\mathrm{d}v_y = \mathrm{e}^{-v^2/2}\mathrm{d}v_x\mathrm{d}v_y = \mathrm{e}^{-v^2/2}v\mathrm{d}v\mathrm{d}\varphi = P(v)\mathrm{d}v\mathrm{d}\varphi, \tag{B.11}$$

so the norm v of the velocity is distributed according to

$$P(y) = y\mathrm{e}^{-y^2/2}. \tag{B.12}$$

From (B.10) we find that the function f is defined by

$$f^{-1}(y) = -\mathrm{e}^{-y^2/2} + \text{Const.} = x \tag{B.13}$$

so that

$$y = f(x) = \sqrt{-2\ln(\text{Const.} - x)}. \tag{B.14}$$

Because x lies between 0 and 1, and y between 0 and ∞, we find for the constant the value 1, and a substitution $x \to 1 - x$ (preserving the interval [0,1] of allowed values for x) enables us to write

$$y = f(x) = \sqrt{-2\ln(x)}. \tag{B.15}$$

This method is very efficient since each random number generated by the uniform generator yields a nonuniformly distributed random number.

From (B.11), we see that it is possible to generate Gaussian random numbers starting from a distribution (B.12), since we can consider the Maxwell distribution as a distribution for the generation of two independent Gaussian random numbers $v_x = v\cos\varphi$, $v_y = v\sin\varphi$. From this it is clear that by generating two random numbers, one being the value v with a distribution according to (B.12) and another being the value φ with a uniform distribution between 0 and 2π, we can construct

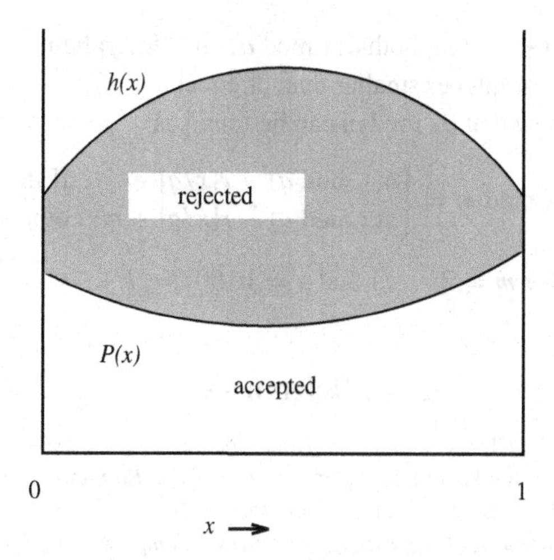

Figure B.1. The von Neumann method for generating nonuniform random numbers.

two numbers v_x and v_y which are both distributed according to a Gaussian. This is called the *Box–Müller method.*

If we cannot find a primitive function for P, we must use a different method. A method by Von Neumann uses at least two uniform random numbers to generate a single nonuniform one. Suppose we want to have a distribution $P(x)$ for x lying between a and b. We start by constructing a generator whose distribution h satisfies $h(x) > \alpha P(x)$ on the interval $[a, b]$. A simple choice is of course the uniform generator, but it is efficient to have a function h with a shape roughly similar to that of P. Now for every x generated by the h-generator, we generate a second random number y uniformly between 0 and 1 and check if y is smaller or larger than $\alpha P(x)/h(x)$. If y is smaller, then x is accepted as the next random number and if y is larger than $\alpha P(x)/h(x)$, it is rejected. The procedure is represented in Figure B.1. Clearly, it is important to have as few rejections as possible, which can be achieved by choosing h such that $\alpha P(x)/h(x)$ is as close as possible to 1 for x between a and b.

Exercises

B.1 Schrage's method [5] enables us to carry out the transformation $ax_n \bmod m$ as it occurs in the modulo random number generator without causing overflow, even when ax_n is greater than the computer's word size. It works as follows. Suppose that for a given a and m, we have two numbers q and r satisfying

$$m = aq + r \quad \text{with } 0 \leq r < q.$$

(a) Show that if $0 < x < m$, both $a(x \bmod q)$ and $r[x/q]$ lie in the range $0, \ldots, m-1$ ($[z]$ is the largest integer smaller than or equal to z).

(b) Using this, show that $ax \bmod m$ can be found as:

$$ax \bmod m = \begin{cases} a(x \bmod q) - r[x/q] & \text{if this is } \geq 0 \\ a(x \bmod q) - r[x/q] + m & \text{otherwise.} \end{cases}$$

(c) Find q and r for $m = 2^{31} - 1$ and $a = 16\,807 (= 7^5)$.

References

[1] N. A. Frigerio and N. Clarck, *Trans. Am. Nucl. Soc.*, **22** (1975), 283–4.

[2] N. A. Frigerio, N. Clarck, and S. Tyler, *Toward Truly Random Random Numbers*, Report, ANL/ES-26 Part 4. Argonne National Laboratory, 1978.

[3] D. E. Knuth, *Seminumerical Algorithms. The Art of Computer Programming*, vol. 2. Reading, MA, Addison-Wesley, 1981.

[4] S. W. Park and K. W. Miller, 'Random number generators: good ones are hard to find,' *Commun. ACM*, **31** (1988), 1192–201.

[5] L. Schrage, 'A more portable random number generator,' *ACM Trans. Math. Software*, **5** (1979), 132–8.

[6] W. H. Press, S. A. Teukolsky, W. T. Vetterling, and B. P. Flannery, *Numerical Recipes*, 2nd edn. Cambridge, Cambridge University Press, 1992.

[7] P. l'Ecuyer, 'Efficient and portable combined random number generators,' *Commun. ACM*, **31** (1988), 742–9 and 774.

[8] G. Marsaglia, 'Random numbers fall mainly in the planes,' *Proc. Nat. Acad. Sci.*, **61** (1968), 25–8.

[9] A. M. Ferrenberg, D. P. Landau, and Y. J. Wong, 'Monte Carlo simulations: hidden errors from "good" random number generators,' *Phys. Rev. Lett.*, **69** (1992), 3382–4.

[10] R. C. Tausworthe, 'Random numbers generated by linear recurrence modulo 2,' *Math. Comp.*, **19** (1965), 201–9.

[11] S. Kirkpatrick and E. P. Stoll, 'A very fast shift-register sequence random number generator,' *J. Comp. Phys.*, **40** (1981), 517–26.

Index